**7** th Edition

# Statistics
## Principles and Methods

**Richard A. Johnson**
University of Wisconsin at Madison

**Gouri K. Bhattacharyya**

WILEY

| PUBLISHER | Laurie Rosatone |
| ACQUISITIONS EDITOR | Joanna Dingle |
| PROJECT EDITOR | Ellen Keohane |
| ASSISTANT EDITOR | Jacqueline Sinacori |
| SENIOR EDITORIAL ASSISTANT | Courtney Welsh |
| SENIOR CONTENT MANAGER | Karoline Luciano |
| SENIOR PRODUCTION EDITOR | Kerry Weinstein |
| MARKETING MANAGER | Melanie Kurkjian |
| SENIOR PRODUCT DESIGNER | Tom Kulesa |
| EDITORIAL OPERATIONS MANAGER | Melissa Edwards |
| MEDIA SPECIALIST | Laura Abrams |
| SENIOR PHOTO EDITOR | Lisa Gee |
| SENIOR DESIGNER | Madelyn Lesure |
| COVER PHOTO | Joseph Van Os/Stockbyte/Getty Images |

This book was set in 10/12 Berling by Laserwords Private Limited, India.

ISBN 978-0-470-90411-4

Printed in the United States of America

10 9 8 7 6 5 4 3

# Preface

## THE NATURE OF THE BOOK

Statistics is the science of learning from data. It encompasses making measurements, reaching conclusions, and communicating the amount of related uncertainty. Its principles help determine what conclusions are reliable and, more generally, provide the navigational guidance for making advances in science and addressing societal issues. Statistics—the subject of data analysis and data-based reasoning—is necessarily playing a vital role in virtually all professions. Some familiarity with this subject is now an essential component of any college education. Yet, pressures to accommodate a growing list of academic requirements often necessitate that this exposure be brief. Keeping these conditions in mind, this book provides students with a first exposure to the powerful ideas of modern statistics. It presents the key statistical concepts and the most commonly applied methods of statistical analysis. It is accessible to freshmen and sophomores from a wide range of disciplines because we have avoided most mathematical derivations. They can pose a stumbling block to learning the essentials in a short period of time.

This book does not require students to have a strong background in mathematics. It is intended for those who seek to learn the basic ideas of statistics and their application in a variety of practical settings. The core material of this book is common to almost all first courses in statistics and is designed to be covered well within a one-semester course in introductory statistics for freshmen to seniors. It is supplemented with some additional special-topics chapters.

## ORIENTATION

The topics treated in this text are, by and large, the ones typically covered in an introductory statistics course. They span three major areas: (i) descriptive statistics, which deals with summarization and description of data; (ii) ideas of probability and an understanding of the manner in which sample-to-sample variation influences our conclusions; and (iii) a collection of statistical methods for analyzing the types of data that are of common occurrence. However, it is the treatment of these topics that makes the text distinctive. Throughout, we have endeavored to give clear and concise explanations of the concepts and important statistical terminology and methods. By means of good motivation, sound explanations, and an abundance of illustrations given in real-world contexts, it emphasizes more than just a superficial understanding.

Each statistical concept or method is motivated by first presenting it in an interesting real-life setting and then focusing on an example to further elaborate important aspects and to illustrate its usefulness. The subsequent discussion is not only limited to showing how a method works but includes an explanation of the why. Even without recourse to mathematics, we are able to make the reader aware of possible pitfalls in the statistical analysis. Students can gain a proper appreciation of statistics only when they are provided with a careful explanation of the underlying logic. Without this understanding, a learning of elementary statistics is bound to be rote and transient.

When describing the various methods of statistical analysis, the reader is continually reminded that the validity of a statistical inference is contingent upon certain model assumptions. Misleading conclusions may result when these assumptions are violated. We think that the teaching of statistics, even at an introductory level, should not be limited to the prescription of methods. Students should be encouraged to develop a critical attitude in applying the methods and to be cautious when interpreting the results. This attitude is especially important in the study of relationship among variables, which is perhaps the most widely used (and also abused) area of statistics. In addition to discussing inference procedures in this context, we have particularly stressed critical examination of the model assumptions and careful interpretation of the conclusions.

To allow an instructor to avoid just teaching formulas and to concentrate on statistical reasoning, this text contains an abundance of data sets in the Data Bank, examples, and exercises. For instance, rather than just declare that a population is normal, the instructor can guide students to an understanding of the importance of checking the validity of the normal assumption and also identifying outliers.

Student projects are becoming an integral part of many first courses in statistics. Our larger data sets provide the basis for numerous projects while the examples help point students toward appropriate statistical analyses.

## SPECIAL FEATURES

1. **Crucial elements are boxed** to highlight important concepts and methods. These boxes provide an ongoing summary of the important items essential for learning statistics. At the end of each chapter, all of its **key ideas and formulas** are summarized.

2. **A rich collection of examples and exercises** is included. These are drawn from a large variety of real-life settings. In fact, many data sets stem from genuine experiments, surveys, or reports.

3. **Exercises** are provided at the end of each major section. They give the reader the opportunity to practice the ideas just learned. Occasionally,

exercises supplement some points raised in the text. A larger collection of exercises appears at the end of a chapter. The starred problems are relatively difficult and suited to the more mathematically competent student.

4. **Using Statistics Wisely**, a feature at the end of each chapter, provides important guidelines for the appropriate use of the statistical procedures presented in the chapter.

5. **Statistics in Context** sections, in the first four chapters and Chapter 7, each describe an important statistical application where a statistical approach to understanding variation is vital. These extended examples reveal, early on in the course, the value of understanding the subject of statistics.

6. *P*–**values** are emphasized in examples concerning tests of hypotheses. Graphs giving the relevant normal or *t* density curve, rejection region, and *P*–value are presented.

7. **Regression analysis** is a primary statistical technique so we provide a more thorough coverage than is usual at this level. The basics of regression are introduced in Chapter 11, whereas Chapter 12 stretches the discussion to several issues of practical importance. These include methods of model checking, handling nonlinear relations, and multiple regression analysis. Complex formulas and calculations are judiciously replaced by computer output so the main ideas can be learned and appreciated with a minimum of stress.

8. **Integrated Technology**, at the end of most chapters, details the steps for using MINITAB, EXCEL, and TI-84/-83 calculator. With this presentation available, discussion can center on interpreting the computer output discussed in the text. Software packages remove much of the drudgery of hand calculation and they allow students to work with larger data sets where patterns are more pronounced. Some computer exercises are included in all chapters where relevant.

9. **Convenient Electronic Data Bank** at the end of the book contains a substantial collection of data. These data sets, together with numerous others throughout the book, allow for considerable flexibility in the choice between concept-orientated and applications-orientated exercises. The Data Bank and the other larger data sets are available for download on the accompanying Web site located at www.wiley.com/college/johnson.

10. **Technical Appendix A** presents a few statistical facts of a mathematical nature. These are separated from the main text so that the instructor can include or exclude them.

## ABOUT THE SEVENTH EDITION

The seventh edition of *STATISTICS—Principles and Methods* maintains the objectives and level of presentation of the earlier editions. The goals are the developing (i) of an understanding of the reasonings by which findings from sample data can be extended to general conclusions and (ii) a familiarity with some basic statistical methods. There are numerous data sets and computer outputs which give an appreciation of the role of the computer in modern data analysis.

Clear and concise explanations introduce the concepts and important statistical terminology and methods. Real-life settings motivate the statistical ideas and well organized discussions proceed to cover statistical methods with heavy emphasis on examples. The seventh edition enhances these special features. The major improvements are:

New Examples A substantial number of new examples, almost all based on real data, are included. Being based on variety of interesting current activities and issues, they expose the student to the widening sphere of application of statistical methods. The majority are in the core chapters together with Chapter 11 on regression.

Rearrangement and Re-writing of Material A re-working of the material on making inferences from two samples or matched pairs, in Chapter 10, includes placing more exercises in its earlier sections. This simplifies the presentation and makes the selection of topics easier for the instructor.

More Emphasis on $P$–values. There is more discussion and several new graphs illustrating $P$–values in examples that involve testing hypotheses. Several new graphs pertain to the $t$ and $\chi^2$ distributions.

Bayes' Theorem Several new examples and exercises give this subject more prominence.

More Data-Based Exercises Many new exercises are keyed to new data-based examples in the text. Several new data sets are also presented in the exercises. Other exercises are based on the polar bear data that is new to the data bank. The abundance of real data sets allows an instructor to address such issues as the assumption of normality and outliers.

Poisson distribution Responding to requests by some users, Section 6 of Chapter 5 now introduces the Poisson distribution. Several examples illustrate its application and the discussion covers rare events and the approximation to the binomial. Key parts can be covered in part of one lecture. Table 3, Appendix B, gives the cumulative Poisson probabilities.

New Exercises Helping Understanding of Concepts Many new exercises provide practice on understanding the concepts while others address understanding related calculations. These new exercises, that augment the already rich collection of exercises, are placed in real-life settings to help promote a greater appreciation of the wide span of applicability of statistical methods.

## ORGANIZATION

This book is organized into fifteen chapters, an optional technical appendix (Appendix A), and a collection of tables (Appendix B). Although designed for a one-semester or a two-quarter course, it is enriched with ample additional material to allow the instructor some choices of topics. Beyond Chapter 1, which sets the theme of statistics and distinguishes population and sample, the subject matter could be classified as follows:

| Topic | Chapter |
|---|---|
| Descriptive study of data | 2, 3 |
| Probability and distributions | 4, 5, 6 |
| Sampling variability | 7 |
| Core ideas and methods of statistical inference | 8, 9, 10 |
| Special topics of statistical inference | 11, 12, 13, 14, 15 |

We regard Chapters 1 to 10 as constituting the core material of an introductory statistics course, with the exception of the starred sections in Chapter 6. Although this material is substantial enough for a one-semester course, many instructors may wish to eliminate some sections in order to cover the basics of regression analysis in Chapter 11. This is most conveniently done by initially skipping Chapter 3 and then taking up only those portions that are linked to Chapter 11. Also, instead of a thorough coverage of probability that is provided in Chapter 4, the later sections of that chapter may receive a lighter coverage.

# SUPPLEMENTS

**Instructor's Solution Manual.** (ISBN 978-1-118-61734-2) This manual contains complete solutions to all exercises. The ISM can also be downloaded by instructors from the book companion Web site.

**Test Bank.** (Available on the accompanying website: www.wiley.com/college/johnson) Contains a large number of additional questions for each chapter.

**Student Solutions Manual.** (ISBN 978-1-118-61631-4) This manual contains complete solutions to all odd-numbered exercises.

**Electronic Data Bank.** (Available on the accompanying website: www.wiley.com/college/johnson) Contains interesting data sets used in the text but that can be used to perform additional analyses with statistical software packages.

**WileyPLUS.** This powerful online tool provides a completely integrated suite of teaching and learning resources in one easy-to-use website. *WileyPLUS* offers an online assessment system with full gradebook capabilities and algorithmically generated skill building questions. This online teaching and learning environment also integrates the entire digital textbook. To view a demo of *WileyPLUS*, contact your local Wiley Sales Representative or visit: www.wiley.com/college/wileyplus.

# ACKNOWLEDGMENTS

We thank Minitab (State College, Pa.) and the SAS Institute (Cary, N.C.) for permission to include commands and output from their software packages. A special thanks to K. T. Wu and Kam Tsui for many helpful suggestions and comments on earlier editions. We also thank all those who have contributed the data sets which enrich the presentation and all those who reviewed the previous editions. The following people gave their careful attention to this edition:

Ron Barry, University of Alaska Fairbanks
Robert Bekes, Santa Clara University
Irene Bloom, Arizona State University
Mingxiang Chen, North Carolina A&T State University
Casey T. Cremins, University of Maryland, College Park
Lisa B. Crespo, University of New Orleans
Christiana Drake, University of California, Davis
Jonathan Eaton, Mississippi State University
Harry Dean Johnson Jr., Washington State University
Analisa M. Flores, University of California, Riverside
David T. Flynn, University of North Dakota
Kathy Gray, California State University, Chico
Wei-Min Huang, Lehigh University
Ilhan M. Izmirli, George Mason University
Yaw Kyei, North Carolina A&T State University
Wei Lin, Ohio University
Jun Masamune, Penn State Altoona
Tatjana Miljkovic, North Dakota State University
Dan Ostrov, Santa Clara University
Rachel Rader, Ohio Northern University
Laurence D. Robinson, Ohio Northern University
Alla Sikorskii, Michigan State University
Robert L. Sims, George Mason University
James Stamey, Baylor University
Natalia Stepanova, Carleton University
Kenneth Strazzeri, George Mason University
Quoc-Phong Vu, Ohio University
George Petrov Yanev, The University of Texas-Pan American

Richard A. Johnson
Gouri K. Bhattacharyya

# WileyPLUS

**WileyPLUS is a research-based online environment for effective teaching and learning.**

*WileyPLUS* builds students' confidence because it takes the guesswork out of studying by providing students with a clear roadmap:

- what to do
- how to do it
- if they did it right

It offers interactive resources along with a complete digital textbook that help students learn more. With *WileyPLUS*, students take more initiative so you'll have greater impact on their achievement in the classroom and beyond.

Now available for

**Bb**
Blackboard

# WileyPLUS

## ALL THE HELP, RESOURCES, AND PERSONAL SUPPORT YOU AND YOUR STUDENTS NEED!

### www.wileyplus.com/resources

**1st DAY OF CLASS ... AND BEYOND!**

2-Minute Tutorials and all of the resources you and your students need to get started

**WileyPLUS**

**Student Partner Program**

Student support from an experienced student user

**Wiley Faculty Network**

Collaborate with your colleagues, find a mentor, attend virtual and live events, and view resources

**www.WhereFacultyConnect.com**

**WileyPLUS**

**Quick Start**

Pre-loaded, ready-to-use assignments and presentations created by subject matter experts

Technical Support 24/7 FAQs, online chat, and phone support

**www.wileyplus.com/support**

© Courtney Keating/iStockphoto

Your *WileyPLUS* Account Manager, providing personal training and support

# Contents

# 1

# Introduction to Statistics

John W. McDonough/SportsIllustrated/Getty Images

Hometown fans attending today's game are but a sample of the population of all local football fans. A self-selected sample may not be entirely representative of the population on issues such as ticket price increases.

## Surveys Provide Information About the Population

| What is your favorite spectator sport? | |
|---|---|
| Football | 40.5% |
| Baseball | 14.0% |
| Basketball | 10.8% |
| Other | 34.7% |

College and professional sports are combined in our summary.[1] Clearly, football is the most popular spectator sport. Actually, the National Football League by itself is more popular than baseball.

For many years, baseball was most popular according to similar surveys. Surveys, repeated at different times, can detect trends in opinion.

# 1. THE SUBJECT AND SCOPE OF STATISTICS

Statistical reasoning allows us to learn from observations that exhibit considerable variation and to make good decisions in the presence of uncertainty. These are, of course, the conditions that dominate our modern world. To introduce this subject, we first develop a definition of statistics as a subject and then give examples that are typically encountered in our everyday lives.

## 1.1 WHAT IS STATISTICS?

Historically, the word statistics originated from the Latin word "status," meaning "state." For a long time, it was identified solely with charts and displays of prevailing economic and demographic conditions. Governments require these statistics to underpin their setting of tax policies and the raising of armies. A major segment of today's population still thinks of statistics as synonymous with forbidding arrays of numbers and myriad graphs like those that predominate government reports. But this view is no longer valid. Gigantic advances during the twentieth century have propelled statistics to recognized importance as a field of data-based reasoning.

What, then, are the role and principal objectives of statistics as a scientific discipline? Stretching well beyond the confines of data display, statistics deals with collecting informative data, interpreting these data, and drawing conclusions about a phenomenon under study. The scope of this subject naturally extends to all processes of acquiring knowledge that involve fact finding by collecting and examining data. Opinion polls (surveys of households to study sociological, economic, or health-related issues), agricultural field experiments (with new seeds, pesticides, or farming equipment), clinical studies of vaccines, and cloud seeding for artificial rain production are just a few examples. The principles and methodology of statistics are useful in answering questions such as, What kind and how much data need to be collected? How should we organize and interpret the data? How can we analyze the data and draw conclusions? How do we assess the strength of the conclusions and gauge their uncertainty?

[1] These percentages are between those obtained by the ESPN Sports Poll, a service of TNS, published in 2009 and a Harris Poll published in 2011.

**Statistics** as a subject provides a body of principles and methodology for designing the process of data collection, summarizing and interpreting the data, and drawing conclusions or generalities.

## 1.2 STATISTICS IN OUR EVERYDAY LIFE

Fact finding by collecting and interpreting data is not confined to professional researchers. All of us must be able to review and interpret numerical facts and figures to better understand issues of environmental protection, the state of the economy, or the performance of competing football teams. In our daily lives, we often learn by doing an implicit analysis of factual information. Statistical reasoning will provide a solid basis for improving this step in the learning process.

We are all familiar to some extent with reports in the news media on important statistics.

*Employment.*   A key to providing timely unemployment numbers is choosing to use a sample rather than attempting a complete enumeration. Monthly, as part of the Current Population Survey, the Bureau of Census collects information about employment status from a sample of about 60,000 households. Households are contacted on a rotating basis with three-fourths of the sample remaining the same for any two consecutive months.

The survey data are analyzed by the Bureau of Labor Statistics, which reports monthly unemployment rates. □

*Gallup Poll.*   This, the best known of the national polls, produces estimates of the percentage of popular vote for each candidate based on interviews with a minimum of 1000 adults per day. Beginning several months before the presidential election, results are regularly published. These reports help predict winners and track changes in voter preferences. □

Our sources of factual information range from individual experience to reports in news media, government records, and articles in professional journals. As consumers of these reports, citizens need some idea of statistical reasoning to properly interpret the data and evaluate the conclusions. Statistical reasoning provides criteria for determining which conclusions are supported by the data and which are not. The credibility of conclusions also depends greatly on the use of statistical methods at the data collection stage. Statistics provides a key ingredient for any systematic approach to improve any type of process from manufacturing to service.

*A/B Testing Approach to Online.*   Many e-commerce companies are currently implementing this strategy for improving their web sites to produce substantially better results. This strategy is an updated method that takes full advantage of the real-time collection of hundreds or more consumer actions each day for the purpose of comparing two versions of a web page or paths to a web page.

Today, many of the largest companies are using and refining this technique to improve their internet sales. For instance, one particular product is selected and then one element of its web page is changed. A picture may be added, an existing picture may be made larger or smaller, or the picture itself may be changed. For instance, a picture of a barbecue grill may be replaced by people grilling at a picnic.

Next, a fraction of the incoming traffic to the product page is directed to the modified page. Each day, for a week or so, the fraction of visitors to the product page who purchase is recorded both for the original page and for the modified page. The two cases give rise to the terminology A and B. If the new page is the more successful of the two, it becomes the new standard and an additional change is tested for improvement. Because of high traffic at the site, even seemingly marginal improvements in drop-off or purchase rates will create substantial additional sales.

This technique was employed, in its early stages of development, during the 2012 presidential campaign to substantially increase donations at one party's web site.

***Quality and Productivity Improvement.***    In the past 35 years, the United States has faced increasing competition in the world marketplace. An international revolution in quality and productivity improvement has heightened the pressure on the U.S. economy. The ideas and teaching of W. Edwards Deming helped rejuvenate Japan's industry in the late 1940s and 1950s. In the 1980s and 1990s, Deming stressed to American executives that, in order to survive, they must mobilize their workforce to make a continuing commitment

Andrew Sacks/Stone/Getty Images

**Statistical reasoning can guide the purposeful collection and analysis of data toward the continuous improvement of any process.**

to quality improvement. His ideas have also been applied to government. The city of Madison, WI, implemented quality improvement projects in the police department and in bus repair and scheduling. In each case, the project goal was better service at less cost. Treating citizens as the customers of government services, the first step was to collect information from them in order to identify situations that needed improvement. One end result was the strategic placement of a new police substation and a subsequent increase in the number of foot patrol persons to interact with the community.

Once a candidate project is selected for improvement, data are collected to assess the current status and then more data are collected after making changes. At this stage, statistical skills in the collection and presentation of summaries are not only valuable but necessary for all participants.

In an industrial setting, statistical training for all employees—production line and office workers, supervisors, and managers—is vital to the quality transformation of American industry.                                                   □

# 2.  STATISTICS IN AID OF SCIENTIFIC INQUIRY

The phrase **scientific inquiry** refers to a systematic process of learning. A scientist sets the goal of an investigation, collects relevant factual information (or data), analyzes the data, draws conclusions, and decides further courses of action. We briefly outline a few illustrative scenarios.

*Training Programs.*    Training or teaching programs designed for a specific type of clientele (college students, industrial workers, minority groups, physically handicapped people, developmentally challenged children, etc.) are continually monitored, evaluated, and modified to improve their usefulness to society. To learn about the comparative effectiveness of different programs, it is essential to collect data on the achievement or growth of skill of the trainees at the completion of each program.                                             □

*Monitoring Advertising Claims.*    The public is constantly bombarded with commercials that claim the superiority of one product brand in comparison to others. When such comparisons are founded on sound experimental evidence, they serve to educate the consumer. Not infrequently, however, misleading advertising claims are made due to insufficient experimentation, faulty analysis of data, or even blatant manipulation of experimental results. Government agencies and consumer groups must be prepared to verify the comparative quality of products by using adequate data collection procedures and proper methods of statistical analysis.                                                        □

*Plant Breeding.*    To increase food production, agricultural scientists develop new hybrids by cross-fertilizing different plant species. Promising new strains need to be compared with the current best ones. Their relative productivity is assessed by planting some of each variety at a number of sites. Yields are recorded and then analyzed for apparent differences. The strains may also be compared on the basis of disease resistance or fertilizer requirements. □

bjonesphotography/iStockphoto

**Statistically designed experiments are needed to document the advantages of the new hybrid versus the old species.**

***Genomics.***   This century's most exciting scientific advances are occurring in biology and genetics. Scientists can now study the genome, or sum total of all of a living organism's genes. The human DNA sequence is now known along with the DNA sequences of hundreds of other organisms.

A primary goal of many studies is to identify the specific genes and related genetic states that give rise to complex traits (e.g., diabetes, heart disease, cancer). New instruments for measuring genes and their products are continually being developed to measure thousands of genes. Due to the impact of the disease and the availability of human tumor specimens, many early studies focused on human cancer. Significant advances have already improved cancer classification, knowledge of cancer biology, and prognostic prediction. A hallmark example of prognostic prediction is Mammaprint approved by the FDA in 2007. This, the first approved genomic test, classifies a breast cancer patient as low or high risk for recurrence.

This is clearly only the beginning. Typically, genomics experiments feature the simultaneous measurement of a great number of responses. As more and more data are collected, there is a growing need for novel statistical methods for analyzing data and thereby addressing critical scientific questions. Statisticians and other computational scientists are playing a major role in these efforts to better human health. ☐

Factual information is crucial to any investigation. The branch of statistics called **experimental design** can guide the investigator in planning the manner and extent of data collection.

After the data are collected, statistical methods are available that summarize and describe the prominent features of data. These are commonly known as **descriptive statistics**. Today, a major thrust of the subject is the evaluation of information present in data and the assessment of the new learning gained from this information. This is the area of **inferential statistics** and its associated methods are known as the methods of **statistical inference**.

It must be realized that a scientific investigation is typically a process of trial and error. Rarely, if ever, can a phenomenon be completely understood or a theory perfected by means of a single, definitive experiment. It is too much to expect to get it all right in one shot. Even after his first success with the light bulb, Thomas Edison had to continue to experiment with numerous materials for the filament before it was perfected.

Data obtained from an experiment provide new knowledge. This knowledge often suggests a revision of an existing theory, and this itself may require further investigation through more experiments and analysis of data. The box below captures the vital point that a scientific process of learning is essentially iterative in nature.

---

### The Conjecture-Experiment-Analysis Learning Cycle

### Learning to Swim

|  | **1st cycle** | **2nd cycle** | **3rd Cycle** |
|---|---|---|---|
| **Conjecture:** | Watch others swim | Try inflatable swim aid | Need instruction |
| **Deduction:** | Be comfortable in deep water | Keep head above water | Able to swim |
| **Experiment:** | Enter deep end of pool | Float with sport tube | Attend a two week class |
| **Analysis:** | Sink and swallow water | Better but really not swimming | Can now float and swim |

---

# 3. TWO BASIC CONCEPTS—POPULATION AND SAMPLE

In the preceding sections, we cite a few examples of situations where evaluation of factual information is essential for acquiring new knowledge. These examples, drawn from widely differing fields, have limited descriptions of the scope and objectives of the studies. However, a few common characteristics are readily discernible.

First, in order to acquire new knowledge, relevant data must be collected. Second, some amount of variability in the data is unavoidable even though observations are made under the same or closely similar conditions. For instance, the treatment for an allergy may provide long-lasting relief for some individuals whereas it may bring only transient relief or even none at all to others. Likewise, it is unrealistic to expect that college freshmen whose high school records are alike would perform equally well in college. Nature does not follow such a rigid law.

A third notable feature is that access to a complete set of data is either physically impossible or from a practical standpoint not feasible. When data are obtained from laboratory experiments or field trials, no matter how much experimentation has been performed, more can always be done. In public opinion or consumer expenditure studies, a complete body of information would emerge only if data were gathered from every individual in the nation—undoubtedly a monumental if not impossible task. To collect an exhaustive set of data related to the damage sustained by all cars of a particular model under collision at a specified speed, every car of that model coming off the production lines would have to be subjected to a collision! Thus, the limitations of time, resources, and facilities, and sometimes the destructive nature of the testing, mean that we must work with incomplete information—the data that are actually collected in the course of an experimental study.

The preceding discussions highlight a distinction between the data set that is actually acquired through the process of observation and the vast collection of all potential observations that can be conceived in a given context. The statistical name for the former is **sample**; for the latter, it is **population**, or **statistical population**. To further elucidate these concepts, we observe that each measurement in a data set originates from a distinct source that may be a patient, tree, farm, household, or some other entity depending on the object of a study. The source of each measurement is called a **sampling unit**, or simply, a **unit**.

To emphasize population as the entire collection of units, we refer to it as the **population of units**.

> A **unit** is a single entity, usually a person or an object, whose characteristics are of interest.
>
> The **population of units** is the complete collection of units about which information is sought.

**TABLE 1**   Populations, Units, and Variables

| Population | Unit | Variables/Characteristics |
|---|---|---|
| Registered voters in your state | Voter | Political party<br>Voted or not in last election<br>Age<br>Sex<br>Conservative/liberal |
| All rental apartments near campus | Apartment | Rent<br>Size in square feet<br>Number of bedrooms<br>Number of bathrooms<br>TV and Internet connections |
| All campus fast food restaurants | Restaurant | Number of employees<br>Seating capacity<br>Hiring/not hiring |
| All computers owned by students at your school | Computer | Speed of processor<br>Size of hard disk<br>Speed of Internet connection<br>Screen size |

There is another aspect to any population and that is the value, for each unit, of a characteristic or variable of interest. There can be several characteristics of interest for a given population of units, as indicated in Table 1.

For a given variable or characteristic of interest, we call the collection of values, evaluated for every unit in the population, the **statistical population** or just the **population**. We refer to the collection of units as the **population of units** when there is a need to differentiate it from the collection of values.

> A statistical **population** is the set of measurements (or record of some qualitative trait) corresponding to the entire collection of units about which information is sought.

The population represents the target of an investigation. We learn about the population by taking a sample from the population. A **sample** or **sample data set** then consists of measurements recorded for those units that are actually observed. It constitutes a part of a far larger collection about which we wish to make inferences—the set of measurements that would result if all the units in the population could be observed.

A **sample** from a statistical population is the subset of measurements that are actually collected in the course of an investigation.

**Example 1**    Identifying the Population and Sample

Questions concerning the effect on health of two or fewer cups of coffee a day are still largely unresolved. Current studies seek to find physiological changes that could prove harmful. An article carried the headline CAFFEINE DECREASES CEREBRAL BLOOD FLOW. It describes a study[2] that establishes a physiological side effect—a substantial decrease in cerebral blood flow for persons drinking two to three cups of coffee daily.

The cerebral blood flow was measured twice on each of 20 subjects. It was measured once after taking an oral dose of caffeine equivalent to two to three cups of coffee and then, on another day, after taking a look-alike dose but without caffeine. The order of the two tests was random and subjects were not told which dose they received. The measured decrease in cerebral blood flow was significant.

Identify the population and sample.

SOLUTION    As the article implies, the conclusion should apply to you and me. The population could well be the potential decreases in cerebral blood flow for all adults living in the United States. It might even apply to all the decreases in blood flow for all caffeine users in the world, although cultural customs may vary the type of caffeine consumption from coffee breaks to tea time to kola nut chewing.

The sample consists of the decreases in blood flow for the 20 subjects who agreed to participate in the study.

**Example 2**    A Misleading Sample

A host of a radio music show announces that she wants to know which singer is the favorite among city residents. Listeners are asked to call in and name their favorite singer.

Identify the population and sample. Comment on how to get a sample that is more representative of the city's population.

SOLUTION    The population is the collection of singer preferences of all city residents and the purported goal is to learn who is the favorite singer. Because it would be nearly impossible to question all the residents in a large city, one must necessarily settle for taking a sample.

Having residents make a local call is certainly a low-cost method of getting a sample. The sample would then consist of the singers named by each person

---

[2]A. Field et al. "Dietary caffeine consumption and withdrawal: Confounding variables in quantitative cerebral perfusion studies?" *Radiology* **227** (2003), pp. 129–135.

who calls the radio station. Unfortunately, with this selection procedure, the sample is not very representative of the responses from all city residents. Those who listen to the particular radio station are already a special subgroup with similar listening tastes. Furthermore, those listeners who take the time and effort to call are usually those who feel strongest about their opinions. The resulting responses could well be much stronger in favor of a particular country western or rock singer than is the case for preference among the total population of city residents or even those who listen to the station.

If the purpose of asking the question is really to determine the favorite singer of the city's residents, we have to proceed otherwise. One procedure commonly employed is a phone survey where the phone numbers are chosen at random. For instance, one can imagine that the numbers 0, 1, 2, 3, 4, 5, 6, 7, 8, and 9 are written on separate pieces of paper and placed in a hat. Slips are then drawn one at a time and replaced between drawings. Later, we will see that computers can mimic this selection quickly and easily. Four draws will produce a random telephone number within a three-digit exchange. Telephone numbers chosen in this manner will certainly produce a much more representative sample than the self-selected sample of persons who call the station.

Self-selected samples consisting of responses to call-in or write-in requests will, in general, not be representative of the population. They arise primarily from subjects who feel strongly about the issue in question. To their credit, many TV news and entertainment programs now state that their call-in polls are nonscientific and merely reflect the opinions of those persons who responded.

## 3.1  USING A RANDOM NUMBER TABLE TO SELECT A SAMPLE

The choice of which population units to include in a sample must be impartial and objective. When the total number of units is finite, the name or number of each population unit could be written on a separate slip of paper and the slips placed in a box. Slips could be drawn one at a time without replacement and the corresponding units selected as the sample of units. Unfortunately, this simple and intuitive procedure is cumbersome to implement. Also, it is difficult to mix the slips well enough to ensure impartiality.

Alternatively, a better method is to take 10 identical marbles, number them 0 through 9, and place them in an urn. After shuffling, select 1 marble. After replacing the marble, shuffle and draw again. Continuing in this way, we create a sequence of random digits. Each digit has an equal chance of appearing in any given position, all pairs have the same chance of appearing in any two given positions, and so on. Further, any digit or collection of digits is unrelated to any other disjoint subset of digits. For convenience of use, these digits can be placed in a table called a **random number table**.

The digits in Table 1 of Appendix B were actually generated using computer software that closely mimics the drawing of marbles. A portion of this table is shown here as Table 2.

**TABLE 2**   Random Digits: A Portion of Table 1, Appendix B

| Row | | | | | | | | | | |
|-----|------|------|------|------|------|------|------|------|------|------|
| 1 | 0695 | 7741 | 8254 | 4297 | 0000 | 5277 | 6563 | 9265 | 1023 | 5925 |
| 2 | 0437 | 5434 | 8503 | 3928 | 6979 | 9393 | 8936 | 9088 | 5744 | 4790 |
| 3 | 6242 | 2998 | 0205 | 5469 | 3365 | 7950 | 7256 | 3716 | 8385 | 0253 |
| 4 | 7090 | 4074 | 1257 | 7175 | 3310 | 0712 | 4748 | 4226 | 0604 | 3804 |
| 5 | 0683 | 6999 | 4828 | 7888 | 0087 | 9288 | 7855 | 2678 | 3315 | 6718 |
| 6 | 7013 | 4300 | 3768 | 2572 | 6473 | 2411 | 6285 | 0069 | 5422 | 6175 |
| 7 | 8808 | 2786 | 5369 | 9571 | 3412 | 2465 | 6419 | 3990 | 0294 | 0896 |
| 8 | 9876 | 3602 | 5812 | 0124 | 1997 | 6445 | 3176 | 2682 | 1259 | 1728 |
| 9 | 1873 | 1065 | 8976 | 1295 | 9434 | 3178 | 0602 | 0732 | 6616 | 7972 |
| 10 | 2581 | 3075 | 4622 | 2974 | 7069 | 5605 | 0420 | 2949 | 4387 | 7679 |
| 11 | 3785 | 6401 | 0504 | 5077 | 7132 | 4135 | 4646 | 3834 | 6753 | 1593 |
| 12 | 8626 | 4017 | 1544 | 4202 | 8986 | 1432 | 2810 | 2418 | 8052 | 2710 |
| 13 | 6253 | 0726 | 9483 | 6753 | 4732 | 2284 | 0421 | 3010 | 7885 | 8436 |
| 14 | 0113 | 4546 | 2212 | 9829 | 2351 | 1370 | 2707 | 3329 | 6574 | 7002 |
| 15 | 4646 | 6474 | 9983 | 8738 | 1603 | 8671 | 0489 | 9588 | 3309 | 5860 |

To obtain a random sample of units from a population of size $N$, we first number the units from 1 to $N$. Then numbers are read from the table of random digits until enough different numbers in the appropriate range are selected.

**Example 3**   **Using the Table of Random Digits to Select Items for a Price Check**

One week, the advertisement for a large grocery store contains 72 special sale items. Five items will be selected with the intention of comparing the sales price with the scan price at the checkout counter. Select the five items at random to avoid partiality.

SOLUTION   The 72 sale items are first numbered from 1 to 72. Since the population size $N = 72$ has two digits, we will select random digits two at a time from Table 2. Arbitrarily, we decide to start in row 7 and columns 19 and 20. Starting with the two digits in columns 19 and 20 and reading down, we obtain

$$12 \quad 97 \quad 34 \quad 69 \quad 32 \quad 86 \quad 32 \quad 51$$

We ignore 97 and 86 because they are larger than the population size 72. We also ignore any number when it appears a second time as 32 does here. Consequently, the sale items numbered

$$12 \quad 34 \quad 69 \quad 32 \quad 51$$

are selected for the price check.

**Example 4**    Selecting a Sample by Random Digit Dialing

A major Internet service provider wants to learn about the proportion of people in one target area who are aware of its latest product. Suppose there is a single three-digit telephone exchange that covers the target area. Use Table 1, in Appendix B, to select six telephone numbers for a phone survey.

SOLUTION    We arbitrarily decide to start at row 31 and columns 25 to 28. Proceeding upward, we obtain

$$7566 \quad 0766 \quad 1619 \quad 9320 \quad 1307 \quad 6435$$

Together with the three-digit exchange, these six numbers form the phone numbers to call in the survey. Every phone number, listed or unlisted, has the same chance of being selected. The same holds for every pair, every triplet, and so on. Commercial phones may have to be discarded and another four digits selected. If there are two exchanges in the area, separate selections could be done for each exchange.

For large sample sizes, it is better to use computer-generated random digits or even computer-dialed random phone numbers.

Data collected with a clear-cut purpose in mind are very different from **anecdotal data**. Most of us have heard people say they won money at a casino, but certainly most people cannot win most of the time as casinos are not in the business of giving away money. People tend to tell good things about themselves. In a similar vein, some drivers' lives are saved when they are thrown free of car wrecks because they were not wearing seat belts. Although such stories are told and retold, you must remember that there is really no opportunity to hear from those who would have lived if they had worn their seat belts. Anecdotal information is usually repeated because it has some striking feature that may not be representative of the mass of cases in the population. Consequently, it is not apt to provide reliable answers to questions.

## Exercises

**1.1** A survey of 451 men revealed that 144 men, or 31.9%, wait until Valentine's day or the day before to purchase flowers. Identify a statistical population and the sample.

**1.2** A magazine article reports that, for the year 2011, insurance claims for jewelry exceeded those for electronics. The insurance covers lost, damaged, or stolen items.[3] Identify a statistical population and the sample for claims concerning jewelry.

**1.3** Twenty college students were asked for their number of close friends; persons who showed sympathy when needed and helped in hard times. The average number reported was just over 2. Identify a statistical population and the sample.

[3]D. Gorski, "Property/casualty claims report," *Best Review*, May 2012, p. 26.

**1.4** The article *What Makes a Great Tweet* concludes that only 36% of tweets are worth reading. A total of 4,220 tweets were rated in this study.[4] Identify a statistical population and the sample.

**1.5** Among 40 adults 20–25 years old questioned about stress levels, 16 responded they are more stressed now than last year. Identify a statistical population and the sample.

**1.6** Rap music was the favorite genre of 6 out of 56 students from a large midwestern university. Identify a statistical population and the sample.

**1.7** About 30% of the persons who visited first aid stations at rave parties in 2008 reported they had a substance-related problem.[5] Should these persons be considered a random sample of party participants with regard to amount of drug use?

**1.8** Which of the following are anecdotal and which are based on a sample?

(a) Ellie told her friends that she is saving $31 a month because she changed to a prepaid cell phone.

(b) One morning, among a large number of coffee shop patrons, the orders of 47 coffee drinks were recorded and 11 of these were espresso-based drinks.

(c) Out of 200 college students at a large university, over half thought about food more than 17 times a day.

**1.9** Which of the following are anecdotal and which are based on a sample?

(a) Seventy-five text messages were sent one day during a lecture by students in a large class.

(b) Erik says he gets the best bargains at online auctions by bidding on items whose termination is early in the morning.

(c) Out of 4,837 college age students in California, 3769 admitted they used a cell phone to talk or text while driving.

**1.10** Twelve bicycles are available for use at the student union. Use Table 1, Appendix B, to select 3 of them for you and two of your friends to ride today.

**1.11** At the last minute, 6 tickets have become available for a big football game. Use Table 1, Appendix B, to select the recipients from among 89 interested students.

# 4. THE PURPOSEFUL COLLECTION OF DATA

Many poor decisions are made, in both business and everyday activities, because people fail to understand and account for variability. Certainly, the purchasing habits of one person may not represent those of the population, or the reaction of one mouse to a potentially toxic chemical compound may not represent that of a large population of mice. However, despite diversity among the purchasing habits of individuals, we can obtain accurate information about the purchasing habits of the population by collecting data on a large number of persons. By the same token, much can be learned about the toxicity of a chemical if many mice are exposed.

Just making the decision to collect data is a key step to answer a question, to provide the basis for taking action, or to improve a process. Once that decision has been made, an important next step is to develop a **statement of purpose** that is both specific and unambiguous. If the subject of the study is public transportation being behind schedule, you must carefully specify what is meant by late. Is it 1 minute, 5 minutes, or more than 10 minutes behind scheduled

[4]P. Andre, M. Bernstein, and K. Luther, "What makes a great tweet," *Harvard Business Review*, May 2012, p. 36.
[5]PLoS ONE—www.plosone.org Dec. 2011, Vol. **6**, issue 12, e29620.

times that should result in calling a bus or commuter train late? Words like soft or uncomfortable in a statement are even harder to quantify. One common approach, for a quality like comfort, is to ask passengers to rate the ride on public transportation on the five-point scale

where the numbers 1 through 5 are attached to the scale, with 1 for very uncomfortable and so on through 5 for very comfortable.

We might conclude that the ride is comfortable if the majority of persons in the sample check either of the top two boxes.

**Example 5**  A Clear Statement of Purpose Concerning Water Quality

Each day, a city must sample the lake water in and around a swimming beach to determine if the water is safe for swimming. During late summer, the primary difficulty is algae growth and the safe limit has been set in terms of water clarity.

SOLUTION  The problem is already well defined so the statement of purpose is straight-forward.

> ***PURPOSE:***  Determine whether or not the water clarity at the beach is below the safe limit.

The city will take measurements of clarity at three separated locations. In Chapter 8, we will learn how to decide if the water is safe despite the variation in the three sample values.

The overall purpose can be quite general but a specific statement of purpose is required at each step to guide the collection of data. For instance:

> ***GENERAL PURPOSE:***  Design a data collection and monitoring program at a completely automated plant that handles radioactive materials.

One issue is to ensure that the production plant will shut down quickly if materials start accumulating anywhere along the production line. More specifically, the weight of materials could be measured at critical positions. A quick shutdown will be implemented if any of these exceed a safe limit. For this step, a statement of purpose could be:

> ***PURPOSE:***  Implement a fast shutdown if the weight at any critical position exceeds 1.2 kilograms.

The safe limit 1.2 kilograms should be obtained from experts; preferably it would be a consensus of expert opinion.

There still remain statistical issues of how many critical positions to choose and how often to measure the weight. These are followed with questions on how to analyze data and specify a rule for implementing a fast shutdown.

A clearly specified statement of purpose will guide the choice of what data to collect and help ensure that it will be relevant to the purpose. Without a clearly specified purpose, or terms unambiguously defined, much effort can be wasted in collecting data that will not answer the question of interest.

## Exercises

**1.12**  What is wrong with this statement of purpose?

> **PURPOSE:**  *Determine whether or not, over the course of the semester, the campus bus reaches your stop at the scheduled time.*

Give an improved statement of purpose.

**1.13**  What is wrong with this statement of purpose?

> **PURPOSE:**  *Determine if a new style wireless mouse is comfortable.*

Give an improved statement of purpose.

**1.14**  Give a statement of purpose for a study to determine the favorite campus area restaurant.

**1.15**  Give a statement of purpose for determining the amount of time it takes to make hotel reservations in San Francisco using the internet.

## STATISTICS IN CONTEXT

A primary health facility became aware that sometimes it was taking too long to return patients' phone calls. Patients phone in with requests for information. These requests are then turned over to doctors or nurses who collect the information and return the call. The overall objective is to understand the current procedure and then improve on it. A good first step is to find how long it takes to return calls under the current procedure. Variation in times from call to call is expected, so the purpose of the initial investigation is to benchmark the variability with the current procedure by collecting a sample of times.

> **PURPOSE:**  Obtain a reference or benchmark for the current procedure by collecting a sample of times to return a patient's call under the current procedure.

For a sample of incoming calls collected during the week, the time received is noted along with the request. When the return call is complete, the elapsed time, in minutes, is recorded. Each of these times is represented as a dot in Figure 1. Notice that over one-third of the calls took over 120 minutes, or over two hours, to return. This is a long time to wait for information if it concerns a child with a high fever or an adult with acute symptoms. If the purpose is to determine what proportion of calls took too long to return, we need to agree on a more precise definition of "too long" in terms of number of minutes. Instead, these data clearly indicate that the process needs improvement and the next step is to proceed in that direction.

Figure 1   Time in minutes to return call.

In any context, to pursue potential improvements of a process, one needs to focus more closely on particulars. Three questions

### When   Where   Who

should always be asked before gathering further data. More specifically, data should be sought that will answer the following questions.

**When** do the difficulties arise? Is it during certain hours, certain days of the week or month, or in coincidence with some other activities?

**Where** do the difficulties arise? Try to identify the locations of bottlenecks and unnecessary delays.

**Who** was performing the activity and who was supervising? The idea is not to pin blame, but to understand the roles of participants with the goal of making improvements.

It is often helpful to construct a **cause-and-effect diagram** or fishbone **diagram**. The main centerline represents the problem or the effect. A somewhat simplified fishbone chart is shown in Figure 2 for the *where* question regarding the location of delays when returning patients' phone calls. The main centerline represents the problem: Where are delays occurring? Calls come to the reception desk, but when these lines are busy, the calls go directly to nurses on the third or fourth floor. The main diagonal arms in Figure 2 represent the floors and the smaller horizontal lines more specific locations on the floor where the delay could occur. For instance, the horizontal line representing a delay in retrieving a patient's medical record connects to the second floor diagonal line. The resulting

Figure 2   A cause-and-effect diagram for the location of delays.

figure resembles the skeleton of a fish. Consideration of the diagram can help guide the choice of what new data to collect.

Fortunately, the quality team conducting this study had already given preliminary consideration to the *When, Where,* and *Who* questions and recorded not only the time of day but also the day and person receiving the call. That is, their current data gave them a start on determining if the time to return calls depends on when or where the call is received.

Although we go no further with this application here, the quality team next developed more detailed diagrams to study the flow of paper between the time the call is received and when it is returned. They then identified bottlenecks in the flow of information that were removed and the process was improved. In later chapters, you will learn how to compare and display data from two locations or old and new processes, but the key idea emphasized here is the purposeful collection of relevant data.

## Exercises

**1.16** How many of the calls in the initial data set took over 125 minutes to answer? How many over 90 minutes?

**1.17** According to the cause-and-effect diagram on page 17, where are the possible delays on the first floor?

**1.18** Refer to the cause-and-effect diagram on page 17. The workers have now noticed that a delay could occur:

(a) On the fourth floor at the pharmacy

(b) On the third floor at the practitioners' station

Redraw the diagram and include this added information.

## 5. OBJECTIVES OF STATISTICS

The subject of statistics provides the methodology to make inferences about the population from the collection and analysis of sample data. These methods enable one to derive plausible generalizations and then assess the extent of uncertainty underlying these generalizations. Statistical concepts are also essential during the planning stage of an investigation when decisions must be made as to the mode and extent of the sampling process.

The major objectives of statistics are:
1. To make inferences about a population from an analysis of information contained in sample data. This includes assessments of the extent of uncertainty involved in these inferences.
2. To design the process and the extent of sampling so that the observations form a basis for drawing valid inferences.

The design of the sampling process is an important step. A good design for the process of data collection permits efficient inferences to be made, often with a straightforward analysis. Unfortunately, even the most sophisticated methods of data analysis cannot, in themselves, salvage much information from data that are produced by a poorly planned experiment or survey.

The early use of statistics in the compilation and passive presentation of data has been largely superseded by the modern role of providing analytical tools with which data can be efficiently gathered, understood, and interpreted. Statistical concepts and methods make it possible to draw valid conclusions about the population on the basis of a sample. Given its extended goal, the subject of statistics has penetrated all fields of human endeavor in which the evaluation of information must be grounded in data-based evidence.

The basic statistical concepts and methods described in this book form the core in all areas of application. We present examples drawn from a wide range of applications to help develop an appreciation of various statistical methods, their potential uses, and their vulnerabilities to misuse.

## USING STATISTICS WISELY

1. Compose a clear statement of purpose and use it to help decide which variables to observe.

2. Carefully define the population of interest.

3. Whenever possible, select samples using a random device or random number table.

4. Do not unquestionably accept conclusions based on self-selected samples.

5. Remember that conclusions reached in TV, magazine, or newspaper reports might not be as obvious as reported. When reading or listening to reports, you must be aware that the advocate, often a politician or advertiser, may only be presenting statistics that emphasize positive features.

## KEY IDEAS

Before gathering data, on a characteristic of interest, identify a **unit** or **sampling unit**. This is usually a person or object. The **population of units** is the complete collection of units. In statistics we concentrate on the collection of values of the characteristic, or record of a qualitative trait, evaluated for each unit in the population. We call this the **statistical population** or just the **population**.

A **sample** or **sample data set** from the population is the subset of measurements that are actually collected.

**Statistics** is a body of principles that helps to first design the process and extent of sampling and then guides the making of inferences about the population (**inferential statistics**). **Descriptive statistics** help summarize the sample. Procedures for **statistical inference** allow us to make generalizations about the population from the information in the sample.

A **statement of purpose** is a key step in designing the data collection process.

## REVIEW EXERCISES

**1.19** A newspaper headline reads,

U.S. TEENS TRUST, FEAR THEIR PEERS

and the article explains that a telephone poll was conducted of 1055 persons 13 to 17 years old. Identify a statistical population and the sample.

**1.20** Consider the population of all students at your college. You want to learn about total monthly entertainment expenses for a student.

(a) Specify the population unit.

(b) Specify the variable of interest.

(c) Specify the statistical population.

**1.21** Consider the population of persons living in Chicago. You want to learn about the proportion of eligible voters who are registered to vote.

(a) Specify the population unit.

(b) Specify the variable of interest.

(c) Specify the statistical population.

**1.22** A student is asked to estimate the mean height of all male students on campus. She decides to use the heights of members of the basketball team because they are conveniently printed in the game program.

(a) Identify the statistical population and the sample.

(b) Comment on the selection of the sample.

(c) How should a sample of males be selected?

**1.23** Psychologists[6] asked 46 golfers, after they played a round, to estimate the diameter of the hole on the green by visually selecting one of nine holes cut in a board.

(a) Specify the population unit.

(b) Specify the statistical population and sample.

**1.24** A phone survey in 2008[7] of 1010 adults included a response to the number of leisure hours per week.

Identify the population unit, statistical population, and sample.

**1.25** It is often easy to put off doing an unpleasant task. At a Web site,[8] persons can take a test and receive a score that determines if they have a serious problem with procrastination. Should the scores from people who take this test online be considered a random sample from the general population? Explain your reasoning.

**1.26** A magazine that features the latest electronics and computer software for homes enclosed a short questionnaire on a postcard. Readers were asked to answer questions concerning their use and ownership of various software and hardware products, and to then send the card to the publisher. A summary of the results appeared in a later issue of the magazine that used the data to make statements such as 40% of readers have purchased program X. Identify a population and sample and comment on the representativeness of the sample. Are readers who have not purchased any new products mentioned in the questionnaire as likely to respond as those who have purchased?

**1.27** Each year a local weekly newspaper gives out "Best of the City" awards in categories such as restaurant, deli, pastry shop, and so on. Readers are asked to fill in their favorites on a form at the web site of the weekly paper. The establishment receiving the most votes is declared the winner in its category. Identify the population and sample and comment on the representativeness of the sample.

**1.28** Which of the following are anecdotal and which are based on sample?

(a) Out of 200 students questioned, 40 admitted they lied regularly.

(b) Bobbie says the produce at Market W is the freshest in the city.

---

[6] J. Witt et al., "Putting to a bigger hole: Golf performance relates to perceived size," *Psychonomic Bulletin and Review* **15**(3) (2008), pp. 581–586.

[7] Harris Interactive telephone survey (October 16–19, 2008).

[8] www.mindtools.com (2012) Are you a procrastinator? (online). [Accessed November 9, 2012].

(c) Out of 50 persons interviewed at a shopping mall, 18 had made a purchase that day.

**1.29** Which of the following are anecdotal and which are based on a sample?

(a) Tom says he gets the best prices on electronics at the www.bestelc.com Internet site.

(b) Out of 22 students, 6 had multiple credit cards.

(c) Among 55 people checking in at the airport, 12 were going to destinations outside of the continental United States.

**1.30** What is wrong with this statement of purpose?

> **PURPOSE:** *Determine if a newly designed roller-ball pen is comfortable to hold when writing.*

Give an improved statement of purpose.

**1.31** What is wrong with this statement of purpose?

> **PURPOSE:** *Determine if it takes too long to get cash from the automated teller machine during the lunch hour.*

Give an improved statement of purpose.

**1.32** Two-person sailboats are available for use at the university dock. Your group is large enough to need three of them. Use Table 1, Appendix B, to select your three boats from among the 10 that are in a line along the shore.

**1.33** Thirty-five classrooms on campus are equiped for multimedia instruction. Use Table 1, Appendix B, to select 4 of these classrooms to visit and check whether or not the instructor is using the equipment during that day's first hour lecture.

**1.34** Fifty band members would like to ride the band bus to an out-of-town game. However, there is room for only 44. Use Table 1, Appendix B, to select the 44 persons who will go. Determine how to make your selection by taking only a few two-digit selections.

**1.35** Eight young students need mentors. Of these, there are three whom you enjoy being with while you are indifferent about the others. Two of the students will be randomly assigned to you. Label the students you like by 0, 1, and 2 and the others by 3, 4, 5, 6, and 7. Then, the process of assigning two students at random is equivalent to choosing two different digits from the table of random digits and ignoring any 8 or 9. Repeat the experiment of assigning two students 20 times by using the table of random digits. Record the pairs of digits you draw for each experiment.

(a) What is the proportion of the 20 experiments that give two students that you like?

(b) What is the proportion of the 20 experiments that give one of the students you like and one other?

(c) What is the proportion of the 20 experiments that give none of the students you like?

**1.36** The United States Environmental Protection Agency[9] reports that in 2010, each American generated 4.43 pounds of solid waste a day.

(a) Does this mean every single American produces the same amount of garbage? What do you think this statement means?

(b) Was the number 4.43 obtained from a sample? Explain.

(c) How would you select a sample?

**1.37** As a very extreme case of self-selection, imagine a five-foot-high solid wood fence surrounding a collection of Great Danes and Miniature Poodles. You want to estimate the proportion of Great Danes inside and decide to collect your sample by observing the first seven dogs to jump high enough to be seen above the fence.

(a) Explain how this is a self-selected sample that is, of course, very misleading.

(b) How is this sample selection procedure like a call-in election poll?

[9]www.epa.gov/epawaste/nonhaz/municipal/index.htm. [Accessed 7/23/2012].

# 2

# Organization and Description of Data

Bill Barley/SuperStock

## Acid Rain Is Killing Our Lakes

Acid precipitation is linked to the disappearance of sport fish and other organisms from lakes. Sources of air pollution, including automobile emissions and the burning of fossil fuels, add to the natural acidity of precipitation.

The Wisconsin Department of Natural Resources initiated a precipitation monitoring program with the goal of developing appropriate air pollution controls to reduce the problem. The acidity of the first 50 rains monitored, measured on a pH scale from 1 (very acidic) to 7 (basic), are summarized by the histogram.

Histogram of acid rain data.

Notice that all the rains are more acidic than normal rain, which has a pH of 5.6. (As a comparison, apples are about pH 3 and milk is about pH 6.)

Researchers in Canada have established that lake water with a pH below 5.6 may severely affect the reproduction of game fish. More research will undoubtedly improve our understanding of the acid rain problem and lead, it is hoped, to an improved environment.

In Chapter 1, we cited several examples of situations where the collection of data by experimentation or observation is essential to acquiring new knowledge. A data set may range in complexity from a few entries to hundreds or even thousands of them. Each entry corresponds to the observation of a specified characteristic of a sampling unit. For example, a nutritionist may provide an experimental diet to 30 undernourished children and record their weight gains after two months. Here, children are the sampling units, and the data set would consist of 30 measurements of weight gains. Once the data are collected, a primary step is organizing the information and extracting a descriptive summary that highlights its salient features. In this chapter, we learn how to organize and describe a set of data by means of tables, graphs, and calculation of some numerical summary measures.□

## 1.  MAIN TYPES OF DATA

Before introducing methods of describing data, we first distinguish between the two basic types:

1.   Qualitative or categorical data
2.   Numerical, quantitative, or measurement data

When the characteristic under study concerns a qualitative trait that is only classified in categories and not numerically measured, the resulting data are called categorical data. Hair color (blond, brown, red, black), employment status (employed, unemployed), and blood type ( O, A, B, AB ) are but some examples. If, on the other hand, the characteristic is measured on a numerical scale, the resulting data consist of a set of numbers and are called measurement data. We will use the term **numerical-valued variable** or just **variable** to refer to a characteristic that is measured on a numerical scale. The word "variable" signifies that the measurements vary over different sampling units. In this terminology, observations of a numerical-valued variable yield measurement data. A few examples of numerical-valued variables are the shoe size of an adult male, daily number of traffic fatalities in a state, intensity of an earthquake, height of a 1-year-old pine seedling, the time spent in line at an automated teller, and the number of offspring in an animal litter.

Although in all these examples the stated characteristic can be numerically measured, a close scrutiny reveals two distinct types of underlying scale of measurement. Shoe sizes are numbers such as 6, $6\frac{1}{2}$, 7, $7\frac{1}{2}$, . . . , which proceed in steps of $\frac{1}{2}$. The count of traffic fatalities can only be an integer and so is the number of offspring in an animal litter. These are examples of **discrete variables**. The name **discrete** draws from the fact that the scale is made up of distinct numbers with gaps in between. On the other hand, some variables such as height, weight, and survival time can ideally take any value in an interval. Since the measurement scale does not have gaps, such variables are called **continuous**.

We must admit that a truly continuous scale of measurement is an idealization. Measurements actually recorded in a data set are always rounded either for the sake of simplicity or because the measuring device has a limited accuracy. Still, even though weights may be recorded in the nearest pounds or time recorded in the whole hours, their actual values occur on a continuous scale so the data are referred to as continuous. Counts are inherently discrete and treated as such, provided that they take relatively few distinct values (e.g., the number of children in a family or the number of traffic violations of a driver). But when a count spans a wide range of values, it is often treated as a continuous variable. For example, the count of white blood cells, number of insects in a colony, and number of shares of stock traded per day are strictly discrete, but for practical purposes, they are viewed as continuous.

**Example 1**     The Various Types of Data

Consider the characteristics

(a)  Height of a skyscraper.

(b)  Favorite primary color.

(c)  Number of visits to a dentist last year.

In each case, determine if the resulting data are categorical, discrete, or continuous.

SOLUTION
(a) Height is continuous. The numerical value is typically recorded to nearest foot, inch, or fraction of inch.

(b) Primary color is categorical. There are three choices: red, green, and blue.

(c) Number of visits is discrete. The numerical value is a count and it is usually low.

A summary description of categorical data is discussed in Section 2.1. The remainder of this chapter is devoted to a descriptive study of measurement data, both discrete and continuous. As in the case of summarization and commentary on a long, wordy document, it is difficult to prescribe concrete steps for summary descriptions that work well for all types of measurement data. However, a few important aspects that deserve special attention are outlined here to provide general guidelines for this process.

---

### Describing a Data Set of Measurements

1. Summarization and description of the overall pattern.
   (a) Presentation of tables and graphs.
   (b) Noting important features of the graphed data including symmetry or departures from it.
   (c) Scanning the graphed data to detect any observations that seem to stick far out from the major mass of the data—the outliers.
2. Computation of numerical measures.
   (a) A typical or representative value that indicates the center of the data.
   (b) The amount of spread or variation present in the data.

---

## 2. DESCRIBING DATA BY TABLES AND GRAPHS

### 2.1 CATEGORICAL DATA

When a qualitative trait is observed for a sample of units, each observation is recorded as a member of one of several categories. Such data are readily organized in the form of a frequency table that shows the counts (**frequencies**) of the individual categories. Our understanding of the data is further enhanced by calculation of the proportion (also called **relative frequency**) of observations in each category.

$$\text{Relative frequency of a category} = \frac{\text{Frequency in the category}}{\text{Total number of observations}}$$

**Example 2**    Calculating Relative Frequencies to Summarize an Opinion Poll

A campus press polled a sample of 280 undergraduate students in order to study student attitude toward a proposed change in the dormitory regulations. Each student was to respond as support, oppose, or neutral in regard to the issue. The numbers were 152 support, 77 neutral, and 51 opposed. Tabulate the results and calculate the relative frequencies for the three response categories.

SOLUTION    Table 1 records the frequencies in the second column, and the relative frequencies are calculated in the third column. The relative frequencies show that about 54% of the polled students supported the change, 18% opposed, and 28% were neutral.

**TABLE 1**    Summary Results of an Opinion Poll

| Responses | Frequency | Relative Frequency |
|-----------|-----------|--------------------|
| Support | 152 | $\dfrac{152}{280} = .543$ |
| Neutral | 77 | $\dfrac{77}{280} = .275$ |
| Oppose | 51 | $\dfrac{51}{280} = .182$ |
| Total | 280 | 1.000 |

**Remark:**  The relative frequencies provide the most relevant information as to the pattern of the data. One should also state the sample size, which serves as an indicator of the credibility of the relative frequencies. (More on this in Chapter 8.)

Categorical data are often presented graphically as a **pie chart** in which the segments of a circle exhibit the relative frequencies of the categories. To obtain the angle for any category, we multiply the relative frequency by 360 degrees, which corresponds to the complete circle. Although laying out the angles by hand can be tedious, many software packages generate the chart with a single command. Figure 1 presents a pie chart for the data in Example 1.

When questions arise that need answering but the decision makers lack precise knowledge of the state of nature or the full ramifications of their decisions, the best procedure is often to collect more data. In the context of quality improvement, if a problem is recognized, the first step is to collect data on the magnitude and possible causes. This information is most effectively communicated through graphical presentations.

A **Pareto diagram** is a powerful graphical technique for displaying events according to their frequency. According to Pareto's empirical law, any collection

Figure 1   Pie chart of student opinion on change in dormitory regulations.

of events consists of only a few that are major in that they are the ones that occur most of the time.

Figure 2 gives a Pareto diagram for the type of defects found in a day's production of facial tissues. The bars correspond to different causes so they do not touch.

Figure 2   Pareto diagram of facial tissue defects.

The cumulative frequency is 22 for the first cause and $22 + 15 = 37$ for the first and second causes combined. This illustrates Pareto's rule, with two of the causes being responsible for 37 out of 50, or 74%, of the defects.

**Example 3**   A Pareto Diagram Clarifies Circumstances Needing Improvement

Graduate students in a counseling course were asked to choose one of their personal habits that needed improvement. In order to reduce the effect of this habit, they were asked to first gather data on the frequency of the occurrence and the circumstances. One student collected the following frequency data on fingernail biting over a two-week period.

| Frequency | Activity |
|-----------|----------|
| 58 | Watching television |
| 21 | Reading newspaper |
| 14 | Talking on phone |
| 7 | Driving a car |
| 3 | Grocery shopping |
| 12 | Other |

Make a Pareto diagram showing the relationship between nail biting and type of activity.

SOLUTION    The cumulative frequencies are 58, 58 + 21 = 79, and so on, out of 115. The Pareto diagram is shown in Figure 3, where watching TV accounts for 50.4% of the instances.

Figure 3    Pareto diagram for nail biting example.

The next step for this person would be to try and find a substitute for nail biting while watching television.

## 2.2 DISCRETE DATA

We next consider summary descriptions of measurement data and begin our discussion with discrete measurement scales. As explained in Section 1, a data set is identified as discrete when the underlying scale is discrete and the distinct values observed are not too numerous.

Similar to our description of categorical data, the information in a discrete data set can be summarized in a frequency table, or frequency distribution that includes a calculation of the relative frequencies. In place of the qualitative categories, we now list the distinct numerical measurements that appear in the data set and then count their frequencies.

**Example 4**    Creating a Frequency Distribution

Retail stores experience their heaviest returns on December 26 and December 27 each year. Most are gifts that, for some reason, did not please the recipient. The number of items returned, by a sample of 30 persons at a large discount department store, are observed and the data of Table 2 are obtained. Determine the frequency distribution.

**TABLE 2**  Number of Items Returned

| | | | | | | | | | |
|---|---|---|---|---|---|---|---|---|---|
| 1 | 4 | 3 | 2 | 3 | 4 | 5 | 1 | 2 | 1 |
| 2 | 5 | 1 | 4 | 2 | 1 | 3 | 2 | 4 | 1 |
| 2 | 3 | 2 | 3 | 2 | 1 | 4 | 3 | 2 | 5 |

SOLUTION    The **frequency distribution** of these data is presented in Table 3. The values are paired with the frequency and the calculated relative frequency.

**TABLE 3**  Frequency Distribution for Number ($x$) of Items Returned

| Value $x$ | Frequency | Relative Frequency |
|:---:|:---:|:---:|
| 1 | 7 | .233 |
| 2 | 9 | .300 |
| 3 | 6 | .200 |
| 4 | 5 | .167 |
| 5 | 3 | .100 |
| Total | 30 | 1.000 |

The frequency distribution of a discrete variable can be presented pictorially by drawing either lines or rectangles to represent the relative frequencies. First, the distinct values of the variable are located on the horizontal axis. For a **line diagram**, we draw a vertical line at each value and make the height of the line equal to the relative frequency. A **histogram** employs vertical rectangles instead of lines. These rectangles are centered at the values and their areas represent relative frequencies. Typically, the values proceed in equal steps so the rectangles are all of the same width and their heights are proportional to the relative frequencies as well as frequencies. Unlike Pareto charts, the bars of a histogram do touch each other.

Figure 4($a$) shows the line diagram and 4($b$) the histogram of the frequency distribution of Table 3.

Figure 4 Graphic display of the frequency distribution of data in Table 3.

## 2.3 DATA ON A CONTINUOUS VARIABLE

What tabular and graphical presentations are appropriate for data sets that contain numerical measurements on a virtually continuous scale? In contrast with the discrete case, a data set of measurements on a continuous variable may contain many distinct values. Then, a table or plot of all distinct values and their frequencies will not provide a condensed or informative summary of the data.

The two main graphical methods used to display a data set of measurements are the **dot diagram** and the **histogram**. Dot diagrams are employed when there are relatively few observations (say, less than 20 or 25); histograms are used with a larger number of observations.

### Dot Diagram

When the data consist of a small set of numbers, they can be graphically represented by drawing a line with a scale covering the range of values of the measurements. Individual measurements are plotted above this line as prominent dots. The resulting diagram is called a **dot diagram**.

**Example 5** A Dot Diagram Reveals an Unusual Observation

The number of days the first six heart transplant patients at Stanford survived after their operations were 15, 3, 46, 623, 126, 64. Make a dot diagram.

SOLUTION These survival times extended from 3 to 623 days. Drawing a line segment from 0 to 700, we can plot the data as shown in Figure 5. This dot diagram shows a cluster of small survival times and a single, rather large value.

Figure 5 Dot diagram for the heart transplant data.

## Paying Attention

First-grade teachers allot a portion of each day to mathematics. An educator, concerned about how students utilize this time, selected 24 students and observed them for a total of 20 minutes spread over several days. The number of minutes, out of 20, that the student was not on task was recorded (courtesy of T. Romberg). These lack-of-attention times are graphically portrayed in the dot diagram in Figure 6. The student with 13 out of 20 minutes off-task stands out enough to merit further consideration. Is this a student who finds the subject too difficult or might it be a very bright child who is bored?

Britt Erlanson/The Image Bank/Getty Images

Paying attention in class. Observations on 24 first-grade students.

Figure 6   Time not concentrating on the mathematics assignment (out of 20 minutes).

## Frequency Distribution on Intervals

When the data consist of a large number of measurements, a dot diagram may be quite tedious to construct. More seriously, overcrowding of the dots causes them to overlap and mar the clarity of the diagram. In such cases, it is convenient to condense the data by grouping the observations according to intervals and recording the frequencies of the intervals. Unlike a discrete frequency distribution, where grouping naturally takes place on points, here we use intervals of values. The main steps in this process are outlined as follows.

---

**Constructing a Frequency Distribution
for a Continuous Variable**

1. Find the minimum and the maximum values in the data set.
2. Choose intervals or cells of equal length that cover the range between the minimum and the maximum without overlapping. These are called class intervals, and their endpoints class boundaries.
3. Count the number of observations in the data that belong to each class interval. The count in each class is the class frequency or cell frequency.
4. Calculate the relative frequency of each class by dividing the class frequency by the total number of observations in the data:

$$\text{Relative frequency} = \frac{\text{Class frequency}}{\text{Total number of observations}}$$

---

The choice of the number and position of the class intervals is primarily a matter of judgment guided by the following considerations. The number of classes usually ranges from 5 to 15, depending on the number of observations in the data. Grouping the observations sacrifices information concerning how the observations are distributed within each cell. With too few cells, the loss of information is serious. On the other hand, if one chooses too many cells and the data set is relatively small, the frequencies from one cell to the next would jump up and down in a chaotic manner and no overall pattern would emerge. As an initial step, frequencies may be determined with a large number of intervals that can later be combined as desired in order to obtain a smooth pattern of the distribution.

Computers conveniently order data from smallest to largest so that the observations in any cell can easily be counted. The construction of a frequency distribution is illustrated in Example 6.

**Example 6**  Creating a Frequency Distribution for Hours of Sleep

Students require different amounts of sleep. A sample of 59 students at a large midwest university reported the following hours of sleep the previous night.

**TABLE 4**   Hours of Sleep for Fifty-Nine Students

| | | | | | | | | | |
|---|---|---|---|---|---|---|---|---|---|
| 4.5 | 4.7 | 5.0 | 5.0 | 5.3 | 5.5 | 5.5 | 5.7 | 5.7 | 5.7 |
| 6.0 | 6.0 | 6.0 | 6.0 | 6.3 | 6.3 | 6.3 | 6.5 | 6.5 | 6.5 |
| 6.7 | 6.7 | 6.7 | 6.7 | 7.0 | 7.0 | 7.0 | 7.0 | 7.3 | 7.3 |
| 7.3 | 7.3 | 7.5 | 7.5 | 7.5 | 7.5 | 7.7 | 7.7 | 7.7 | 7.7 |
| 8.0 | 8.0 | 8.0 | 8.0 | 8.3 | 8.3 | 8.3 | 8.5 | 8.5 | 8.5 |
| 8.5 | 8.7 | 8.7 | 9.0 | 9.0 | 9.0 | 9.3 | 9.3 | 10.0 | |

Construct a frequency distribution of the sleep data.

SOLUTION    To construct a frequency distribution, we first notice that the minimum hours of sleep in Table 4 is 4.5 and the maximum is 10.0. We choose class intervals of length 1.2 hours as a matter of convenience.

The selection of class boundaries is a bit of fussy work. Because the data have one decimal place, we could add a second decimal to avoid the possibility of any observation falling exactly on the boundary. For example, we could end the first class interval at 5.45. Alternatively, and more neatly, we could write 4.3–5.5 and make the **endpoint convention** that the left-hand end point is included but not the right.

The first interval contains 5 observations so its frequency is 5 and its relative frequency is $\frac{5}{59}$ = .085. Table 5 gives the frequency distribution. The relative frequencies add to 1, as they should (up to rounding error) for any frequency distribution. We see, for instance, that just about one-third of the students .271 + .051 = .322 got 7.9 hours or more of sleep.

**TABLE 5**   Frequency Distribution for Hours of Sleep Data (left endpoints included but right endpoints excluded)

| Class Interval | Frequency | Relative Frequency |
|:---:|:---:|:---:|
| 4.3–5.5 | 5 | $\frac{5}{59}$ = .085 |
| 5.5–6.7 | 15 | $\frac{15}{59}$ = .254 |
| 6.7–7.9 | 20 | $\frac{20}{59}$ = .339 |
| 7.9–9.1 | 16 | $\frac{16}{59}$ = .271 |
| 9.1–10.3 | 3 | $\frac{3}{59}$ = .051 |
| Total | 59 | 1.000 |

*Remark:* The rule requiring equal class intervals is inconvenient when the data are spread over a wide range but are highly concentrated in a small part of the range with relatively few numbers elsewhere. Using smaller intervals where the data are highly concentrated and larger intervals where the data are sparse helps to reduce the loss of information due to grouping.

In every application involving an **endpoint convention**, it is important that you clearly state which endpoint is included and which is excluded. This information should be presented in the title or in a footnote of any frequency distribution.

### Histogram

A frequency distribution can be graphically presented as a histogram. To draw a histogram, we first mark the class intervals on the horizontal axis. On each interval, we then draw a vertical rectangle whose **area represents the relative frequency**—that is, the proportion of the observations occurring in that class interval.

To create rectangles whose area is equal to relative frequency, use the rule

$$\text{Height} = \frac{\text{Relative frequency}}{\text{Width of interval}}$$

The total area of all rectangles equals 1, the sum of the relative frequencies.

$$\text{The total area of a histogram is } 1.$$

The histogram for Table 5 is shown in Figure 7. For example, the rectangle drawn on the class interval 4.3–5.5 has area = .071 × 1.2 = .085, which is the relative frequency of this class. Actually, we determined the height .071 as

$$\text{Height} = \frac{\text{Relative frequency}}{\text{Width of interval}} = \frac{.085}{1.2} = .071$$

The units on the vertical axis can be viewed as relative frequencies per unit of the horizontal scale. For instance, .071 is the relative frequency per hour for the interval 4.3–5.5.

Visually, we note that the rectangle having largest area, or most frequent class interval, is 6.7–7.9. Also, proportion .085 + .254 = .339 of the students slept less than 6.7 hours.

*Remark:* When all class intervals have equal widths, the heights of the rectangles are proportional to the relative frequencies that the areas represent. The formal calculation of height, as area divided by the width,

Figure 7   Histogram of the sleep data of Tables 4 and 5.
Sample size = 59.

is then redundant. Instead, one can mark the vertical scale according to the relative frequencies—that is, make the heights of the rectangles equal to the relative frequencies. The resulting picture also makes the areas represent the relative frequencies if we read the vertical scale as if it is in units of the class interval. This leeway when plotting the histogram is not permitted in the case of unequal class intervals.

Figure 8 shows one ingenious way of displaying two histograms for comparison. In spite of their complicated shapes, their back-to-back plot as a "pyramid" allows

Figure 8   Population pyramid (histograms) of the male and female age distributions in the United States in 2010. (*Source:* U.S. Bureau of the Census.)

for easy visual comparison. Females are the clear majority in the last age groups of the male and female age distributions.

## Stem-and-Leaf Display

A **stem-and-leaf display** provides a more efficient variant of the histogram for displaying data, especially when the observations are two-digit numbers. This plot is obtained by sorting the observations into rows according to their leading digit. To make a stem-and-leaf display:

1. List the digits 0 through 9 in a column and draw a vertical line. These correspond to the leading digit.
2. For each observation, record its second digit to the right of this vertical line in the row where the first digit appears.
3. Finally, arrange the second digits in each row so they are in increasing order.

The stem-and-leaf display for the data of Table 6 is shown in Table 7.

**TABLE 6**   Examination Scores of 50 Students

| | | | | | | | | |
|---|---|---|---|---|---|---|---|---|
| 75 | 98 | 42 | 75 | 84 | 87 | 65 | 59 | 63 |
| 86 | 78 | 37 | 99 | 66 | 90 | 79 | 80 | 89 |
| 68 | 57 | 95 | 55 | 79 | 88 | 76 | 60 | 77 |
| 49 | 92 | 83 | 71 | 78 | 53 | 81 | 77 | 58 |
| 93 | 85 | 70 | 62 | 80 | 74 | 69 | 90 | 62 |
| 84 | 64 | 73 | 48 | 72 | | | | |

**TABLE 7**   Stem-and-Leaf Display for the Examination Scores

| | |
|---|---|
| 0 | |
| 1 | |
| 2 | |
| 3 | 7 |
| 4 | 289 |
| 5 | 35789 |
| 6 | 022345689 |
| 7 | 01234556778899 |
| 8 | 00134456789 |
| 9 | 0023589 |

In the stem-and-leaf display, the column of first digits to the left of the vertical line is viewed as the stem, and the second digits as the leaves. Viewed sidewise, it looks like a histogram with a cell width equal to 10. However, it is more informative than a histogram because the actual data points are retained. In fact, every observation can be recovered exactly from this stem-and-leaf display.

A stem-and-leaf display retains all the information in the leading digits of the data. When the leaf unit $= .01$, $3.5 \mid 0\ 2\ 3\ 7\ 8$ presents the data 3.50, 3.52,

3.53, 3.57, and 3.58. Leaves may also be two-digit at times. When the first leaf digit = .01, .4 | 07 13 82 90 presents the data .407, .413, .482, and .490.

Further variants of the stem-and-leaf display are described in Exercises 2.25 and 2.26. This versatile display is one of the most applicable techniques of exploratory data analysis.

When the sample size is small or moderate, no information is lost with the stem-and-leaf diagram because you can see every data point. The major disadvantage is that, when the sample size is large, diagrams with hundreds of numbers in a row cannot be constructed in a legible manner.

## Exercises

**2.1** Cities must find better ways to dispose of solid waste. According to the Environmental Protection Agency, the composition of the 254 million tons of solid municipal waste created in 2010 was

| | |
|---|---|
| Paper and paperboard | 28.5% |
| Food waste | 13.9% |
| Yard waste | 13.4% |
| Plastics | 12.4% |
| Metals | 9.0% |
| Other materials | |

(a) Determine the percentage of other materials in the solid waste. This category includes glass, wood, rubber, and so on.

(b) Create a Pareto chart.

(c) What percentage of the total solid waste is paper or paperboard? What percentage is from the top two categories? What percentage is from the top five categories?

**2.2** Recorded here are the blood types of 40 persons who have volunteered to donate blood at a plasma center. Summarize the data in a frequency table. Include calculations of the relative frequencies.

| | | | | | | | | | |
|---|---|---|---|---|---|---|---|---|---|
| O | O | A | B | A | O | A | A | A | O |
| B | O | B | O | O | A | O | O | A | A |
| A | A | AB | A | B | A | A | O | O | A |
| O | O | A | A | A | O | A | O | O | AB |

**2.3** A student at the University of Wisconsin surveyed 40 students in her dorm concerning their participation in extracurricular activities during the past week. The data on number of activities are

| | | | | | | | | | | | |
|---|---|---|---|---|---|---|---|---|---|---|---|
| 1 | 5 | 0 | 1 | 4 | 3 | 0 | 2 | 1 | 6 | 1 | 1 | 0 | 0 |
| 2 | 0 | 0 | 3 | 1 | 2 | 1 | 2 | 2 | 2 | 2 | 2 | 1 | 0 |
| 2 | 2 | 3 | 4 | 2 | 7 | 2 | 2 | 3 | 3 | 1 | 1 | | |

Present these data in a frequency table and in a relative frequency bar chart.

**2.4** The number of automobile accidents reported per month helps to identify intersections that require improvement. The number of crashes per month reported at an intersection near a university campus in Madison, Wisconsin, are

| | | | | | | | | | | |
|---|---|---|---|---|---|---|---|---|---|---|
| 1 | 3 | 3 | 3 | 2 | 2 | 3 | 1 | 2 | 4 | 1 | 4 |
| 1 | 3 | 1 | 1 | 1 | 0 | 1 | 2 | 2 | 5 | 5 | 2 |
| 5 | 5 | 4 | 3 | 3 | 6 | 1 | 2 | 3 | 2 | 4 | 3 |
| 4 | 4 | 3 | 5 | 3 | 3 | 3 | 5 | 1 | 5 | 5 | 3 |
| 4 | 2 | 2 | 0 | 0 | 1 | 4 | 1 | 0 | 2 | 0 | |

Present these data in a frequency table and in a relative frequency bar chart.

**2.5** The following table shows how workers in one department get to work.

| Mode of Transportation | Frequency |
|---|---|
| Drive alone | 25 |
| Car pool | 3 |
| Ride bus | 7 |
| Other | 5 |

(a) Calculate the relative frequency of each mode of transportation.

(b) Construct a pie chart.

**2.6** Of the $207 million raised by a major university's fund drive, $117 million came from individuals and bequests, $24 million from industry and business, and $66 million from foundations and associations. Present this information in the form of a pie chart.

**2.7** Data from one campus dorm on the number of burglaries are collected each week of the semester. These data are to be grouped into the classes 0–1, 2–3, 3–5, 6 or more. Both endpoints included. Explain where a difficulty might arise.

**2.8** Data from one campus dorm, on the number of complaints about the dorm food are collected each week of the semester. These weekly counts are to be grouped into the classes 0–1, 2–3, 4–5, 7 or more. Both endpoints are included. Explain where a difficulty might arise.

**2.9** A sample of persons will each be asked to give the number of their close friends. The responses are to be grouped into the following classes: 0, 1–3, 3–5, 6 or more. Left endpoint is included. Explain where difficulties might arise.

**2.10** The weights of the players on the university football team (to the nearest pound) are to be grouped into the following classes: 160–175, 175–190, 190–205, 205–220, 220–235, 235 or more. The left endpoint is included but not the right endpoint. Explain where difficulties might arise.

**2.11** On flights from San Francisco to Chicago, the number of empty seats are to be grouped into the following classes: 0–4, 5–9, 10–14, 15–19, more than 19.

Is it possible to determine from this frequency distribution the exact number of flights on which there were:

(a) Fewer than 10 empty seats?

(b) More than 14 empty seats?

(c) At least 5 empty seats?

(d) Exactly 9 empty seats?

(e) Between 5 and 15 empty seats inclusively?

**2.12** A major West Coast power company surveyed 50 customers who were asked to respond to the statement, "People should rely mainly on themselves to solve problems caused by power outages" with one of the following responses.

1. Definitely agree.
2. Somewhat agree.
3. Somewhat disagree.
4. Definitely disagree.

The responses are as follows:

```
4 2 1 3 3 2 4 2 1 1 2 2 2 2 1 3 4
1 4 4 1 3 2 4 1 4 3 3 1 1 1 2 1 1
4 4 4 4 4 1 2 2 2 4 4 4 1 3 4 2
```

Construct a frequency table.

**2.13** A sample of 50 departing airline passengers at the main check-in counter produced the following number of bags checked through to final destinations.

```
0 1 2 2 1 2 1 2 3 0 1 0
1 1 0 1 3 0 1 2 1 1 1 2
1 2 2 1 2 0 0 2 2 1 1 1
1 1 1 1 2 0 1 3 0 1 2 1
1 3
```

(a) Make a relative frequency line diagram.

(b) Comment on the pattern.

(c) What proportion of passengers who check in at the main counter fail to check any bags?

**2.14** A person with asthma took measurements by blowing into a peak-flow meter on seven consecutive days.

```
429   425   471   422   432   444   454
```

Display the data in a dot diagram.

**2.15** Before microwave ovens are sold, the manufacturer must check to ensure that the radiation coming through the door is below a specified safe limit. The amounts of radiation leakage ($mW/cm^2$)

with the door closed from 25 ovens are as follows
(courtesy of John Cryer):

| | | | | | | |
|---|---|---|---|---|---|---|
| 15 | 9 | 18 | 10 | 5 | 12 | 8 |
| 5 | 8 | 10 | 7 | 2 | 1 | |
| 5 | 3 | 5 | 15 | 10 | 15 | |
| 9 | 8 | 18 | 1 | 2 | 11 | |

Display the data in a dot diagram.

**2.16** A campus area merchant recorded the number of
bad checks received per month, for five months

$$4 \quad 5 \quad 4 \quad 7 \quad 6$$

Display the data in a dot diagram.

**2.17** The city of Madison regularly checks the water
quality at swimming beaches located on area lakes.
The concentration of fecal coliforms, in number of
colony forming units (CFU) per 100 ml of water, was
measured on fifteen days during the summer at one
beach.

| | | | | | | | |
|---|---|---|---|---|---|---|---|
| 180 | 1600 | 90 | 140 | 50 | 260 | 400 | 90 |
| 380 | 110 | 10 | 60 | 20 | 340 | 80 | |

(a) Make a dot diagram.

(b) Comment on the pattern and any unusual features.

(c) The city closes any swimming beach if a count is
over 1350. What proportion of days, among the
fifteen, was this beach closed?

**2.18** Tornadoes kill many people every year in the
United States. The yearly number of lives lost during
the 50 years 1962 through 2011 are summarized in
the following table.

| Number of Deaths | Frequency |
|---|---|
| 24 or less | 3 |
| 25–49 | 18 |
| 50–74 | 14 |
| 75–99 | 6 |
| 100–149 | 5 |
| 150–199 | 1 |
| 200 or more | 3 |
| Total | 50 |

(a) Calculate the relative frequency for the intervals
[ 0, 25 ), [ 25, 50 ) and so on where the right-
hand endpoint is excluded. Take the last interval
to be [ 200, 600 ).

(b) Plot the relative frequency histogram. (*Hint*:
Since the intervals have unequal widths, make
the height of each rectangle equal to the relative frequency divided by the width of the
interval.)

(c) What proportion of the years had 49 or fewer
deaths due to tornadoes?

(d) Comment on the shape of the distribution.

**2.19** A zoologist collected wild lizards in the
Southwestern United States. Thirty lizards from
the genus *Phrynosoma* were placed on a treadmill
and their speed measured. The recorded speed
(meters/second) is the fastest time to run a half
meter (courtesy of K. Bonine).

| | | | | | | | |
|---|---|---|---|---|---|---|---|
| 1.28 | 1.36 | 1.24 | 2.47 | 1.94 | 2.52 | 2.67 | 1.29 |
| 1.56 | 2.66 | 2.17 | 1.57 | 2.10 | 2.54 | 1.63 | 2.11 |
| 2.57 | 1.72 | 0.76 | 1.02 | 1.78 | 0.50 | 1.49 | 1.57 |
| 1.04 | 1.92 | 1.55 | 1.78 | 1.70 | 1.20 | | |

(a) Construct a frequency distribution using the
class intervals 0.45–0.90, 0.90–1.35, and
so on, with the endpoint convention that
the left endpoint is included and the right
endpoint is excluded. Calculate the relative
frequencies.

(b) Make a histogram.

**2.20** The United States Geological Survey maintains
data on large earthquakes including those of magnitude greater than 6.0 in California. Through 2010,
the ordered magnitudes of the 57 quakes are

| | | | | | | | | | | |
|---|---|---|---|---|---|---|---|---|---|---|
| 6.1 | 6.1 | 6.1 | 6.1 | 6.1 | 6.2 | 6.2 | 6.2 | 6.2 | 6.3 | 6.3 |
| 6.3 | 6.4 | 6.4 | 6.4 | 6.4 | 6.4 | 6.4 | 6.4 | 6.5 | 6.5 | 6.5 |
| 6.5 | 6.5 | 6.5 | 6.6 | 6.6 | 6.6 | 6.7 | 6.7 | 6.7 | 6.8 | 6.8 |
| 6.8 | 6.8 | 6.8 | 6.9 | 6.9 | 7.0 | 7.0 | 7.0 | 7.1 | 7.1 | 7.1 |
| 7.2 | 7.2 | 7.2 | 7.2 | 7.2 | 7.3 | 7.3 | 7.3 | 7.3 | 7.4 | 7.8 |
| 7.8 | 7.9 | | | | | | | | | |

Construct a histogram using equal-length intervals
starting with ( 6.0, 6.3 ] where the right-hand endpoint is included but not the left-hand endpoint.

**2.21** Referring to Exercise 2.20, construct a density histogram using the intervals ( 6.0, 6.3 ], ( 6.3, 6.6 ], ( 6.6, 6.9 ], ( 6.9, 7.2 ], and ( 7.2, 7.9 ].

**2.22** The following data represent the scores of 40 students on a college qualification test (courtesy of R. W. Johnson).

```
162  171  138  145  144  126  145  162  174  178
167   98  161  152  182  136  165  137  133  143
184  166  115  115   95  190  119  144  176  135
194  147  160  158  178  162  131  106  157  154
```

Make a stem-and-leaf display.

**2.23** A federal government study of the oil reserves in Elk Hills, CA, included a study of the amount of iron present in the oil.

| Amount of Iron (percent ash) | | | | |
|----|----|----|----|----|
| 20 | 18 | 25 | 26 | 17 |
| 14 | 20 | 14 | 18 | 15 |
| 22 | 15 | 17 | 25 | 22 |
| 12 | 52 | 27 | 24 | 41 |
| 34 | 20 | 17 | 20 | 19 |
| 20 | 16 | 20 | 15 | 34 |
| 22 | 29 | 29 | 34 | 27 |
| 13 |  6 | 24 | 47 | 32 |
| 12 | 17 | 36 | 35 | 41 |
| 36 | 32 | 46 | 30 | 51 |

Make a stem-and-leaf display.

**2.24** The following is a stem-and-leaf display with two-digit leaves. (The leading leaf digit = 10.0.)

```
1 |
2 | 46   68   93
3 | 19   44   71   82   97
4 | 05   26   43   90
5 | 04   68
6 | 13
```

List the corresponding measurements.

**2.25** If there are too many leaves on some stems in a stem-and-leaf display, we might double the number of stems. The leaves 0–4 could hang on one stem and 5–9 on the repeated stem. For the observations

```
193  198  200  202  203  203  205  205  206  207
207  208  212  213  214  217  219  220  222  226  237
```

we would get the **double-stem display**

```
19 | 3
19 | 8
20 | 0233
20 | 556778
21 | 234
21 | 79
22 | 02
22 | 6
23 |
23 | 7
```

Construct a double-stem display with one-digit leaves for the data of Exercise 2.23.

**2.26** If the double-stem display still has too few stems, we may wish to construct a stem-and-leaf display with a separate stem to hold leaves 0 and 1, 2 and 3, 4 and 5, 6 and 7, and a stem to hold 8 and 9. The resulting stem-and-leaf display is called a **five-stem display**. The following is a five-digit stem-and-leaf display. (Leaf unit = 1.0)

```
1 | 8
2 | 001
2 | 2233
2 | 444555
2 | 667
2 | 9
3 | 0
```

List the corresponding measurements.

**2.27** The following table lists values of the Consumer Price Index for 24 selected areas both for 2007 and 2011. Construct a five-stem display for the consumer price index in 2011.

|              | 2007 | 2011 |
| ------------ | ---- | ---- |
| Anchorage    | 181  | 202  |
| Atlanta      | 198  | 209  |
| Boston       | 227  | 244  |
| Chicago      | 198  | 219  |
| Cincinnati   | 188  | 211  |
| Cleveland    | 186  | 211  |
| Dallas       | 195  | 208  |
| Denver       | 194  | 220  |
| Detroit      | 195  | 212  |
| Honolulu     | 219  | 244  |
| Houston      | 182  | 200  |
| Kansas City  | 186  | 214  |

|               | 2007 | 2011 |
| ------------- | ---- | ---- |
| Los Angeles   | 210  | 232  |
| Miami         | 210  | 231  |
| Milwaukee     | 198  | 217  |
| Minneapolis   | 195  | 219  |
| New York      | 221  | 248  |
| Philadelphia  | 216  | 233  |
| Pittsburgh    | 194  | 225  |
| Portland      | 203  | 225  |
| St. Louis     | 192  | 210  |
| San Diego     | 218  | 253  |
| San Francisco | 211  | 233  |
| Seattle       | 210  | 233  |

# 3.  MEASURES OF CENTER

The graphic procedures described in Section 2 help us to visualize the pattern of a data set of measurements. To obtain a more objective summary description and a comparison of data sets, we must go one step further and obtain numerical values for the location or center of the data and the amount of variability present. Because data are normally obtained by sampling from a large population, our discussion of numerical measures is restricted to data arising in this context. Moreover, when the population is finite and completely sampled, the same arithmetic operations can be carried out to obtain numerical measures for the population.

To effectively present the ideas and associated calculations, it is convenient to represent a data set by symbols to prevent the discussion from becoming anchored to a specific set of numbers. A data set consists of a number of measurements that are symbolically represented by $x_1, x_2, \ldots, x_n$. The last subscript $n$ denotes the number of measurements in the data, and $x_1, x_2, \ldots$ represent the first observation, the second observation, and so on. For instance, a data set consisting of the five measurements 2.1, 3.2, 4.1, 5.6, and 3.7 is represented in symbols by $x_1, x_2, x_3, x_4, x_5$, where $x_1 = 2.1$, $x_2 = 3.2$, $x_3 = 4.1$, $x_4 = 5.6$, and $x_5 = 3.7$.

The most important aspect of studying the distribution of a sample of measurements is locating the position of a central value about which the measurements are distributed. The two most commonly used indicators of center are the **mean** and the **median**.

The **mean,** or **average**, of a set of measurements is the sum of the measurements divided by their number. For instance, the mean of the five measurements

2.1, 3.2, 4.1, 5.6, and 3.7 is

$$\frac{2.1 + 3.2 + 4.1 + 5.6 + 3.7}{5} = \frac{18.7}{5} = 3.74$$

To state this idea in general terms, we use symbols. If a sample consists of $n$ measurements $x_1, x_2, \ldots, x_n$, the mean of the sample is

$$\frac{x_1 + x_2 + \cdots + x_n}{n} = \frac{\text{sum of the } n \text{ measurements}}{n}$$

The notation $\bar{x}$, read $x$ bar, represents a sample mean. To further simplify the writing of a sum, the Greek capital letter $\sum$ (sigma) is used as a statistical shorthand. With this symbol:

> The sum $x_1 + x_2 + \cdots + x_n$ is denoted as $\sum_{i=1}^{n} x_i$.
>
> Read this as "the sum of all $x_i$ with $i$ ranging from 1 to $n$."

For example, $\sum_{i=1}^{5} x_i$ represents the sum $x_1 + x_2 + x_3 + x_4 + x_5$.

**Remark:**   When the number of terms being summed is understood from the context, we often simplify to $\sum x_i$, instead of $\sum_{i=1}^{n} x_i$. Some further operations with the $\sum$ notation are discussed in Appendix A1.
We are now ready to formally define the sample mean.

> The **sample mean** of a set of $n$ measurements $x_1, x_2, \ldots, x_n$ is the sum of these measurements divided by $n$. The sample mean is denoted by $\bar{x}$.
>
> $$\bar{x} = \frac{\sum_{i=1}^{n} x_i}{n} \quad \text{or} \quad \frac{\sum x_i}{n}$$

According to the concept of "average," the mean represents a center of a data set. If we picture the dot diagram of a data set as a thin weightless horizontal bar on which balls of equal size and weight are placed at the positions of the data points, then the mean $\bar{x}$ represents the point on which the bar will balance. The computation of the sample mean and its physical interpretation are illustrated in Example 7.

**Example 7**    Calculating and Interpreting the Sample Mean

The birth weights in pounds of five babies born one day in the same hospital are 9.2, 6.4, 10.5, 8.1, and 7.8. Obtain the sample mean and create a dot diagram.

SOLUTION    The mean birth weight for these data is

$$\bar{x} = \frac{9.2 + 6.4 + 10.5 + 8.1 + 7.8}{5} = \frac{42.0}{5} = 8.4 \text{ pounds}$$

The dot diagram of the data appears in Figure 9, where the sample mean (marked by $\triangle$) is the balancing point or center of the picture.

Figure 9    Dot diagram and the sample mean for the birth-weight data.

Another measure of center is the middle value.

> The sample median of a set of $n$ measurements $x_1, \ldots, x_n$ is the middle value when the measurements are arranged from smallest to largest.

Roughly speaking, the median is the value that divides the data into two equal halves. In other words, 50% of the data lie below the median and 50% above it. If $n$ is an odd number, there is a unique middle value and it is the median. If $n$ is an even number, there are two middle values and the median is defined as their average. For instance, the ordered data 3, 5, 7, 8 have two middle values 5 and 7, so the median = $(5 + 7)/2 = 6$.

**Example 8**    Calculating the Sample Median

Find the median of the birth-weight data given in Example 7.

SOLUTION    The measurements, ordered from smallest to largest, are

$$6.4 \quad 7.8 \quad \boxed{8.1} \quad 9.2 \quad 10.5$$

The middle value is 8.1, and the median is therefore 8.1 pounds.

**Example 9**    Choosing between the Mean and Median

Calculate the median of the survival times given in Example 5. Also calculate the mean and compare.

SOLUTION    To find the median, first we order the data. The ordered values are

$$3 \quad 15 \quad 46 \quad 64 \quad 126 \quad 623$$

There are two middle values, so

$$\text{Median} = \frac{46 + 64}{2} = 55 \text{ days}$$

The sample mean is

$$\bar{x} = \frac{3 + 15 + 46 + 64 + 126 + 623}{6} = \frac{877}{6} = 146.2 \text{ days}$$

Note that one large survival time greatly inflates the mean. Only 1 out of the 6 patients survived longer than $\bar{x} = 146.2$ days. Here the median of 55 days appears to be a better indicator of the center than the mean.

Example 9 demonstrates that the median is not affected by a few very small or very large observations, whereas the presence of such extremes can have a considerable effect on the mean. For extremely asymmetrical distributions, the median is likely to be a more sensible measure of center than the mean. That is why government reports on income distribution quote the median income as a summary, rather than the mean. A relatively small number of very highly paid persons can have a great effect on the mean salary.

If the number of observations is quite large ( greater than, say, 25 or 30 ), it is sometimes useful to extend the notion of the median and divide the ordered data set into quarters. Just as the point for division into halves is called the median, the points for division into quarters are called **quartiles**. The points of division into more general fractions are called **percentiles**.

> The sample 100 $p$-th percentile is a value such that after the data are ordered from smallest to largest, at least 100$p$% of the observations are at or below this value and at least 100 $( 1 - p )$% are at or above this value.

If we take $p = .5$, the above conceptual description of the sample $100(.5) = 50$th percentile specifies that at least half the observations are equal or smaller and at least half are equal or larger. If we take $p = .25$, the sample $100(.25) = 25$th percentile has proportion one-fourth of the observations that are the same or smaller and proportion three-fourths that are the same or larger.

We adopt the convention of taking an observed value for the sample percentile except when two adjacent values satisfy the definition, in which case their average is taken as the percentile. This coincides with the way the median is defined when the sample size is even. When all values in an interval satisfy the definition of a percentile, the particular convention used to locate a point in the interval does not appreciably alter the results in large data sets, except perhaps for the determination of extreme percentiles (those before the 5th or after the 95th percentile).

The following operating rule will simplify the calculation of the sample percentile.

## Calculating the Sample 100p-th Percentile

1. Order the data from smallest to largest.
2. Determine the product ( *sample size* ) × ( *proportion* ) = $np$.

If $np$ is not an integer, round it up to the next integer and find the corresponding ordered value.

If $np$ is an integer, say $k$, calculate the average of the $k$th and $(k + 1)$st ordered values.

The quartiles are simply the 25th, 50th, and 75th percentiles.

## Sample Quartiles

Lower (first) quartile          $Q_1$ = 25th percentile
Second quartile (or median)     $Q_2$ = 50th percentile
Upper (third) quartile          $Q_3$ = 75th percentile

**Example 10**  Calculating Quartiles to Summarize Length of Phone Calls

An administrator wanted to study the utilization of long-distance telephone service by a department. One variable of interest is the length, in minutes, of long-distance calls made during one month. There were 38 calls that resulted in a connection. The lengths of calls, already ordered from smallest to largest, are presented in Table 8. Locate the quartiles and also determine the 90th percentile.

**TABLE 8**  The Lengths of Long-Distance Phone Calls in Minutes

| | | | | | | | | | |
|---|---|---|---|---|---|---|---|---|---|
| 1.6 | 1.7 | 1.8 | 1.8 | 1.9 | 2.1 | 2.5 | 3.0 | 3.0 | 4.4 |
| 4.5 | 4.5 | 5.9 | 7.1 | 7.4 | 7.5 | 7.7 | 8.6 | 9.3 | 9.5 |
| 12.7 | 15.3 | 15.5 | 15.9 | 15.9 | 16.1 | 16.5 | 17.3 | 17.5 | 19.0 |
| 19.4 | 22.5 | 23.5 | 24.0 | 31.7 | 32.8 | 43.5 | 53.3 | | |

SOLUTION  To determine the first quartile, we take $p = .25$ and calculate the product $38 \times .25 = 9.5$. Because 9.5 is not an integer, we take the next largest integer, 10. In Table 8, we see that the 10th ordered observation is 4.4 so the first quartile is $Q_1 = 4.4$ minutes.

We confirm that this observation has 10 values *at or below* it and 29 values *at or above* so that it does satisfy the conceptual definition of the first quartile.

For the median, we take $p = .5$ and calculate $38 \times .5 = 19$. Because this is an integer, we average the 19th and 20th smallest observations to obtain the median, $(9.3 + 9.5)/2 = 9.4$ minutes.

Next, to determine the third quartile, we take $p = .75$ and calculate $38 \times .75 = 28.5$. The next largest integer is 29, so the 29th ordered observation is the third quartile $Q_3 = 17.5$ minutes. More simply, we could mimic the calculation of the first quartile but now count down 10 observations starting with the largest value.

For the 90th percentile, we determine $38 \times .90 = 34.2$, which we increase to 35. The 90th percentile is 31.7 minutes. Only 10% of calls last 31.7 minutes or longer.

## Exercises

**2.28** Calculate the mean and median for each of the following data sets.

(a) 3   7   4   11   5

(b) 4   2   8   4   2

**2.29** Calculate the mean and median for each of the following data sets.

(a) 2   5   1   4   3

(b) 26   30   38   32   26   31

(c) −3   0   −2   −1   2   −3   0

**2.30** The height that bread rises may be one indicator of how light it will be. As a first step, before modifying her existing recipe, a student cook measured the raise height ( cm ) on eight occasions:

6.3   6.9   5.7   5.4   5.6   5.5   6.6   6.5

Find the mean and median of the raised heights.

**2.31** With reference to the water quality in Exercise 2.17:

(a) Find the sample mean.

(b) Does the sample mean or the median give a better indication of the water quality of a "typical" day? Why?

**2.32** The monthly income in dollars for seven sales persons at a car dealership are

2450   2275   2425   4700   2650   2350   2475

(a) Calculate the mean and median salary.

(b) Which of the two is preferable as a measure of center and why?

**2.33** Records show that in Las Vegas, NV, the normal daily maximum temperature (°F) for each month starting in January is

56   62   68   77   87   99   105   102   95   82   66   57

Verify that the mean of these figures is 79.67. Comment on the claim that the daily maximum temperature in Las Vegas averages a pleasant 79.67.

**2.34** A sample of six university students responded to the question, "How much time, in minutes, did you spend on the social network site yesterday?" One student never used the site and was dropped from the study.

100   45   60   130   30

Calculate the sample mean time for the five students that use the site.

**2.35** With reference to the radiation leakage data given in Exercise 2.15:

(a) Calculate the sample mean.

(b) Which gives a better indication of the amount of radiation leakage, the sample mean or the median?

**2.36** In 2012 there were nine skyscrapers in the world over 300 meters tall. The ordered values of height are

366   381   442   452   484   492   508   601   828

The tallest is in Dubai.

(a) Calculate the sample mean height for the nine skyscrapers.

(b) Drop the largest value and recalculate the mean. Comment on the effect of dropping one very large observation.

**2.37** The weights ( oz ) of nineteen babies born in Madison, Wisconsin, are summarized in the computer output

**Descriptive Statistics: Weight**

| Variable | N | Mean | Median | StDev |
|----------|----|--------|--------|-------|
| Weight | 19 | 118.05 | 117.00 | 15.47 |

Locate two measures of center tendency, or location, and interpret the values.

**2.38** With reference to the extracurricular activities data in Exercise 2.3, obtain the

(a) Sample mean.

(b) Sample median.

(c) Comment on the effect of a large observation.

**2.39** With reference to the number of returns in Example 4, obtain the sample (a) mean and (b) median.

**2.40** Old Faithful, the most famous geyser in Yellowstone Park, had the following durations (measured in seconds) in six consecutive eruptions:

240    248    113    268    117    253

(a) Find the sample median.

(b) Find the sample mean.

**2.41** Loss of calcium is a serious problem for older women. To investigate the amount of loss, a researcher measured the initial amount of bone mineral content in the radius bone of the dominant hand of elderly women and then the amount remaining after one year. The differences, representing the loss of bone mineral content, are given in the following table (courtesy of E. Smith).

| | | | | |
|----|----|----|----|----|
| 8 | 7 | 13 | 3 | 6 |
| 4 | 8 | 6 | 3 | 4 |
| 0 | 1 | 11 | 7 | 1 |
| 8 | 6 | 12 | 13 | 10 |
| 9 | 11 | 3 | 2 | 9 |
| 7 | 1 | 16 | 3 | 2 |
| 10 | 15 | 2 | 5 | 8 |
| 17 | 8 | 2 | 5 | 5 |

(a) Find the sample mean.

(b) Does the sample mean or the median give a better indication of the amount of mineral loss?

**2.42** Physical education researchers interested in the development of the overarm throw measured the horizontal velocity of a thrown ball at the time of release. The results for first-grade children (in feet/second) (courtesy of L. Halverson and M. Roberton) are

Males

54.2  39.6  52.3  48.4  35.9  30.4  25.2  45.4  48.9
48.9  45.8  44.0  52.5  48.3  59.9  51.7  38.6  39.1
49.9  38.3

Females

30.3  43.0  25.7  26.7  27.3  31.9  53.7  32.9  19.4
23.7  23.3  23.3  37.8  39.5  33.5  30.4  28.5

(a) Find the sample median for males.

(b) Find the sample median for females.

(c) Find the sample median for the combined set of males and females.

**2.43**  On opening day one season, 10 major league base-ball games were played and they lasted the following numbers of minutes.

167  211  187  176  170  158  198  218  145  232

Find the sample median.

**2.44**  If you were to use the data on the length of major league baseball games in Exercise 2.43 to estimate the total amount of digital memory needed to film another 10 major league baseball games, which is the more meaningful description, the sample mean or the sample median? Explain.

**2.45**  The following measurements of the diameters (in feet) of Indian mounds in southern Wisconsin were gathered by examining reports in the *Wisconsin Archaeologist* (courtesy of J. Williams).

22  24  24  30  22  20  28  30  24  34  36  15  37

(a)  Create a dot diagram.

(b)  Calculate the mean and median and then mark these on the dot diagram.

(c)  Calculate the quartiles.

**2.46**  With reference to Exercise 2.3, calculate the quartiles.

**2.47**  Refer to the data of college qualification test scores given in Exercise 2.22.

(a)  Find the median.

(b)  Find $Q_1$ and $Q_3$.

**2.48**  A large mail-order firm employs numerous persons to take phone orders. Computers on which orders are entered also automatically collect data on phone activity. One variable useful for planning staffing levels is the number of calls per shift handled by each employee. From the data collected on 25 workers, calls per shift were (courtesy of Land's End)

| | | | | | | | | |
|---|---|---|---|---|---|---|---|---|
| 118 | 118 | 57 | 92 | 127 | 109 | 96 | 68 | 73 |
| 69 | 106 | 91 | 93 | 94 | 102 | 105 | 100 | 104 |
| 80 | 50 | 96 | 82 | 72 | 108 | 73 | | |

Calculate the sample mean.

**2.49**  With reference to Exercise 2.48, calculate the quartiles.

**2.50**  The speedy lizard data, from Exercise 2.19, are

| | | | | | | | |
|---|---|---|---|---|---|---|---|
| 1.28 | 1.36 | 1.24 | 2.47 | 1.94 | 2.52 | 2.67 | 1.29 |
| 1.56 | 2.66 | 2.17 | 1.57 | 2.10 | 2.54 | 1.63 | 2.11 |
| 2.57 | 1.72 | 0.76 | 1.02 | 1.78 | 0.50 | 1.49 | 1.57 |
| 1.04 | 1.92 | 1.55 | 1.78 | 1.70 | 1.20 | | |

(a)  Find the sample median, first quartile, and third quartile.

(b)  Find the sample 90th percentile.

**2.51**  With reference to the water quality data in Exercise 2.17:

(a)  Find the sample median, first quartile, and third quartile.

(b)  Find the sample 90th percentile.

**2.52**  *Some properties of the mean and median.*

1.  If a fixed number $c$ is added to all measurements in a data set, then the mean of the new measurements is

    $c +$ ( the original mean ).

2.  If all measurements in a data set are multiplied by a fixed number $d$, then the mean of the new measurements is

    $d \times$ ( the original mean ).

(a)  Verify these properties for the data set

    4    8    8    7    9    6

    taking $c = 4$ in property ( 1 ) and $d = 2$ in ( 2 ).

(b)  The same properties also hold for the median. Verify these for the data set and the numbers $c$ and $d$ given in part (a).

**2.53**  On a day, the noon temperature measurements (in °F) reported by five weather stations in a state were

    74    80    76    76    73

(a)  Find the mean and median temperature in °F.

(b)  The Celsius (°C) scale is related to the Farenheit (°F) scale by $C = \frac{5}{9}( F - 32 )$. What are the mean and median temperatures in °C? (Answer without converting each temperature measurement to °C. Use the properties stated in Exercise 2.52.)

**2.54** Given here are the mean and median salaries of machinists employed by two competing companies *A* and *B*.

|  | Company | |
|---|---|---|
|  | *A* | *B* |
| Mean salary | $80,000 | $75,500 |
| Median salary | $66,000 | $69,000 |

Assume that the salaries are set in accordance with job competence and the overall quality of workers is about the same in the two companies.

(a) Which company offers a better prospect to a machinist having superior ability? Explain your answer.

(b) Where can a medium-quality machinist expect to earn more? Explain your answer.

**2.55** Refer to the alligator data in Table D.11 of the Data Bank. Using the data on testosterone $x_4$ for male alligators:

(a) Make separate dot plots for the Lake Apopka and Lake Woodruff alligators.

(b) Calculate the sample means for each group.

(c) Do the concentrations of testosterone appear to differ between the two groups? What does this suggest the contamination has done to male alligators in the Lake Apopka habitat?

**2.56** Refer to the alligator data in Table D.11 of the Data Bank. Using the data on testosterone $x_4$ from Lake Apopka:

(a) Make separate dot plots for the male and female alligators.

(b) Calculate the sample means for each group.

(c) Do the concentrations of testosterone appear to differ between the two groups? We would expect differences. What does your graph suggest the contamination has done to alligators in the Lake Apopka habitat?

# 4. MEASURES OF VARIATION

Besides locating the center of the data, any descriptive study of data must numerically measure the extent of variation around the center. Two data sets may exhibit similar positions of center but may be remarkably different with respect to variability. For example, the dots in Figure 10*b* are more scattered than those in Figure 10*a*.

Figure 10    Dot diagrams with similar center values but different amounts of variation.

Because the sample mean $\bar{x}$ is a measure of center, the variation of the individual data points about this center is reflected in their deviation from the mean

$$\text{Deviation} = \text{Observation} - (\text{Sample mean})$$

$$= x - \bar{x}$$

For instance, the data set 3, 5, 7, 7, 8 has mean $\bar{x} = (3 + 5 + 7 + 7 + 8)/5 = 30/5 = 6$, so the deviations are calculated by subtracting 6 from each observation. See Table 9.

**TABLE 9**  Calculation of Deviations

| Observation $x$ | Deviation $x - \bar{x}$ |
|:---:|:---:|
| 3 | −3 |
| 5 | −1 |
| 7 | 1 |
| 7 | 1 |
| 8 | 2 |

One might feel that the average of the deviations would provide a numerical measure of spread. However, some deviations are positive and some negative, and the total of the positive deviations exactly cancels the total of the negative ones. In the foregoing example, we see that the positive deviations add to 4 and the negative ones add to −4, so the total deviation is 0. With a little reflection on the definition of the sample mean, the reader will realize that this was not just an accident. For any data set, the total deviation is 0 (for a formal proof of this fact, see Appendix A1).

$$\sum (\text{Deviations}) = \sum (x_i - \bar{x}) = 0$$

To obtain a measure of spread, we must eliminate the signs of the deviations before averaging. One way of removing the interference of signs is to square the numbers. A measure of spread, called the **sample variance**, is constructed by adding the squared deviations and dividing the total by the number of observations minus one.

Sample variance of $n$ observations:

$$s^2 = \frac{\text{sum of squared deviations}}{n - 1}$$

$$= \frac{\sum\limits_{i=1}^{n} (x_i - \bar{x})^2}{n - 1}$$

**Example 11**  Calculating Sample Variance

Calculate the sample variance of the data   3   5   7   7   8.

SOLUTION   For this data set, $n = 5$. To find the variance, we first calculate the mean, then the deviations and the squared deviations. See Table 10.

**TABLE 10**   Calculation of Variance

| Observation $x$ | Deviation $x - \bar{x}$ | $(\text{Deviation})^2$ $(x - \bar{x})^2$ |
|:---:|:---:|:---:|
| 3 | −3 | 9 |
| 5 | −1 | 1 |
| 7 | 1 | 1 |
| 7 | 1 | 1 |
| 8 | 2 | 4 |
| Total     30 | 0 | 16 |

$$\sum x \qquad \sum (x - \bar{x}) \qquad \sum (x - \bar{x})^2$$

$$\bar{x} = \frac{30}{5} = 6$$

$$\text{Sample variance } s^2 = \frac{16}{5 - 1} = 4$$

**Remark:**   Although the sample variance is conceptualized as the average squared deviation, notice that the divisor is $n - 1$ rather than $n$. The divisor, $n - 1$, is called the degrees of freedom[1] associated with $s^2$.

---

[1] The deviations add to 0 so a specification of any $n - 1$ deviations allows us to recover the one that is left out. For instance, the first four deviations in Example 11 add to −2, so to make the total 0, the last one must be +2, as it really is. In the definition of $s^2$, the divisor $n - 1$ represents the number of deviations that can be viewed as free quantities.

Because the variance involves a sum of squares, its unit is the square of the unit in which the measurements are expressed. For example, if the data pertain to measurements of weight in pounds, the variance is expressed in (pounds)$^2$. To obtain a measure of variability in the same unit as the data, we take the positive square root of the variance, called the **sample standard deviation**. The standard deviation rather than the variance serves as a basic measure of variability.

---

### Sample Standard Deviation

$$s = \sqrt{\text{Variance}} = \sqrt{\frac{\sum_{i=1}^{n} (x_i - \bar{x})^2}{n - 1}}$$

---

**Example 12**    Calculating the Sample Standard Deviation

Calculate the standard deviation for the data of Example 11.

SOLUTION    We already calculated the variance $s^2 = 4$ so the standard deviation is $s = \sqrt{4} = 2$.

To show that a larger spread of the data does indeed result in a larger numerical value of the standard deviation, we consider another data set in Example 13.

**Example 13**    Using Standard Deviations to Compare Variation in Two Data Sets

Calculate the standard deviation for the data 1, 4, 5, 9, 11. Plot the dot diagram of this data set and also the data set of Example 11.

SOLUTION    The standard deviation is calculated in Table 11. The dot diagrams, given in Figure 11, show that the data points of Example 11 have less spread than those of Example 13. This visual comparison is confirmed by a smaller value of $s$ for the first data set.

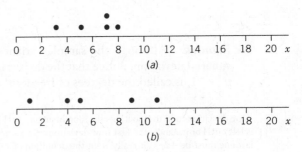

Figure 11   Dot diagrams of two data sets.

**TABLE 11**   Calculation of $s$

| $x$ | $(x - \bar{x})$ | $(x - \bar{x})^2$ |
|---|---|---|
| 1 | $-5$ | 25 |
| 4 | $-2$ | 4 |
| 5 | $-1$ | 1 |
| 9 | 3 | 9 |
| 11 | 5 | 25 |
| Total   30 | 0 | 64 |

$$\bar{x} = 6 \qquad\qquad s^2 = \frac{64}{4} = 16$$

$$s = \sqrt{16} = 4$$

An alternative formula for the sample variance

$$s^2 = \frac{1}{n-1} \left[ \sum x_i^2 - \frac{(\sum x_i)^2}{n} \right]$$

does not require the calculation of the individual deviations. In hand calculation, the use of this alternative formula often reduces the arithmetic work, especially when $\bar{x}$ turns out to be a number with many decimal places. The equivalence of the two formulas is shown in Appendix A1.2.

**Example 14**   Calculating Sample Variance Using the Alternative Formula

In a psychological experiment a stimulating signal of fixed intensity is used on six experimental subjects. Their reaction times, recorded in seconds, are 4, 2, 3, 3, 6, 3. Calculate the standard deviation for the data by using the alternative formula.

SOLUTION   These calculations can be conveniently carried out in tabular form:

| $x$ | $x^2$ |
|---|---|
| 4 | 16 |
| 2 | 4 |
| 3 | 9 |
| 3 | 9 |
| 6 | 36 |
| 3 | 9 |
| Total   21 | 83 |
| $= \sum x$ | $= \sum x^2$ |

$$s^2 = \frac{1}{n-1}\left[\sum x^2 - \frac{(\sum x)^2}{n}\right] = \frac{83 - (21)^2/6}{5} = \frac{83 - 73.5}{5}$$

$$= \frac{9.5}{5} = 1.9$$

$$s = \sqrt{1.9} = 1.38 \text{ seconds}$$

The reader may do the calculations with the first formula and verify that the same result is obtained.

**Sample z-score**
The sample *z scale*, (or *standard scale*), measures the position of a value relative to the sample mean in units of the standard deviation. Specifically,

$$\text{z-score of a measurement} = \frac{\text{Measurement} - \bar{x}}{s}$$

**Example 15**  Litter Size of Lions in the Wild

Lions typically have babies in twos and threes but sometimes four or five. Many do not survive until age one. To protect the very young, the mother will take the babies away from the pride for the first 6 weeks.

The sizes of eight litters born to one pride of lions are (courtesy of Martina Trinkel)

$$3 \quad 5 \quad 3 \quad 3 \quad 2 \quad 3 \quad 3 \quad 1$$

(a)  Find the sample mean, variance, and standard deviation.

(b)  Calculate the z-score for a liter of size 2.

SOLUTION  (a)  We calculate

$$\bar{x} = \frac{3 + 5 + 3 + 3 + 2 + 3 + 3 + 1}{8} = \frac{23}{8} = 2.88 \quad \text{cubs}$$

and, to avoid the rounding error in the mean, we use the alternative formula

$$s^2 = \frac{\sum_{i=1}^{n} x_i^2 - \left(\sum_{i=1}^{n} x_i\right)^2/n}{n-1}$$

$$= \frac{3^2 + 5^2 + 3^2 + 3^2 + 2^2 + 3^2 + 3^2 + 1^2 - (23)^2/8}{8-1}$$

$$= 1.268$$

The sample standard deviation $s = \sqrt{1.268} = 1.126$ cubs.

(b)  The z-score for the value 2 is $(2 - 2.88)/1.126 = -.78$ so it is .78 standard deviations below the sample mean number of cubs.

In Example 13, the data set with visibly larger variation yields a larger numerical value of $s$. The issue there surrounds a comparison between different data sets. In the context of a single data set, can we relate the numerical value of $s$ to the physical closeness of the data points to the center $\bar{x}$? To this end, we view one standard deviation as a benchmark distance from the mean $\bar{x}$. For bell-shaped distributions, an empirical rule relates the standard deviation to the proportion of the data that lie in an interval around $\bar{x}$.

---

**Empirical Guidelines for Symmetric Bell-Shaped Distributions**

| Approximately | 68% | of the data lie within $\bar{x} \pm s$ |
|---|---|---|
| | 95% | of the data lie within $\bar{x} \pm 2s$ |
| | 99.7% | of the data lie within $\bar{x} \pm 3s$ |

---

**Example 16**   Comparing the Sleep Data with the Empirical Guidelines

Examine the 59 hours of sleep in Table 4 in the context of the empirical guideline.

SOLUTION   Using a computer (see, for instance, Exercise 2.124), we obtain

$$\bar{x} = 7.18 \qquad s = 1.28 \qquad 2s = 2(1.28) = 2.56$$

The interval $7.18 - 1.28 = 5.90$ to $8.46 = 7.18 + 1.28$ contains 40 observations or $100(40/59) = 67.8\%$ of the data.

Going two standard deviations either side of $\bar{x}$ results in the interval

$$7.18 - 2.56 = 4.62 \quad \text{to} \quad 9.74 = 7.18 + 2.56$$

By actual count, all the observations except 4.5 and 10.0 fall in this interval. We find that $57/59 = .966$, or 96.6% of the observations lie within two standard deviations of $\bar{x}$.

The empirical guidelines suggest 68% and 95% respectively, so they are close.

**Other Measures of Variation**

Another measure of variation that is sometimes employed is

---

Sample range = Largest observation − Smallest observation

---

The range gives the length of the interval spanned by the observations.

**Example 17**   Calculating the Sample Range

Calculate the range for the hours of sleep data given in Example 6.

SOLUTION   The data given in Table 4 contained

$$\text{Smallest observation} = 4.5$$

$$\text{Largest observation} = 10.0$$

Therefore, the length of the interval covered by these observations is

$$\text{Sample range} = 10.0 - 4.5 = 5.5 \text{ hours}$$

As a measure of spread, the range has two attractive features: It is extremely simple to compute and interpret. However, it suffers from the serious disadvantage that it is much too sensitive to the existence of a single very large or very small observation in the data set. Also, it ignores the information present in the scatter of the intermediate points.

To circumvent the problem of using a measure that may be thrown far off the mark by one or two wild or unusual observations, a compromise is made by measuring the interval between the first and third quartiles.

$$\text{Sample interquartile range} = \text{Third quartile} - \text{First quartile}$$

The sample interquartile range represents the length of the interval covered by the center half of the observations. This measure of the amount of variation is not disturbed if a small fraction of the observations are very large or very small. The sample interquartile range is usually quoted in government reports on income and other distributions that have long tails in one direction, in preference to standard deviation as the measure of spread.

**Example 18**   Calculating the Interquartile Range

Calculate the sample interquartile range for the length of long distance phone call data given in Table 8.

SOLUTION   In Example 10, the quartiles were found to be $Q_1 = 4.4$ and $Q_3 = 17.5$. Therefore,

$$\text{Sample interquartile range} = Q_3 - Q_1$$

$$= 17.5 - 4.4$$

$$= 13.1 \text{ minutes}$$

Nearly 50% of the middle calls are within an interval of length 13.1 minutes.

## Boxplots

A recently created graphic display, called a **boxplot**, highlights the summary information in the quartiles. Begin with the

**Five-number summary:**    minimum, $Q_1$, $Q_2$, $Q_3$, maximum.

The center half of the data, from the first to the third quartile, is represented by a rectangle (box) with the median indicated by a bar. A line extends from $Q_3$ to the maximum value and another from $Q_1$ to the minimum. Figure 12 gives the boxplot for the length of phone call data in Table 8. The long line to the right is a consequence of the largest value, 53.3 minutes, and, to some extent, the second largest value, 43.5 minutes.

Figure 12   Boxplot of the length of phone call data in Table 8.

Boxplots are particularly effective for displaying several samples alongside each other for the purpose of visual comparison.

Figure 13, on page 58, displays the amount of reflected light in the near-infrared band as recorded by satellite when flying over forest areas and urban areas, respectively. Because high readings tend to correspond to forest and low readings to urban areas, the readings have proven useful in classifying unknown areas.

**Example 19**    Comparing Boxplots for the Reflected Light Data

Refer to the boxplots for the amount of reflected near-infrared light in Figure 13.

    (a)   Do forests or urban areas produce the largest median reflected light?

    (b)   Which has the largest interquartile range, IQR?

SOLUTION

    (a)   It is clear that forests have the largest median. Its median is over 10 units higher than that of urban areas.

    (b)   The height of the box is the IQR. The IQR of the forest data is over twice that of the IQR for urban areas.

Figure 13    Boxplots of near-infrared light
reflected from forest and urban areas.

One common variation of the boxplot is to modify the rule that the lines always extend to the minimum and the maximum. Rather, they extend to the most extreme observation that is within one and a half times the interquartile range. Values outside of this range are shown individually. Most software programs use this convention.

**Example 20**    A Modified Boxplot—Possible Outliers are Detached

Construct a modified boxplot for the phone call data and compare with the boxplot in Example 18.

SOLUTION    From Example 10 and Example 18, $Q_1 = 4.4$, $Q_3 = 17.5$ and the IQR $= $. $13.1$ so $1.5 \times$ IQR $= 1.5 \times 13.1 = 19.65$. The smallest observation is closer than this to $Q_1 = 4.4$ but

$$\text{maximum} - Q_3 = 53.3 - 17.5 = 35.8 \qquad \text{exceeds}$$

$$19.65 = 1.5 \times \text{IQR}$$

As shown in Figure 14, the line to right extends to 32.8, the most extreme observation within 19.65 units of $Q_3$. Observations larger than this are shown individually.

This modified boxplot does suggest two possible outliers that are not obvious in the simple boxplot shown in Figure 12.

Figure 14   Modified boxplot reveals the presence of two very long phone calls.

## Exercises

**2.57** For the data set

$$8 \quad 3 \quad 4$$

(a) Calculate the deviations $(x - \bar{x})$ and check to see that they add up to 0.

(b) Calculate the sample variance and the standard deviation.

**2.58** Repeat (a) and (b) of Exercise 2.57 for the data set

$$3 \quad 8 \quad 1$$

**2.59** For the data set   8   6   14   4

(a) Calculate the deviations $(x - \bar{x})$ and check to see that they add up to 0.

(b) Calculate the variance and the standard deviation.

**2.60** Repeat (a) and (b) of Exercise 2.59 for the data set

$$2.5 \quad 1.7 \quad 2.1 \quad 1.5 \quad 1.7$$

**2.61** For the data of Exercise 2.57, calculate $s^2$ by using the alternative formula.

**2.62** A sample of six university students responded to the question "How much time, in minutes, did you spend on the social network site yesterday?" One student never used the site and was dropped from the study.

$$100 \quad 45 \quad 60 \quad 130 \quad 30$$

Calculate, using these five values,

(a) The sample variance using the alternative formula.

(b) The sample standard deviation.

(c) Including a sixth value 0 for the student who never uses the site, re-calculate the sample standard deviation. Comment on the change in value.

**2.63** For each data set, calculate $s^2$.

(a) 1   4   3   2   2

(b) −2   1   −1   −3   0   −2

(c) 9   8   8   9   8   8   9

**2.64** In 2012 there were nine skyscrapers in the world over 300 meters tall. The ordered values of height are

366  381  442  452  484  492  508  601  828

where the largest is in Dubai.

(a) Calculate the sample variance for the nine skyscrapers.

(b) Calculate the sample standard deviation.

(c) Drop the largest value and recalculate the sample standard deviation. Comment on the change in value.

**2.65** Find the standard deviation of the measurements of diameters given in Exercise 2.45.

**2.66** A campus area merchant recorded the number of bad checks received per month, for five months

$$4 \quad 5 \quad 4 \quad 7 \quad 6$$

Calculate:

(a) The sample variance.

(b) The sample standard deviation.

**2.67** The city of Madison regularly checks the quality of water at swimming beaches located on area lakes. Fifteen times the concentration of fecal coliforms, in number of colony forming units (CFU) per 100 ml of water, was measured during the summer at one beach.

| 180 | 1600 | 90 | 140 | 50 | 260 | 400 | 90 |
|-----|------|----|-----|----|-----|-----|----|
| 380 | 110 | 10 | 60 | 20 | 340 | 80 | |

(a) Calculate the sample variance.

(b) Calculate the sample standard deviation.

(c) One day, the water quality was bad—the reading was 1600 CFU—and the beach was closed. Drop this value and calculate the sample standard deviation for the days where the water quality was suitable for swimming. Comment on the change.

**2.68** With reference to the radiation leakage data given in Exercise 2.15, calculate:

(a) The sample variance.

(b) The sample standard deviation.

**2.69** With reference to the data on the length of 10 major league baseball games in Exercise 2.43:

(a) Find the sample mean.

(b) Find the sample variance.

(c) Find the sample standard deviation.

**2.70** With reference to checked bags in Exercise 2.13,

(a) Find the sample mean.

(b) Find the sample standard deviation.

**2.71** The performance time (minutes) for Beethoven's Ninth Symphony is obtained from each of seven compact discs.

66.9   66.2   71.0   68.6   65.4   68.4   71.9

(a) Find the sample median.

(b) Find the sample mean.

(c) Find the sample standard deviation.

**2.72** Recent crime reports on the number of aggravated assaults at each of the 18 largest universities reporting for the year are summarized in the computer output.

| Variable | N | Mean | StDev |
|----------|---|------|-------|
| AggAsult | 18 | 5.44 | 4.89 |

| Minimum | Q1 | Median | Q3 | Maximum |
|---------|-----|--------|------|---------|
| 0.00 | 2.00 | 4.00 | 8.25 | 17.00 |

(a) Locate a measure of variation and also calculate the sample variance.

(b) Calculate the interquartile range and interpret this value.

(c) Give a value for a standard deviation that would correspond to greater variation in the numbers of aggravated assaults.

**2.73** The weights (oz) of nineteen babies born in Madison Wisconsin are summarized in the computer output.

**Descriptive Statistics: Weight**

| Variable | N | Mean | Median | StDev |
|----------|---|------|--------|-------|
| Weight | 19 | 118.05 | 117.00 | 15.47 |

| Variable | Minimum | Maximum | Q1 | Q3 |
|----------|---------|---------|------|------|
| Weight | 89.00 | 144.00 | 106.00 | 131.00 |

(a) Locate a measure of variation and also calculate the sample variance.

(b) Calculate the interquartile range and interpret this value.

(c) Give a value for a standard deviation that would correspond to smaller variation in the weights.

**2.74** *Some properties of the standard deviation.*

1. If a fixed number $c$ is added to all measurements in a data set, the deviations $(x - \bar{x})$ remain unchanged (see Exercise 2.52). Consequently, $s^2$ and $s$ remain unchanged.

2. If all measurements in a data set are multiplied by a fixed number $d$, the deviations $(x - \bar{x})$ get multiplied by $d$. Consequently, $s^2$ gets multiplied by $d^2$, and $s$ by $|d|$. (*Note*: The standard deviation is never negative.)

Verify these properties for the data set

$$5 \quad 9 \quad 9 \quad 8 \quad 10 \quad 7$$

taking $c = 4$ in property (1) and $d = 2$ in (2).

**2.75** For the data set of Exercise 2.22, calculate the interquartile range.

**2.76** For the extracurricular data of Exercise 2.3, calculate the interquartile range.

**2.77** Should you be surprised if the range is larger than twice the interquartile range? Explain.

**2.78** Calculations with the test scores data of Exercise 2.22 give $\bar{x} = 150.125$ and $s = 24.677$.

(a) Find the proportion of the observations in the intervals $\bar{x} \pm 2s$ and $\bar{x} \pm 3s$.

(b) Compare your findings in part (a) with those suggested by the empirical guidelines for bell-shaped distributions.

**2.79** Refer to the data on bone mineral content in Exercise 2.41.

(a) Calculate $\bar{x}$ and $s$.

(b) Find the proportion of the observations that are in the intervals $\bar{x} \pm s$, $\bar{x} \pm 2s$, and $\bar{x} \pm 3s$.

(c) Compare the results of part (b) with the empirical guidelines.

**2.80** Refer to the data on lizards in Exercise 2.19.

(a) Calculate $\bar{x}$ and $s$.

(b) Find the proportion of the observations that are in the intervals $\bar{x} \pm s$, $\bar{x} \pm 2s$, and $\bar{x} \pm 3s$.

(c) Compare the results of part (b) with the empirical guidelines.

**2.81** Refer to the data on number of returns in Example 4.

(a) Calculate $\bar{x}$ and $s$.

(b) Find the proportions of the observations that are in the intervals $\bar{x} \pm s$, $\bar{x} \pm 2s$, and $\bar{x} \pm 3s$.

(c) Compare the results of part (b) with the empirical guidelines.

**2.82** Comparisons using sample z scores.

$$z\text{-value of a measurement} = \frac{\text{Measurement} - \bar{x}}{s}$$

When two measurements originate from different sources, converting them to the z scale helps to draw a sensible interpretation of their relative magnitudes. For instance, suppose a student scored 65 in a math course and 72 in a history course. These (raw) scores tell little about the student's performance. If the class averages and standard deviations were $\bar{x} = 60, s = 20$ in math and $\bar{x} = 78, s = 10$ in history, this student's

$$z\text{ score in math} = \frac{65 - 60}{20} = .25$$

$$z\text{ score in history} = \frac{72 - 78}{10} = -.60$$

Thus, the student was .25 standard deviations above the average in math and .6 standard deviations below the average in history.

(a) If $\bar{x} = 490$ and $s = 120$, find the z scores of 350 and 620.

(b) For a z score of 2.4, what is the raw score if $\bar{x} = 210$ and $s = 50$?

**2.83** The weights (oz) of nineteen babies born in Madison Wisconsin are summarized in the computer output.

**Descriptive Statistics: Weight**

| Variable | N | Mean | Median | StDev |
|----------|-----|--------|--------|-------|
| Weight | 19 | 118.05 | 117.00 | 15.47 |

Obtain the z score for a baby weighing

(a) 102 oz

(b) 144 oz

**2.84** Two cities provided the following information on public school teachers' salaries.

| | Minimum | $Q_1$ | Median | $Q_3$ | Maximum |
|--------|---------|--------|--------|--------|---------|
| City A | 38,400 | 44,000 | 48,300 | 50,400 | 56,300 |
| City B | 39,600 | 46,500 | 51,200 | 55,700 | 61,800 |

(a) Construct a boxplot for the salaries in City A.

(b) Construct a boxplot, on the same graph, for the salaries in City B.

(c) Are there larger differences at the lower or the higher salary levels? Explain.

**2.85** Refer to the data on throwing speed in Exercise 2.42. Make separate boxplots to compare males and females.

**2.86** Refer to Exercise 2.27 and the data on the consumer price index for various cities. Find the increase, for each city, by subtracting the 2007 value from the 2011 value.

(a) Obtain the five-number summary: minimum, $Q_1$, $Q_2$, $Q_3$, and maximum. Which city had the largest increase? Were there any decreases?

(b) Make a boxplot of the increases.

**2.87** Refer to Exercise 2.27 and the data on the consumer price index for various cities. Find the increase, for each city, by subtracting the 2007 value from the 2011 value.

(a) Find the sample mean and standard deviation of these differences.

(b) What proportion of the increases lie between $\bar{x} \pm 2s$?

**2.88** Refer to Example 6 and the data on hours of sleep.

(a) Obtain the five-number summary: minimum, $Q_1$, $Q_2$, $Q_3$, and maximum.

(b) Make a boxplot of the hours of sleep.

**2.89** Refer to Exercise 2.3 and the data on extracurricular activities. Find the sample mean and standard deviation.

**2.90** Presidents also take midterms! After two years of the president's term, members of Congress are up for election. The following table gives the number of net seats lost, by the party of the president, in the House of Representatives starting from Eisenhower's time.

Net House Seats Lost in Midterm Elections

| | | |
|---|---|---|
| 1958 | Eisenhower (R) | 48 |
| 1962 | Kennedy (D) | 4 |
| 1966 | Johnson (D) | 47 |
| 1970 | Nixon (R) | 12 |
| 1974 | Nixon/Ford (R) | 48 |
| 1978 | Carter (D) | 15 |
| 1982 | Reagan (R) | 26 |
| 1986 | Reagan (R) | 5 |
| 1990 | Bush (R) | 8 |
| 1994 | Clinton (D) | 52 |
| 1998 | Clinton (D) | −5 (gain) |
| 2002 | Bush (R) | −8 |
| 2006 | Bush (R) | 30 |
| 2010 | Obama(D) | 63 |

For the data on the number of House seats lost:

(a) Calculate the sample mean.

(b) Calculate the standard deviation.

(c) Make a dot plot.

(d) What is one striking feature of these data that could be useful in predicting future midterm election results? (*Hint*: Would you have expected more elections to result in net gains?)

**2.91** With reference to Exercise 2.90:

(a) Calculate the median number of lost House seats.

(b) Find the maximum and minimum losses and identify these with a president.

(c) Determine the range for the number of House seats lost.

# 5. CHECKING THE STABILITY OF THE OBSERVATIONS OVER TIME

The calculations for the sample mean and sample variance treat all the observations alike. The presumption is that there are no apparent trends in data over time and there are no unusual observations. Another way of saying this is that the process producing the observations is in statistical control. The concept of statistical control allows for variability in the observations but requires that the pattern of variability be the same over time. Variability should not increase or decrease with time and the center of the pattern should not change.

To check on the stability of the observations over time, observations should be plotted versus time, or at least the order in which they were taken. The resulting plot is called a **time plot** or sometimes a **time series plot**.

**Example 21**   A Time Plot of Overtime Hours

The Madison Police Department charts several important variables, one of which is the number of overtime hours due to extraordinary events. These events would include murders, major robberies, and so forth. Although any one event is not very predictable, there is some constancy when data are grouped into six-month periods.

The values of overtime hours for extraordinary events for eight recent years, beginning with 2200, 875, . . . , through 1223, are

| 2200 | 875 | 957 | 1758 | 868 | 398 | 1603 | 523 |
| 2034 | 1136 | 5326 | 1658 | 1945 | 344 | 807 | 1223 |

Is the extraordinary event overtime hours process in control? Construct a time plot and comment.

SOLUTION   The time plot is shown in Figure 15. There does not appear to be any trend, but there is one large value of 5326 hours.

Figure 15   Time plot of extraordinary event hours versus time order.

**Example 22**   A Time Plot of the Yen/Dollar Exchange Rate

The exchange rate between the United States and Japan can be stated as the number of yen that can be purchased with $1. Although this rate changes daily, we quote the official value for the year:

| Year | 1986 | 1987 | 1988 | 1989 | 1990 | 1991 | 1992 | 1993 | 1994 | 1995 | 1996 | 1997 |
|------|------|------|------|------|------|------|------|------|------|------|------|------|
| Exchange rate | 168.4 | 144.6 | 128.2 | 138.1 | 145.0 | 134.6 | 126.8 | 111.1 | 102.2 | 94.0 | 108.8 | 121.1 |

| 1998 | 1999 | 2000 | 2001 | 2002 | 2003 | 2004 | 2005 | 2006 | 2007 | 2008 | 2009 | 2010 | 2011 |
|------|------|------|------|------|------|------|------|------|------|------|------|------|------|
| 130.9 | 113.7 | 107.8 | 121.6 | 125.2 | 115.9 | 108.2 | 110.1 | 116.3 | 117.8 | 103.4 | 93.6 | 87.8 | 79.7 |

Is this exchange rate in statistical control? Make a time plot and comment.

SOLUTION   The time plot is shown in Figure 16. There is a general downhill trend over most of the time period so the exchange rate is definitely not in statistical control. A dollar has purchased fewer and fewer yen over the years. It is the downward trend that is the primary feature and a cause for serious concern with regard to trade deficits.

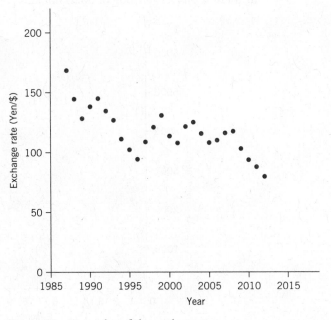

Figure 16   Time plot of the exchange rate.

Manufacturers need to monitor critical dimensions, temperatures, and other variables so that, although they vary, the variation is kept small enough so that the quality of the final product is maintained. A graphical approach, called a control chart, is recommended for this purpose because it allows for the visual inspection of outliers and trends. It adds a centerline and control limits to the time plot to help identify unusual observations.

To construct a control chart:

1.  Plot the observations versus time order.
2.  Add a solid centerline at the level of the sample mean $\bar{x}$.
3.  Add dashed lines for the control limits at $\bar{x} - 2s$ and $\bar{x} + 2s$.

According to the empirical rule, if the process is in statistical control so the observations are stable over time, only about 5% of the observations will fall outside of the control limits. In many applications, manufacturers use $\bar{x} - 3s$ and $\bar{x} + 3s$ as the control limits.

The upper and lower control limits on the control charts help to identify unusually low or unusually high observations. It allows us to distinguish between typical variation and variation that is especially large and could be due to special or assignable causes. Any time an observation falls outside of the control limits, an effort should be made to search for the reason.

**Example 23**    A Control Chart for Overtime Hours

Manufacturing processes are not the only ones that can benefit from control charting. Refer to the data in Example 21 on the number of overtime hours for police due to extraordinary events. Is the extraordinary event overtime hours process in control? Construct a control chart and comment.

Figure 17    Control chart for extraordinary event overtime hours.

SOLUTION    A computer calculation gives $\bar{x} = 1478$ and $s = 1183$ so the centerline is drawn at the sample mean 1478 and the upper control limit is $\bar{x} + 2s = 1478 + 2 \times 1183 = 3844$. The lower control limit is negative; we replace it by 0. Figure 17 gives the resulting control chart for extraordinary event overtime hours.

There is no discernible trend, but one point does exceed the upper control limit. By checking more detailed records, it was learned that the point outside of the control limits occurred when protests took place on campus in response to the bombing of a foreign capital. These events required city police to serve 1773 extraordinary overtime hours in just one 2-week period and 683 in the next period. That is, there was really one exceptional event, or special cause, that could be identified with the unusual point.

The one large value, 5326 hours, not only affects the centerline by inflating the mean, but it also increases the variance and that raises the upper control limit. In Exercise 2.98, you are asked to remove this outlier and redo the control chart.

# Exercises

**2.92** Make a time plot of the mail order phone call data in Exercise 2.48 and comment on the statistical control.

**2.93** A city department has introduced a quality improvement program and has allowed employees to get credit for overtime hours when attending meetings of their quality groups. The total number of overtime meeting hours for each of the 26 pay periods in one year by row were

| 30 | 215 | 162 | 97 | 194 | 163 | 60 | 41 | 100 |
|----|-----|-----|-----|-----|-----|-----|-----|-----|
| 43 | 96 | 69 | 80 | 42 | 162 | 75 | 95 | 65 |
| 57 | 131 | 54 | 114 | 64 | 114 | 38 | 140 | |

Make a time plot of the overtime meeting hours data.

**2.94** Make a control chart for the data referred to in Exercise 2.92 and comment.

**2.95** Make a control chart for the data in Exercise 2.93 and comment.

**2.96** The exchange rate between the United States and Canada can be stated as the number of Canadian dollars that can be purchased with $1. The official values for the year are

| Year | 1995 | 1996 | 1997 | 1998 | 1999 | 2000 |
|------|------|------|------|------|------|------|
| Exchange rate | 1.37 | 1.36 | 1.38 | 1.48 | 1.49 | 1.48 |

| | 2001 | 2002 | 2003 | 2004 | 2005 | 2006 |
|--|------|------|------|------|------|------|
| | 1.55 | 1.57 | 1.40 | 1.30 | 1.21 | 1.13 |

| | 2007 | 2008 | 2009 | 2010 | 2011 |
|--|------|------|------|------|------|
| | 1.07 | 1.07 | 1.14 | 1.03 | .99 |

Is this exchange rate in statistical control? Make a time plot and comment.

**2.97** Make a control chart of the data in Exercise 2.96 and comment.

**2.98** Make a control chart for the extraordinary event overtime data in Example 23 after removing the outlier identified in that example. You need to recalculate the mean and standard deviation.

# 6.  MORE ON GRAPHICS

The importance of graphing your data cannot be overemphasized. If a feature you expect to see is not present in the plots, statistical analyses will be of no avail. Moreover, creative graphics can often highlight features in the data and even give new insights.

The devastation of Napoleon's Grand Army during his ill-fated attempt to capture Russia was vividly depicted by Charles Minard. The 422,000 troops that entered Russia near Kaunas are shown as a wide (blue) river flowing toward Moscow and the retreating army as a small (blue) stream. The width of the band indicates the size of the army at each location on the map. Even the simplified version of the original graphic, appearing in Figure 18, dramatically conveys the losses that reduced the army of 422,000 men to 10,000 returning members. The temperature scale at the bottom, pertaining to the retreat, helps to explain the loss of life, including the incident where thousands died trying to cross the Berezina River in subzero temperatures. (For a copy of Minard's more detailed map and additional discussion, see E. R. Tufte, *The Visual Display of Quantitative Information*, Cheshire, CT: Graphics Press, 1983.)

Figure 18    The demise of Napoleon's army in Russia, 1812–1813, based on Charles Minard.

Another informative graphic, made possible with modern software, is the display of the frequency of serious tornadoes rated F3 to F5 in Figure 19. This figure presents a ten year average, by state. A primary "tornado alley" is clearly evident.

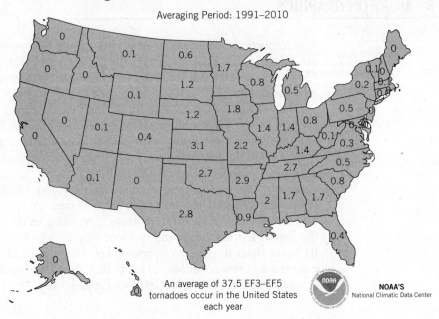

Figure 19  Yearly number of strong tornadoes averaged over 10 years. (*Source*: National Climatic Data Center, U.S. Department of Commerce.)

Dave Carpenter/www.CartoonStock.com

**Graphs can give a vivid overall picture.**

# STATISTICS IN CONTEXT

The importance of visually inspecting data cannot be overemphasized. We present a mini-case study (courtesy of Don Ermer) that shows the importance of first appropriately plotting and then monitoring manufacturing data. This statistical application concerns a ceramic part used in a popular brand of coffee makers. To make this ceramic part, a mixture of clay, water, and oil is poured into the cavity between two dies of a pressing machine. After pressing but before the part is dried to a hardened state, critical dimensions are measured. The depth of a slot is of interest here.

Sources of variation in the slot depth abound: the natural uncontrolled variation in the clay–water–oil mixture, the condition of the press, differences in operators, and so on. Some variation in the depth of the slot is inevitable. Even so, for the part to fit when assembled, the slot depth needs to be controlled within certain limits.

Every half hour during the first shift, slot depth is measured on three ceramic parts selected from production. Table 12 gives the data obtained on a Friday. The sample mean for the first sample of 218, 217, and 219 (thousandths of inch) is $(218 + 217 + 219)/3 = 654/3 = 218$, and so on.

**TABLE 12**   Slot Depth (thousandths of an inch)

| Time | 7:00 | 7:30 | 8:00 | 8:30 | 9:00 | 9:30 | 10:00 | |
|------|------|------|------|------|------|------|-------|---|
| 1 | 218 | 218 | 216 | 217 | 218 | 218 | 219 | |
| 2 | 217 | 218 | 218 | 220 | 219 | 217 | 219 | |
| 3 | 219 | 217 | 219 | 221 | 216 | 217 | 218 | |
| SUM | 654 | 653 | 653 | 658 | 653 | 652 | 656 | |
| $\bar{x}$ | 218.0 | 217.7 | 217.7 | 219.3 | 217.7 | 217.3 | 218.7 | |

| Time | 10:30 | 11:00 | 11:30 | 12:30 | 1:00 | 1:30 | 2:00 | 2:30 |
|------|-------|-------|-------|-------|------|------|------|------|
| 1 | 216 | 216 | 218 | 219 | 217 | 219 | 217 | 215 |
| 2 | 219 | 218 | 219 | 220 | 220 | 219 | 220 | 215 |
| 3 | 218 | 217 | 220 | 221 | 216 | 220 | 218 | 214 |
| SUM | 653 | 651 | 657 | 660 | 653 | 658 | 655 | 644 |
| $\bar{x}$ | 217.7 | 217.0 | 219.0 | 220.0 | 217.7 | 219.3 | 218.3 | 214.7 |

An x-bar chart will indicate when changes have occurred and there is a need for corrective actions. Because there are 3 slot measurements at each time, it is the 15 sample means that are plotted versus time order. We will take the centerline to be the mean of the 15 sample means, or

$$\text{Centerline: } \bar{\bar{x}} = \frac{218.0 + \cdots + 214.7}{15} = 218.0$$

When each plotted mean is based on several observations, the variance can be estimated by combining the variances from each sample. The first sample has variance $s_1^2 = [(218 - 218)^2 + (217 - 218)^2 + (219 - 218)^2]/(3 - 1) = 1.000$ and so on. The details are not important, but a computer calculation of the variance used to set control limits first determines the average of the 15 individual sample variances,

$$\frac{1.000 + 0.333 + \cdots + 0.333}{15} = 1.58$$

and, for reasons given in Chapter 7, divides by 3 to give the variance of a single sample mean. That is, $1.58/3 = .527$ is the appropriate $s^2$. The control limits are set at three times the estimated standard deviation $s = \sqrt{.527} = .726$, or $3 \times .726 = 2.2$ units from the centerline.

$$\text{Lower control limit: LCL} = 218.0 - 2.2 = 215.8$$

$$\text{Upper control limit: UCL} = 218.0 + 2.2 = 220.2$$

The x-bar chart is shown in Figure 20. What does the chart tell us?

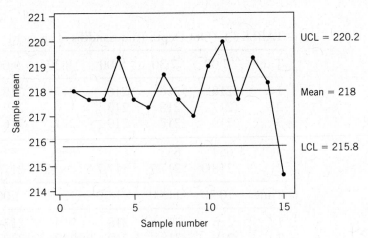

Figure 20   X-bar chart for depth.

The x-bar chart shows that the process was stable throughout the day and no points were out of control except the last sample. It was then that an unfortunate oversight occurred. Because it was near the end of her shift and the start of the weekend, the operator did not report the out-of-control value to either the setup person or the foreman. She knew the setup person was already cleaning up for the end of the shift and that the foreman was likely thinking about going across the street to the Legion Bar for some refreshments as soon as the shift ended. The operator did not want to ruin anyone's weekend plans so she kept quiet.

When the pressing machine was started up on Monday morning, one of the dies broke. The cost of the die was over a thousand dollars. But, when a customer was called and told there would be a delay in delivering the ceramic parts, he canceled the order. Certainly the loss of a customer is an even more expensive item.

Later, it was concluded that the clay had likely dried and stuck to the die leading to the break. A problem was predicted by the chart on Friday. Although the chart correctly indicated a problem at that time, someone had to act for the monitoring procedure to work.

## USING STATISTICS WISELY

1. As a first step, always graph the data as a dot diagram or histogram to assess the overall pattern of data.

2. When comparing histograms based on different class intervals, be sure to create histograms whose height is relative frequency divided by width of interval.

3. Calculate summary statistics to describe the data set. Always determine the sample mean and standard deviation. The five-number summary

   minimum    first quartile    median    third quartile    maximum

   provides an additional summary when the sample sizes are moderately large. It helps describe cases where the dot diagram or histogram has a single long tail.

4. Use the median to describe the center when a small sample contains an extreme observation. The sample median is not influenced by a few very large or very small observations that may even be incorrectly recorded.

5. Do not routinely calculate summary statistics without identifying unusual observations (outliers) that may have undue influence on the value of a summary statistic.

## KEY IDEAS AND FORMULAS

**Qualitative data** refer to frequency counts in categories. These are summarized by calculating the

$$\text{Relative frequency} = \frac{\text{Frequency}}{\text{Total number of observations}}$$

for the individual categories.

The term **numerical-valued variable** or just **variable** refers to a characteristic that varies over units and is measured on a numerical scale. **Discrete variables** are usually counts and all discrete variables have gaps in their scale of measurement. **Continuous variables**, like height or weight, can conceptually take any value in an interval. Data resulting from measurements of a variable are either **discrete** or **continuous** data.

For a **discrete data** set, the **frequency** is the count of the number of observations having a distinct value. The **relative frequency** is the proportion of sample units having this property.

$$\text{Relative frequency} = \frac{\text{Frequency}}{\text{Total number of observations}}$$

Summarize a discrete data set by a **frequency distribution** that lists the distinct data points and their corresponding relative frequencies. Either a **line diagram** or a **histogram** can be used for a graphical display.

Continuous measurement data should be graphed as a **dot diagram** when the data set is small, say, fewer than 20 or 25 observations. Larger data sets are first summarized in a **frequency table**. This is constructed by grouping the observations in **class intervals**, preferably of equal lengths. The class intervals are nonoverlapping and cover the range of the data set from smallest to largest. We recommend specifying an **endpoint convention** that tells which of the **class boundaries**, or endpoints of the class intervals, to include and which to exclude from each class interval. A list of the class intervals along with the corresponding relative frequencies provides a **frequency distribution** that can graphically be displayed as a **histogram**. The histogram is constructed to have total area 1, equal to total relative frequency. That is, for each class interval, we draw a rectangle whose **area represents the relative frequency** of the class interval.

A **stem-and-leaf display** is another effective means of display when the data set is not too large. It is more informative than a histogram because it retains the individual observations in each class interval instead of lumping them into a frequency count. Two variants are the **double-stem display** and **five-stem display**.

**Pareto diagrams** display events according to their frequency in order to highlight the most important few that occur most of the time.

A summary of measurement data (discrete or continuous) should also include numerical measures of center and spread.

Two important measures of center are

$$\textbf{Sample mean} \qquad \bar{x} = \frac{\sum x}{n}$$

**Sample median** = middle most value of the ordered data set

The **quartiles** and, more generally, **percentiles** are other useful locators of the distribution of a data set. The second quartile is the same as the median. The **sample quartiles** divide the **ordered data** into nearly four equal parts. The $100p$ -th percentile has at least proportion $p$ at or below and proportion $1 - p$ at or above.

The amount of **variation**, or **spread**, of a data set is measured by the **sample standard deviation** $s$. The **sample variance** $s^2$ is given by

$$s^2 = \frac{\sum (x - \bar{x})^2}{n - 1}$$

Also, $s^2 = \dfrac{1}{n-1}\left[\sum x^2 - \dfrac{(\sum x)^2}{n}\right]$ (convenient for hand calculation)

$$\text{Sample standard deviation } s = +\sqrt{s^2}$$

$$z\text{-score} = \dfrac{x - \bar{x}}{s} \text{ for a value } x.$$

The standard deviation indicates the amount of spread of the data points around the mean $\bar{x}$. If the histogram appears **symmetric and bell-shaped**, then the interval

| | |
|---|---|
| $\bar{x} \pm s$ | includes approximately 68% of the data |
| $\bar{x} \pm 2s$ | includes approximately 95% of the data |
| $\bar{x} \pm 3s$ | includes approximately 99.7% of the data |

Two other measures of variation are

**Sample range** $=$ Largest observation $-$ Smallest observation

and

**Sample interquartile range** $=$ Third quartile $-$ First quartile

The **five-number summary**, namely, the median, the first and third quartiles, the smallest observation, and the largest observation, together serve as useful indicators of the distribution of a data set. These are displayed in a **boxplot**.

# TECHNOLOGY

Creating graphs and computing statistical summaries have become considerably easier because of recent developments in software. In all our professional applications of statistics, we begin by entering the data in a worksheet. We then read aloud from the worksheet while another person checks against the original. This procedure has eliminated many errors in data entry and allowed us to proceed knowing that the computer software is using the correct data.

In the technology sections of this text, we give the essential details for using MINITAB, EXCEL, and the TI-84/83 Plus graphing calculator. The first two use the worksheet format.

## MINITAB

The MINITAB screen is split with the bottom part being the worksheet. In the example here, we have typed sleep in the top entry and then in row 1 of the first column we have typed the first number of hours of sleep in Table 4. The rest of the hours of sleep are typed in the other cells in column 1.

Alternatively, the data sets in the book are stored as MINITAB worksheets on the book's Web site. Go to www.wiley.com/college/johnson and click on this book. For instance, the hours of sleep are in C2T4.mtw, indicating Table 4 of Chapter 2. To open this worksheet:

Under **File** choose **Open Worksheet** and go to the drive that contains the downloaded files.

Click on the MINITAB folder and then click on the file name C2T4. Click **OK**.

This will activate the worksheet in MINITAB and you do not have to manually enter the numbers.

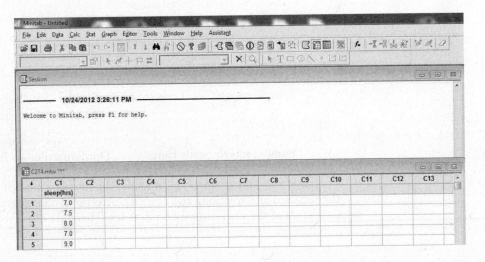

The summary statistics can be obtained by pulling down the menu in the top bar under **Stat**, then choosing **Basic Statistics** and then **Graphic Summary**. More specifically,

**Data** in C1

**Dialog box:**

**Stat** > **Basic Statistics** > **Graphical Summary**.

Type C1 in **Variables**. Click **OK**.

MINITAB uses a slightly different definition of the quartiles and their values may slightly differ from those calculated by the method in this book.

MINITAB will also create a histogram. With the data in C1,

**Dialog box:**

**Graph** > **Histogram**.

Select *Simple*. Click **OK**.

Type C1 in **Graph variables**. Click **OK**.

MINITAB will also create boxplots, dot plots, and stem-and-leaf displays. With the data in C1,

**Dialog box:**

**Graph > Dotplot.**

Select *Simple*. Click **OK**.

Type C1 in **Graph variables**. Click **OK**.

produces a dot plot. To obtain a boxplot, replace the first step by

**Graph > Boxplot.**

and to obtain a stem-and-leaf display, replace the first two steps by the single step

**Graph > Stem-and-Leaf.**

Clicking on **Labels** before the last **OK** will allow you to put titles on your graph.

**EXCEL**

Begin with the data in column A. For the hours of sleep from Table 4, the spread sheet is

Alternatively, the data sets in the book are stored in EXCEL workbooks on the book's web site. Go to www.wiley.com/college/johnson and click on this book. For instance, the hours of sleep are in C2T4.xls, indicating Table 4 of Chapter 2. Go to the drive containing the downloaded files and click on the EXCEL folder and then on the file name C2T4. The workbook having the data from Table 4 will then open.

Most of the statistical procedures we will use start with

Click the **Data** tab and then **Data Analysis**.

If **Data Analysis** is not listed, then it must be added once. To do so, select the **File** tab then **Options** and **Add-Ins**. Under **Add-Ins**, select **Analysis Toolpak** and **click OK**. After **Manage Excel Add-Ins**. Click **OK**.

To obtain the summary statistics,

Select **the Data tab**, then **Data Analysis**, and then **Descriptive Statistics**.

Click **OK**. Place cursor in the **Input Range** window and use the mouse to highlight the data in column A. Check **Summary Statistics** and click **OK**.

### TI-84/83 PLUS

Press **STAT**, select **1:Edit**, and then press **ENTER**. Enter the data in $L_1$ by typing 4.5 and then pressing **ENTER**, typing 4.7 and then pressing **ENTER** and so on.

With the data in $L_1$ press **STAT**, highlight **Calc**, and select **1 : 1 – Var Stats**. Then press **ENTER**.

## REVIEW EXERCISES

**2.99** Recorded here are the numbers of civilians employed in the United States by major occupation groups for the years 2010 and 2000. (*Source: Statistical Abstract of the United States, 2012.*)

(a) For each year, calculate the relative frequencies of the occupation groups.

(b) Comment on changes in the occupation pattern between 2000 and 2010.

|  | Number of Workers in Millions | |
|---|---|---|
|  | 2010 | 2000 |
| Goods producing | 17.8 | 24.6 |
| Service (private) | 112.1 | 107.1 |
| Government | 22.5 | 20.8 |
| Total | 152.4 | 152.5 |

**2.100** Table 13, on page 77, gives data collected from the students attending an elementary statistics course at the University of Wisconsin. These data include sex, height, number of years in college, and the general area of intended major [Humanities (H);

Social Science (S); Biological Science (B); Physical Science (P)].

(a) Summarize the data of "intended major" in a frequency table.

(b) Summarize the data of "year in college" in a frequency table and draw either a line diagram or a histogram.

**2.101** Referring to Exercise 2.100, plot the dot diagrams of heights separately for the male and female students and compare.

**2.102** Refer to the data on power outages in Table D.1 in the Data Bank. Make a Pareto chart for the cause of the outage.

**2.103** The dollar amounts claimed by businessmen for their lunches are to be grouped into the following classes: 0–5, 5–10, 10–15, 15–20, 20 or more. The left endpoint is included. Is it possible to determine from this frequency distribution the exact number of lunches for which the amount claimed was:

(a) Less than 15?

(b) 10 or more?

(c) 30 or more?

**2.104** Mung bean sprouts are more widely used in Asian cooking than the beans themselves. To study their growth, an experimenter presoaked some beans until they sprouted about 1 millimeter. Five were randomly selected and placed in individual petri dishes. After 96 hours, their lengths (mm)

$$143 \quad 131 \quad 101 \quad 143 \quad 111$$

were obtained. Find the mean and standard deviation.

**2.105** The weights of twenty adult grizzly bears captured and released are summarized in the computer output

**Descriptive Statistics: bearwt**

| Variable | N | Mean | Median | StDev |
|----------|-----|-------|--------|-------|
| Bearwt | 20 | 227.4 | 232.5 | 82.7 |

(a) Locate two measures of center tendency, or location, and interpret the values.

(b) Locate the standard deviation.

(c) Calculate the $z$ score for a grizzly bear that weighs 320 pounds.

**TABLE 13**  Class Data

| Student No. | Sex | Height in Inches | Year in College | Intended Major | Student No. | Sex | Height in Inches | Year in College | Intended Major |
|-------------|-----|------------------|-----------------|----------------|-------------|-----|------------------|-----------------|----------------|
| 1 | F | 67 | 3 | S | 26 | M | 67 | 1 | B |
| 2 | M | 72 | 3 | P | 27 | M | 68 | 3 | P |
| 3 | M | 70 | 4 | S | 28 | M | 72 | 4 | B |
| 4 | M | 70 | 1 | B | 29 | F | 68 | 3 | P |
| 5 | F | 61 | 4 | P | 30 | F | 66 | 2 | B |
| 6 | F | 66 | 3 | B | 31 | F | 65 | 2 | B |
| 7 | M | 71 | 3 | H | 32 | M | 64 | 4 | B |
| 8 | M | 67 | 4 | B | 33 | M | 72 | 1 | H |
| 9 | M | 65 | 3 | S | 34 | M | 67 | 4 | B |
| 10 | F | 67 | 3 | B | 35 | M | 73 | 3 | S |
| 11 | M | 74 | 3 | H | 36 | F | 71 | 4 | B |
| 12 | M | 68 | 3 | S | 37 | M | 71 | 3 | B |
| 13 | M | 74 | 2 | P | 38 | M | 69 | 2 | S |
| 14 | F | 64 | 4 | P | 39 | F | 69 | 4 | P |
| 15 | M | 69 | 3 | S | 40 | M | 74 | 4 | S |
| 16 | M | 64 | 3 | B | 41 | M | 73 | 3 | B |
| 17 | M | 72 | 4 | P | 42 | M | 68 | 3 | B |
| 18 | M | 71 | 3 | B | 43 | F | 66 | 2 | S |
| 19 | F | 67 | 2 | S | 44 | M | 73 | 2 | P |
| 20 | M | 70 | 4 | S | 45 | M | 73 | 2 | S |
| 21 | M | 66 | 4 | S | 46 | M | 67 | 4 | S |
| 22 | F | 67 | 2 | B | 47 | F | 62 | 3 | S |
| 23 | M | 68 | 4 | S | 48 | M | 68 | 2 | B |
| 24 | M | 71 | 3 | H | 49 | M | 71 | 3 | S |
| 25 | M | 75 | 1 | S | | | | | |

**2.106** The stem-and-leaf display given here shows the final examination scores of students in a sociology course. (Leaf unit = 1.0)

| Stem-and-Leaf Display of Scores | |
|---|---|
| 2 | 57 |
| 3 | 244 |
| 4 | 1179 |
| 5 | 03368 |
| 6 | 012447 |
| 7 | 223556899 |
| 8 | 00457 |
| 9 | 0036 |

(a) Find the median score.

(b) Find the quartiles $Q_1$ and $Q_3$.

(c) What proportion of the students scored below 70? 80 and over?

**2.107** The following are the numbers of passengers on the minibus tour of Hollywood.

9  12  10  11  11  7  12   6  11  4  10  10  11  9  10
7  10   8   8   9 8   9  11   9  8   6  10   6  8  11

(a) Find the sample median.

(b) Find the sample mean.

(c) Find the sample variance.

**2.108** The following table shows the age at inauguration of each U.S. president.

| Name | Age at Inauguration |
|---|---|
| 1. Washington | 57 |
| 2. J. Adams | 61 |
| 3. Jefferson | 57 |
| 4. Madison | 57 |
| 5. Monroe | 58 |

| Name | Age at Inauguration |
|---|---|
| 6. J. Q. Adams | 57 |
| 7. Jackson | 61 |
| 8. Van Buren | 54 |
| 9. W. H. Harrison | 68 |
| 10. Tyler | 51 |
| 11. Polk | 49 |
| 12. Taylor | 64 |
| 13. Fillmore | 50 |
| 14. Pierce | 48 |
| 15. Buchanan | 65 |
| 16. Lincoln | 52 |
| 17. A. Johnson | 56 |
| 18. Grant | 46 |
| 19. Hayes | 54 |
| 20. Garfield | 49 |
| 21. Arthur | 50 |
| 22. Cleveland | 47 |
| 23. B. Harrison | 55 |
| 24. Cleveland | 55 |
| 25. McKinley | 54 |
| 26. T. Roosevelt | 42 |
| 27. Taft | 51 |
| 28. Wilson | 56 |
| 29. Harding | 55 |
| 30. Coolidge | 51 |
| 31. Hoover | 54 |
| 32. F. D. Roosevelt | 51 |
| 33. Truman | 60 |
| 34. Eisenhower | 62 |
| 35. Kennedy | 43 |
| 36. L. Johnson | 55 |
| 37. Nixon | 56 |
| 38. Ford | 61 |
| 39. Carter | 52 |
| 40. Reagan | 69 |
| 41. G. Bush | 64 |
| 42. Clinton | 46 |
| 43. G. W. Bush | 54 |
| 44. Obama | 47 |

(a) Make a stem-and-leaf display with a double stem.

(b) Find the median, $Q_1$ and $Q_3$.

**2.109**

(a) Calculate $\bar{x}$ and $s$ for the data 6, 8, 4, 9, 8.

(b) Consider the data set 106, 108, 104, 109, 108, which is obtained by adding 100 to each number given in part (a). Use your results of part (a) and the properties stated in Exercises 2.52 and 2.74 to obtain the $\bar{x}$ and $s$ for this modified data set. Verify your results by direct calculations with this new data set.

(c) Consider the data set $-18, -24, -12, -27, -24$, which is obtained by multiplying each number of part (a) by $-3$. Repeat the problem given in part (b) for this new data set.

**2.110** The boxplot and summary statistics for data from 48 student on the time spent, in minutes, at a social network site the previous day are

```
Descriptive Statistics: Social Network(m)

Variable            N    Mean   StDev
Social Network(m)  48   35.96   29.11

Minimum  Q1    Median  Q3    Maximum
  1.00  15.00  26.00  45.0   130.00
```

(a) From the boxplot, determine whether or not the distribution of values is nearly symmetric, has a long tail to the left, or a long tail to the right.

(b) Calculate the interquartile range and then multiply by 1.5 to determine the maximum possible length of the right-hand whisker.

(c) Identify the approximate values of the two outliers suggested by the boxplot.

**2.111** In a genetic study, a regular food was placed in each of 20 vials and the number of flies of a particular genotype feeding on each vial recorded. The counts of flies were also recorded for another set of 20 vials that contained grape juice. The following data sets were obtained (courtesy of C. Denniston and J. Mitchell).

**No. of Flies (Regular Food)**

| 15 | 20 | 31 | 16 | 22 | 22 | 23 | 33 | 38 | 28 |
|----|----|----|----|----|----|----|----|----|----|
| 25 | 20 | 21 | 23 | 29 | 26 | 40 | 20 | 19 | 31 |

**No. of Flies (Grape Juice)**

| 6 | 19 | 0 | 2 | 11 | 12 | 13 | 12 | 5 | 16 |
|---|----|---|---|----|----|----|----|---|----|
| 2 | 7 | 13 | 20 | 18 | 19 | 19 | 9 | 9 | 9 |

(a) Plot separate dot diagrams for the two data sets.

(b) Make a visual comparison of the two distributions with respect to their centers and spreads.

(c) Calculate $\bar{x}$ and $s$ for each data set.

**2.112** The data below were obtained from a detailed record of purchases of toothpaste over several years (courtesy of A. Banerjee). The usage times (in weeks) per ounce of toothpaste for a household taken from a consumer panel were

.74 .45 .80 .95 .84 .82 .78 .82 .89 .75 .76 .81
.85 .75 .89 .76 .89 .99 .71 .77 .55 .85 .77 .87

(a) Plot a dot diagram of the data.

(b) Find the relative frequency of the usage times that do not exceed .80.

(c) Calculate the mean and the standard deviation.

(d) Calculate the median and the quartiles.

**2.113** To study how first-grade students utilize their time when assigned to a math task, a researcher observes 24 students and records their times off-task out of 20 minutes. The dotplot appears on page 31. (courtesy of T. Romberg).

**Times Off-Task (minutes)**

| 4 | 0 | 2 | 2 | 4 | 1 |
|----|---|----|---|---|----|
| 4 | 6 | 9 | 7 | 2 | 7 |
| 5 | 4 | 13 | 7 | 7 | 10 |
| 10 | 0 | 5 | 3 | 9 | 8 |

For this data set, find:

(a) Mean and standard deviation.

(b) Median.

(c) Range.

**2.114** The following summary statistics were obtained from a data set.

$$\bar{x} = 80.5 \quad \text{Median} = 84.0$$
$$s = 10.5 \quad Q_1 = 75.5$$
$$Q_3 = 96.0$$

Approximately what proportion of the observations are:

(a) Below 96.0?

(b) Above 84.0?

(c) In the interval 59.5 to 101.5?

(d) In the interval 75.5 to 96.0?

(e) In the interval 49.0 to 112.0?

State which of your answers are based on the assumption of a bell-shaped distribution.

**2.115** The 50 measurements of acid rain in Wisconsin, whose histogram is given on page 23 are

| | | | | | |
|---|---|---|---|---|---|
| 3.58 | 3.80 | 4.01 | 4.01 | 4.05 | 4.05 |
| 4.12 | 4.18 | 4.20 | 4.21 | 4.27 | 4.28 |
| 4.30 | 4.32 | 4.33 | 4.35 | 4.35 | 4.41 |
| 4.42 | 4.45 | 4.45 | 4.50 | 4.50 | 4.50 |
| 4.50 | 4.51 | 4.52 | 4.52 | 4.52 | 4.57 |
| 4.58 | 4.60 | 4.61 | 4.61 | 4.62 | 4.62 |
| 4.65 | 4.70 | 4.70 | 4.70 | 4.70 | 4.72 |
| 4.78 | 4.78 | 4.80 | 5.07 | 5.20 | 5.26 |
| 5.41 | 5.48 | | | | |

(a) Calculate the median and quartiles.

(b) Find the 90th percentile.

(c) Determine the mean and standard deviation.

(d) Display the data in the form of a boxplot.

**2.116** Refer to Exercise 2.115.

(a) Determine the intervals $\bar{x} \pm s$, $\bar{x} \pm 2s$, and $\bar{x} \pm 3s$.

(b) What proportions of the measurements lie in those intervals?

(c) Compare your findings with the empirical guidelines for bell-shaped distributions.

**2.117** Refer to the earthquake size data in Exercise 2.20.

(a) Calculate the median and quartiles.

(b) Calculate the mean and standard deviation.

(c) Display the data in the form of a boxplot.

**2.118** The Dow Jones average provides an indication of overall market level. The changes in this average, from year end, from 1969 to 1970 through 2010 to 2011 are summarized in the following frequency table, where the left-hand endpoint is excluded.

| Yearly Changes in the Dow Jones Average | |
|---|---|
| Change in DJ Average | Frequency |
| $(-4500, -1600]$ | 1 |
| $(-1600, -850]$ | 1 |
| $(-850, -250]$ | 2 |
| $(-250, 0]$ | 8 |
| $(0, 250]$ | 13 |
| $(250, 850]$ | 7 |
| $(850, 1650]$ | 6 |
| $(1650, 2450]$ | 4 |
| Total | 42 |

(a) Calculate the relative frequency for the intervals.

(b) Plot the relative frequency histogram. (*Hint:* Since the intervals have unequal widths, make the height of each rectangle equal to the relative frequency divided by the width of the interval.)

(c) What proportion of the changes are not positive?

(d) Comment on the location and shape of the distribution.

**2.119** The winning times of the men's 400-meter freestyle swimming in the Olympics (1964 to 2012) appear in the following table.

| Winning Times in Minutes and Seconds | |
| --- | --- |
| Year | Time |
| 1964 | 4:12.2 |
| 1968 | 4:09.0 |
| 1972 | 4:00.27 |
| 1976 | 3:51.93 |
| 1980 | 3:51.31 |
| 1984 | 3:51.23 |
| 1988 | 3:46.95 |
| 1992 | 3:45.00 |
| 1996 | 3:47.97 |
| 2000 | 3:40.59 |
| 2004 | 3:43.10 |
| 2008 | 3:41.86 |
| 2012 | 3:40.14 |

(a) Draw a dot diagram and label the points according to time order.

(b) Explain why it is not reasonable to group the data into a frequency distribution.

**2.120** The **mode** of a collection of observations is defined as the observed value with largest relative frequency. The mode is sometimes used as a center value. There can be more than one mode in a data set. Find the mode for the checked bag data given in Exercise 2.13.

**2.121** Lightning causes many deaths each year in the United States. The yearly number of deaths for 53 years, 1959 through 2011 are, by rows,

183 129 149 153 165 129 149 110 88 129 131 122 122
94 124 102 91 74 98 88 63 74 66 77 77 67
74 68 88 68 67 74 73 41 43 69 85 53 42
44 46 51 44 51 43 32 38 47 45 27 34 29
26

Obtain the mean and standard deviation.

**2.122** With reference to the lightning data in Exercise 2.121,

(a) Make a time plot of the data.

(b) Comment on the appropriateness of presenting the mean and standard deviation as summaries.

### The Following Exercises Require a Computer.

Calculations of the descriptive statistics such as $\bar{x}$ and $s$ are increasingly tedious with larger data sets. Current computer software programs alleviate the drudgery of hand calculations. Use MINITAB or some other software package program.

**2.123** Find $\bar{x}$ and $s$ for:

(a) The lizard data in Exercise 2.19.

(b) The acid rain data in Exercise 2.115.

**2.124** Lumber intended for building houses and other structures must be monitored for strength. The

Figure 21   MINITAB output for Exercise 2.124.

measurement of strength (pounds per square inch) for 61 specimens of Southern Pine (*Source*: U.S. Forest Products Laboratory) yielded

| | | | | | |
|------|------|------|------|------|------|
| 4001 | 3927 | 3048 | 4298 | 4000 | 3445 |
| 4949 | 3530 | 3075 | 4012 | 3797 | 3550 |
| 4027 | 3571 | 3738 | 5157 | 3598 | 4749 |
| 4263 | 3894 | 4262 | 4232 | 3852 | 4256 |
| 3271 | 4315 | 3078 | 3607 | 3889 | 3147 |
| 3421 | 3531 | 3987 | 4120 | 4349 | 4071 |
| 3686 | 3332 | 3285 | 3739 | 3544 | |
| 4103 | 3401 | 3601 | 3717 | 4846 | |
| 5005 | 3991 | 2866 | 3561 | 4003 | |
| 4387 | 3510 | 2884 | 3819 | 3173 | |
| 3470 | 3340 | 3214 | 3670 | 3694 | |

Using MINITAB, the sequence of choices

---

**Data**(in 1.124.txt):

Strength

**Dialog box:**

**Stat > Basic Statistics > Graphical Summary.**

Type *Strength* in **Variables.** *Click* **OK.**

---

produces a rather complete summary of the data. It includes the output on page 81 in Figure 21.

(a) Use this output to identify a departure from a bell-shaped pattern.

(b) MINITAB uses a slightly different scheme to determine the first and third quartiles, but the difference is not of practical importance with large samples. Calculate the first quartile using the definition in this book and compare with the value in the output.

**2.125** Find $\bar{x}$ and $s$ for the data set in Table 4.

**2.126** Find $\bar{x}$ and $s$ for the final times to run 1.5 miles in Table D.5 in the Data Bank.

**2.127** The SAS computer software package produced the output in the next column. Compare the mean and standard deviation with that of the MINITAB output in Exercise 2.124. Which output gives more digits?

The UNIVARIATE Procedure
**Variable: Strength**

Moments

| | |
|-----------------|-------------|
| N | 61 |
| Mean | 3800.95082 |
| Sum Observations | 231858 |
| Std Deviation | 513.541087 |
| Variance | 263724.448 |

| | |
|----------------------|-------------|
| Range | 2291 |
| Interquartile Range | 601.00000 |

| Quantile | Estimate |
|-------------|----------|
| 100% Max | 5157 |
| 99% | 5157 |
| 95% | 4846 |
| 90% | 4349 |
| 75% Q3 | 4071 |
| 50% Median | 3738 |
| 25% Q1 | 3470 |
| 10% | 3173 |
| 5% | 3075 |
| 1% | 2866 |
| 0% Min | 2866 |

Extreme Observations

| ----Lowest---- | | ----Highest--- | |
|-------|-----|-------|-----|
| Value | Obs | Value | Obs |
| 2866 | 49 | 4749 | 18 |
| 2884 | 54 | 4846 | 46 |
| 3048 | 3 | 4949 | 7 |
| 3075 | 9 | 5005 | 47 |
| 3078 | 27 | 5157 | 16 |

Figure 22   SAS output for Exercise 2.127.

**2.128** The salmon fisheries support a primary industry in Alaska and their management is of high priority. Salmon are born in freshwater rivers and streams but then swim out into the ocean for a few years before returning to spawn and die. In order to identify the origins of mature fish, researchers studied growth rings on their scales. The growth the first year in freshwater is measured by the width of the growth rings for that period of life. The growth ring for the first year in the ocean environment will give an indication of growth for that period. A set of these measurements are given in Table D.7 in the Data Bank.

(a) Describe the freshwater growth for males by making a histogram and calculating the mean, standard deviation, and quartiles.

(b) Describe the freshwater growth for females by making a histogram and calculating the mean, standard deviation, and quartiles.

(c) Construct boxplots to compare the growth of males and females.

**2.129** Refer to the alligator data in Table D.11 of the Data Bank. Using the data on $x_5$ for thirty-seven alligators:

(a) Make a histogram.

(b) Obtain the sample mean and standard deviation.

**2.130** Refer to Exercise 2.129.

(a) Obtain the quartiles.

(b) Obtain the 90th percentile. How many of the alligators above the 90th percentile are female?

**2.131** Refer to the data on malt extract in Table D.8 of the Data Bank.

(a) Obtain sample mean and standard deviation.

(b) Obtain quartiles.

(c) Check conformity with the empirical rule.

# 3

# Descriptive Study of Bivariate Data

Courtesy of Wisconsin Department of
Natural Resources

## Ozone-Temperature Association and Air Quality

Ozone is a primary component of smog and breathing ground-level ozone can produce serious health effects. High daily ozone concentrations are associated with increases in asthma attacks and also mortality especially among the elderly. A broad segment of the population experiences respiratory problems and allergy type symptoms when temperature and ozone are both at high levels.

The Environmental Protection Agency has set safe standards and ozone alerts are issued on days when high levels of ozone are expected. During the summers of 2011 and 2012, the Wisconsin Department of Natural Resources recorded the daily maximum temperature and ozone (parts per billion) for Milwaukee County. A plot, based on the readings from 23 Wednesdays, reveals a relationship.

Days with high maximum temperature tend to have high ozone. Other days have small values of both temperature and ozone. This indicates a positive association between maximum temperature and ozone.

Chapter 2 explains how to organize and summarize data concerning a single variable. Observations on *two or more* variables are often recorded for the individual sampling units; the height and weight of individuals, or the number of goals scored by and against a team. By studying such **bivariate or multivariate** data, one typically wishes to discover if any relationships exist between the variables, how strong the relationships appear to be, and whether one variable of primary interest can be effectively predicted from information on the values of the other variables. To illustrate the concepts, we restrict our attention to the simplest case where only two characteristics are observed on the individual sampling units. Some examples are:

Gender and the type of occupation of college graduates.
Smoking habit and lung capacity of adult males.
Average daily carbohydrate intake and protein intake of 10-year-old children.
The age of an aircraft and the time required for repairs.

The two characteristics observed may both be qualitative traits, both numerical variables, or one of each kind. For brevity, we will only deal with situations where the characteristics observed are either both categorical or both numerical. Summarization of bivariate categorical data is discussed in Section 1. Sections 3, 4, and 5 are concerned with bivariate measurement data and treat such issues as graphical presentations, examination of relationship, and prediction of one variable from another.

## 1. SUMMARIZATION OF BIVARIATE CATEGORICAL DATA

When two traits are observed for the individual sampling units and each trait is recorded in some qualitative categories, the resulting data can be summarized in the form of a two-way frequency table. The categories for one trait are marked along the left margin, those for the other along the upper margin, and the frequency counts recorded in the cells. The total frequency for any row is given in the right-hand margin and those for any column given at the bottom margin. Both are called **marginal totals**.

Data in this summary form are commonly called **cross-classified** or **cross-tabulated data**. In statistical terminology, they are also called **contingency tables**.

**Example 1**    Calculation of Relative Frequencies Aids Interpretation

Four hundred undergraduates were surveyed concerning their part-time work during the semester. The number of hours worked last week was categorized as: worked 10 or less hours, worked more than 10 hours, or did not work. The students were also categorized as underclassman or upperclassman. The cross-tabulated frequency counts are presented as Table 1.

**TABLE 1**    Cross-Tabulated Frequency Counts of Work Hours

|  | No Job | Work 10 Hours or Less | Work More Than 10 Hours | Total |
|---|---|---|---|---|
| Underclassman | 132 | 28 | 20 | 180 |
| Upperclassman | 124 | 44 | 52 | 220 |
| Total | 256 | 72 | 72 | 400 |

The entries of this table are self-explanatory. For instance, of the 400 students polled, 180 were underclassman. Among these, 132 did not work, 28 worked 10 hours or less, and 20 worked more than 10 hours. To gain further understanding of how the responses are distributed, calculate the relative frequencies.

SOLUTION    For this purpose, we divide each cell frequency by the sample size 400. The relative frequencies, for instance $44 / 400 = .11$, are shown in Table 2.

**TABLE 2**  Relative Frequencies for the Data of Table 1

|  | No Job | Work 10 Hours or Less | Work More Than 10 Hours | Total |
|---|---|---|---|---|
| Underclassman | .33 | .07 | .05 | .45 |
| Upperclassman | .31 | .11 | .13 | .55 |
| Total | .64 | .18 | .18 | 1.00 |

Depending on the specific context of a cross-tabulation, one may also wish to examine the cell frequencies relative to a marginal total. In Example 1, you may wish to compare the pattern of part-time work for underclassmen with that of the upperclassman. This is accomplished by calculating the relative frequencies separately for the two groups. For instance, 44 / 220 = .200, as Table 3 shows.

**TABLE 3**  Relative Frequencies by Class

|  | No Job | Work 10 Hours or Less | Work More Than 10 Hours | Total |
|---|---|---|---|---|
| Underclassman | .733 | .156 | .111 | 1.000 |
| Upperclassman | .564 | .200 | .236 | 1.000 |

From the calculations presented in Table 3, it appears that a larger proportion of upperclassmen hold part-time jobs and that they tend to work more than 10 hours a week.

Now the pertinent question is: Can these observed differences be explained by chance or are there real differences in the pattern of part-time work between the populations of the two classes? We pursue this aspect of statistical inference in Chapter 13.

## 1.1  SIMPSON'S PARADOX

Quite surprising and misleading conclusions can occur when data from different sources are combined into a single table. We illustrate this reversal of implications with graduate school admissions data.

**Example 2**  Combining Tables Can Produce Misleading Summaries

We consider graduate school admissions at a large midwestern university but, to simplify, we use only two departments as the whole school. We are interested in comparing admission rates by gender and obtain the data in Table 4 for the school.

**TABLE 4**    School Admission Rates

|        | Admit | Not Admit | Total Applicants |
|--------|-------|-----------|------------------|
| Male   | 233   | 324       | 557              |
| Female | 88    | 194       | 282              |
| Total  | 321   | 518       | 839              |

Does there appear to be a gender bias?

SOLUTION    It is clear from these admissions statistics that the proportion of males admitted, 233 / 557 = .418, is greater than the proportion of females admitted, 88 / 282 = .312.

Does this imply some type of discrimination? Not necessarily. By checking the admission records, we were able to further categorize the cases according to department in Table 5. Table 4 is the aggregate of these two sets of data.

**TABLE 5**    Admission Rates by Department

| Mechanical Engineering | | | | History | | |
|--------|-------|--------------|-------|--------|-------|--------------|-------|
|        | Admit | Not Admit | Total |        | Admit | Not Admit | Total |
| Male   | 151   | 35        | 186   | Male   | 82    | 289       | 371   |
| Female | 16    | 2         | 18    | Female | 72    | 192       | 264   |
| Total  | 167   | 37        | 204   | Total  | 154   | 481       | 635   |

One of the two departments, mechanical engineering, has mostly male applicants. Even so, the proportion of males admitted, 151 / 186 = .812, is smaller than the proportion of females admitted, 16 / 18 = .889. The same is true for the history department where the proportion of males admitted, 82 / 371 = .221, is again smaller than the proportion of females admitted, 72 / 264 = .273. When the data are studied department by department, the reverse but correct conclusion holds; females have a higher admission rate in both cases!

To obtain the correct interpretation, these data need to be presented as the full three-way table of gender-admission action-department as given above. If department is ignored as in Table 4, and the data aggregated across this variable, "department" can act as an unrecorded or lurking variable. In this example, it has reversed the direction of possible gender bias and led to the erroneous conclusion that males have a higher admission rate than females.

The reversal of the comparison, such as in Example 2, when data are combined from several groups is called **Simpson's paradox**.

When data from several sources are aggregated into a single table, there is always the danger that unreported variables may cause a reversal of the findings. In practical applications, there is not always agreement on how much effort to expend following up on unreported variables. When comparing two medical treatments, the results often need to be adjusted for the age, gender, and sometimes current health of the subjects and other variables.

## Exercises

**3.1** Nausea from air sickness affects some travelers. A drug company, wanting to establish the effectiveness of its motion sickness pill, randomly gives either its pill or a look-alike sugar pill (placebo) to 200 passengers.

| | Degree of Nausea | | | | |
|---|---|---|---|---|---|
| | None | Slight | Moderate | Severe | Total |
| Pill | 43 | 36 | 18 | 3 | 100 |
| Placebo | 19 | 33 | 36 | 12 | 100 |
| Total | | | | | |

(a) Complete the marginal totals.

(b) Calculate the relative frequencies separately for each row.

(c) Comment on any apparent differences in response between the pill and the placebo.

**3.2** Breakfast cereals from three leading manufacturers can be classified either above average or below average in sugar content. Data for ten cereals from each manufacturer are given below:

| | Below Average | Above Average | Total |
|---|---|---|---|
| General Mills | 2 | 8 | 10 |
| Kellogg's | 3 | 7 | 10 |
| Quaker | 4 | 6 | 10 |
| Total | | | |

(a) Complete the marginal totals.

(b) Calculate the relative frequencies separately for each row.

**3.3** At the conclusion of one semester, a sample of 250 juniors was questioned about how long they studied for each of their final exams. Students were also classified as social, biological, or physical science majors.

Number of Hours of Study for Each Final

| Major | 10 or less | More than 10 |
|---|---|---|
| Biological | 30 | 45 |
| Physical | 15 | 35 |
| Social | 65 | 60 |

Compare the times studying for finals by calculating the relative frequencies.

**3.4** A survey was conducted to study the attitudes of the faculty, academic staff, and students in regard to a proposed measure for reducing the heating and air-conditioning expenses on campus.

| | Favor | Indifferent | Opposed | Total |
|---|---|---|---|---|
| Faculty | 36 | 42 | 122 | 200 |
| Academic staff | 44 | 77 | 129 | 250 |
| Students | 106 | 178 | 116 | 400 |

Compare the attitude patterns of the three groups by computing the relative frequencies.

**3.5** Groundwater from 19 wells was classified as low or high in alkalinity and low or high in dissolved iron. There were 9 wells with high alkalinity, 7 that were high in iron, and 5 that were high in both.

(a) Based on these data, complete the following two-way frequency table.

(b) Calculate the relative frequencies of the cells.

(c) Calculate the relative frequencies separately for each row.

|  | Iron | |
|---|---|---|
| Alkalinity | Low | High |
| Low | | |
| High | | |

**3.6** Interviews with 150 persons engaged in a stressful occupation revealed that 57 were alcoholics, 64 were mentally depressed, and 42 were both.

(a) Based on these records, complete the following two-way frequency table.

(b) Calculate the relative frequencies.

|  | Alcoholic | Not Alcoholic | Total |
|---|---|---|---|
| Depressed | | | |
| Not depressed | | | |
| Total | | | |

**3.7** Cross-tabulate the "Class data" of Exercise 2.100 according to gender (M, F) and the general areas of intended major (H, S, B, P). Calculate the relative frequencies.

**3.8** A psychologist interested in obese children gathered data on a group of children and their parents.

|  | Child | |
|---|---|---|
| Parent | Obese | Not Obese |
| At least one obese | 12 | 24 |
| Neither obese | 8 | 36 |

(a) Calculate the marginal totals.

(b) Convert the frequencies to relative frequencies.

(c) Calculate the relative frequencies separately for each row.

**3.9** Typically, there is a gender unbalance among tenured faculty, especially in the sciences. At a large university, tenured faculty members in two departments, English and Computer Science, were categorized according to gender.

|  | Male | Female |
|---|---|---|
| English | 23 | 19 |
| Computer Science | 27 | 5 |

(a) Calculate relative frequencies separately for each row.

(b) Comment on major differences in the patterns for the two rows.

**3.10** A large research hospital and a community hospital are located in your area. The surgery records for the two hospitals are:

|  | Died | Survived | Total |
|---|---|---|---|
| Research hospital | 90 | 2110 | 2200 |
| Community hospital | 23 | 677 | 700 |

The outcome is "survived" if the patient lives at least six weeks.

(a) Calculate the proportion of patients that survive surgery at each of the hospitals.

(b) Which hospital do these data suggest you should choose for surgery?

**3.11** Refer to Exercise 3.10. Not all surgery cases, even of the same type, are equally serious. Large research hospitals tend to get the most serious surgery cases, whereas community hospitals tend to get more of the routine cases. Suppose that patients can be classified as being in either "Good" or "Poor" condition and the outcomes of surgery are as shown in the table on the next page.

(a) Calculate the proportions that survive for each hospital and each condition.

(b) From these data, which hospital would you choose if you were in good condition? If you were in bad condition?

(c) Compare your answer with that to Exercise 3.10. Explain this reversal as an example of Simpson's paradox and identify the lurking variable in Exercise 3.10.

Survival Rates by Condition

Good Condition

| | Died | Survived | Total |
|---|---|---|---|
| Research hospital | 15 | 685 | 700 |
| Community hospital | 16 | 584 | 600 |
| Total | 31 | 1269 | 1300 |

Poor Condition

| | Died | Survived | Total |
|---|---|---|---|
| Research hospital | 75 | 1425 | 1500 |
| Community hospital | 7 | 93 | 100 |
| Total | 82 | 1518 | 1600 |

## 2.  A DESIGNED EXPERIMENT FOR MAKING A COMPARISON

We regularly encounter claims that, as a group, smokers have worse health records than nonsmokers with respect to one disease or another or that a new medical treatment is better than the former one. Properly designed experiments can often provide data that are so conclusive that a comparison is clear-cut. An example of a comparative study illustrates the design issue and how to conduct an experiment.

During the early development of a medicated skin patch to help smokers break the habit, a test was conducted with 112 volunteers. The experimenters wanted to avoid errorneous conclusions caused by the so-called placebo effect when a treatment, with no therapeutic value, is administered to a subject but their symptoms improve anyway. One explanation is the positive thinking of subjects having high expectations of getting better and who believe the real treatment will work. Consequently, half the volunteers were given an unmedicated skin patch. The data will consist of a count of the number of persons who are abstinent at the end of the study.

*Purpose:* To determine the effectiveness of a medicated nicotine patch for smoking cessation based on the end-of-therapy numbers of abstinent persons in the medicated and unmedicated groups.

What is involved in comparing two approaches or methods for doing something? First the subjects, or experimental units, must be assigned to the two groups in such a manner that neither method is favored. One approach is to list the subjects' names on a piece of paper, cut the paper into strips, each with one name on it, and then draw one at a time until half the names are drawn. Ideally, we like to have groups of equal size, so if there is an odd number of subjects, draw just over one-half. These subjects are assigned to the first approach. The other subjects are assigned to the second approach. This step, called **random assignment**, helps guarantee a valid comparison. Any subject likely to respond positively has the same chance of supporting the first approach as supporting the second approach. When subjects cannot be randomly assigned, we will never know if observed differences in the number of abstinent smokers is due to the approaches or some other variables associated with the particular persons assigned to the two groups.

Subjects are not told whether their randomly assigned group is the medicated or unmedicated (placebo) group. As with many medical trials, this is a **double blind trial**. That is, the medical staff in contact with the patients are also kept unaware of which patients are getting the treated patch and which are not. At the end of the study, the number of persons in each group who are abstinent and who are smoking is recorded.

The data[1] collected from this experiment are summarized in Table 6.

**TABLE 6**   Quitting Smoking

|  | Abstinent | Smoking | |
|---|---|---|---|
| Medicated patch | 21 | 36 | 57 |
| Unmedicated patch | 11 | 44 | 55 |
|  | 32 | 80 | 112 |

The proportion abstinent is $21/57 = .368$ for the medicated skin patch group and only $11/55 = .200$ for the control. The medicated patch seems to work. Later, in Chapter 10, we verify that the difference $.368 - .200 = .168$ is greater than can be explained by chance variation.

In any application where the subjects might learn from the subjects before them, it would be a poor idea to perform all the trials for treatment 1 and then all those for treatment 2. Learning or other uncontrolled variables must not be given the opportunity to systematically affect the experiment. We could number the subjects 1 to 112 and write each of these numbers on a separate slip of paper. The 112 slips of paper should be mixed and then drawn one at a time to determine the sequence in which the trials are conducted.

Researchers continue to investigate the effectiveness of patches. One study presents evidence against the effectiveness of patches.[2]

[1]M. Fiore, S. Kenford, D. Jorenby, D. Wetter, S. Smith, and T. Baker. "Two studies of the clinical effectiveness of the nicotine patch with different counseling treatments." *Chest* **105** (1994), pp. 524–533.
[2]H. Albert et al., "A prospective cohort study challenging the effectiveness of population-based medical intervention for smoking cessation", *Tobacco Control*, January 2012, doi:10.1136/tobaccocontrol-2011-050129.

# Exercises

**3.12** With reference to the quit-smoking experiment, suppose two new subjects are available. Explain how you would assign one subject to receive the placebo and one to receive the medicated patch.

**3.13** With reference to the quit-smoking experiment:

(a) Suppose the placebo trials were ignored and you were only told that 21 of 57 were abstinent after using the medicated patches. Would this now appear to be stronger evidence in favor of the patches?

(b) Explain why the placebo trials provide a more valid reference for results of the medicated patch trials.

# 3. SCATTER DIAGRAM OF BIVARIATE MEASUREMENT DATA

How do we describe data sets concerning two variables, each measured on a numerical scale? For ease of reference, we will label one variable $x$ and the other $y$. Thus, two numerical observations $(x, y)$ are recorded for each sampling unit. These observations are *paired* in the sense that an $(x, y)$ pair arises from the same sampling unit. An $x$ observation from one pair and an $x$ or $y$ from another are unrelated. For $n$ sampling units, we can write the measurement pairs as

$$(x_1, y_1), (x_2, y_2), \ldots, (x_n, y_n)$$

The set of $x$ measurements alone, if we disregard the $y$ measurements, constitutes a data set for one variable. The methods of Chapter 2 provide graphical presentations of the measurements and the mean, standard deviation, and other quantities. Likewise, the $y$ measurements can be studied disregarding the $x$ measurements. However, a major purpose of collecting bivariate data is to answer such questions as:

Are the variables related?

What form of relationship is indicated by the data?

Can we quantify the strength of their relationship?

Can we predict one variable from the other?

Studying either the $x$ measurements by themselves or the $y$ measurements by themselves would not help answer these questions.

An important first step in studying the relationship between two variables is to graph the data. To this end, the variable $x$ is marked along the horizontal axis and $y$ on the vertical axis. The pairs $(x, y)$ of observations are then plotted as dots on the graph. The resulting diagram is called a scatter diagram or scatter plot. By looking at the scatter diagram, a visual impression can be formed about the relation between the variables. For instance, we can observe whether the points band around a line or a curve or if they form a patternless cluster.

**Example 3**    A Scatter Diagram Provides a Visual Display of a Relationship
Recorded in Table 7 are the data of

$$x = \text{Undergraduate GPA}$$

and

$$y = \text{Score in the Graduate Management Aptitude Test (GMAT)}$$

for applicants seeking admission to an MBA program.
Construct a scatter diagram.

**TABLE 7**  Data of Undergraduate GPA
$x$ and GMAT Score $y$

| $x$ | $y$ | $x$ | $y$ | $x$ | $y$ |
|------|------|------|------|------|------|
| 3.63 | 447 | 2.36 | 399 | 2.80 | 444 |
| 3.59 | 588 | 2.36 | 482 | 3.13 | 416 |
| 3.30 | 563 | 2.66 | 420 | 3.01 | 471 |
| 3.40 | 553 | 2.68 | 414 | 2.79 | 490 |
| 3.50 | 572 | 2.48 | 533 | 2.89 | 431 |
| 3.78 | 591 | 2.46 | 509 | 2.91 | 446 |
| 3.44 | 692 | 2.63 | 504 | 2.75 | 546 |
| 3.48 | 528 | 2.44 | 336 | 2.73 | 467 |
| 3.47 | 552 | 2.13 | 408 | 3.12 | 463 |
| 3.35 | 520 | 2.41 | 469 | 3.08 | 440 |
| 3.39 | 543 | 2.55 | 538 | 3.03 | 419 |
|      |     |      |      | 3.00 | 509 |

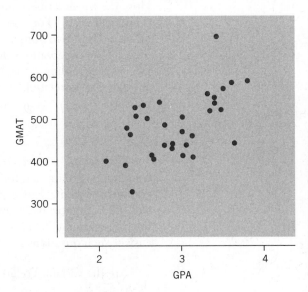

Figure 1    Scatter diagram of applicants' scores.

SOLUTION    The scatter diagram is plotted in Figure 1. The southwest-to-northeast pattern
of the points indicates a positive relation between $x$ and $y$. That is, the
applicants with a high GPA tend to have a high GMAT. Evidently, the relation
is far from a perfect mathematical relation.

When the two measurements are made on two or more groups, visual
comparisons between groups are made by plotting the points on the same scatter
plot. A different symbol is used for each group. The resulting graph is called a
multiple scatter plot where "multiple" refers to groups.

**Example 4**    A Multiple Scatter Diagram for Visually Comparing Relationships

Concern was raised by environmentalists that spills of contaminants were affecting wildlife in and around an adjacent lake. Estrogenic contaminants in the environment can have grave consequences on the ability of living things to reproduce. Researchers examined the reproductive development of young male alligators hatched from eggs taken from around (1) Lake Apopka, the lake which was contaminated, and (2) Lake Woodruff, which acted as a control. The contaminants were thought to influence sex steroid concentrations. The concentrations of two steroids, estradiol and testosterone, were determined by radioimmunoassay.

**Lake Apopka**

| Estradiol | 38 | 23 | 53 | 37 | 30 |
|---|---|---|---|---|---|
| Testosterone | 22 | 24 | 8 | 6 | 7 |

**Lake Woodruff**

| Estradiol | 29 | 64 | 19 | 36 | 27 | 16 | 15 | 72 | 85 |
|---|---|---|---|---|---|---|---|---|---|
| Testosterone | 47 | 20 | 60 | 75 | 12 | 54 | 33 | 53 | 100 |

(a) Make a scatter diagram of the two concentrations for the Lake Apopka alligators.

(b) Create a multiple scatter diagram by adding to the same plot the pairs of concentrations for the Lake Woodruff male alligators. Use a different symbol for the two lakes.

(c) Comment on any major differences between the two groups of male alligators.

SOLUTION    (a) Figure 2a gives the scatter diagram for the Lake Apopka alligators.

(a) Scatter diagram for Lake Apopka

Figure 2a    Scatter diagram. $A$ = Lake Apopka.

(b) Multiple scatter diagram

Figure 2b    Scatter diagram. $B$ = Lake Woodruff.

(b)  Figure 2*b* is the multiple scatter diagram with the points for Lake Woodruff marked as *B*.

(c)  The most prominent feature of the data is that the male alligators from the contaminated lake have, generally, much lower levels of testosterone than those from the nearly pollution-free control lake. (The *A*'s are at the bottom third of the multiple scatter diagram.) Low testosterone levels in males have grave consequences regarding reproduction.

# 4.  THE CORRELATION COEFFICIENT—A MEASURE OF LINEAR RELATION

The scatter diagram provides a visual impression of the nature of the relation between the $x$ and $y$ values in a bivariate data set. In a great many cases, the points appear to band around a straight line. Our visual impression of the closeness of the scatter to a linear relation can be quantified by calculating a numerical measure, called the **correlation coefficient**.

The correlation coefficient, denoted by $r$, is a measure of strength of the linear relation between the $x$ and $y$ variables. Before introducing its formula, we outline some important features of the correlation coefficient and discuss the manner in which it serves to measure the strength of a linear relation.

1.  The value of $r$ is always between $-1$ and $+1$.

2.  The magnitude of $r$ indicates the strength of a linear relation, whereas its sign indicates the direction. More specifically,

$r > 0$ if the pattern of $(x, y)$ values is a band that runs from lower left to upper right.

$r < 0$ if the pattern of $(x, y)$ values is a band that runs from upper left to lower right.

$r = +1$ if all $(x, y)$ values lie exactly on a straight line with a positive slope (perfect positive linear relation).

$r = -1$ if all $(x, y)$ values lie exactly on a straight line with a negative slope (perfect negative linear relation).

A value of $r$ close to $+1$ or $-1$, represents a strong linear relation.

3.  A value of $r$ close to zero means that the linear association is very weak.

The correlation coefficient is close to zero when there is no visible pattern of relation; that is, the $y$ values do not change in any direction as the $x$ values change. A value of $r$ near zero could also happen because the points band around a curve that is far from linear. After all, $r$ measures linear association, and a markedly bent curve is far from linear.

Figure 3 shows the correspondence between the appearance of a scatter diagram and the value of $r$. Observe that $(e)$ and $(f)$ correspond to situations

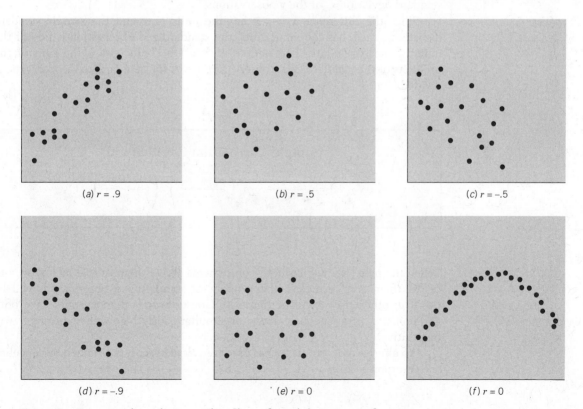

(a) $r = .9$    (b) $r = .5$    (c) $r = -.5$

(d) $r = -.9$    (e) $r = 0$    (f) $r = 0$

Figure 3   Correspondence between the values of $r$ and the amount of scatter.

where $r = 0$. The zero correlation in ($e$) is due to the absence of any relation between $x$ and $y$, whereas in ($f$) it is due to a relation following a curve that is far from linear.

## 4.1 CALCULATION OF $r$

The sample correlation coefficient quantifies the association between two numerically valued characteristics. It is calculated from $n$ pairs of observations on the two characteristics

$$(x_1, y_1), (x_2, y_2), \ldots, (x_n, y_n)$$

The correlation coefficient is best interpreted in terms of the standardized observations, or sample z values

$$\frac{\text{Observation} - \text{Sample mean}}{\text{Sample standard deviation}} = \frac{x_i - \bar{x}}{s_x}$$

where $s_x = \sqrt{\sum_{i=1}^{n}(x_i - \bar{x})^2 / (n - 1)}$ and the subscript $x$ on $s$ distinguishes the sample standard deviation of the $x$ observations from the sample standard deviation $s_y$ of the $y$ observations.

Since the difference $x_i - \bar{x}$ has the units of $x$ and the sample standard deviation $s_x$ also has the same units, the standardized observation is free of the units of measurements. The sample correlation coefficient is the sum of the products of the standardized $x$ observation times the standardized $y$ observations divided by $n - 1$.

---

### Sample Correlation Coefficient

$$r = \frac{1}{n - 1} \sum_{i=1}^{n} \left( \frac{x_i - \bar{x}}{s_x} \right) \left( \frac{y_i - \bar{y}}{s_y} \right)$$

---

When the pair $(x_i, y_i)$ has both components above their sample means or both below their sample means, the product of standardized observations will be positive; otherwise it will be negative. Consequently, if most pairs have both components simultaneously above or simultaneously below their means, $r$ will be positive.

An alternative formula for $r$ is used for calculation. It is obtained by canceling the common term $n - 1$.

### Calculation Formula for the Sample Correlation Coefficient

$$r = \frac{S_{xy}}{\sqrt{S_{xx}}\,\sqrt{S_{yy}}}$$

where

$$S_{xy} = \sum (x - \bar{x})(y - \bar{y})$$

$$S_{xx} = \sum (x - \bar{x})^2 \qquad S_{yy} = \sum (y - \bar{y})^2$$

The quantities $S_{xx}$ and $S_{yy}$ are the sums of squared deviations of the $x$ observations and the $y$ observations, respectively. $S_{xy}$ is the sum of cross products of the $x$ deviations with the $y$ deviations. This formula is examined in more detail in Chapter 11.

**Example 5**   Calculation of Sample Correlation

Calculate $r$ for the $n = 4$ pairs of observations

$$(2, 5) \ (1, 3) \ (5, 6) \ (0, 2)$$

SOLUTION   We first determine the mean $\bar{x}$ and deviations $x - \bar{x}$ and then $\bar{y}$ and the deviations $y - \bar{y}$. See Table 8.

**TABLE 8**   Calculation of $r$

| $x$ | $y$ | $x - \bar{x}$ | $y - \bar{y}$ | $(x - \bar{x})^2$ | $(y - \bar{y})^2$ | $(x - \bar{x})(y - \bar{y})$ |
|---|---|---|---|---|---|---|
| 2 | 5 | 0 | 1 | 0 | 1 | 0 |
| 1 | 3 | −1 | −1 | 1 | 1 | 1 |
| 5 | 6 | 3 | 2 | 9 | 4 | 6 |
| 0 | 2 | −2 | −2 | 4 | 4 | 4 |
| Total  8 | 16 | 0 | 0 | 14 | 10 | 11 |
| $\bar{x} = 2$ | $\bar{y} = 4$ | | | $S_{xx}$ | $S_{yy}$ | $S_{xy}$ |

Consequently,

$$r = \frac{S_{xy}}{\sqrt{S_{xx}}\sqrt{S_{yy}}} = \frac{11}{\sqrt{14}\sqrt{10}} = .930$$

The value .930 is large and it implies a strong association where both $x$ and $y$ tend to be small or both tend to be large.

It is sometimes convenient, when using hand-held calculators, to evaluate $r$ using the alternative formulas for $S_{xx}$, $S_{yy}$, and $S_{xy}$ (see Appendix A1.2).

$$S_{xx} = \sum x^2 - \frac{(\sum x)^2}{n} \qquad S_{yy} = \sum y^2 - \frac{(\sum y)^2}{n}$$

$$S_{xy} = \sum xy - \frac{(\sum x)(\sum y)}{n}$$

This calculation is illustrated in Table 9.

TABLE 9    Alternate Calculation of $r$

|  | $x$ | $y$ | $x^2$ | $y^2$ | $xy$ |
|---|---|---|---|---|---|
|  | 2 | 5 | 4 | 25 | 10 |
|  | 1 | 3 | 1 | 9 | 3 |
|  | 5 | 6 | 25 | 36 | 30 |
|  | 0 | 2 | 0 | 4 | 0 |
| Total | 8 | 16 | 30 | 74 | 43 |
|  | $\sum x$ | $\sum y$ | $\sum x^2$ | $\sum y^2$ | $\sum xy$ |

$$r = \frac{43 - \dfrac{8 \times 16}{4}}{\sqrt{30 - \dfrac{8^2}{4}}\sqrt{74 - \dfrac{(16)^2}{4}}} = .930$$

We remind the reader that $r$ measures the closeness of the pattern of scatter to a line. Figure 3f on page 97 presents a strong relationship between $x$ and $y$, but one that is not linear. The small value of $r$ for these data does not properly reflect the strength of the relation. Clearly $r$ is not an appropriate summary of a curved pattern. Another situation where the sample correlation coefficient $r$ is not appropriate occurs when the scatter plot breaks into two clusters. Faced with separate clusters as depicted in Figure 4, it is best to try and determine the underlying cause. It may be that a part of the sample has come from one population and a part from another.

Figure 4    $r$ is not appropriate—samples from two populations.

## 4.2 CORRELATION AND CAUSATION

Data analysts often jump to unjustified conclusions by mistaking an observed correlation for a cause-and-effect relationship. A high sample correlation coefficient does not necessarily signify a causal relation between two variables. A classic example concerns an observed high positive correlation between the number of storks sighted and the number of births in a European city. It is hoped no one would use this evidence to conclude that storks bring babies or, worse yet, that killing storks would control population growth.

The observation that two variables tend to simultaneously vary in a certain direction does not imply the presence of a direct relationship between them. If we record the monthly number of homicides $x$ and the monthly number of religious meetings $y$ for several cities of widely varying sizes, the data will probably indicate a high positive correlation. It is the fluctuation of a third variable (namely, the city population) that causes $x$ and $y$ to vary in the same direction, despite the fact that $x$ and $y$ may be unrelated or even negatively related. Picturesquely, the third variable, which in this example is actually causing the observed correlation between crime and religious meetings, is referred to as a **lurking variable**. The false correlation that it produces is called a **spurious correlation**. It is more a matter of common sense than of statistical reasoning to determine if an observed correlation has a practical interpretation or is spurious.

> An observed correlation between two variables may be **spurious**. That is, it may be caused by the influence of a third variable.

When using the correlation coefficient as a measure of relationship, we must be careful to avoid the possibility that a lurking variable is affecting any of the variables under consideration.

**Example 6**    Spurious Correlation Caused by Lurking Variables

Figure 5 gives a scatter diagram of the number of persons in prison, under state or federal jurisdiction, and the number of cell phone subscribers in each of 10 years. Both variables are measured in millions (see Exercise 3.29).

Figure 5   Scatter diagram reveals pattern of strong positive correlation.

This plot exhibits a pattern of strong positive correlation; the numerical value $r = .983$. Would restricting the number of cell phones result in fewer persons in prison?

SOLUTION   The scatter diagram reveals a strong positive correlation, but common sense suggests there is no cause-and-effect relation to tie an increase in the number of cell phone subscribers to an increase in the prison population. Realistically, the two variables should not have a causal relationship.

In Figure 6 we have repeated the scatter diagram but have labeled each point according to the year. For example, 03 stands for 2003. The years increase exactly in the same order as the points from lower left to upper right in the scatter diagram. More things change over time or from year to year. Year is just a stand-in, or proxy, for all of them. Since the population of the United States increased over these years, population size could be one lurking variable.

Figure 6   Scatter diagram pattern has strong relation to year.

Once the year notation is added to the graph, it is clear that other variables are leading to the observed correlation. A graph, with time order added, can often help discredit false claims of causal relations. If the order of the years had been scrambled, we could not discredit the suggestion of a causal relation.

### Lurking Variables

The Insurance Institute for Highway Safety 2011 report announced the safest and unsafest 2005–2008 car models for the period 2006 to 2009 in terms of fewest fatalities per one million registered vehicle years. The death rates, shown in parentheses, are given in terms of one million cars that are registered for the year.

| Lowest Fatality Rates | | Highest Fatality Rates | |
|---|---|---|---|
| Mercedes E-class 4-door AWD | (0) | Nissan 350Z | (143) |
| Mercedes E-class 4-door | (12) | Chevrolet Aveo 4-door | (119) |
| Saab 9–3 4-door | (16) | Chevrolet Cobalt 4-door | (117) |
| Honda Accord 4-door | (19) | Kia Spectra | (102) |
| Accura 3.2 TL | (21) | Chevrolet Malibu Classic 4-door | (99) |
| Accura RL | (21) | Hyundai Tiburon 2-door | (96) |
| | | Nissan Versa 4-door | (96) |

Although it must be acknowledged there is truth in the statement that large cars are generally safer than small cars, there is a big lurking variable here—the driver. How often does the teenager cruise in the luxury car? There is a strong correlation between the age of the driver and the type of car driven and also between the age of the driver and driver behavior.

**WARNING.** Don't confuse the existence of a high correlation between two variables with the situation where one is the cause of the other. Recall Example 6, where the number of cell phones and the number of persons in prison have a high correlation. There is no commonsense connection—no causal relation.

A high correlation between two variables can sometimes be caused when there is a third, "lurking" variable that strongly influences both of them.

**Example 7**    A Very Expensive Mistake of Confusing Association and Causation

Toward the end of 2006, a drug company made an exciting announcement about a new type of pill for treating heart disease.[3] They had already invested $1 billion for development and $90 billion to expand manufacturing facilities. The drug had recently entered the last phase of clinical trials that is required before FDA approval. The science seemed sound. High levels of cholesterol is consistently linked to heart disease and accumulations in arterial walls leads to blood vessels clogged with fat. Existing drugs lower the level of so-called bad cholesterol (LDL). The new drug was designed to block the conversion of good cholesterol (HDL) to LDL and thereby reduce the amount of the bad.

[3] Based on J. Lehrer, "Trials and errors: Why science is failing us," *Wired*, (Jan. 2012), pp. 102–109.

Amazingly, two days later the company announced that their clinical trial was being terminated. The drug was actually resulting in higher rates of chest pain and heart failure along with a 60 percent increase in mortality. The week of the announcement, the value of the company dropped by $21 billion. Comment on the mistake of confusing association with causation.

SOLUTION    Certain steps seemed to verify that cholesterol was associated with blockage of blood vessels but the causal connection of the blocking conversion step to mortality was only assumed. Unfortunately, not only is the relationship not causal, but the drug caused an increase in mortality. The monetary loss is over $100 billion.

Figure 7    A lesson in spurious correlation.

# Exercises

**3.14** Would you expect a positive, negative, or nearly zero correlation for each of the following? Give reasons for your answers.

(a) The physical fitness of a dog and the physical fitness of the owner.

(b) For each person, the number of songs downloaded from the Internet last month and the number of hours listening to MP3 format music.

(c) For a student, the number of friends listed on their personal Internet sites and the number of hours they are active on the Internet.

**3.15** In each of the following instances, would you expect a positive, negative, or zero correlation?

(a) Number of salespersons and total dollar sales for real estate firms.

(b) Total payroll and percent of wins of national league baseball teams.

(c) The amount spent on a week of TV advertising and sales of a cola.

(d) Age of adults and their ability to maintain a strenuous exercise program.

**3.16** Data since 2000 reveals a moderately strong positive correlation between the federal debt and attendance at National Football League games. Would restricting the number of persons attending games reduce the national debt? Explain your answer.

**3.17** If the value of $r$ is small, can we conclude that there is not a strong relationship between the two variables?

**3.18** For the data set

| $x$ | 1 | 2 | 7 | 4 | 6 |
|---|---|---|---|---|---|
| $y$ | 5 | 4 | 1 | 3 | 2 |

(a) Construct a scatter diagram.

(b) Guess the sign and value of the correlation coefficient.

(c) Calculate the correlation coefficient.

**3.19** Refer to the alligator data in Table D.11 of the Data Bank. Using the data on $x_3$ and $x_4$ for male and female alligators from Lake Apopka:

(a) Make a scatter diagram of the pairs of concentrations for the male alligators. Calculate the sample correlation coefficient.

(b) Create a multiple scatter diagram by adding, on the same plot, the pairs of concentrations for the female alligators. Use a different symbol for females. Calculate the sample correlation coefficient for this latter group.

(c) Comment on any major differences between the male and female alligators.

**3.20** (a) Construct scatter diagrams of the data sets

(i)
| $x$ | 0 | 4 | 2 | 6 | 3 |
|---|---|---|---|---|---|
| $y$ | 4 | 6 | 2 | 8 | 5 |

(ii)
| $x$ | 0 | 4 | 2 | 6 | 3 |
|---|---|---|---|---|---|
| $y$ | 8 | 2 | 5 | 4 | 6 |

(b) Calculate $r$ for the data set (i).

(c) Guess the value of $r$ for the data set (ii) and then calculate $r$. (Note: The $x$ and $y$ values are the same for both sets, but they are paired differently in the two cases.)

**3.21** Match the following values of $r$ with the correct diagrams (Figure 8).

(a) $r = -.3$

(b) $r = .1$

(c) $r = .9$

**3.22** Is the correlation in Figure 9 about (a) .1, (b) .5, (c) .9, or (d) −.7?

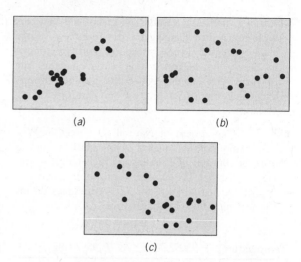

(a)          (b)

(c)

Figure 8    Scatter diagrams for Exercise 3.21.

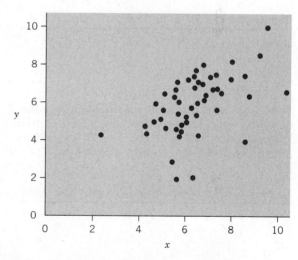

Figure 9    Scatter diagram for Exercise 3.22.

**3.23** Calculations from a data set of $n = 36$ pairs of $(x, y)$ values have provided the following results.

$$\sum (x - \bar{x})^2 = 530.7 \quad \sum (y - \bar{y})^2 = 235.4$$

$$\sum (x - \bar{x})(y - \bar{y}) = -204.3$$

Obtain the correlation coefficient.

**3.24** Over the years, a traffic officer noticed that cars with fuzzy dice hanging on the rear-view mirror always seemed to be speeding. Perhaps tongue in cheek, he suggested that outlawing the sale of fuzzy dice would reduce the number of cars exceeding the speed limit. Comment on lurking variables.

**3.25** The Department of Natural Resources in Wisconsin monitors air quality in the state. Ozone is a major component of smog and high temperatures also contribute. The summer maximum daily ozone (ppm) and temperature for 23 Wednesdays during the summer of 2011 and 2012 are

| Temperature | 79 93 62 75 69 78 70 98 74 82 77 84 |
|---|---|
| max Ozone | 56 69 43 53 43 44 26 75 42 49 36 60 |

| Temperature | 91 70 69 93 85 93 102 82 81 80 96 |
|---|---|
| max Ozone | 48 40 41 58 56 87 76 63 48 45 79 |

Calculate *r*.

**3.26** A zoologist collected 20 wild lizards in the southwestern United States. After measuring their total length (mm), they were placed on a treadmill and their speed (m/sec) recorded.

| Speed | 1.28 | 1.36 | 1.24 | 2.47 | 1.94 | 2.52 | 2.67 |
|---|---|---|---|---|---|---|---|
| Length | 179 | 157 | 169 | 146 | 143 | 131 | 159 |

| Speed | 1.29 | 1.56 | 2.66 | 2.17 | 1.57 | 2.10 | 2.54 |
|---|---|---|---|---|---|---|---|
| Length | 142 | 141 | 130 | 142 | 116 | 130 | 140 |

| Speed | 1.63 | 2.11 | 2.57 | 1.72 | 0.76 | 1.02 |
|---|---|---|---|---|---|---|
| Length | 138 | 137 | 134 | 114 | 90 | 114 |

(a) Create a scatter plot. Comment on any unusual observations.

(b) Calculate the sample correlation coefficient.

**3.27** An ongoing study of wolves is being conducted at the Yukon-Charley Rivers National Preserve. Table D.9 in the Data Bank gives the physical characteristics of wolves that were captured.

Males

| Body length (cm) | 134 | 143 | 148 | 127 | 136 | 146 |
|---|---|---|---|---|---|---|
| Weight (lb) | 71 | 93 | 101 | 84 | 88 | 117 |

| Body length (cm) | 142 | 139 | 140 | 133 | 123 |
|---|---|---|---|---|---|
| Weight (lb) | 86 | 86 | 93 | 86 | 106 |

(a) Plot length versus weight for the male wolves. From your visual inspection, estimate the value of the correlation coefficient.

(b) Calculate the sample correlation coefficient for male wolves.

(c) Create a multiple scatter diagram by adding the points for female wolves from Table D.9 to your plot in part (a). Do the patterns of correlation for males and females appear to be similar or different?

**3.28** An ongoing study of wolves is being conducted at the Yukon-Charley Rivers National Preserve. Table D.9 in the Data Bank gives the physical characteristics of wolves that were captured.

Females

| Body length (cm) | 123 | 129 | 143 | 124 | 125 | 122 | 125 | 122 |
|---|---|---|---|---|---|---|---|---|
| Weight (lb) | 57 | 84 | 90 | 71 | 71 | 77 | 68 | 73 |

(a) Plot length versus weight for the female wolves. From your visual inspection, estimate the value of the correlation coefficient.

(b) Calculate the sample correlation coefficient for female wolves.

**3.29** Refer to Example 6 concerning spurious correlation. Replace number of cell phone subscribers with the number of registered motorcycles in millions.

**TABLE 10**  Variables Showing Spurious Correlation

| Inmates (mil) | Cell phones (mil) | Motorcycles (mil) | Year |
|------|------|------|------|
| 1.33 | 109.4 | 4.3 | 2000 |
| 1.35 | 128.4 | 4.9 | 2001 |
| 1.38 | 140.8 | 5.0 | 2002 |
| 1.41 | 158.7 | 5.4 | 2003 |
| 1.43 | 187.1 | 5.8 | 2004 |
| 1.53 | 207.9 | 6.0 | 2005 |
| 1.57 | 233.0 | 6.6 | 2006 |
| 1.60 | 255.4 | 7.1 | 2007 |
| 1.61 | 270.3 | 7.7 | 2008 |
| 1.61 | 285.6 | 7.9 | 2009 |

(a) Create a scatter diagram and identify the kind of association.

(b) Comment on possible lurking variables.

**3.30** *A further property of* $r$. Suppose all $x$ measurements are changed to $x' = ax + b$ and all $y$ measurements to $y' = cy + d$, where $a$, $b$, $c$, and $d$ are fixed numbers ($a \neq 0$, $c \neq 0$). Then the correlation coefficient remains unchanged if $a$ and $c$ have the same signs; it changes sign but not numerical value if $a$ and $c$ are of opposite signs.

This property of $r$ can be verified along the lines of Exercise 2.74 in Chapter 2. In particular, the deviations $x - \bar{x}$ change to $a(x - \bar{x})$ and the deviations $y - \bar{y}$ change to $c(y - \bar{y})$. Consequently, $\sqrt{S_{xx}}$, $\sqrt{S_{yy}}$, and $S_{xy}$ change to $|a|\sqrt{S_{xx}}$, $|c|\sqrt{S_{yy}}$, and $acS_{xy}$, respectively (recall that we must take the positive square root of a sum of squares of the deviations). Therefore, $r$ changes to

$$\frac{ac}{|a||c|}r = \begin{cases} r \text{ if } a \text{ and } c \text{ have the same signs} \\ -r \text{ if } a \text{ and } c \text{ have opposite signs} \end{cases}$$

(a) For a numerical verification of this property of $r$, consider the data of Exercise 3.18. Change the $x$ and $y$ measurements according to

$$x' = 2x - 3$$
$$y' = -y + 10$$

Calculate $r$ from the ($x'$, $y'$) measurements and compare with the result of Exercise 3.18.

(b) Suppose from a data set of height measurements in inches and weight measurements in pounds, the value of $r$ is found to be .86. What would the value of $r$ be if the heights were measured in centimeters and weights in kilograms?

**3.31** The amount of municipal solid waste created has become a major problem. According to the Environmental Protection Agency, the yearly amounts (millions of tons) are:

| Year | 1960 | 1970 | 1980 | 1990 | 2000 | 2010 |
|------|------|------|------|------|------|------|
| Garbage (millions of tons) | 88 | 121 | 152 | 208 | 243 | 250 |
| Population (millions) | 179 | 203 | 227 | 249 | 282 | 309 |

(a) Plot the amount of garbage (millions of tons) versus year.

(b) Visually, does there appear to be a strong correlation? Explain.

(c) Give one possible lurking variable.

**3.32** Refer to the data on garbage in Exercise 3.31.

(a) Plot the amount of garbage (millions of tons) versus population (millions).

(b) Does there appear to be a strong correlation? Explain.

(c) How does your interpretation of the association differ from that in Exercise 3.31, parts (b) and (c)?

**3.33** Refer to the data on garbage in Exercise 3.31.

(a) Replace year by (year − 1960). Calculate the correlation coefficient between (year − 1960) and amount of garbage in millions of tons.

(b) Based on your calculation in part (a), what is the correlation between the year itself and the amount of garbage? Explain.

**3.34** Refer to the data on garbage in Exercise 3.31.

(a) Calculate the correlation coefficient between the amount of garbage in millions of tons and the population size in millions.

(b) Based on your calculation in part (a), give the correlation coefficient between the amount of garbage in pounds and population size in number of persons. Explain your answer. [*Hint*: Recall that there are 2000 pounds in a ton so ( number of pounds ) = 2000 × ( number of tons ).]

# 5.  PREDICTION OF ONE VARIABLE FROM ANOTHER (LINEAR REGRESSION)

A study of the relation between two variables is often motivated by a need to predict one from the other. The administrator of a job training program may wish to study the relation between the duration of training and the score of the trainee on a subsequent skill test. A forester may wish to estimate the timber volume of a tree from the measurement of the trunk diameter a few feet above the ground. A medical technologist may be interested in predicting the blood alcohol measurement from the read-out of a newly devised breath analyzer.

In such contexts as these, the **predictor** or **input** variable is denoted by $x$, and the **response** or **output** variable is labeled $y$. The object is to find the nature of relation between $x$ and $y$ from experimental data and use the relation to predict the response variable $y$ from the input variable $x$. Naturally, the first step in such a study is to plot and examine the scatter diagram. If a linear relation emerges, the calculation of the numerical value of $r$ will confirm the strength of the linear relation. Its value indicates how effectively $y$ can be predicted from $x$ by fitting a straight line to the data.

A line is determined by two constants as illustrated in Figure 10: its height above the origin (**intercept**) and the amount that $y$ increases whenever $x$ is increased by one unit (**slope**).

Chapter 11 explains an objective method of best fitting a straight line, called the **method of least squares**. This best fitting line, or least squares line, is close to the points graphed in the scatter plot in terms of minimizing the average amount of vertical distance.

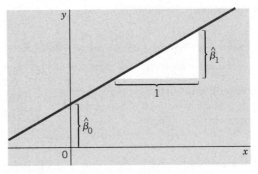

Figure 10   The line $\widehat{y} = \widehat{\beta}_0 + \widehat{\beta}_1 x$.

---

**Equation of the Line Fitted by Least Squares**

$$\hat{y} = \hat{\beta}_0 + \hat{\beta}_1 x$$

where

$$\text{Slope } \hat{\beta}_1 = \frac{S_{xy}}{S_{xx}} = \frac{\sum (x - \bar{x})(y - \bar{y})}{\sum (x - \bar{x})^2}$$

$$\text{Intercept } \hat{\beta}_0 = \bar{y} - \hat{\beta}_1 \bar{x}$$

---

Besides the sample mean $\bar{x}$ and $\bar{y}$, the fitted line involves the sum of the squared deviations of the $x$ observations, $S_{xx}$, and the sum of the cross products of the $x$ observations and the $y$ deviations, $S_{xy}$. The formulas are examined in more detail in Chapter 11.

**Example 8**  Calculation of the Line Fitted by Least Squares

A chemist wishes to study the relation between the drying time of a paint and the concentration of a base solvent that facilitates a smooth application. The data of concentration (grams) $x$ and the observed drying times (minutes) $y$ are recorded in the first two columns of Table 11. Plot the data, calculate $r$, and obtain the fitted line.

**TABLE 11**  Data of Concentration $x$ and Drying Time $y$ (in minutes) and the Basic Calculations

| Concentration $x$ (g) | Drying Time $y$ (min) | $x^2$ | $y^2$ | $xy$ |
|:---:|:---:|:---:|:---:|:---:|
| 0 | 1 | 0 | 1 | 0 |
| 1 | 5 | 1 | 25 | 5 |
| 2 | 3 | 4 | 9 | 6 |
| 3 | 9 | 9 | 81 | 27 |
| 4 | 7 | 16 | 49 | 28 |
| Total  10 | 25 | 30 | 165 | 66 |

SOLUTION  The scatter diagram in Figure 11 gives the appearance of a linear relation. To calculate $r$ and determine the equation of the fitted line, we first calculate the basic quantities $\bar{x}$, $\bar{y}$, $S_{xx}$, $S_{yy}$, and $S_{xy}$ using the totals in Table 11.

Figure 11   Scatter diagram.

$$\bar{x} = \frac{10}{5} = 2 \quad \bar{y} = \frac{25}{5} = 5$$

$$S_{xx} = 30 - \frac{(10)^2}{5} = 10$$

$$S_{yy} = 165 - \frac{(25)^2}{5} = 40$$

$$S_{xy} = 66 - \frac{10 \times 25}{5} = 16$$

$$r = \frac{16}{\sqrt{40 \times 10}} = \frac{16}{20} = .8$$

$$\hat{\beta}_1 = \frac{16}{10} = 1.6$$

$$\hat{\beta}_0 = 5 - (1.6)2 = 1.8$$

The equation of the fitted line is

$$\hat{y} = 1.8 + 1.6x$$

The estimated slope 1.6 tells us that one additional gram of solvent results in an increase of 1.6 minutes in average drying time. The fitted line is shown on the scatter diagram in Figure 11.

If we are to predict the drying time $y$ corresponding to the concentration 2.5, we substitute $x = 2.5$ in our prediction equation and get the result.

At $x = 2.5$, the predicted drying time $= 1.8 + 1.6(2.5) = 5.8$ minutes. Graphically, this amounts to reading the ordinate of the fitted line at $x = 2.5$.

Software programs greatly simplify the calculation and plotting of the fitted line. The MINITAB commands for Example 8, with column 1 labeled $x$ and column 2, $y$

**Data:**

C1: 0  1  2  3  4
C2: 1  5  3  9  7

**Stat > Regression > Fitted Line Plot.**
Type C2 in **Response**. Type C1 in **Predictors**.
Under **Type of Regression Model** choose **Linear**. Click **OK**.

produce the output shown in Figure 12.

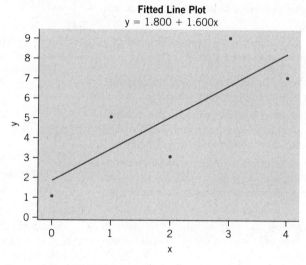

Figure 12    MINITAB output for fitted line in Example 8.

The sample correlation $r$ was introduced as a measure of association between two variables. When $r$ is near 1 or $-1$, points in the scatter plot are closely clustered about a straight line and the association is high. In these circumstances, the value of one variable can be accurately predicted from the value of the other. Said another way, when the value of $r^2$ is near 1, we can predict the value of $y$ from its corresponding $x$ value. In all cases, the slope of the least squares line $\widehat{\beta}_1$ and the sample correlation $r$ are related since $\widehat{\beta}_1 = r\sqrt{S_{yy}}/\sqrt{S_{xx}}$. If the sample correlation is positive, then the slope of the least squares line is positive. Otherwise, both are negative or both zero.

Here we have only outlined the basic ideas concerning the prediction of one variable from another in the context of a linear relation. Chapter 11 expands on these ideas and treats the associated statistical inference problems.

## Exercises

**3.35** Plot the line $y = 2 + 3x$ on graph paper by locating points for $x = 1$ and $x = 4$. What is its intercept? Its slope?

**3.36** Plot the line $y = 6 - 2x$ on graph paper by locating the points for $x = 0$ and $x = 3$. What is its intercept? Its slope?

**3.37** A store manager has determined that the monthly profit $y$ realized from selling a particular brand of car battery is given by

$$y = 10x - 155$$

where $x$ denotes the number of these batteries sold in a month.

(a) If 41 batteries were sold in a month, what was the profit?

(b) At least how many batteries must be sold in a month in order to make a profit?

**3.38** Identify the predictor variable $x$ and the response variable $y$ in each of the following situations.

(a) A training director wishes to study the relationship between the duration of training for new recruits and their performance in a skilled job.

(b) The aim of a study is to relate the carbon monoxide level in blood samples from smokers with the average number of cigarettes they smoke per day.

(c) An agronomist wishes to investigate the growth rate of a fungus in relation to the level of humidity in the environment.

(d) A market analyst wishes to relate the expenditures incurred in promoting a product in test markets and the subsequent amount of product sales.

**3.39** Given these five pairs of $(x, y)$ values

| $x$ | 1 | 2 | 3 | 4 | 5 |
|---|---|---|---|---|---|
| $y$ | 1 | 2.2 | 2.6 | 3.4 | 3.9 |

(a) Plot the points on graph paper.

(b) From a visual inspection, draw a straight line that appears to fit the data well.

(c) Compute the least squares estimates $\hat{\beta}_0$ and $\hat{\beta}_1$ and draw the fitted line.

**3.40** For the data set

| $x$ | 1 | 2 | 7 | 4 | 6 |
|---|---|---|---|---|---|
| $y$ | 6 | 5 | 2 | 4 | 3 |

(a) Construct a scatter diagram.

(b) From a visual inspection, draw a straight line that appears to fit the data well.

(c) Calculate the least squares estimates and draw the least squares fitted line on your plot.

**3.41** In an experiment to study the relation between the time waiting in line, $y$ (minutes), to get to the head of the checkout line at her favorite grocery store and the number of persons ahead in line, $x$, a student collected the following statistics:

$$n = 9 \qquad \Sigma x = 19 \qquad \Sigma y = 39.9$$
$$S_{xx} = 9.4 \qquad S_{yy} = 17.8 \qquad S_{xy} = 10.2$$

(a) Find the equation of the least squares fitted line.

(b) Using the fitted line, predict the time waiting in line when 3 persons are already in line.

**3.42** Wolves used to range over much of Michigan, Minnesota, and Wisconsin. They were reintroduced several years ago, but counts over the winter showed that the populations are expanding rapidly over the past few years.

Total Number of Wolves in Wisconsin

| Year | 1998 | 1999 | 2000 | 2001 | 2002 | 2003 |
|---|---|---|---|---|---|---|
| No. wolves | 178 | 204 | 248 | 257 | 327 | 325 |

| Year | 2004 | 2005 | 2006 | 2007 | 2008 | 2009 | 2010 |
|---|---|---|---|---|---|---|---|
| No. wolves | 373 | 436 | 467 | 546 | 549 | 637 | 782 |

(a) Plot the number of wolves versus the year the count was taken.

(b) Fit a least squares line to summarize the growth. To simplify the calculation, code 1998 as $x = 1$, 1999 as $x = 2$, and so on.

(c) Does your fitted straight line summarize the growth in the wolf population over this period of time? If so, what numerical value summarizes the change in population size from one winter to the next?

| Year | 1960 | 1970 | 1980 | 1990 | 2000 | 2010 |
|------|------|------|------|------|------|------|
| Garbage (million tons) | 88 | 121 | 152 | 208 | 243 | 250 |
| Population (millions) | 179 | 203 | 227 | 249 | 282 | 309 |

**3.43** The amount of municipal solid waste created has become a major problem. According to the Environmental Protection Agency, the yearly amount (millions of tons) are:

(a) Plot the amount of garbage (millions of tons) versus population (millions).

(b) Fit a straight line using $x =$ population in millions.

(c) According to the fitted line, how much garbage is created by a typical person?

## USING STATISTICS WISELY

1. To study the association between two variables, you need to collect the pair of values obtained from each unit in the sample. There is no information about association in the summaries of the observations on individual variables.

2. To study association when both variables are categorical, cross-tabulate the frequencies in a two-way table. Calculate relative frequencies based on the total number.

3. To look for association between any pair of variables whose values are numerical, create a scatter diagram and look for a pattern of association.

4. Never confuse a strong association with a causal relationship. The relation may be due to a lurking variable.

5. Remember that the correlation coefficient measures the clustering of points about a straight line. It is not appropriate for a relationship on a curve or disjoint groups of points.

6. Before using a fitted line to predict one variable from another, create a scatter plot and inspect the pattern to see if a straight-line relationship is appropriate.

## KEY IDEAS AND FORMULAS

Comparative trials often have a **placebo**, or inactive treatment, which serves as a control. This eliminates from the comparison a placebo effect where some subjects in the control group responded positively to an ineffective treatment because their expectation to improve is so strong. An experiment has **double blind** trials where neither the subject nor the person administering the treatments knows which treatment is given.

A **random assignment** of treatments helps prevent uncontrolled sources of variation from systematically influencing the responses.

Data on two traits can be summarized in a two-way table of frequencies where the categories for one trait are in the left margin and categories for the other trait along the upper margin. These are said to be cross-classified or cross-tabulated data and the summary tables of frequencies are called contingency tables.

The combining of two contingency tables that pertain to the same two traits, but are based on very different populations, can lead to very misleading conclusions if the two tables are combined. This is called Simpson's paradox when there is a third variable that strongly influences the other two.

A scatter plot or scatter diagram shows all the values $(x_i, y_i)$ of a pair of variables as points in two dimensions. This plot can be visually inspected for the strength of association between the two variables.

The correlation coefficient $r$ measures how closely the scatter approximates a straight-line pattern.

A positive value of correlation indicates a tendency of large values of $x$ to occur with large values of $y$, and also for small values of both to occur together.

A negative value of correlation indicates a tendency of large values of $x$ to occur with small values of $y$ and vice versa.

A high correlation does not necessarily imply a causal relation.

In fact, a high value of correlation between two variables may be spurious. That is, the two variables may not be connected but the apparent correlation is caused by a third lurking variable that strongly influences both of the original two variables.

A least squares fit of a straight line helps describe the relation of the response or output variable $y$ to the preditor or input variable $x$.

A $y$ value may be predicted for a known $x$ value by reading from the fitted line $\widehat{y} = \widehat{\beta}_0 + \widehat{\beta}_1 x$.

For pairs of measurements $(x, y)$

$$\text{Sample correlation} \quad r = \frac{S_{xy}}{\sqrt{S_{xx}}\sqrt{S_{yy}}}$$

where

$$S_{xx} = \sum (x - \bar{x})^2, \; S_{yy} = \sum (y - \bar{y})^2, \; \text{and}$$
$$S_{xy} = \sum (x - \bar{x})(y - \bar{y})$$

$$\text{Fitted line} \quad \widehat{y} = \widehat{\beta}_0 + \widehat{\beta}_1 x$$

where

$$\text{Slope} \quad \widehat{\beta}_1 = \frac{S_{xy}}{S_{xx}} \quad \text{and} \quad \text{Intercept} \quad \widehat{\beta}_0 = \bar{y} - \widehat{\beta}_1 \bar{x}$$

### *Fitting a straight line and calculating the correlation coefficient*

## MINITAB

### *Fitting a straight line*

Begin with the values for the predictor variable $x$ in C1 and the response variable $y$ in C2.

> **Stat** > **Regression** > **Fitted Line Plot**.
> Type C2 in **Response**. Type C1 in **Predictors**.
> Click **OK**.

To calculate the correlation coefficient, start as above with data in C1 and C2.

> **Stat** > **Basic Statistics** > **Correlation**.
> Type C1 C2 in **Variables**. Click **OK**.

## EXCEL

### *Fitting a straight line*

Begin with the values of the predictor variable in column A and the values of the response variable in column B.

> Highlight the data and click on the **INSERT** tab. Click on **Recommended Charts** and choose **XY (Scatter)**. Click **OK**.
>
> Click on **Add Chart** element and choose **Trendline** and then **Linear**. Click OK.
>
> To calculate the correlation coefficient, begin with the first variable in column A and the second variable in column B.
>
> Click on a blank cell. Select **Formulas** and then **More Functions**.
>
> Select **Statistical** and then **PEARSON**.
>
> **Highlight** the data in Column A for **Array1** and **Highlight** the data in Column B for **Array2**. Then, click **OK**.

## TI-84/83 Plus

### *Fitting a straight line*

Press **STAT** then **1: Edit**. Enter the values of the predictor variable in $L_1$ and those of the response variable in $L_2$.

> Select **STAT** then **Calc** and then **4: LinReg ( ax + b )**.
> With **LinReg** on the Home screen press **Enter**.
>
> The calculator will return the intercept $a$, slope $b$, correlation coefficient $r$. (If $r$ is not shown, go to the 2nd **O: CATALOG** and select **DiagnosticON**. Press ENTER twice. Then go back to **LinREg**.)

**3.44** Applicants for welfare are allowed an appeals process when they feel they have been unfairly treated. At the hearing, the applicant may choose self-representation or representation by an attorney. The appeal may result in an increase, decrease, or no change in benefit recommendation. Court records of 320 appeals cases provided the data below. Calculate the relative frequencies for each row and compare the patterns of the appeals decisions between the two types of representation.

| Type of Representation | Amount of Aid | | | Total |
|---|---|---|---|---|
| | Increased | Unchanged | Decreased | |
| Self | 59 | 108 | 17 | |
| Attorney | 70 | 63 | 3 | |
| Total | | | | |

**3.45** Sugar content (g) and carbohydrate content (g) are obtained from the package of the breakfast cereals referred to in Exercise 3.2.

| General Mills | | Kellogg's | | Quaker | |
|---|---|---|---|---|---|
| Sugar | Carb. | Sugar | Carb. | Sugar | Carb. |
| 9 | 2 | 8 | 1 | 9 | 3 |
| 9 | 1 | 2 | 3 | 13 | 5 |
| 1 | 2 | 8 | 0 | 1 | 2 |
| 9 | 1 | 10 | 10 | 0 | 0 |
| 6 | 1 | 12 | 10 | 2 | 12 |
| 9 | 2 | 15 | 7 | 9 | 3 |
| 9 | 3 | 12 | 10 | 6 | 2 |
| 10 | 3 | 3 | 0 | 11 | 2 |
| 10 | 1 | 12 | 5 | 12 | 2 |
| 8 | 1 | 4 | 0 | 8 | 2 |

(a) Calculate the sample mean carbohydrates for all 30 cereals.

(b) Construct a table like the one in Exercise 3.2 but using carbohydrates rather than sugar.

(c) Calculate the relative frequencies separately for each row. Comment on any pattern.

**3.46** Concern over the quality of eyeglasses purchased over the internet led to a study where many glasses were individually ordered from several different online vendors.[4] Among the lenses with only antireflective coating, 61 had a thickness at the center greater than 1.9 mm and 31 were thinner. Of the 61 thicker lenses, 12 failed impact testing while 18 of the 31 thinner lenses failed.

(a) Create a cross-classifed frequency table with thick/thin lenses and pass/fail as the two traits.

(b) Calculate relative frequencies based on the total number 92.

(c) Comment on any apparent pattern in the relative frequencies and what it means to this problem.

**3.47** A dealer's recent records of 80 truck sales provided the following frequency information on size of truck and type of drive.

| Truck Size | 2-Wheel Drive | 4-Wheel Drive |
|---|---|---|
| Small | 12 | 23 |
| Full | 20 | 25 |

(a) Determine the marginal totals.

(b) Obtain the table of relative frequencies.

(c) Calculate the relative frequencies separately for each row.

(d) Does there appear to be a difference in the choice of drive for purchasers of small- and full-size trucks?

**3.48** A high-risk group of 1083 male volunteers was included in a major clinical trial for testing a new vaccine for type B hepatitis. The vaccine was given to 549 persons randomly selected from the group, and the others were injected with a neutral substance (placebo). Eleven of the vaccinated people and 70 of the nonvaccinated ones later got the disease.

(a) Present these data in the following two-way frequency table.

[4]K. Citek, et al. "Safety and compliance of prescription spectacles ordered by the public via the Internet," *Optometry* **82** (2011), pp. 549–555.

(b) Compare the rates of incidence of hepatitis between the two subgroups.

|  | Hepatitis | No Hepatitis | Total |
|---|---|---|---|
| Vaccinated |  |  |  |
| Not vaccinated |  |  |  |
| Total |  |  |  |

**3.49** Would you expect a positive, negative, or nearly zero correlation for each of the following? Give reasons for your answers.

(a) The time a student spends playing computer games each week and the time they spend talking with friends in a group.

(b) The number of finals taken by undergraduates and their number of hours of sleep during finals week.

(c) A person's height and the number of movies he or she watched last month.

(d) The temperature at a baseball game and beer sales.

**3.50** Examine each of the following situations and state whether you would expect to find a high correlation between the variables. Give reasons why an observed correlation cannot be interpreted as a direct relationship between the variables and indicate at least one possible lurking variable.

(a) The correlation between the number of Internet users and truck sales in cities of varying sizes.

(b) The correlation between yearly sales of cell phones and number of new credit cards issued over the past 10 years.

(c) Correlation between the concentration $x$ of air pollutants and the number of riders $y$ on public transportation facilities when the data are collected from several cities that vary greatly in size.

**3.51** Forest fires in the Western states are responsible for large losses. It was pointed out that the losses are greatest when many firefighters are present and least when there are only a few. Is this convincing evidence that losses can be cut by restricting the number of firefighters sent to battle the flames? Comment.

**3.52** The tar yield of cigarettes is often assayed by the following method: A motorized smoking machine takes a two-second puff once every minute until a fixed butt length remains. The total tar yield is determined by laboratory analysis of the pool of smoke taken by the machine. Of course, the process is repeated on several cigarettes of a brand to determine the average tar yield. Given here are the data of average tar yield and the average number of puffs for six brands of filter cigarettes.

| Average tar (milligrams) | 12.2 | 14.3 | 15.7 | 12.6 | 13.5 | 14.0 |
|---|---|---|---|---|---|---|
| Average no. of puffs | 8.5 | 9.9 | 10.7 | 9.0 | 9.3 | 9.5 |

(a) Plot the scatter diagram.

(b) Calculate $r$.

> *Remark:* Fewer puffs taken by the smoking machine means a faster burn time. The amount of tar inhaled by a human smoker depends largely on how often the smoker puffs.

**3.53** As part of a study of the psychobiological correlates of success in athletes, the following measurements (courtesy of W. Morgan) are obtained from members of the U.S. Olympic wrestling team.

| Anger $x$ | 6 | 7 | 5 | 21 | 13 | 5 | 13 | 14 |
|---|---|---|---|---|---|---|---|---|
| Vigor $y$ | 28 | 23 | 29 | 22 | 20 | 19 | 28 | 19 |

(a) Plot the scatter diagram.

(b) Calculate $r$.

(c) Obtain the least squares line.

(d) Predict the vigor score $y$ when the anger score is $x = 8$.

**3.54** Given the following $(x, y)$ values

| $x$ | 0 | 2 | 5 | 4 | 1 | 6 |
|---|---|---|---|---|---|---|
| $y$ | 5 | 4 | 4 | 2 | 7 | 2 |

(a) Make a scatter plot.

(b) Calculate $r$.

**3.55** Given these five pairs of values

| $x$ | 0 | 3 | 5 | 8 | 9 |
|---|---|---|---|---|---|
| $y$ | 1 | 2 | 4 | 3 | 5 |

(a) Plot the scatter diagram.

(b) From a visual inspection, draw a straight line that appears to fit the data well.

(c) Compute the least squares estimates $\widehat{\beta}_0$, $\widehat{\beta}_1$ and draw the fitted line.

**3.56** For samples collected concerning the following pairs of variables, decide whether you should make a scatter plot or create a contingency table.

(a) The amount earned and the number of weeks worked during the last summer vacation.

(b) Home ownership (own/rent) and having own bedroom (no/yes) during the freshman year of high school.

(c) Number of days that groceries were purchased and the number of days gas was purchased, in the past week.

**3.57** Identify the predictor variable $x$ and the response variable $y$ in each of the following situations.

(a) The state highway department wants to study the relationship between road roughness and a car's gas consumption.

(b) A concession salesperson at football games wants to relate yearly total fall sales to the number of games the home team wins.

(c) A sociologist wants to investigate the number of weekends a college student goes home in relation to the trip distance.

**3.58** The Following Exercises Require a Computer In Figure 12, we have illustrated the output from MINITAB commands for fitting a straight line. To create the scatter plot, without the fitted line, choose:

**Graph > Scatter plot**. Choose **Simple**. Click **OK**. Type C2 in **Y variables** and C1 in **X variables**. Click **OK**.

Use MINITAB (or another package program) to obtain the scatter diagram, correlation coefficient, and regression line for:

(a) The GPA and GMAT scores data of Table 7 in Example 3.

(b) The temperature $x$ and ozone $y$ data in Exercise 3.25.

**3.59** For fitting body length to weight for all wolves given in Table D.9 in the Data Bank, use MINITAB or some other computer package to obtain:

(a) The scatter diagram.

(b) The correlation coefficient.

(c) The regression line.

**3.60** Use MINITAB or some other computer package to obtain the scatter diagram, correlation coefficient, and regression line of:

(a) The final on the initial times to row given in Table D.4 in the Data Bank.

(b) Drop one unusual pair and repeat part (a). Comment on any major differences.

**3.61** A director of student counseling is interested in the relationship between the numerical score $x$ and the social science score $y$ on college qualification tests. The following data (courtesy of R. W. Johnson) are recorded.

| $x$ | 41 | 39 | 53 | 67 | 61 | 67 |
|---|---|---|---|---|---|---|
| $y$ | 29 | 19 | 30 | 27 | 28 | 27 |

| $x$ | 46 | 50 | 55 | 72 | 63 | 59 |
|---|---|---|---|---|---|---|
| $y$ | 22 | 29 | 24 | 33 | 25 | 20 |

| $x$ | 53 | 62 | 65 | 48 | 32 | 64 |
|---|---|---|---|---|---|---|
| $y$ | 28 | 22 | 27 | 22 | 27 | 28 |

| $x$ | 59 | 54 | 52 | 64 | 51 | 62 |
|---|---|---|---|---|---|---|
| $y$ | 30 | 29 | 21 | 36 | 20 | 29 |

| $x$ | 56 | 38 | 52 | 40 | 65 | 61 |
|---|---|---|---|---|---|---|
| $y$ | 34 | 21 | 25 | 24 | 32 | 29 |

| $x$ | 64 | 64 | 53 | 51 | 58 | 65 |
|---|---|---|---|---|---|---|
| $y$ | 27 | 26 | 24 | 25 | 34 | 28 |

(a) Plot the scatter diagram.

(b) Calculate $r$.

# 4

# Probability

## Uncertainty of Weather Forecasts

Today's forecast: Increasing cloudiness with a 25% chance of snow.

Huntstock/Stockbyte/Getty Images

Probabilities express the chance of events that cannot be predicted with certainty. Even unlikely events sometimes occur.

In Chapter 1, we introduced the notions of *sample* and *statistical population* in the context of investigations where the outcomes exhibit variation. Although complete knowledge of the statistical population remains the target of an investigation, we typically only have the partial information contained in a sample. Chapter 2 focused on some methods for describing the salient features of a data set by graphical presentations and calculation of the mean, standard deviation, and other summary statistics. However, when the data set represents a sample from a statistical population, its description is only a preliminary part of a statistical analysis. Our major goal is to make generalizations or inferences about the target population on the basis of information obtained from the sample data. It is the subject of probability that is essential for understanding the reasoning that leads to such generalizations.

In everyday conversations, we all use expressions of the kind:

"Most likely our team will win this Saturday."
"It is unlikely that the weekend will be cold."
"I have a 50–50 chance of getting a summer job at the camp."

The phrases "most likely," "probable," "quite likely," and so on are used qualitatively to indicate the chance that an event will occur. Probability, as a subject, provides a means of quantifying uncertainty. In general terms, the probability of an event is a numerical value that gauges how likely it is that the event will occur. We assign probability on a scale from 0 to 1 with a very low value indicating extremely unlikely, a value close to 1 indicating very likely, and the intermediate values interpreted accordingly. A full appreciation for the concept of a numerical measure of uncertainty and its role in statistical inference can be gained only after the concept has been pursued to a reasonable extent. We can, however, preview the role of probability in one kind of statistical reasoning.

*Suppose it has been observed that in 50% of the cases a certain type of muscular pain goes away by itself. A hypnotist claims that her method is effective in relieving the pain. For experimental evidence, she hypnotizes 15 patients and 12 get relief from the pain. Does this demonstrate that hypnotism is effective in stopping the pain?*

*Let us scrutinize the claim from a statistical point of view. If indeed the method had nothing to offer, there could still be a 50–50 chance that a patient is cured. Observing 12 cures out of 15 amounts to obtaining 12 heads in 15 tosses of a coin. We will see later that the probability of at least 12 heads in 15 tosses of a fair coin is .018, indicating that the event is not likely to happen. Thus, if we tentatively assume the model (or hypothesis) that the method is ineffective, 12 or more cures are very unlikely. Rather than agree that an unlikely event has occurred, we instead conclude that the experimental evidence strongly supports the hypnotist's claim.*

This kind of reasoning, called *testing a statistical hypothesis*, is explored in greater detail later. For now, we are concerned with introducing the ideas that guide the assigning of numerical values for probabilities.

# 1.  PROBABILITY OF AN EVENT

The probability of an event is viewed as a numerical measure of the chance that the event will occur. The idea is naturally relevant to situations where the outcome of an experiment or observation exhibits variation.

Although we have already used the terms "experiment" and "event," a more specific explanation is now in order. In the present context, the term experiment is not limited to the studies conducted in a laboratory. Rather, it is used in a broad sense to include any operation of data collection or observation where the outcomes are subject to variation. Rolling a die, drawing a card from a shuffled deck, sampling a number of customers for an opinion survey, and quality inspection of items from a production line are just a few examples.

> An **experiment** is the process of observing a phenomenon that has variation in its outcomes.

Before attempting to assign probabilities, it is essential to consider all the eventualities of the experiment. Pertinent to their description, we introduce the following terminologies and explain them through examples.

> The **sample space** associated with an experiment is the collection of all possible distinct outcomes of the experiment.
>
> Each outcome is called an **elementary outcome**, a **simple event**, or an **element of the sample space**.
>
> An **event** is the set of elementary outcomes possessing a designated feature.

The elementary outcomes, which together comprise the sample space, constitute the ultimate breakdown of the potential results of an experiment. For instance, in rolling a die, the elementary outcomes are the points 1, 2, 3, 4, 5, and 6, which together constitute the sample space. The outcome of a football game would be either a win, loss, or tie for the home team. Each time the experiment is performed, one and only one elementary outcome can occur. A sample space can be specified by either listing all the elementary outcomes, using convenient symbols to identify them, or making a descriptive statement that characterizes the entire collection. For general discussion, we denote:

The sample space by $\mathcal{S}$

The elementary outcomes by $e_1, e_2, e_3, \ldots$

Events by $A, B,$ and so on.

In specific applications, the elementary outcomes may be given other labels that provide a more vivid identification.

> An event $A$ occurs when any one of the elementary outcomes in $A$ occurs.

**Example 1**    A Tree Diagram and Events for Coin Tossing

Toss a coin twice and record the outcome head ($H$) or tail ($T$) for each toss. Let $A$ denote the event of getting exactly one head and $B$ the event of getting no heads at all. List the sample space and give the compositions of $A$ and $B$.

SOLUTION    For two tosses of a coin, the elementary outcomes can be conveniently identified by means of a **tree diagram**.

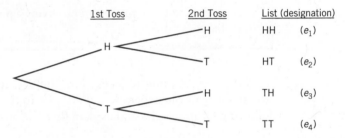

| 1st Toss | 2nd Toss | List (designation) | |
|---|---|---|---|
| H | H | HH | $(e_1)$ |
| | T | HT | $(e_2)$ |
| T | H | TH | $(e_3)$ |
| | T | TT | $(e_4)$ |

The sample space can then be listed as $S = \{\text{HH, HT, TH, TT}\}$. With the designation given above, we can also write

$$S = \{e_1, e_2, e_3, e_4\}.$$

The order in which the elements of $S$ are listed is inconsequential. It is the collection that matters.

Consider the event $A$ of getting exactly one head. Scanning the above list, we see that only the elements HT ($e_2$) and TH ($e_3$) satisfy this requirement. Therefore, the event $A$ has the composition

$$A = \{e_2, e_3\}$$

which is, of course, a subset of $S$. The event $B$ of getting no heads at all consists of the single element $e_4$ so $B = \{e_4\}$. That is, $B$ is a simple event as well as an event. The term "event" is a general term that includes simple events.

**Example 2**    A Sample Space and an Event Based on a Count

On a Saturday afternoon, 135 customers are observed during check-out and the number paying by card, credit or debit, is recorded. Identify (a) the sample space and (b) the event that more than 50% of purchases are made with a card.

SOLUTION

(a) Since the number of customers who purchase with a card could be any of the numbers 0, 1, 2, . . . , 135, the sample space can be listed simply as

$$S = \{0, 1, 2, \ldots, 135\}$$

Using the notation $e$ for elementary outcome, one can also describe this sample space as $S = \{e_0, e_1, e_2, \ldots, e_{135}\}$

(b) Let $A$ stand for the event that more than 50% of the customers purchase with a card. Calculating $.5 \times 135 = 67.5$, we identify

$$A = \{68, 69, \ldots, 135\}$$

Both Examples 1 and 2 illustrate sample spaces that have a finite number of elements. There are also sample spaces with infinitely many elements. For instance, suppose a gambler at a casino will continue pulling the handle of a slot machine until he hits the first jackpot. The conceivable number of attempts does not have a natural upper limit so the list never terminates. That is, $S = \{1, 2, 3, \ldots\}$ has an infinite number of elements. However, we notice that the elements can be arranged one after another in a sequence. An infinite sample space where such an arrangement is possible is called "countably infinite." Either of these two types of sample spaces is called a **discrete sample space**.

Another type of infinite sample space is also important. Suppose a car with a full tank of gasoline is driven until its fuel runs out and the distance traveled recorded. Since distance is measured on a continuous scale, any nonnegative number is a possible outcome. Denoting the distance traveled by $d$, we can describe this sample space as $S = \{d; \ d \geq 0\}$, that is, the set of all real numbers greater than or equal to zero. Here the elements of $S$ form a continuum and cannot be arranged in a sequence. Any $S$ that is an interval is called a **continuous sample space**.

To avoid unnecessary complications, we develop the basic principles of probability in the context of finite sample spaces. We first elaborate on the notion of the probability of an event as a numerical measure of the chance that it will occur. The most intuitive interpretation of this quantification is to consider the fraction of times the event would occur in many repeated trials of the experiment.

---

The probability of an event is a numerical value that represents the proportion of times the event is expected to occur when the experiment is repeated many times under identical conditions.

The probability of event $A$ is denoted by $P(A)$.

---

Since a proportion must lie between 0 and 1, the probability of an event is a number between 0 and 1. To explore a few other important properties of probability, let us refer to the experiment in Example 1 of tossing a coin twice. The event $A$ of getting exactly one head consists of the elementary outcomes HT ( $e_2$ ) and TH ( $e_3$ ). Consequently, $A$ occurs if either of these occurs. Because

$$\begin{bmatrix} \text{Proportion of times} \\ A \text{ occurs} \end{bmatrix} = \begin{bmatrix} \text{Proportion of times} \\ \text{HT occurs} \end{bmatrix} + \begin{bmatrix} \text{Proportion of times} \\ \text{TH occurs} \end{bmatrix}$$

the number that we assign as $P(A)$ must be the sum of the two numbers $P(\text{HT})$ and $P(\text{TH})$. Guided by this example, we state some general properties of probability.

**The probability of an event is the sum of the probabilities assigned to all the elementary outcomes contained in the event.**

Next, since the sample space $S$ includes all conceivable outcomes, in every trial of the experiment some element of $S$ must occur. Viewed as an event, $S$ is certain to occur, and therefore its probability is 1.

**The sum of the probabilities of all the elements of $S$ must be 1.**

An assessment of the probabilities of events and their consequences can help to guide decisions.

In summary:

---

Probability must satisfy:

1. $0 \le P(A) \le 1$ for all events $A$

2. $P(A) = \sum_{\text{all } e \text{ in } A} P(e)$

3. $P(S) = \sum_{\text{all } e \text{ in } S} P(e) = 1$

---

We have deduced these basic properties of probability by reasoning from the definition that the probability of an event is the proportion of times the event is expected to occur in many repeated trials of the experiment.

## Exercises

**4.1** Match the proposed probability of $A$ with the appropriate verbal description. (More than one description may apply.)

| Probability | | Verbal Description |
|---|---|---|
| (a) | .03 | (i)   No chance of happening |
| (b) | .96 | (ii)  Very likely to happen |
| (c) | 2.0 | (iii) As much chance of occurring as not |
| (d) | −.1 | (iv)  Very little chance of happening |
| (e) | .3 | (v)   May occur but by no means certain |
| (f) | 0 | (vi)  An incorrect assignment |
| (g) | .5 | |

**4.2** For each numerical value assigned to the probability of an event, identify the verbal statements that are appropriate.

(a) 1.1   (b) $\dfrac{1}{1.1}$   (c) $\dfrac{1}{2}$

(d) $\dfrac{43}{47}$   (e) $\dfrac{1}{89}$   (f) 1.0

Verbal statements: (i) cannot be a probability, (ii) the event is very unlikely to happen, (iii) 50–50 chance of happening, (iv) sure to happen, (v) more likely to happen than not.

**4.3** Identify the statement that best describes each $P(A)$.

(a) $P(A) = .04$   (i) $P(A)$ is incorrect.
(b) $P(A) = .36$   (ii) $A$ rarely occurs.
(c) $P(A) = -.5$   (iii) $A$ occurs moderately often.

**4.4** Construct a sample space for each of the following experiments.

(a) Someone claims to be able to taste the difference between the same brand of bottled, tap, and canned draft beer. A glass of each is poured and given to the subject in an unknown order. The subject is asked to identify the contents of each glass. The number of correct identifications will be recorded.

(b) Record the number of traffic fatalities in a state next year.

(c) Observe the length of time a new digital video recorder will continue to work satisfactorily without service.

Which of these sample spaces are discrete and which are continuous?

**4.5** Identify these events in Exercise 4.4.

(a) Not more than one correct identification.

(b) Less accidents than last year.
(*Note*: If you don't know last year's value, use 345.)

(c) Longer than the 90-day warranty but less than 425.4 days.

**4.6** When bidding on two projects, the president and vice president of a construction company make the following probability assessments for winning the contracts.

| President | Vice President |
|---|---|
| $P(\text{win none}) = .1$ | $P(\text{win none}) = .1$ |
| $P(\text{win only one}) = .5$ | $P(\text{win Project 1}) = .4$ |
| $P(\text{win both}) = .4$ | $P(\text{win Project 2}) = .2$ |
| | $P(\text{win both}) = .3$ |

For both cases, examine whether or not the probability assignment is permissible.

**4.7** Bob, John, Linda, and Sue are the finalists in the campus bowling tournament. The winner and the first runner-up will be sent to a statewide competition.

(a) List the sample space concerning the outcomes of the local tournament.

(b) Give the composition of each of the following events.

$A = $ Linda wins the local tournament

$B = $ Bob does not go to the state tournament

**4.8** Consider the following experiment: A coin will be tossed twice. If both tosses show heads, the experiment will stop. If one head is obtained in the two tosses, the coin will be tossed one more time, and in the case of both tails in the two tosses, the coin will be tossed two more times.

(a) Make a tree diagram and list the sample space.

(b) Give the composition of the following events.

$A = [\text{Two heads}] \quad B = [\text{Two tails}]$

**4.9** There are four elementary outcomes in a sample space. If $P(e_1) = .3$, $P(e_2) = .4$, and $P(e_3) = .2$, what is the probability of $e_4$?

**4.10** Suppose $\mathcal{S} = \{e_1, e_2, e_3\}$. If the simple events $e_1$, $e_2$, and $e_3$ are all equally likely, what are the numerical values $P(e_1)$, $P(e_2)$, and $P(e_3)$?

**4.11** The sample space for the response of a single person's attitude toward a political issue consists of the three elementary outcomes $e_1 = \{\text{Unfavorable}\}$, $e_2 = \{\text{Favorable}\}$, and $e_3 = \{\text{Undecided}\}$. Are the following assignments of probabilities permissible?

(a) $P(e_1) = .7$, $P(e_2) = .2$, $P(e_3) = .1$

(b) $P(e_1) = .3$, $P(e_2) = .3$, $P(e_3) = .3$

(c) $P(e_1) = .5$, $P(e_2) = .5$, $P(e_3) = 0$

**4.12** A campus organization will select one day of the week for an end-of-year picnic. Assume that the weekdays, Monday through Friday, are equally likely and that each weekend day, Saturday and Sunday, is twice as likely as a weekday to be selected.

(a) Assign probabilities to the seven outcomes.

(b) Find the probability a weekday will be selected.

**4.13** The month in which the year's highest temperature occurs in a city has probabilities in the ratio $1:3:6:10$ for May, June, July, and August, respectively. Find the probability that the highest temperature occurs in either May or June.

**4.14** *Probability and odds.* The probability of an event is often expressed in terms of odds. Specifically, when we say that the odds are $k$ to $m$ that an event will occur, we mean that the probability of the event is $k / (k + m)$. For instance, "the odds are 4 to 1 that candidate Jones will win" means that $P(\text{Jones will win}) = \frac{4}{5} = .8$. Express the following statements in terms of probability. (Bookmakers give the odds of not winning.)

(a) The odds are 3 to 1 that there will be good weather tomorrow.

(b) The odds are 8 to 3 that the city council will delay the funding of a new sports arena.

# 2. METHODS OF ASSIGNING PROBABILITY

An assignment of probabilities to all the events in a sample space determines a probability model. In order to be a valid probability model, the probability assignment must satisfy the properties 1, 2, and 3 stated in the previous section.

Any assignment of numbers $P(e_i)$ to the elementary outcomes will satisfy the three conditions of probability provided these numbers are nonnegative and their sum over all the outcomes $e_i$ in $S$ is 1. However, to be of any practical import, the probability assigned to an event must also be in agreement with the concept of probability as the proportion of times the event is expected to occur. Here we discuss the implementation of this concept in two important situations.

## 2.1  EQUALLY LIKELY ELEMENTARY OUTCOMES—
## THE UNIFORM PROBABILITY MODEL

Often, the description of an experiment ensures that each elementary outcome is as likely to occur as any other. For example, consider the experiment of rolling a fair die and recording the top face. The sample space can be listed as

$$S = \{e_1, e_2, e_3, e_4, e_5, e_6\}$$

where $e_1$ stands for the elementary outcome of getting the face 1, and similarly, $e_2, \ldots, e_6$. Without actually rolling a die, we can deduce the probabilities. Because a fair die is a symmetric cube, each of its six faces is as likely to appear as any other. In other words, each face is expected to occur one-sixth of the time. The probability assignments should therefore be

$$P(e_1) = P(e_2) = \cdots = P(e_6) = \frac{1}{6}$$

and any other assignment would contradict the statement that the die is fair. We say that rolling a fair die conforms to a **uniform probability model** because the total probability 1 is evenly apportioned to all the elementary outcomes.

What is the probability of getting a number higher than 4? Letting $A$ denote this event, we have the composition $A = \{e_5, e_6\}$, so

$$P(A) = P(e_5) + P(e_6) = \frac{1}{6} + \frac{1}{6} = \frac{1}{3}$$

When the elementary outcomes are modeled as equally likely, we have a uniform probability model. If there are $k$ elementary outcomes in $S$, each is assigned the probability of $1/k$.

An event $A$ consisting of $m$ elementary outcomes is then assigned

$$P(A) = \frac{m}{k} = \frac{\text{No. of elementary outcomes in } A}{\text{No. of elementary outcomes in } S}$$

Gregor Mendel, pioneer geneticist, perceived a pattern in the characteristics of generations of pea plants and conceived a theory of heredity to explain them. According to Mendel, inherited characteristics are transmitted from one generation to another by genes. Genes occur in pairs and the offspring obtain their pair by taking one gene from each parent. A simple uniform probability model lies at the heart of Mendel's explanation of the selection mechanism.

One experiment that illustrated Mendel's theory consists of cross fertilizing a pure strain of red flowers with a pure strain of white flowers. This produces hybrids having one gene of each type that are pink-flowered. Crossing these hybrids leads to one of four possible gene pairs. Under Mendel's laws, these four are equally likely. Consequently, $P[\text{Pink}] = \frac{1}{2}$ and $P[\text{White}] = P[\text{Red}] = \frac{1}{4}$. (Compare with the experiment of tossing two coins.)

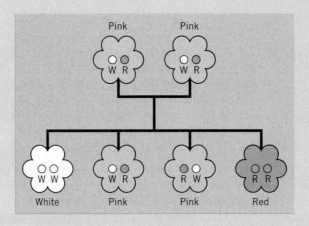

An experiment carried out by Correns, one of Mendel's followers, resulted in the frequencies 141, 291, and 132 for the white, pink, and red flowers, respectively. These numbers are nearly in the ratio 1 : 2 : 1. (*Source*: W. Johannsen, *Elements of the Precise Theory of Heredity*, Jena: G. Fischer, 1909.)

**Example 3**    The Uniform Probability Model for Tossing a Fair Coin

Find the probability of getting exactly one head in two tosses of a fair coin.

SOLUTION    As listed in Example 1, there are four elementary outcomes in the sample space: $S = \{\text{HH, HT, TH, TT}\}$. The very concept of a fair coin implies that the four elementary outcomes in $S$ are equally likely. We therefore assign the probability $\frac{1}{4}$ to each of them. The event $A = [\text{One head}]$ has two elementary outcomes—namely, HT and TH. Hence, $P(A) = \frac{2}{4} = .5$.

**Example 4**    Random Selection and the Uniform Probability Model

Suppose that among 50 students in a class, 42 are right-handed and 8 left-handed. If one student is randomly selected from the class, what is the probability that the selected student is left-handed?

SOLUTION    The intuitive notion of random selection is that each student is as likely to be selected as any other. If we view the selection of each individual student as an elementary outcome, the sample space consists of 50 $e$'s of which 8 are in the event "left-handed." Consequently, $P[\text{Left-handed}] = \frac{8}{50} = .16$.

> *Note:*   Considering that the selected student will be either left-handed ($L$) or right-handed ($R$), we can write the sample space as $S = \{L, R\}$, but we should be aware that the two elements $L$ and $R$ are not equally likely.

## 2.2 PROBABILITY AS THE LONG-RUN RELATIVE FREQUENCY

In many situations, it is not possible to construct a sample space where the elementary outcomes are equally likely. If one corner of a die is cut off, it would be unreasonable to assume that the faces remain equally likely and the assignments of probability to various faces can no longer be made by deductive reasoning. When speaking of the probability (or risk) that a man will die in his thirties, one may choose to identify the occurrence of death at each decade or even each year of age as an elementary outcome. However, no sound reasoning can be provided in favor of a uniform probability model. In fact, from extensive mortality studies, demographers have found considerable disparity in the risk of death for different age groups.

When the assumption of equally likely elementary outcomes is not tenable, how do we assess the probability of an event? The only recourse is to repeat the experiment many times and observe the proportion of times the event occurs. Letting $N$ denote the number of repetitions (or trials) of an experiment, we set

$$\text{Relative frequency of event } A \text{ in } N \text{ trials} = \frac{\text{No. of times } A \text{ occurs in } N \text{ trials}}{N}$$

For instance, let $A$ be the event of getting a 6 when rolling a die. If the die is rolled 100 times and 6 comes up 23 times, the observed relative frequency of $A$ would be $\frac{23}{100} = .23$. In the next 100 tosses, 6 may come up 18 times. Collecting these two sets together, we have $N = 200$ trials with the observed relative frequency

$$\frac{23 + 18}{200} = \frac{41}{200} = .205$$

Imagine that this process is continued by recording the results from more and more tosses of the die and updating the calculations of relative frequency.

Whatever the experiment, the relative frequency exhibits a remarkable behavior as the number of repetitions becomes large. Figure 1 shows a typical

(a) Relative frequency versus number of trials. First 1–50.   (b) Relative frequency versus number of trials. First 500 trials.

Figure 1   Stabilization of relative frequency.

plot of the relative frequency of an event $A$ versus the number $N$ of trials of the experiment. We see that the relative frequencies fluctuate as $N$ changes, but the fluctuations become damped with increasing $N$. Two persons separately performing the same experiment $N$ times are not going to get exactly the same graph. However, the numerical value at which the relative frequency stabilizes, in the long run, will be the same. This concept, called **long-run stability of relative frequency**, is illustrated in Figure 1*b*.

Figure 1*a* graphically displays the considerable fluctuations present in the relative frequency as the number of trials increases from 1 to 50. Figure 1*b* displays the relative frequencies for the first 500 trials. In Figure 1*b*, the stabilization of relative frequency is evident, although the results for the first 50 trials are a little hard to discern in this view.

---

### Probability as Long-Run Relative Frequency

We define $P(A)$, the probability of an event $A$, as the value to which the relative frequency stabilizes with increasing number of trials.

Although we will never know $P(A)$ exactly, it can be estimated accurately by repeating the experiment many times.

---

The property of the long-run stabilization of relative frequencies is based on the findings of experimenters in many fields who have undertaken the strain of studying the behavior of the relative frequencies under prolonged repetitions of their experiments. French gamblers, who provided much of the early impetus for the study of probability, performed experiments tossing dice and coins, drawing cards, and playing other games of chance thousands and thousands of times. They observed the stabilization property of relative frequency and applied this knowledge to achieve an understanding of the uncertainty involved in these games.

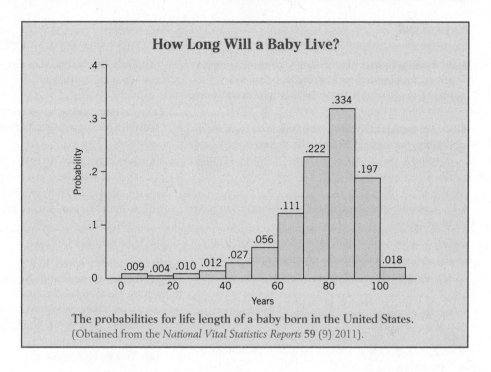

The probabilities for life length of a baby born in the United States.
(Obtained from the *National Vital Statistics Reports* 59 (9) 2011).

Demographers have compiled and studied volumes of mortality data to examine the relative frequency of the occurrence of such events as death in particular age groups. In each context, the relative frequencies were found to stabilize at specific numerical values as the number of cases studied increased. Life and accident insurance companies actually depend on the stability property of relative frequencies.

As another example of an idealized model, consider the assignment of probabilities to the day of the week a child will be born. We may tentatively assume the simple model that all seven days of the week are equally likely. Each day is then assigned the probability $\frac{1}{7}$. If $A$ denotes the event of a birth on the weekend (Saturday or Sunday), our model leads to the probability $P(A) = \frac{2}{7}$. The plausibility of the uniform probability model can only be ascertained from an extensive set of birth records.

Each newborn in the United States can be considered as a trial of the experiment where the day of birth determines whether or not the event $A$ occurs. One year[1], the outcomes for 4130.7 thousand newborns constitute a very large number of replications.

The resulting proportion of babies born on either Saturday or Sunday is $\frac{811.5}{4130.7} = .196$. This is quite far from the value $\frac{2}{7} = .286$ predicted by the uniform model. The difference $.285 - .196 = .089$ is much larger than would ordinarily occur by chance. A reasonable explanation is the increasing prevalence of elective induction of labor, which is mostly performed on weekdays.

[1] *National Vital Statistics Reports* **60** (1) (November 3, 2011).

## *Exercises*

**4.15** Refer to the day of birth data in the preceding text above. Assuming conditions are the same today, estimate the probability that a baby will be born during a weekday. That is, not on Saturday or Sunday.

**4.16** Among 41,131 turkey permit holders for a recent hunting season in Wisconsin, 8845 harvested a bird. Assuming conditions are the same today, estimate the probability that a turkey will be harvested for a single permit.

**4.17** Consider the experiment of tossing a fair coin three times.

(a) List the sample space by drawing a tree diagram.

(b) Assign probabilities to the elementary outcomes.

(c) Find the probability of getting exactly one head.

**4.18** A letter is chosen at random from the word "TEAM." What is the probability that it is a vowel?

**4.19** A stack contains eight tickets numbered 1, 1, 2, 2, 2, 3, 3, 3. One ticket will be drawn at random and its number will be noted.

(a) List the sample space and assign probabilities to the elementary outcomes.

(b) What is the probability of drawing an odd-numbered ticket?

**4.20** Suppose you are eating at a pizza parlor with two friends. You have agreed to the following rule to decide who will pay the bill. Each person will toss a coin. The person who gets a result that is different from the other two will pay the bill. If all three tosses yield the same result, the bill will be shared by all. Find the probability that:

(a) Only you will have to pay.

(b) All three will share.

**4.21** A white and a colored die are tossed. The possible outcomes are shown in the illustration below.

(a) Identify the events $A = [\text{Sum} = 6]$, $B = [\text{Sum} = 7]$, $C = [\text{Sum is even}]$, $D = [\text{Same number on each die}]$.

(b) If both die are "fair," assign probability to each elementary outcome.

(c) Obtain $P(A)$, $P(B)$, $P(C)$, $P(D)$.

**4.22** A roulette wheel has 34 slots, 2 of which are green, 16 are red, and 16 are black. A successful bet on black or red doubles the money, whereas one on green fetches 30 times as much. If you play the game once by betting $5 on the black, what is the probability that:

(a) You will lose your $5?

(b) You will win $5?

**\*4.23** One part of a quiz consists of two multiple-choice questions with the suggested answers: True (T), False (F), or Insufficient Data to Answer (I). An unprepared student randomly marks one of the three answers to each question.

(a) Make a tree diagram to list the sample space, that is, all possible pairs of answers the student might mark.

(b) What is the probability of exactly one correct answer?

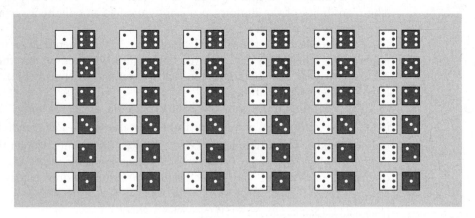

**4.24** Based on the data of the Center for Health Statistics, the 2009 birth rates in 50 states are grouped in the following frequency table.

| Birth rate (per thousand) | 10–12 | 12–14 | 14–16 |
|---|---|---|---|
| No. of states | 4 | 12 | 29 |

| Birth rate (per thousand) | 16–18 | 18 and over | Total |
|---|---|---|---|
| No. of states | 4 | 1 | 50 |

(Endpoint convention: Lower point is included, upper is not.)

If one state is selected at random, what is the probability that the birth rate there is:

(a) Under 14?

(b) Under 18 but not under 12?

(c) 16 or over?

**4.25** Fifteen persons reporting to a Red Cross center one day are typed for blood, and the following counts are found:

| Blood group | O | A | B | AB | Total |
|---|---|---|---|---|---|
| No. of persons | 3 | 5 | 6 | 1 | 15 |

If one person is randomly selected, what is the probability that this person's blood group is:

(a) AB?

(b) Either A or B?

(c) Not O?

**4.26** Friends will be called, one after another, and asked to go on a weekend trip with you. You will call until one agrees to go (A) or four friends are asked, whichever occurs first. List the sample space for the number of calls.

**4.27** Campers arriving at a summer camp will be asked one after another whether they have protection against Lyme disease (Y) or not (N). The inspection will continue until one camper is found to be not protected or until five campers are checked, whichever occurs first. List the sample space for the number of inspections.

**4.28** (a) Consider the simplistic model that human births are evenly distributed over the 12 calendar months. If a person is randomly selected, say, from a phone directory, what is the probability that his or her birthday would be in November or December?

(b) The following record shows a classification of births (thousands) for 2011 in the United States. Calculate the relative frequency of births for each month and comment on the plausibility of the uniform probability model.

| Jan. | 321 | July | 346 |
|---|---|---|---|
| Feb. | 299 | Aug. | 360 |
| March | 331 | Sept. | 346 |
| April | 314 | Oct. | 329 |
| May | 327 | Nov. | 322 |
| June | 338 | Dec. | 328 |
| | | Total | 3961 |

**4.29** A government agency will randomly select one of the 14 paper mills in a state to investigate its compliance with federal safety standards. Suppose, unknown to the agency, 9 of these mills are in compliance, 3 are borderline cases, and 2 are in gross violation.

(a) Formulate the sample space in such a way that a uniform probablility model holds.

(b) Find the probability that a gross violator will be detected.

**4.30** A plant geneticist crosses two parent strains, each with gene pairs of type $aA$. An offspring receives one gene from each parent.

(a) Construct the sample space for the genetic type of the offspring.

(b) Assign probabilities assuming that the selection of genes is random.

(c) If $A$ is dominant and the $aa$ offspring are short while all the others are tall, find $P$[short offspring].

**4.31** Explain why the long-run relative frequency interpretation of probability does not apply to the following situations.

(a) The proportion of days when the home loan rate at your bank is above its value at the start of the year.

(b) The proportion of cars that do not meet emission standards if the data are collected from service stations where the mechanics have been asked to check emissions while attending to other requested services.

**4.32** A local bookstore intended to award three gift certificates in the amounts $100, $50, and $25 to the first, second, and third customer to identify a mystery author. Unfortunately, a careless clerk in charge of mailing forgot the order and just randomly placed the gift certificates in the already addressed envelopes.

(a) List the sample space using $F$, $S$, and $T$ for the three persons.

(b) State the compositions of the events

$A = $ [ exactly one certificate is sent to the correct person ]

$B = $ [ all of the certificates are sent to incorrect persons ]

**4.33** Refer to Exercise 4.32.

(a) Assign probabilities to the elementary outcomes.

(b) Find $P(A)$ and $P(B)$.

**4.34** Refer to Exercise 4.28. Using relative frequencies to estimate probabilities, find which 3 consecutive months have the lowest probability of a new birth.

## 3. EVENT OPERATIONS AND TWO LAWS OF PROBABILITY

To make probability calculations that support generalizations from the actual sample to the complete population, we need to calculate probabilities of combined events. For example, airlines need probabilities to assess whether the number of no shows for a flight is unusually large or unusually low.

Recall that the probability of an event $A$ is the sum of the probabilities of all the elementary outcomes that are in $A$. It often turns out, however, that the event of interest has a complex structure that requires tedious enumeration of its elementary outcomes. On the other hand, this event may be related to other events that can be handled more easily. The purpose of this section is to first introduce the three most basic event relations: **complement, union,** and **intersection.** These event relations will then motivate some laws of probability.

The event operations are conveniently described in graphical terms. We first represent the sample space as a collection of points in a diagram, each identified with a specific elementary outcome. The geometric pattern of the plotted points is irrelevant. What is important is that each point is clearly tagged to indicate which elementary outcome it represents and to watch that no elementary outcome is missed or duplicated in the diagram. To represent an event $A$, identify the points that correspond to the elementary outcomes in $A$, enclose them in a boundary line, and attach the tag $A$. This representation, called a **Venn diagram,** is illustrated in Figure 2.

**Example 5**   Venn Diagram for Coin Tossing

Make a Venn diagram for the experiment of tossing a coin twice and indicate the following events.

$A$ :   Tail at the second toss.

$B$ :   At least one head.

SOLUTION    Here the sample space is $S = \{$ HH, HT, TH, TT $\}$, and the two events have the compositions $A = \{$ HT, TT $\}$, $B = \{$ HH, HT, TH $\}$. Figure 2 shows the Venn diagram.

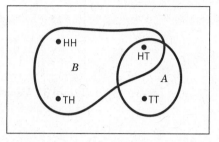

Figure 2    Venn diagram of the events in Example 5.

**Example 6**    A Venn Diagram for the Selection of Puppies

Four young lab puppies from different litters are available for a new method of training.

| Dog | Sex | Age (weeks) |
|-----|-----|-------------|
| 1 | M | 10 |
| 2 | M | 15 |
| 3 | F | 10 |
| 4 | F | 10 |

Two dogs will be selected by lottery to receive the training. Considering all possible choices of two puppies, make a Venn diagram and show the following events.

$A:$    The selected dogs are of the same sex.

$B:$    The selected dogs are of the same age.

SOLUTION    Here the elementary outcomes are the possible choices of a pair of numbers from $\{$ 1, 2, 3, 4 $\}$. These pairs are listed and labeled as $e_1$, $e_2$, $e_3$, $e_4$, $e_5$, $e_6$ for ease of reference.

$$\{1, 2\} \ (e_1) \qquad \{2, 3\} \ (e_4)$$
$$\{1, 3\} \ (e_2) \qquad \{2, 4\} \ (e_5)$$
$$\{1, 4\} \ (e_3) \qquad \{3, 4\} \ (e_6)$$

The pair $\{$ 1, 2 $\}$ has both puppies of the same sex, and so does the pair $\{$ 3, 4 $\}$. Consequently, $A = \{e_1, e_6\}$. Those with the same ages are $\{$ 1, 3 $\}$, $\{$ 1, 4 $\}$, and $\{$ 3, 4 $\}$, so $B = \{e_2, e_3, e_6\}$. Figure 3 shows the Venn diagram.

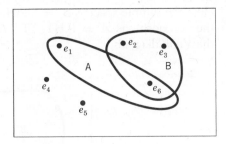

Figure 3   Venn diagram of the events
in Example 6.

We now proceed to define the three basic event operations and introduce the corresponding symbols.

The **complement** of an event $A$, denoted by $\bar{A}$, is the set of all elementary outcomes that are not in $A$. The occurrence of $\bar{A}$ means that $A$ *does not occur.*

The **union** of two events $A$ and $B$, denoted by $A \cup B$, is the set of all elementary outcomes that are in $A$, $B$, or both. The occurrence of $A \cup B$ means that *either A or B or both occur.*

The **intersection** of two events $A$ and $B$, denoted by $AB$, is the set of all elementary outcomes that are in $A$ and $B$. The occurrence of $AB$ means that *both A and B occur.*

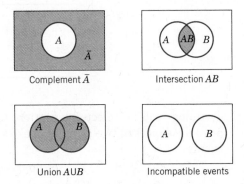

Note that $A \cup B$ is a larger set containing $A$ as well as $B$, whereas $AB$ is the common part of the sets $A$ and $B$. Also it is evident from the definitions that $A \cup B$ and $B \cup A$ represent the same event, while $AB$ and $BA$ are both expressions for the intersection of $A$ and $B$. The operations of union and intersection can be extended to more than two events. For instance, $A \cup B \cup C$ stands for the set of all points that are in *at least one* of $A$, $B$, and $C$, whereas $ABC$ represents the *simultaneous occurrence* of all three events.

Two events $A$ and $B$ are called **incompatible** or **mutually exclusive** if their intersection $AB$ is empty. Because incompatible events have no elementary outcomes in common, they cannot occur simultaneously.

**Example 7**    Determining the Composition of Events Defined by Complement, Union, or Intersection

Refer to the experiment in Example 6 of selecting two puppies out of four. Let $A = $ [ Same sex ], $B = $ [ Same age ], and $C = $ [ Different sexes ]. Give the compositions of the events

$$C, \ \bar{A}, \ A \cup B, \ AB, \ BC$$

SOLUTION    We know from Example 6 that $A = \{e_1, e_6\}$ and $B = \{e_2, e_3, e_6\}$. The pairs consisting of different sexes are $\{1, 3\}, \{1, 4\}, \{2, 3\}$, and $\{2, 4\}$, so $C = \{e_2, e_3, e_4, e_5\}$. The event $\bar{A}$ is the same as the event $C$. Employing the definitions of union and intersection, we obtain

$$A \cup B = \{e_1, e_2, e_3, e_6\}$$
$$AB = \{e_6\}$$
$$BC = \{e_2, e_3\}$$

Let us now examine how probabilities behave as the operations of complementation, union, and intersection are applied to events. It would be worthwhile for the reader to review the properties of probability listed in Section 1. In particular, recall that $P(A)$ is the sum of probabilities of the elementary outcomes that are in $A$, and $P(S) = 1$.

First, let us examine how $P(\bar{A})$ is related to $P(A)$. The sum $P(A) + P(\bar{A})$ is the sum of the probabilities of all elementary outcomes that are in $A$ plus the sum of the probabilities of elementary outcomes not in $A$. Together, these two sets comprise $S$ and we must have $P(S) = 1$. Consequently, $P(A) + P(\bar{A}) = 1$, and we arrive at the following law.

---

### Law of Complement

$$P(A) = 1 - P(\bar{A})$$

---

This law or formula is useful in calculating $P(A)$ when $\bar{A}$ is of a simpler form than $A$ so that $P(\bar{A})$ is easier to calculate.

Turning to the operation of union, recall that $A \cup B$ is composed of points (or elementary outcomes) that are in $A$, in $B$, or in both $A$ and $B$. Consequently, $P(A \cup B)$ is the sum of the probabilities assigned to these elementary outcomes, each probability taken *just once*. Now, the sum $P(A) + P(B)$ includes contributions from all these points, but it double counts those in the region $AB$ (see the figure of $A \cup B$). To adjust for this double counting, we must therefore subtract $P(AB)$ from $P(A) + P(B)$. This results in the following law.

---

### Addition Law

$$P(A \cup B) = P(A) + P(B) - P(AB)$$

---

If the events $A$ and $B$ are incompatible, their intersection $AB$ is empty, so $P(AB) = 0$, and we obtain

---

### Special Addition Law for Incompatible Events

$$P(A \cup B) = P(A) + P(B)$$

---

The addition law expresses the probability of a larger event $A \cup B$ in terms of the probabilities of the smaller events $A$, $B$, and $AB$. Some applications of these two laws are given in the following examples.

**Example 8**    Using the Law of Complement for Probability

A child is presented with three word-association problems. With each problem, two answers are suggested—one is correct and the other wrong. If the child has no understanding of the words whatsoever and answers the problems by guessing, what is the probability of getting at least one correct answer?

SOLUTION    Let us denote a correct answer by C and a wrong answer by W. The elementary outcomes can be conveniently enumerated by means of a tree diagram.

| Problem 1 | Problem 2 | Problem 3 | Elementary outcomes |
|---|---|---|---|
| | | C | CCC |
| | C | W | CCW |
| C | | C | CWC |
| | W | W | CWW |
| | | C | WCC |
| | C | W | WCW |
| W | | C | WWC |
| | W | W | WWW |

There are 8 elementary outcomes in the sample space and, because they are equally likely, each has the probability $\frac{1}{8}$. Let $A$ denote the event of getting at least one correct answer. Scanning our list, we see that $A$ contains 7 elementary outcomes, all except WWW. Our direct calculation yields $P(A) = \frac{7}{8}$.

Now let us see how this probability calculation could be considerably simplified. First, making a complete list of the sample space is not necessary. Since the elementary outcomes are equally likely, we need only determine that there are a total of 8 elements in $S$. How can we obtain this count without making a list? Note that an outcome is represented by three letters. There are 2 choices for each letter—namely, C or W. We then have $2 \times 2 \times 2 = 8$

ways of filling the three slots. The tree diagram explains this multiplication rule of counting. Evidently, the event $A$ contains many elementary outcomes. On the other hand, $\bar{A}$ is the event of getting all answers wrong. It consists of the single elementary outcome WWW, so $P(\bar{A}) = \frac{1}{8}$. According to the law of complement,

$$P(A) = 1 - P(\bar{A})$$
$$= 1 - \frac{1}{8} = \frac{7}{8}$$

**Example 9**  Using the Addition Law for Probability

Refer to Example 6 where two puppies are selected from four by lottery. What is the probability that the selected puppies are either of the same sex or the same age?

SOLUTION  In Example 6, we already enumerated the six elementary outcomes that comprise the sample space. The lottery selection makes all choices equally likely and the uniform probability model applies. The two events of interest are

$$A = [\text{Same sex}] = \{e_1, e_6\}$$
$$B = [\text{Same age}] = \{e_2, e_3, e_6\}$$

Because $A$ consists of two elementary outcomes and $B$ consists of three,

$$P(A) = \frac{2}{6} \quad \text{and} \quad P(B) = \frac{3}{6}$$

Here we are to calculate $P(A \cup B)$. To employ the addition law, we also need to calculate $P(AB)$ where $AB = \{e_6\}$, so $P(AB) = \frac{1}{6}$. Therefore,

$$P(A \cup B) = P(A) + P(B) - P(AB)$$
$$= \frac{2}{6} + \frac{3}{6} - \frac{1}{6} = \frac{4}{6} = \frac{2}{3}$$

which is confirmed by the observation that $A \cup B = \{e_1, e_2, e_3, e_6\}$ indeed has four outcomes.

**Example 10**  Determining Probabilities from Those Given in a Venn Diagram

The accompanying Venn diagram shows three events $A$, $B$, and $C$ and also the probabilities of the various intersections. [For instance, $P(AB) = .07$, $P(A\bar{B}) = .13$.] Determine:

(a)  $P(A)$
(b)  $P(B\bar{C})$
(c)  $P(A \cup B)$

SOLUTION    To calculate a probability, first identify the set in the Venn diagram. Then add the probabilities of those intersections that together comprise the stated event. We obtain

(a)  $P(A) = .13 + .07 = .20$
(b)  $P(B\bar{C}) = .10 + .07 = .17$
(c)  $P(A \cup B) = .13 + .07 + .10 + .15 = .45$

**Example 11**    Expressing Relations between Events in Set Notation

Refer to Example 10. Express the following events in set notation and find their probabilities.

(a)  Both $B$ and $C$ occur.
(b)  $C$ occurs and $B$ does not.
(c)  Exactly one of the three events occurs.

SOLUTION    The stated events and their probabilities are

(a)  $BC$    $P(BC) = .15$
(b)  $\bar{B}C$    $P(\bar{B}C) = .28$
(c)  $(A\bar{B}\bar{C}) \cup (\bar{A}B\bar{C}) \cup (\bar{A}\bar{B}C)$
     The probability $= .13 + .10 + .28 = .51$

## Exercises

**4.35** Next week, a student worker will be assigned one morning of kitchen duty. The sample space has seven elementary outcomes $e_1, e_2, \ldots, e_7$ where $e_1$ represents Sunday, $e_2$ Monday, and so on. Two events are given as $A = \{e_4, e_5, e_6, e_7\}$ and $B = \{e_1, e_6, e_7\}$.

(a) Draw a Venn diagram and show the events $A$ and $B$.

(b) Determine the composition of the following events: (i) $AB$ (ii) $\bar{B}$ (iii) $A\bar{B}$ (iv) $A \cup B$.

**4.36** A sample space consists of 8 elementary outcomes with the following probabilities.

$$P(e_1) = .07 \quad P(e_2) = P(e_3) = P(e_4) = .11$$
$$P(e_5) = P(e_6) = P(e_7) = P(e_8) = .15$$

Three events are given as
$A = \{e_1, e_2, e_5, e_6, e_7\},$
$B = \{e_2, e_3, e_6, e_7\},$ and $C = \{e_6, e_8\}$.

(a) Draw a Venn diagram and show these events.

(b) Give the composition and determine the probability of    (i) $\bar{B}$    (ii) $BC$    (iii) $A \cup C$    (iv) $\bar{A} \cup C$.

**4.37** Refer to Exercise 4.36. Corresponding to each verbal description given here, write the event in set notation, give its composition, and find its probability.

(a)  $C$ does not occur.
(b)  Both $A$ and $B$ occur.
(c)  $A$ occurs and $B$ does not occur.
(d)  Neither $A$ nor $C$ occurs.

**4.38** Suppose you have had interviews for summer jobs at a grocery store, a discount store, and a movie theater. Let $G$, $D$, and $M$ denote the events of your getting an offer from the grocery store, the discount store, and the movie theater, respectively. Express the following events in set notation.

(a) You get offers from the discount store and the movie theater.

(b) You get offers from the discount store and the movie theater but fail to get an offer from the grocery store.

(c) You do not get offers from the grocery store and the movie theater.

**4.39** Four applicants are interviewed for an administrative position with an environmental lobby. They have the following characteristics.

1. Psychology major, male, GPA 3.5
2. Chemistry major, female, GPA 3.3
3. Journalism major, female, GPA 3.7
4. Mathematics major, male, GPA 3.8

One of the candidates will be hired.

(a) Draw a Venn diagram and exhibit these events:

$A$ = [ A social science major is hired ]
$B$ = [ The GPA of the selected candidate is higher than 3.6 ]
$C$ = [ A male candidate is hired ]

(b) Give the composition of the events $A \cup B$ and $AB$.

**4.40** For the experiment of Exercise 4.39, give a verbal description of each of the following events and also state the composition of the event.

(a) $\bar{C}$
(b) $C\bar{A}$
(c) $A \cup \bar{C}$

**4.41** A sample space consists of 9 elementary outcomes $e_1, e_2, \ldots, e_9$ whose probabilities are

$P(e_1) = P(e_2) = .04 \quad P(e_3) = P(e_4) = P(e_5) = .2$
$P(e_6) = P(e_7) = .1 \quad P(e_8) = P(e_9) = .06$

Suppose

$A = \{e_1, e_5, e_8\}, B = \{e_2, e_5, e_8, e_9\}.$

(a) Calculate $P(A)$, $P(B)$, and $P(AB)$.
(b) Using the addition law of probability, calculate $P(A \cup B)$.
(c) List the composition of the event $A \cup B$ and calculate $P(A \cup B)$ by adding the probabilities of the elementary outcomes.

(d) Calculate $P(\bar{B})$ from $P(B)$ and also by listing the composition of $\bar{B}$.

**4.42** Refer to Exercise 4.35. Suppose the elementary outcomes are assigned these probabilities.

$P(e_1) = P(e_2) = P(e_3) = .15$

$P(e_4) = P(e_5) = .06 \quad P(e_6) = .2 \quad P(e_7) = .23$

(a) Find $P(A)$, $P(B)$, and $P(AB)$.
(b) Employing the laws of probability and the results of part (a), calculate $P(\bar{A})$ and $P(A \cup B)$.
(c) Verify your answers to part (b) by adding the probabilities of the elementary outcomes in each of $\bar{A}$ and $A \cup B$.

**4.43** Consider the two events.

$$A = [\text{Obese}] \quad B = [\text{Male}]$$

for persons in the age group 20–39 years old. A survey taken in early 2009–2010 by the National Center for Health Statistics, suggests the probabilities

$$P(A) = .36 \quad P(B) = .50 \quad P(AB) = .18$$

for a randomly selected person.

(a) Enter these probabilities in the following table.
(b) Determine the probabilities of $A\bar{B}$, $\bar{A}B$, and $\bar{A}\bar{B}$ and fill in the table.

|       | $B$ | $\bar{B}$ |
|-------|-----|-----------|
| $A$   |     |           |
| $\bar{A}$ |     |           |

**4.44** Refer to Exercise 4.43. Express the following events in set notation and find their probabilities.

(a) $B$ occurs and $A$ does not occur.
(b) Neither $A$ nor $B$ occurs.
(c) Either $A$ occurs or $B$ does not occur.

**4.45** Consider the two events.

$A$ = [ 5 or more alcoholic drinks in one day last year ]

$B$ = [ Female ]

A 2008 survey taken by the National Center for Health Statistics, suggests the probabilities

$P(B) = .50 \quad P(A\bar{B}) = .23 \quad P(AB) = .14$

for a randomly selected person in the age group 18–24 years old. The following table shows the probabilities concerning A and B.

|        | B    | $\bar{B}$ |   |
|--------|------|------|---|
| A      | .14  | .23  |   |
| $\bar{A}$ |      |      |   |
|        | .50  |      |   |

(a)  Determine the missing entries.

(b)  What is the probability that A does not occur and B does occur?

(c)  Find the probability that either A or B occurs.

(d)  Find the probability that one of these events occurs and the other does not.

**4.46**  If $P(A) = .3$ and $P(B) = .8$, can A and B be mutually exclusive? Why or why not?

**4.47**  From the probabilities shown in this Venn diagram, determine the probabilities of the following events.

(a)  A does not occur.

(b)  A occurs and B does not occur.

(c)  Exactly one of the events A and B occurs.

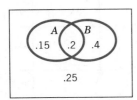

**4.48**  In a class of 32 seniors and graduate students, 20 are men and 12 are graduate students of whom 8 are women. If a student is randomly selected from this class, what is the probability that the selected student is (a) a senior? (b) a male graduate student?

**4.49**  Of 18 fast food restaurants in a city, 7 are in violation of sanitary standards, 8 are in violation of safety standards, and 4 are in violation of both. If a fast food restaurant is chosen at random, what is the probability that it is in compliance with both safety and sanitary standards?

**4.50**  Given that the probability that A occurs is .3, the probability that B does not occur is .6, and the probability that either A or B occurs is .5, find:

(a)  The probability that A does not occur.

(b)  The probability that both A and B occur.

(c)  The probability that A occurs and B does not occur.

**4.51**  The medical records of the male diabetic patients reporting to a clinic during one year provide the following percentages.

|                  | Light Case | | Serious Case | |
|------------------|------------|------|--------------|------|
|                  | Diabetes in Parents | | Diabetes in Parents | |
| Age of Patient   | Yes  | No   | Yes  | No   |
| Below 40         | 15   | 10   | 8    | 2    |
| Above 40         | 15   | 20   | 20   | 10   |

Suppose a patient is chosen at random from this group, and the events A, B, and C are defined as follows.

$$A = [\text{He has a serious case}]$$
$$B = [\text{He is below 40}]$$
$$C = [\text{His parents are diabetic}]$$

(a)  Find the probabilities $P(A)$, $P(B)$, $P(BC)$, $P(ABC)$.

(b)  Describe the following events verbally and find their probabilities: (i) $\bar{A}\bar{B}$  (ii) $\bar{A} \cup \bar{C}$  (iii) $\bar{A}B\bar{C}$.

**4.52**  The following frequency table shows the classification of 58 landfills in a state according to their concentration of the three hazardous chemicals arsenic, barium, and mercury.

|         | Barium | | | |
|---------|--------|------|------|------|
|         | High | | Low | |
|         | Mercury | | Mercury | |
| Arsenic | High | Low | High | Low |
| High    | 1    | 3   | 5    | 9   |
| Low     | 4    | 8   | 10   | 18  |

If a landfill is selected at random, find the probability that it has:

(a) A high concentration of barium.

(b) A high concentration of mercury and low concentrations of both arsenic and barium.

(c) High concentrations of any two of the chemicals and low concentration of the third.

(d) A high concentration of any one of the chemicals and low concentrations of the other two.

**4.53** A bank rewards its employees by giving awards to any employee who is cited by a customer for giving special service. Each award consists of two gift certificates contained in a sealed envelope. Each envelope contains certificates for one of the five following combinations of items.

1. Dinner and box of candy.

2. Round of golf and flowers.

3. Lunch and flowers.

4. Box of candy and lunch.

5. Music CD and lunch.

(a) An employee, cited twice for service, first selects one envelope from a collection of five and then the second from the full collection of five choices. List the sample space and assign probabilities to the simple events.

(b) State the compositions of the events

$$A = \{\text{The employee gets flowers}\}$$
$$B = \{\text{The employee gets lunch}\}$$
$$AB = \{\text{The employee gets flowers and lunch}\}$$

and give their probabilities.

# 4. CONDITIONAL PROBABILITY AND INDEPENDENCE

The probability of an event $A$ must often be updated after information is obtained as to whether or not a related event $B$ has taken place. Information about some aspect of the experimental results may therefore necessitate a revision of the probability of an event concerning some other aspect of the results. The revised probability of $A$ when it is known that $B$ has occurred is called the **conditional probability of $A$ given $B$** and is denoted by $P(A \mid B)$. To illustrate how such modification is made, we consider an example that will lead us to the formula for conditional probability.

**Example 12** Conditional Probability of Using Alternative Medicine Given Body Weight

Complementary alternative medicine (CAM), including acupuncture, yoga, and massage has become more popular. By combining information in two tables,[2] we obtain information concerning use of CAM in the past year and weight class based on body mass index. The proportions in the various categories appear in Table 1.

(a) What is the probability that a person selected at random from this population will have used complementary and alternative medicine in the past year?

[2] *Statistical Abstract of the United States*, Table 203 (2009) and *National Health Statistics Reports*, Table 7 **12** (December 10, 2008).

**TABLE 1**   Body Weight and Complementary and Alternative Medicine

|        | Underweight | Healthy weight | Overweight | Obese | Total |
|--------|-------------|----------------|------------|-------|-------|
| CAM    | .01         | .13            | .12        | .12   | .38   |
| No CAM | .02         | .19            | .21        | .20   | .62   |
| Total  | .03         | .32            | .33        | .32   | 1.00  |

(b)  A person selected at random is found to be overweight. What is the probability that this person used complementary and alternative medicine in the past year?

SOLUTION   Let $A$ denote the event that a person used CAM, and let $B$ denote the event that a person is overweight.

(a)  Because 38% of the people used CAM and the individual is selected at random, we conclude that $P(A) = .38$. This is the unconditional probability of $A$.

(b)  When we are given the information that the selected person is overweight, the categories in the first, second, and fourth columns of Table 1 are not relevant to this person. The third column shows that among the subgroup of overweight persons, the proportion using CAM is $.12/.33$. Therefore, given the information that the person is in this subgroup, the probability that he or she used CAM

$$P(A|B) = \frac{.12}{.33} = .364$$

Noting that $P(AB) = .12$ and $P(B) = .33$, we have derived $P(A|B)$ by taking the ratio $P(AB)/P(B)$. In other words, $P(A|B)$ is the proportion of the population having the characteristic $A$ among all those having the characteristic $B$.

---

The conditional probability of $A$ given $B$ is denoted by $P(A|B)$ and defined by the formula

$$P(A|B) = \frac{P(AB)}{P(B)}$$

Equivalently, this formula can be written

$$P(AB) = P(B)P(A|B)$$

This latter version is called the **multiplication law of probability**.

Similarly, the conditional probability of B given A can be expressed

$$P(B|A) = \frac{P(AB)}{P(A)}$$

which gives the relation $P(AB) = P(A) P(B|A)$. Thus, the multiplication law of probability states that the conditional probability of an event multiplied by the probability of the conditioning event gives the probability of the intersection.

The multiplication law can be used in one of two ways, depending on convenience. When it is easy to compute $P(A)$ and $P(AB)$ directly, these values can be used to compute $P(A|B)$, as in Example 12. On the other hand, if it is easy to calculate $P(B)$ and $P(A|B)$ directly, these values can be used to compute $P(AB)$.

**Example 13**    Conditional Probability of Survival

Refer to the box "How Long Will a Baby Live?" in Section 2.2. It shows the probabilities of death within 10-year age groups.

    (a)  What is the probability that a newborn child will survive beyond age 90?

    (b)  What is the probability that a person who has just turned 80 will survive beyond age 90?

SOLUTION    (a)  Let A denote the event "Survive beyond 90." Adding the probabilities of death in the age groups 90–100 and beyond, we find

$$P(A) = .197 + .018 = .215$$

    (b)  Letting B denote the event "Survive beyond 80," we see that the required probability is the conditional probability $P(A|B)$. Because $AB = A$, $P(A) = .215$, and

$$P(B) = .334 + .197 + .018 = .549$$

we obtain

$$P(A|B) = \frac{P(AB)}{P(B)} = \frac{.215}{.549} = .392$$

**Example 14**    Using the Multiplication Law of Probability

There are 25 pens in a container on your desk. Among them, 20 write well but 5 have defective ink cartridges. You select 2 pens to take to a business appointment. Calculate the probability that:

    (a)  Both pens are defective.

    (b)  One pen is defective but the other writes well.

SOLUTION  We use the symbols D for "defective" and G for "writes well," and attach subscripts to identify the order of the selection. For instance, $G_1 D_2$ represents the event that the first pen checked writes well and the second is defective.

(a)  Here the problem is to calculate $P(D_1D_2)$. Evidently, $D_1D_2$ is the intersection of the two events $D_1$ and $D_2$. Using the multiplication law, we write

$$P(D_1D_2) = P(D_1) P(D_2 | D_1)$$

In order to calculate $P(D_1)$, we need only consider selecting one pen at random from 20 good and 5 defective pens. Clearly, $P(D_1) = \frac{5}{25}$. The next step is to calculate $P(D_2 | D_1)$. Given that $D_1$ has occurred, there remains 20 good and 4 defective pens at the time the second selection is made. Therefore, the conditional probability of $D_2$ given $D_1$ is $P(D_2 | D_1) = \frac{4}{24}$. Multiplying these two probabilities, we get

$$P(\text{both defective}) = P(D_1D_2) = \frac{5}{25} \times \frac{4}{24} = \frac{1}{30} = .033$$

(b)  The event [exactly one defective] is the union of the two incompatible events $G_1D_2$ and $D_1G_2$. The probability of each of these can be calculated by the multiplication law as in part (a). Specifically,

$$P(G_1D_2) = P(G_1) P(D_2 | G_1) = \frac{20}{25} \times \frac{5}{24} = \frac{1}{6}$$

$$P(D_1G_2) = P(D_1) P(G_2 | D_1) = \frac{5}{25} \times \frac{20}{24} = \frac{1}{6}$$

The required probability is $P(G_1D_2) + P(D_1G_2) = \frac{2}{6} = $ .333.

**Remark:**  In solving the problems of Example 14, we have avoided listing the sample space corresponding to the selection of two pens from a collection of 25. A judicious use of the multiplication law has made it possible to focus attention on one draw at a time, thus simplifying the probability calculations.

A situation that merits special attention occurs when the conditional probability $P(A | B)$ turns out to be the same as the unconditional probability $P(A)$. Information about the occurrence of $B$ then has no bearing on the assessment of the probability of $A$. Therefore, when we have the equality $P(A | B) = P(A)$, we say the events $A$ and $B$ are independent.

---

Two events $A$ and $B$ are independent if
$$P(A | B) = P(A)$$

Equivalent conditions are
$$P(B | A) = P(B)$$

or
$$P(AB) = P(A) P(B)$$

The last form follows by recalling that $P(A|B) = P(AB)/P(B)$, so that the condition $P(A|B) = P(A)$ is equivalent to

$$P(AB) = P(A)P(B)$$

which may be used as an alternative definition of independence. The other equivalent form is obtained from

$$P(B|A) = \frac{P(AB)}{P(A)} = \frac{P(A)P(B)}{P(A)} = P(B)$$

The form $P(AB) = P(A)P(B)$ shows that the definition of independence is symmetric in $A$ and $B$.

**Example 15**   Demonstrating Dependence between Use of CAM and Overweight

Are the two events $A = [$ used $CAM]$ and $B = [$ Overweight $]$ independent for the population in Example 12?

SOLUTION   Referring to that example, we have

$$P(A) = .38$$

$$P(A|B) = \frac{P(AB)}{P(B)} = \frac{.12}{.33} = .364$$

Because these two probabilities are different, the two events $A$ and $B$ are dependent.

*Caution:*   Do not confuse the terms "incompatible events" and "independent events." We say $A$ and $B$ are incompatible when their intersection $AB$ is empty, so $P(AB) = 0$. On the other hand, if $A$ and $B$ are independent, $P(AB) = P(A)P(B)$. Both these properties cannot hold as long as $A$ and $B$ have nonzero probabilities.

We introduced the condition of independence in the context of checking a given assignment of probability to see if $P(A|B) = P(A)$. A second use of this condition is in the assignment of probability when the experiment consists of two physically unrelated parts. When events $A$ and $B$ refer to unrelated parts of an experiment, $AB$ is assigned the probability $P(AB) = P(A)P(B)$.

**Example 16**   Using Independence to Assign Probability

Engineers use the term "reliability" as an alternative name for the probability that a device does not fail. Suppose a mechanical system consists of two components that function independently. From extensive testing, it is known that component 1 has reliability .98 and component 2 has reliability .95. If the system can function only if both components function, what is the reliability of the system?

SOLUTION    Consider the events

$$A_1: \quad \text{Component 1 functions}$$

$$A_2: \quad \text{Component 2 functions}$$

$$S: \quad \text{System functions}$$

Here we have the event relation $S = A_1 A_2$. Given that the components operate independently, we take the events $A_1$ and $A_2$ to be independent. Consequently, the multiplication law assigns

$$P(S) = P(A_1) P(A_2) = .98 \times .95 = .931$$

and the system reliability is .931.

In this system, the components are said to be connected in series, and the system is called a series system. A two-battery flashlight is an example. The conventional diagram for a series system is shown in the illustration:

---

**Example 17**   Independence and Assigning Probabilities When Sampling with Replacement

In the context of Example 14, suppose that a box contains 25 cards identifying the pens and their ability to write. One card is drawn at random. It is returned to the box and then another card is drawn at random. What is the probability that both draws produce pens that will not write?

SOLUTION    As before, we use the letter $D$ for defective and $G$ for a pen that writes. By returning the first card to the box, the contents of the box remain unchanged. Hence, with each draw, $P(D) = \frac{5}{25}$, and the results of the two draws are independent. Instead of working with conditional probability as we did in Example 14, we can use the property of independence to calculate

$$P(D_1 D_2) = P(D_1) P(D_2) = \frac{5}{25} \times \frac{5}{25} = .04$$

**Remark 1:**   Evidently, this method of probability calculation extends to any number of draws if after each draw the selected card is returned to the box. For instance, the probability that the first draw produces a $D$ and the next two draws produce $G$'s is

$$P(D_1 G_2 G_3) = \frac{5}{25} \times \frac{20}{25} \times \frac{20}{25} = .128$$

**Remark 2:**   Sampling with replacement is seldom used in practice, but it serves as a conceptual frame for simple probability calculations when a problem concerns sampling from a large population. For example, consider drawing 3 cards from a box containing 2500 cards, of which 2000 are $G$'s and 500 $D$'s. Whether or not a selected card is returned to the box before the next draw makes little difference in the probabilities. The model of independence serves as a reasonable approximation.

The connection between dependent trials and the size of the population merits further emphasis.

**Example 18**   Dependence and Sampling without Replacement

If the outcome of a single trial of any experiment is restricted to just two possible outcomes, it can be modeled as drawing a single ball from an urn containing only red (R) and white (W) balls. In the previous example, these two possible outcomes were good and defective. Consider taking a sample of size 3, without replacement, from each of two populations:

1.   Small population where the urn contains 7 $W$ and 3 $R$.
2.   Large population where the urn contains 7000 $W$ and 3000 $R$.

Compare with a sample of size 3 generated from a spinner having a probability .7 of white, where $R$ or $W$ is determined by a separate spin for each trial.

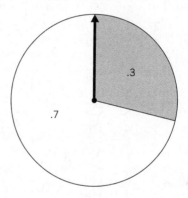

(a)   Calculate the probability assigned to each possible sample.
(b)   Let $D$ = [ At least one $W$ ]. Calculate the probability of $D$.

SOLUTION

(a)   We write $RWR$ for the outcome where the first and third draws are $R$ and the second is $W$. Applying the general multiplication rule $P(ABC) = P(A) P(B|A) P(C|AB)$, when sampling the small population, we get

$$P(RWR) = P(R) P(W|R) P(R|RW) = \frac{3}{10} \times \frac{7}{9} \times \frac{2}{8} = \frac{42}{720}$$

For the larger population,

$$P(RWR) = \frac{3000}{10,000} \times \frac{7000}{9999} \times \frac{2999}{9998} \approx (.3) \times (.7) \times (.3) = (.3)^2(.7)$$

When the population size is large, the assumption of independence produces a very good approximation.

**TABLE 2**    A Comparison of Finite Populations and the Spinner Model

| Draw 3 balls without replacement | Small population | Large population | Spinner |
|---|---|---|---|

$$P(ABC) =$$
$$P(A)\,P(B\,|\,A)\,P(C\,|\,AB)$$

$$P(ABC) \approx$$
$$P(A)\,P(B)\,P(C)$$

$$P(ABC) =$$
$$P(A)\,P(B)\,P(C)$$

| Outcome | Not independent | Approximately independent | Independent |
|---|---|---|---|
| RRR | $\dfrac{3}{10} \times \dfrac{2}{9} \times \dfrac{1}{8} = \dfrac{6}{720}$ | $\dfrac{3000}{10{,}000} \times \dfrac{2999}{9999} \times \dfrac{2998}{9998} \approx (.3)(.3)(.3)$ | $(.3)(.3)(.3)$ |
| RRW | $\dfrac{3}{10} \times \dfrac{2}{9} \times \dfrac{7}{8} = \dfrac{42}{720}$ | $\approx (.3)(.3)(.7)$ | $(.3)^2(.7)$ |
| RWR | $\dfrac{3}{10} \times \dfrac{7}{9} \times \dfrac{2}{8} = \dfrac{42}{720}$ | $\approx (.3)(.7)(.3)$ | $(.3)^2(.7)$ |
| WRR | $\dfrac{7}{10} \times \dfrac{3}{9} \times \dfrac{2}{8} = \dfrac{42}{720}$ | $\approx (.7)(.3)(.3)$ | $(.3)^2(.7)$ |
| RWW | $\dfrac{3}{10} \times \dfrac{7}{9} \times \dfrac{6}{8} = \dfrac{126}{720}$ | $\approx (.3)(.7)(.7)$ | $(.3)(.7)^2$ |
| WRW | $\dfrac{7}{10} \times \dfrac{3}{9} \times \dfrac{6}{8} = \dfrac{126}{720}$ | $\approx (.7)(.3)(.7)$ | $(.3)(.7)^2$ |
| WWR | $\dfrac{7}{10} \times \dfrac{6}{9} \times \dfrac{3}{8} = \dfrac{126}{720}$ | $\approx (.7)(.7)(.3)$ | $(.3)(.7)^2$ |
| WWW | $\dfrac{7}{10} \times \dfrac{6}{9} \times \dfrac{5}{8} = \dfrac{210}{720}$ | $\approx (.7)(.7)(.7)$ | $(.7)^3$ |
| $P(D) = 1 - P(D) = 1 - \dfrac{6}{720}$ | | $\approx 1 - (.3)^3$ | $= 1 - (.3)^3$ |
| If $A = 1^{st}$ is $R$, then | $B = 2^{nd}$ is $R$  $ABC = \{RRR\} = D$ | $C = 3^{rd}$ is $R$  $D = \{RRW,\ RWR,\ WRR,\ RWW,\ WRW,\ WWR,\ WWW\}$ | $D = $ at least one $W$ |

Under the spinner model, the probability of $R$ is .3 for the first trial and this probability is the same for all trials. A spinner is a classic representation of a device with no memory, so that the outcome of the current trial is independent of the outcomes of all the previous trials. According to the product rule for independence, we assign

$$P(RWR) = (.3) \times (.7) \times (.3)$$

Notice that the spinner model is equivalent to sampling with replacement from either of the two finite populations.

The results for all eight possible samples are shown in Table 2.

(b) The event $D$ is complicated, whereas $\bar{D} = \{RRR\}$, a single outcome. By the law of the complement,

$$P(D) = 1 - P(\bar{D}) = 1 - \frac{3}{10} \times \frac{2}{9} \times \frac{1}{8} = 1 - \frac{6}{720}$$

In the second case, $P(D)$ is approximately $1 - (.3) \times (.3) \times (.3)$ and this answer is exact for the spinner model.

Table 2 summarizes sampling from a small finite population, a large but finite population, and the spinner model. Dependence does matter when sampling without replacement from a small population.

## Exercises

**4.54**  A person is randomly selected from persons working in your state. Consider the two events

$A = $ [ Earned over $70,000 last year ]
$B = $ [ College graduate ]

Given that the person is a college graduate, would you expect the probability of $A$ to be larger, the same, or smaller than the unconditional probability $P(A)$? Explain your answer. Are $A$ and $B$ independent according to your reasoning?

**4.55**  A person is randomly selected from persons working in your state. Consider the two events

$A = $ [ Lawyer ]
$B = $ [ Driving a new luxury car ]

Given that the person selected drives a new luxury car, would you expect the probability of $A$ to be larger, the same, or smaller than the unconditional probability $P(A)$? Explain your answer. Are $A$ and $B$ independent according to your reasoning?

**4.56**  Refer to Exercise 4.43. Find

(a) The conditional probability that $B$ occurs given that $A$ occurs.

(b) The conditional probability that $B$ does not occur given that $A$ occurs.

(c) The conditional probability that $B$ occurs given that $A$ does not occur.

**4.57**  The following data relate to the proportions in a population of drivers.

$A = $ Defensive driver training last year
$B = $ Accident in current year

The probabilities are given in the accompanying Venn diagram. Find $P(B|A)$. Are $A$ and $B$ independent?

**4.58**  A witness identified a suspect as having blond hair. However, when tested under lighting conditions similar to those on the night of the incident, it was concluded that the probability of correctly identifying blond hair was .90 and the probability of being incorrect about nonblond hair was .2. Suppose that .4 is the proportion of persons with blond hair that live in the area. Let $A = $ [ Identified as blond ] and $B = $ [ Person is blond ].

(a) Fill in the four probabilities, and the marginal totals, in the table.

(b) Given that this person was identified as blond, find the probability of being blond.

|   | B | $\bar{B}$ |
|---|---|---|
| A |   |   |
| $\bar{A}$ |   |   |

**4.59**  For two events $A$ and $B$, the following probabilities are given.

$$P(A) = .4 \quad P(B) = .25 \quad P(A|B) = .7$$

Use the appropriate laws of probability to calculate

(a)  $P(\bar{A})$

(b)  $P(AB)$

(c)  $P(A \cup B)$

**4.60**  Records of student patients at a dentist's office concerning fear of visiting the dentist suggest the following proportions.

|   | School | | |
|---|---|---|---|
|   | Elementary | Middle | High |
| Fear | .12 | .08 | .05 |
| Do not fear | .28 | .25 | .22 |

For a student selected at random, consider the events

$$A = [\,\text{Fear}\,] \quad M = [\,\text{Middle school}\,]$$

(a)  Find the probabilities

$$P(A) \quad P(AM)$$
$$P(M) \quad P(A \cup M)$$

(b)  Are $A$ and $M$ independent?

**4.61**  An urn contains two green balls and three red balls. Suppose two balls will be drawn at random one after another and *without replacement* (i.e., the first ball drawn is *not* returned to the urn before the second one is drawn).

(a)  Find the probabilities of the events

$$A = [\,\text{Green ball appears in the first draw}\,]$$

$$B = [\,\text{Green ball appears in the second draw}\,]$$

(b)  Are the two events independent? Why or why not?

(c)  Now suppose two balls will be drawn *with replacement* (i.e., the first ball drawn will be returned to the urn before the second draw). Repeat parts (a) and (b).

**4.62**  One type of passenger airplane has four engines but it can still fly with only one. Suppose each engine has probability .002 of failing during a flight.

(a)  Calculate the probability of all four engines failing during the same flight. Assume independence.

(b)  One accident did occur when all four engines failed. The subsequence investigation revealed that a mechanic made the same mistake servicing all four engines. Explain why your probability calculation in part (a) does not apply here.

 Because of this incident, airlines are now required to have different mechanics for each engine.

(c)  Suggest another factor, other than having service performed by the same mechanic, that would violate the assumption of independence.

**4.63**  An insurance company's records of 12,299 automobile insurance policies revealed that 2073 made a claim. Among insured drivers under age 25, there were 1032 claims out of 5192 policies (courtesy of J. Hickman). Assuming the claim history is valid for the current year, determine the probability that

(a)  An insured driver will make a claim.

(b)  An insured driver under age 25 will make a claim.

(c)  Comment on the need to charge more for insurance for a person under age 25.

**4.64**  If the probability of running out of gas is .03 and the probability the electronic starting system will not work is .01,

(a)  What is the probability that there will be enough gas and that the starting system will work? Assume the two events are independent.

(b)  When may independence be a poor assumption?

**4.65**  Suppose $P(A) = .6$ and $P(B) = .22$.

(a)  Determine $P(A \cup B)$ if $A$ and $B$ are independent.

(b)  Determine $P(A \cup B)$ if $A$ and $B$ are mutually exclusive.

(c)  Find $P(A|\bar{B})$ if $A$ and $B$ are mutually exclusive.

**4.66** Refer to Exercise 4.49.

(a) If a fast food restaurant selected at random is found to comply with safety standards, what is the probability that it violates sanitary standards?

(b) If a restaurant selected at random is found to violate at least one of the two standards, what is the probability that it complies with safety standards?

**4.67** Of 20 rats in a cage, 12 are males and 9 are infected with a virus that causes hemorrhagic fever. Of the 12 male rats, 7 are infected with the virus. One rat is randomly selected from the cage.

(a) If the selected rat is found to be infected, what is the probability that it is a female?

(b) If the selected rat is found to be a male, what is the probability that it is infected?

(c) Are the events "the selected rat is infected" and "the selected rat is male" independent? Why or why not?

**4.68** A restaurant critic goes to a place twice. If she has an unsatisfactory experience during both visits, she will go once more. Otherwise she will make only the two visits.

Assuming that the results for different visits are independent and that the probability of a satisfactory experience in any one visit is .8

(a) Assign probabilities to each outcome.

(b) Find the probability of at least two unsatisfactory visits.

(c) Find the conditional probability of at least one satisfactory visit given at least one unsatisfactory visit.

**4.69** Of three events, $A$, $B$, and $C$, suppose events $A$ and $B$ are independent and events $B$ and $C$ are mutually exclusive. Their probabilities are $P(A) = .7$, $P(B) = .2$, and $P(C) = .3$. Express the following events in set notation and calculate their probabilities.

(a) Both $B$ and $C$ occur.

(b) At least one of $A$ and $B$ occurs.

(c) $B$ does not occur.

(d) All three events occur.

**4.70** Approximately 40% of the Wisconsin population have type O blood. If 4 persons are selected at random to be donors, find $P[\text{at least one type O}]$.

**4.71** The primary cooling unit in a nuclear power plant has reliability .999. There is also a back-up cooling

unit to substitute for the primary unit when it fails. The reliability of the back-up unit is .910. Find the reliability of the cooling system of the power plant. Assume independence.

**4.72** An accountant screens large batches of bills according to the following sampling inspection plan. She inspects 4 bills chosen at random from each batch and passes the batch if, among the 4, none is irregular. Find the probability that a batch will be passed if, in fact:

(a) 5% of its bills are irregular.

(b) 20% of its bills are irregular.

**4.73** In 80% of the cases an electronic scanner is successful in detecting flaws in a material specimen. Three material specimens containing flaws will be tested with the scanner. Assume that the tests are independent.

(a) List the sample space and assign probabilities to the simple events.

(b) Find the probability that the scanner is successful in at least two of the three cases.

**4.74** Refer to Exercise 4.52. Given that a landfill selected at random is found to have a high concentration of mercury, what is the probability that its concentration is:

(a) High in barium?

(b) Low in both arsenic and barium?

(c) High in either arsenic or barium?

**4.75** Of the patients reporting to a clinic with the symptoms of sore throat and fever, 25% have strep throat, 40% have an allergy, and 10% have both.

(a) What is the probability that a patient selected at random has strep throat, an allergy, or both?

(b) Are the events "strep throat" and "allergy" independent?

**\*4.76** Consider tossing two fair coins and the events

$A$:   Head in the first toss

$B$:   Head in the second toss

$C$:   Both heads or both tails in the two tosses

(a) Verify that the property of independence holds for all event pairs.

(b) Show that $P(ABC)$ is different from the product $P(A)\,P(B)\,P(C)$. (This illustrates the fact that pairwise independence does not ensure complete independence.)

# 5. BAYES' THEOREM

Bayes' theorem expresses the conditional probability $P(B|A)$ in terms of $P(A|B)$ and the two unconditional probabilities $P(A)$ and $P(B)$. It is a rule about probabilities that expresses how an evaluation of a probability should be rationally updated to account for observing the event $A$. Although we do not pursue the wide reaching consequences and applications to probabilities that are subjectively evaluated, below we do introduce the common terms *prior probability* and *posterior probability* for the before and after values.

To begin, we first show how the multiplication rule of probability leads to a result called the rule of total probability. An event $A$ can occur either when an event $B$ occurs or when it does not occur. That is, $A$ can be written as the disjoint union of $AB$ and $A\bar{B}$. Consequently,

$$P(A) = P(AB) + P(A\bar{B})$$

Using the multiplication rule of probability, we obtain the rule of total probability.

---

**Rule of Total Probability**

$$P(A) = P(A|B)P(B) + P(A|\bar{B})P(\bar{B})$$

---

**Example 19**   Rule of Total Probability and Uncertainty with Medical Tests

Let $A$ be the event that a person tests positive for a serious virus and $B$ be the event that the person actually has the virus. Suppose that the virus is present in 1.4% of the population. Because medical tests are sometimes incorrect, we model uncertainty by assigning probability. Suppose the conditional probability that the test is positive, given that the person has the virus, is $.995 = P(A|B)$. Also, suppose that $.02 = P(A|\bar{B})$ is the conditional probability that a person not having the virus tests positive; a false positive.

Determine the probability that a person will test positive, $P(A)$.

SOLUTION   We are given $P(B) = .014$ so $1 - P(B) = P(\bar{B}) = .986$. Then

$$P(A) = P(A|B)P(B) + P(A|\bar{B})P(\bar{B})$$
$$= .995 \times .014 + .02 \times .986 = .034$$

The same reasoning prevails if there are three events $B_1$, $B_2$, and $B_3$ that are mutually exclusive and whose union is the whole sample space as illustrated in Figure 4. Then, $A$ is the union of the mutually exclusive events $AB_1$, $AB_2$, and $AB_3$ and the rule of total probability becomes

$$P(A) = P(A|B_1)P(B_1) + P(A|B_2)P(B_2) + P(A|B_3)P(B_3)$$

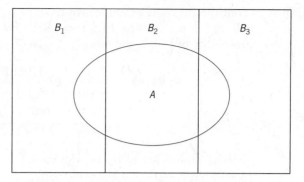

Figure 4    Event $A$ and mutually exclusive events $B_1$, $B_2$, and $B_3$ with $B_1 \cup B_2 \cup B_3 = S$.

Suppose the two events $A$ and $B$ can occur together and, before observing either, we know the probability $P(B)$ so $P(\bar{B}) = 1 - P(B)$ is also known. We call these two probabilities the prior probabilities since they represent the probabilities associated with $B$ and $\bar{B}$ before we know the status of event $A$ or any other event. When we also know the two conditional probabilities $P(A \mid B)$ and $P(A \mid \bar{B})$, the probability of $B$ can be updated when we observe the status of $A$.

Once we know $A$ has occurred, the updated or posterior probability of $B$ is given by the conditional probability

$$P(B \mid A) = \frac{P(AB)}{P(A)}$$

The numerator can be written as $P(A B) = P(A \mid B) P(B)$ by the multiplication rule and the denominator $P(A) = P(A \mid B) P(B) + P(A \mid \bar{B}) P(\bar{B})$ by the rule of total probability. Substituting these two alternate expressions into the formula for conditional probability, we obtain Bayes' Theorem.

---

### Bayes' Theorem

$$P(B \mid A) = \frac{P(A \mid B) P(B)}{P(A \mid B) P(B) + P(A \mid \bar{B}) P(\bar{B})}$$

The posterior probability of $\bar{B}$ is then $P(\bar{B} \mid A) = 1 - P(B \mid A)$

---

**Example 20**    Bayes' Theorem and the Uncertainty with Medical Tests

Refer to Example 19 where $A$ is the event that a person tests positive for a serious virus and $B$ is the event that the person actually has the virus.

Suppose a person tests positive. Use Bayes' Theorem to update the probability that the person actually has the virus. That is, determine the posterior probability $P(B \mid A)$.

SOLUTION   From the previous example, we have $P(A \mid B) = .995$, $P(A \mid \bar{B}) = .02$, and $P(B) = .014$ so $P(\bar{B}) = .986$. By Bayes' Theorem, the posterior probability of having the virus is

$$P(B \mid A) = \frac{P(A \mid B) P(B)}{P(A \mid B) P(B) + P(A \mid \bar{B}) P(\bar{B})}$$

$$= \frac{.995 \times .014}{.995 \times .014 + .02 \times .986} = .414$$

The probability of having the virus has increased dramatically from .014 to .414 but it is still less than one-half.

Without using Bayes' Theorem, most people would think that the probability they have the virus is much higher. Bayes' Theorem adjusts for the fact that the virus is rare.

Bayes' Theorem plays an integral part in most schemes for filtering spam. It gives a probability that a chance message is really spam given the presence of certain words.

**Example 21**   Identifying Spam Using Bayes' Theorem

A first step toward identifying spam is to create a list of words that are more likely to appear in spam than in normal messages. For instance, there are words like buy or the brand name of an enhancement drug that are more likely to occur in spam messages than in normal messages. Suppose a specified list of words is available and that your database of 1000 messages contains 350 that are spam. Among the spam messages, 280 contain words in the list. Of the 650 normal messages, only 65 contain words in the list.

Obtain the probability that a message is spam given that the message contains words in the list.

SOLUTION   Let $A = $ [ message contains words in list ] be the event a message is identified as spam and let $B = $ [ message is spam ]. Here we use the observed relative frequencies from the database as approximations to the probabilities.

$$P(B) = \frac{350}{1000} = .35 \qquad P(\bar{B}) = 1 - .35 = .65$$

$$P(A \mid B) = \frac{280}{350} = .80 \qquad P(A \mid \bar{B}) = \frac{65}{650} = .10$$

Bayes' Theorem expresses the probability of being spam, given that a message is identified as spam, as

$$P(B \mid A) = \frac{P(A \mid B) P(B)}{P(A \mid B) P(B) + P(A \mid \bar{B}) P(\bar{B})}$$

The updated, or posterior probability,

$$P(A|B) = \frac{.80 \times .35}{.80 \times .35 + .10 \times .65} = \frac{.28}{.345} = .812$$

This posterior probability of being spam is calculated after it is observed that the message is identified as spam. Because this probability is quite large, we suspect that this message is spam. When $P(A) = .35$, or 35% of the incoming messages are spam, we likely would want the spam filter to remove this message.

Existing programs to filter spam learn and improve as you mark spam or not spam on your incoming messages.

When there are three events $B_1$, $B_2$, and $B_3$ that are mutually exclusive and whose union is the whole sample space, as in Figure 4, Bayes' Theorem becomes

$$P(B_1|A) = \frac{P(A|B_1)P(B_1)}{P(A|B_1)P(B_1) + P(A|B_2)P(B_2) + P(A|B_3)P(B_3)}$$

**Example 22**    Discovering Which of Three Suppliers Sent a Defective Part

A manufacturer of tablets encountered a shortage of LED screens. Shipments from the usual supplier had 4% defective screens. Because of the shortage, the manufacturer had to obtain 20% of the screens from a second supplier who sent 7% defective screens, and 15% from a third supplier who sent 10% defective screens. Find the

(a) Probability that an incoming LED screen will be defective.

(b) Probability that a defective LED screen was sent by the second supplier.

SOLUTION    Here we have three possible suppliers so we define one event for each.

$$B_1 = [\text{Supplier 1}] \qquad B_2 = [\text{Supplier 2}] \qquad B_3 = [\text{Supplier 3}]$$

Let $A = [\text{Defective LED Screen}]$. We are given that $P(B_2) = .20$ and $P(B_3) = .15$ so $P(B_1) = 1 - .20 - .15 = .65$. Also, $P(A|B_1) = .04$, $P(A|B_2) = .07$ and $P(A|B_3) = .10$.

(a) The law of total probability states that

$$P(A) = P(A|B_1)P(B_1) + P(A|B_2)P(B_2) + P(A|B_3)P(B_3)$$

so

$$P(A) = .04 \times .65 + .07 \times .20 + .10 \times .15 = .055$$

(b) Next, by Bayes' Theorem for three classes,

$$P(B_i|A) = \frac{P(A|B_i)P(B_i)}{P(A|B_1)P(B_1) + P(A|B_2)P(B_2) + P(A|B_3)P(B_3)} \qquad \text{for } i = 1, 2, 3$$

where the denominator was just calculated by the law of total probability. We find

$$P(B_2 | A) = \frac{.07 \times .20}{.055} = .255$$

Although Supplier 2 provides only 20% of the screens, this source produces just over 25% of the defective ones.

## Exercises

**4.77** *Imperfect clinical test.* Repeat Example 20 but change $P(A | \bar{B})$ to .03.

**4.78** Carol and Karl both solve difficult computer problems that come to the student desk. Carol makes 60% of the repairs and Karl 40%. However, Carol's repairs are incomplete 4% of the time and Karl's are incomplete 6% of the time.

(a) Determine the probability that a repair is incomplete.

(b) If a repair is found to be incomplete, what is the probability that the repair was made by Karl?

**4.79** In a county, men constitute 60% of the labor force. The rates of unemployment are 5.1% and 4.3% among males and females, respectively.

(a) In the context, of selecting a worker at random from the country labor force, state what probabilities the foregoing percentages represent. (Use symbols such as $M$ for male, $E$ for employed.)

(b) What is the overall rate of unemployment in the county?

(c) If a worker selected at random is found to be unemployed, what is the probability that the worker is a woman?

**4.80** Refer to Example 21 concerning spam but now suppose that a new list is available. Among the 1000 messages, the 350 spam messages have 310 that contain words on this list and the 650 normal messages have 70 that contain words on the list.

(a) Obtain the probability that a message is spam given that the message contains certain words.

(b) Would you prefer the list here or the one in Example 21? Why?

**4.81** In professional tennis, the server often tries to ace her opponent on the first serve. If the first serve is out of bounds, the server tries to be more accurate with the second and last serve. From records of women's tennis at Wimbledon[3] the probability is .61 that the first serve is in bounds. When the first serve is in, the server wins the point with probability .62. When a second serve is required, the probability is .86 that it is in bounds but the probability drops to .54 of ultimately winning the point. For a typical player, with these probabilities, determine

(a) The probability of winning the point when serving.

(b) If you only know that she wins the point, what is the probability she won on the first serve?

**4.82** A large firm has 85% of its service calls made by a contractor, and 10 of these calls result in customer complaints. The other 15% of the service calls are made by their own employees, and these calls have a 5% complaint rate. Find the

(a) Probability of receiving a complaint.

(b) Probability that the complaint was from a customer serviced by the contractor.

**4.83** Refer to Example 22 where a manufacturer had difficulty getting enough LED screens. Now suppose, because of the shortage, the manufacturer had to obtain 30% of the screens from the second supplier and 15% from the third supplier. Find the

(a) Probability that a LED screen will be defective.

(b) Probability that a defective LED screen was sent by the second supplier.

---

[3] J. Magnus and F. Klaassen, "On the advantage of serving first in a tennis set: Four years at Wimbledon." *The Statistician,* 48(2), 1999.

# 6.  RANDOM SAMPLING FROM A FINITE POPULATION

In our earlier examples of probability calculations, we have used the phrase "randomly selected" to mean that all possible selections are equally likely. It usually is not difficult to enumerate all the elementary outcomes when both the population size and sample size are small numbers. With larger numbers, making a list of all the possible choices becomes a tedious job. However, a counting rule is available that enables us to solve many probability problems.

We begin with an example where the population size and the sample size are both small numbers so all possible samples can be conveniently listed.

**Example 23**    Selecting a Random Sample of Size 2 from a Population of Size 5

There are five qualified applicants for two editorial positions on a college newspaper. Two of these applicants are men and three women. If the positions are filled by randomly selecting two of the five applicants, what is the probability that neither of the men is selected?

SOLUTION    Suppose the three women applicants are identified as $a$, $b$, and $c$ and the two men as $d$ and $e$. Two members are selected at random from the population:

$$\underbrace{\{a, \quad b, \quad c,}_{\text{women}} \quad \underbrace{d, \quad e\}}_{\text{men}}$$

The possible samples may be listed as

$$
\begin{array}{cccc}
\{a, b\} & \{b, c\} & \{c, d\} & \{d, e\} \\
\{a, c\} & \{b, d\} & \{c, e\} & \\
\{a, d\} & \{b, e\} & & \\
\{a, e\} & & &
\end{array}
$$

As the list shows, our sample space has 10 elementary outcomes. The notion of random selection entails that these are all equally likely, so each is assigned the probability $\frac{1}{10}$. Let $A$ represent the event that two women are selected. Scanning our list, we see that $A$ consists of the three elementary outcomes

$$\{a, b\} \quad \{a, c\} \quad \{b, c\}$$

Consequently,

$$P(A) = \frac{\text{No. of elements in } A}{\text{No. of elements in } S} = \frac{3}{10} = .3$$

Note that our probability calculation in Example 23 only requires knowledge of the two counts: the number of elements in $S$ and the number of elements in

A. Can we arrive at these counts without formally listing the sample space? An important counting rule comes to our aid. (See Appendix A.2)

---

### The Rule of Combinations

**Notation:** The number of possible choices of $r$ objects from a group of $N$ distinct objects is denoted by $\binom{N}{r}$, which reads as "$N$ choose $r$."

**Formula:**

$$\binom{N}{r} = \frac{N \times (N-1) \times \cdots \times (N-r+1)}{r \times (r-1) \times \cdots \times 2 \times 1}$$

---

More specifically, the numerator of the formula $\binom{N}{r}$ is the product of $r$ consecutive integers starting with $N$ and proceeding downward. The denominator is also the product of $r$ consecutive integers, but starting with $r$ and proceeding down to 1.

To motivate the formula, let us consider the number of possible choices (or collections) of three letters from the seven letters $\{\, a, b, c, d, e, f, g \,\}$. This count is denoted by $\binom{7}{3}$.

It is easier to arrive at a formula for the number of ordered selections. The first choice can be any of the 7 letters, the second can be any of the remaining 6, and the third can be any of the remaining 5. Thinking in terms of a tree diagram (without actually drawing one), we arrive at the following count.

**The number of ordered selections of 3 letters from 7 is given by the product $7 \times 6 \times 5$.**

Next, note that a particular collection, say $\{\, a, b, c \,\}$, can produce $3 \times 2 \times 1$ orderings, as one can also verify by a tree diagram. The $\binom{7}{3}$ number of collections, each producing $3 \times 2 \times 1$ orderings, generates a total of $\binom{7}{3} \times 3 \times 2 \times 1$ orderings. Because this count must equal $7 \times 6 \times 5$, we get

$$\binom{7}{3} = \frac{7 \times 6 \times 5}{3 \times 2 \times 1}$$

This explains the formula of $\binom{N}{r}$ for the case $N = 7$ and $r = 3$.

Although not immediately apparent, there is a certain symmetry in the counts $\binom{N}{r}$. The process of selecting $r$ objects is the same as choosing $N - r$

objects to leave behind. Because every choice of $r$ objects corresponds to a choice of $N - r$ objects,

$$\binom{N}{r} = \binom{N}{N-r}$$

This relation often simplifies calculations. Since $\binom{N}{N} = 1$, we take $\binom{N}{0} = 1$.

**Example 24**    Evaluating Some Combinations

Calculate the values of $\binom{5}{2}$, $\binom{15}{4}$, and $\binom{15}{11}$.

SOLUTION

$$\binom{5}{2} = \frac{5 \times 4}{2 \times 1} = 10 \qquad \binom{15}{4} = \frac{15 \times 14 \times 13 \times 12}{4 \times 3 \times 2 \times 1} = 1365$$

Using the relation $\binom{N}{r} = \binom{N}{N-r}$, we have

$$\binom{15}{11} = \binom{15}{4} = 1365$$

**Example 25**    Calculating a Probability Using Combinations

Refer to Example 23 concerning a random selection of two persons from a group of two men and three women. Calculate the required probability without listing the sample space.

SOLUTION    The number of ways two persons can be selected out of five is given by

$$\binom{5}{2} = \frac{5 \times 4}{2 \times 1} = 10$$

Random selection means that the 10 outcomes are equally likely. Next, we are to count the outcomes that are favorable to the event $A$ that both selected persons are women. Two women can be selected out of three in

$$\binom{3}{2} = \frac{3 \times 2}{2 \times 1} = 3 \text{ ways}$$

Taking the ratio, we obtain the result

$$P(A) = \frac{3}{10} = .3$$

**Example 26** Number of Sequences Having $x$ Heads in $N$ Flips of a Coin

The model of coin flips applies in many settings where each flip, or trial, has only two possible outcomes such as boy/girl or defective/not defective. We use the familiar coin flip terminology when counting the number of outcomes with a specified number of heads, boys, or defectives.

(a) A coin is tossed 7 times. Show that the number of outcomes with

   (i) Exactly 3 heads is $\binom{7}{3} = 35$

   (ii) Exactly $x$ heads $\binom{7}{x}$

(b) A coin is tossed $N$ times. Show that the number of outcomes with exactly $x$ heads is $\binom{N}{x}$

SOLUTION

(a) When a coin is tossed 7 times, an outcome can be expressed as a sequence of heads (H) and tails (T). There are seven choices of either an H or T and this amounts to placing one of the two symbols in each of seven positions.

— — — — — — —

For a sequence to have exactly 3 H's, and also $7 - 3 = 4$ T's, an H must be placed in exactly 3 of the positions. These can be chosen in

$$\binom{7}{3} = \frac{7 \times 6 \times 5}{3 \times 2 \times 1} = 35 \text{ ways.}$$

If 3 is replaced by $x$ heads, the reasoning is similar and we obtain $\binom{7}{x}$. Notice that $x = 0$ results in $\binom{7}{0} = 1$ according to our convention above.

(b) Here, there are $N$ places to fill in with exactly $x$ H's and the argument in part (a) about choosing exactly $x$ places for an H gives the result $\binom{N}{x}$.

**Example 27** Probabilities of Being Selected Under Random Selection

After some initial challenges, there remain 16 potential jurors of which 10 are male and 6 female. The defense attorney can dismiss 4 additional persons on the basis of answers to her questions.

(a) How many ways can the 4 additional jurors be selected for dismissal?

(b) How many selections are possible that result in 1 male and 3 females being dismissed?

(c) If the selection process were random, what is the probability that 1 male and 3 females would be dismissed?

SOLUTION  (a)  According to the counting rule $\binom{N}{r}$, the number of ways 4 jurors can be selected out of 16 is

$$\binom{16}{4} = \frac{16 \times 15 \times 14 \times 13}{4 \times 3 \times 2 \times 1} = 1820$$

(b)  One male can be chosen from 10 in $\binom{10}{1} = 10$ ways. Also, 3 females can be chosen from 6 in

$$\binom{6}{3} = \frac{6 \times 5 \times 4}{3 \times 2 \times 1} = 20 \text{ ways}$$

Each of the 10 choices of a male can accompany each of the 20 choices of 3 females. Reasoning from the tree diagram, we conclude that the number of possible samples with the stated composition is

$$\binom{10}{1} \times \binom{6}{3} = 10 \times 20 = 200$$

(c)  Random sampling requires that the 1820 possible samples are all equally likely. Of these, 200 are favorable to the event $A = [1 \text{ male and 3 females}]$. Consequently,

$$P(A) = \frac{200}{1820} = .110$$

The notion of a random sample from a finite population is crucial to statistical inference. In order to generalize from a sample to the population, it is imperative that the sampling process be impartial. This criterion is evidently met if we allow the selection process to be such that all possible samples are given equal opportunity to be selected. This is precisely the idea behind the term **random sampling**, and a formal definition can be phrased as follows.

> A sample of size $n$ selected from a population of $N$ distinct objects is said to be a **random sample** if each collection of size $n$ has the same probability $1 / \binom{N}{n}$ of being selected.

Note that this is a conceptual rather than an operational definition of a random sample. On the surface, it might seem that a haphazard selection by the experimenter would result in a random sample. Unfortunately, a seemingly haphazard selection may have hidden bias. For instance, when asked to name a random integer between 1 and 9, more persons respond with 7 than any other number. Also, odd numbers are more popular than even numbers. Therefore, the selection of objects must be entrusted to some device that cannot think. The selection process must be mechanized to make it truly haphazard!

To accomplish the goal of a random selection, one may make a card for each of the N members of the population, shuffle, and then draw n cards. This method is easy to understand but awkward to apply to large populations. It is best to use random numbers as described in Chapter 1. Random numbers are conveniently generated on a computer (see Chapter 4 Technology section).

Earlier in this chapter, we stated that probability constitutes the major vehicle of statistical inference. In the context of random sampling from a population, the tools of probability enable us to gauge the likelihood of various potential outcomes of the sampling process. Ingrained in our probability calculations lies the artificial assumption that the composition of the population is *known*. The route of statistical inference is exactly in the opposite direction, as depicted in Figure 5. It is the composition of the population that is unknown while we have at hand the observations (data) resulting from a random sample. Our object of inference is to ascertain what compositions (or models) of the population are compatible with the observed sample data. We view a model as plausible unless probability calculations based on this model make the sample outcome seem unlikely.

Figure 5    Probability versus statistical inference.

## Exercises

**4.84** Evaluate:

(a) $\binom{6}{3}$    (b) $\binom{10}{4}$    (c) $\binom{22}{2}$

(d) $\binom{22}{20}$    (e) $\binom{30}{3}$    (f) $\binom{30}{27}$

**4.85** Of 10 available candidates for membership in a university committee, 6 are men and 4 are women. The committee is to consist of 4 persons.

(a) How many different selections of the committee are possible?

(b) How many selections are possible if the committee must have 2 men and 2 women?

**4.86** If a coin is tossed 11 times, the outcome can be recorded as an 11-character sequence of $H$'s and $T$'s according to the results of the successive tosses. In differently, in how many ways can one choose 4 positions out of 11 to put the letter *H*?)

**4.87** A psychologist will select 5 preschool children from a class of 11 students in order to try out new abuse awareness material.

(a) How many different selections are possible?

(b) Suppose 4 of the 11 children are males. If the 5 selected children were to consist of 2 males and 3 females, how many different selections are possible?

**4.88** Out of 12 people applying for an assembly job, 3 cannot do the work. Suppose two persons will be hired.

(a) How many distinct pairs are possible?

(b) In how many of the pairs will 0 or 1 person not be able to do the work?

(c) If two persons are chosen in a random manner, what is the probability that neither will be able to do the job?

**4.89** After a preliminary screening, the list of qualified jurors consists of 10 males and 7 females. The 5 jurors the judge selects from this list are all males. Did the selection process seem to discriminate against females? Answer this by computing the probability of having no female members in the jury if the selection is random.

**4.90** Suppose you participate in a lottery conducted by a local store to give away four prizes. Each customer is allowed to place 2 cards in the barrel. Suppose the barrel contains 5000 cards from which the 4 winning cards will be chosen at random. What is the probability that at least one of your cards will be drawn?

**4.91** A batch of 20 used automobile alternators contains 4 defectives. If 3 alternators are sampled at random, find the probability of the event

(a) $A =$ [ None of the defectives appear ]

(b) $B =$ [ Exactly two defectives appear ]

**4.92** *Ordered sampling versus unordered sampling.* Refer to Exercise 4.91. Suppose the sampling of 3 alternators is done by randomly choosing one after another and without replacement. The event $A$ can then be described as $G_1 G_2 G_3$, where G denotes "good" and the suffixes refer to the order of the draws. Use the method of Example 14 to calculate

how many ways can there be 4 *H*'s and 7 *T*'s? (Put $P(A)$ and $P(B)$. Verify that you get the same results as in Exercise 4.91.

This illustrates the following fact: To arrive at a random sample, we may randomly draw one object at a time without replacement and then disregard the order of the draws.

**4.93** A college senior is selected at random from each state. Next, one senior is selected at random from the group of 50. Does this procedure produce a senior selected at random from those in the United States?

**4.94** An instructor will choose 3 problems from a set of 7 containing 3 hard and 4 easy problems. If the selection is made at random, what is the probability that only the hard problems are chosen?

**4.95** Nine agricultural plots for an experiment are laid out in a square grid as shown. Three plots are to be selected at random.

(a) Find the probability that all 3 are in the same row.

(b) Find the probability that all 3 are in different rows.

| 1 | 2 | 3 |
|---|---|---|
| 4 | 5 | 6 |
| 7 | 8 | 9 |

**4.96** In one area of an orchard, there are 17 trees, of which 10 are bushy and 7 lean. If 4 trees are randomly selected for testing a new spray, what is the probability that exactly 2 bushy trees are selected?

**\*4.97** Referring to Exercise 4.96, now suppose that the trees are located in two rows: Row *A* has 8 trees of which 4 are bushy, and row *B* has 9 trees of which 6 are bushy. Two trees are to be randomly selected from each row for testing the spray, and the selections are independent for the two rows.

(a) Find the probability that the trees selected in row *A* are both bushy and those selected in row *B* are both lean.

(b) Find the probability that of the total of 4 trees selected in the manner described above, exactly 2 are bushy.

**4.98** Are the following methods of selection likely to produce a random sample of 5 students from your school? Explain.

(a) Pick 5 students throwing flying discs on the mall.

(b) Pick 5 students who are studying in the library on Friday night.

(c) Select 5 students sitting near you in your statistics course.

**4.99** An advertisement seeking volunteers for a clinical research draws 11 respondents. Of these respondents, 5 are below age 30 and 6 are over 30. The researcher will randomly select 4 persons to assign to a particular treatment regimen.

(a) How many selections are possible?

(b) What is the probability exactly 3 of the selected persons are below age 30?

**\*4.100** Refer to Exercise 4.99, and further suppose that the 5 respondents who are below 30 consist of 2 males and 3 females, whereas those above 30 consist of 4 males and 2 females. Now, the researcher wants to randomly select 2 males and 2 females to be assigned to the treatment regimen. (The random selections from the different sexes are, of course, independent.)

(a) How many selections are possible?

(b) What is the probability that both selected males are over 30 and both selected females are under 30?

**4.101** A box of tulip bulbs contains six bulbs that produce yellow flowers and five bulbs that produce red flowers. Four bulbs are to be randomly selected without replacement. Find the probability that:

(a) Exactly two of the selected bulbs produce red flowers.

(b) At least two of the selected bulbs produce red flowers.

(c) All four selected bulbs produce flowers of an identical color.

**4.102** A file cabinet has 8 student folders arranged alphabetically according to last name. Three files are selected at random.

(a) How many different selections are possible?

(b) Find the probability that the selected folders are all adjacent.
(*Hint*: Enumerate the selections of adjacent folders.)

## STATISTICS IN CONTEXT

Over 22,000 man-made objects are above our heads in low earth orbits. Beginning in 1957, when the Russians launched Sputnik into space, a great number of satellites and even a space station have been placed in orbits around the earth. However, because there is still some air, even at those heights, all objects in orbit continually slow down and eventually make a fiery re-entry into our atmosphere. Based on records of a large number of objects, 10 to 40% of the objects can hit the ground as charred remains. How is the probability determined that your city will be hit by a specific piece of space junk?[4]

Let $A$ be the event that charred remains do hit the earth. It is necessary for experts to specify the probability of $A$ using knowledge of the size and composition of the object as well as its re-entry angle.

For purposes of illustration, we take $P(A) = .30$. Next, let $B$ be the event that your city is hit, given that some remains reach the earth. The probability $P(B|A)$ will depend on both the size of your city and its location relative to the current orbit of the object. Suppose that, based on their calculations, the experts conclude that $P(B|A) = .0002$. Then, by the multiplication law, the probability the object will hit your city is

$$P(BA) = P(A)P(B|A) = .30 \times .0002 = .00006$$

or 6 out of 100,000 times.

[4]D. Mosher, "Space Junk—How the FAA estimates the risk of a satellite hitting you." *Wired*, Nov. 2011, p. 48.

The FAA must actually address the chances of hitting you or other people. Their calculations are more complicated because they include the size of the remains that hit the ground and the population density in the area of concern. At the 2011 re-entry of a bus sized object, NASA put the probability of hitting someone at 1 / 3200. Through 2012, no person has ever been killed by space debris.

## USING STATISTICS WISELY

1. Begin by creating a sample space $S$ that specifies all possible outcomes for the experiment.

2. Always assign probabilities to events that satisfy the axioms of probability. In the discrete case, the possible outcomes can be arranged in a sequence. The axioms are then automatically satisfied when probability $P(e)$ is assigned to the elementary outcome $e$, where

$$0 \le P(e) \quad \text{and} \quad \sum_{\text{all } e \text{ in } S} P(e) = 1$$

and then the probability of any even $A$ is defined as

$$P(A) = \sum_{\text{all } e \text{ in } A} P(e)$$

3. Always use the rules of probability when combining the probabilities of events.

4. Do not confuse independent events with mutually exclusive events. When $A$ and $B$ are mutually exclusive, only one of them can occur. Their intersection is empty and so has probability 0.

5. Do not assign probability to $AB$ according to the special product rule

$$P(AB) = P(A)P(B)$$

unless the conditions for independence hold. Independence may be plausible when the events $A$ and $B$ pertain to physically unrelated parts of a large system and there are no common causes that jointly affect the occurrence of both events.

## KEY IDEAS AND FORMULAS

An **experiment** is any process of observing a phenomenon that has variation in its outcomes. Each possible outcome is called an **elementary outcome**, a **simple event**, or an **element of the sample space**. The **sample space** is the collection of all of these outcomes. A **discrete sample** space has outcomes that can be arranged in a, possibly infinite, sequence. In contrast, a **continuous sample space** is an interval of possible outcomes.

A **tree diagram**, with separate sets of branches for each stage of an experiment, can help identify the elementary outcomes.

If an experiment is repeated a large number of times, experimentally we observe that the relative frequency of an event $A$

$$\frac{\text{Number of times } A \text{ occurs}}{\text{Number of time experiment is performed}}$$

will stabilize at a numerical value. This **long-run stability of relative frequency** motivates us to assign a number $P(A)$ between 0 and 1 as the probability of the event $A$. In the other direction, we can approximate the probability of any event by repeating an experiment many times.

When the **sample space** is **discrete, probability** is then expressed as any assignment of nonnegative numbers to the elementary outcomes so that probability 1 is assigned to the whole sample space. The probability model of an experiment is described by:

1. The sample space, a list or statement of all possible distinct outcomes.
2. Assignment of probabilities to all the elementary outcomes. $P(e) \geq 0$ and $\sum P(e) = 1$, where the sum extends over all $e$ in $S$.

The **probability of an event** $A$ is the sum of the probabilities of all the elementary outcomes that are in $A$.

$$P(A) = \sum_{\text{all } e \text{ in } A} P(e)$$

A uniform probability model holds when all the elementary outcomes in $S$ are equiprobable. With a **uniform probability model**,

$$P(A) = \frac{\text{No. of } e \text{ in } A}{\text{No. of } e \text{ in } S}$$

In all cases, $P(A)$, viewed as the long-run relative frequency of $A$, can be approximately determined by repeating the experiment a large number of times.

Elementary outcomes and events can be portrayed in a **Venn diagram**. The event operations **union, intersection**, and **complement** can be depicted as well as the result of combining several operations.

The three basic laws of **probability** are

**Law of complement**   $P(A) = 1 - P(\bar{A})$

**Addition law**   $P(A \cup B) = P(A) + P(B) - P(AB)$

**Multiplication law**   $P(AB) = P(B)P(A|B)$

These are useful in probability calculations when events are formed with the operations of complement, union, and intersection.

Two events are **incompatible** or **mutually exclusive** if their intersection is empty. In that case we have the **special addition law for incompatible events**

$$P(A \cup B) = P(A) + P(B)$$

The concept of conditional probability is useful to determine how the probability of an event $A$ must be revised when another event $B$ has occurred. It forms the basis of the multiplication law of probability and the notion of **independence** of events.

Conditional probability of $A$ given $B$

$$P(A \mid B) = \frac{P(AB)}{P(B)}$$

Two events $A$ and $B$ are said to be **independent** if $P(A \mid B) = P(A)$. An equivalent condition for independence is that $P(AB) = P(A) P(B)$.

**Rule of Total Probability**

$$P(A) = P(A \mid B) P(B) + P(A \mid \bar{B}) P(\bar{B})$$

**Bayes' Theorem**

$$P(B \mid A) = \frac{P(A \mid B) P(B)}{P(A \mid B) P(B) + P(A \mid \bar{B}) P(\bar{B})}$$

The notion of **random sampling** is formalized by requiring that all possible samples are equally likely to be selected. The rule of combinations facilitates the calculation of probabilities in the context of random sampling from $N$ distinct units.

**Rule of Combinations**

$$\binom{N}{r} = \frac{N \times (N - 1) \times \ldots \times (N - r + 1)}{r \times (r - 1) \times \ldots \times 2 \times 1}$$

# TECHNOLOGY

### *Generating random digits*

#### MINITAB

The following commands illustrate the generation of 5 random digits between 1 and 237 inclusive. As with random-digit tables, it is possible to get repeated values. It is prudent to generate a few more digits than you need in order to get enough unique numbers.

**Dialog box:**

**Calc > Random Data > Integer.** Type **5** in **Generate.** Type *C1* in **Store,** *1* in **Minimum,** and *237* in **Maximum.** Click **OK.**

#### EXCEL

The following commands illustrate the generation of 5 random digits between 1 and 237 inclusive. As with random-digit tables, it is possible to get repeated values.

Click the **Data** tab then **Data Analysis**,[5] and then **Random Number Generation**. Click **OK**. Type *1* in **Number of Variables**, *5* in **Number of Random Numbers**.

Select *Uniform* for **Distribution**, type *1* for **Between** and *238* after. (238 is 1 larger than the desired limit 237)

Type (any positive number) *743* in **Random Seed**.

Under **Output options**, choose **Output Range** and type *A1*.

Click **OK**.

The random numbers appear in the first column of the spreadsheet. You just ignore the decimal part of each entry to obtain random digits between 1 and 237 inclusive.

[5]If **Data Analysis** is not on tools menu, see directions for adding in Chapter 2, Technology EXCEL section.

### TI-84/83 PLUS

The following commands show the generation of 5 random digits between 1 and 237 inclusive. As with random-digit tables, it is possible to get repeated values.

Enter any nonzero four digit number on the Home screen.

Press the **STO** → button. Press the **MATH** button.

Select the **PRB** menu and then select **1: rand**.

From the Home screen press **ENTER**.

Press the **MATH** button. Select the **PRB** menu and then **5: randInt(**.

With **randInt(** on the Home screen, enter 1,237,5) so that the screen reads randInt (1,237,5)

Press **ENTER** to obtain the 5 random digits.

## REVIEW EXERCISES

**4.103** Describe the sample space for each of the following experiments.

(a) The number of different words used in a sentence containing 24 words.

(b) The air pressure (psi) in the right front tire of a car.

(c) In a survey, 50 students are asked to respond "yes" or "no" to the question "Do you hold at least a part-time job while attending school?" Only the number answering "yes" will be recorded.

(d) The time a TV satellite remains in operation.

**4.104** For the experiments in Exercise 4.103, which sample spaces are discrete and which are continuous?

**4.105** Identify these events in the corresponding parts of Exercise 4.103.

(a) More than 22 words.

(b) Air pressure less than or equal to 28 psi.

(c) At most 25% hold jobs.

(d) Less than 500.5 days.

**4.106** Examine each of these probability assignments and state what makes it improper.

(a) Concerning tomorrow's weather,

$$P(\text{Rain}) = .4$$
$$P(\text{Cloudy but no rain}) = .5$$
$$P(\text{Sunny}) = .2$$

(b) Concerning your passing of the statistics course,

$$P(\text{Pass}) = 1.1 \qquad P(\text{Fail}) = .1$$

(c) Concerning your grades in statistics and economics courses,

$$P(A \text{ in statistics}) = .3$$
$$P(A \text{ in economics}) = .6$$
$$P(A's \text{ in both statistics and economics}) = .4$$

**4.107** A driver is stopped for erratic driving, and the alcohol content of his blood is checked. Specify the sample space and the event $A = [\text{level exceeds legal limit}]$ if the legal limit is .10%.

**4.108** The Wimbledon men's tennis championship ends when one player wins three sets.

(a) How many elementary outcomes end in three sets? In four?

*(b) If the players are evenly matched, what is the probability that the tennis match ends in four sets?

**4.109** There are four tickets numbered 1, 2, 3, and 4. Suppose a two-digit number will be formed by first drawing one ticket at random and then drawing a second ticket at random from the remaining three. (For instance, if the first ticket drawn shows 3 and the second shows 1, the number recorded is 31.) List the sample space and determine the following probabilities.

(a) An even number.

(b) A number larger than 20.

(c) A number between 22 and 30.

**4.110** To compare two varieties of wheat, say, $a$ and $b$, a field trial will be conducted on four square plots located in two rows and two columns. Each variety will be planted on two of these plots.

Plot arrangement

(a) List all possible assignments for variety $a$.

(b) If the assignments are made completely at random, find the probability that the plots receiving variety $a$ are:
  (i) In the same column.
  (ii) In different rows and different columns.

**4.111** Refer to Exercise 4.110. Instead of a completely random choice, suppose a plot is chosen at random from each row and assigned to variety $a$. Find the probability that the plots receiving $a$ are in the same column.

**4.112** Chevalier de Méré, a French nobleman of the seventeenth century, reasoned that in a single throw of a fair die, $P(1) = \frac{1}{6}$, so in two throws, $P(1 \text{ appears at least once}) = \frac{1}{6} + \frac{1}{6} = \frac{1}{3}$. What is wrong with the above reasoning? Use the sample space of Exercise 4.21 to obtain the correct answer.

**4.113** Does the uniform model apply to the following observations? Explain.

(a) Day of week on which the most persons depart by airplane from Chicago.

(b) Day of week on which the monthly low temperature occurs.

(c) Day of week on which the maximum amount of ozone is recorded.

(d) Month of year when a department store has the maximum sales revenues.

**4.114** A three-digit number is formed by arranging the digits 1, 2, and 5 in a random order.

(a) List the sample space.

(b) Find the probability of getting a number less than 400.

(c) What is the probability that an even number is obtained?

**4.115** In checking the conditions of a used car, let $A$ denote the event that the car has a faulty transmission, $B$ the event that it has faulty brakes, and $C$ the event that it has a faulty exhaust system. Describe in words what the following events represent:

(a) $A \cup B$

(b) $ABC$

(c) $\overline{ABC}$

(d) $\overline{A} \cup \overline{B}$

**4.116** Express the following statements in the notations of the event operations.

(a) $A$ occurs and $B$ does not.

(b) Neither $A$ nor $B$ occurs.

(c) Exactly one of the events $A$ and $B$ occurs.

**4.117** Suppose each of the numbers .1, .3, and .5 represents the probability of one of the events $A$, $AB$, and $A \cup B$. Connect the probabilities to the appropriate events.

**4.118** From the probabilities exhibited in this Venn diagram, find $P(\overline{A})$, $P(AB)$, $P(B \cup C)$, and $P(BC)$.

**4.119** Using event relations, express the following events in terms of the three events $A$, $B$, and $C$.

(a) All three events occur.

(b) At least one of the three events occurs.

(c) $A$ and $B$ occur and $C$ does not.

(d) Only $B$ occurs.

**4.120** Let the three events $A$, $B$, and $C$ represent the cases that a randomly selected student is good at answering multiple choice, essay, and mathematical questions, respectively. The probabilities of the

various intersections are given in the accompanying table. [For instance, $P(AB\overline{C}) = .10$.]

|   | $B$ | | $\overline{B}$ | |
|---|---|---|---|---|
|   | $C$ | $\overline{C}$ | $C$ | $\overline{C}$ |
| $A$ | .05 | .10 | .08 | .14 |
| $\overline{A}$ | .20 | .15 | .18 | .10 |

(a) Draw a Venn diagram, identify the intersections, and mark the probabilities.

(b) Determine the probabilities

$$P(AB) \qquad P(A\overline{C}) \qquad P(C)$$

(c) Fill in the accompanying probability table concerning the events $A$ and $B$.

|   | $B$ | $\overline{B}$ |
|---|---|---|
| $A$ |   |   |
| $\overline{A}$ |   |   |
|   |   |   |

**4.121** Referring to Exercise 4.120 describe each event in the context of the problem and calculate the probabilities of the following events.

(a) Both $B$ and $C$ occur.

(b) Either $B$ or $C$ occurs.

(c) $B$ occurs and $C$ does not occur.

(d) Only one of the three events $A$, $B$, and $C$ occurs.

**4.122** Concerning three events $A$, $B$, and $C$, the following probabilities are specified.

$$P(A) = .51 \quad P(AB) = .17 \quad P(ABC) = .12$$
$$P(B) = .45 \quad P(BC) = .20$$
$$P(C) = .50 \quad P(AC) = .33$$

Draw a Venn diagram and determine the probabilities of all the intersections that appear in the diagram. Also, make a probability table like the one given in Exercise 4.120

**4.123** Referring to Exercise 4.122 find the probability that:

(a) $B$ occurs and $C$ does not occur.

(b) At least one of the events $A$ and $B$ occurs.

(c) Exactly two of the events $A$, $B$, and C occur.

**4.124** Suppose a fair die has its even-numbered faces painted red and the odd-numbered faces are white. Consider the experiment of rolling the die once and the events

$$A = [\,2 \text{ or } 3 \text{ shows up}\,]$$
$$B = [\,\text{Red face shows up}\,]$$

Find the following probabilities:

(a) $P(A)$     (c) $P(AB)$     (e) $P(A \cup B)$

(b) $P(B)$     (d) $P(A \mid B)$

**4.125** Given $P(AB) = .4$ and $P(B) = .5$, find $P(A \mid B)$. If, further, $P(A) = .8$, are $A$ and $B$ independent?

**4.126** Suppose three events $A$, $B$, and C are such that $B$ and C are mutually exclusive and

$$P(A) = .6 \quad P(B) = .3 \quad P(C) = .25$$
$$P(A \mid B) = \frac{2}{3} \quad P(\bar{A}C) = .1$$

(a) Show the events in a Venn diagram.

(b) Determine the probabilities of all the intersections and mark them in the Venn diagram.

(c) Find the probability that only one of the three events occurs.

**4.127** Refer to Exercise 4.126 for each pair of events given below, determine whether or not the events are independent.

(a) $A$, C

(b) $A\bar{B}$, C

**4.128** Let $A$ be the event that a person is a binge drinker and $B$ be the event that the person is female. For a person selected at random in the United States, the probabilities are[6]

$$P(B) = .50 \quad P(A \mid B) = .11 \quad P(A \mid \bar{B}) = .23$$

(a) Express in words, in the context of this problem, the third probability statement.

(b) Determine the probability that the person selected is a binge drinker.

(c) If the person selected is found to be a binge drinker, what is the probability of being female?

**4.129** Refer to the probability table given in Exercise 4.120 concerning three events $A$, $B$, and C.

(a) Find the conditional probability of $A$ given that $B$ does not occur.

(b) Find the conditional probability of B given that both $A$ and C occur.

(c) Determine whether or not the events $A$ and C are independent.

**4.130** Mr. Hope, a character apprehended by Sherlock Holmes, was driven by revenge to commit two murders. He presented two seemingly identical pills, one containing a deadly poison, to an adversary who selected one while Mr. Hope took the other. The entire procedure was then to be repeated with the second victim. Mr. Hope felt that Providence would protect him, but what is the probability of the success of his endeavor?

**4.131** Refer to the Statistic in Context discussion about space junk. Change the probability of hitting to

(a) .4 and repeat the calculation of probability for hitting your city.

(b) .1 and repeat the calculation of probability for hitting your city.

**4.132** A mail order firm sends 70% of its orders by a parcel service that delivers late only 2% of the time. It sends the other 30% by a combination bus service and local transportation that delivers late 7% of the time. Find the

(a) The probability that an order will be delivered late.

(b) The probability that an order delivered late was sent by the parcel service.

**4.133** Identity theft is a growing problem in the United States. According to a Federal Trade Commission Report about 280,000 identity complaints were filed for 2011. Among the 43.2 million persons in the 20–29 year old age group, 56,689 complaints were filed. The 20–29 year old age group makes up proportion .139 of the total population. Use the relative frequencies to approximate the probability, that for the current year

(a) A person in the 20−29 age group files an identity theft complaint.

(b) A person not in the 20−29 age group files an identity theft complaint. Comment on your answers to parts (a) and (b).

(c) A random person will file an identity theft complaint.

(d) If a complaint is filed, what is the probability it is by someone in the 20−29 age group.

**4.134** Three production lines contribute to the total pool of a company's product. Line 1 provides 20% to the pool and 10% of its products are defective; Line 2 provides 50% to the pool and 5% of its products are defective; Line 3 contributes 30% to the pool and 6% of its products are defective.

(a) What percent of the items in the pool are defective?

(b) Suppose an item is randomly selected from the pool and found to be defective. What is the probability that it came from Line 1?

**4.135** In an optical sensory experiment, a subject shows a fast response ($F$), a delayed response ($D$), or no response at all ($N$). The experiment will be performed on two subjects.

(a) Using a tree diagram, list the sample space.

(b) Suppose, for each subject, $P(F) = .4$, $P(D) = .3$, $P(N) = .3$, and the responses of different subjects are independent.
  (i) Assign probabilities to the elementary outcomes.
  (ii) Find the probability that at least one of the subjects shows a fast response.
  (iii) Find the probability that both of the subjects respond.

**4.136** Four upper level undergraduate students are available to serve on a committee.

| Student | Gender | Year in school |
|---------|--------|----------------|
| 1 | M | Junior |
| 2 | M | Senior |
| 3 | F | Junior |
| 4 | F | Senior |

Two students will be selected at random to serve on the committee. Let

$A$ - The students selected are of the same gender.
$B$ - The students selected are the same year in school.

(a) Make a Venn diagram showing the outcomes and the two events.

(b) Find the probability of $A \cup B$.

(c) Are $A$ and $B$ independent? Explain why or why not.

(d) Find the probability of $\bar{A}B$.

**4.137** A local moving company owns 11 trucks. Three are randomly selected for compliance with emission standards and all are found to be noncompliant. The company argues that these are the only three that do not meet the standards. Calculate the probability that, if only three are noncompliant, all three would be in the sample. Comment on the veracity of the company's claim.

**4.138** In all of William Shakespeare's works, he used 884,647 different words. Of these, 14,376 appeared only once and 4343 appeared twice. If one word is randomly selected from a list of these 884,647 different words:

(a) What is the probability that the selected word appears only once?

(b) What is the probability that the selected word appears exactly twice?

(c) What is the probability that the selected word appears more than twice?

(d) Suppose that, instead of randomly selecting a word from the list of different words, you randomly select a word from a book of Shakespeare's complete works by selecting a page, line, and word number from a random-digit table. Is the probability of selecting a word that appears exactly once larger, smaller, or the same as your answer to part (a)?

**4.139** An IRS agent receives a batch of 18 tax returns that were flagged by computer for possible tax evasions. Suppose, unknown to the agent, 7 of these returns have illegal deductions and the other 11 are in good standing. If the agent randomly selects 4 of these returns for audit, what is the probability that:

# Probability Distributions

John Terence Turner/Photodisc/Getty Images

## Rescue Service on a Lake

Student sailors and other boaters on Lake Mendota are protected by a boating rescue service. The relative frequencies from a long record for summer days leads to an approximate distribution of the number of rescues per day.

(a) None of the returns that contain illegal deductions are selected?

(b) At least 2 have illegal deductions?

**\*4.140 Polya's urn scheme.** An urn contains 4 red and 6 green balls. One ball is drawn at random and its color is observed. The ball is then returned to the urn, and 3 new balls of the same color are added to the urn. A second ball is then randomly drawn from the urn that now contains 13 balls.

(a) List all outcomes of this experiment (use symbols such as $R_1 G_2$ to denote the outcome of the first ball red and the second green).

(b) What is the probability that the first ball drawn is green?

(c) What is the conditional probability of getting a red ball in the second draw given that a green ball appears in the first?

(d) What is the (unconditional) probability of getting a green ball in the second draw?

**\*4.141 Birthdays.** It is somewhat surprising to learn the probability that 2 persons in a class share the same birthday. As an approximation, assume that the 365 days are equally likely birthdays.

(a) What is the probability that least 2 have the same birth soning associated with a tree there are $365 \times 365 \times 36$ day outcomes. Of these, 365 correspond to no common birthd

(b) Generálize the above reasoning to Show that

$$P[\text{No common birthday}] =$$
$$\frac{365 \times 364 \times \cdots \times (365 - N + }{(365)^N}$$

Some numerical values are

| $N$ | 5 | 9 | 18 | 22 | 23 |
|---|---|---|---|---|---|
| $P[\text{No common birthday}]$ | .973 | .905 | .653 | .524 | .493 |

We see that with $N = 23$ persons, the probability is greater than $\frac{1}{2}$ that at least two share a common birthday.)

The distribution describes the randomness of daily rescue activity. For instance, on any given day, the most probable number of rescues is 2. The distribution can be the basis for decisions concerning manpower and the need for additional rescue boats.

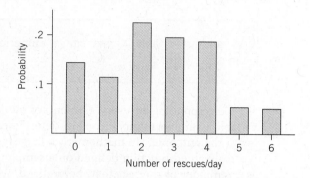

A prescription for the probability model of an experiment contains two basic ingredients: the sample space and the assignment of probability to each elementary outcome. In Chapter 4, we encountered several examples where the elementary outcomes had only qualitative descriptions rather than numerical values. For instance, with two tosses of a coin, the outcomes HH, HT, TH, and TT are pairs of letters that identify the occurrences of heads or tails. If a new vaccine is studied for the possible side effects of nausea, the response of each subject may be severe, moderate, or no feeling of nausea. These are qualitative outcomes rather than measurements on a numerical scale.

Often, the outcomes of an experiment are numerical values: for example, the daily number of burglaries in a city, the hourly wages of students on summer jobs, and scores on a college placement examination. Even in the former situation where the elementary outcomes are only qualitatively described, interest frequently centers on some related numerical aspects. If a new vaccine is tested on 100 individuals, the information relevant for an evaluation of the vaccine may be the numbers of responses in the categories—severe, moderate, or no nausea. The detailed record of 100 responses can be dispensed with once we have extracted this summary. Likewise, for an opinion poll conducted on 500 residents to determine support for a proposed city ordinance, the information of particular interest is how many residents are in favor of the ordinance, and how many are opposed. In these examples, the individual observations are not numerical, yet a numerical summary of a collection of observations forms the natural basis for drawing inferences. In this chapter, we concentrate on the numerical aspects of experimental outcomes.

## 1. RANDOM VARIABLES

Focusing our attention on the numerical features of the outcomes, we introduce the idea of a random variable.

> A random variable $X$ associates a numerical value with each outcome of an experiment.

Corresponding to every elementary outcome of an experiment, a random variable assumes a numerical value, determined from some characteristic pertaining to the outcome. (In mathematical language, we say that a random variable $X$ is a real-valued function defined on a sample space.) The word "random" serves as a reminder of the fact that, beforehand, we do not know the outcome of an experiment or its associated value of $X$.

In general, we use capital letters like $X$ or $Y$ to denote random variables and lowercase letters like $x$ or $y$ to denote their specific values. For instance $X$ represents the outcome of rolling a die before we roll so the outcome is uncertain. Then $x$ represents the actual outcome of the roll so $x = 3$ if the result is three dots.

**Example 1**    The Number of Heads as a Random Variable

Consider $X$ to be the number of heads obtained in three tosses of a coin.
  (a)  List the numerical values of $X$ and the corresponding elementary outcomes.
  (b)  Identify all of the elementary outcomes that produce the value $x$.

SOLUTION    (a)  First, $X$ is a variable since the number of heads in three tosses of a coin can have any of the values $x = 0$, 1, 2, or 3. Second, this variable is random in the sense that the value that would occur in a given instance cannot be predicted with certainty. We can, though, make a list of the elementary outcomes and the associated values of $X$.

| Outcome | Value of $X$ |
|---------|--------------|
| HHH | 3 |
| HHT | 2 |
| HTH | 2 |
| HTT | 1 |
| THH | 2 |
| THT | 1 |
| TTH | 1 |
| TTT | 0 |

Note that, for each elementary outcome there is only one value of $X$. However, several elementary outcomes may yield the same value.

(b) Scanning our list, we now identify the events (i.e., the collections of the elementary outcomes) that correspond to the distinct values of $X$.

| Numerical Value of $X$ as an Event | Composition of the Event |
|---|---|
| $[X = 0] = \{TTT\}$ | |
| $[X = 1] = \{HTT, THT, TTH\}$ | |
| $[X = 2] = \{HHT, HTH, THH\}$ | |
| $[X = 3] = \{HHH\}$ | |

Guided by this example, we observe the following general facts.

**The events corresponding to the distinct values of $X$ are incompatible.**

**The union of these events is the entire sample space.**

Typically, the possible values of a random variable $X$ can be determined directly from the description of the random variable without listing the sample space. However, to assign probabilities to these values, treated as events, it is sometimes helpful to refer to the sample space.

**Example 2**    A Random Variable That Is a Count with a Finite Maximum Value

Fifty cars are entered in a 100-mile road race. Let $X$ be the number of cars that actually finish the race. Here $X$ could conceivably take any of the values $0, 1, \ldots 50$.

**Example 3**    A Random Variable That Is a Count with No Upper Limit

Once a week, a student buys a single lottery ticket. Let $X$ be the number of tickets she purchases before she wins at least $1000 on a ticket. The possible values of $X$ are then $1, 2, 3, \ldots$, where the list never terminates.

A random variable is said to be **discrete** if it has either a finite number of values or infinitely many values that can be arranged in a sequence. All the preceding examples are of this type. On the other hand, if a random variable represents some measurement on a continuous scale and is therefore capable of assuming all values in an interval, it is called a **continuous** random variable. Of course, any measuring device has a limited accuracy and, therefore, a continuous scale must be interpreted as an abstraction. Some examples of continuous random variables are the height of an adult male, the daily milk yield of a holstein, and the survival time of a patient following a heart attack.

Probability distributions of discrete random variables are explored in this chapter. As we shall see, the developments stem directly from the concepts of probability introduced in Chapter 4. A somewhat different outlook is involved in the process of conceptualizing the distribution of a continuous random variable. Details for the continuous case are postponed until Chapter 6.

# Exercises

**5.1** Identify each of the following as a discrete or continuous random variable.

(a) Number of empty seats on a flight from Atlanta to London.

(b) Yearly low temperature in your city.

(c) The height of the highest wave on the north shore of Hawaii next winter.

(d) Time it takes for a plumber to fix a bathroom faucet.

(e) Number of cars ticketed for illegal parking on campus today.

**5.2** Identify the variable as a discrete or a continuous random variable in parts (a)–(e).

(a) The loss of weight following a diet program.

(b) The magnitude of an earthquake as measured on the open-ended Richter scale.

(c) The seating capacity of a roller coaster.

(d) The number of cars sold at a dealership on one day.

(e) The amount of electricity produced at a wind farm next Friday.

**5.3** Two of the integers { 1, 3, 5, 6, 7 } are chosen at random without replacement. Let $X$ denote the difference = largest − smallest.

(a) List all choices and the corresponding values of $X$.

(b) List the distinct values of $X$ and determine their probabilities.

**5.4** The three finalists for an award are $A$, $B$, and $C$. They are rated by two judges. Each judge assigns the ratings 1 for best, 2 for intermediate, and 3 for worst. Let $X$ denote the total score for finalist $A$ (the sum of the ratings received from the two judges).

(a) List all pairs of ratings that finalist $A$ can receive.

(b) List the distinct values of $X$.

**5.5** Refer to Exercise 5.4. Suppose instead there are two finalists $A$ and $B$ and four judges. Each judge assigns the ratings 1 for the best and 2 for the worst finalists.

(a) List all possible assignments of ratings to finalist $A$ by the four judges.

(b) List the distinct values of $X$, the total score of $A$.

**5.6** Two brands of beverages, $B$ and $M$, are popular with students. The owner of one campus establishment observes sales and, for each of three weekends, records which brand has the highest sales. List the possible outcomes, and for each outcome record the number of weekends $X$ that the sales of $B$ are highest. (Assume there are no ties.)

**5.7** Each week a grocery shopper buys either canned (C) or bottled (B) soft drinks. The type of soft drink purchased in 3 consecutive weeks is to be recorded.

(a) List the sample space.

(b) If a different type of soft drink is purchased than in the previous week, we say that there is a switch. Let $X$ denote the number of switches. Determine the value of $X$ for each elementary outcome. (*Example*: For BBB, $X = 0$; for BCB, $X = 2$.)

**5.8** A child psychologist interested in how friends are selected studies groups of three children. For one group, Ann, Barb, and Carol, each is asked which of the other two she likes best.

(a) Make a list of the outcomes. (Use $A$, $B$, and $C$ to denote the three children.)

(b) Let $X$ be the number of times Carol is chosen. List the values of $X$.

## 2. PROBABILITY DISTRIBUTION OF A DISCRETE RANDOM VARIABLE

The list of possible values of a random variable $X$ makes us aware of all the eventualities of an experiment as far as the realization of $X$ is concerned. By employing the concepts of probability, we can ascertain the chances of observing the various values. To this end, we introduce the notion of a probability distribution.

> The **probability distribution** or, simply the **distribution**, of a discrete random variable $X$ is a list of the distinct numerical values of $X$ along with their associated probabilities.
> Often, a formula can be used in place of a detailed list.

**Example 4**    The Probability Distribution for Tossing a Fair Coin

If $X$ represents the number of heads obtained in three tosses of a fair coin, find the probability distribution of $X$.

SOLUTION    In Example 1, we have already listed the eight elementary outcomes and the associated values of $X$. The distinct values of $X$ are $x = 0$, 1, 2, and 3. We now calculate their probabilities.

The model of a fair coin entails that the eight elementary outcomes are equally likely, so each is assigned the probability $\frac{1}{8}$. The event $[X = 0]$ has the single outcome TTT, so its probability is $\frac{1}{8}$. Similarly, the probabilities of $[X = 1]$, $[X = 2]$, and $[X = 3]$ are found to be $\frac{3}{8}$, $\frac{3}{8}$, and $\frac{1}{8}$, respectively. Collecting these results, we obtain the probability distribution of $X$ displayed in Table 1.

**TABLE 1**    The Probability Distribution of $X$, the Number of Heads in Three Tosses of a Coin

| $X$ Takes Value | Probability |
|:---:|:---:|
| 0 | $\frac{1}{8}$ |
| 1 | $\frac{3}{8}$ |
| 2 | $\frac{3}{8}$ |
| 3 | $\frac{1}{8}$ |
| Total | 1 |

For general discussion, we use the notation $x_1$, $x_2$, and so on, to designate the distinct values of a random variable $X$. The probability that a particular value $x_i$ occurs is denoted by $f(x_i)$. As in Example 4, if $X$ can take $k$ possible values $x_1, \ldots x_k$ with the corresponding probabilities $f(x_1), \ldots f(x_k)$, the probability distribution of $X$ can be displayed in the format of Table 2. Since the quantities $f(x_i)$ represent probabilities, they must all be numbers between 0 and 1. Furthermore, when summed over all possible values of $X$, these probabilities must add up to 1.

**TABLE 2**  Form of a Discrete Probability Distribution

| Value $x$ | Probability $f(x)$ |
|:---:|:---:|
| $x_1$ | $f(x_1)$ |
| $x_2$ | $f(x_2)$ |
| . | . |
| . | . |
| . | . |
| $x_k$ | $f(x_k)$ |
| Total | 1 |

The probability distribution of a discrete random variable $X$ is described as the function

$$f(x_i) = P[X = x_i]$$

which gives the probability for each value and satisfies:

1.  $0 \leq f(x_i) \leq 1$ for each value $x_i$ of $X$

2.  $\displaystyle\sum_{i=1}^{k} f(x_i) = 1$

A probability distribution or the probability function describes the manner in which the total probability 1 gets apportioned to the individual values of the random variable.

A graphical presentation of a probability distribution helps reveal any pattern in the distribution of probabilities. Is there symmetry about some value or a long tail to one side? Is the distribution peaked with a few values having high probabilities or is it uniform?

We consider a display similar in form to a relative frequency histogram, discussed in Chapter 2. It will also facilitate the building of the concept of a continuous distribution. To draw a probability histogram, we first mark the values of $X$ on the horizontal axis. With each value $x_i$ as center, a vertical rectangle is drawn whose area equals the probability $f(x_i)$. The probability histogram for the distribution of Example 4 is shown in Figure 1.

Figure 1   The probability histogram of $X$, the number of heads in three tosses of a coin.

**Example 5**   Probability Distribution for News Source Preference

Suppose 60% of the students at a large university prefer getting their daily news from the Internet as opposed to television. These are the only two choices. Four students are randomly selected. Let $X$ be the number of students sampled who prefer news from the Internet. Obtain the probability distribution of $X$ and plot the probability histogram.

SOLUTION   Because each student prefers either the Internet (I) or television (T) for news, the number of elementary outcomes concerning a sample of four students is $2 \times 2 \times 2 \times 2 = 16$. These can be conveniently enumerated in the scheme of Example 8, Chapter 4, called a tree diagram. However, we list them here according to the count $X$.

| $X = 0$ | $X = 1$ | $X = 2$ | $X = 3$ | $X = 4$ |
|---------|---------|---------|---------|---------|
| TTTT    | TTTI    | TTII    | TIII    | IIII    |
|         | TTIT    | TITI    | ITII    |         |
|         | TITT    | TIIT    | IITI    |         |
|         | ITTT    | ITTI    | IIIT    |         |
|         |         | ITIT    |         |         |
|         |         | IITT    |         |         |

Our objective here is to calculate the probability of each value of $X$. To this end, we first reflect on the assignment of probabilities to the elementary outcomes. For one student selected at random, we obviously have $P(I) = .6$ and $P(T) = .4$ because 60% of the students prefer Internet news. Moreover, as the population is vast while the sample size is very small, the observations on four students can, for all practical purposes, be treated as independent. That is, knowledge that the first student selected prefers Internet news does not change the probability that the second will prefer Internet news and so on.

Invoking independence and the multiplication law of probability, we calculate $P(TTTT) = .4 \times .4 \times .4 \times .4 = .0256$ so $P(X = 0) = .0256$. The event $[X = 1]$ has four elementary outcomes, each containing three T's and one I. Since $P(TTTI) = (.4)^3 \times (.6) = .0384$ and the same result holds for each of these 4 elementary outcomes, we get $P[X = 1] = 4 \times .0384 = .1536$. In the same manner,

$$P[X = 2] = 6 \times (.4)^2 \times (.6)^2 = .3456$$

$$P[X = 3] = 4 \times (.4) \times (.6)^3 = .3456$$

$$P[X = 4] = (.6)^4 = .1296$$

Collecting these results, we obtain the probability distribution of X presented in Table 3 and the probability histogram plotted in Figure 2.

**TABLE 3**    The Probability Distribution of X in Example 5

| $x$ | $f(x)$ |
|---|---|
| 0 | .0256 |
| 1 | .1536 |
| 2 | .3456 |
| 3 | .3456 |
| 4 | .1296 |
| Total | 1.0000 |

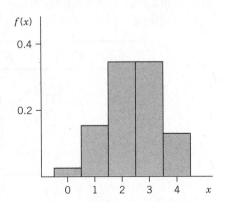

Figure 2    Probability histogram for news source preference.

At this point, we digress briefly for an explanation of the role of probability distributions in statistical inference. To calculate the probabilities associated with the values of a random variable, we require a full knowledge of the uncertainties of the experimental outcomes. For instance, when X represents some numerical characteristic of a random sample from a population, we assume a known composition of the population in order that the distribution of $X$ can be calculated numerically. In Example 5, the chances of observing the various values of $X$ were calculated under the assumption that the proportion of all students who prefer Internet news was .6. Ordinarily, in practical applications, this population quantity would be unknown to us. Suppose the letter $p$ stands for this unknown proportion of students who prefer Internet news. Statistical inference attempts to determine the values of $p$ that are deemed plausible in light of the value of $X$ actually observed in a sample. To fix ideas, suppose all four of the students sampled prefer Internet news. Based on this observation, is .6 a plausible value of $p$? Table 3 shows that if $p$ were indeed .6, the chance of observing the extreme value $X = 0$ is only .0256. This very low probability casts doubt on the hypothesis that $p = .6$. This kind of statistical reasoning is explored further in later chapters.

The probability distributions in Examples 4 and 5 were obtained by first assigning probabilities to the elementary outcomes using a process of logical deduction. When this cannot be done, one must turn to an empirical determination of the distribution. This involves repeating the experiment a large number of times and using the relative frequencies of the various values of $X$ as approximations of the corresponding probabilities.

**Example 6**    **A Probability Distribution Based on an Empirical Study**

The rate of ownership of smart devices—smart phone, tablet or PC—is rapidly increasing among students.

Let $X$ denote the number of smart devices a college senior owns. From a survey of 400 college seniors, suppose the frequency distribution of Table 4 was observed. Approximate the probability distribution of $X$.

**TABLE 4**    Frequency Distribution of the Number $X$ of Smart Devices

| Smart Devices $(x)$ | Frequency | Relative Frequency[a] |
|:---:|:---:|:---:|
| 0 | 61 | .15 |
| 1 | 153 | .38 |
| 2 | 106 | .27 |
| 3 | 56 | .14 |
| 4 | 24 | .06 |
| Total | 400 | 1.00 |

[a]Rounded to second decimal.

SOLUTION    Viewing the relative frequencies as empirical estimates of the probabilities, we have essentially obtained an approximate determination of the probability distribution of $X$. The true probability distribution would emerge if a vast number (ideally, the entire population) of seniors were surveyed.

The reader should bear in mind an important distinction between a relative frequency distribution and the probability distribution. The former is a sample-based entity and is therefore susceptible to variation on different occasions of sampling. By contrast, the probability distribution is a stable entity that refers to the entire population. It is a theoretical construct that serves as a model for describing the variation in the population.

The probability distribution of X can be used to calculate the probabilities of events defined in terms of X.

**Example 7**    Determining the Probability of an Event Determined from a Probability Distribution

Table 5 describes the number of homework assignments due next week for a randomly selected set of students taking at least 14 credits. Determine the probability that (a) X is equal to or larger than 2 and (b) X is less than or equal to 2.

**TABLE 5**    A Probability Distribution for Number of Homework Assignments Due Next Week

| Value $x$ | Probability $f(x)$ |
|:---------:|:------------------:|
| 0 | .02 |
| 1 | .23 |
| 2 | .40 |
| 3 | .25 |
| 4 | .10 |

SOLUTION

(a)  The event $[X \geq 2]$ is composed of $[X = 2]$, $[X = 3]$, and $[X = 4]$. Thus,

$$P[X \geq 2] = f(2) + f(3) + f(4)$$
$$= .40 + .25 + .10 = .75$$

(b)  Similarly, we also calculate

$$P[X \leq 2] = f(0) + f(1) + f(2)$$
$$= .02 + .23 + .40 = .65$$

## Exercises

**5.9**  Faced with a tight deadline on two major projects, you decide to hire two of the five available persons to help complete the work. They have 2, 3, 5, 3, and 2 years experience, respectively. Since their references are very similar, you decide to select two of these workers at random. Let X denote the sum of their years experience. Obtain the probability distribution of X.

**5.10**  Refer to Exercise 5.9 but let X denote the maximum years experience among the two persons selected.

(a) List all choices and the corresponding values of X.

(b) List the distinct values of X.

(c) Obtain the probability distribution of X.

**5.11** Let the random variable X represent the sum of the points in two tosses of a die.

(a) List the possible values of X.

(b) For each value of X, list the corresponding elementary outcomes.

(c) Obtain the probability distribution of X.

**5.12** Examine if the following are legitimate probability distributions.

| (a) | | (b) | |
|---|---|---|---|
| $\dot{x}$ | $f(x)$ | $x$ | $f(x)$ |
| −1 | .3 | 1 | .2 |
| 2 | .5 | 3 | .4 |
| 7 | .2 | 4 | .3 |
| 9 | .1 | 6 | .1 |

| (c) | | (d) | |
|---|---|---|---|
| $x$ | $f(x)$ | $x$ | $f(x)$ |
| −2 | .30 | 0 | .3 |
| 0 | .40 | 1 | −.1 |
| 2 | .30 | 2 | .8 |
| 4 | 0 | | |

**5.13** For each case, list the values of X and $f(x)$ and examine if the specification represents a probability distribution. If it does not, state what properties are violated.

(a) $f(x) = \frac{1}{6}(x - 1)$ for $x = 1, 2, 3, 4$

(b) $f(x) = \frac{1}{3}(x - 3)$ for $x = 2, 3, 4, 5$

(c) $f(x) = \frac{1}{20}(2x+4)$ for $x = -2, -1, 0, 1, 2$

(d) $f(x) = \frac{8}{15}\frac{1}{2^x}$ for $x = 0, 1, 2, 3$

**5.14** The probability distribution of X is given by the function

$$f(x) = \frac{1}{30}\binom{5}{x} \quad \text{for} \quad x = 1, 2, 3, 4$$

Find

(a) $P[X = 3]$

(b) $P[X \text{ is even}]$.

**5.15** Refer to Exercise 5.7. Suppose that for each purchase $P(B) = \frac{1}{2}$ and the decisions in different weeks are independent. Assign probabilities to the elementary outcomes and obtain the distribution of X.

**5.16** Refer to Exercise 5.8. Assuming each choice is equally likely, determine the probability distribution of X.

**5.17** Market researchers are concerned if people who view a commercial remember the product. They often make phone surveys two hours after a commercial is shown. Suppose that 20% of the people who watch one commercial remember the product two hours later. Four persons are randomly selected from those who viewed the commercial. Let X denote the number in the sample who remember the product. Obtain the probability distribution of X.

**5.18** New video games are rated, by editors, at various Web sites (e.g., www.gamespot.com). You are equally interested in five games that received editors' ratings of

$$10 \quad 10 \quad 10 \quad 9 \quad 9$$

on a ten point scale. Suppose you decide to randomly choose two games to purchase at this time. Let X denote the sum of the ratings on the two games purchased. List the possible values of X and determine the probability distribution.

**5.19** Suppose, for a loaded die, the probabilities of the faces

$$1 \quad 2 \quad 3 \quad 4 \quad 5 \quad 6$$

are in the ratios $3:1:1:1:1:3$. Let X denote the number appearing on a single roll of the die.

(a) Determine the probability distribution of X.

(b) What is the probability of getting an even number?

**5.20** A surprise quiz contains three multiple-choice questions: Question 1 has four suggested answers, Question 2 has three, and Question 3 has two. A completely unprepared student decides to choose the answers at random. Let X denote the number of questions the student answers correctly.

(a) List the possible values of X.

(b) Find the probability distribution of X.

(c) Find $P[\text{At least 1 correct}] = P[X \geq 1]$.

(d) Plot the probability histogram.

**5.21** A probability distribution is partially given in the following table with the additional information that the even values of X are equally likely. Determine the missing entries in the table.

| x | f(x) |
|---|------|
| 1 | .1 |
| 2 | |
| 3 | 0 |
| 4 | |
| 5 | .3 |
| 6 | |

**5.22** Consider the following setting of a random selection: A box contains 100 cards, of which 25 are numbered 1, 35 are numbered 2, 30 are numbered 3, 10 are numbered 4. One card will be drawn from the box and its number X observed. Give the probability distribution of X.

**5.23** Two probability distributions are shown in the following tables. For each case, describe a specific setting of random selection (like the one given in Exercise 5.22) that yields the given probability distribution.

(a)

| x | f(x) |
|---|------|
| 2 | .32 |
| 4 | .44 |
| 6 | .24 |

(b)

| x | f(x) |
|----|------|
| -2 | 3/4 |
| 0 | 4/14 |
| 4 | 5/14 |
| 5 | 2/14 |

**5.24** In a study of the life length of a species of mice, 120 newborn mice are observed. The numbers staying alive past the first, second, third, and fourth years are 106, 72, 25, and 0, respectively. Let X denote the life length (in discrete units of whole years) of this species of mice. Using these data, make an empirical determination of the probability distribution of X.

**5.25** Use the approximate probability distribution in Example 6 to calculate
(a) $P[X \leq 3]$
(b) $P(X \geq 2]$
(c) $P[2 \leq X \leq 3]$

**5.26** Of eight candidates seeking three positions at a counseling center, five have degrees in social science

and three do not. If three candidates are selected at random, find the probability distribution of X, the number having social science degrees among the selected persons.

**5.27** Based on recent records, the manager of a car painting center has determined the following probability distribution for the number of customers per day.

| x | f(x) |
|---|------|
| 0 | .05 |
| 1 | .20 |
| 2 | .30 |
| 3 | .25 |
| 4 | .15 |
| 5 | .05 |

(a) If the center has the capacity to serve two customers per day, what is the probability that one or more customers will be turned away on a given day?

(b) What is the probability that the center's capacity will not be fully utilized on a day?

(c) By how much must the capacity be increased so the probability of turning a customer away is no more than .10?

**5.28** Among cable TV customers, let X denote the number of television sets in a single-family residential dwelling. From an examination of the subscription records of 361 residences in a city, the frequency distribution is obtained.

| No. of Television Sets (x) | No. of Residences (Frequency) |
|----------------------------|-------------------------------|
| 1 | 64 |
| 2 | 161 |
| 3 | 89 |
| 4 | 47 |
| Total | 361 |

(a) Based on these data, obtain an approximate determination of the probability distribution of X.

(b) Why is this regarded as an approximation?

(c) Plot the probability histogram.

# 3. MEAN (EXPECTED VALUE) AND STANDARD DEVIATION OF A PROBABILITY DISTRIBUTION

We now introduce a numerical measure for the center of a probability distribution and another for its spread. In Chapter 2, we discussed the concepts of mean, as a measure of the center of a data set, and standard deviation, as a measure of spread. Because probability distributions are theoretical models in which the probabilities can be viewed as long-run relative frequencies, the sample measures of center and spread have their population counterparts.

To motivate their definitions, we first refer to the calculation of the mean of a data set. Suppose a die is tossed 20 times and the following data obtained.

$$4, \quad 3, \quad 4, \quad 2, \quad 5, \quad 1, \quad 6, \quad 6, \quad 5, \quad 2$$
$$2, \quad 6, \quad 5, \quad 4, \quad 6, \quad 2, \quad 1, \quad 6, \quad 2, \quad 4$$

The mean of these observations, called the sample mean, is calculated as

$$\bar{x} = \frac{\text{Sum of the observations}}{\text{Sample size}} = \frac{76}{20} = 3.8$$

Alternatively, we can first count the frequency of each point and use the relative frequencies to calculate the mean as

$$\bar{x} = 1 \left( \frac{2}{20} \right) + 2 \left( \frac{5}{20} \right) + 3 \left( \frac{1}{20} \right) + 4 \left( \frac{4}{20} \right) + 5 \left( \frac{3}{20} \right)$$
$$+ 6 \left( \frac{5}{20} \right) = 3.8$$

This second calculation illustrates the formula

$$\text{Sample mean } \bar{x} = \sum (\text{Value} \times \text{Relative frequency})$$

Rather than stopping with 20 tosses, if we imagine a very large number of tosses of a die, the relative frequencies will approach the probabilities, each of which is $\frac{1}{6}$ for a fair die. The mean of the (infinite) collection of tosses of a fair die should then be calculated as

$$1 \left( \frac{1}{6} \right) + 2 \left( \frac{1}{6} \right) + \cdots + 6 \left( \frac{1}{6} \right) = \sum (\text{Value} \times \text{Probability}) = 3.5$$

Motivated by this example and the stability of long-run relative frequency, it is then natural to define the mean of a random variable $X$ or its probability distribution as

$$\sum (\text{Value} \times \text{Probability}) \quad \text{or} \quad \sum x_i f(x_i)$$

where $x_i$'s denote the distinct values of $X$. The mean of a probability distribution is also called the population mean for the variable $X$ and is denoted by the Greek letter $\mu$.

The mean of a random variable $X$ is also called its expected value or expectation and, alternatively, denoted by $E(X)$. That is, the mean $\mu$ and expected value $E(X)$ are the same quantity and will be used interchangeably.

---

The mean of $X$ or population mean

$$E(X) = \mu$$

$$= \sum (\text{Value} \times \text{Probability}) = \sum x_i f(x_i)$$

Here the sum extends over all the distinct values $x_i$ of $X$.

---

**Example 8** Calculating the Population Mean Number of Heads

With $X$ denoting the number of heads in three tosses of a fair coin, calculate the mean of $X$.

SOLUTION The probability distribution of $X$ was recorded in Table 1. From the calculations exhibited in Table 6 we find that the mean is 1.5.

**TABLE 6** Mean of the Distribution of Table 1

| $x$ | $f(x)$ | $xf(x)$ |
|-----|--------|---------|
| 0 | $\dfrac{1}{8}$ | 0 |
| 1 | $\dfrac{3}{8}$ | $\dfrac{3}{8}$ |
| 2 | $\dfrac{3}{8}$ | $\dfrac{6}{8}$ |
| 3 | $\dfrac{1}{8}$ | $\dfrac{3}{8}$ |
| Total | 1 | $\dfrac{12}{8} = 1.5 = \mu$ |

The mean of a probability distribution has a physical interpretation. If a metal sheet is cut in the shape of the probability histogram, then $\mu$ represents the point on the base at which the sheet will balance. For instance, the mean $\mu = 1.5$ calculated in Example 8 is exactly at the center of mass for the distribution depicted in Figure 1. Because the amount of probability corresponds to the amount of mass in a bar, we interpret the balance point $\mu$ as the center of the probability distribution.

Like many concepts of probability, the idea of the mean or expectation originated from studies of gambling. When $X$ refers to the financial gain in a game of chance, such as playing poker or participating in a state lottery, the name "expected gain" is more appealing than "mean gain." In the realm of statistics, both the names "mean" and "expected value" are widely used.

**Example 9**   Expected Value—Setting a Premium

A trip insurance policy pays $3000 to the customer in case of a loss due to theft or damage on a five-day trip. If the risk of such a loss is assessed to be 1 in 200, what is the expected cost, per customer, to cover?

SOLUTION   The probability that the company will be liable to pay $3000 to a customer is $\frac{1}{200}$ = .005. Therefore, the probability distribution of $X$, the payment per customer, is as follows.

| Payment $x$ | Probability $f(x)$ |
|:---:|:---:|
| $   0 | .995 |
| $3000 | .005 |

We calculate

$$E(X) = 0 \times .995 + 3000 \times .005$$

$$= \$15.00$$

The company's expected cost per customer is $15.00 and, therefore, a premium equal to this amount is viewed as the fair premium. If this premium is charged and no other costs are involved, then the company will neither make a profit nor lose money in the long run. In practice, the premium is set at a higher price because it must include administrative costs and intended profit.

---

### No Casino Game Has a Positive Expected Profit

Each year, thousands of visitors come to casinos to gamble. Although all count on being lucky and a few indeed return with a smiling face, most leave the casino with a light purse. But, what should a gambler's expectation be?

Consider a simple bet on the red of a roulette wheel that has 18 red, 18 black, and 2 green slots. This bet is at even money so a $10 wager on red has an expected profit of

$$E(\text{Profit}) = (10) \left( \frac{18}{38} \right) + (-10) \left( \frac{20}{38} \right) = -.526$$

The negative expected profit says we expect to lose an average of 52.6¢ on every $10 bet. Over a long series of bets, the relative frequency of winning will approach the probability $\frac{18}{38}$ and that of losing will approach $\frac{20}{38}$, so a player will lose a substantial amount of money.

Other bets against the house have a similar negative expected profit. How else could a casino stay in business?

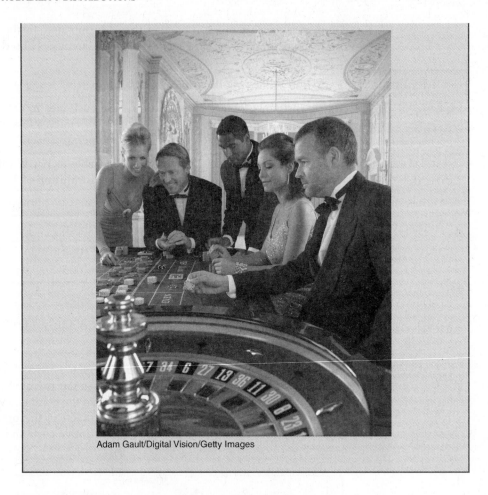

Adam Gault/Digital Vision/Getty Images

The concept of expected value also leads to a numerical measure for the spread of a probability distribution—namely, the standard deviation. When we define the standard deviation of a probability distribution, the reasoning parallels that for the standard deviation discussed in Chapter 2.

Because the mean $\mu$ is the center of the distribution of $X$, we express variation of $X$ in terms of the deviation $X - \mu$. We define the variance of $X$ as the expected value of the squared deviation $(X - \mu)^2$. To calculate this expected value, we note that

| $(X - \mu)^2$ Takes Value | With Probability |
|---|---|
| $(x_1 - \mu)^2$ | $f(x_1)$ |
| $(x_2 - \mu)^2$ | $f(x_2)$ |
| . | . |
| . | . |
| . | . |
| $(x_k - \mu)^2$ | $f(x_k)$ |

The expected value of $(X - \mu)^2$ is obtained by multiplying each value $(x_i - \mu)^2$ by the probability $f(x_i)$ and then summing these products. This motivates the definition:

$$\text{Variance of } X = \sum (\text{Deviation})^2 \times (\text{Probability})$$
$$= \sum (x_i - \mu)^2 f(x_i)$$

The variance of $X$ is abbreviated as $\text{Var}(X)$ and is also denoted by $\sigma^2$. The standard deviation of $X$ is the positive square root of the variance and is denoted by $\text{sd}(X)$ or $\sigma$ (a Greek lowercase sigma.)

The variance of $X$ is also called the population variance and $\sigma$ denotes the population standard deviation.

---

**Variance and Standard Deviation of $X$**

$$\sigma^2 = \text{Var}(X) = \sum (x_i - \mu)^2 f(x_i)$$
$$\sigma = \text{sd}(X) = +\sqrt{\text{Var}(X)}$$

---

**Example 10**    Calculating a Population Variance and Standard Deviation

Calculate the variance and the standard deviation of the distribution of $X$ that appears in the left two columns of Table 7.

SOLUTION    We calculate the mean $\mu$, the deviations $x - \mu$, $(x - \mu)^2$, and finally $(x - \mu)^2 f(x)$. The details are shown in Table 7.

$$\text{Var}(X) = \sigma^2 = 1.2$$
$$\text{sd}(X) = \sigma = \sqrt{1.2} = 1.095$$

**TABLE 7    Calculation of Variance and Standard Deviation**

| $x$ | $f(x)$ | $x f(x)$ | $(x - \mu)$ | $(x - \mu)^2$ | $(x - \mu)^2 f(x)$ |
|-----|--------|----------|-------------|---------------|---------------------|
| 0 | .1 | 0 | $-2$ | 4 | .4 |
| 1 | .2 | .2 | $-1$ | 1 | .2 |
| 2 | .4 | .8 | 0 | 0 | 0 |
| 3 | .2 | .6 | 1 | 1 | .2 |
| 4 | .1 | .4 | 2 | 4 | .4 |
| Total | 1.0 | $2.0 = \mu$ | | | $1.2 = \sigma^2$ |

An alternative formula for $\sigma^2$ often simplifies the numerical work (see Appendix A2.2).

---

**Alternative Formula for Hand Calculation**

$$\sigma^2 = \sum x_i^2 f(x_i) - \mu^2$$

---

**Example 11**    Alternative Calculation of Variance

We illustrate the alternative formula for $\sigma^2$ using the probability distribution in Example 10. See Table 8.

**TABLE 8**    Calculation of Variance by the Alternative Formula

| $x$ | $f(x)$ | $xf(x)$ | $x^2f(x)$ |
|:---:|:---:|:---:|:---:|
| 0 | .1 | .0 | .0 |
| 1 | .2 | .2 | .2 |
| 2 | .4 | .8 | 1.6 |
| 3 | .2 | .6 | 1.8 |
| 4 | .1 | .4 | 1.6 |
| Total | 1.0 | 2.0 | 5.2 |
|  |  | $= \mu$ | $= \sum x^2 f(x)$ |

$$\sigma^2 = 5.2 - (2.0)^2$$
$$= 1.2$$
$$\sigma = \sqrt{1.2} = 1.095$$

The standard deviation $\sigma$, rather than $\sigma^2$, is the appropriate measure of spread. Its unit is the same as that of $X$. For instance, if $X$ refers to income in dollars, $\sigma$ will have the unit dollar, whereas $\sigma^2$ has the rather artificial unit $(\text{dollar})^2$.

## Exercises

**5.29**  Given the following probability distribution concerning the number of Web sites visited almost every day:

(a)  Construct the probability histogram.

(b)  Find $E(X)$, $\sigma^2$, and $\sigma$.

| $x$ | $f(x)$ |
|:---:|:---:|
| 1 | .1 |
| 2 | .2 |
| 3 | .3 |
| 4 | .4 |

**5.30** A wait person proposed a distribution for the number of meals served on a two-for-one deal.

| x | f(x) |
|---|------|
| 2 | .3 |
| 4 | .5 |
| 6 | .1 |
| 8 | .1 |

Find the mean and standard deviation.

**5.31** In bidding for a remodeling project, a carpenter determines that he will have a net profit of $5000 if he gets the contract and a net loss of $86 if his bid fails. If the probability of his getting the contract is .2, calculate his expected return.

**5.32** A book club announces a sweepstakes in order to attract new subscribers. The prizes and the corresponding chances are listed here (typically, the prizes are listed in bold print in an advertisement flyer while the chances are entered in fine print or not mentioned at all).

| Prize | Chance |
|-------|--------|
| $50,000 | 1 in one million |
| $ 5,000 | 1 in 250,000 |
| $100 | 1 in 5,000 |
| $20 | 1 in 500 |

Suppose you have just mailed in a sweepstakes ticket and X stands for your winnings.

(a) List the probability distribution of X. (*Caution*: What is not listed is the chance of winning nothing, but you can figure that out from the given information.)

(b) Calculate your expected winnings.

**5.33** Calculate the mean and standard deviation for the probability distribution of Example 5.

**5.34** Referring to Exercise 5.27, find the mean and standard deviation of the number of customers.

**5.35** A construction company submits bids for two projects. Listed here are the profit and the probability of winning each project. Assume that the outcomes of the two bids are independent.

| | Profit | Chance of Winning Bid |
|---|--------|----------------------|
| Project A | $175,000 | .50 |
| Project B | $220,000 | .65 |

(a) List the possible outcomes (win/not win) for the two projects and find their probabilities.

(b) Let X denote the company's total profit out of the two contracts. Determine the probability distribution of X.

(c) If it costs the company $2000 for preparatory surveys and paperwork for the two bids, what is the expected net profit?

**5.36** Refer to Exercise 5.35, but suppose that the projects are scheduled consecutively with A in the first year and B in the second year. The company's chance of winning project A is still .50. Instead of the assumption of independence, now assume that if the company wins project A, its chance of winning B becomes .80 due to a boost of its image, whereas its chance drops to .40 in case it fails to win A. Under this premise, do parts (a)–(c).

**5.37** Upon examination of the claims records of 280 policy holders over a period of five years, an insurance company makes an empirical determination of the probability distribution of X = number of claims in five years.

| x | f(x) |
|---|------|
| 0 | .315 |
| 1 | .289 |
| 2 | .201 |
| 3 | .114 |
| 4 | .063 |
| 5 | .012 |
| 6 | .006 |

(a) Calculate the expected value of X.

(b) Calculate the standard deviation of X.

**5.38** Suppose the probability distribution of a random variable X is given by the function

$$f(x) = \frac{12}{25} \cdot \frac{1}{x} \quad \text{for} \quad x = 1, 2, 3, 4$$

Calculate the mean and standard deviation of this distribution.

**5.39** The probability distribution of a random variable $X$ is given by the function

$$f(x) = \frac{1}{84}\binom{5}{x}\binom{4}{3-x} \quad \text{for} \quad x = 0, 1, 2, 3$$

(a) Calculate the numerical probabilities and list the distribution.

(b) Calculate the mean and standard deviation of $X$.

**5.40** Given here are the probability distributions of two random variables $X$ and $Y$.

| $x$ | $f(x)$ | $y$ | $f(y)$ |
|-----|--------|-----|--------|
| 1 | .1 | 0 | .2 |
| 2 | .3 | 2 | .4 |
| 3 | .4 | 4 | .3 |
| 4 | .2 | 6 | .1 |

(a) From the $X$ distribution, determine the distribution of the random variable $8 - 2X$ and verify that it coincides with the $Y$ distribution. (Hence, identify $Y = 8 - 2X$.)

(b) Calculate the mean and standard deviation of $X$ (call these $\mu_X$ and $\sigma_X$, respectively).

(c) From the $Y$ distribution, calculate the mean and standard deviation of $Y$ (call these $\mu_Y$ and $\sigma_Y$, respectively).

(d) If $Y = a + bX$, then according to theory, we must have the relations $\mu_Y = a + b\mu_X$ and $\sigma_Y = |b|\sigma_X$. Verify these relations from your results in parts (b) and (c).

**5.41** A communications systems salesperson contacts four businesses during a week. Each contact can result in either a sale, with probability .3, or no

sale, with probability .7. Assume that the results for customer contacts are independent.

(a) List the elementary outcomes and assign probabilities.

(b) If $X$ denotes the number of communications systems sold during the week, obtain the probability distribution of $X$.

(c) Calculate the expected value of $X$.

**5.42** Refer to Exercise 5.41. Suppose the communications systems are priced at $5000, and let $Y$ denote the salesman's total sales (in dollars) during a week.

(a) Give the probability distribution of $Y$.

(b) Calculate $E(Y)$ and see that it is the same as $5000 \times E(X)$.

**5.43** Definition: The median of a distribution is the value $m_0$ of the random variable such that $P[X \leq m_0] \geq .5$ and $P[X \geq m_0] \geq .5$. In other words, the probability at or below $m_0$ is at least .5, and the probability at or above $m_0$ is at least .5. Find the median of the distribution given in Exercise 5.29.

**5.44** Given the two probability distributions

| $x$ | $f(x)$ | $y$ | $f(y)$ |
|-----|--------|-----|--------|
| 1 | .2 | 0 | .1 |
| 2 | .6 | 1 | .2 |
| 3 | .2 | 2 | .4 |
|   |    | 3 | .2 |
|   |    | 4 | .1 |

(a) Construct probability histograms. Which distribution has a larger spread?

(b) Verify that both distributions have the same mean.

(c) Compare the two standard deviations.

# 4. SUCCESSES AND FAILURES—BERNOULLI TRIALS

Often, an experiment can have only two possible outcomes. Example 5 concerned individual students who either preferred Internet or television news. The proportion of the population that preferred Internet news was .60. Also, only two outcomes are possible for a single trial in the scenarios of Examples 1 and 2. In all these circumstances, a simple probability model can be developed for the chance variation in the outcomes. Moreover, the population proportion need not be known as in the previous examples. Instead, the probability distribution will involve this unknown population proportion as a parameter.

Sampling situations where the elements of a population have a dichotomy abound in virtually all walks of life. A few examples are:

Inspect a specified number of items coming off a production line and count the number of defectives.

Survey a sample of voters and observe how many favor an increase of public spending on welfare.

Analyze the blood specimens of a number of rodents and count how many carry a particular viral infection.

Examine the case histories of a number of births and count how many involved delivery by Cesarean section.

Selecting a single element of the population is envisioned as a trial of the (sampling) experiment, so that each trial can result in one of two possible outcomes. Our ultimate goal is to develop a probability model for the number of outcomes in one category when repeated trials are performed.

### Boy or Girl?

©Blaine Harrington/Age Fotostock

A model for the potential sex of a newborn is the assignment of probability to each of the two outcomes. For most applications, .5 is assigned to "male" but extensive official statistics establish that the probability is actually about .52.

An organization of the key terminologies, concerning the successive repetitions of an experiment, is now in order. We call each repetition by the simpler name—a trial. Furthermore, the two possible outcomes of a trial are now assigned the technical names success (S) and failure (F) just to emphasize the point that they are the only two possible results. These names bear no connotation of success or failure in real life. Customarily, the outcome of primary interest in a study is labeled success (even if it is a disastrous event). In a study of the rate of unemployment, the status of being unemployed may be attributed the statistical name success!

Further conditions on the repeated trials are necessary in order to arrive at our intended probability distribution. Repeated trials that obey these conditions are called Bernoulli trials after the Swiss mathematician Jacob Bernoulli.

Perhaps the simplest example of Bernoulli trials is the prototype model of tossing a coin, where the occurrences *head* and *tail* can be labeled S and F, respectively. For a fair coin, we assign probability $p = \frac{1}{2}$ to success and $q = \frac{1}{2}$ to failure.

---

### Bernoulli Trials

1. Each trial yields one of two outcomes, technically called success $(S)$ and failure $(F)$.

2. For each trial, the probability of success $P(S)$ is the same and is denoted by $p = P(S)$. The probability of failure is then $P(F) = 1 - p$ for each trial and is denoted by $q$, so that $p + q = 1$.

3. Trials are independent. The probability of success in a trial remains unchanged given the outcomes of all the other trials.

---

**Example 12**    Sampling from a Population with Two Categories of Elements

Consider a lot (population) of items in which each item can be classified as either defective or nondefective. Suppose that a lot consists of 15 items, of which 5 are defective and 10 are nondefective.

Do the conditions for Bernoulli trials apply when sampling (1) with replacement and (2) without replacement?

SOLUTION    1.  *Sampling with replacement.* An item is drawn at random (i.e., in a manner that all items in the lot are equally likely to be selected). The quality of the item is recorded and it is returned to the lot before the next drawing. The conditions for Bernoulli trials are satisfied. If the occurrence of a defective is labeled S, we have $P(S) = \frac{5}{15}$.

2.  *Sampling without replacement.* In situation (2), suppose that 3 items are drawn one at a time but without replacement. Then the condition concerning the independence of trials is violated. For the first drawing, $P(S) = \frac{5}{15}$.

If the first draw produces S, the lot then consists of 14 items, 4 of which are defective. Given this information about the result of the first draw, the conditional probability of obtaining an S on the second draw is then $\frac{4}{14} \neq \frac{5}{15}$, which establishes the lack of independence.

This violation of the condition of independence loses its thrust when the population is vast and only a small fraction of it is sampled. Consider sampling 3 items without replacement from a lot of 1500 items, 500 of which are defective. With $S_1$ denoting the occurrence of an S in the first draw and $S_2$ that in the second, we have

$$P(S_1) = \frac{500}{1500} = \frac{5}{15}$$

and

$$P(S_2 \mid S_1) = \frac{499}{1499}$$

For most practical purposes, the latter fraction can be approximated by $\frac{5}{15}$. Strictly speaking, there has been a violation of the independence of trials, but it is to such a negligible extent that the model of Bernoulli trials can be assumed as a good approximation.

Example 12 illustrates the important points:

> If elements are sampled from a dichotomous population at random and with replacement, the conditions for Bernoulli trials are satisfied.
>     When the sampling is made without replacement, the condition of the independence of trials is violated. However, if the population is large and only a small fraction of it (less than 10%, as a rule of thumb) is sampled, the effect of this violation is negligible and the model of the Bernoulli trials can be taken as a good approximation.

Example 13 further illustrates the kinds of approximations that are sometimes employed when using the model of the Bernoulli trials.

**Example 13**    Testing a New Antibiotic—Bernoulli Trials?

Suppose that a newly developed antibiotic is to be tried on 10 patients who have a certain disease and the possible outcomes in each case are cure (S) or no cure (F).

Comment on the applicability of the Bernoulli trial model.

SOLUTION    Each patient has a distinct physical condition and genetic constitution that cannot be perfectly matched by any other patient. Therefore, strictly speaking, it may not be possible to regard the trials made on 10 different patients as 10 repetitions of an experiment under identical conditions, as the definition of Bernoulli trials demands.

We must remember that the conditions of a probability model are abstractions that help to realistically simplify the complex mechanism governing the outcomes of an experiment. Identification with Bernoulli trials in such situations is to be viewed as an approximation of the real world, and its merit rests on how successfully the model explains chance variations in the outcomes.

A key step toward improving the approximation is selecting the patients at random. In this case, it is reasonable to treat the outcomes as identically distributed.

# Exercises

**5.45** Is the model of Bernoulli trials plausible in each of the following situations? Discuss in what manner (if any) a serious violation of the assumptions can occur.

(a) Seven friends go to a blockbuster movie and each is asked whether the movie was excellent.

(b) A musical aptitude test is given to 10 students and the times to complete the test are recorded.

(c) Items coming off an assembly line are inspected and classified as defective or nondefective.

(d) Going room by room in your dorm and recording whether or not the internet connection is down.

**5.46** In each case, examine whether or not repetitions of the stated experiment conform to the model of Bernoulli trials. Where the model is appropriate, determine the numerical value of $p$ or indicate how it can be determined.

(a) Roll a fair die and observe the number that shows up.

(b) Roll a fair die and observe whether or not the number 6 shows up.

(c) Roll two fair dice and observe the total of the points that show up.

(d) Roll two fair dice and observe whether or not both show the same number.

(e) Roll a loaded die and observe whether or not the number 6 shows up.

**5.47** A jar contains 25 candies of which 6 are brown, 12 are yellow, and 7 are of other colors. Consider 4 successive draws of 1 candy at random from the jar and suppose the appearance of a yellow candy is the event of interest. For each of the following situations, state whether or not the model of Bernoulli trials is reasonable, and if so, determine the numerical value of $p$.

(a) After each draw, the selected candy is returned to the jar.

(b) After each draw, the selected candy is not returned to the jar.

(c) After each draw, the selected candy is returned to the jar and one new candy of the same color is added in the jar.

**5.48** Refer to Exercise 5.47 and suppose instead that the mix consists of 2500 candies, of which 600 are brown, 1200 are yellow, and 700 are of other colors. Repeat parts (a)–(c) of Exercise 5.47 in this setting.

**5.49** From four agricultural plots, two are selected at random for a pesticide treatment. The other two plots serve as controls. For each plot, denote by S the event that it is treated with the pesticide. Consider the assignment of treatment or control to a single plot as a trial.

(a) Is $P(S)$ the same for all trials? If so, what is the numerical value of $P(S)$?

(b) Are the trials independent? Why or why not?

**5.50** Refer to Exercise 5.49. Now suppose for each plot a fair coin is tossed. If a head shows up, the plot is treated; otherwise, it is a control. With this manner of treatment allocation, answer parts (a) and (b).

**5.51** A market researcher intends to study the consumer preference between regular and decaffeinated coffee. Examine the plausibility of the model of Bernoulli trials in the following situations.

(a) One hundred consumers are randomly selected and each is asked to report the types of coffee (regular or decaffeinated) purchased in the five most recent occasions. If we consider each purchase as a trial, this inquiry deals with 500 trials.

(b) Five hundred consumers are randomly selected and each is asked about the most recent purchase of coffee. Here again the inquiry deals with 500 trials.

**5.52** A backpacking party carries three emergency signal flares, each of which lights with a probability of .98. Assuming that the flares operate independently, find:

(a) The probability that at least one flare lights.

(b) The probability that exactly two flares light.

**5.53** Consider Bernoulli trials with success probability $p = .3$.

(a) Find the probability that four trials result in all failures.

(b) Given that the first four trials result in all failures, what is the conditional probability that the next four trials are all successes?

(c) Find the probability that the first success occurs in the fourth trial.

**5.54** According to numbers provided by the Bureau of Labor Statistics, the probability that three unrelated new businesses will last for five years is .216. What is the probability that all three have failed before that time? Assume that the conditions for Bernoulli trials hold.

**5.55** According to a recent survey, about 40% of adults reported that the most helpful method to achieve a goal is to share their goal with friends and ask them for support. Let $S$ be the event that an adult feels sharing with friends is the most helpful method to achieve a goal. If a random sample of five adults is selected,

(a) Find the probability of the sequence *SFFSF*.

(b) Find the probability of exactly 2 $S$'s.

**5.56** A graphic designer makes a presentation to clients and this results in sales of her services in one-fourth of the cases. Assuming the results for different clients are independent

(a) Find the probability that exactly 3 of the next 4 presentations result in sales.

(b) Find the probability that none of the presentations result in a sale.

**5.57** An animal either dies (D) or survives (S) in the course of a surgical experiment. The experiment is to be performed first with two animals. If both survive, no further trials are to be made. If exactly one animal survives, one more animal is to undergo the experiment. Finally, if both animals die, two additional animals are to be tried.

(a) List the sample space.

(b) Assume that the trials are independent and the probability of survival in each trial is $\frac{1}{4}$. Assign probabilities to the elementary outcomes.

(c) Let $X$ denote the number of survivors. Obtain the probability distribution of $X$ by referring to part (b).

**5.58** The accompanying table shows the percentages of residents in a large community when classified according to gender and presence of a particular allergy.

|        | Allergy | |
|--------|---------|--------|
|        | Present | Absent |
| Male   | 16      | 36     |
| Female | 9       | 39     |

Suppose that the selection of a person is considered a trial and the presence of the allergy is considered a success. For each case, identify the numerical value of $p$ and find the required probability.

(a) Four persons are selected at random. What is the probability that none has the allergy?

(b) Four males are selected at random. What is the probability that none has the allergy?

(c) Two males and two females are selected at random. What is the probability that none has the allergy?

## 5. THE BINOMIAL DISTRIBUTION

A basic distribution models chance variation in any experiment consisting of a fixed number of Bernoulli trials. The random variable X of interest is the frequency count of one of the categories. Previously, its distribution was calculated under the assumption that the population proportion is known. For instance, the probability distribution of Table 3, from Example 5, resulted from the specification that 60% of the population of students prefer news from the Internet. In a practical situation, however, the population proportion is usually an unknown quantity. When this is so, the probability distribution of X cannot be numerically determined. However, we will see that it is possible to construct a model for the probability distribution of X that contains the unknown population proportion as a **parameter**. The probability model serves as the major vehicle of drawing inferences about the population from observations of the random variable X.

> A probability model is an assumed form of the probability distribution that describes the chance behavior for a random variable X.
>
> Probabilities are expressed in terms of relevant population quantities, called the parameters.

Consider a **fixed number** $n$ of Bernoulli trials with the success probability $p$ in each trial. The number of successes obtained in $n$ trials is a random variable that we denote by X. The probability distribution of this random variable X is called a binomial distribution.

The binomial distribution depends on the two quantities $n$ and $p$. For instance, the distribution appearing in Table 1 is precisely the binomial distribution with $n = 3$ and $p = .5$, whereas that in Table 3 is the binomial distribution with $n = 4$ and $p = .6$.

> ### The Binomial Distribution
>
> Denote
>
> $$n = \text{a fixed number of Bernoulli trials}$$
> $$p = \text{the probability of success in each trial}$$
> $$X = \text{the ( random ) number of successes in } n \text{ trials}$$
>
> The random variable X is called a **binomial random variable**. Its distribution is called a **binomial distribution**.

A review of the developments in Example 5 will help motivate a formula for the general binomial distribution.

**Example 14**   Example 5 Revisited—An Example of the Binomial Distribution

The random variable $X$ represents the number of students who prefer news from the Internet among a random sample of $n = 4$ students from a large university. Instead of the numerical value .6, we now denote the population proportion of students who prefer Internet news by the symbol $p$. Obtain the binomial distribution of $X$.

SOLUTION   We relabel the outcome "Internet" as a success (S) and "not Internet" as a failure (F). The elementary outcomes of sampling 4 students, the associated probabilities, and the value of $X$ are listed as follows.

| FFFF | SFFF | SSFF | SSSF | SSSS |
|------|------|------|------|------|
|      | FSFF | SFSF | SSFS |      |
|      | FFSF | SFFS | SFSS |      |
|      | FFFS | FSSF | FSSS |      |
|      |      | FSFS |      |      |
|      |      | FFSS |      |      |

| Value of $X$ | 0 | 1 | 2 | 3 | 4 |
|---|---|---|---|---|---|
| Probability of each outcome | $q^4$ | $p q^3$ | $p^2 q^2$ | $p^3 q$ | $p^4$ |
| Number of outcomes | $1$ $= \binom{4}{0}$ | $4$ $= \binom{4}{1}$ | $6$ $= \binom{4}{2}$ | $4$ $= \binom{4}{3}$ | $1$ $= \binom{4}{4}$ |

Because the population of students at a large university is vast, the trials can be treated as independent. Also, for an individual trial, $P(S) = p$ and $P(F) = q = 1 - p$. The event $[X = 0]$ has one outcome, FFFF, whose probability is

$$P[X = 0] = P(FFFF) = q \times q \times q \times q = q^4$$

To arrive at an expression for $P[X = 1]$, we consider the outcomes listed in the second column. The probability of SFFF is

$$P(SFFF) = p \times q \times q \times q = p q^3$$

and the same result holds for every outcome in this column. There are 4 outcomes so we obtain $P[X = 1] = 4 p q^3$. The factor 4 is the number of outcomes with one S and three F's. Even without making a complete list of the outcomes, we can obtain this count. Every outcome has 4 places and the 1 place where S occurs can be selected from the total of 4 in $\binom{4}{1} = 4$ ways, while the remaining 3 places must be filled with an F. Continuing in the same line

**TABLE 9**   Binomial Distribution with $n = 4$ Trials

| Value $x$ | 0 | 1 | 2 | 3 | 4 |
|---|---|---|---|---|---|
| Probability $f(x)$ | $\binom{4}{0} p^0 q^4$ | $\binom{4}{1} p^1 q^3$ | $\binom{4}{2} p^2 q^2$ | $\binom{4}{3} p^3 q^1$ | $\binom{4}{4} p^4 q^0$ |

of reasoning, we see that the value $X = 2$ occurs with $\binom{4}{2} = 6$ outcomes, each of which has a probability of $p^2 q^2$. Therefore $P [X = 2] = \binom{4}{2} p^2 q^2$. After we work out the remaining terms, the binomial distribution with $n = 4$ trials can be presented as in Table 9.

It would be instructive for the reader to verify that the numerical probabilities appearing in Table 3 are obtained by substituting $p = .6$ and $q = .4$.

Extending the reasoning of Example 14 to the case of a general number $n$ of Bernoulli trials, we observe that there are $\binom{n}{x}$ outcomes that have exactly $x$ successes and $n - x$ failures. The probability of every such outcome is $p^x q^{n-x}$. Therefore,

$$f(x) = P[X = x] = \binom{n}{x} p^x q^{n-x} \qquad \text{for} \quad x = 0, 1, \ldots n$$

is the formula for the binomial probability distribution with $n$ trials.

---

The binomial distribution with $n$ trials and success probability $p$ is described by the function

$$f(x) = P[X = x] = \binom{n}{x} p^x (1 - p)^{n-x}$$

for the possible values $x = 0, 1, \ldots n$.

---

**Example 15**   The Binomial Distribution and Genetics

According to the Mendelian theory of inherited characteristics, a cross fertilization of related species of red- and white-flowered plants produces a generation whose offspring contain 25% red-flowered plants. Suppose that a horticulturist wishes to cross 5 pairs of the cross-fertilized species. Of the resulting 5 offspring, what is the probability that:

(a)   There will be no red-flowered plants?

(b)   There will be 4 or more red-flowered plants?

SOLUTION    Because the trials are conducted on different parent plants, it is natural to assume that they are independent. Let the random variable $X$ denote the number of red-flowered plants among the 5 offspring. If we identify the occurrence of a red as a success S, the Mendelian theory specifies that $P(S) = p = \frac{1}{4}$, and hence $X$ has a binomial distribution with $n = 5$ and $p = .25$. The required probabilities are therefore

(a)  $P[X = 0] = f(0) = (.75)^5 = .237$

(b)  $P[X \geq 4] = f(4) + f(5) = \binom{5}{4}(.25)^4(.75)^1$

$$+ \binom{5}{5}(.25)^5(.75)^0 = .015 + .001 = .016$$

To illustrate the manner in which the values of $p$ influence the shape of the binomial distribution, the probability histograms for three binomial distributions with $n = 6$ and $p$ values of .5, .3, and .7, respectively, are presented in Figure 3. When $p = .5$, the binomial distribution is symmetric with the highest probability occurring at the center (see Figure 3a).

For values of $p$ smaller than .5, more probability is shifted toward the smaller values of $x$ and the distribution has a longer tail to the right. Figure 3b, where the binomial histogram for $p = .3$ is plotted, illustrates this tendency. On the other hand, Figure 3c with $p = .7$ illustrates the opposite tendency: The value of $p$ is higher than .5, more probability mass is shifted toward higher values of $x$, and the distribution has a longer tail to the left. Considering the histograms in Figures 3b and 3c, we note that the value of $p$ in one histogram is the same as the value of $q$ in the other. The probabilities in one histogram are exactly the same as those in the other, but their order is reversed. This illustrates a general property of the binomial distribution: When $p$ and $q$ are interchanged, the distribution of probabilities is reversed. For instance, with $p = .3$, $f(1) = .302$ is the probability of 1 success or, equivalently 5 failures where $1 - .3 = .7$ is the probability of failure. Reversing the labels success/failure establishes the reversal.

Figure 3    Binomial distributions for $n = 6$.

## 5.1  HOW TO USE THE BINOMIAL TABLE (APPENDIX B, TABLE 2)

Although the binomial distribution is easily evaluated on a computer and some hand calculators, we provide a short table in Appendix B, Table 2. It covers selected sample sizes $n$ ranging from 1 to 25 and several values of $p$. For a given pair $(n, p)$, the table entry corresponding to each $c$ represents the cumulative probability $P[X \leq c] = \sum_{x=0}^{c} f(x)$, as is explained in the following scheme.

<table>
<tr><td colspan="2" align="center">The Binomial Distribution</td><td></td><td colspan="2" align="center">Appendix B, Table 2 Provides</td></tr>
<tr><td>Value $x$</td><td>Probability $f(x)$</td><td>$c$</td><td colspan="2">Table Entry $\sum_{x=0}^{c} f(x) = P[X \leq c]$</td></tr>
<tr><td align="center">0</td><td align="center">$f(0)$</td><td align="center">0</td><td colspan="2">$f(0)$</td></tr>
<tr><td align="center">1</td><td align="center">$f(1)$</td><td align="center">1</td><td colspan="2">$f(0) + f(1)$</td></tr>
<tr><td align="center">2</td><td align="center">$f(2)$</td><td align="center">2</td><td colspan="2">$f(0) + f(1) + f(2)$</td></tr>
<tr><td align="center">.</td><td align="center">.</td><td align="center">.</td><td colspan="2">.</td></tr>
<tr><td align="center">.</td><td align="center">.</td><td align="center">.</td><td colspan="2">.</td></tr>
<tr><td align="center">.</td><td align="center">.</td><td align="center">.</td><td colspan="2">.</td></tr>
<tr><td align="center">$n$</td><td align="center">$f(n)$</td><td align="center">$n$</td><td colspan="2">1.000</td></tr>
<tr><td align="center">Total</td><td align="center">1</td><td></td><td colspan="2"></td></tr>
</table>

The probability of an individual value $x$ can be obtained from this table by a subtraction of two consecutive entries. For example,

$$P[X = 2] = f(2) = \left(\begin{array}{c}\text{table entry at}\\ c = 2\end{array}\right) - \left(\begin{array}{c}\text{table entry at}\\ c = 1\end{array}\right)$$

**Example 16**   Binomial Distribution for the Number Cured

Suppose it is known that a new treatment is successful in curing a muscular pain in 50% of the cases. If it is tried on 15 patients, find the probability that:

(a)  At most 6 will be cured.

(b)  The number cured will be no fewer than 6 and no more than 10.

(c)  Twelve or more will be cured.

SOLUTION   Designating the cure of a patient by S and assuming that the results for individual patients are independent, we note that the binomial distribution with $n = 15$ and $p = .5$ is appropriate for $X = $ number of patients who are cured. To compute the required probabilities, we consult the binomial table for $n = 15$ and $p = .5$.

(a)  $P[X \leq 6] = .304$, which is directly obtained by reading from the row $c = 6$.

(b)  We are to calculate, as illustrated in Figure 4,

$$P[6 \leq X \leq 10] = f(6) + f(7) + f(8) + f(9) + f(10)$$

$$= \sum_{x=6}^{10} f(x)$$

The table entry corresponding to $c = 10$ gives

$$P[X \leq 10] = \sum_{x=0}^{10} f(x) = .941$$

and the entry corresponding to $c = 5$ yields

$$P[X \leq 5] = \sum_{x=0}^{5} f(x) = .151$$

Because their difference represents the sum $\sum_{x=6}^{10} f(x)$, we obtain

$$P[6 \leq X \leq 10] = P[X \leq 10] - P[X \leq 5]$$

$$= .941 - .151$$

$$= .790$$

(c)  To find $P[X \geq 12]$, we use the law of complement:

$$P[X \geq 12] = 1 - P[X \leq 11]$$

$$= 1 - .982$$

$$= .018$$

Figure 4  Probability line diagrams illustrate the calculation of $P[6 \leq X \leq 10]$ as a difference of two tabled cumulative probabilities.

Note that $[X < 12]$ is the same event as $[X \leq 11]$.

(*An Aside*: Refer to our "muscular pain" example at the beginning of Chapter 4. The mystery surrounding the numerical probability .018 is now resolved.)

## 5.2 THE MEAN AND STANDARD DEVIATION OF THE BINOMIAL DISTRIBUTION

Although we already have a general formula that gives the binomial probabilities for any $n$ and $p$, in later chapters we will need to know the mean and the standard deviation of the binomial distribution. The expression $np$ for the mean is apparent from the following intuitive reasoning: If a fair coin is tossed 100 times, the expected number of heads is $100 \times \frac{1}{2} = 50$. Likewise, if the probability of an event is $p$, then in $n$ trials the event is expected to happen $np$ times. The mean and standard deviation are obtained in Appendix A3.4.

The binomial distribution with $n$ trials and success probability $p$ has

$$\text{Mean} = np$$

$$\text{Variance} = npq \ (\text{Recall: } q = 1 - p)$$

$$\text{sd} = \sqrt{npq}$$

**Example 17**  Calculating the Population Mean and Standard Deviation of a Binomial Distribution

For the binomial distribution with $n = 3$ and $p = .5$, calculate the mean and the standard deviation.

SOLUTION   Employing the formulas, we obtain

$$\text{Mean} = np = 3 \times .5 = 1.5$$

$$\text{sd} = \sqrt{npq} = \sqrt{3 \times .5 \times .5} = \sqrt{.75} = .866$$

The mean agrees with the results of Example 8. The reader may wish to check the standard deviation by numerical calculations using the definition of $\sigma$.

**Example 18**  Binomial Distribution and Proportion of Sports Articles Written by Robots

Algorithms, based on artificial intelligence, now write sport stories and financial reports. One company started by doing reports on sports that usually received very little coverage. Its algorithm, fed with box scores and play-by-play information, created over one million online reports of little league baseball games in 2012. Expanding their initial work with women's sports, they also write many stories on the Big Ten Network site[1] including frequent updates during football games.

Suppose that 18 new stories are scheduled to appear on a Web site next week and that, typically, 60 percent are written by the algorithm as opposed to sportswriters.

(a) Find the probability that 6 or fewer will be written by the algorithm.

(b) Obtain the mean and standard deviation of the number written by the algorithm.

SOLUTION    Let $X$ be the number of so-called robot written stories among the $n = 18$ that appear. The number of trials is fixed. As long as the editor makes the decision to include or exclude an article based solely on the article alone, we can treat the trials as independent. A binomial distribution with $n = 18$ and $p = .60$ appears reasonable for the number of robot written articles (successes).

(a) From the binomial table, we obtain

$$P[X \leq 6] = .020$$

If the proportion $p$ does equal .60, only 2 in 100 times would we observe 6 or fewer of the 18 articles written by a robot.

(b) By the formulas for mean and standard deviation,

$$\text{Mean} = np = 18 \times .60 = 10.8$$

$$\text{sd} = \sqrt{npq} = \sqrt{18 \times .6 \times .4} = \sqrt{4.32} = 2.08$$

## Exercises

**5.59** For each situation, state whether or not a binomial distribution holds for the random variable X. Also, identify the numerical values of $n$ and $p$ when a binomial distribution holds.

(a) A fair die is rolled 10 times, and X denotes the number of times 6 shows up.

(b) A fair die is rolled until 6 appears, and X denotes the number of rolls.

(c) In a jar, there are ten marbles, of which four are numbered 1, three are numbered 2, two are numbered 3, and one is numbered 4. Three marbles are drawn at random, one after another and with replacement, and X denotes the count of the selected marbles that are numbered either 1 or 2.

(d) The same experiment as described in part (c), but now X denotes the sum of the numbers on the selected marbles.

**5.60** Construct a tree diagram for three Bernoulli trials. Attach probabilities in terms of $p$ and $q$ to each outcome and then table the binomial distribution for $n = 3$.

**5.61** In each case, find the probability of $x$ successes in $n$ Bernoulli trials with success probability $p$ for each trial.

(a) $x = 2$    $n = 3$    $p = .35$

(b) $x = 3$    $n = 6$    $p = .25$

(c) $x = 2$    $n = 6$    $p = .65$

[1]S. Levy, "The rise of the robot reporter," *Wired*, May, 2012 pp. 132–137.

**5.62** (a) Plot the probability histograms for the binomial distributions for $n = 5$ and $p$ equal to .2, .5, and .8.

(b) Locate the means.

(c) Find $P[X \geq 4]$ for each of the three cases.

**5.63** An interior designer makes a presentation to potential clients and this results in sales of her services in 35% of the cases. Let $X$ denote the number of sales in the next four presentations. Assuming the results for different clients are independent, calculate the probabilities $f(x) = P[X = x]$ for $x = 0, 1, \ldots 4$ and find

(a) $P[X \leq 3]$

(b) $P[X \geq 3]$

(c) $P[X = 2 \text{ or } 4]$

**5.64** Refer to Exercise 5.63. What is the most probable value of $X$ (called the **mode** of a distribution)?

**5.65** About 75% of dog owners buy holiday presents for their dogs.[2] Suppose $n = 4$ dog owners are randomly selected. Find the probability of:

(a) Three or more buy their dog holiday presents.

(b) At most three buy their dog holiday presents.

(c) Find the expected number of persons, in the sample, who buy their dog holiday presents.

**5.66** According to a recent survey, outside of their own family members, 26% of adult Americans have no close friend to confide in. If this is the prevailing probability today, find the probability that in a random sample of $n = 5$ adults

(a) Two or more have no close friend.

(b) At most two have no close friend.

(c) Find the expected number of persons who have no close friend.

**5.67** Suppose 15% of the trees in a forest have severe leaf damage from air pollution. If 5 trees are selected at random, find the probability of:

(a) Three of the selected trees have severe leaf damage.

(b) No more than two have severe leaf damage.

**5.68** Rh-positive blood appears in 85% of the white population in the United States. If 8 people are sampled at random from that population, find the probability that:

(a) At least 6 of them have Rh-positive blood.

(b) At most 3 of them have Rh-negative blood, that is, an absence of Rh positive.

**5.69** Using the binomial table, find the probability of:

(a) Four successes in 13 trials when $p = .3$.

(b) Eight failures in 13 trials when $p = .7$.

(c) Eight successes in 13 trials when $p = .3$.

Explain why you get identical answers in parts (b) and (c).

**5.70** According to the U.S. Census Bureau, in 2011 about 10% of persons between 25 and 39 years old live alone. For a random sample of size $n$, use the binomial table to find the probability of

(a) 1 or fewer persons living alone when $n = 12$.

(b) 2 or more persons living alone when $n = 12$.

(c) Find the expected number when $n = 12$.

(d) 1 or fewer persons living alone when $n = 20$.

**5.71** About 30% of adults say that reading is a favorite leisure activity. Let success be the outcome that reading is a favorite leisure activity. Find the probability that:

(a) More than 5 trials are needed in order to obtain 3 successes. (*Hint*: In other words, the event is at most 2 successes in 5 trials.)

(b) More than 9 trials are needed in order to obtain 5 successes.

**5.72** A survey report states that 70% of adult women visit their doctors for a physical examination at least once in two years. If 20 adult women are randomly selected, find the probability that:

(a) Fewer than 14 of them have had a physical examination in the past two years.

(b) At least 17 of them have had a physical examination in the past two years.

---

[2]75% is between the two 2008 survey results obtained by the American Kennel Club and Harris Interactive Poll.

**5.73** Calculate the mean and standard deviation of the binomial distribution using the formulas in mean $= np$  sd $= \sqrt{np(1-p)}$

(a) Exercise 5.65 if $n$ is changed to 20.

(b) Exercise 5.70 when $n = 20$.

(c) Exercise 5.71 when $n = 40$.

**5.74** (a) For the binomial distribution with $n = 3$ and $p = .6$, list the probability distribution $(x, f(x))$ in a table.

(b) From this table, calculate the mean and standard deviation by using the methods of Section 3.

(c) Check your results with the formulas mean $= np$, sd $= \sqrt{npq}$.

**5.75** Suppose that 20% of the college seniors support an increase in federal funding for care of the elderly. If 20 college seniors are randomly selected, what is the probability that at most 3 of them support the increased funding?

**5.76** Referring to Exercise 5.75, find:

(a) The expected number of college seniors, in a random sample of 20, supporting the increased funding.

(b) The probability that the number of sampled college seniors supporting the increased funding equals the expected number.

**5.77** According to the Centers for Disease Control 50% of practicing physicians are in the specialty area of primary care. Assuming that the same rate prevails, find the mean and standard deviation of the number of physicians specializing in primary care out of a current random selection of 545 medical graduates.

**5.78** According to the Mendelian theory of inheritance of genes, offspring of a dihybrid cross of peas could be any of the four types: round-yellow ($RY$), wrinkled-yellow ($WY$), round-green ($RG$) and wrinkled-green ($WG$), and their probabilities are in the ratio $9:3:3:1$.

(a) If $X$ denotes the number of $RY$ offspring from 130 such crosses, find the mean and standard deviation of $X$.

(b) If $Y$ denotes the number of $WG$ offspring from 85 such crosses, find the mean and standard deviation of $Y$.

**5.79** The following table (see Exercise 5.58) shows the percentages of residents in a large community when classified according to gender and presence of a particular allergy. For each part below, find the mean and standard deviation of the specified random variable.

|        | Allergy |        |
|--------|---------|--------|
|        | Present | Absent |
| Male   | 16      | 36     |
| Female | 9       | 39     |

(a) $X$ stands for the number of persons having the allergy in a random sample of 40 persons.

(b) $Y$ stands for the number of males having the allergy in a random sample of 40 males.

(c) $Z$ stands for the number of females not having the allergy in a random sample of 40 females.

## The Following Exercise Requires a Computer

**5.80** Many computer packages produce binomial probabilities. We illustrate the MINITAB commands for obtaining the binomial probabilities with $n = 5$ and $p = .33$. The probabilities $P[X = x]$ are obtained by first setting 0, 1, 2, 3, 4, 5 in C1 and then selecting:

Calc > Probability distributions > Binomial.

Type 5 in **Number of trials**

and .33 in **Probability**.

Enter C1 in **Input constant**. Click **OK**.

which produces the output.

**Probability Density Function**

Binomial with n = 5 and p = 0.33

```
x          P(X=x)
0        0.135013
1        0.332493
2        0.327531
3        0.161321
4        0.039728
5        0.003914
```

To obtain the cumulative probabilities $P(X \leq x)$, click **Cumulative probability** instead of **Probability**. The resulting output is shown on the right.

Using the computer, calculate

(a) $P[X \leq 8]$ and $P[X = 8]$ when $p = .67$ and $n = 12$

(b) $P[10 \leq X \leq .15]$ when $p = .43$ and $n = 35$

**Cumulative Distribution Function**

Binomial with n = 5 and p = 0.33

| x | P(X<=x) |
|---|---------|
| 0 | 0.13501 |
| 1 | 0.46751 |
| 2 | 0.79504 |
| 3 | 0.95636 |
| 4 | 0.99609 |
| 5 | 1.00000 |

Poor Linus. Chance did not even favor him with half correct.
Reprinted by permission.

## THE BINOMIAL DISTRIBUTION IN CONTEXT

Requests for credit cards must be processed to determine if the applicant meets certain financial standards. In many instances, such as when the applicant already has a long-term good credit record, only a short review is required. Usually this consists of a credit score assigned on the basis of the answers to questions on the application and then a computerized check of credit records. Many other cases require a full review with manual checks of information to determine the credit worthiness of the applicant.

Each week, a large financial institution selects a sample of 20 incoming applications and counts the number requiring full review. From data collected over several weeks, it is observed that about 40% of the applications require full review. If we take this long-run relative frequency as the probability, what is an unusually large number of full reviews and what is an unusually small number?

Let $X$ be the number in the sample that require a full review. From the binomial table, with $n = 20$ and $p = .4$, we get

$$P[X \leq 3] = .016$$

$$P[X \geq 13] = 1 - P[X \leq 12] = 1 - .979 = .021$$

Taken together, the probability of $X$ being 3 or less or 13 or more is .037, so those values should occur less than four times in 100 samples. That is, they could be considered unusual. In Exercise 5.81, you will be asked to show that either including 4 or including 12 will lead to a combined probability greater than .05. That is, the large and small values should then occur more than 1 in 20 times. For many people, this would be too frequent to be considered rare or unusual.

For the count $X$, we expect $np = 20 \times .4 = 8$ applications in the sample to require a full review. The standard deviation of this count is $\sqrt{20(.4)(1 - .4)} = 2.191$. Alternatively, when $n$ is moderate or large, we could describe unusual as being two or more standard deviations from the mean. A value at least two standard deviations, or $2(2.191) = 4.382$, above the mean of 8 must be 13 or more. A value 2 or more standard deviations below the mean must be 3 or less. These values correspond exactly to the values above that which we called unusual. In other cases, the two standard deviations approach provides a reasonable and widely used approximation.

### *p* Charts for Trend

A series of sample proportions should be visually inspected for trend. A graph called a $p$ chart helps identify times when the population proportion has changed from its long-time stable value. Because many sample proportions will be graphed, it is customary to set control limits at 3 rather than 2 standard deviations.

When $p_0$ is the expected or long-run proportion, we obtain a lower control limit by dividing the lower limit for $X$, $np_0 - 3\sqrt{np_0(1 - p_0)}$, by the sample size. Doing the same with the upper bound, we obtain an upper control limit.

| Lower control limit | Upper control limit |
|:---:|:---:|
| $p_0 - 3\sqrt{\dfrac{p_0(1 - p_0)}{n}}$ | $p_0 + 3\sqrt{\dfrac{p_0(1 - p_0)}{n}}$ |

In the context of credit applications that require a full review, $.4 = p_0$, so the control limits are

$$p_0 - 3\sqrt{\frac{p_0(1 - p_0)}{n}} = .4 - 3\sqrt{\frac{.4(1 - .4)}{20}} = .4 - 3(.110) = .07$$

$$p_0 + 3\sqrt{\frac{p_0(1 - p_0)}{n}} = .4 + 3\sqrt{\frac{.4(1 - .4)}{20}} = .4 + 3(.110) = .73$$

The **centerline** is drawn as a solid line at the expected or long-run proportion $p_0 = .4$, and the two control limits each at a distance of three standard deviations of the sample proportion from the centerline are also drawn as horizontal lines as in Figure 5. Sample proportions that fall outside of the control limits are considered unusual and should result in a search for a cause that may include a change in the mix of type of persons requesting credit cards.

The number of applications requiring full review out of the 20 in the sample were recorded for 19 weeks:

11  7  8  4  9  10  4  8  8  7  10  6  9  10  7  7  6  9  10

After converting to sample proportions by dividing by 20, we can graph the points in a *p* chart as in Figure 5. All the points are in control, and the financial institution does not appear to have reached the point where the mix of applicants includes more marginal cases.

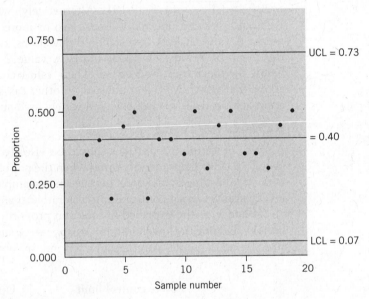

Figure 5   A *p* chart for the proportion of applications requiring a full review.

# Exercises

**5.81** Refer to the credit card application approval process on page 213 where unusual values are defined.

(a) Show that if 4 is included as an unusual value, then the probability $P[\,X \leq 4 \text{ or } X \geq 13\,]$ is greater than .05.

(b) Show that if 12 is included as an unusual value, then the probability $P[\,X \leq 3 \text{ or } X \geq 12\,]$ is greater than .05.

**5.82** Refer to the credit card application approval process on pages 213 and 214.

(a) Make a $p$ chart using the centerline and control limits calculated for $p_0 = .4$.

(b) Suppose the next five weeks bring 12, 10, 15, 11, and 16 applications requiring full review. Graph the corresponding proportions on your $p$ chart.

(c) Identify any weeks where the chart signals "out of control."

**5.83** Several fast food restaurants advertise quarter-pound hamburgers. This could be interpreted as meaning half the hamburgers made have an uncooked weight of at least a quarter-pound and half have a weight that is less. An inspector checks 20 uncooked hamburgers at each restaurant.

(a) Make a $p$ chart using the centerline and control limits calculated for $p_0 = .5$.

(b) Suppose that five restaurants have 8, 11, 7, 15, and 10 underweight hamburgers in samples of size 20. Graph the corresponding proportions on your $p$ chart.

(c) Identify any restaurants where the chart signals "out of control."

**5.84** Refer to Exercise 5.83.

(a) What are the unusual values for the number of underweight hamburgers in the sample if they correspond to proportions outside of the control limits of the $p$ chart?

(b) Use the binomial table to find the probability of observing one of these unusual values.

# 6. THE POISSON DISTRIBUTION AND RARE EVENTS

The Poisson distribution is an important probability distribution for describing the number of times an event $S$ randomly occurs in one unit of time or one unit of space. In a unit of time, each instant of time can be regarded as a potential trial in which the event may or may not occur. Although there are conceivably an infinite number of trials, usually only a few or moderate number of events take place.

For example, we could count the number of persons who use a campus money machine tomorrow during the half hour beginning at noon. Each user is an occurrence of the event $S$. As a second example, we could take a liter of lake water and count the number of zoo-plankton. The liter of water may be regarded as consisting of a large number of micro-volumes each of which may or may not contain a zoo-plankton. There are virtually an infinite number of trials, but the rate or average count per liter is only a moderate number. These examples show us that there may not be a clear cut definition of a trial, either in time or space, and that the count has no natural upper limit.

The Poisson distribution is determined by the average, or expected, rate of occurrence in a unit of time. In the context of a binomial distribution, the probability $p$ of $S$ occurring is required to be the same for all trials. For the Poisson distribution, it is the rate that is the same for all intervals of the same length.

To obtain a Poisson distribution, the event $S$ not only must occur randomly in time but also obey the following conditions stated as postulates.

(a) *Independence:* The number of times $S$ occurs in any interval of time is independent of the number of occurrences of $S$ in any other disjoint time interval.

(b) *Lack of clustering:* The chance of two or more occurrences happening simultaneously is negligible and can be assumed to be 0.

(c) *Rate:* The average number of occurrences per unit time is a constant denoted by $m$. This rate does not change over time.

Let $X$ be the random variable representing the number of times $S$ occurs in a unit time interval. Under these three postulates, the probability distribution of $X$ is given by the formula:

---

The Poisson distribution for

$$X = \text{the number of times } S \text{ occurs in a unit time interval,}$$

when the rate is $m$, has probability distribution

$$f(x) = P[X = x] = \frac{e^{-m} m^x}{x!}$$

for possible values $x = 0, 1, 2, \ldots$. Here $e$ is approximately equal to 2.71828.

$$\text{Mean} = m, \quad \text{Variance} = m$$

---

This distribution is called the **Poisson distribution with parameter** $m$. Here $m$ is the rate or the average number of times that $S$ is expected to occur. A somewhat remarkable feature is that the mean and variance both equal $m$.

There is a different Poisson distribution for different values $m$. They are all asymmetrical. If $m$ is an integer $f(m - 1) = f(m)$ and each is larger than any other probability. If $m$ is not an integer, the largest probability is assigned to the integer part of $m$. When $0 < m < 1$, the probability of 0 is largest and the probabilities $f(x)$ decrease as the value $x$ increases as illustrated in Figure 6(a) for $m = .6$. The distribution for $m = 3$, representing $m > 1$, has a more typical behavior. As illustrated in Figure 6 (b) the probabilities $f(x)$ increase to $f(2) = f(3)$ and then decrease as $x$ increases. This distribution has a long right-hand tail. As $m$ becomes large, the distribution becomes approximately symmetric.

## 6.1  HOW TO USE THE POISSON TABLE (APPENDIX B, TABLE 3)

We provide a short table of the Poisson cumulative distribution in Appendix B, Table 3, but still recommend the use of computer software or an appropriate hand-held calculator. For each fixed $m$, the cumulative probabilities $P[X \leq c] = \sum_{x=0}^{c} f(x)$ become a row of Table 3. The probability of an individual value $x$ is obtained by subtracting the row entry in column $c - 1$ from the entry in column $c$.

Figure 6   Two Poisson distributions (a) $m = .6$ and (b) $m = 3$.

**Example 19**   Poisson Distribution as a Model for Random Internet Interruptions

The manager of a department computing system states that the rate of interruptions to the internet service is .4 per month. Use the Poisson distribution to find the probability of

    (a)   Zero interruptions in one month.

    (b)   One or more interruptions in one month.

    (c)   Exactly two interruptions in 3 consecutive months.

    (d)   At least one but less than four interruptions in 3 consecutive months.

SOLUTION   Interruptions to the network occur randomly and the conditions for the Poisson distribution initially appear reasonable. Unlike the binomial case, there is no $q$ corresponding to a failure and there is no choice of a fixed Bernoulli trial here because one can always work with smaller intervals. We have $m = .4$ for the expected number of interruptions (successes) in one month. Let $X$ be the number of interruptions observed.

(a)  By the formula for Poisson probabilities $f(0) = e^{-m} = e^{-.4} = .670$. Over many months, there will be no interruptions to the internet in about two-thirds of them.

Using Table 3, and the fact that 0 is the smallest possible value, $f(0) = P[X = 0]$ is also the cumulative probability $P[X \leq 0] = .670$.

(b)  $P[X \geq 1] = 1 - P[X = 0] = 1 - .670 = .330$ by part (a).

(c)  A rate of .4 per month is equivalent to $3(.4) = 1.2$ during a three month period. From Table 3, with $m = 1.2$, we find

$$f(2) = P[X \leq 2] - P[X \leq 0] = .879 - .663 = .216$$

There will be exactly two interruptions in about one-fifth of the nonoverlapping three month periods.

(d)  Because the event $[1 \leq X < 4]$ is the same event as $[1 \leq X \leq 3]$,

$$P[1 \leq X < 4] = P[X \leq 3] - P[X \leq 0] = .966 - .301 = .665$$

Although the number of interruptions in a three-month period vary by chance, two-thirds of the time it will be 1, 2, or 3.

---

**Example 20**    A Poisson Distribution for the Number of Zoo-Plankton

Environmental scientists, concerned about the effects of releasing warm water from a power plant's cooling system into a Great Lake, decided to sample many organisms both inside and outside of a warm water plume. One choice was the zoo-plankton Cyclops. The scientists take a sample of water of size 100 cc and count the number of Cyclops. The expected number is 1.6 per 100 cc.

Use the Poisson distribution to find the probability of

(a)  Two Cyclops in a 100 cc sample.

(b)  Less than or equal 5 but more than one in a 100 cc sample.

(c)  Exactly 4 Cyclops in a sample of size 200 cc.

(d)  Three or more Cyclops in a sample of size 200 cc.

SOLUTION    The Cyclops occur randomly within the sample volume of water. The conditions for the Poisson distribution tentatively appear reasonable. Again there is no clear choice of a fixed Bernoulli trial, because one can always work with smaller volumes, and no probability $q$ corresponding to not observing a Cyclops (failure).

We have $m = 1.6$ for the expected number of Cyclops in one 100 cc sample. Let $X$ be the number of Cyclops observed.

(a)  By the formula for Poisson probabilities

$$f(2) = \frac{e^{-m} m^2}{2!} = \frac{e^{-1.6}(1.6)^2}{2} = .258$$

Using Table 3, we subtract the appropriate entries in the row where $m = 1.6$,

$$f(2) = P[X \leq 2] - P[X \leq 1] = .783 - .525 = .258$$

About 1 in 4 samples will contain exactly 2 Cyclops.

(b) $[1 < X \leq 5]$ is the same event as $[2 \leq X \leq 5]$,

$$P[1 < X \leq 5] = P[X \leq 5] - P[X \leq 1] = .994 - .525 = .469$$

Almost half the samples will contain between 2 and 4 Cyclops.

(c) A rate of 1.6 per 100 cc is equivalent to $2(1.6) = 3.2$ in a 200 cc sample. From Table 3, with $m = 3.2$, we find

$$f(4) = P[X \leq 4] - P[X \leq 3] = .781 - .603 = .178$$

There will be exactly 4 Cyclops in about one in 6 of the 200 cc samples.

(d) By the law of the complement,

$$P[X \geq 3] = 1 - P[X \leq 2] = 1 - .380 = .620$$

A little less than two-thirds of the 200 cc samples will contain at least 3 Cyclops.

## 6.2 POISSON APPROXIMATION TO BINOMIAL FOR *p* SMALL AND *n* LARGE

A binomial distribution with very large $n$, very small $p$, and $np = m$ of moderate magnitude is well approximated by the Poisson probability. For the individual probabilities $P[X = x]$,

$$\text{Binomial: } f(x) = \binom{n}{x} p^x (1 - p)^{n-x} \approx \frac{e^{-m} m^x}{x!} = f(x) \text{: Poisson}$$

for $x = 1, 2, \ldots n$. It is in the context of this approximation to the binomial, where $p$ is very small, that the Poisson distribution has been called a distribution for rare events. The Poisson distribution does, of course, apply to large values of $m$ but not as approximation to the binomial probability.

The quality of this approximation is illustrated in Table 10 on page 220 where $m = np = 2$ is fixed and $n$ successively increased and $p$ decreased.

The approximation improves with an increasing number of trials $n$. However, with modern statistical software it is possible to get accurate binomial probabilities for almost any $n$ and that is what we recommend.

**Example 21**   The Poisson Approximation to Binomial

The rate of red-green color blindness among females in the United States is .4 percent so $p = .004$. A random sample of $n = 1000$ females is observed. Use the Poisson approximation to the binomial to evaluate the probability of

**TABLE 10**    The Poisson Approximation to the Binomial
Distribution

| $n$ | $p$ | Binomial X = 0 | Probability X = 4 |
|---|---|---|---|
| 10 | .2 | .1074 | .0881 |
| 20 | .1 | .1216 | .0898 |
| 50 | .04 | .1299 | .0902 |
| 100 | .02 | .1326 | .0902 |
| Poisson Probability with $m = 2$ | | .1353 | .0902 |

(a)  Exactly two are red-green color blind.

(b)  Two or fewer are red-green color blind.

(c)  Compare the exact and approximate probabilities.

SOLUTION

(a)  The mean of the binomial is $np = 1000 \times .004 = 4$ and so we approximate with the Poisson distribution having the same mean $m = 4$.

$$f(2) = P[X = 2] = \frac{e^{-m} m^x}{x!} = \frac{e^{-4} 4^2}{2} = .1465$$

The value 2 would occur in about 1 of 7 samples.

(b)  The probability of 2 or fewer is

$$f(0) + f(1) + f(2) = \frac{e^{-4}}{0!} + \frac{e^{-4} 4}{1!} + \frac{e^{-4} 4^2}{2}$$

$$= .0183 + .0733 + .1465 = .2382$$

About one-fourth of the time, samples of size 1000 will contain 2 or fewer females with red-green color blindness.

(c)  A computer calculation, see the Technology section, gives exact binomial probabilities $f(2) = P[X = 2]$ and $P[\le 2]$. The probabilities are close to the approximate values but the exact are preferable.

```
Binomial with n = 1000 and p = 0.004
x        P(X=x)                 x        P(X<=x)
2     0.146378                  2        0.237516
```

The mean and variance of the Poisson distribution can be compared to those of the binomial (see Table 11).

**TABLE 11**   Mean and Variance of the Poisson and
Binomial Distribution

|  | Poisson ($m$) | Binomial ($n$, $p$) |
|---|---|---|
| Mean | $m$ | $np$ |
| Variance | $m$ | $npq$ |

Because the Poisson distribution approximates the binomial distribution when $np = m$ is moderate and $p$ is quite small, $q$ is very close to 1. In this case, $npq$ is very close to $np$, which provides an intuitive justification of why the mean and variance of the Poisson distribution are equal.

## Exercises

**5.85** A Poisson distribution with $m = 5.5$ governs the daily number of insurance claims handled by an adjuster. Determine the probability that she handles

(a) 6 claims tomorrow.

(b) 12 claims in the next two days.

(c) 3 or fewer tomorrow.

**5.86** The number of weekly breakdowns of a department's computing system is a random variable having a Poisson distribution with $m = .4$. What is the probability the computer will run for two consecutive weeks without a breakdown?

**5.87** A Poisson distribution with $m = 1.2$ governs the number of fireflies that flash in a 10 second period during an August evening in the yard. Determine the probability that:

(a) 2 fireflies flash.

(b) Less than or equal 3 flash.

(c) 6 fireflies flash in a 30 second period.

**5.88** A Poisson distribution with $m = 2.5$ governs the yearly number of tornado touchdowns in a central area of the state. Find the probabilities $P[X \leq 1]$ and $P[X \geq 1]$.

**5.89** For health reasons, homes need to be inspected for radon gas that decays and produces alpha particles. One device counts the number of alpha particles that hit its detector. To a good approximation, in one area, the count for the next week follows a Poisson distribution with mean 3. Determine

(a) The probability of 5 next week.

(b) The variance and standard deviation.

(c) The probability of 12 in three weeks.

**5.90** Records over several years show that the probability is .00003 that a car will have a flat tire while driving through a long tunnel. About 16,000 cars use the tunnel each week. Determine the probability that 2 or more cars will have a flat next week

(a) Using the Poisson approximation to the binomial.

(b) Using computer software to obtain the exact result.

**5.91** According to a Federal Trade Commission Report, about .13 % of the persons in the 20–29 year old age group filed identity theft complaints during 2011. From a random sample of $n = 500$ persons 20–29 years old, let $X$ denote the number who filed an identity theft complaint last year. Assuming that the proportion .13% prevails, use the Poisson distribution to approximate the probability of:

(a) $X = 0$

(b) $X \geq 1$

(c) $1 \leq X \leq 3$

## USING STATISTICS WISELY

1. The assignment of a value to each possible outcome, which creates a random variable, should quantify an important feature of the outcomes.

2. Describe the chance behavior of a discrete random variable $X$ by its probability distribution

$$f(x) = P[X = x] \text{ for each possible value } x$$

3. Summarize a probability distribution, or the random variable, by its

$$\text{Mean: } \mu = \sum_{\text{all } x} x \cdot f(x)$$

$$\text{Variance: } \sigma^2 = \sum_{\text{all } x} (x - \mu)^2 \cdot f(x)$$

4. If the use of Bernoulli trials is reasonable, probabilities concerning the number of successes in $n$ Bernoulli trials can be calculated using the formula for the binomial distribution

$$f(x) = \binom{n}{x} p^x (1 - p)^{n-x} \quad \text{for} \quad x = 0, 1, \ldots n$$

having mean $np$ and variance $np(1 - p)$.

5. Never use the formula $np(1 - p)$ for the variance of a count of successes in $n$ trials without first checking that the conditions for Bernoulli trials, independent trials with the same probability of success for each trial, hold. If the conditions are satisfied, then the binomial distribution is appropriate.

6. The Poisson distribution may be a reasonable model if the rate is constant in time, the counts for nonoverlapping intervals are independent, and two successes essentially cannot occur at the same time. However, if a long series of observations are taken over time, you should check constant rate by comparing the number of successes in the first half of the data with the number of successes in the second half.

## KEY IDEAS AND FORMULAS

The outcomes of an experiment are quantified by assigning each of them a numerical value related to a characteristic of interest. The rule for assigning the numerical value is called a **random variable** $X$.

A random variable having a finite number of values, or a sequence of values like a count, is called **discrete**. If a random variable can take any value in an interval, it is called a **continuous** random variable.

The **probability distribution** of $X$, or simply **distribution**, describes the manner in which probability is distributed over the possible values of $X$. Specifically, it is a list or formula giving the pairs $x$ and $f(x) = P[X = x]$.

A probability distribution serves as a model for explaining variation in a population.

A **probability histogram** graphically displays a discrete distribution using bars whose area equals the probability.

A probability distribution has a

$$\text{Mean} \quad \mu = \sum (\text{Value} \times \text{Probability}) = \sum x f(x)$$

which is interpreted as the **population mean**. This quantity is also called the **expected value** $E(X)$. Although $X$ is a variable, $E(X)$ is a constant.

The **population variance** is

$$\sigma^2 = E(X - \mu)^2 = \sum (x - \mu)^2 f(x)$$

The **standard deviation** $\sigma$ is the positive square root of variance. The standard deviation is a measure of the spread or variation of the population.

**Bernoulli trials** are defined by the characteristics: (1) two possible outcomes, **success** (S) or **failure** (F) for each trial; (2) a constant probability of success; and (3) independence of trials.

Sampling from a finite population without replacement violates the requirement of independence. If the population is large and the sample size small, the trials can be treated as independent for all practical purposes.

The number of successes $X$ in a fixed number of Bernoulli trials is called a **binomial random variable**. Its probability distribution, called the **binomial distribution**, is given by

$$f(x) = \binom{n}{x} p^x q^{n-x} \text{ for } x = 0, \ldots, n$$

where $n =$ number of trials, $p =$ probability of success in each trial, and $q = 1 - p$. The **binomial distribution** has

$$\textbf{Mean} = np$$

$$\textbf{Standard deviation} = \sqrt{npq}$$

The number of successes in a unit of time or space can be modeled as a **Poisson random variable**, if the appropriate conditions hold. Its probability distribution, called the **Poisson distribution**, is given by

$$P[X \leq x] = \frac{e^{-m} m^x}{x!}$$

where $m$ is the rate or expected number of successes in one unit of time or space. The **Poisson distribution** has

$$\textbf{Mean} = m$$

$$\textbf{Standard Deviation} = \sqrt{m}$$

A **probability model** is an assumed model for the probability distribution of a random variable. Probabilities are expressed in terms of **parameters** that are features of the population. For example, the binomial distribution is a probability model and the proportion $p$ is a parameter.

A **$p$ chart** displays sample proportions to reveal trends or changes in the population proportion over time.

## TECHNOLOGY

### *Calculating the binomial probabilities*
### $P[X = x]$ and $P[X \leq x]$

### MINITAB

### *Calculating the binomial probability*

$$P[X = x] = \binom{n}{x} p^x (1 - p)^{n-x}$$

The following commands illustrate the calculation of $P[X = 14]$ under the binomial distribution having $n = 21$ and $p = .57$:

Dialog box:

**Calc** > **Probability distributions** > **Binomial**. Select **Probability**.

Type *21* in **Number of Trials** and *.57* in **Event Probability**.

Select **Input constant**, and enter *14*. Click **OK**.

Alternatively to get all of the binomial probabilities for $n = 21$, you can first

Enter 0, 1, 2, . . . 21 in C1. Then, instead of clicking **Input constant**,

**Enter** C1 **in Input column and** C2 **in Optional storage.** Click **OK**.

### *Calculating the cumulative binomial probability*

$$P[X \leq c] = \sum_{x=0}^{c} \binom{n}{x} p^x (1 - p)^{n-x}$$

Follow the same steps as in the calculation of the probability of single terms except:

Select **Cumulative probability** instead of **Probability**.

### EXCEL
### *Calculating the binomial probability*

$$P[X = x] = \binom{n}{x} p^x (1 - p)^{n-x}$$

We illustrate the calculation of $P[X = 14]$ under the binomial distribution having $n = 21$ and $p = .57$.

With the cursor in a blank cell, select the **Formulas** tab and then **More Functions**. Choose **Statistical** and then **BINOM.DIST**.

Type *14* in **Number_s**, *21* in **Trials**, *.57* in **Probability_s**, and *FALSE* in **Cumulative**. Click **OK**.

### *Calculating the cumulative binomial probability*

$$P[X \leq c] = \sum_{x=0}^{c} \binom{n}{x} p^x (1 - p)^{n-x}$$

Follow the same steps as in the calculation of $P[X = x]$ except:

Type *TRUE* in **Cumulative**.

### TI-84/-83 PLUS

#### *Calculating the binomial probability*

$$P[X = x] = \binom{n}{x} p^x (1 - p)^{n-x}$$

The following commands illustrate the calculation of $P[X = 14]$ under the binomial distribution having $n = 21$ and $p = .57$:

Press $2^{nd}$ and **DISTR**.

Select[3] **0:binompdf(** and press **ENTER**.

With **binompdf(** on the home screen, type *21, .57, 14)* to give: binompdf( 21, .57, 14). Press **ENTER**.

#### *Calculating the cumulative binomial probability*

$$P[X \le c] = \sum_{x=0}^{c} \binom{n}{x} p^x (1 - p)^{n-x}$$

Follow the same steps as in the calculation of $P[X = x]$ except replace the second step by

Select **A:binomcdf(** and press **ENTER**.

### MINITAB

#### *Calculating the Poisson probability*

$$P[X = x] = \frac{e^{-m} m^x}{x!}$$

The following steps illustrate the calculation of $P[X = 3]$ under the Poisson distribution having $m = 2.37$.

> **Dialog box:**

> **Calc > Probability Distribution > Poisson.**

> Select **Probability**. Type 2.37 in **Mean**.

> Select **Input constant** and type 3. Click **OK**.

#### *Calculating the cumulative Poisson probability*

$$P[X \le c] = \sum_{x=0}^{c} \frac{e^{-m} m^x}{x!}$$

Follow the same steps as in the calculation of $P[X = x]$ except:

> Select **Cumulative probability** instead of **Probability**.

### EXCEL

#### *Calculating the Poisson probability $P[X = x] = \dfrac{e^{-m} m^x}{x!}$*

The following steps illustrate the calculation of $P[X = 3]$ under the Poisson distribution having $m = 2.37$.

---

[3]The options O and A are different for some TI-84 calculators.

With the cursor in a blank cell, select the **Formulas** tab and then **More Functions**. Choose **Statistical** and then **POISSON.DIST**.

Type 3 in *x*, 2.37 in **Mean** and *FALSE* in **Cumulative**. Click **OK**.

### Calculating the cumulative Poisson probability

$$P\,[\,X\,\leq\,c\,]\;=\;\sum_{x=0}^{c}\,\frac{e^{-m}\,m^{x}}{x!}$$

Follow the same steps as in the calculation of a single term except:

Type *TRUE* in **Cumulative**.

### TI-84/-83 PLUS
### Calculating the Poisson probability $P\,[\,X\,=\,x\,]\;=\;\dfrac{e^{-m}m^{x}}{x\,!}$

The following steps illustrate the calculation of $P[\,X\,=\,3\,]$ under the Poisson distribution having $m\,=\,2.37$.

2nd**DIST** to read the **DISTR** menu. Select A: **Poissonpdf** and press **ENTER**. With **Poissonpdf(** on the home screen, Type 3, 2.37 to give Poissonpdf( 3, 2.37 )

Press **ENTER**.

### Calculating the cumulative Poisson probability

$$P\,[\,X\,\leq\,c\,]\;=\;\sum_{x=0}^{c}\,\frac{e^{-m}\,m^{x}}{x!}$$

Follow the same steps as in the calculation of a single term except:

Choose **Poissoncdf(** instead of **Poissonpdf(**

## REVIEW EXERCISES

**5.92**   Let X denote the difference (no. of heads − no. of tails) in three tosses of a coin.

(a)  List the possible values of X.

(b)  List the elementary outcomes associated with each value of X.

**5.93**   A large science department at the university made a list of its top 25 juniors and top 30 seniors. The first list contains 8 females and the second 10 females. One person is randomly selected from each list and the two selections are independent. Let X denote the number of females selected.

(a)  For each possible value of X, identify the elementary outcomes associated with that value.

(b)  Determine the probability distribution of X.

*****5.94**   Refer to Exercise 5.93 but now suppose the sampling is done in two stages: First a list is selected at random and then, from that list, two persons are selected at random without replacement. Let Y denote the number of females in the sample.

(a)  List the elementary outcomes concerning the possible selections of this and the possible compositions of the sample (use a tree diagram). Find their probabilities. (*Hint*: Use conditional probability and the multiplication rule.)

(b)  Determine the probability distribution of Y.

**5.95** A list of the world's 10 largest companies, in terms of value, contains 3 from the United States, 4 from China, and 1 each from Australia, Brazil, and the United Kingdom. A potential investor randomly selects 3 of the companies to research further.

(a) Find the probability distribution of $X$, the number of United States companies selected.

(b) Determine the mean and variance of $X$.

**5.96** Refer to the monthly intersection accident data in Exercise 2.4. Considering an even longer record leads to a distribution for $X$ = number of accidents in a month.

| Value $x$ | Probability $f(x)$ |
|---|---|
| 0 | .08 |
| 1 | .20 |
| 2 | .19 |
| 3 | .24 |
| 4 | .14 |
| 5 | .13 |
| 6 | .02 |

(a) Calculate $E(X)$.

(b) Calculate sd$(X)$.

(c) Draw the probability histogram and locate the mean.

**5.97** The following distribution has been proposed for the number of times a student will eat a gourmet restaurant dinner next week.

| $x$ | $f(x)$ |
|---|---|
| 0 | .3 |
| 1 | .4 |
| 2 | .3 |

(a) Calculate the mean and variance.

(b) Plot the probability histogram and locate $\mu$.

**5.98** Refer to Exercise 5.96.

(a) List the $x$ values that lie in the interval $\mu - \sigma$ to $\mu + \sigma$ and calculate $P[\mu - \sigma \le X \le \mu + \sigma]$.

(b) List the $x$ values that lie in the interval $\mu - 2\sigma$ to $\mu + 2\sigma$ and calculate $P[\mu - 2\sigma \le X \le \mu + 2\sigma]$.

**5.99** A student buys a lottery ticket for $1. For every 1000 tickets sold, two bicycles are to be given away in a drawing.

(a) What is the probability that the student will win a bicycle?

(b) If each bicycle is worth $200, determine the student's expected gain.

**5.100** In the finals of a match play golf tournament, the winner will get $90,000 and the loser $15,000. Find the expected winnings of player B if (a) the two finalists are evenly matched and (b) player B has probability .8 of winning.

**5.101** The number of overnight emergency calls $X$ to the answering service of a heating and air conditioning firm have the probabilities .05, .1, .15, .35, .20, and .15 for 0, 1, 2, 3, 4, and 5 calls, respectively.

(a) Find the probability of fewer than 3 calls.

(b) Determine $E(X)$ and sd$(X)$.

**5.102** Suppose the number of parking tickets $X$ issued during a police officer's shift has the probability distribution

| $x$ | 0 | 1 | 2 | 3 | 4 |
|---|---|---|---|---|---|
| $f(x)$ | 0.13 | 0.14 | 0.43 | 0.20 | 0.10 |

(a) Find the mean and standard deviation of the number of parking tickets issued.

(b) Let $A = [X \le 2]$ and $B = [X \ge 1]$. Find $P(A \mid B) = P(X \le 2 | X \ge 1)$.

(c) Suppose the numbers of tickets issued on different days are independent. What is the probability that, over the next five days, no parking tickets will be issued on exactly one of the days?

**5.103** A botany student is asked to match the popular names of three house plants with their obscure botanical names. Suppose the student never heard of these names and is trying to match by sheer guess. Let X denote the number of correct matches.

(a) Obtain the probability distribution of X.

(b) What is the expected number of matches?

**5.104** The number of days, X, that it takes the post office to deliver a letter between City A and City B has the probability distribution

| x | f(x) |
|---|---|
| 3 | .5 |
| 4 | .3 |
| 5 | .2 |

Find:

(a) The expected number of days.

(b) The standard deviation of the number of days.

**5.105** A roulette wheel has 38 slots, of which 18 are red, 18 black, and 2 green. A gambler will play three times, each time betting $5 on red. The gambler gets $10 if red occurs and loses the bet otherwise. Let X denote the net gain of the gambler in 3 plays (for instance, if he loses all three times, then $X = -15$).

(a) Obtain the probability distribution of X.

(b) Calculate the expected value of X.

(c) Will the expected net gain be different if the gambler alternates his bets between red and black? Why or why not?

**5.106** Suppose that X can take the values 0, 1, 2, 3, and 4, and the probability distribution of X is incompletely specified by the function

$$f(x) = \frac{1}{4}\left(\frac{3}{4}\right)^x \quad \text{for} \quad x = 0, 1, 2, 3$$

Find (a) $f(4)$ (b) $P[X \geq 2]$ (c) $E(X)$ and (d) sd($X$).

**5.107** *The cumulative probabilities for a distribution.* A probability distribution can also be described by a function that gives the accumulated probability at or below each value of X. Specifically,

$$F(c) = P[X \leq c] = \sum_{x \leq c} f(x)$$

Cumulative distribution function at $c$ = Sum of probabilities of all values $x \leq c$

| x | f(x) | F(x) |
|---|---|---|
| 1 | .07 | .07 |
| 2 | .12 | .19 |
| 3 | .25 | |
| 4 | .30 | |
| 5 | .16 | |
| 6 | .10 | |

For the probability distribution given here, we calculate

$$F(1) = P[X \leq 1] = f(1) = .07$$
$$F(2) = P[X \leq 2] = f(1) + f(2) = .19$$

(a) Complete the $F(x)$ column in this table.

(b) Now cover the $f(x)$ column with a strip of paper. From the $F(x)$ values, reconstruct the probability function $f(x)$.

[ *Hint*: $f(x) = F(x) - F(x-1)$. ]

**5.108** Let the random variable Y denote the proportion of times a head occurs in three tosses of a coin, that is, $Y = $ (No. of heads in 3 tosses)/3.

(a) Obtain the probability distribution Y.

(b) Draw the probability histogram.

(c) Calculate the $E(Y)$ and sd($Y$).

**5.109** Is the model of Bernoulli trials plausible in each of the following situations? Identify any serious violations of the conditions.

(a) A beginning golfer tries 40 putts, 5 feet from the hole, while practicing on the putting green.

(b) Persons applying for a driver's license will be recorded as writing left- or right-handed.

(c) For each person taking a seat at a lunch counter, observe the time it takes to be served.

(d) Each day of the first week in April is recorded as being either clear or cloudy.

(e) Cars selected at random will or will not pass state safety inspection.

**5.110** Give an example (different from those appearing in Exercise 5.109) of repeated trials with two possible outcomes where:

(a) The model of Bernoulli trials is reasonable.

(b) The condition of independence is violated.

(c) The condition of equal $P(S)$ is violated.

**5.111** If the probability of having a male child is .5, find the probability that the third child is the first son.

**5.112** A basketball team scores 35% of the times it gets the ball. Find the probability that the first basket occurs on its third possession. (Assume independence.)

**5.113** If in three Bernoulli trials the probability that the first two trials are both failures is $\frac{9}{49}$, what is the probability that the first two are successes and the third is a failure?

**5.114** The proportion of people having the blood type O in a large southern city is .4. For two randomly selected donors:

(a) Find the probability of at least one type O.

(b) Find the expected number of type O.

(c) Repeat parts (a) and (b) if there are three donors.

**5.115** A viral infection is spread by contact with an infected person. Let the probability that a healthy person gets the infection in one contact be $p = .4$.

(a) An infected person has contact with six healthy persons. Specify the distribution of $X =$ No. of persons who contract the infection.

(b) Find $P[X \leq 3]$, $P[X = 0]$, and $E(X)$.

**5.116** The probability that a voter will believe a rumor about a politician is .3. If 20 voters are told individually:

(a) Find the probability that none of the 20 believes the rumor.

(b) Find the probability that seven or more believe the rumor.

(c) Determine the mean and standard deviation of the number who believe the rumor.

**5.117** National safety statistics suggest that about 33% of the persons treated in an emergency room because of moped accidents are under 16 years of age. Suppose you count the number of persons under 16 among the next 14 moped accident victims to come to the emergency room.

(a) Find the mean of $X$.

(b) Find the standard deviation of $X$.

(c) Find the probability that the first injured person is under 16 years old and the second is at least 16 years old.

**5.118** For each situation, state if a binomial distribution is reasonable for the random variable $X$. Justify your answer.

(a) A multiple-choice examination consists of 10 problems, each of which has 5 suggested answers. A student marks answers by pure guesses (i.e., one answer is chosen at random out of the 5), and $X$ denotes the number of marked answers that are wrong.

(b) A multiple-choice examination has two parts: Part 1 has 8 problems, each with 5 suggested answers, and Part 2 has 10 problems, each with 4 suggested answers. A student marks answers by pure guesses, and $X$ denotes the total number of problems that the student correctly answers.

(c) Twenty-five married couples are interviewed about exercise, and $X$ denotes the number of persons (out of the 50 people interviewed) who exercise regularly.

**5.119** A school newspaper claims that 70% of the students support its view on a campus issue. A random sample of 20 students is taken, and 10 students agree with the newspaper. Find $P[10 \text{ or less agree}]$ if 70% support the view and comment on the plausibility of the claim.

**5.120** At one large midwest university, about 40% of the college seniors have a social science major. Fourteen seniors are selected at random. Let $X$ denote the number that have a social science major. Determine

(a) $P[3 \leq X \leq 9]$     (d) $E(X)$

(b) $P[3 < X \leq 9]$     (e) $sd(X)$

(c) $P[3 < X < 9]$

**5.121** Refer to the population of social science majors in Exercise 5.120 but change the sample size to $n = 5$. Using the binomial table,

(a) List the probability distribution.

(b) Plot the probability histogram.

(c) Calculate $E(X)$ and Var$(X)$ from the entries in the list from part (a).

(d) Calculate $E(X) = np$ and Var$(X) = npq$ and compare your answer with part (c).

**5.122** For a binomial distribution with $p = .15$, find the smallest number $n$ such that 1 success is more probable than no successes in $n$ trials.

**5.123** Only 30% of the people in a large city feel that its mass transit system is adequate. If 20 persons are selected at random, find the probability that 10 or more will feel that the system is adequate. Find the probability that exactly 10 will feel that the system is adequate.

**5.124** A sociologist feels that only half of the high school seniors capable of graduating from college go to college. Of 17 high school seniors who have the ability to graduate from college, find the probability that 12 or more will go to college if the sociologist is correct. Assume that the seniors will make their decisions independently. Also find the expected number.

**5.125** Jones claims to have extrasensory perception (ESP). In order to test the claim, a psychologist shows Jones five cards that carry different pictures. Then Jones is blindfolded and the psychologist selects one card and asks Jones to identify the picture. This process is repeated 16 times. Suppose, in reality, that Jones has no ESP but responds by sheer guesses.

(a) What is the probability that the identifications are correct at most 8 times?

(b) What is the probability that the identifications are wrong at least 10 times?

(c) Find the expected value and standard deviation of the number of correct identifications.

**5.126** A manufacturer of luxury cars is experiencing some difficulties with a new metallic paint. During the final check at the end of assembly line, let $X =$ be the number of painting imperfections per car. This random variable is described by a Poisson distribution mean 2.6. Determine the probability of:

(a) 0 imperfection on any given car.

(b) 1 imperfection on any given car.

(c) 3 imperfections on any given car.

(d) 5 or more imperfections on any given car.

**5.127** A Poisson distribution describes $X =$ the number of bad checks a mega-store receives each day. If this store receives an average of 2 bad checks per day,

(a) Find the mean and variance.

What are the probabilities it will receive

(b) 3 bad checks in a given day?

(c) 5 bad checks over any 2 consecutive days?

(d) 10 or less in three consecutive days.

**5.128** Among the library's books purchased in the past three years, 1.2 % cannot be located. A random sample of 150 books is selected from a much larger number purchased in the last three years. Approximate the probability of

(a) Exactly 2 cannot be located.

(b) 2 or more cannot be located.

**5.129** An inspector samples bags of potato chips to see if they fall short of the weight, 14 ounces, printed on the bag. Samples of 20 bags are selected and the number with weight less than 14 ounces is recorded.

(a) Make a $p$ chart using the centerline and control limits calculated for $p_0 = .5$.

(b) Suppose that samples from ten different days have

11  8  14  10  13  12  7  14  10  13

underweight bags. Graph the corresponding proportions on your $p$ chart.

(c) Identify any days where the chart signals "out of control."

**5.130** *Geometric distribution.* Instead of performing a fixed number of Bernoulli trials, an experimenter performs trials until the first success occurs. The number of successes is now fixed at 1, but the number of trials $Y$ is now random. It can assume any of the values 1, 2, 3, and so on with no upper limit.

(a) Show that $f(y) = q^{y-1}p$ for $y = 1, 2, \ldots$

(b) Find the probability of 3 or fewer trials when $p = .5$.

# 6

# The Normal Distribution

altrendo nature/Stockbyte/Getty Images Inc.

## Bell-Shaped Distribution of Heights of Red Pine Seedlings

Trees are a renewable resource that is continually studied to both monitor the current status and improve this valuable natural resource. One researcher followed the growth of red pine seedlings.

The heights (mm) of 1456 three-year-old seedlings are summarized in the histogram. This histogram suggests a distribution with a single peak and which falls off in a symmetric manner. The histogram of the heights of adult males, or of adult females, has a similar pattern. A bell-shaped distribution is likely to be appropriate for the size of many things in nature.

Height in millimeters

# 1. PROBABILITY MODEL FOR A CONTINUOUS RANDOM VARIABLE

Up to this point, we have limited our discussion to probability distributions of discrete random variables. Recall that a discrete random variable takes on only some isolated values, usually integers representing a count. We now turn our attention to the probability distribution of a continuous random variable—one that can ideally assume any value in an interval. Variables measured on an under-lying continuous scale, such as weight, strength, life length, and temperature, have this feature.

Just as probability is conceived as the long-run relative frequency, the idea of a continuous probability distribution draws from the relative frequency histogram for a large number of measurements. The reader may wish to review Section 2.3 of Chapter 2 where grouping of data in class intervals and construction of a relative frequency histogram were discussed. We have remarked that with an increasing number of observations in a data set, histograms can be constructed with class intervals having smaller widths. We now pursue this point in order to motivate the idea of a continuous probability distribution. To focus the discussion let us consider that the weight $X$ of a newborn baby is the continuous random variable of our interest. How do we conceptualize the probability distribution of $X$? Initially, suppose that the birth weights of 100 babies are recorded, the data grouped in class intervals of 1 pound, and the relative frequency histogram in Figure 1a on page 234, is obtained. Recall that a relative frequency histogram has the following properties:

1. The total area under the histogram is 1.
2. For two points $a$ and $b$ such that each is a boundary point of some class, the relative frequency of measurements in the interval $a$ to $b$ is the **area** under the histogram above this interval.

For example, Figure 1a shows that the interval 7.5 to 9.5 pounds contains a proportion . 28 + .25 = .53 of the 100 measurements.

Next, we suppose that the number of measurements is increased to 5000 and they are grouped in class intervals of .25 pound. The resulting relative frequency histogram appears in Figure 1b. This is a refinement of the histogram in Figure 1a in that it is constructed from a larger set of observations and exhibits relative frequencies for finer class intervals. (Narrowing the class interval without increasing the number of observations would obscure the overall shape of the distribution.) The refined histogram in Figure 1b again has the properties 1 and 2 stated above.

By proceeding in this manner, even further refinements of relative frequency histograms can be imagined with larger numbers of observations and smaller class intervals. In pursuing this conceptual argument, we ignore the difficulty that the accuracy of the measuring device is limited. In the course of refining the histograms, the jumps between consecutive rectangles tend to dampen out, and the top of the histogram approximates the shape of a smooth curve, as illustrated in Figure 1c. Because probability is interpreted as long-run relative frequency, the curve obtained as the limiting form of the relative frequency histograms represents the manner in which the total probability 1 is distributed over the interval of possible values of the random variable $X$. This curve is called the **probability density curve** of the continuous random variable $X$. The mathematical function $f(x)$ whose graph produces this curve is called the **probability density function** of the continuous random variable $X$.

The properties 1 and 2 that we stated earlier for a relative frequency histogram are shared by a probability density curve that is, after all, conceived as a limiting smoothed form of a histogram. Also, since a histogram can never protrude below the $x$ axis, we have the further fact that $f(x)$ is nonnegative for all $x$.

(a)

(b)

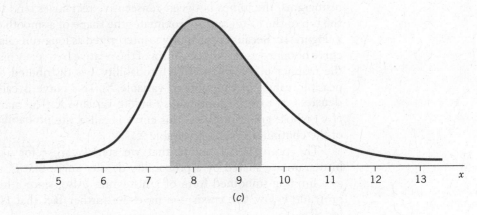

(c)

Figure 1    Probability density curve viewed as a limiting form of relative frequency histograms.

The probability density function $f(x)$ describes the distribution of probability for a continuous random variable. It has the properties:

1.   The total area under the probability density curve is 1.
2.   $P[a \leq X \leq b]$ = area under the probability density curve between $a$ and $b$.
3.   $f(x) \geq 0$ for all $x$.

Although the total area is 1, $f(x)$ can be greater than 1.

   Unlike the description of a discrete probability distribution, the probability density $f(x)$ for a continuous random variable does not represent the probability that the random variable will exactly equal the value $x$, or the event $[X = x]$ Instead, a probability density function relates the probability of an interval $[a, b]$ to the area under the curve in a strip over this interval. A single point $x$, being an interval with a width of 0, supports 0 area, so $P[X = x] = 0$.

With a continuous random variable, the probability that $X = x$ is **always** 0. It is only meaningful to speak about the probability that $X$ lies in an interval.

   The deduction that the probability at every single point is zero needs some clarification. In the birth-weight example, the statement $P[X = 8.5 \text{ pounds}] = 0$ probably seems shocking. Does this statement mean that no child can have a birth weight of 8.5 pounds? To resolve this paradox, we need to recognize that the accuracy of every measuring device is limited, so that here the number 8.5 is actually indistinguishable from all numbers in an interval surrounding it, say, $[8.495, 8.505]$. Thus, the question really concerns the probability of an interval surrounding 8.5, and the area under the curve is no longer 0.

   When determining the probability of an interval $a$ to $b$, we need not be concerned if either or both endpoints are included in the interval. Since the probabilities of $X = a$ and $X = b$ are both equal to 0,

$$P[a \leq X \leq b] = P[a < X \leq b] = P[a \leq X < b]$$
$$= P[a < X < b]$$

In contrast, these probabilities may not be equal for a discrete distribution.

   Fortunately, for important distributions, areas have been extensively tabulated. In most tables, the entire area to the left of each point is tabulated. To obtain the probabilities of other intervals, we must apply the following rules.

$$P[\,a\ <\ X\ <\ b\,]\ =\ (\text{Area to left of } b) - (\text{Area to the left of } a)$$

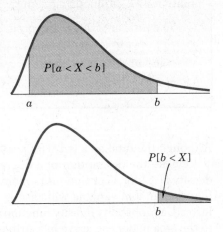

$$P[\,b\ <\ X\,]\ =\ 1 - (\text{Area to left of } b)$$

## 1.1 SPECIFICATION OF A PROBABILITY MODEL

A probability model for a continuous random variable is specified by giving the mathematical form of the probability density function. If a fairly large number of observations of a continuous random variable are available, we may try to approximate the top of the staircase silhouette of the relative frequency histogram by a mathematical curve.

In the absence of a large data set, we may tentatively assume a reasonable model that may have been suggested by data from a similar source. Of course, any model obtained in this way must be closely scrutinized to verify that it conforms to the data at hand. Section 6 addresses this issue.

## 1.2 FEATURES OF A CONTINUOUS DISTRIBUTION

As is true for relative frequency histograms, the probability density curves of continuous random variables could possess a wide variety of shapes. A few of these are illustrated in Figure 2. Many statisticians use the term skewed for a long tail in one direction.

A continuous random variable $X$ also has a mean, or expected value $E(X)$, as well as a variance and a standard deviation. Their interpretations are the same as in the case of discrete random variables, but their formal definitions involve integral calculus and are therefore not pursued here. However, it is instructive to see in Figure 3 that the mean $\mu = E(X)$ marks the balance point of the probability mass. The median, another measure of center, is the value of $X$ that divides the area under the curve into halves each with probability 0.5.

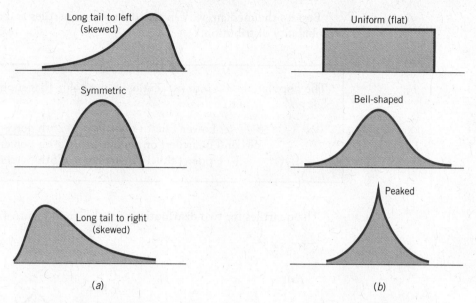

Figure 2    Different shapes of probability density curves. (*a*) Symmetry and deviations from symmetry. (*b*) Different peakedness.

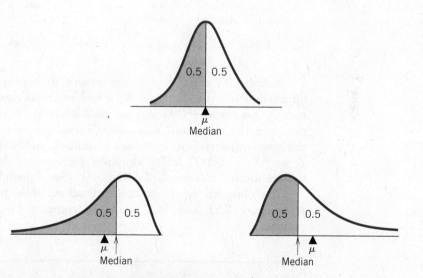

Figure 3    Mean as the balance point and median as the point of equal division of the probability mass.

Besides the median, we can also define the quartiles and other percentiles of a probability distribution.

The population 100 $p$-th percentile is an $x$ value that supports area $p$ to its left and $1 - p$ to its right.

$$
\begin{aligned}
\textbf{Lower (first) quartile} &= \text{25th percentile} \\
\textbf{Second quartile (or median)} &= \text{50th percentile} \\
\textbf{Upper (third) quartile} &= \text{75th percentile}
\end{aligned}
$$

The quartiles for two distributions are shown in Figure 4.

Figure 4   Quartiles of two continuous distributions.

Statisticians often find it convenient to convert random variables to a dimensionless scale. Suppose $X$, a real estate salesperson's commission for a month, has mean \$6000 and standard deviation \$500. Subtracting the mean produces the deviation $X - 6000$ measured in dollars. Then, dividing by the standard deviation, expressed in dollars, yields the dimensionless variable $Z = (X - 6000) / 500$. Moreover, the standardized variable $Z$ can be shown to have mean 0 and standard deviation 1. (See Appendix A3.1 for details.)

The observed values of standardized variables provide a convenient way to compare SAT and ACT scores or compare heights of male and female partners.

The standardized variable

$$
Z = \frac{X - \mu}{\sigma} = \frac{\text{Variable} - \text{Mean}}{\text{Standard deviation}}
$$

has mean 0 and sd 1.

# Exercises

**6.1** Which of the functions sketched below could be a probability density function for a continuous random variable? Why or why not?

(a)

(b)

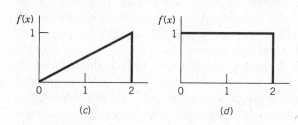

(c)

(d)

**6.2** Determine the following probabilities from the curve $f(x)$ diagrammed in Exercise 6.1(a).

(a) $P[0 < X < .5]$

(b) $P[.5 < X < 1]$

(c) $P[1.0 < X < 2]$

(d) $P[X = 1]$

**6.3** For the curve $f(x)$ graphed in Exercise 6.1(c), which of the two intervals $[0 < X < .5]$ or $[1.5 < X < 2]$ is assigned a higher probability?

**6.4** Determine the median and the quartiles for the probability distribution depicted in Exercise 6.1(a).

**6.5** Determine the median and the quartiles for the curve depicted in Exercise 6.1(c).

**6.6** Determine the 15th percentile of the curve in Exercise 6.1(a).

**6.7** If a student is more likely to be late than on time for the 1:20 PM history class:

(a) Determine if the median of the student's arrival time distribution is earlier than, equal to, or later than 1:20 PM.

(b) On the basis of the given information, can you determine if the mean of the student's arrival time distribution is earlier than, equal to, or later than 1:20 PM? Comment.

**6.8** Which of the distributions in Figure 3 are compatible with the following statements?

(a) The first test was too easy because over half the class scored above the mean.

(b) In spite of recent large increases in salary, half of the professional football players still make less than the average salary.

**6.9** Find the standardized variable $Z$ if $X$ has

(a) Mean 15 and standard deviation 4.

(b) Mean 61 and standard deviation 9.

(c) Mean 161 and variance 25.

**6.10** Males 20 to 29 years old have a mean height of 69.5 inches with a standard deviation of 3.1 inches. Females 20 to 29 years old have a mean height of 64.4 inches with a standard deviation of 3.0 inches. (Based on *Statistical Abstract of the U.S. 2012*, Table 209.)

(a) Find the standardized variable for the heights of males.

(b) Find the standardized variable for the heights of females.

(c) For a 68-inch-tall person, find the value of the standardized variable for males.

(d) For a 68-inch-tall person, find the value of the standardized variable for females. Compare your answer with part (c) and comment.

**6.11** Two pieces of wood need to be glued together. After applying glue to both pieces, they must be clamped to obtain the best results. The required clamping time in any application is a random variable. Find the standardized variable $Z$, for the clamping time $X$, when using a manufacturers'

(a) Wood glue which has mean 25 minutes and standard deviation 3 minutes.

(b) General purpose glue which has mean 90 minutes and standard deviation 12 minutes.

## 2.  THE NORMAL DISTRIBUTION—ITS GENERAL FEATURES

The normal distribution, which may already be familiar to some readers as the curve with the bell shape, is sometimes associated with the names of Pierre Laplace and Carl Gauss, who figured prominently in its historical development. Gauss derived the normal distribution mathematically as the probability distribution of the error of measurements, which he called the "normal law of errors." Subsequently, astronomers, physicists, and, somewhat later, data collectors in a wide variety of fields found that their histograms exhibited the common feature of first rising gradually in height to a maximum and then decreasing in a symmetric manner. Although the normal curve is not unique in exhibiting this form, it has been found to provide a reasonable approximation in a great many situations. Unfortunately, at one time during the early stages of the development of statistics, it had many overzealous admirers. Apparently, they felt that all real-life data must conform to the bell-shaped normal curve, or otherwise, the process of data collection should be suspect. It is in this context that the distribution became known as the **normal distribution**. However, scrutiny of data has often revealed inadequacies of the normal distribution. In fact, the universality of the normal distribution is only a myth, and examples of quite nonnormal distributions abound in virtually every field of study. Still, the normal distribution plays a central role in statistics, and inference procedures derived from it have wide applicability and form the backbone of current methods of statistical analysis.

Although we are speaking of the importance of the normal distribution, our remarks really apply to a whole class of distributions having bell-shaped densities. There is a normal distribution for each value of its mean $\mu$ and its standard deviation $\sigma$.

A few details of the normal curve merit special attention. The curve is symmetric about its mean $\mu$, which locates the peak of the bell (see Figure 5). The interval running one standard deviation in each direction from $\mu$ has a probability of .683, the interval from $\mu - 2\sigma$ to $\mu + 2\sigma$ has a probability of .954, and

Does the $\bar{x}$ ranch have average beef?

the interval from $\mu - 3\sigma$ *to* $\mu + 3\sigma$ has a probability of .997. It is these probabilities that give rise to the empirical rule stated in Chapter 2. The curve never reaches 0 for any value of $x$, but because the tail areas outside $(\mu - 3\sigma, \mu + 3\sigma)$ are very small, we usually terminate the graph at these points.

A normal distribution has a bell-shaped density[1] as shown in Figure 5. It has

$$\text{Mean} = \mu$$
$$\text{Standard deviation} = \sigma$$

Figure 5   Normal distribution.

The probability of the interval extending

One sd on each side of the mean:  $P[\mu - \sigma < X < \mu + \sigma] = .683$

Two sd on each side of the mean:  $P[\mu - 2\sigma < X < \mu + 2\sigma] = .954$

Three sd on each side of the mean: $P[\mu - 3\sigma < X < \mu + 3\sigma] = .997$

[1]The formula, which need not concern us, is

$$f(x) = \frac{1}{\sqrt{2\pi}\sigma} e^{-\frac{1}{2}\left(\frac{x-\mu}{\sigma}\right)^2} \qquad \text{for} \qquad -\infty < x < \infty$$

where $\pi$ is the area of a circle having unit radius, or approximately 3.1416, and $e$ is approximately 2.7183.

**Notation**

The normal distribution with a mean of $\mu$ and a standard deviation of $\sigma$ is denoted by $N(\mu, \sigma)$.

Interpreting the parameters, we can see in Figure 6 that a change of mean from $\mu_1$ to a larger value $\mu_2$ merely slides the bell-shaped curve along the axis until a new center is established at $\mu_2$. There is no change in the shape of the curve.

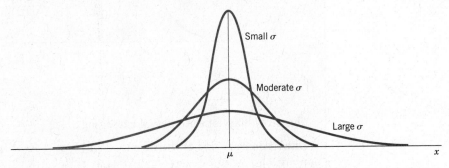

Figure 6   Two normal distributions with different means, with $\mu_1$ less than $\mu_2$, but the same standard deviation.

A different value for the standard deviation results in a different maximum height of the curve and changes the amount of the area in any fixed interval about $\mu$ (see Figure 7). The position of the center does not change if only $\sigma$ is changed.

Figure 7   Decreasing $\sigma$ increases the maximum height and the concentration of probability about $\mu$.

## 3. THE STANDARD NORMAL DISTRIBUTION

The particular normal distribution that has a mean of 0 and a standard deviation of 1 is called the standard normal distribution. It is customary to denote the standard normal variable by $Z$. The standard normal curve is illustrated in Figure 8.

Figure 8    The standard normal curve.

---

The standard normal distribution has a bell-shaped density with

Mean $\mu = 0$
Standard deviation $\sigma = 1$

The standard normal distribution is denoted by $N(0, 1)$.

---

## 3.1  FINDING PROBABILITIES FROM AREAS UNDER THE CURVE

### Use of the Standard Normal Table (Appendix B, Table 4)

The standard normal table in the appendix gives the area to the left of a specified value of $z$ as

$$P[Z \leq z] = \text{Area under curve to the left of } z$$

For the probability of an interval $[a, b]$,

$$P[a \leq Z \leq b] = [\text{Area to left of } b] - [\text{Area to left of } a]$$

The following properties can be observed from the symmetry of the standard normal curve about 0 as exhibited in Figure 9.

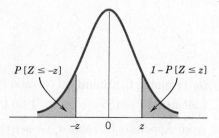

Figure 9    Equal normal tail probabilities.

1. $P[Z \leq 0] = .5$
2. $P[Z \leq -z] = 1 - P[Z \leq z] = P[Z \geq z]$

---

**Example 1**  Determining Standard Normal Probabilities for Tail Events

Find $P[Z \leq 1.37]$ and $P[Z > 1.37]$

SOLUTION  From the normal table, we see that the probability or area to the left of 1.37 is .9147. (See Table 1.) Consequently, $P[Z \leq 1.37] = .9147$. Moreover, because $[Z > 1.37]$ is the complement of $[Z \leq 1.37]$,

$$P[Z > 1.37] = 1 - P[Z \leq 1.37] = 1 - .9147 = .0853$$

as we can see in Figure 10. An alternative method is to use symmetry to show that $P[Z > 1.37] = P[Z < -1.37]$, which can be obtained directly from the normal table.

**TABLE 1**  How to Read from Appendix B, Table 4 for $z = 1.37 = 1.3 + .07$

| $z$ | .00 | . . . | .07 | . . . |
|-----|-----|-------|-----|-------|
| .0  |     |       |     |       |
| .   |     |       |     |       |
| .   |     |       |     |       |
| 1.3 |     |       | .9147 |     |
| .   |     |       |     |       |
| .   |     |       |     |       |
| .   |     |       |     |       |

Figure 10  An upper tail normal probability.

---

**Example 2**  Determining the Standard Normal Probability of an Interval

Calculate $P[-.155 < Z < 1.60]$.

SOLUTION  From Appendix B, Table 4, we see that

$$P[Z \leq 1.60] = \text{Area to left of } 1.60 = .9452$$

We interpolate[2] between the entries for $-.15$ and $-.16$ to obtain

$$P[\,Z \leq -.155\,] = \text{Area to left of } -.155 = .4384$$

Therefore,

$$P[\,-.155 < Z < 1.60\,] = .9452 - .4384 = .5068$$

which is the shaded area in Figure 11.

.5068

−.155

Figure 11   Normal probability of an interval.

**Example 3**   Determining the Standard Normal Probability Outside of an Interval

Find $P[\,Z < -1.9 \text{ or } Z > 2.1\,]$.

SOLUTION   The two events $[\,Z < -1.9\,]$ and $[\,Z > 2.1\,]$ are incompatible, so we add their probabilities:

$$P[\,Z < -1.9 \text{ or } Z > 2.1\,] = P[\,Z < -1.9\,] + P[\,Z > 2.1\,]$$

As indicated in Figure 12, $P[\,Z > 2.1\,]$ is the area to the right of 2.1, which is $1 - [\text{Area to left of } 2.1\,] = 1 - .9821 = .0179$. The normal table gives $P[\,Z < -1.9\,] = .0287$ directly. Adding these two quantities, we get

$$P[\,Z < -1.9 \text{ or } Z > 2.1\,] = .0287 + .0179 = .0466$$

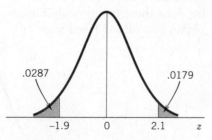

.0287    .0179

−1.9    0    2.1

Figure 12   Normal probabilities for Example 3.

[2]Since $z = -.155$ is halfway between $-.15$ and $-.16$, the interpolated value is halfway between the table entries .4404 and .4364. The result is .4384. We actually used computer software (see Technology section) to get .4384. You may want to just eyeball a value between two entries in the table.

### 3.2 FINDING PERCENTILES AND OTHER VALUES FOR Z

Given a probability, we can read the standard normal table in reverse to obtain a value for $z$.

**Example 4**   Determining an Upper Percentile of the Standard Normal Distribution

Locate the value of $z$ that satisfies $P[\, Z \, > \, z\,] \; = \; .025$

SOLUTION   If we use the property that the total area is 1, the area to the left of $z$ must be $1 \, - \, .0250 \, = \, .9750$. The marginal value with the tabular entry .9750 is $z \, = \, 1.96$ (diagrammed in Figure 13).

| $z$ | .00 | $\ldots$ | .06 | $\ldots$ |
|-----|-----|----------|------|----------|
| .0  |     |          |      |          |
| $\cdot$ |     |      |      |          |
| 1.9 |     |          | .9750 |         |
| $\cdot$ |     |      |      |          |
| $\cdot$ |     |      |      |          |

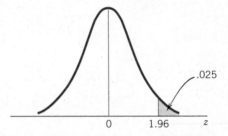

Figure 13   $P[\, Z \, > \, 1.96\,] \; = \; .025.$

**Example 5**   Determining $z$ for Given Equal Tail Areas

Obtain the value of $z$ for which $P[\, -z \, \leq \, Z \, \leq \, z\,] \; = \; .90.$

SOLUTION   We observe from the symmetry of the curve that

$$P[\, Z \, < \, -z\,] \; = \; P[\, Z \, > \, z\,] \; = \; .05$$

From the normal table, we see that $z \, = \, 1.65$ gives $P[\, Z \, < \, -1.65\,] \, = \, .0495$ and $z \, = \, 1.64$ gives $P[\, Z \, < \, -1.64\,] \; = \; .0505$. Because .05 is halfway between these two probabilities, we interpolate between the two $z$ values to obtain $z \, = \, 1.645$ (see Figure 14).

Figure 14   $P[\, Z \, < \, -1.645 \text{ or } Z \, > \, 1.645\,] \; = \; .10.$

*Suggestion:*   The preceding examples illustrate the usefulness of a sketch to depict an area under the standard normal curve. A correct diagram shows how to combine the left-side areas given in the normal table.

## Exercises

**6.12** Find the area under the standard normal curve to the left of

(a) $z = 1.26$          (b) $z = .26$

(c) $z = -1.71$         (d) $z = -2.43$

**6.13** Find the area under the standard normal curve to the left of

(a) $z = .63$           (b) $z = 1.03$

(c) $z = -1.03$         (d) $z = -1.35$

**6.14** Find the area under the standard normal curve to the right of

(a) $z = 1.16$

(b) $z = .64$

(c) $z = -1.71$

(d) $z = -1.525$ (interpolate)

**6.15** Find the area under the standard normal curve to the right of

(a) $z = .63$

(b) $z = 2.63$

(c) $z = -1.23$

(d) $z = 1.635$ (interpolate)

**6.16** Find the area under the standard normal curve over the interval

(a) $z = -.75$ to $z = .75$

(b) $z = -1.09$ to $z = 1.09$

(c) $z = .32$ to $z = 2.65$

(d) $z = -.745$ to $z = 1.244$ (interpolate)

**6.17** Find the area under the standard normal curve over the interval

(a) $z = -.44$ to $z = .44$

(b) $z = -1.33$ to $z = 1.33$

(c) $z = .40$ to $z = 2.03$

(d) $z = 1.405$ to $z = 2.306$ (interpolate)

**6.18** Identify the $z$ values in the following diagrams of the standard normal distribution (interpolate, as needed).

**6.19** Identify the $z$ values in the following diagrams of the standard normal distribution (interpolate, as needed).

**6.20** For a standard normal random variable $Z$, find
  (a) $P[Z < .62]$
  (b) $P[Z < -.62]$
  (c) $P[Z > 1.59]$
  (d) $P[Z > -1.59]$
  (e) $P[-1.3 < Z < 2.61]$
  (f) $P[.08 < Z < .8]$
  (g) $P[-1.62 < Z < -.34]$
  (h) $P[|Z| < 1.65]$

**6.21** Find the $z$ value in each of the following cases.
  (a) $P[Z < z] = .1762$
  (b) $P[Z > z] = .10$
  (c) $P[-z < Z < z] = .954$
  (d) $P[-.6 < Z < z] = .50$

**6.22** Find the quartiles of the standard normal distribution.

**6.23** Find
  (a) $P[Z < .33]$.
  (b) The 33rd percentile of the standard normal distribution.
  (c) $P[Z < .97]$.
  (d) The 97th percentile of the standard normal distribution.

**6.24** Find
  (a) $P[Z < .46]$.
  (b) The 46th percentile of the standard normal distribution.
  (c) $P[Z < .85]$.
  (d) The 85th percentile of the standard normal distribution.

## 4. PROBABILITY CALCULATIONS WITH NORMAL DISTRIBUTIONS

Fortunately, no new tables are required for probability calculations regarding the general normal distribution. Any normal distribution can be set in correspondence to the standard normal by the following relation.

> If $X$ is distributed as $N(\mu, \sigma)$, then the standardized variable
> $$Z = \frac{X - \mu}{\sigma}$$
> has the standard normal distribution.

This property of the normal distribution allows us to cast a probability problem concerning $X$ into one concerning $Z$. To find the probability that $X$ lies in a given interval, convert the interval to the $z$ scale and then calculate the probability by using the standard normal table (Appendix B, Table 4).

**Example 6**    Converting a Normal Probability to a Standard Normal Probability

Given that $X$ has the normal distribution $N(60, 4)$, find $P[55 \leq X \leq 63]$.

SOLUTION    Here, the standardized variable is $Z = \dfrac{X - 60}{4}$. The distribution of $X$ is shown in Figure 15, where the distribution of $Z$ and the $z$ scale are also displayed below the $x$ scale. In particular,

$$x = 55 \text{ gives } z = \frac{55 - 60}{4} = -1.25$$

$$x = 63 \text{ gives } z = \frac{63 - 60}{4} = .75$$

Figure 15    Converting to the z scale.

Therefore,

$$P[\, 55 \leq X \leq 63 \,] = P[\, -1.25 \leq Z \leq .75 \,]$$

Using the normal table, we find $P[\, Z \leq .75 \,] = .7734$ and $P[\, Z \leq -1.25 \,] = .1056$, so the required probability is $.7734 - .1056 = .6678$.

The working steps employed in Example 6 can be formalized into the rule:

If $X$ is distributed as $N(\mu, \sigma)$, then

$$P[\, a \leq X \leq b \,] = P\left[\, \frac{a - \mu}{\sigma} \leq Z \leq \frac{b - \mu}{\sigma} \,\right]$$

where $Z$ has the standard normal distribution.

**Example 7**    Probabilities Concerning Calories in a Lunch Salad

The number of calories in a salad on the lunch menu is normally distributed with mean $= 200$ and sd $= 5$. Find the probability that the salad you select will contain:

(a)  More than 208 calories.

(b)  Between 190 and 200 calories.

SOLUTION    Letting $X$ denote the number of calories in the salad, we have the standardized variable

$$Z = \frac{X - 200}{5}$$

(a) The $z$ value corresponding to $x = 208$ is

$$z = \frac{208 - 200}{5} = 1.6$$

Therefore,

$$P[X > 208] = P[Z > 1.6]$$
$$= 1 - P[Z \leq 1.6]$$
$$= 1 - .9452 = .0548$$

(b) The $z$ values corresponding to $x = 190$ and $x = 200$ are

$$\frac{190 - 200}{5} = -2.0 \quad \text{and} \quad \frac{200 - 200}{5} = 0$$

respectively. We calculate

$$P[190 \leq X \leq 200] = P[-2.0 \leq Z \leq 0]$$
$$= .5 - .0228 = .4772$$

**Example 8**    Determining a Percentile of a Normal Population

The hours of sleep data in Example 6, Chapter 2, suggest that the population of hours of sleep can be modeled as a normal distribution with mean $= 7.2$ hours and sd $= 1.3$ hours.

(a) Determine the probability assigned to sleeping less than 6.5 hours.

(b) Find the 70th percentile of the distribution for hours of sleep.

SOLUTION    If we denote the hours of sleep by $X$, the standardized score

$$Z = \frac{X - 7.2}{1.3}$$

is distributed as $N(0, 1)$.

(a) The $z$ score corresponding to 6.5 is

$$z = \frac{6.5 - 7.2}{1.3} = -.538$$

So, interpolating,

$$P[X < 6.5] = P[Z < -.538] = .295$$

Thus, 29.5%, or about 30%, of the students sleep less than 6.5 hours. In other words, 6.5 hours nearly locates the 30th percentile.

(b)  We first find the 70th percentile in the $z$ scale and then convert it to the $x$ scale. From the standard normal table, we interpolate to find

$$P[\, Z \, \leq \, .524 \,] \, = \, .70$$

The standardized score $z \, = \, .524$ corresponds to

$$x \, = \, 7.2 \, + \, 1.3 (\, .524 \,)$$
$$= \, 7.88$$

Therefore, the 70th percentile score is about 7.88 or nearly eight hours.

## Exercises

**6.25**  Records suggest that the normal distribution with mean 50 and standard deviation 9 is a plausible model for a measurement of the amount of suspended solids (ppm) in river water. Find

(a)  $P[\, X \, < \, 46.4 \,]$

(b)  $P[\, X \, \leq \, 57.2 \,]$

(c)  $P[\, X \, > \, 57.2 \,]$

(d)  $P[\, X \, > \, 60.8 \,]$

(e)  $P[\, 33.8 \, \leq \, X \, \leq \, 64.4 \,]$

(f)  $P[\, 52.5 \, \leq \, X \, \leq \, 60.9 \,]$

**6.26**  Data suggests that the normal distribution with mean 13.0 and standard deviation 2.4 is a plausible model for the length (feet) of adult anaconda snakes. Find

(a)  $P[\, X \, < \, 10.4 \,]$

(b)  $P[\, X \, \leq \, 17.8 \,]$

(c)  $P[\, X \, > \, 17.8 \,]$

(d)  $P[\, X \, > \, 16.72 \,]$

(e)  $P[\, 10.24 \, \leq \, X \, \leq \, 18.4 \,]$

(f)  $P[\, 14.8 \, \leq \, X \, \leq \, 17.2 \,]$

**6.27**  Referring to Exercise 6.25, find $b$ such that

(a)  $P[\, X \, < \, b \,] \, = \, .975$

(b)  $P[\, X \, > \, b \,] \, = \, .025$

(c)  $P[\, X \, < \, b \,] \, = \, .305$

**6.28**  Referring to Exercise 6.26, find $b$ such that

(a)  $P[\, X \, < \, b \,] \, = \, .7995$

(b)  $P[\, X \, > \, b \,] \, = \, .002$

(c)  $P[\, X \, < \, b \,] \, = \, .015$

**6.29**  Scores on a certain nationwide college entrance examination follow a normal distribution with a mean of 500 and a standard deviation of 100. Find the probability that a student will score:

(a)  Over 650.

(b)  Less than 250.

(c)  Between 325 and 675.

**6.30**  It is reasonable to model the distribution that produced the College Qualification Test (CQT) data in Exercise 2.22 as a normal distribution with mean 150 and standard deviation 25.4.

(a)  What is the probability of a new student scoring above 195?

(b)  What score has probability .15 of being exceeded.

**6.31**  According to the children's growth chart that doctors use as a reference, the heights of two-year-old boys are nearly normally distributed with a mean of 34.5 inches and a standard deviation of 1.4 inches. If a two-year-old boy is selected at random, what is the probability that he will be between 32.5 and 36.5 inches tall?

**6.32**  The time it takes a symphony orchestra to play Beethoven's *Ninth Symphony* has a normal distribution with a mean of 64.3 minutes and a standard deviation of 1.15 minutes. The next time it is played, what is the probability that it will take between 62.5 and 67.7 minutes?

**6.33**  The weights of apples served at a restaurant are normally distributed with a mean of 5 ounces and standard deviation of 1.2 ounces. What is the probability that the next person served will be given an apple that weighs less than 4 ounces?

**6.34** The diameter of hail hitting the ground during a storm is normally distributed with a mean of .5 inch and a standard deviation of .1 inch. What is the probability that:

(a) A hailstone picked up at random will have a diameter greater than .71 inch?

(b) Two hailstones picked up in a row will have diameters greater than .6 inch? (Assume independence of the two diameters.)

(c) By the end of the storm, what proportion of the hailstones would have had diameters greater than .71 inch?

**6.35** Refer to Exercise 6.10 where, according to current U.S. Census Bureau data, the heights of 20- to 29-year-old women can be well approximated by a normal distribution with mean 64.4 inches and standard deviation 3.0 inches.

(a) What is the probability that the height of a randomly selected woman 20 to 29 years old exceeds 70 inches?

(b) What is the probability that the height of a randomly selected woman 20 to 29 years old is less than or equal to 60 inches?

**6.36** Suppose the contents of bottles of water coming off a production line have a normal distribution with mean 9.1 ounces and standard deviation .1 ounce.

(a) If every bottle is labeled 9 ounces, what proportion of the bottles contain less than the labeled amount?

(b) If only 2.5% of the bottles exceed weight $w$, what is the value of $w$?

**6.37** The time for an emergency medical squad to arrive at the sports center at the edge of town is distributed as a normal variable with $\mu = 17$ minutes and $\sigma = 3$ minutes.

(a) Determine the probability that the time to arrive is:
   (i) More than 22 minutes.
   (ii) Between 13 and 21 minutes.
   (iii) Between 15.5 and 18.5 minutes.

(b) Which arrival period of duration 1 minute is assigned the highest probability by the normal distribution?

**6.38** It is reasonable to model the distribution that produced the lizards' speed test in Exercise 2.19 as normal distribution with mean 1.7 m/s and standard deviation .57 m/s. Find the probability that the speed of a new lizard

(a) Will exceed 2.5 m/s.

(b) Will be less than 1.5.

(c) Will lie between 1.5 and 2.5.

# 5.   THE NORMAL APPROXIMATION TO THE BINOMIAL

The binomial distribution, introduced in Chapter 5, pertains to the number of successes $X$ in $n$ independent trials of an experiment. When the success probability $p$ is not too near 0 or 1 and the number of trials is large, the normal distribution serves as a good approximation to the binomial probabilities. Bypassing the mathematical proof, we concentrate on illustrating the manner in which this approximation works.

Figure 16 presents the binomial distribution for the number of trials $n$ being 5, 12, and 25 when $p = .4$ Notice how the distribution begins to assume the distinctive bell shape for increasing $n$. Even though the binomial distributions with $p = .4$ are not symmetric, the lack of symmetry becomes negligible for large $n$.

Figure 16    The binomial distributions for $p = .4$ and $n = 5, 12, 25$.

Figure 17 presents the binomial distribution with $p = .4$ but with $n$ increased to 40. The normal distribution having the same mean $\mu = np = 40 \times .4 = 16$ and variance $\sigma^2 = np(1 - p) = 40 \times .4 \times .6 = 9.6$ is also shown. The approximation is quite good. The approximation illustrated in Figure 17 provides the clue on how to approximate the binomial probability

$$P[X = x] = \binom{n}{x} p^x (1 - p)^{n - x}$$

Figure 17   The binomial distribution for $p = .4$ and $n = 40$ along with the normal density having the same mean 16 and standard deviation 3.10.

by a normal probability. The normal probability assigned to a single value $x$ is zero. However, as shown in Figure 18, the probability assigned to the interval $x - \frac{1}{2}$ to $x + \frac{1}{2}$ is the appropriate comparison. The addition and subtraction of $\frac{1}{2}$ is called the **continuity correction**.

Figure 18   Idea of continuity correction.

The idea of the continuity correction is to approximate the rectangle with area $\binom{n}{x} p^x (1 - p)^{n-x}$ by the area under a normal curve. For $n = 15$ and $p = .4$, the binomial distribution assigns

$$P[X = 7] = .787 - .610 = .177$$

Recall from Chapter 5 that the binomial distribution has

$$\text{Mean} = np = 15(.4) = 6$$
$$\text{sd} = \sqrt{np(1 - p)} = \sqrt{15(.4)(.6)} = 1.897$$

To obtain an approximation, we select the normal distribution with the same mean, $\mu = 6$, and same $\sigma = 1.897$. The normal approximation is then the probability assigned to the interval $7 - \frac{1}{2}$ to $7 + \frac{1}{2}$.

$$P[\, 6.5 < X < 7.5 \,] = P \left[ \frac{6.5 - 6}{1.897} < \frac{X - 6}{1.897} < \frac{7.5 - 6}{1.897} \right]$$

$$\approx P[\, .264 < Z < .791 \,] = .7855 - .6041 = .1814$$

Of course $n = 15$ is small, so the approximation .1814 differs somewhat from the exact value .177. However, the accuracy of the approximation increases with the number of trials $n$.

The normal approximation to the binomial applies when $n$ is large and the success probability $p$ is not too close to 0 or 1. The binomial probability of $[\, a \leq X \leq b \,]$ is approximated by the normal probability of $\left[ a - \frac{1}{2} \leq X \leq b + \frac{1}{2} \right]$.

---

### The Normal Approximation to the Binomial

When $np$ and $n(1 - p)$ are both large,[3] say, greater than 15, the binomial distribution is well approximated by the normal distribution having mean $= np$ and sd $= \sqrt{np(1 - p)}$. That is,

$$Z = \frac{X - np}{\sqrt{np(1 - p)}} \qquad \text{is approximately } N(0, 1)$$

---

**Example 9**   Normal Approximation to the Binomial

In a large scale statewide survey concerning television viewing by children, about 40% of babies a few months old were reported to watch TV regularly. In a future random sample of 150 babies in this age group, let $X$ be the number who regularly watch TV. Approximate the probability that

(a)   $X$ is between 52 and 71 both inclusive.

(b)   $X$ is greater than 67.

SOLUTION

(a)   Because the population is large, $X$ has a binomial distribution with $p = .4$. To obtain a normal approximation, we first calculate the mean and standard deviation of $X$. Since $n = 150$,

$$\text{Mean} = np = 150(.4) = 60$$

$$\text{sd} = \sqrt{np(1 - p)} = \sqrt{150(.4)(.6)} = \sqrt{36} = 6$$

---

[3]To be consistent with the rule in Chapter 13 some suggest using the normal approximation when $np$ is greater than 5. The exact calculation of binomial probabilities using statistical software is always preferable.

The standardized variable is

$$Z = \frac{X - 60}{6}$$

The event $[\, 52 \leq X \leq 71 \,]$ includes both endpoints. The appropriate continuity correction is to *subtract* $\frac{1}{2}$ from the lower end and *add* $\frac{1}{2}$ to the upper end. We then approximate

$$P[\, 51.5 \leq X \leq 71.5 \,] = P\left[ \frac{51.5 - 60}{6} \leq \frac{X - 60}{6} \leq \frac{71.5 - 60}{6} \right]$$

$$\approx P[\, -1.417 \leq Z \leq 1.917 \,]$$

where $Z$ is standard normal. From the normal table, we interpolate

$$P[\, -1.417 \leq Z \leq 1.917 \,] = .9724 - .0782 = .8942$$

and approximate $P[\, 52 \leq X \leq 71 \,]$ by the normal probability .8942.

(b) For $[\, X > 67 \,]$ we reason that 67 is not included so that $[\, X \geq 67 + .5 \,]$ or $[\, X \geq 67.5 \,]$ is the event of interest:

$$P[\, X \geq 67.5 \,] = P\left[ \frac{X - 60}{6} \geq \frac{67.5 - 60}{6} \right]$$

$$\approx P[\, Z \geq 1.25 \,] = 1 - .8944$$

The normal approximation to the binomial gives $P[\, X > 67 \,] \approx$ .1056.

**Example 10**  Normal Probability Approximation to the Prevalence of Identity Fraud

Smartphone users have a higher risk, than average, of being victims of identity fraud. In 2011, it was reported that about 7% of smartphone users[4] were victims. Over half of the victims did not lock their devices with a password.

Suppose that the same rate of theft holds today. Find the probability, that out of a random sample of 250 smartphone users, the number that are victims of identity fraud in the previous year will be

(a)  10 or fewer.

(b)  19 or more.

SOLUTION  Let $X$ be the number of smartphone users, in the sample, who were victims of identity fraud last year. Under the assumption that the proportion remains

---

[4]A Javelin Strategy and Research report: *2012 Identity Fraud Report*.

at .07, the distribution of $X$ is modeled by the binomial distribution with $n = 250$ and $p = .07$. Since $n$ is large and

$$np = 250 \times .07 = 17.5, \qquad \sqrt{np(1 - p)} = \sqrt{17.5 \times .93} = 4.034$$

The binomial distribution is approximately, $N(17.5, 4.034)$.

(a) Because $X$ is a count, using the continuity correction, we have

$$P[X \leq 10] = P\left[Z \leq \frac{10.5 - 17.5}{4.034}\right]$$

$$= P[Z \leq -1.859] = .0314$$

In just over 3 in 100 times would a sample of size 250 smartphone users contain 10 or fewer victims of identity fraud.

(b) Because $X$ is a count, the event $[X \geq 19]$ is the same as $[X \geq 18.5]$. Using the continuity correction,

$$P[X \geq 19] = P\left[Z \geq \frac{18.5 - 17.5}{4.034}\right]$$

$$= 1 - .5979 = .4021$$

About 40% of the samples of size 250 would contain 19 or more fraud victims.

**Remark:** If the object is to calculate binomial probabilities, today the best practice is to evaluate them directly using an established statistical computing package. The numerical details need not concern us. However, the fact that

$$\frac{X - np}{\sqrt{np(1 - p)}} \qquad \text{is approximately normal}$$

when $np$ and $n(1 - p)$ are both large remains important. We will use it in later chapters when discussing inferences about proportions. Because the continuity correction will not be crucial, we will drop it for the sake of simplicity. Beyond this chapter, we employ the normal approximation but *without* the continuity correction.

## Exercises

**6.39** Let the number of successes $X$ have a binomial distribution with $n = 25$ and $p = .6$

(a) Find the exact probabilities of each of the following:

$$X = 17 \quad 11 \leq X \leq 18 \quad 11 < X < 18$$

(b) Apply the normal approximation to each situation in part (a).

**6.40** Let the number of successes $X$ have a binomial distribution with $n = 25$ and $p = .4$

(a) Find the exact probability of each of the following:

$$X = 11 \qquad 6 \le X \le 12 \qquad 6 < X < 12$$

(b) Apply the normal approximation to each situation in part (a).

**6.41** A study by the National Endowment of the Arts revealed that 19.7% of adults age 18–24 played a musical instrument in the past 12 months. Suppose that is still the current rate. What is the normal approximation to the probability, that in a random sample of 100 adults age 18–24, the number who played a musical instrument in the last 12 months will be

(a) 16 or fewer.

(b) more than 26.

(c) between 15 and 29 inclusive.

**6.42** A recent study reported that 54% of the adults in the United States drink at least one cup of coffee a day. Suppose that this is still the current rate. What is the normal approximation to the probability that, in a random sample of 1000 adults, the number that drink at least one cup a day will be

(a) less than 519.

(b) Equal to or more than 556.

**6.43** State whether or not the normal approximation to the binomial is appropriate in each of the following situations.

(a) $n = 90, \quad p = .24$

(b) $n = 100, \quad p = .03$

(c) $n = 120, \quad p = .98$

(d) $n = 61, \quad p = .40$

**6.44** State whether or not the normal approximation to the binomial is appropriate in each of the following situations.

(a) $n = 500, \quad p = .23$

(b) $n = 10, \quad p = .40$

(c) $n = 300, \quad p = .02$

(d) $n = 150, \quad p = .97$

(e) $n = 100, \quad p = .71$

**6.45** Copy Figure 16 and add the standard score scale $z = (x - np)/\sqrt{np(1 - p)}$ underneath the $x$-axis for $n = 5, 12, 25$. Notice how the distributions center on zero and most of the probability lies between $z = -2$ and $z = 2$.

**6.46** The median age of residents of the United States is 37.2 years. If a survey of 200 residents is taken, approximate the probability that at least 110 will be under 37.2 years of age.

**6.47** The unemployment rate in a city is 7.9%. A sample of 300 persons is selected from the labor force. Approximate the probability that

(a) Less than 18 unemployed persons are in the sample.

(b) More than 30 unemployed persons are in the sample.

**6.48** A survey reports that 96% of the people think that violence has increased in the past five years. Out of a random sample of 50 persons, 48 expressed the opinion that citizens have become more violent in the past five years. Does the normal approximation seem appropriate for $X$ = the number of persons who expressed the opinion that citizens have become more violent in the past five years? Explain.

**6.49** According to the U.S. Statistical Abstract 2012, about 27.2% of persons age 18–24 participated in charity work in the past year. Among a sample of 64 persons in this age group, find the probability that 20 or more participated in charity work.

**6.50** The weekly amount spent by a small company for in-state travel has approximately a normal distribution with mean $1450 and standard deviation $220. What is the probability that the actual expenses will exceed $1560 in 20 or more weeks during the next year?

**6.51** With reference to Exercise 6.50, calculate the probability that the actual expenses would exceed $1500 for between 18 and 24 weeks, inclusive during the next year.

**6.52** In a large midwestern university, 30% of the students live in apartments. If 200 students are randomly selected, find the probability that the number of them living in apartments will be between 55 and 70 inclusive.

**6.53** According to a study of mobility, 26% of U.S. residents in the age group 20 to 24 years moved to different housing in 2010 from where they lived in 2009. (Based on *Statistical Abstract of the U.S. 2012* Table 30.) If the same percentage holds today, give the approximate probability that in a random sample of 100 residents 20 to 24 years old, there will be 34 or more persons who have moved in the past year.

**6.54** Suppose that 20% of the trees in a forest are infested with a certain type of parasite.

(a) What is the probability that, in a random sample of 300 trees, the number of trees having the parasite will be between 49 and 71 inclusive?

(b) After sampling 300 trees, suppose that 72 trees are found to have the parasite. Does this provide strong evidence that the population proportion is higher than 20%? Base your answer on $P[X \geq 72]$ when 20% are infested.

# *6.  CHECKING THE PLAUSIBILITY OF A NORMAL MODEL

Does a normal distribution serve as a reasonable model for the population that produced the sample? One reason for our interest in this question is that many commonly used statistical procedures require the population to be nearly normal. If a normal distribution is tentatively assumed to be a plausible model, the investigator must still check this assumption once the sample data are obtained.

Although they involve subjective judgment, graphical procedures prove most helpful in detecting serious departures from normality. Histograms can be inspected for lack of symmetry. The thickness of the tails can be checked for conformance with the normal by comparing the proportions of observations in the intervals $(\bar{x} - s, \bar{x} + s)$, $(\bar{x} - 2s, \bar{x} + 2s)$, and $(\bar{x} - 3s, \bar{x} + 3s)$ with those suggested by the empirical guidelines for the bell-shaped (normal) distribution.

A more effective way to check the plausibility of a normal model is to construct a special graph, called a **normal-scores plot**, of the sample data. In order to describe this method, we first explain the meaning of normal scores, indicate how the plot is constructed, and then explain how to interpret the plot. For an easy explanation of the ideas, we work with a small sample size. In practical applications, at least 15 or 20 observations are needed to detect a meaningful pattern in the plot.

The term **normal scores** refers to an idealized sample from the standard normal distribution—namely, the $z$ values that divide the standard normal distribution into equal-probability intervals. For purposes of discussion, suppose the sample size is $n = 4$. Figure 19 shows the standard normal distribution where four points are located on the $z$ axis so the distribution is divided into five segments of equal probability $\frac{1}{5} = .2$ These four points, denoted by $m_1$, $m_2$, $m_3$, and $m_4$, are precisely the normal scores for a sample of size $n = 4$. Using Appendix B, Table 4, we find that

$$m_1 = -.84$$
$$m_2 = -.25$$
$$m_3 = .25$$
$$m_4 = .84$$

Figure 19   The $N(0, 1)$ distribution and the normal scores for $n = 4$.

A normal-scores plot allows us to visually assess how well a sample mimics the idealized normal sample. To construct a normal-scores plot:

1.    Order the sample data from smallest to largest.
2.    Obtain the normal scores.
3.    Pair the $i$th largest observation with the $i$th largest normal score and plot the pairs in a graph.

**Example 11**    Making a Normal-Scores Plot for Sample Size 4

Suppose a random sample of size 4 has produced the observations 68, 82, 44, and 75. Construct a normal-scores plot.

SOLUTION    The ordered observations and the normal scores are shown in Table 2, and the normal-scores plot of the data is given in Figure 20.

**TABLE 2**   Normal Scores

| Normal Scores | Ordered Sample |
|---|---|
| $m_1 = -.84$ | 44 |
| $m_2 = -.25$ | 68 |
| $m_3 = .25$ | 75 |
| $m_4 = .84$ | 82 |

Figure 20   Normal-scores plot of Table 2 data.

## *6.1 INTERPRETATION OF THE PLOT

How does the normal-scores plot of the data help in checking normality? To explain the main idea, we continue our discussion with the data of Example 11. Let $\mu$ and $\sigma$ denote the mean and standard deviation of the population from which the sample was obtained. The normal scores that are the idealized $z$ observations can then be converted to the $x$ scale by the usual relation $x = \mu + \sigma z$. The actual $x$ observations and the corresponding idealized observations are given in Table 3. If the population were indeed normal, we would expect the two columns of Table 3 to be close. In other words, a plot of the observed $x$ values versus the normal scores would produce a straight-line pattern, where the intercept of the line would indicate the value of $\mu$ and the slope of the line would indicate $\sigma$.

**TABLE 3**  Idealized Sample

| Observed x Values | Idealized x Values |
|---|---|
| 44 | $\mu + \sigma\, m_1$ |
| 68 | $\mu + \sigma\, m_2$ |
| 75 | $\mu + \sigma\, m_3$ |
| 82 | $\mu + \sigma\, m_4$ |

> A straight-line pattern in a normal-scores plot supports the plausibility of a normal model. A curve appearance indicates a departure from normality.

A normal-scores plot is easily obtained using software packages. The next example illustrates a straight-line pattern consistent with the assumption that the population is normal.

**Example 12**  A Normal-Scores Plot that Looks Normal

Consider the data on the growth of salmon in Table D.7 of the Data Bank. We plot the freshwater growth of female salmon measured in terms of width of growth rings in hundredths of an inch. Create a normal-scores plot of these data and comment on the pattern.

SOLUTION  We use MINITAB software as described in Exercise 6.55 to make the normal-scores plot in Figure 21 on page 262.

Notice that the plot conforms quite well to the straight-line pattern expected for normal observations. The assumption of a normal distribution seems quite reasonable.

Figure 21  A normal-scores plot of the female salmon growth in freshwater.

The MINITAB package that produced Figure 21 uses one of the many slight variants of the normal scores above but the plots are similar if the sample size is greater than 20.

## Exercises

### The Following Exercise Require a Computer

**6.55** *Normal-scores plot.* Use a computer program to make a normal-scores plot for the volume of timber data in Table 4. (Courtesy of Professor Alan Ek) Comment on the departure from normality displayed by the normal-scores plot.

**TABLE 4**  Volume of Timber in Cords

| | | | | |
|---|---|---|---|---|
| 39.3 | 14.8 | 6.3 | .9 | 6.5 |
| 3.5 | 8.3 | 10.0 | 1.3 | 7.1 |
| 6.0 | 17.1 | 16.8 | .7 | 7.9 |
| 2.7 | 26.2 | 24.3 | 17.7 | 3.2 |
| 7.4 | 6.6 | 5.2 | 8.3 | 5.9 |
| 3.5 | 8.3 | 44.8 | 8.3 | 13.4 |
| 19.4 | 19.0 | 14.1 | 1.9 | 12.0 |
| 19.7 | 10.3 | 3.4 | 16.7 | 4.3 |
| 1.0 | 7.6 | 28.3 | 26.2 | 31.7 |
| 8.7 | 18.9 | 3.4 | 10.0 | |

We illustrate a normal-scores plot using MINITAB. With the data set in column 1, the MINITAB commands

> **Calc > Calculator.**
>
> Type C2 in **Store**. Type *NSCOR(C1)* in **Expression**. Click **OK**.
>
> **Graph > Scatterplot**. Select *Simple*. Type C1 under **Y** variables and C2 under **X** variables. Click **OK**.

will create a normal-scores plot for the observations in C1. (MINITAB uses a variant of the normal scores, $m_i$, that we defined.)

**\*6.56** Use MINITAB or another package program to make a normal-scores plot of the malt extract data in Table D.8 of the Data Bank.

**\*6.57** Use MINITAB or another package program to make a normal-scores plot of the College Qualification Test (CQT) data in Exercise 2.22. Comment on the pattern.

**\*6.58** Use MINITAB or another package program to make a normal-scores plot of the lizards' speed test in Exercise 2.19. Comment on the pattern.

## \*7.  TRANSFORMING OBSERVATIONS TO ATTAIN NEAR NORMALITY

A valid application of many powerful techniques of statistical inference, especially those suited to small or moderate samples, requires that the population distribution be reasonably close to normal. When the sample measurements appear to have been taken from a population that departs drastically from normality, an appropriate conversion to a new variable may bring the distribution close to normal. Efficient techniques can then be safely applied to the converted data, whereas their application to the original data would have been questionable. Inferential methods requiring the assumption of normality are discussed in later chapters. The goal of our discussion here is to show how a transformation can improve the approximation to a normal distribution.

There is no rule for determining the best transformation in a given situation. For any data set that does not have a symmetric histogram, we consider a variety of transformations.

---

### Some Useful Transformations

| Make large values larger: | Make large values smaller: |
|---|---|
| $x^3$, $x^2$ | $\sqrt{x}$, $\sqrt[4]{x}$, $\log_e x$, $\frac{1}{x}$ |

---

You may recall that $\log_e x$ is the natural logarithm. Fortunately, computers easily calculate and order the transformed values, so that several transformations in a list can be quickly tested. Note, however, that the observations must be positive if we intend to use $\sqrt{x}$, $\sqrt[4]{x}$, and $\log_e x$.

We search for a good transformation by visually comparing graphs of the data with those typical of normal populations. If the data set contains a few numbers that appear to be detached far to the right, $\sqrt{x}$, $\sqrt[4]{x}$, and $\log_e x$, or negative powers that would pull these stragglers closer to the other data points should be considered.

---

**Example 13**  A Transformation to Improve Normality

A forester records the volume of timber, measured in cords, for 49 plots selected in a large forest. The data are given in Table 4 and the corresponding histogram appears in Figure 22*a* on page 265. The histogram exhibits a long tail to the right, so it is reasonable to consider the transformations $\sqrt{x}$, $\sqrt[3]{x}$, $\log_e x$, and $1/x$. Transform the data to near normality.

SOLUTION    The most satisfactory result, obtained by taking the square root twice,

$$\text{Transformed data} = \sqrt[4]{\text{Volume}}$$

is illustrated in Table 5 and Figure 22b. The latter histogram more nearly resembles a symmetric bell-shaped pattern expected for normal populations.

**TABLE 5    The Transformed Data $\sqrt[4]{\text{Volume}}$**

| | | | | |
|---|---|---|---|---|
| 2.50 | 1.96 | 1.58 | .97 | 1.60 |
| 1.37 | 1.70 | 1.78 | 1.07 | 1.63 |
| 1.57 | 2.03 | 2.02 | .91 | 1.68 |
| 1.28 | 2.26 | 2.22 | 2.05 | 1.34 |
| 1.64 | 1.60 | 1.51 | 1.70 | 1.56 |
| 1.37 | 1.70 | 2.59 | 1.70 | 1.91 |
| 2.10 | 2.09 | 1.94 | 1.17 | 1.86 |
| 2.11 | 1.79 | 1.36 | 2.02 | 1.44 |
| 1.00 | 1.66 | 2.31 | 2.26 | 2.37 |
| 1.72 | 2.09 | 1.36 | 1.78 | |

Tony Baggett/iStockphoto

HOW MUCH TIMBER IS IN THIS FOREST?
The volume of timber available for making lumber can only be estimated by sampling the number of trees in randomly selected plots within the forest. The distribution of tree size must also be taken into account.

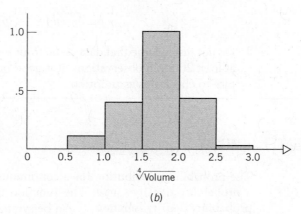

Figure 22   An illustration of the transformation technique. (*a*) Histogram of timber volume. (*b*) Histogram of $\sqrt[4]{\text{Volume}}$.

## Exercises

### The Following Exercises Require a Computer.

*6.59 *Transformations and normal-scores plots*.
The MINITAB computer language makes it possible to easily transform data. With the data already set in column 1, the commands

Dialog box:

**Calc > Calculator**. Type C2 in **Store**. Type *LOGE*(C1) in **Expression**. Click **OK**.

**Calc > Calculator**. Type C3 in **Store**. Type *SQRT*(C1) in **Expression**. Click **OK**.

**Calc > Calculator**. Type C4 in **Store**. Type *SQRT*(C3) in **Expression**. Click **OK**.

will place the natural logarithm $\log_e x$ in C2, $\sqrt{x}$ in C3, and $x^{1/4}$ in C4. Normal-scores plots can then be constructed as in Exercise 6.55.

Refer to the lightning data in Exercise 2.121 of Chapter 2. Make a normal-scores plot of the

(a) Original observations.

(b) Fourth root of the original observations.

(c) Comment on the quality of the approximation by a normal distribution in both cases.

*6.60 Refer to the volume of timber data in Example 13.

(a) Make a normal-scores plot of the original data.

(b) Make a normal-scores plot of the fourth root of the data.

(c) Compare the two plots and comment.

## USING STATISTICS WISELY

1. A sketch of the bell-shaped normal curve and the area of interest can prevent blunders when determining probabilities and percentiles.

2. Never apply the normal approximation to the binomial, treating

$$Z = \frac{X - np}{\sqrt{np(1 - p)}}$$

as standard normal, when the expected number of successes (or failures) is too small. That is, when either

$$np \quad \text{or} \quad n(1 - p) \text{ is 15 or less}$$

3. Do not just assume that data come from a normal distribution. When there are at least 20 to 25 observations, it is good practice to construct a normal-scores plot to check this assumption.

## KEY IDEAS AND FORMULAS

The probability distribution for a continuous random variable $X$ is specified by a probability density curve. The function that specifies this curve is called a probability density function. It can be symmetric about the mean of $X$ or it can be skewed, meaning that it has a long tail to either the left or the right.

The probability that $X$ lies in an interval from $a$ to $b$ is determined by the area under the probability density curve between $a$ and $b$. The total area under the curve is 1, and the curve is never negative.

The population 100 $p$-th percentile is an $x$ value that has probability $p$ to its left and probability $1 - p$ to its right.

When $X$ has mean $\mu$ and standard deviation $\sigma$, the standardized variable

$$Z = \frac{X - \mu}{\sigma}$$

has mean 0 and standard deviation 1.

The normal distribution has a symmetric bell-shaped curve centered at the mean. The intervals extending one, two, and three standard deviations around the mean contain the probabilities .683, .954, and .997, respectively.

If $X$ is normally distributed with mean $\mu$ and standard deviation $\sigma$, then

$$Z = \frac{X - \mu}{\sigma}$$

has the standard normal distribution.

When the number of trials $n$ is large and the success probability $p$ is not too near 0 or 1, the binomial distribution is well approximated by a normal distribution with mean $np$ and sd $= \sqrt{np(1 - p)}$. Specifically, the probabilities for a binomial variable $X$ can be approximately calculated by treating

$$Z = \frac{X - np}{\sqrt{np(1 - p)}}$$

as standard normal. For a moderate number of trials $n$, the approximation is improved by appropriately adjusting by $\frac{1}{2}$ called a **continuity correction**.

The **normal scores** are an ideal sample from a standard normal distribution. Plotting each ordered observation versus the corresponding normal score creates a **normal-scores plot**, which provides a visual check for possible departures from a normal distribution.

Transformation of the measurement scale often helps to convert a long-tailed distribution to one that resembles a normal distribution.

## TECHNOLOGY

### *Probability and percentiles for the standard normal and general normal distribution*

#### MINITAB

MINITAB uses the same steps for calculations with the standard normal and cases of other means and standard deviations. We illustrate with the calculation of $P[\, X \, \leq \, 8 \,]$ when $X$ is normal with mean 5 and standard deviation 12.5.

> **Calc > Probability Distributions > Normal.**
>
> Select **Cumulative probability**. Type *5* in **Mean**.
>
> Type *12.5* in **Standard deviation**.
>
> Select **Input Constant** and type *8*. Click **OK**.

The default settings Mean 0 and Standard deviation 1 simplify the steps for obtaining standard normal probabilities.

The inverse problem of finding $b$ so that $P[\, X \, \leq \, b \,] = a$, where $a$ is a specified probability, is illustrated with finding $b$ so that $P[\, X \, \leq \, b \,] = .9700$ when $X$ is normal with mean 5 and standard deviation 12.5.

> **Calc > Probability Distributions > Normal.**
>
> Select **Inverse cummulative probability**. Type *5* in **Mean**.
>
> Type *12.5* in **Standard deviation**.
>
> Select **Input Constant** and type *.9700*. Click **OK**.

#### EXCEL

EXCEL uses the function **NORM.S.DIST** for standard normal probabilities and **NORM.DIST** for a general normal distribution. We illustrate with the calculation of $P[\, X \, \leq \, 8 \,]$ when $X$ is normal with mean 5 and standard deviation 12.5.

Click the **Formulas** tab and then **More Functions**.

Choose **Statistical** and then **NORM.DIST**. Click **OK**.

Type 8 in **X**, 5 in **Mean**, *12.5* in **Standard_dev** and *TRUE* in **Cumulative**. Click **OK**.

The inverse problem of finding $b$ so that $P[X \le b] = a$, where $a$ is a specified probability, is illustrated with finding $b$ so that $P[X \le b] = .9700$ when $X$ is normal with mean 5 and standard deviation 12.5.

Click the **Formulas** tab and then **More Functions**.

Choose **Statistical**, and then **NORM.INV**. Click **OK**.

Type *.9700* in **Probability**, 5 in **Mean**, *12.5* in **Standard_dev**. Click **OK**.

To solve the standard normal inverse problem replace **NORM.INV** by **NORM.S.INV**.

### TI-84/-83 PLUS

We illustrate with the calculation of $P[X \le 8]$ when $X$ is normal with mean 5 and standard deviation 12.5.

In the Home screen, press **2$^{nd}$ VARS**

From the **DISTR** menu, select **2: Normalcdf(**.

Type entries to obtain **Normalcdf(−1E99, 8, 5, 12.5)**, where $(-)\ 10^{99}$ produces the entry $-1E99$.

Then press **ENTER**.

The inverse problem of finding $b$ so that $P[X \le b] = a$, where $a$ is a specified probability, is illustrated with finding $b$ so that $P[X \le b] = .9700$ when $X$ is normal with mean 5 and standard deviation 12.5.

In the Home screen, press **2$^{nd}$ VARS**.

From the **DISTR** mean, select **3: InvNorm(**.

Type entries to obtain **InvNorm( .9700, 5, 12.5 )**.

Then press **ENTER**.

---

## REVIEW EXERCISES

**6.61** Determine (a) the median and (b) the quartiles for the distribution shown in the following illustration.

**6.62** For $X$ having the density in Exercise 6.61, find (a) $P[X > .7]$ (b) $P[.5 \le X \le .7]$ and (c) $P[.5 < X < .7]$

**6.63** In the context of the height of red pine seedlings presented at the front of the chapter, describe the reasoning that leads from a histogram to the concept of a probability density curve. (Think of successive histograms based on 100 heights, 500 heights, 1456 heights, and then an unlimited number.)

**6.64** For a standard normal random variable $Z$, find

(a) $P[Z < 1.56]$

(b) $P[Z > 1.245]$

(c) $P[.61 < Z < 1.92]$

(d) $P[-1.47 < Z < 1.055]$

**6.65** For the standard normal distribution, find the value $z$ such that

(a) Area to its left is .0838.

(b) Area to its left is .047.

(c) Area to its right is .2611.

(d) Area to its right is .12.

**6.66** Find the 20th, 40th, 60th, and 80th percentiles of the standard normal distribution.

**6.67** If $Z$ is a standard normal random variable, what is the probability that

(a) $Z$ exceeds .62?

(b) $Z$ lies in the interval $(-1.40, 1.40)$?

(c) $|Z|$ exceeds 3.0?

(d) $|Z|$ is less than 2.0?

**6.68** According to Example 12, a normal distribution with mean 115 and standard deviation 22 hundredths of an inch describes variation in female salmon growth in freshwater.

(a) If a newly caught female salmon has growth 108, what is the corresponding standardized score?

(b) If a standardized score is $-.8$, what is the growth measurement?

(c) Find the interval of standardized scores corresponding to the growth measurements 105 to 128.

(d) Find the interval of growth measurements corresponding to the standardized scores of $-1.5$ to 1.5.

**6.69** The bell-shaped histogram for the heights of three-year-old red pine seedlings on page 232 is consistent with the assumption of a normal distribution having mean $= 280$ and sd $= 58$ millimeters. Let $X$ denote the height, at three years of age, of the next red pine that will be measured. Find

(a) $P[X < 337]$

(b) $P[X < 240]$

(c) $P[X > 230]$

(d) $P[X > 90]$

(e) $P[235 < X < 335]$

(f) $P[305 < X < 405]$

**6.70** If $X$ has a normal distribution with $\mu = 100$ and $\sigma = 5$, find $b$ such that

(a) $P[X < b] = .6700$

(b) $P[X > b] = .0110$

(c) $P[|X - 100| < b] = .966$

**6.71** Suppose that a student's verbal score $X$ from next year's Graduate Record Exam can be considered an observation from a normal population having mean 499 and standard deviation 120. Find

(a) $P[X > 600]$

(b) 90th percentile of the distribution.

(c) Probability that the student scores below 400.

**6.72** The lifting capacities of a class of industrial workers are normally distributed with mean 65 pounds and standard deviation 8 pounds. What proportion of these workers can lift an 80-pound load?

**6.73** The bonding strength of a drop of plastic glue is normally distributed with mean 100 pounds and standard deviation 8 pounds. A broken plastic strip is repaired with a drop of this glue and then subjected to a test load of 90 pounds. What is the probability that the bonding will fail?

**6.74** *Grading on a curve.* The scores on an examination are normally distributed with mean $\mu = 70$ and standard deviation $\sigma = 8$. Suppose that the instructor decides to assign letter grades according to the following scheme (left endpoint included).

| Scores | Grade |
| --- | --- |
| Less than 58 | F |
| 58 to 66 | D |
| 66 to 74 | C |
| 74 to 82 | B |
| 82 and above | A |

Find the percentage of students in each grade category.

**6.75**  Suppose the duration of trouble-free operation of a new robotic vacuum cleaner is normally distributed with mean 750 days and standard deviation 100 days.

(a) What is the probability that the vacuum cleaner will work for at least two years without trouble?

(b) The company wishes to set the warranty period so that no more than 10% of the vacuum cleaners would need repair services while under warranty. How long a warranty period must be set?

**6.76**  Suppose the amount of a popular sport drink in bottles leaving the filling machine has a normal distribution with mean 101.5 milliliters (ml) and standard deviation 1.6 ml.

(a) If the bottles are labeled 100 ml, what proportion of the bottles contain less than the labeled amount.

(b) If only 5% of the bottles have contents that exceed a specified amount $v$, what is the value of $v$?

**6.77**  Suppose the amount of sun block lotion in plastic bottles leaving a filling machine has a normal distribution. The bottles are labeled 300 milliliters (ml) but the actual mean is 302 ml and the standard deviation is 2 ml.

(a) What is the probability that an individual bottle will contain less than 299 ml?

(b) If only 5% of the bottles have contents that exceed a specified amount $v$, what is the value of $v$?

**\*6.78**  *A property of the normal distribution.* Suppose the random variable $X$ is normally distributed with mean $\mu$ and standard deviation $\sigma$. If $Y$ is a linear function of $X$—that is, $Y = a + bX$, where $a$ and $b$ are constants—then $Y$ is also normally distributed with

$$\text{Mean} = a + b\mu$$

$$\text{sd} = |b|\sigma$$

For instance, if $X$ is distributed as $N(25, 2)$ and $Y = 7 - 3X$, then the distribution of $Y$ is normal with Mean $= 7 - 3(25) = -68$ and sd $= |-3| \times 2 = 6$.

(a) At the "low" setting of a water heater, the temperature $X$ of water is normally distributed with Mean $= 102°F$ and sd $= 4°F$. If $Y$ refers to the temperature measurement in the centigrade

scale, that is, $Y = \frac{5}{9}(X - 32)$, what is the distribution of $Y$?

(b) Referring to part (a), find the probability of $[35 \leq Y \leq 42]$.

> **Remark:**  The relation between a general normal and the standard normal is only a special case of this property. Specifically, the standardized variable $Z$ is the linear function.
>
> $$Z = \frac{X - \mu}{\sigma} = -\frac{\mu}{\sigma} + \frac{1}{\sigma}X$$
>
> where $Z$ has
>
> $$\text{Mean} = -\frac{\mu}{\sigma} + \frac{1}{\sigma}\mu = 0$$
>
> $$\text{sd} = \frac{1}{\sigma}\sigma = 1$$

**6.79**  Let $X$ denote the number of successes in $n$ Bernoulli trials with a success probability of $p$.

(a) Find the exact probabilities of each of the following:
(i)   $X \leq 7$ when $n = 25$, $p = .4$
(ii)  $11 \leq X \leq 16$ when $n = 20$, $p = .7$
(iii) $X \geq 9$ when $n = 16$, $p = .5$

(b) Use a normal approximation for each situation in part (a).

**6.80**  It is known from past experience that 7% of the tax bills are paid late. If 20,000 tax bills are sent out, approximate the probability that:

(a) Less than 1350 are paid late.

(b) 1480 or more are paid late.

**6.81**  A particular program, say, program $A$, previously drew 30% of the television audience. To determine if a recent rescheduling of the programs on a competing channel has adversely affected the audience of program $A$, a random sample of 400 viewers are asked whether or not they currently watch this program.

(a) If the percentage of viewers watching program $A$ has not changed, what is the probability that fewer than 103 out of a sample of 400 will be found to watch the program?

(b) If the number of viewers of the program is actually found to be less than 103, will this strongly support the suspicion that the population percentage has dropped? Use your calculation from part (a).

**6.82**  The number of successes $X$ has a binomial distribution. State whether or not the normal approximation is appropriate in each of the following situations:

(a)  $n = 400$, $p = .23$
(b)  $n = 20$, $p = .03$
(c)  $n = 90$, $p = .98$.

**6.83**  Because 10% of the reservation holders are "no-shows," a U.S. airline sells 400 tickets for a flight that can accommodate 370 passengers.

(a)  Find the approximate probability that one or more reservation holders will not be accommodated on the flight.

(b)  Find the approximate probability of fewer than 350 passengers on the flight.

**6.84**  On a Saturday afternoon, 147 customers are observed during check-out and the number paying by card, credit or debit, is recorded. Records from the store suggest that 43% of customers pay by card. Approximate the probability that:

(a)  More than 60 will pay by card.

(b)  Between 60 and 70, inclusive, will pay by card.

**6.85**  In all of William Shakespeare's works, he used 884,647[5] different words. Of these, 14,376 appeared only once. In 1985 a 429-word poem was discovered that may have been written by Shakespeare. To keep the probability calculations simple, assume that the choices between a new word and one from the list of 884,647 are independent for each of the 429 words. Approximate the probability that a new word will not be on the list, by the relative frequency of words used once.

(a)  Find the expected number of new words in the poem.

(b)  Use the normal approximation to the binomial to determine the probability of finding 12 or more new words in the poem. Use the continuity correction.

(c)  Use the normal approximation to the binomial to determine the probability of finding 2 or fewer new words in the poem. Use the continuity correction.

(d)  Use the normal approximation to the binomial to determine the probability of finding more than 2 but less than 12 new words in the poem. On the basis of your answer, decide if 9 = actual number of new words not in the list is consistent with Shakespeare having written the poem or if it contradicts this claim.

## The Following Exercises Require a Computer.

**\*6.86**  Refering to Exercise 6.55, use MINITAB or another package program to make a normal-scores plot of the computer anxiety scores in Table D.4 of the Data Bank.

**6.87**  Refer to Exercise 6.59 and to the yearly deaths due to lightening in Exercise 2.121. Make a normal-scores plot for the square root of the number of deaths. Comment on the agreement with the ideal straight-line pattern.

[5]See R. Thisted and B. Efron. "Did Shakespeare write a newlydiscovered poem?" *Biometrika*, **74**, (1987) pp. 445–455.

# 7

# Variation in Repeated Samples—Sampling Distributions

1 The Sampling Distribution of a Statistic
2 Distribution of the Sample Mean and the Central Limit Theorem

Image Source/Getty Images

Bowlers are well aware that their three-game averages are less variable than their single-game scores. Sample means always have less variability than individual observations.

## Bowling Averages

A bowler records individual game scores and the average for a three-game series.

At the heart of statistics lies the ideas of inference. They enable the investigator to argue from the particular observations in a sample to the general case. These generalizations are founded on an understanding of the manner in which variation in the population is transmitted, by sampling, to variation in statistics like the sample mean. This key concept is the subject of this chapter.

Typically, we are interested in learning about some numerical feature of the population, such as the proportion possessing a stated characteristic, the mean and standard deviation of the population, or some other numerical measure of center or variability.

> A numerical feature of a population is called a **parameter**.

The true value of a population parameter is an unknown constant. It can be correctly determined only by a complete study of the population. The concepts of statistical inference come into play whenever this is impossible or not practically feasible.

If we only have access to a sample from the population, our inferences about a parameter must then rest on an appropriate sample-based quantity. Whereas a parameter refers to some numerical characteristic of the population, a sample-based quantity is called a **statistic**.

> A **statistic** is a numerical valued function of the sample observations.

For example, the sample mean

$$\overline{X} = \frac{X_1 + \cdots + X_n}{n}$$

is a statistic because its numerical value can be computed once the sample data, consisting of the values of $X_1, \ldots, X_n$, are available. Likewise, the sample median and the sample standard deviation are also sample-based quantities so each is a statistic. Note that every statistic is a random variable. A sample-based quantity (statistic) must serve as our source of information about the value of a parameter. Three points are crucial:

1. Because a sample is only a part of the population, the numerical value of a statistic cannot be expected to give us the exact value of the parameter.

2. The observed value of a statistic depends on the particular sample that happens to be selected.

3. There will be some variability in the values of a statistic over different occasions of sampling.

A brief example helps illustrate these important points. Suppose an urban planner wishes to study the average commuting distance of workers from their home to their principal place of work. Here the statistical population consists of the commuting distances of all the workers in the city. The mean of this finite but vast and unrecorded set of numbers is called the population mean, which we denote by $\mu$. We want to learn about the parameter $\mu$ by collecting data from a sample of workers. Suppose 80 workers are randomly selected and the (sample) mean of their commuting distances is found to be $\bar{x} = 8.3$ miles. Evidently, the population mean $\mu$ cannot be claimed to be exactly 8.3 miles. If one were to observe another random sample of 80 workers, would the sample mean again be 8.3 miles? Obviously, we do not expect the two results to be identical. Because the commuting distances do vary in the population of workers, the sample mean would also vary on different occasions of sampling. In practice, we observe only one sample and correspondingly a single value of the sample mean such as $x = 8.3$. However, it is the idea of the variability of the $\bar{x}$ values in repeated sampling that contains the clue to determining how precisely we can hope to determine $\mu$ from the information on $\bar{x}$.

# 1. THE SAMPLING DISTRIBUTION OF A STATISTIC

A key concept is that the value of the sample mean, or any other statistic, varies as different samples are collected. Because any statistic, the sample mean in particular, varies from sample to sample, it is a random variable and has its own probability distribution. The variability of the statistic, in repeated sampling, is described by this probability distribution.

> The probability distribution of a statistic is called its **sampling distribution**.

The qualifier "sampling" indicates that the distribution is conceived in the context of repeated sampling from a population. We often drop the qualifier and simply say the **distribution of a statistic**.

Although in any given situation, we are limited to one sample and the corresponding single value for a statistic, over repeated samples from a population the statistic varies and has a sampling distribution. The sampling distribution of a statistic is determined from the distribution $f(x)$ that governs the population, and it also depends on the sample size $n$. Let us see how the distribution of $\overline{X}$ can be determined in a simple situation where the sample size is 2 and the population consists of 3 units.

**Example 1**    Illustration of a Sampling Distribution

A population consists of three housing units, where the value of $X$, the number of rooms for rent in each unit, is shown in the illustration.

Consider drawing a random sample of size 2 with replacement. That is, we select a unit at random, put it back, and then select another unit at random. Denote by $X_1$ and $X_2$ the observation of $X$ obtained in the first and second drawing, respectively. Find the sampling distribution of $\overline{X} = (X_1 + X_2)/2$.

SOLUTION    The population distribution of $X$ is given in Table 1, which simply formalizes the fact that each of the $X$ values 2, 3, and 4 occurs in $\frac{1}{3}$ of the population of the housing units.

TABLE 1    The Population Distribution

| $x$ | $f(x)$ |
|-----|--------|
| 2 | $\dfrac{1}{3}$ |
| 3 | $\dfrac{1}{3}$ |
| 4 | $\dfrac{1}{3}$ |

Because each unit is equally likely to be selected, the observation $X_1$ from the first drawing has the same distribution as given in Table 1. Since the sampling is with replacement, the second observation $X_2$ also has this same distribution.

The possible samples $(x_1, x_2)$ of size 2 and the corresponding values of $\overline{X}$ are

| $(x_1, x_2)$ | (2, 2) | (2, 3) | (2, 4) | (3, 2) | (3, 3) | (3, 4) | (4, 2) | (4, 3) | (4, 4) |
|---|---|---|---|---|---|---|---|---|---|
| $\overline{x} = \dfrac{x_1 + x_2}{2}$ | 2 | 2.5 | 3 | 2.5 | 3 | 3.5 | 3 | 3.5 | 4 |

The nine possible samples are equally likely so, for instance, $P[\overline{X} = 2.5] = \frac{2}{9}$. Continuing in this manner, we obtain the distribution of $\overline{X}$, which is given in Table 2.

This sampling distribution pertains to repeated selection of random samples of size 2 with replacement. It tells us that if the random sampling is repeated a large number of times, then in about $\frac{1}{9}$, or 11%, of the cases, the sample mean would be 2, and in $\frac{2}{9}$, or 22%, of the cases, it would be 2.5, and so on.

**TABLE 2**   The Probability Distribution of
$$\overline{X} = (X_1 + X_2)/2$$

| Value of $\overline{X}$ | Probability |
|:---:|:---:|
| 2 | $\dfrac{1}{9}$ |
| 2.5 | $\dfrac{2}{9}$ |
| 3 | $\dfrac{3}{9}$ |
| 3.5 | $\dfrac{2}{9}$ |
| 4 | $\dfrac{1}{9}$ |

Figure 1 shows the probability histograms of the distributions in Tables 1 and 2.

(a) Population distribution.

(b) Sampling distribution of $\overline{X} = (X_1 + X_2)/2$.

Figure 1   Idea of a sampling distribution.

In the context of Example 1, suppose instead the population consists of 300 housing units, of which 100 units have 2 rooms, 100 units have 3 rooms, and 100 units have 4 rooms for rent. When we sample two units from this large population, it makes little difference whether or not we replace the unit after the first selection. Each observation still has the same probability distribution—namely, $P[X = 2] = P[X = 3] = P[X = 4] = \frac{1}{3}$, which characterizes the population.

When the population is very large and the sample size relatively small, it is inconsequential whether or not a unit is replaced before the next unit is selected. Under these conditions, too, we refer to the observations as a random sample.

Before its value becomes available, each observation is modeled as a random variable. What are the key conditions required for a sample to be random?

> The observations $X_1, X_2, \ldots, X_n$ are a **random sample of size $n$ from the population distribution** if they result from independent selections and each observation, namely $X_1, X_2$, through $X_n$, has the same distribution as the population.

More concisely, under the independence and same distribution conditions, we refer to the observations as a **random sample**.

Because of variation in the population, the random sample will vary and so will $\overline{X}$, the sample median, or any other statistic.

The next example further explores sampling distributions by focusing first on the role of independence in constructing a sampling distribution and then emphasizing that this distribution too has a mean and variance.

**Example 2**   The Sample Mean and Median Each Have a Sampling Distribution

Given a characteristic $X$, a large population is described by the probability distribution

| $x$ | $f(x)$ |
|-----|--------|
| 0   | .2     |
| 3   | .3     |
| 12  | .5     |

Let $X_1, X_2, X_3$ be a random sample of size 3 from this distribution.

(a)   List all the possible samples and determine their probabilities.

(b)   Determine the sampling distribution of the sample mean.

(c)   Determine the sampling distribution of the sample median.

SOLUTION   (a)   Because we have a random sample, each of the three observations $X_1, X_2, X_3$ has the same distribution as the population and they are independent. So, the sample 0, 3, 0 has probability $(.2) \times (.3) \times (.2) = 0.12$. The calculations for all $3 \times 3 \times 3 = 27$ possible samples are given in Table 3.

**TABLE 3**  Sampling Distributions

Population
Distribution

| $x$ | $f(x)$ |
|-----|--------|
| 0   | .2     |
| 3   | .3     |
| 12  | .5     |

Population mean: $E(X) = 0(.2) + 3(.3) + 12(.5) = 6.9$
$= \mu$

Population variance: $\text{Var}(X) = 0^2(.2) + 3^2(.3)$
$+12^2(.5) - 6.9^2$
$= 27.09 = \sigma^2$

|     | Possible Samples | | | Sample Mean | Sample Median | Probability |
|-----|-------|-------|-------|-------------|---------------|-------------|
|     | $x_1$ | $x_2$ | $x_3$ | $\bar{x}$ | $m$ |  |
| 1   | 0  | 0  | 0  | 0  | 0  | $(.2)(.2)(.2) = .008$ |
| 2   | 0  | 0  | 3  | 1  | 0  | $(.2)(.2)(.3) = .012$ |
| 3   | 0  | 0  | 12 | 4  | 0  | $(.2)(.2)(.5) = .020$ |
| 4   | 0  | 3  | 0  | 1  | 0  | $(.2)(.3)(.2) = .012$ |
| 5   | 0  | 3  | 3  | 2  | 3  | $(.2)(.3)(.3) = .018$ |
| 6   | 0  | 3  | 12 | 5  | 3  | $(.2)(.3)(.5) = .030$ |
| 7   | 0  | 12 | 0  | 4  | 0  | $(.2)(.5)(.2) = .020$ |
| 8   | 0  | 12 | 3  | 5  | 3  | $(.2)(.5)(.3) = .030$ |
| 9   | 0  | 12 | 12 | 8  | 12 | $(.2)(.5)(.5) = .050$ |
| 10  | 3  | 0  | 0  | 1  | 0  | $(.3)(.2)(.2) = .012$ |
| 11  | 3  | 0  | 3  | 2  | 3  | $(.3)(.2)(.3) = .018$ |
| 12  | 3  | 0  | 12 | 5  | 3  | $(.3)(.2)(.5) = .030$ |
| 13  | 3  | 3  | 0  | 2  | 3  | $(.3)(.3)(.2) = .018$ |
| 14  | 3  | 3  | 3  | 3  | 3  | $(.3)(.3)(.3) = .027$ |
| 15  | 3  | 3  | 12 | 6  | 3  | $(.3)(.3)(.5) = .045$ |
| 16  | 3  | 12 | 0  | 5  | 3  | $(.3)(.5)(.2) = .030$ |
| 17  | 3  | 12 | 3  | 6  | 3  | $(.3)(.5)(.3) = .045$ |
| 18  | 3  | 12 | 12 | 9  | 12 | $(.3)(.5)(.5) = .075$ |
| 19  | 12 | 0  | 0  | 4  | 0  | $(.5)(.2)(.2) = .020$ |
| 20  | 12 | 0  | 3  | 5  | 3  | $(.5)(.2)(.3) = .030$ |
| 21  | 12 | 0  | 12 | 8  | 12 | $(.5)(.2)(.5) = .050$ |
| 22  | 12 | 3  | 0  | 5  | 3  | $(.5)(.3)(.2) = .030$ |
| 23  | 12 | 3  | 3  | 6  | 3  | $(.5)(.3)(.3) = .045$ |
| 24  | 12 | 3  | 12 | 9  | 12 | $(.5)(.3)(.5) = .075$ |
| 25  | 12 | 12 | 0  | 8  | 12 | $(.5)(.5)(.2) = .050$ |
| 26  | 12 | 12 | 3  | 9  | 12 | $(.5)(.5)(.3) = .075$ |
| 27  | 12 | 12 | 12 | 12 | 12 | $(.5)(.5)(.5) = .125$ |
|     |    |    |    |    |    | Total = 1.000 |

*(Continued)*

**TABLE 3**    *(Cont.)*

Sampling Distribution of $\overline{X}$

| $\bar{x}$ | $f(\bar{x})$ |
|---|---|
| 0 | .008 |
| 1 | $.036 = .012 + .012 + .012$ |
| 2 | $.054 = .018 + .018 + .018$ |
| 3 | .027 |
| 4 | $.060 = .020 + .020 + .020$ |
| 5 | $.180 = .030 + .030 + .030 + .030 + .030 + .030$ |
| 6 | $.135 = .045 + .045 + .045$ |
| 8 | $.150 = .050 + .050 + .050$ |
| 9 | $.225 = .075 + .075 + .075$ |
| 12 | $.125 = .125$ |

$$E(\overline{X}) = \sum \bar{x} f(\bar{x}) = 0(.008) + 1(.036) + 2(.054) + 3(.027)$$
$$+ 4(.060) + 5(.180) + 6(.135)$$
$$+ 8(.150) + 9(.225) + 12(.125)$$
$$= 6.9 \text{ same as } E(X), \text{ pop.mean}$$

$$\operatorname{Var}(\overline{X}) = \sum \bar{x}^2 f(\bar{x}) - \mu^2 = 0^2(.008) + 1^2(.036) + 2^2(.054) + 3^2(.027)$$
$$+ 4^2(.060) + 5^2(.180) + 6^2(.135)$$
$$+ 8^2(.150) + 9^2(.225) + 12^2(.125) - (6.9)^2$$
$$= 9.03 = \frac{27.09}{3} = \frac{\sigma^2}{3}$$

$\operatorname{Var}(\overline{X})$ is one-third of the population variance.

Sampling Distribution of the Median $m$

| $m$ | $f(m)$ |
|---|---|
| 0 | $.104 = .008 + .012 + .020 + .012$ <br> $+ .020 + .012 + .020$ |
| 3 | $.396 = .018 + .030 + .030 + .018 + .030$ <br> $+ .018 + .027 + .045 + .030 + .045$ <br> $+ .030 + .030 + .045$ |
| 12 | $.500 = .050 + .075 + .050 + .075$ <br> $+ .050 + .075 + .125$ |

Mean of the distribution of sample median
$$= 0(.104) + 3(.396) + 12(.500) = 7.188 \neq 6.9 = \mu$$
Different from the mean of the population distribution.

Variance of the distribution of sample median
$$= 0^2(.104) + 3^2(.396) + 12^2(.500) - (7.188)^2 = 23.897$$
[*not* one-third of the population variance 27.09]

(b) The probabilities of all samples giving the same value $\bar{x}$ are added to obtain the sampling distribution on the second page of Table 3.

(c) The calculations and sampling distribution of the median are also given on the second page of Table 3.

To illustrate the idea of a sampling distribution, we considered simple populations with only three possible values and small sample sizes $n = 2$ and $n = 3$. The calculation gets more tedious and extensive when a population has many values of $X$ and $n$ is large. However, the procedure remains the same. Once the population and sample size are specified:

1. List all possible samples of size $n$.

2. Calculate the value of the statistic for each sample.

3. List the distinct values of the statistic obtained in step 2. Calculate the corresponding probabilities by identifying all the samples that yield the same value of the statistic.

We leave the more complicated cases to statisticians who can sometimes use additional mathematical methods to derive exact sampling distributions.

Instead of a precise determination, one can turn to the computer in order to approximate a sampling distribution. The idea is to program the computer to actually draw a random sample and calculate the statistic. This procedure is then repeated a large number of times and a relative frequency histogram constructed from the values of the statistic. The resulting histogram is an approximation to the sampling distribution. This approximation is used in Example 6.

## Exercises

**7.1** Identify each of the following as either a parameter or a statistic.

(a) Population standard deviation.

(b) Sample interquartile range.

(c) Population 20th percentile.

(d) Sample first quartile.

(e) Sample median.

**7.2** Identify the parameter, statistic, and population when they appear in each of the following statements.

(a) During a recent year, forty-one different movies received the distinction of generating the most box office revenue for a weekend.

(b) A survey of 400 minority persons living in Chicago revealed that 41 were out of work.

(c) Out of a sample of 100 dog owners who applied for dog licenses in northern Wisconsin, 18 had a Labrador retriever.

**7.3** Data obtained from asking the wrong questions at the wrong time or in the wrong place can lead to misleading summary statistics. Explain why the following collection procedures are likely to produce useless data.

(a) To evaluate the number of students who are employed at least part time, the investigator interviews students who are taking an evening class.

(b) To study the pattern of spending of persons earning the minimum wage, a survey is taken during the first three weeks of December.

**7.4** Explain why the following collection procedures are likely to produce data that fail to yield the desired information.

(a) To evaluate public opinion about a new global trade agreement, an interviewer asks persons, "Do you feel that this unfair trade agreement should be canceled?"

(b) To determine how eighth-grade girls feel about having boys in the classroom, a random sample from a private girls' school is polled.

**7.5** From the set of numbers { 3, 5, 7 }, a random sample of size 2 is selected with replacement.

(a) List all possible samples and evaluate $\bar{x}$ for each.

(b) Determine the sampling distribution of $\overline{X}$.

**7.6** A random sample of size 2 is selected, with replacement, from the set of numbers { 0, 2, 4 }.

(a) List all possible samples and evaluate $\bar{x}$ and $s^2$ for each.

(b) Determine the sampling distribution of $\overline{X}$.

(c) Determine the sampling distribution of $s^2$.

**7.7** A bride-to-be asks a prospective wedding photographer to show a sample of her work. She provides ten pictures. Should the bride-to-be consider this a random sample of the quality of pictures she will get? Comment.

**7.8** To determine the time a cashier spends on a customer in the express lane, the manager decides to record the time to check-out for the customer who is being served at 10 past the hour, 20 past the hour, and so on. Will measurements collected in this manner be a random sample of the times a cashier spends on a customer?

**7.9** *Using a physical device to generate random samples.* Using a die, generate a sample and evaluate the statistic. Then repeat many times and obtain an estimate of the sampling distribution. In particular, investigate the sampling distribution of the median for a sample of size 3 from the population distribution.

| $x$ | $f(x)$ |
|-----|--------|
| 1 | 2 / 6 |
| 2 | 3 / 6 |
| 4 | 1 / 6 |

(a) Roll the die. Assign $X = 1$ if 1 or 2 dots show. Complete the assignment of values so that $X$ has the population distribution when the die is fair.

(b) Roll the die two more times and obtain the median of the three observed values of X.

(c) Repeat to obtain a total of 25 samples of size 3. Calculate the relative frequencies, among the 75 values, of 1, 2, and 4. Compare with the population probabilities and explain why they should be close.

(d) Obtain the 25 values of the sample median and create a frequency table. Explain how the distribution in this table approximates the actual sampling distribution. It is easy to see how this approach extends to any sample size.

**7.10** Referring to Exercise 7.9, use a die to generate samples of size 3. Investigate the sampling distribution of the number of times a value 1 occurs in a sample of size 3.

(a) Roll the die and assign $X = 1$ if 1 dot shows and $X = 0$, otherwise. Repeat until you obtain a total of 25 samples of size 3. Calculate the relative frequencies, among the 75 values, of 1. Compare with the population probabilities and explain why they should be close.

(b) Let a random variable $Y$ equal 1 if at least one value of $X$ in the sample is 1. Set $Y$ equal to 0 otherwise. Then the relative frequency of $[Y = 1]$ is an estimate of $P[Y = 1]$, the probability of at least one value 1 in the sample of size 3. Give your estimate.

## 2. DISTRIBUTION OF THE SAMPLE MEAN AND THE CENTRAL LIMIT THEOREM

Statistical inference about the population mean is of prime practical importance. Inferences about this parameter are based on the sample mean

$$\overline{X} = \frac{X_1 + X_2 + \cdots + X_n}{n}$$

and its sampling distribution. Consequently, we now explore the basic properties of the sampling distribution of $\overline{X}$ and explain the role of the normal distribution as a useful approximation.

In particular, we want to relate the sampling distribution of $\overline{X}$ to the population from which the random sample was selected. We denote the parameters of the population by

---

Population mean $= \mu$

Population standard deviation $= \sigma$

---

The sampling distribution of $\overline{X}$ also has a mean $E(\overline{X})$ and a standard deviation sd $(\overline{X})$ These can be expressed in terms of the population mean $\mu$ and standard deviation $\sigma$. (The interested reader can consult Appendix A4 for details.)

---

**Mean and Standard Deviation of $\overline{X}$**

The distribution of the sample mean, based on a random sample of size $n$, has

$$E(\overline{X}) = \mu \quad (= \text{Population mean})$$

$$\text{Var}(\overline{X}) = \frac{\sigma^2}{n} \quad \left( = \frac{\text{Population variance}}{\text{Sample size}} \right)$$

$$\text{sd}(\overline{X}) = \frac{\sigma}{\sqrt{n}} \quad \left( = \frac{\text{Population standard deviation}}{\sqrt{\text{Sample size}}} \right)$$

---

The first result shows that the distribution of $\overline{X}$ is centered at the population mean $\mu$ in the sense that expectation serves as a measure of center of a distribution. The last result states that the standard deviation of $\overline{X}$ equals the population standard deviation divided by the square root of the sample size. That is, the variability of the sample mean is governed by the two factors: the population variability $\sigma$ and the sample size $n$. Large variability in the population induces large variability in $\overline{X}$ thus making the sample information about $\mu$ less dependable. However, this can be countered by choosing $n$ large. For instance, with $n = 100$, the standard deviation of $\overline{X}$ is $\sigma / \sqrt{100} = \sigma / 10$, a tenth of the population standard deviation. With increasing sample size, the standard deviation $\sigma / \sqrt{n}$ decreases and the distribution of $\overline{X}$ tends to become more concentrated around the population mean $\mu$.

**Example 3**  The Mean and Variance of the Sampling Distribution of $\bar{X}$

Calculate the mean and standard deviation for the population distribution given in Table 1 and for the distribution of $\bar{X}$ given in Table 2. Verify the relations $E(\bar{X}) = \mu$ and $sd(\bar{X}) = \sigma / \sqrt{n}$.

SOLUTION   The calculations are performed in Table 4.

**TABLE 4**  Mean and Variance of $\bar{X} = (X_1 + X_2)/2$

| Population Distribution | | | | Distribution of $\bar{X} = (X_1 + X_2)/2$ | | | |
|---|---|---|---|---|---|---|---|
| $x$ | $f(x)$ | $xf(x)$ | $x^2 f(x)$ | $\bar{x}$ | $f(\bar{x})$ | $\bar{x}f(\bar{x})$ | $\bar{x}^2 f(\bar{x})$ |
| 2 | $\frac{1}{3}$ | $\frac{2}{3}$ | $\frac{4}{3}$ | 2 | $\frac{1}{9}$ | $\frac{2}{9}$ | $\frac{4}{9}$ |
| 3 | $\frac{1}{3}$ | $\frac{3}{3}$ | $\frac{9}{3}$ | 2.5 | $\frac{2}{9}$ | $\frac{5}{9}$ | $\frac{12.5}{9}$ |
| 4 | $\frac{1}{3}$ | $\frac{4}{3}$ | $\frac{16}{3}$ | 3 | $\frac{3}{9}$ | $\frac{9}{9}$ | $\frac{27}{9}$ |
| | | | | 3.5 | $\frac{2}{9}$ | $\frac{7}{9}$ | $\frac{24.5}{9}$ |
| Total | 1 | 3 | $\frac{29}{3}$ | 4 | $\frac{1}{9}$ | $\frac{4}{9}$ | $\frac{16}{9}$ |
| | | | | Total | 1 | 3 | $\frac{84}{9}$ |

$$\mu = 3$$

$$\sigma^2 = \frac{29}{3} - (3)^2 = \frac{2}{3}$$

$$E(\bar{X}) = 3 = \mu$$

$$Var(\bar{X}) = \frac{84}{9} - (3)^2 = \frac{1}{3}$$

By direct calculation, $sd(\bar{X}) = 1/\sqrt{3}$. This is confirmed by the relation

$$sd(\bar{X}) = \frac{\sigma}{\sqrt{n}} = \sqrt{\frac{2}{3}} / \sqrt{2} = \frac{1}{\sqrt{3}}$$

We now state two important results concerning the shape of the sampling distribution of $\bar{X}$. The first result gives the exact form of the distribution of $\bar{X}$ when the population distribution is normal.

---

### $\bar{X}$ Is Normal When Sampling from a Normal Population

In random sampling from a **normal** population with mean $\mu$ and standard deviation $\sigma$, the sample mean $\bar{X}$ has the normal distribution with mean $\mu$ and standard deviation $\sigma / \sqrt{n}$.

---

**Example 4**   Determining Probabilities Concerning $\bar{X}$—Normal Populations

The weight of a pepperoni and cheese pizza from a local provider is a random variable whose distribution is normal with mean 16 ounces and standard deviation 1 ounce. You intend to purchase four pepperoni and cheese pizzas. What is the probability that:

(a)   The average weight of the four pizzas will be greater than 17.1 ounces?

(b)   The total weight of the four pizzas will not exceed 61.0 ounces?

SOLUTION   Because the population is normal, the distribution of the sample mean $\bar{X} = (X_1 + X_2 + X_3 + X_4)/4$ is exactly normal with mean 16 ounces and standard deviation $1 / \sqrt{4} = .5$ ounce.

(a)   Since $\bar{X}$ is $N(16, .5)$

$$P[\bar{X} > 17.1] = P\left[\frac{\bar{X} - 16}{.5} > \frac{17.1 - 16}{.5}\right]$$

$$= P[Z > 2.20] = 1 - .9861 = .0139$$

Only rarely, just over one time in a hundred purchases of four pizzas, would the average weight exceed 17.1 ounces.

(b)   The event that the total weight $X_1 + X_2 + X_3 + X_4 = 4\bar{X}$ does not exceed 61.0 ounces is the same event that the average weight $\bar{X}$ is less than or equal to $61.0/4 = 15.25$. Consequently

$$P[X_1 + X_2 + X_3 + X_4 \leq 61.0] = P[\bar{X} \leq 15.25]$$

$$= P\left[\frac{\bar{X} - 16}{.5} \leq \frac{15.25 - 16}{.5}\right]$$

$$= P[Z \leq -1.50] = .0668$$

Only about seven times in one hundred purchases would the total weight be less than 61.0 ounces.

---

When sampling from a nonnormal population, the distribution of $\bar{X}$ depends on the particular form of the population distribution that prevails. A surprising result, known as the **central limit theorem**, states that when the sample size

$n$ is large, the distribution $\overline{X}$ is approximately normal, regardless of the shape of the population distribution. In practice, the normal approximation is usually adequate when $n$ is greater than 30.

---

### Central Limit Theorem

Whatever the population, the distribution of $\overline{X}$ is approximately normal when $n$ is large

In random sampling from an arbitrary population with mean $\mu$ and standard deviation $\sigma$, when $n$ is large, the distribution of $\overline{X}$ is approximately normal with mean $\mu$ and standard deviation $\sigma / \sqrt{n}$. Consequently,

$$Z = \frac{\overline{X} - \mu}{\sigma / \sqrt{n}} \quad \text{is approximately } N(0, 1)$$

---

As our rule of thumb, we treat **$n$ greater than 30 as large**.

Whether the population distribution is continuous, discrete, symmetric, or asymmetric, the central limit theorem asserts that as long as the population variance is finite, the distribution of the sample mean $\overline{X}$ is nearly normal if the sample size is large. In this sense, the normal distribution plays a central role in the development of statistical procedures.

**Example 5**  Probability Calculations for $\overline{X}$—Based on a Large Sample of Activities

Extensive data, including that in Exercise 2.3, suggest that the number of extracurricular activities per week can be modeled as a distribution with mean 1.9 and standard deviation 1.6.

(a)  If a random sample of size 41 is selected, what is the probability that the sample mean will lie between 1.7 and 2.1?

(b)  With a sample size of 100, what is the probability that the sample mean will lie between 1.7 and 2.1?

SOLUTION  (a)  The discrete distribution for the number of activities has $\mu = 1.9$ and $\sigma = 1.6$. Since $n = 41$ is large, the central limit theorem tells us that the distribution of $\overline{X}$ is approximately normal with

$$\text{Mean} = \mu = 1.9$$

$$\text{Standard deviation} = \frac{\sigma}{\sqrt{n}} = \frac{1.6}{\sqrt{41}} = .250$$

To calculate $P[\,1.7 < \overline{X} < 2.1\,]$, we convert to the standardized variable

$$Z = \frac{\overline{X} - \mu}{\sigma / \sqrt{n}} = \frac{\overline{X} - 1.9}{.250}$$

The $z$ values corresponding to 1.7 and 2.1 are

$$\frac{1.7 - 1.9}{.250} = -.8 \quad \text{and} \quad \frac{2.1 - 1.9}{.250} = .8$$

$$
\begin{array}{cccccc}
& 1.4 & 1.65 & 1.9 & 2.15 & 2.4 & \overline{X} \\
\hline
& -2 & -1 & 0 & 1 & 2 & Z
\end{array}
$$

Consequently,

$$P[\, 1.7 \; < \; \overline{X} \; < \; 2.1\,] \; = \; P[\, -.8 \; < \; Z \; < \; .8\,]$$

$$= \; .7881 \; - \; .2119 \; (\,\text{using the normal table}\,)$$

$$= \; .5762$$

More than half of the time, the sample mean will fall in the interval (1.7, 2.1).

(b)  We now have $n = 100$, so $\sigma / \sqrt{n} = 1.6 / \sqrt{100} = .16$, and

$$Z = \frac{\overline{X} - 1.9}{.16}$$

Therefore,

$$P[\, 1.7 \; < \; \overline{X} \; < \; 2.1\,] \; = \; P\left[\, \frac{1.7 - 1.9}{.16} \; < \; Z \; < \; \frac{2.1 - 1.9}{.16}\,\right]$$

$$= \; P[\, -1.25 \; < \; Z \; < \; 1.25\,]$$

$$= \; .8944 \; - \; .1056$$

$$= \; .7888$$

Note that the interval (1.7, 2.1) is centered at $\mu = 1.9$. The probability that $\overline{X}$ will lie in this interval is larger for $n = 100$ than for $n = 41$.

Although a proof of the central limit theorem requires higher mathematics, we can empirically demonstrate how this result works.

**Example 6**    Demonstrating the Central Limit Theorem

Consider a population having a discrete uniform distribution that places a probability of .1 on each of the integers 0, 1, . . . , 9. This may be an appropriate model for the distribution of the last digit in telephone numbers or the first overflow digit in computer calculations. The line diagram of this distribution appears in Figure 2. The population has $\mu = 4.5$ and $\sigma = 2.872$.

Take 100 samples of size 5, calculate $\bar{x}$ for each, and make a histogram to approximate the sampling distribution of $\overline{X}$.

Figure 2    Uniform distribution on the integers 0, 1, . . . , 9.

SOLUTION    By means of a computer, 100 random samples of size 5 were generated from this distribution, and $\bar{x}$ was computed for each sample. The results of this repeated random sampling are presented in Table 5. The relative frequency histogram in Figure 3 is constructed from the 100 observed values of $\bar{x}$ Although the population distribution (Figure 2) is far from normal, the top of the histogram of the $\bar{x}$ values (Figure 3) has the appearance of a bell-shaped curve, even for the small sample size of 5. For larger sample sizes, the normal distribution would give an even closer approximation.

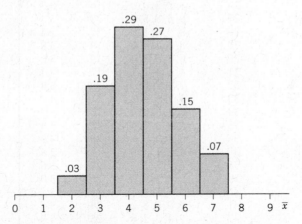

Figure 3    Relative frequency histogram of the $\bar{x}$ values recorded in Table 5.

Calculating from the 100 simulated $\bar{x}$ values in Table 5, we find the sample mean and standard deviation to be 4.54 and 1.215, respectively. These are in close agreement with the theoretical values for the mean and standard deviation of $\overline{X}$: $\mu = 4.5$ and $\sigma / \sqrt{n} = 2.872 / \sqrt{5} = 1.284$.

**TABLE 5**   Samples of Size 5 from a Discrete Uniform Distribution

| Sample Number | Observations | Sum | Mean $\bar{x}$ | Sample Number | Observations | Sum | Mean $\bar{x}$ |
|---|---|---|---|---|---|---|---|
| 1 | 4, 7, 9, 0, 6 | 26 | 5.2 | 51 | 4, 7, 3, 8, 8 | 30 | 6.0 |
| 2 | 7, 3, 7, 7, 4 | 28 | 5.6 | 52 | 2, 0, 3, 3, 2 | 10 | 2.0 |
| 3 | 0, 4, 6, 9, 2 | 21 | 4.2 | 53 | 4, 4, 2, 6, 3 | 19 | 3.8 |
| 4 | 7, 6, 1, 9, 1 | 24 | 4.8 | 54 | 1, 6, 4, 0, 6 | 17 | 3.4 |
| 5 | 9, 0, 2, 9, 4 | 24 | 4.8 | 55 | 2, 4, 5, 8, 9 | 28 | 5.6 |
| 6 | 9, 4, 9, 4, 2 | 28 | 5.6 | 56 | 1, 5, 5, 4, 0 | 15 | 3.0 |
| 7 | 7, 4, 2, 1, 6 | 20 | 4.0 | 57 | 3, 7, 5, 4, 3 | 22 | 4.4 |
| 8 | 4, 4, 7, 7, 9 | 31 | 6.2 | 58 | 3, 7, 0, 7, 6 | 23 | 4.6 |
| 9 | 8, 7, 6, 0, 5 | 26 | 5.2 | 59 | 4, 8, 9, 5, 9 | 35 | 7.0 |
| 10 | 7, 9, 1, 0, 6 | 23 | 4.6 | 60 | 6, 7, 8, 2, 9 | 32 | 6.4 |
| 11 | 1, 3, 6, 5, 7 | 22 | 4.4 | 61 | 7, 3, 6, 3, 6 | 25 | 5.0 |
| 12 | 3, 7, 5, 3, 2 | 20 | 4.0 | 62 | 7, 4, 6, 0, 1 | 18 | 3.6 |
| 13 | 5, 6, 6, 5, 0 | 22 | 4.4 | 63 | 7, 9, 9, 7, 5 | 37 | 7.4 |
| 14 | 9, 9, 6, 4, 1 | 29 | 5.8 | 64 | 8, 0, 6, 2, 7 | 23 | 4.6 |
| 15 | 0, 0, 9, 5, 7 | 21 | 4.2 | 65 | 6, 5, 3, 6, 2 | 22 | 4.4 |
| 16 | 4, 9, 1, 1, 6 | 21 | 4.2 | 66 | 5, 0, 5, 2, 9 | 21 | 4.2 |
| 17 | 9, 4, 1, 1, 4 | 19 | 3.8 | 67 | 2, 9, 4, 9, 1 | 25 | 5.0 |
| 18 | 6, 4, 2, 7, 3 | 22 | 4.4 | 68 | 9, 5, 2, 2, 6 | 24 | 4.8 |
| 19 | 9, 4, 4, 1, 8 | 26 | 5.2 | 69 | 0, 1, 4, 4, 4 | 13 | 2.6 |
| 20 | 8, 4, 6, 8, 3 | 29 | 5.8 | 70 | 5, 4, 0, 5, 2 | 16 | 3.2 |
| 21 | 5, 2, 2, 6, 1 | 16 | 3.2 | 71 | 1, 1, 4, 2, 0 | 8 | 1.6 |
| 22 | 2, 2, 9, 1, 0 | 14 | 2.8 | 72 | 9, 5, 4, 5, 9 | 32 | 6.4 |
| 23 | 1, 4, 5, 8, 8 | 26 | 5.2 | 73 | 7, 1, 6, 6, 9 | 29 | 5.8 |
| 24 | 8, 1, 6, 3, 7 | 25 | 5.0 | 74 | 3, 5, 0, 0, 5 | 13 | 2.6 |
| 25 | 1, 2, 0, 9, 6 | 18 | 3.6 | 75 | 3, 7, 7, 3, 5 | 25 | 5.0 |
| 26 | 8, 5, 3, 0, 0 | 16 | 3.2 | 76 | 7, 4, 7, 6, 2 | 26 | 5.2 |
| 27 | 9, 5, 8, 5, 0 | 27 | 5.4 | 77 | 8, 1, 0, 9, 1 | 19 | 3.8 |
| 28 | 8, 9, 1, 1, 8 | 27 | 5.4 | 78 | 6, 4, 7, 9, 3 | 29 | 5.8 |
| 29 | 8, 0, 7, 4, 0 | 19 | 3.8 | 79 | 7, 7, 6, 9, 7 | 36 | 7.2 |
| 30 | 6, 5, 5, 3, 0 | 19 | 3.8 | 80 | 9, 4, 2, 9, 9 | 33 | 6.6 |
| 31 | 4, 6, 4, 2, 1 | 17 | 3.4 | 81 | 3, 3, 3, 3, 3 | 15 | 3.0 |
| 32 | 7, 8, 3, 6, 5 | 29 | 5.8 | 82 | 8, 7, 7, 0, 3 | 25 | 5.0 |
| 33 | 4, 2, 8, 5, 2 | 21 | 4.2 | 83 | 5, 3, 2, 1, 1 | 12 | 2.4 |
| 34 | 7, 1, 9, 0, 9 | 26 | 5.2 | 84 | 0, 4, 5, 2, 6 | 17 | 3.4 |
| 35 | 5, 8, 4, 1, 4 | 22 | 4.4 | 85 | 3, 7, 5, 4, 1 | 20 | 4.0 |
| 36 | 6, 4, 4, 5, 1 | 20 | 4.0 | 86 | 7, 4, 5, 9, 8 | 33 | 6.6 |
| 37 | 4, 2, 1, 1, 6 | 14 | 2.8 | 87 | 3, 2, 9, 0, 5 | 19 | 3.8 |
| 38 | 4, 7, 5, 5, 7 | 28 | 5.6 | 88 | 4, 6, 6, 3, 3 | 22 | 4.4 |
| 39 | 9, 0, 5, 9, 2 | 25 | 5.0 | 89 | 1, 0, 9, 3, 7 | 20 | 4.0 |
| 40 | 3, 1, 5, 4, 5 | 18 | 3.6 | 90 | 2, 9, 6, 8, 5 | 30 | 6.0 |
| 41 | 9, 8, 6, 3, 2 | 28 | 5.6 | 91 | 4, 8, 0, 7, 6 | 25 | 5.0 |
| 42 | 9, 4, 2, 2, 8 | 25 | 5.0 | 92 | 5, 6, 7, 6, 3 | 27 | 5.4 |
| 43 | 8, 4, 7, 2, 2 | 23 | 4.6 | 93 | 3, 6, 2, 5, 6 | 22 | 4.4 |
| 44 | 0, 7, 3, 4, 9 | 23 | 4.6 | 94 | 0, 1, 1, 8, 4 | 14 | 2.8 |
| 45 | 0, 2, 7, 5, 2 | 16 | 3.2 | 95 | 3, 6, 6, 4, 5 | 24 | 4.8 |
| 46 | 7, 1, 9, 9, 9 | 35 | 7.0 | 96 | 9, 2, 9, 8, 6 | 34 | 6.8 |
| 47 | 4, 0, 5, 9, 4 | 22 | 4.4 | 97 | 2, 0, 0, 6, 8 | 16 | 3.2 |
| 48 | 5, 8, 6, 3, 3 | 25 | 5.0 | 98 | 0, 4, 5, 0, 5 | 14 | 2.8 |
| 49 | 4, 5, 0, 5, 3 | 17 | 3.4 | 99 | 0, 3, 7, 3, 9 | 22 | 4.4 |
| 50 | 7, 7, 2, 0, 1 | 17 | 3.4 | 100 | 2, 5, 0, 0, 7 | 14 | 2.8 |

It might be interesting for the reader to collect similar samples by reading the last digits of numbers from a telephone directory and then to construct a histogram of the $\bar{x}$ values.

Another graphic example of the central limit theorem appears in Figure 4, where the population distribution represented by the solid curve is a continuous asymmetric distribution with $\mu = 2$ and $\sigma = 1.41$. The distributions of the sample mean $\bar{X}$ for sample sizes $n = 3$ and $n = 10$ are plotted as dashed and dotted curves on the graph. These indicate that with increasing $n$, the distributions become more concentrated around $\mu$ and look more like the normal distribution.

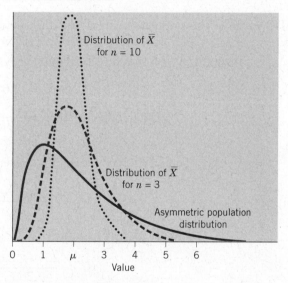

Figure 4   Distributions of $\bar{X}$ for $n = 3$ and $n = 10$ in sampling from an asymmetric population.

---

**Example 7**  Probability Calculations for $\bar{X}$—Number of Items Returned

Retail stores experience their heaviest returns on December 26 and December 27 each year. A small sample of number of items returned is given in Example 4, Chapter 2, but a much larger sample size is required to approximate the probability distribution. Suppose the relative frequencies, determined from a sample of size 300, suggest the probability distribution in Table 6.

This distribution for number of gifts returned has mean 2.61 and standard deviation 1.34 items. Assume the probability distribution in Table 6 still holds for this year.

(a)  If this year, a random sample of size 45 is selected, what is the probability that the sample mean will be greater than 2.9 items?

(b)  Find an upper bound $b$ such that the total number of items returned by 45 customers will be less than $b$ with probability .95.

**TABLE 6** Number X of Items
Returned

| Number Items Returned ($x$) | Probability |
|---|---|
| 1 | .25 |
| 2 | .28 |
| 3 | .20 |
| 4 | .17 |
| 5 | .08 |
| 6 | .02 |

SOLUTION  Here the population mean and the standard deviation are $\mu = 2.61$ and $\sigma = 1.34$, respectively. The sample size $n = 45$ is large, so the central limit theorem ensures that the distribution of $\overline{X}$ is approximately normal with

$$\text{Mean} = 2.61$$
$$\text{Standard deviation} = \frac{\sigma}{\sqrt{n}} = \frac{1.34}{\sqrt{45}} = .200$$

(a)  To find $P[\ \overline{X} > 2.9\ ]$, we convert to the standardized variable

$$Z = \frac{\overline{X} - 2.61}{.200}$$

and calculate the $z$ value

$$\frac{2.9 - 2.61}{.200} = 1.45$$

The required probability is

$$P[\overline{X} > 2.9] = P[Z > 1.45]$$
$$= 1 - P[Z \leq 1.45]$$
$$= 1 - .9265 = .0735$$

The sample mean will exceed 2.9 items in just over 1 in 14 times

(b)  Let $X_i$ denote the number of items returned by the $i$ th person. Then we recognize that the event the total number of items returned is less than or equal $b$, $[\ X_1 + X_2 + \cdots + X_{45} \leq b]$ is the same event as $[\overline{X} < b/45]$.

By the central limit theorem, since $z_{.05} = 1.645$, $b$ must satisfy

$$.95 = P[Z \leq 1.645] = P\left[\frac{\overline{X} - 2.61}{.200} \leq \frac{b/45 - 2.61}{.200}\right]$$

It follows that

$$b = 45 \cdot (1.645 \times .200 + 2.61) = 132.3$$

In the context of this problem, the total must be an integer so, conservatively, we may take 133 items as the bound.

The normal approximation to the binomial distribution, introduced in Chapter 6, is just a special case of the central limit theorem.

**Example 8**    Central Limit Theorem Is Basis for Normal Approximation to Binomial

Let $X$ be distributed as the binomial distribution with sample size $n$ and proportion parameter $p$. Show that, for large $n$, the sample proportion of successes $X / n$ satisfies

(a)    $\dfrac{\left(\dfrac{X}{n} - p\right)}{\sqrt{p(1 - p)}/\sqrt{n}}$    is approximately    $N(0, 1)$

(b)    The normal approximation to $X$ in Chapter 6 follows from the equality

$$\frac{X - np}{\sqrt{np(1 - p)}} = \frac{\left(\dfrac{X}{n} - p\right)}{\sqrt{p(1 - p)}/\sqrt{n}} \quad \text{is approximately} \quad N(0, 1)$$

SOLUTION    The random variable $X_1$ that equals 1 if the first trial is a success and 0 otherwise, has mean and variance

$$\mu = E(X_1) = 0(1 - p) + 1p = p \quad \text{and}$$
$$\sigma^2 = \text{Var}(X_1) = 0^2(1 - p) + 1^2 p - p^2 = p(1 - p)$$

Continuing, for the $i$-th trial, let $X_i$ equal 1 if the $i$th trial is a success and 0 otherwise. The binomial count $X = X_1 + X_2 + \cdots + X_n$ is a sum and $\overline{X} = X/n$

(a)    Applying the central limit theorem to the random sample $X_1, X_2, \ldots, X_n$ where $\mu = p$ and $\sigma = \sqrt{p(1 - p)}$ we have

$$\frac{\left(\dfrac{X}{n} - p\right)}{\sqrt{p(1 - p)}/\sqrt{n}} \quad \text{is approximately} \quad N(0, 1)$$

(b)    Multiplying the statistic by $1 = n/n$,

$$\frac{n}{n} \frac{\left(\dfrac{X}{n} - p\right)}{\sqrt{p(1 - p)}/\sqrt{n}} = \frac{\left(n\dfrac{X}{n} - np\right)}{\sqrt{np(1 - p)}}$$

so the two forms are equivalent.

The condition from Chapter 6, that both $np$ and $n(1 - p)$ are large, ensures that the approximation is reasonably good.

## *Exercises*

**7.11** The population density function and that for the sampling distribution of $\overline{X}$ are shown in Figure 5. Identify which one is the sampling distribution and explain your answer.

Figure 5   Two density functions. Exercise 7.11.

**7.12** The population density function and that for the sampling distribution of $\overline{X}$, for $n = 2$, are shown in Figure 6. Identify which one is the sampling distribution and explain your answer.

Figure 6   Two density functions. Exercise 7.12.

**7.13** Refer to the lightning data in Exercise 2.121 of Chapter 2. One plausible model for the population distribution has mean $\mu = 76$ and standard deviation $\sigma = 35$ deaths per year. Calculate the mean and standard deviation of $\overline{X}$ for a random sample of size (a) 4 and (b) 25.

**7.14** Refer to the data on earthquakes of magnitude greater than 6.0 in Exercise 2.20. The data suggests that one plausible model, for $X =$ magnitude, is a population distribution having mean $\mu = 6.7$ and standard deviation $\sigma = .46$. Calculate the expected value and standard deviation of $\overline{X}$ for a random sample of size (a) 9 and (b) 16.

**7.15** Refer to the monthly intersection accident data in Exercise 5.96. The data suggests that one plausible

model, for $X =$ the number of accidents in one month, is a population distribution having mean $\mu = 2.6$ and variance $\sigma^2 = 2.4$. Determine the standard deviation of $\overline{X}$ for a random sample of size (a) 25, (b) 100, and (c) 400. (d) How does quadrupling the sample size change the standard deviation of $\overline{X}$?

**7.16** Refer to Exercise 7.13. Determine the standard deviation of $\overline{X}$ for a random sample of size (a) 9, (b) 36, and (c) 144. (d) How does quadrupling the sample size change the standard deviation of $\overline{X}$?

**7.17** Using the sampling distribution determined for $\overline{X} = (X_1 + X_2)/2$ in Exercise 7.5, verify that $E[\overline{X}] = \mu$ and $sd(\overline{X}) = \sigma/\sqrt{2}$.

**7.18** Using the sampling distribution determined for $\overline{X} = (X_1 + X_2)/2$ in Exercise 7.6, verify that $E[\overline{X}] = \mu$ and $sd(\overline{X}) = \sigma/\sqrt{2}$.

**7.19** Suppose the number of different computers used by a student last week has distribution

| Value | Probability |
|-------|-------------|
| 0 | .3 |
| 1 | .4 |
| 2 | .3 |

Let $X_1$ and $X_2$ be independent and each have the same distribution as the population.

(a) Determine the missing elements in the table for the sampling distribution of $\overline{X} = (X_1 + X_2)/2$.

| $\bar{x}$ | Probability |
|-----------|-------------|
| 0.0 | |
| 0.5 | |
| 1.0 | .34 |
| 1.5 | .24 |
| 2.0 | |

(b) Find the expected value of $\overline{X}$.

(c) If the sample size is increased to 36, give the mean and variance of $\overline{X}$.

**7.20** As suggested in Example 8, Chapter 6, the population of hours of sleep can be modeled as a normal distribution with mean 7.2 hours and standard deviation 1.3 hours. For a sample of size 4, determine the

(a) Mean of $\overline{X}$.

(b) Standard deviation of $\overline{X}$.

(c) Distribution of $\overline{X}$.

**7.21** According to Example 12, Chapter 6, a normal distribution with mean 115 and standard deviation 22 hundredths of an inch describes the variation in female salmon growth in freshwater. For a sample of size 6, determine the

(a) Mean of $\overline{X}$.

(b) Standard deviation of $\overline{X}$.

(c) Distribution of $\overline{X}$.

**7.22** A population has distribution

| Value | Probability |
|:-----:|:-----------:|
| 0 | .7 |
| 2 | .1 |
| 4 | .2 |

Let $X_1$ and $X_2$ be independent and each have the same distribution as the population.

(a) Determine the missing elements in the table for the sampling distribution of $\overline{X} = (X_1 + X_2)/2$.

| $\overline{x}$ | Probability |
|:-----:|:-----------:|
| 0 | |
| 1 | |
| 2 | .29 |
| 3 | |
| 4 | .04 |

(b) Find the expected value of $\overline{X}$.

(c) If the sample size is increased to 25, give the mean and variance of $\overline{X}$.

**7.23** Suppose the weights of the contents of cans of mixed nuts have a normal distribution with mean 32.4 ounces and standard deviation .4 ounce.

(a) If every can is labeled 32 ounces, what proportion of the cans have contents that weigh less than the labeled amount?

(b) If two packages are randomly selected, specify the mean, standard deviation, and distribution of the average weight of the contents.

(c) If two packages are randomly selected, what is the probability that the average weight is less than 32 ounces?

**7.24** Suppose the amount of a popular sport drink in bottles leaving the filling machine has a normal distribution with mean 101.5 milliliters (ml) and standard deviation 1.6 ml.

(a) If the bottles are labeled 100 ml, what proportion of the bottles contain less than the labeled amount?

(b) If four bottles are randomly selected, find the mean and standard deviation of the average content.

(c) What is the probability that the average content of four bottles is less than 100 ml?

**7.25** The distribution of personal income of full-time retail clerks working in a large eastern city has $\mu = \$51,000$ and $\sigma = \$5000$.

(a) What is the approximate distribution for $\overline{X}$ based on a random sample of 100 persons?

(b) Evaluate $P[\overline{X} > 51,500]$.

**7.26** The result of a recent survey suggests that one plausible population distribution, for $X =$ number of persons with whom an adult discusses important matters, can be modeled as a population having mean $\mu = 2.0$ and standard deviation $\sigma = 2.0$. A random sample of size 100 is obtained.

(a) What can you say about the probability distribution of the sample mean $\overline{X}$?

(b) Find the probability that $\overline{X}$ exceeds 2.3.

**7.27** The lengths of the trout fry in a pond at the fish hatchery are approximately normally distributed with mean 3.4 inches and standard deviation .8 inch. Three dozen fry are netted and their lengths measured.

(a) What is the probability that the sample mean length of the 36 netted trout fry is less than 3.2 inches?

(b) Why might the fish in the net not represent a random sample of trout fry in the pond?

**7.28** The heights of male students at a university have a nearly normal distribution with mean 70 inches and standard deviation 2.8 inches. If 5 male students are randomly selected to make up an intramural basketball team, what is the probability that the heights of the team averages over 72.0 inches?

**7.29** According to the growth chart that doctors use as a reference, the heights of two-year-old boys are normally distributed with mean 34.5 inches and standard deviation 1.3 inches. For a random sample of 6 two-year-old boys, find the probability that the sample mean is between 34.1 and 35.2 inches.

**7.30** The weight of an almond is normally distributed with mean .05 ounce and standard deviation .015 ounce. Find the probability that a package of 100 almonds weighs between 4.8 and 5.3 ounces. That is, find the probability that $\overline{X}$ is between .048 and .053 ounce.

**7.31** Refer to Table 5 on page 288.

(a) Calculate the sample median for each sample.

(b) Construct a frequency table and make a histogram.

(c) Compare the histogram for the median with that given in Figure 3 for the sample mean. Does your comparison suggest that the sampling distribution of the mean or median has the smaller variance?

**7.32** The number of days, X, that it takes the post office to deliver a notarized letter cross-country between City A and City B has the probability distribution

| $x$ | 3 | 4 | 5 |
|---|---|---|---|
| $f(x)$ | .5 | .3 | .2 |

(a) Find the expected number of days and the standard deviation of the number of days.

(b) A company in City A sends a notarized letter to a company in City B with a notarized return receipt request that is to be mailed immediately upon receiving the letter. Find the probability distribution of total number of days from the time the letter is mailed until the return receipt arrives back at the company in City A. Assume the two delivery times have the same distribution and are independent.

(c) A single notarized letter is sent from City A on each of 100 different days. What is the approximate probability that more than 25 of the letters take 5 days to reach City B?

**7.33** The number of complaints per day, X, received by a cable TV distributor has the probability distribution

| $x$ | 0 | 1 | 2 | 3 |
|---|---|---|---|---|
| $f(x)$ | .4 | .3 | .1 | .2 |

(a) Find the expected number of complaints per day.

(b) Find the standard deviation of the number of complaints.

(c) What is the probability distribution of total number of complaints received in two days? Assume the numbers of complaints on different days are independent.

(d) What is the approximate probability that the distributor receives more than 125 complaints in 90 days?

# STATISTICS IN CONTEXT

Troy, a Canadian importer of cut flowers, must effectively deal with uncertainty every day that he is in business. For instance, he must order enough of each kind of flower to supply his regular wholesale customers and yet not have too many left each day. Fresh flowers are no longer fresh on the day after arrival.

Troy purchases his fresh flowers from growers in the United States, Mexico, and Central and South America. Because most of the growers purchase their growing stock and chemicals from the United States, all of the selling prices are quoted in U.S. dollars. On a typical day, he purchases tens of thousands of cut flowers. Troy knows their price in U.S. dollars, but this is not his ultimate cost. Because of a fluctuating exchange rate, he does not know his ultimate cost at the time of purchase.

As with most businesses, Troy takes about a month to pay his bills. He must pay in Canadian dollars, so fluctuations in the Canadian/U.S. exchange rate from the time of purchase to the time the invoice is paid are a major source of uncertainty. Can this uncertainty be quantified and modeled?

The Canadian dollar to U.S. dollar exchange rate equals the number of Canadian dollars that must be paid for each U.S. dollar. Data from several years will provide the basis for a model. As given in Table 7. The exchange rate was 1.0791 on 7/31/2009 and, reading across rows, the rate was .9958 on 12/31/2012.

If the exchange rate is 1.100, one dollar and ten cents Canadian is required to pay for each single U.S. dollar. An invoice for 1000 U.S. dollars costs Troy 1100 Canadian dollars while it costs just 958 Canadian dollars if the rate drops to .9580. The complication is that Troy knows the exchange rate at the time of purchase but he actually pays at the rate prevailing a month later.

It is the change in the exchange rate from time of purchase to payment that creates uncertainty. Although the exchange rate changes every day, we consider the monthly rates. The value of the difference

$$\text{Current month exchange rate } - \text{ Previous month exchange rate}$$

would describe the change in cost, per dollar invoiced, resulting from the one-month delay between purchasing and paying for a shipment of cut flowers. If the rate goes down, Troy makes money. If the rate goes up, he loses money.

The sampling distribution of the one-month difference is the key to understanding the variation in profit due to changes in the Canadian/U.S. exchange rate. Our approach is to approximate this sampling distribution by using the 41 differences calculated from the closing rates in Table 7. These $n = 41$ one-month differences have $\bar{x} = -.0021$ and $s = .0234$.

Figure 7 gives a time plot of these differences for the period 8/31/2009 to 12/31/2012. The 41 differences appear to be stable.

**TABLE 7**   Monthly Canadian to U.S. Dollar Exchange Rates
7/31/2009–12/31/2012

| | | | | | | |
|---|---|---|---|---|---|---|
| 1.0791 | 1.0967 | 1.0719 | 1.0767 | 1.0570 | 1.0461 | 1.0652 |
| 1.0520 | 1.0156 | 1.0112 | 1.0497 | 1.0606 | 1.0293 | 1.0640 |
| 1.0293 | 1.0187 | 1.0266 | 1.0009 | 1.0020 | .9737 | .9717 |
| .9486 | .9688 | .9642 | .9539 | .9783 | 1.0389 | .9932 |
| 1.0199 | 1.0168 | 1.0050 | .9866 | .9990 | .9886 | 1.0349 |
| 1.0190 | 1.0014 | .9862 | .9837 | .9994 | .9931 | .9958 |

**Figure 7**   Time plot of the one-month differences in the Canadian/U.S. exchange rate.

Figure 8 presents a histogram and a normal-scores plot. Neither graph appears to contain a pattern that contradicts the plausibility of a normal sampling distribution. The slight bend in the normal-scores plot does suggest some lack of symmetry but the deviation from normal is not statistically significant.

**Figure 8**   One-month differences in the Canadian/U.S. exchange rate: (a) histogram (b) normal-scores plot.

Figure 9    Normal distribution model of the one-month differences in the Canadian/U.S. exchange rate.

Based on the 41 observed one-month differences, we choose to model the sampling distribution as normal. According to the statistical methods developed in the next chapter, 0 is a plausible value for the mean $\mu$. Also, we round up our estimated value and take $\sigma = .024$.

Figure 9 shows the $N(0, .024)$ distribution we select as our model. Each month, an independent difference is selected from this distribution. The interval from $-2\sigma$ to $2\sigma$ is assigned probability .954 as indicated by the shaded area. The probability outside of this interval is .046. This distribution shows that changes in the exchange rate larger than .072 in absolute value are very unlikely. Values in the interval $-.024$ to $.024$, or within one standard deviation of 0, are assigned probability .6826. The long-run frequency interpretation of probability asserts that the long-run proportion of differences within this last interval will be approximately .6826.

What can be expected for the 60 differences over a five-year period? The binomial distribution with $p = .046$ applies so that the expected number of times the absolute value of the difference will exceed $2\sigma = .048$ is only $.046 \times 60 = 2.76$ times.

We have successfully modeled the uncertainty in the the exchange rate over a one-month period. If Troy paid all of his invoices in exactly one month, this quantifies the variability he faces.

## Exercises

**7.34** Refer to the model for monthly differences in the exchange rate shown in Figure 9. Calculate the mean and variance of the number of times, in two years, that the absolute value of a one-month difference exceeds (a) .048 (b) .072

**7.35** Refer to the Statistics in Context section concerning the flower importer.

(a) Suppose it takes the importer two months to pay his invoices. Proceed by taking the sum of the adjacent one-month differences to obtain the two-month differences. Make a histogram of these differences two months apart. (The two-month differences may not be independent but the histogram is the correct summary of uncertainty for some two-month period in the future.)

(b) Compare your histogram in part (a) with that for the one-month changes. Which is more variable?

## USING STATISTICS WISELY

1. Understand the concept of a sampling distribution. Each observation is the value of a random variable so a sample of $n$ observations varies from one possible sample to another. Consequently, a statistic such as a sample mean varies from one possible sample to another. The probability distribution that describes the chance behavior of the sample mean is called its sampling distribution.

2. When the underlying distribution has mean $\mu$ and variance $\sigma^2$, remember that the sampling distribution of $\overline{X}$ has

$$\text{Mean of } \overline{X} = \mu = \text{Population mean}$$

$$\text{Variance of } \overline{X} = \frac{\sigma^2}{n} = \frac{\text{Population variance}}{n}$$

3. When the underlying distribution is normal with mean $\mu$ and variance $\sigma^2$, calculate exact probabilities for $\overline{X}$ using the normal distributionwith mean $\mu$ and variance $\dfrac{\sigma^2}{n}$.

$$P[\overline{X} \leq b] = P\left[Z \leq \frac{b - \mu}{\sigma / \sqrt{n}}\right]$$

4. Apply the central limit theorem, when the sample size is large, to approximate the sampling distribution of $\overline{X}$ by a normal distribution with mean $\mu$ and variance $\sigma^2 / n$. The probability $P[X \leq b]$ is approximately equal to the standard normal probability $P\left[Z \leq \dfrac{b - \mu}{\sigma / \sqrt{n}}\right]$

5. Do not confuse the population distribution, which describes the variation for a single random variable, with the sampling distribution of a statistic.

6. When the population distribution is noticeably nonnormal, do not try to conclude that the sampling distribution of $\overline{X}$ is normal unless the sample size is at least moderately large, more than 30 by a rule of thumb.

## KEY IDEAS AND FORMULAS

The observations $X_1, X_2, \ldots, X_n$ are a random sample of size $n$ from the population distribution if they result from independent selections and each observation has the same distribution as the population. Under these conditions, we refer to the observations as a random sample.

A parameter is a numerical characteristic of the population. It is a constant, although its value is typically unknown to us. The object of a statistical analysis of sample data is to learn about the parameter.

A numerical characteristic of a sample is called a **statistic**. The value of a statistic varies in repeated sampling.

When random sampling from a population, a statistic is a random variable. The probability distribution of a statistic is called its **sampling distribution**.

The sampling distribution of $\overline{X}$ has mean $\mu$ and standard deviation $\sigma / \sqrt{n}$, where $\mu$ = population mean, $\sigma$ = population standard deviation, and $n$ = sample size.

With increasing $n$, the distribution of $\overline{X}$ is more concentrated around $\mu$.

If the population distribution is normal, $N(\mu, \sigma)$, the distribution of $\overline{X}$ is $N(\mu, \sigma / \sqrt{n})$.

Regardless of the shape of the population distribution, the distribution of $\overline{X}$ is approximately $N(\mu, \sigma / \sqrt{n})$, provided that $n$ is large. This result is called the **central limit theorem**.

## REVIEW EXERCISES

**7.36** The population density function and that for the sampling distribution of $\overline{X}$, for $n = 2$, are shown in Figure 10. Identify which one is the sampling distribution and explain your answer.

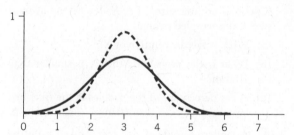

Figure 10   Two density functions. Exercise 7.36.

**7.37** Could the two density functions shown in Figure 11 be a population density function and that for the sampling distribution of $\overline{X}$. Explain why or why not.

Figure 11   Two density functions. Exercise 7.37.

**7.38** A population consists of the four numbers { 0, 2, 4, 6 }. Consider drawing a random sample of size 2 with replacement.

(a) List all possible samples and evaluate $\bar{x}$ for each.

(b) Determine the sampling distribution of $\overline{X}$.

(c) Write down the population distribution and calculate its mean $\mu$ and standard deviation $\sigma$.

(d) Calculate the mean and standard deviation of the sampling distribution of $\overline{X}$, obtained in part (b), and verify that these agree with $\mu$ and $\sigma / 2$, respectively.

**7.39** Refer to Exercise 7.38 and, instead of $\overline{X}$, consider the statistic

Sample range $R$ = Largest observation − Smallest observation

For instance, if the sample observations are (2, 6), the range is $6 - 2 = 4$.

(a) Calculate the sample range for all possible samples.

(b) Determine the sampling distribution of $R$.

**7.40** What sample size is required in order that the standard deviation of $\overline{X}$ be:

(a) $\frac{1}{4}$ of the population standard deviation?

(b) $\frac{1}{7}$ of the population standard deviation?

(c) 12% of the population standard deviation?

**7.41**    A population has distribution

| Value | Probability |
|-------|-------------|
| 1 | .2 |
| 2 | .6 |
| 3 | .2 |

Let $X_1$ and $X_2$ be independent and each have the same distribution as the population.

(a) Determine the missing elements in the table for the sampling distribution of $\overline{X} = (X_1 + X_2)/2$.

| $\overline{x}$ | Probability |
|------|-------------|
| 1.0 | |
| 1.5 | |
| 2.0 | .44 |
| 2.5 | .24 |
| 3.0 | |

(b) Find the expected value of $\overline{X}$.

(c) If the sample size is increased to 81, give the mean and variance of $\overline{X}$.

**7.42**    A population has distribution

| Value | Probability |
|-------|-------------|
| 1 | .6 |
| 3 | .3 |
| 5 | .1 |

Let $X_1$ and $X_2$ be independent and each have the same distribution as the population.

(a) Determine the missing elements in the table for the sampling distribution of $\overline{X} = (X_1 + X_2)/2$.

| $\overline{x}$ | Probability |
|------|-------------|
| 1 | |
| 2 | .36 |
| 3 | .21 |
| 4 | |
| 5 | |

(b) Find the expected value of $\overline{X}$.

(c) If the sample size is increased to 25, give the mean and variance of $\overline{X}$.

**7.43**    Suppose the weights of the contents of cans of mixed nuts have a normal distribution with mean 32.4 ounces and standard deviation .4 ounce. For a random sample of size $n = 9$

(a) What are the mean and standard deviation of $\overline{X}$?

(b) What is the distribution of $\overline{X}$? Is this distribution exact or approximate?

(c) Find the probability that $\overline{X}$ lies between 32.3 and 32.6.

**7.44**    The weights of pears in an orchard are normally distributed with mean .32 pound and standard deviation .08 pound.

(a) If one pear is selected at random, what is the probability that its weight is between .28 and .34 pound?

(b) If $\overline{X}$ denotes the average weight of a random sample of four pears, what is the probability that $\overline{X}$ is between .28 and .34 pound?

**7.45**    Suppose that the size of pebbles in a river bed is normally distributed with mean 12.1 mm and standard deviation 3.2 mm. A random sample of 9 pebbles are measured. Let $\overline{X}$ denote the average size of the sampled pebbles.

(a) What is the distribution of $\overline{X}$?

(b) What is the probability that $\overline{X}$ is smaller than 10 mm?

(c) What percentage of the pebbles in the river bed are of size smaller than 10 mm?

**7.46**    A random sample of size 150 is taken from the population of the ages of juniors enrolled at a large university during one semester. This population has mean 21.1 years and standard deviation 2.6. The population distribution is not normal.

(a) Is it reasonable to assume a normal distribution for the sample mean $\overline{X}$? Why or why not?

(b) Find the probability that $\overline{X}$ lies between 20.85 and 21.54 years.

(c) Find the probability that $\overline{X}$ exceeds 25.91 years.

**7.47**    A company that manufactures car mufflers finds that the labor to set up and run a nearly automatic machine has mean $\mu = 1.9$ hours and $\sigma = 1.2$ hours. For a random sample of 36 runs

(a) Determine the mean and standard deviation of $\overline{X}$.

(b) What can you say about the distribution of $\overline{X}$?

**7.48** Refer to Exercise 7.47. Evaluate

(a) $P[\overline{X} > 2.2]$

(b) $P[1.65 < \overline{X} < 2.25]$

**7.49** Visitors to a popular Internet site rated the newest gaming console on a scale of 1 to 5 stars. The following probability distribution is proposed based on over 1400 individual ratings.

| $x$ | $f(x)$ |
|---|---|
| 1 | .02 |
| 2 | .02 |
| 3 | .04 |
| 4 | .12 |
| 5 | .80 |

(a) For a future random sample of 40 ratings, what are the mean and standard deviation of $\overline{X}$?

(b) What is the distribution of $\overline{X}$? Is this distribution exact or approximate?

(c) Find the probability that $\overline{X}$ lies between 4.6 and 4.8 stars.

**7.50** A special purpose coating must have the proper abrasion. The standard deviation is known to be 21. Consider a random sample of 49 abrasion measurements.

(a) Find the probability that the sample mean $\overline{X}$ lies within 2 units of the population mean—that is, $P[-2 \leq \overline{X} - \mu \leq 2]$.

(b) Find the number $k$ so that

$$P[-k \leq \overline{X} - \mu \leq k] = .90.$$

(c) What is the probability that $\overline{X}$ differs from $\mu$ by more than 4 units?

**7.51** The daily number of kayaks sold, $X$, at a water sports store has the probability distribution

| $x$ | 0 | 1 | 2 |
|---|---|---|---|
| $f(x)$ | .5 | .3 | .2 |

(a) Find the expected number of kayaks sold in a day.

(b) Find the standard deviation of the number of kayaks sold in a day.

(c) Find the probability distribution of the total number of kayaks sold in the next two days. Suppose that the number of sales on different days are independent.

(d) Over the next 64 days, what is the approximate probability that at least 53 kayaks are sold?

(e) How many kayaks should the store order to have approximate probability .95 of meeting the total demand in the next 64 days?

**7.52** Suppose packages of cream cheese coming from an automated processor have weights that are normally distributed. For one day's production run, the mean is 8.2 ounces and the standard deviation is 0.1 ounce.

(a) If the packages of cream cheese are labeled 8 ounces, what proportion of the packages weigh less than the labeled amount?

(b) If only 5% of the packages exceed a specified weight $w$, what is the value of $w$?

(c) Suppose two packages are selected at random from the day's production. What is the probability that the average weight of the two packages is less than 8.3 ounces?

(d) Suppose 5 packages are selected at random from the day's production. What is the probability that at most one package weighs at least 8.3 ounces?

**7.53** Suppose the amount of sun block lotion in plastic bottles leaving a filling machine has a normal distribution. The bottles are labeled 300 milliliters (ml) but the actual mean is 302 ml and the standard deviation is 2 ml.

(a) What is the probability that an individual bottle contains less than 299 ml?

(b) Two bottles can be purchased together in a twin-pack. What is the probability that the mean content of bottles in a twin-pack is less than 299 ml? Assume the contents of the two bottles are independent.

(c) If you purchase two twin-packs of the lotion, what is the probability that only one of the twin-packs has a mean bottle content less than 299 ml?

## CLASS PROJECTS

1. (a) Count the number of occupants $X$ including the driver in each of 20 passing cars. Calculate the mean $\bar{x}$ of your sample.

   (b) Repeat part (a) 10 times.

   (c) Collect the data sets of the individual car counts $x$ from the entire class and construct a relative frequency histogram.

   (d) Collect the $\bar{x}$ values from the entire class (10 from each student) and construct a relative frequency histogram for $\bar{x}$ choosing appropriate class intervals.

   (e) Plot the two relative frequency histograms and comment on the closeness of their shapes to the normal distribution.

2. (a) Collect a sample of size 7 and compute $\bar{x}$ and the sample median.

   (b) Repeat part (a) 30 times.

   (c) Plot dot diagrams for the values of the two statistics in part (a). These plots reflect the individual sampling distributions.

   (d) Compare the amount of variation in $\overline{X}$ and the median.

   In this exercise, you might record weekly soft-drink consumptions, sentence lengths, or hours of sleep for different students.

## COMPUTER PROJECT

1. Conduct a simulation experiment on the computer to verify the central limit theorem. Generate $n = 6$ observations from the continuous distribution that is uniform on 0 to 1. Calculate $\overline{X}$ Repeat 150 times. Make a histogram of the $\overline{X}$ values and a normal-scores plot. Does the distribution of $\overline{X}$ appear to be normal for $n = 6$? You may wish to repeat with $n = 20$.

   If MINITAB is available, you could use the commands

   > **Calc > Random Data > Uniform.**
   > Type *150* after **Generate** and *C1 – C6* in **Store in Column(s)**. Click **OK**.
   > **Calc > Row Statistics**. Click **Mean** and type *C1 – C6* in **Input Variables**.
   > Type C7 in **Store**. Click **OK**.

   The 150 means in C7 can then be summarized using the dialog box sequence described in Chapter 2.

   > **Stat > Basic Statistics > Graphical summary.**
   > Type C7 in **Variables**. Click **OK**.

Any inference about a population parameter involves some uncertainty because it is based on a sample rather than the entire population. To be meaningful, a statistical inference must include a specification of the uncertainty that is determined using the ideas of probability and the sampling distribution of the statistic.

The purpose of an investigation, spelled out in a clear statement of purpose as described in Chapter 1, can reveal the nature of the inference required to answer important questions.

The two most important types of inferences are (1) **estimation of parameter(s)** and (2) **testing of statistical hypotheses**. The true value of a parameter is an unknown constant that can be correctly ascertained only by an exhaustive study of the population, if indeed that were possible. Our objective may be to obtain a guess or an estimate of the unknown true value along with a determination of its accuracy. This type of inference is called **estimation of parameters**. An alternative objective may be to examine whether the sample data support or contradict the investigator's conjecture about the true value of the parameter. This latter type of inference is called **testing of statistical hypotheses**.

**Example 1**  Types of Inference: Point Estimation, Interval Estimation, and Testing Hypotheses

What is the degree of participation in community service? A student at a large midwestern university questioned $n = 40$ students concerning the amount of time they spent doing community service during the past month. The data on times, in hours, are presented in Table 1.

**TABLE 1**  Community Service (hours) in Month

| 0 | 0 | 0 | 0 | 0 | 0 | 0 | 1 | 1 | 1 |
|---|---|---|---|---|---|---|---|---|---|
| 2 | 2 | 2 | 2 | 2 | 3 | 3 | 3 | 3 | 4 |
| 4 | 4 | 4 | 5 | 5 | 5 | 5 | 5 | 5 | 5 |
| 5 | 5 | 6 | 6 | 6 | 8 | 10 | 15 | 20 | 25 |

Employing the ideas of Chapter 2, we can calculate a descriptive summary for this set of measurements.

Sample mean $\bar{x} = 4.55$   Sample standard deviation $s = 5.17$
Sample median $= 4.00$   First quartile $= 1.50$   Third quartile $= 5.00$

However the target of our investigation is not just the particular set of measurements recorded but also concerns the vast population of hours of community service for all students at this university or even similar universities. The population distribution of time is unknown to us. Consequently, parameters such as the population mean $\mu$ and population standard deviation $\sigma$ are also unknown. If we take the view that the 40 observations represent a random sample from the population distribution of times, one goal of this study may be to "learn about $\mu$." More specifically, depending on the purpose of the study, we may wish to do one, two, or all three of the following:

1.  Estimate a single value for the unknown $\mu$ (**point estimation**).
2.  Determine an interval of plausible values for $\mu$ (**interval estimation**).
3.  Decide whether or not the mean time $\mu$ is 2.6 hours, which was found to be the mean time in an earlier study (**testing statistical hypotheses**).

**Example 2**    Inferences about an Unknown Proportion

A university financial official wishes to determine what proportion of last year's graduates, who took out student loans, are very concerned about repaying the loan. She feels, correctly, that this assessment could be made quickly and effectively by sampling a small fraction of those students.

The persons selected are asked if they are very concerned, as opposed to not very concerned, about how long it will take to repay their student loan. Suppose that a random sample of 300 graduates, who did take out student loans, are interviewed and 46 are very concerned about how long it will take them to repay their loan.[1]

A descriptive summary of this finding is provided by

$$\text{Sample proportion very concerned} = \frac{46}{200} = .23$$

Here the target population is the entire collection of last year's graduates that held student loans. The feature of interest, or parameter, is the proportion, $p$, that are very concerned about the length of time it will take them to repay the loan. The value of $p$ is unknown. The sample proportion $= .23$ sheds some light on $p$, but it is subject to some error since it draws only on a part of the population. The financial officer would like to evaluate its margin of error and provide an interval of plausible values of $p$.

Also, the financial official may wish to test the hypothesis that the proportion of very concerned graduates, $p$, for her university is not higher than the value given by a nationwide study.

# 2. POINT ESTIMATION OF A POPULATION MEAN

The object of point estimation is to calculate, from the sample data, a single number that is likely to be close to the unknown value of the parameter. The available information is assumed to be in the form of a random sample $X_1, X_2, \ldots, X_n$ of size $n$ taken from the population. We wish to formulate a

[1]The proportion is the same as that reported in a Harris Interactive survey, done for CouponCabin, in April 2012.

statistic such that its value computed from the sample data would closely reflect the value of the population parameter.

> A statistic intended for estimating a parameter is called a **point estimator**, or simply an **estimator**.
>     The standard deviation of an estimator is called its **standard error**: S.E.

When we estimate a population mean from a random sample, perhaps the most intuitive estimator is the sample mean,

$$\overline{X} = \frac{X_1 + X_2 + \cdots + X_n}{n}$$

For instance, to estimate the mean hours of community service in Example 1, we would naturally compute the mean of the sample measurements. Employing the estimator $\overline{X}$, with the data of Table 1, we get the result $\bar{x} = 4.55$ hours, which we call a **point estimate**, or simply an **estimate of** $\mu$.

Without an assessment of accuracy, a single number quoted as an estimate may not serve a very useful purpose. We must indicate the extent of variability in the distribution of the estimator. The standard deviation, alternatively called the **standard error** of the estimator, provides information about its variability.

In order to study the properties of the sample mean $\overline{X}$ as an estimator of the population mean $\mu$, let us review the results from Chapter 7.

1. $E(\overline{X}) = \mu$.
2. $\text{sd}(\overline{X}) = \dfrac{\sigma}{\sqrt{n}}$ so $\text{S.E.}(\overline{X}) = \dfrac{\sigma}{\sqrt{n}}$.
3. With large $n$, $\overline{X}$ is nearly normally distributed with mean $\mu$ and standard deviation $\sigma / \sqrt{n}$.

The first two results show that the distribution of $\overline{X}$ is centered around $\mu$ and its standard error is $\sigma / \sqrt{n}$, where $\sigma$ is the population standard deviation and $n$ the sample size.

To understand how closely $\overline{X}$ is expected to estimate $\mu$, we now examine the third result, which is depicted in Figure 1. Recall that, in a normal distribution,

Figure 1   Approximate normal distribution of $\overline{X}$.

the interval running two standard deviations on either side of the mean contains probability .954. Thus, prior to sampling, the probability is .954 that the estimator $\overline{X}$ will be at most a distance $2\sigma / \sqrt{n}$ from the true parameter value $\mu$. This probability statement can be rephrased by saying that **when we are estimating $\mu$ by $\overline{X}$, the 95.4% error margin is $2\sigma / \sqrt{n}$**.

Use of the probability .954, which corresponds to the multiplier 2 of the standard error, is by no means universal. The following notation will facilitate our writing of an expression for the $100(1 - \alpha)\%$ error margin, where $1 - \alpha$ denotes the desired high probability such as .95 or .90.

---

**Notation**

$z_{\alpha/2}$ = Upper $\alpha/2$ point of standard normal distribution

That is, the area to the right of $z_{\alpha/2}$ is $\alpha/2$, and the area between $-z_{\alpha/2}$ and $z_{\alpha/2}$ is $1 - \alpha$ (see Figure 2).

---

Figure 2    The notation $z_{\alpha/2}$.

A few values of $z_{\alpha/2}$ obtained from the normal table appear in Table 2 for easy reference. To illustrate the notation, suppose we want to determine the 90% error margin. We then set $1 - \alpha = .90$ so $\alpha/2 = .05$ and, from Table 2, we have $z_{.05} = 1.645$. Therefore, when estimating $\mu$ by $\overline{X}$, the 90% error margin is $1.645\,\sigma / \sqrt{n}$.

**TABLE 2**  Values of $z_{\alpha/2}$

| $1 - \alpha$ | .80 | .85 | .90 | .95 | .99 |
|---|---|---|---|---|---|
| $z_{\alpha/2}$ | 1.28 | 1.44 | 1.645 | 1.96 | 2.58 |

A minor difficulty remains in computing the standard error of $\overline{X}$. The expression involves the unknown population standard deviation $\sigma$, but we can estimate $\sigma$ by the sample standard deviation.

$$S = \sqrt{\frac{\sum_{i=1}^{n} (X_i - \overline{X})^2}{n - 1}}$$

When $n$ is large, the effect of estimating the standard error $\sigma / \sqrt{n}$ by $S / \sqrt{n}$ can be neglected. We now summarize.

---

### Point Estimation of the Mean

Parameter: Population mean $\mu$.

Data:        $X_1, \ldots, X_n$        (a random sample of size $n$)

Estimator: $\overline{X}$        (sample mean)

$$\text{S.E.}(\overline{X}) = \frac{\sigma}{\sqrt{n}} \qquad \text{Estimated S.E.}(\overline{X}) = \frac{S}{\sqrt{n}}$$

For large $n$, the $100(1 - \alpha)\%$ error margin is $z_{\alpha/2}\, \sigma / \sqrt{n}$.
(If $\sigma$ is unknown, use $S$ in place of $\sigma$.)

---

**Example 3**    Point Estimation of the Mean Time of Community Service

From the data of Example 1, consisting of 40 measurements of the time devoted to community service the past month, give a point estimate of the population mean amount of time and state a 95% error margin.

SOLUTION    The sample mean and the standard deviation computed from the 40 measurements in Table 1 are

$$\overline{x} = \frac{\sum x_i}{40} = 4.55$$

$$s = \sqrt{\frac{\sum (x_i - \overline{x})^2}{39}} = \sqrt{26.715} = 5.17$$

To calculate the 95% error margin, we set $1 - \alpha = .95$ so that $\alpha/2 = .025$ and $z_{\alpha/2} = 1.96$. Therefore, the 95% error margin is

$$\frac{1.96\, s}{\sqrt{n}} = \frac{1.96 \times 5.17}{\sqrt{40}} = 1.60 \text{ hours}$$

Our estimate of the population mean time is 4.55 hours per month. We do not expect the population mean to be exactly this value and we attach a 95% error margin of plus and minus 1.60 hours.

**Caution:**    (a) Standard error should not be interpreted as the "typical" error in a problem of estimation as the word "standard" may suggest. For instance, when $\text{S.E.}(\overline{X}) = .3$, we should not think that the error $(\overline{X} - \mu)$ is likely to be .3, but rather, prior to observing the data, the probability is approximately .954 that the error will be within $\pm 2(\text{S.E.}) = \pm .6$.

(b) An estimate and its variability are often reported in either of the forms: estimate $\pm$S.E. or estimate $\pm 2(\text{S.E.})$. In reporting a numerical result such as $53.4 \pm 4.6$, we must specify whether 4.6 represents S.E., 2(S.E.), or some other multiple of the standard error.

## 2.1 DETERMINING THE SAMPLE SIZE

During the planning stage of an investigation, it is important to address the question of sample size. Because sampling is costly and time-consuming, the investigator needs to know beforehand the sample size required to give the desired precision.

In order to determine how large a sample is needed for estimating a population mean, we must specify

$$d = \text{Desired error margin}$$

and

$$1 - \alpha = \text{Probability associated with error margin}$$

Referring to the expression for a $100(1 - \alpha)\%$ error margin, we then equate:

$$z_{\alpha/2} \frac{\sigma}{\sqrt{n}} = d$$

This gives an equation in which $n$ is unknown. Solving for $n$, we obtain

$$n = \left[ \frac{z_{a/2}\,\sigma}{d} \right]^2$$

which determines the required sample size. Of course, the solution is rounded to the next higher integer, because a sample size cannot be fractional.

This determination of sample size is valid provided $n > 30$, so that the normal approximation to $\overline{X}$ is satisfactory.

To be $100\,(1 - \alpha)\%$ sure that the error of estimation $|\overline{X} - \mu|$ does not exceed $d$, the **required sample size** is

$$n = \left[ \frac{z_{\alpha/2}\,\sigma}{d} \right]^2$$

If $\sigma$ is completely unknown, a small-scale preliminary sampling is necessary to obtain an estimate of $\sigma$ to be used in the formula to compute $n$.

**Example 4**    Determining a Sample Size for Collecting Water Samples

A limnologist wishes to estimate the mean phosphate content per unit volume of lake water. It is known from studies in previous years that the standard deviation has a fairly stable value of $\sigma = 4$. How many water samples must the limnologist analyze to be 90% certain that the error of estimation does not exceed 0.8 milligrams?

SOLUTION    Here $\sigma = 4$ and $1 - \alpha = .90$, so $\alpha / 2 = .05$. The upper .05 point of the $N(0, 1)$ distribution is $z_{.05} = 1.645$. The tolerable error is $d = .8$. Computing

$$ n = \left[ \frac{1.645 \times 4}{.8} \right]^2 = 67.65 $$

we round up to determine that the required sample size is $n = 68$.

## Exercises

**8.1**  A researcher wants to estimate $\mu$, the mean number of minutes before a person scores over 200 on a new computer game. She wishes to get estimates, separately, for each of the groups (a) novices, (b) occasional game players, and (c) expert game players. When using $\bar{X}$ to estimate $\mu$ find the (i) standard error and the (ii) $100(1 - \alpha)\%$ error margin for each group.

(a) Novice, $n = 125$, $\sigma = 65$, 98% error margin

(b) Occasional $n = 47$, $\sigma = 38$, 95% error margin

(c) Expert, $n = 6$, $\sigma = 20$, 90% error margin

**8.2**  A researcher asked participants to wear a simple counting device for a week and they counted the number of times that they thought about food.[2] Determine a point estimate of the daily mean number of times a person thinks about food and the $(1 - \alpha)\%$ error margin separately for males and females.

(a) Female    $n = 36$         $\bar{x} = 14.97$
             $s = 9.97$     $1 - \alpha = .95$

(b) Male       $n = 29$         $\bar{x} = 24.49$
             $s = 23.16$    $1 - \alpha = .95$

**8.3**  During 2010, the Wisconsin Department of Natural Resources captured 10 male wolves and 6 female wolves which they weighed, radio collared, and released.

(a) Female $n = 6$ $\bar{x} = 63.6$ pounds
$$ \sum (x_i - \bar{x})^2 = 333.5 $$

(b) Male $n = 10$ $\bar{x} = 79.4$ pounds
$$ \sum (x_i - \bar{x})^2 = 409.4 $$

Obtain the point estimate of $\mu$ and the estimated standard error separately for male and female wolves.

**8.4**  To study the growth of pine trees at an early stage, a nursery worker records 40 measurements of the heights (cm) of one-year-old red pine seedlings. (courtesy of Professor Alan Ek)

| | | | | | | | |
|---|---|---|---|---|---|---|---|
| 2.6 | 1.9 | 1.8 | 1.6 | 1.4 | 2.2 | 1.2 | 1.6 |
| 1.6 | 1.5 | 1.4 | 1.6 | 2.3 | 1.5 | 1.1 | 1.6 |
| 2.0 | 1.5 | 1.7 | 1.5 | 1.6 | 2.1 | 2.8 | 1.0 |
| 1.2 | 1.2 | 1.8 | 1.7 | 0.8 | 1.5 | 2.0 | 2.2 |
| 1.5 | 1.6 | 2.2 | 2.1 | 3.1 | 1.7} | 1.7 | 1.2 |

The summary statistics are

$$ n = 40 \quad \bar{x} = 1.715 \quad s = .475 \text{ centimeter} $$

[2]T. Fisher, Z. Moore, and M. Pittenger, "Sex on the brain?: an examination of frequency of sexual cognitions as a function of gender, erotophilia, and social desirability," *Journal of Sex Research* 49(1), (2012), pp. 69–77.

Find the (a) point estimate of the population mean height $\mu$, (b) standard error, and (c) 98% error margin.

**8.5** A credit company randomly selected 50 contested items and recorded the dollar amount being contested. These contested items had sample mean $\bar{x} = 95.74$ dollars and $s = 24.63$ dollars. Construct a point estimate for the population mean contested amount, $\mu$, and give its 90% error margin.

**8.6** A manager at a power company monitored the employee time required to process high-efficiency lamp bulb rebates. A random sample of 40 applications gave a sample mean time of 3.8 minutes and a standard deviation of 1.2 minutes. Construct a point estimate for the population mean time to process, $\mu$, and give its 90% error margin.

**8.7** When estimating $\mu$ from a large sample, suppose that one has found the 95% error margin of $\overline{X}$ to be 4.2. From this information, determine:

(a) The estimated S.E. of $\overline{X}$.

(b) The 90% error margin.

**8.8** The same owners operate two coffee shops in a large building. One is (a) small and the other (b) large. On any day, the number of customers is only observed for one shop. Determine the point estimate of $\mu$, the mean number of persons served during a weekday morning, and the $100(1 - \alpha)\%$ error margin, separately, for each shop when

(a) Small $n = 65$  $\bar{x} = 103$

$\qquad s = 15$  $1 - \alpha = .95$

(b) Large $n = 45$  $\bar{x} = 260$

$\qquad s = 40$  $1 - \alpha = .975$

**8.9** When estimating the mean of a population, how large a sample is required in order that the 95% error margin be:

(a) $\frac{1}{8}$ of the population standard deviation?

(b) 15% of the population standard deviation?

**8.10** Assume that the standard deviation of the number of violent incidents in one hour of children's shows on television is 3.2. An investigator would like to be 99% sure that the true mean number of violent incidents per hour is estimated within 1.4 incidents. For how many randomly selected hours does she need to count the number of violent incidents?

**8.11** Referring to Exercise 8.5, suppose that the survey of 50 contested items was, in fact, a pilot study intended to give an idea of the population standard deviation. Assuming $\sigma = \$25$, determine the sample size that is needed for estimating the population mean amount contested with a 98% error margin of $5.00.

**8.12** Assume that the standard deviation of the heights of five-year-old boys is 3.5 inches. How many five-year-old boys need to be sampled if we want to be 90% sure that the population mean height is estimated within .5 inch?

**8.13** Let the abbreviation PSLT stand for the percent of the gross family income that goes into paying state and local taxes. Suppose one wants to estimate the mean PSLT for the population of all families in New York City with gross incomes in the range $70,000 to $80,000. If $\sigma = 2.5$, how many such families should be surveyed if one wants to be 90% sure of being able to estimate the true mean PSLT within .5?

**8.14** To estimate $\mu$ with a 90% error margin of 2.9 units, one has determined that the required sample size is 108. What then is the required sample size if one wants the 95% error margin to be 1.8 units?

# 3. CONFIDENCE INTERVAL ESTIMATION OF A POPULATION MEAN

For point estimation, a single number lies in the forefront even though a standard error is attached. Instead, it is often more desirable to produce an interval of values that is likely to contain the true value of the parameter.

Ideally, we would like to be able to collect a sample and then use it to calculate an interval that would definitely contain the true value of the parameter. This goal, however, is not achievable because of sample-to-sample variation. Instead, we insist that, before sampling, the proposed interval will contain the true value with a specified high probability. This probability, called the **level of confidence**, is typically taken as .90, .95, or .99.

To develop this concept, we first confine our attention to the construction of a confidence interval for a population mean $\mu$, assuming that the population is normal and the standard deviation $\sigma$ is *known*. This restriction helps to simplify the initial presentation of the concept of a confidence interval. Later on, we treat the more realistic case where $\sigma$ is also unknown.

A probability statement about $\overline{X}$ based on the normal distribution provides the cornerstone for the development of a confidence interval. From Chapter 7, recall that when the population is normal, the distribution of $\overline{X}$ is also normal. It has mean $\mu$ and standard deviation $\sigma / \sqrt{n}$. Here $\mu$ is unknown, but $\sigma / \sqrt{n}$ is a known number because the sample size $n$ is known and we have assumed that $\sigma$ is known.

The normal table shows that the probability is .95 that a normal random variable will lie within 1.96 standard deviations from its mean. For $\overline{X}$, we then have

$$P \left[ \mu - 1.96 \frac{\sigma}{\sqrt{n}} < \overline{X} < \mu + 1.96 \frac{\sigma}{\sqrt{n}} \right] = .95$$

as shown in Figure 3.

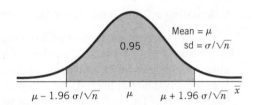

Figure 3   Normal distribution of $\overline{X}$.

Now, the relation

$$\mu - 1.96 \frac{\sigma}{\sqrt{n}} < \overline{X} \qquad \text{is the same as} \qquad \mu < \overline{X} + 1.96 \frac{\sigma}{\sqrt{n}}$$

and

$$\overline{X} < \mu + 1.96 \frac{\sigma}{\sqrt{n}} \qquad \text{is the same as} \qquad \overline{X} - 1.96 \frac{\sigma}{\sqrt{n}} < \mu$$

as we can see by transposing $1.96\sigma / \sqrt{n}$ from one side of an inequality to the other. Therefore, the event

$$\left[ \mu - 1.96 \frac{\sigma}{\sqrt{n}} < \overline{X} < \mu + 1.96 \frac{\sigma}{\sqrt{n}} \right]$$

is equivalent to

$$\left[ \overline{X} - 1.96 \frac{\sigma}{\sqrt{n}} < \mu < \overline{X} + 1.96 \frac{\sigma}{\sqrt{n}} \right]$$

In essence, both events state that the difference $\overline{X} - \mu$ lies between $-1.96\,\sigma\,/\sqrt{n}$ and $1.96\,\sigma\,/\sqrt{n}$. Thus, the probability statement

$$P\left[\mu - 1.96\,\frac{\sigma}{\sqrt{n}} < \overline{X} < \mu + 1.96\,\frac{\sigma}{\sqrt{n}}\right] = .95$$

can also be expressed as

$$P\left[\overline{X} - 1.96\,\frac{\sigma}{\sqrt{n}} < \mu < \overline{X} + 1.96\,\frac{\sigma}{\sqrt{n}}\right] = .95$$

This second form tells us that, before we sample, the random interval from $\overline{X} - 1.96\,\sigma\,/\sqrt{n}$ to $\overline{X} + 1.96\,\sigma\,/\sqrt{n}$ will include the unknown parameter $\mu$ with a probability of .95. Because $\sigma$ is assumed to be known, both the upper and lower endpoints can be computed as soon as the sample data are available. Guided by the above reasonings, we say that the interval

$$\left(\overline{X} - 1.96\,\frac{\sigma}{\sqrt{n}}\,, \quad \overline{X} + 1.96\,\frac{\sigma}{\sqrt{n}}\right)$$

or its realization $(\,\overline{x} - 1.96\,\sigma\,/\sqrt{n}\,, \; \overline{x} + 1.96\,\sigma\,/\sqrt{n}\,)$ is a 95% confidence interval for $\mu$ when the population is normal and $\sigma$ known.

**Example 5** Calculating a Confidence Interval—Normal Population $\sigma$ Known

The daily carbon monoxide (CO) emission from a large production plant is measured on 25 randomly selected weekdays. The production process is always being modified and the current mean value of daily CO emissions $\mu$ is unknown. Data collected over several years confirm that, for each year, the distribution of CO emission is normal with a standard deviation of .8 ton.

Suppose the sample mean is found to be $\overline{x} = 2.7$ tons. Construct a 95% confidence interval for the current daily mean emission $\mu$.

SOLUTION    The population is normal, and the observed value $\overline{x} = 2.7$.

$$\left(2.7 - 1.96\,\frac{.8}{\sqrt{25}}\,, \quad 2.7 + 1.96\,\frac{.8}{\sqrt{25}}\right) = (\,2.39,\,3.01\,)\text{ tons}$$

is a 95% confidence interval for $\mu$. Since $\mu$ is unknown, we cannot determine whether or not $\mu$ lies in this interval.

Referring to the confidence interval obtained in Example 5, we must **not** speak of the probability of the fixed interval (2.39, 3.01) covering the true mean $\mu$. The particular interval $(\,2.39,\,3.01\,)$ either does or does not cover $\mu$, and we will never know which is the case.

We need not always tie our discussion of confidence intervals to the choice of a 95% level of confidence. An investigator may wish to specify a different high probability. We denote this probability by $1 - \alpha$ and speak of a $100(\,1 - \alpha\,)\%$ confidence interval. The only change is to replace 1.96 with $z_{\alpha/2}$, where $z_{\alpha/2}$ denotes the upper $\alpha/2$ point of the standard normal distribution (i.e., the area to the right of $z_{\alpha/2}$ is $\alpha/2$, as shown in Figure 2).

In summary, when the population is normal and $\sigma$ is known, a $100(1 - \alpha)\%$ confidence interval for $\mu$ is given by

$$\left( \overline{X} - z_{\alpha/2} \frac{\sigma}{\sqrt{n}}, \quad \overline{X} + z_{\alpha/2} \frac{\sigma}{\sqrt{n}} \right)$$

## 3.1 INTERPRETATION OF CONFIDENCE INTERVALS

To better understand the meaning of a confidence statement, we use the computer to perform repeated samplings from a normal distribution with $\mu = 100$ and $\sigma = 10$. Ten samples of size 7 are selected, and a 95% confidence interval $\bar{x} \pm 1.96 \times 10/\sqrt{7}$ is computed from each. For the first sample, $\bar{x} = 104.3$ and the interval is $104.3 \pm 7.4$, or 96.9 to 111.7. This and the other intervals are illustrated in Figure 4, where each vertical line segment represents one confidence interval. The midpoint of a line is the observed value of $\overline{X}$ for that particular sample. Also note that all the intervals are of the same length $2 \times 1.96\sigma/\sqrt{n} = 14.8$. Of the 10 intervals shown, 9 cover the true value of $\mu$. This is not surprising, because the specified probability .95 represents the long-run relative frequency of these intervals covering the true $\mu = 100$.

Figure 4    Interpretation of the confidence interval for $\mu$.

Because confidence interval statements are the most useful way to communicate information obtained from a sample, certain aspects of their formulation merit special emphasis. Stated in terms of a 95% confidence interval for $\mu$, these are:

1.   Before we sample, a confidence interval $( \overline{X} - 1.96\sigma/\sqrt{n}, \overline{X} + 1.96 \sigma/\sqrt{n} )$ is a random interval that attempts to cover the true value of the parameter $\mu$.

2.  The probability

$$P\left[ \overline{X} - 1.96 \frac{\sigma}{\sqrt{n}} < \mu < \overline{X} + 1.96 \frac{\sigma}{\sqrt{n}} \right] = .95$$

interpreted as the long-run relative frequency over many repetitions of sampling asserts that about 95% of the intervals will cover $\mu$.

3.  Once $\bar{x}$ is calculated from an observed sample, the interval

$$\left( \bar{x} - 1.96 \frac{\sigma}{\sqrt{n}}, \quad \bar{x} + 1.96 \frac{\sigma}{\sqrt{n}} \right)$$

which is a realization of the random interval, is presented as a 95% confidence interval for $\mu$. A numerical interval having been determined, it is no longer sensible to speak about the probability of its covering a fixed quantity $\mu$.

4.  In any application we never know if the 95% confidence interval covers the unknown mean $\mu$. Relying on the long-run relative frequency of coverage in property 2, we adopt the terminology confidence once the interval is calculated.

At this point, one might protest, "I have only one sample and I am not really interested in repeated sampling." But if the confidence estimation techniques presented in this text are mastered and followed each time a problem of interval estimation arises, then over a lifetime approximately 95% of the intervals will cover the true parameter. Of course, this is contingent on the validity of the assumptions underlying the techniques—independent normal observations here.

## 3.2  LARGE SAMPLE CONFIDENCE INTERVALS FOR $\mu$

Having established the basic concepts underlying confidence interval statements, we now turn to the more realistic situation for which the population standard deviation $\sigma$ is unknown. We require the sample size $n$ to be large in order to dispense with the assumption of a normal population. The central limit theorem then tells us that $\overline{X}$ is nearly normal whatever the form of the population. Referring to the normal distribution of $\overline{X}$ in Figure 5 and the discussion accompanying

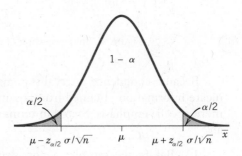

Figure 5    Normal distribution of $\overline{X}$.

Figure 2, we again have the probability statement

$$P\left[\overline{X} - z_{\alpha/2}\frac{\sigma}{\sqrt{n}} < \mu < \overline{X} + z_{\alpha/2}\frac{\sigma}{\sqrt{n}}\right] = 1 - \alpha$$

(Strictly speaking, this probability is approximately $1 - \alpha$ for a nonnormal population.) Even though the interval

$$\left(\overline{X} - z_{\alpha/2}\frac{\sigma}{\sqrt{n}}, \quad \overline{X} + z_{\alpha/2}\frac{\sigma}{\sqrt{n}}\right)$$

will include $\mu$ with the probability $1 - \alpha$, it does not serve as a confidence interval because it involves the unknown quantity $\sigma$. However, because $n$ is large, replacing $\sigma/\sqrt{n}$ with its estimator $S/\sqrt{n}$ does not appreciably affect the probability statement. Summarizing, we find that the large sample confidence interval for $\mu$ has the form

$$\text{Estimate} \pm (z\ \text{Value})(\text{Estimated standard error}).$$

---

### Large Sample Confidence Interval for $\mu$

When $n$ is large, a $100(1 - \alpha)\%$ confidence interval for $\mu$ is given by

$$\left(\overline{X} - z_{\alpha/2}\frac{S}{\sqrt{n}}, \quad \overline{X} + z_{\alpha/2}\frac{S}{\sqrt{n}}\right)$$

where $S$ is the sample standard deviation.

---

**Example 6**  Confidence Interval for the Daily Mean Time at Social Network Sites

Most college students are enamored with social network sites. A sample of 53 students at a large midwestern university answered the question, "How many minutes did you spend on the social network site yesterday?" A total of 5 stated they never use the network and were eliminated from the sample. The $n = 48$ times, in minutes, are

| | | | | | | | |
|---|---|---|---|---|---|---|---|
| 45 | 70 | 77 | 5 | 35 | 32 | 25 | 20 |
| 40 | 15 | 40 | 5 | 20 | 27 | 5 | 90 |
| 45 | 70 | 130 | 30 | 25 | 40 | 120 | 15 |
| 10 | 45 | 25 | 20 | 50 | 1 | 1 | 23 |
| 65 | 30 | 65 | 80 | 20 | 60 | 5 | 10 |
| 25 | 25 | 10 | 40 | 15 | 40 | 20 | 15 |

Compute (a) 90% and (b) 80% confidence intervals for the population mean time per day that college students spend on the social network site.

SOLUTION   A computer calculation produces the summary statistics

| Variable | N | Mean | StDev | Min | Q1 | Median | Q3 | Max |
|----------|---|------|-------|-----|-----|--------|-----|-----|
| Network(m) | 48 | 35.96 | 29.11 | 1 | 15 | 26 | 45 | 130 |

The sample mean does not include students who never use the site but only those that do. Let $\mu$ be the mean time, in minutes, that users spend at the site in one day. A histogram or box plot reveals that these data have a longer tail to the right. However, because the sample size $n = 48$ is large, the normal approximation to the distribution of $\overline{X}$ is appropriate. From the summary statistics, we have

$$\overline{x} = 35.96 \text{ minutes} \quad \text{and} \quad s = 29.11 \text{ minutes}$$

(a) With $1 - \alpha = .90$, we have $\alpha/2 = .05$ and $z_{.05} = 1.645$,

$$1.645\frac{s}{\sqrt{n}} = \frac{1.645 \times 29.11}{\sqrt{48}} = 6.91$$

The 90% confidence interval for the population mean number of minutes spent at a social network site, $\mu$, becomes

$$\left(\overline{x} - 1.645\frac{s}{\sqrt{n}}, \quad \overline{x} + 1.645\frac{s}{\sqrt{n}}\right) = (35.96 - 6.91, 35.96 + 6.91)$$

or, approximately $(29.1, 42.9)$ minutes per day. We can be 90% confident that the mean minutes per day $\mu$ is in the interval 29.1 to 42.9 minutes. We have this confidence because about 90% of the time, random samples of 48 students that use the site will produce intervals $\overline{x} \pm 1.645\, s/\sqrt{n}$ that contain $\mu$.

(b) With $1 - \alpha = .80$, we have $\alpha/2 = .10$ and $z_{.10} = 1.28$,

$$1.28\frac{s}{\sqrt{n}} = \frac{1.28 \times 29.11}{\sqrt{48}} = 5.38$$

The 80% confidence interval for $\mu$ becomes

$$(35.96 - 5.38, 35.96 + 5.38) \qquad \text{or} \qquad (30.6, 41.3)$$

Comparing the two confidence intervals, we note that the 80% confidence interval is shorter than the 90% interval. A shorter interval seems to give a more precise location for $\mu$ but suffers from a lower long-run frequency of being correct.

## 3.3  CONFIDENCE INTERVAL FOR A PARAMETER

The concept of a confidence interval applies to any parameter, not just the mean. It requires that a lower limit $L$ and an upper limit $U$ be computed from the sample data. Then the random interval from $L$ to $U$ must have the specified probability of covering the true value of the parameter. The large sample $100(1 - \alpha)\%$ confidence interval for $\mu$ has

$$L = \overline{X} - z_{\alpha/2} \frac{S}{\sqrt{n}} \qquad U = \overline{X} + z_{\alpha/2} \frac{S}{\sqrt{n}}$$

---

### Definition of a Confidence Interval for a Parameter

An interval $(L, U)$ is a $100(1 - \alpha)\%$ confidence interval for a parameter if

$$P[L < \text{Parameter} < U] = 1 - \alpha$$

and the endpoints $L$ and $U$ are computable from the sample.

---

**Example 7**   A Confidence Interval for the Mean Time to Complete a Test

Madison recruits for the fire department need to complete a timed test that simulates working conditions. It includes placing a ladder against a building, pulling out a section of fire hose, dragging a weighted object, and crawling in a simulated attic environment. The times, in seconds, for recruits to complete the test for Madison firefighter are

| | | | | | | | | | | | |
|---|---|---|---|---|---|---|---|---|---|---|---|
| 425 | 389 | 380 | 421 | 438 | 331 | 368 | 417 | 403 | 416 | 385 | 315 |
| 427 | 417 | 386 | 386 | 378 | 300 | 321 | 286 | 269 | 225 | 268 | 317 |
| 287 | 256 | 334 | 342 | 269 | 226 | 291 | 280 | 221 | 283 | 302 | 308 |
| 296 | 266 | 238 | 286 | 317 | 276 | 254 | 278 | 247 | 336 | 296 | 259 |
| 270 | 302 | 281 | 228 | 317 | 312 | 327 | 288 | 395 | 240 | 264 | 246 |
| 294 | 254 | 222 | 285 | 254 | 264 | 277 | 266 | 228 | 347 | 322 | 232 |
| 365 | 356 | 261 | 293 | 354 | 236 | 285 | 303 | 275 | 403 | 268 | 250 |
| 279 | 400 | 370 | 399 | 438 | 287 | 363 | 350 | 278 | 278 | 234 | 266 |
| 319 | 276 | 291 | 352 | 313 | 262 | 289 | 273 | 317 | 328 | 292 | 279 |
| 289 | 312 | 334 | 294 | 297 | 304 | 240 | 303 | 255 | 305 | 252 | 286 |
| 297 | 353 | 350 | 276 | 333 | 285 | 317 | 296 | 276 | 247 | 339 | 328 |
| 267 | 305 | 291 | 269 | 386 | 264 | 299 | 261 | 284 | 302 | 342 | 304 |
| 336 | 291 | 294 | 323 | 320 | 289 | 339 | 292 | 373 | 410 | 257 | 406 |
| 374 | 268 | | | | | | | | | | |

Obtain a 95% confidence interval for the mean time of recruits who complete the test.

SOLUTION   A computer calculation gives

| | |
|---|---|
| SAMPLE SIZE | 158 |
| MEAN | 307.77 |
| STD DEV | 51.852 |

Since $1 - \alpha = .95$, $\alpha / 2 = .025$, and $z_{.025} = 1.96$, the large sample 95% confidence interval for $\mu$ becomes

$$\left( \bar{x} - z_{\alpha/2} \frac{s}{\sqrt{n}}, \quad \bar{x} + z_{\alpha/2} \frac{s}{\sqrt{n}} \right)$$

$$= \left( 307.77 - 1.96 \frac{51.852}{\sqrt{158}}, \quad 307.77 + 1.96 \frac{51.852}{\sqrt{158}} \right)$$

or

$$( 299.68, 315.86 ) \text{ seconds}$$

When the sample size is large, the sample also contains information on the shape of the distribution that can be elicited by graphical displays. Figure 6 gives the stem-and-leaf display, with the data rounded to two places, accompanied by the boxplot. The confidence interval pertains to the mean of a population with a long right-hand tail.

```
STEM-AND-LEAF of TIME N    158
LEAF UNIT = 10

   2 │ 2222223333
   2 │ 44444555555555
   2 │ 6666666666666666777777777777777
   2 │ 88888888888888889999999999999999
   3 │ 000000000001111111111
   3 │ 222222233333333
   3 │ 444555555
   3 │ 6667777
   3 │ 88888899
   3 │ 00001111
   4 │ 22233
```

Figure 6   A stem-and-leaf display and boxplot give more information about the form of the population.

## Exercises

**8.15** Refer to Example 4, Chapter 2, on page 29. Table 2 records the number of items returned by 30 persons to a large discount department store in late December. The summary statistics are

$$n = 30, \quad \bar{x} = 2.538, \quad s = 1.303$$

Obtain a 98% confidence interval for $\mu$, the population mean number of items returned.

**8.16** A company wants to check the consistency of electronic copies of signatures for consumer credit purchases. A sample of 49 electronic signatures are available from the same customer. One measure of consistency in signing is the total length that the script is outside the signature box. The sample of 49 signatures yielded a mean length of .21 with a standard deviation of .19 centimeter.

Obtain a 99% confidence interval for this customer's population mean length outside the box.

**8.17** Each day of the year, a large sample of cellular phone calls is selected and a 95% confidence interval is calculated for the mean length of all cellular phone calls on that day. Of these 365 confidence intervals, one for each day of the year, approximately how many will cover their respective population means? Explain your reasoning.

**8.18**  A forester measures 100 needles off a pine tree and finds $\bar{x}$ = 3.1 centimeters and $s$ = 0.7 centimeter. She reports that a 95% confidence interval for the mean needle length is

$$3.1 - 1.96\frac{0.7}{\sqrt{100}} \quad \text{to} \quad 3.1 + 1.96\frac{0.7}{\sqrt{100}}$$

$$\text{or} \quad (2.96, 3.24)$$

(a) Is the statement correct?

(b) Does the interval ( 2.96, 3.24 ) cover the true mean? Explain.

**8.19**  In a study on the nutritional qualities of fast foods, the amount of fat was measured for a random sample of 35 hamburgers of a particular restaurant chain. The sample mean and standard deviation were found to be 30.2 and 3.8 grams, respectively. Use these data to construct a 95% confidence interval for the mean fat content in hamburgers served in these restaurants.

**8.20**  In the same study described in Exercise 8.19, the sodium content was also measured for the sampled hamburgers, and the sample mean and standard deviation were 658 and 47 milligrams, respectively. Determine a 98% confidence interval for the true mean sodium content.

**8.21**  An entomologist sprayed 120 adult Melon flies with a specific low concentration of malathion and observed their survival times. The mean and standard deviation were found to be 18.3 and 5.2 days, respectively. Use these data to construct a 99% confidence interval for the true mean survival time.

**8.22**  Students are asked about the number of songs they downloaded from a pay-for-songs Web site the last month. From a random sample of 39 students, the sample mean was 4.7 with a standard deviation of 3.2.

(a) Obtain a 95% confidence interval for $\mu$, the mean number of songs downloaded by the population of all students.

(b) Does $\mu$ lie in your interval obtained in part(a)?

(c) In a long series of repeated experiments, with new samples of 39 students collected for each experiment, what proportion of the resulting confidence intervals will contain the true population mean? Explain your reasoning.

**8.23**  The freshness of produce at a superstore is rated on a scale of 1 to 5 with 5 being very fresh. From a random sample of 49 customers, the average score was 3.8 with a standard deviation of .7.

(a) Obtain a 95% confidence interval for the population mean, $\mu$, the mean score for the distribution of all possible customers.

(b) Does $\mu$ lie in your interval obtained in part(a)? Explain.

(c) In a long series of repeated experiments, with new random samples of 49 customers each day, what proportion of the resulting confidence intervals will contain the true population mean? Explain your reasoning.

**8.24**  Referring to Example 7, where the 158 times to complete the firefighter test have mean 307.77 and standard deviation 51.852, obtain a 99% confidence interval for the mean time of all possible recruits who would complete the test.

**8.25**  Based on a survey of 140 employed persons in a city, the mean and standard deviation of the commuting distances between home and the principal place of business are found to be 8.6 and 4.3 miles, respectively. Determine a 90% confidence interval for the mean commuting distance for the population of all employed persons in the city.

**8.26**  A manager at a power company monitored the employee time required to process high-efficiency lamp bulb rebates. A random sample of 40 applications gave a sample mean time of 3.8 minutes and a standard deviation of 1.2 minutes. Construct a 90% confidence interval for the mean time to process $\mu$.

**8.27**  A credit company randomly selected 50 contested items and recorded the dollar amount being contested. These contested items had a sample mean $\bar{x}$ = 95.74 dollars and $s$ = 24.63 dollars. Construct a 95% confidence interval for the mean amount contested, $\mu$.

**8.28**  In a study to determine whether a certain stimulant produces hyperactivity, 55 mice were injected with 10 micrograms of the stimulant. Afterward, each mouse was given a hyperactivity rating score. The mean score was $\bar{x}$ = 14.9 and $s$ = 2.8. Give a 95% confidence interval for the population mean score $\mu$.

**8.29** Refer to the Statistics in Context section of Chapter 7 concerning monthly changes in the Canadian to U.S. exchange rate. A computer calculation gives $\bar{x} = -.0021$ and $s = .0234$ for the $n = 41$ monthly changes. Find a 95% confidence interval for the mean monthly change.

**8.30** An employee of an on-campus copy center wants to determine the mean number of copies before a cartridge needs to be replaced. She records the life length in thousands of copies for 43 cartridges and obtains

$$n = 43 \quad \bar{x} = 8.12 \quad s = 1.78 \text{ thousand copies}$$

Obtain a 90% confidence interval for the population mean, $\mu$, number of copies in thousands before a cartridge should be replaced.

**8.31** Refer to the 40 height measurements given in Exercise 8.4, which have

$$n = 40 \quad \bar{x} = 1.715 \quad s = .475 \text{ centimeter}$$

Calculate a 99% confidence interval for the population mean height.

**8.32** Radiation measurements on a sample of 65 microwave ovens produced $\bar{x} = .11$ and $s = .06$. Determine a 95% confidence interval for the mean radiation.

**8.33** Refer to the data on the growth of female salmon in the marine environment in Table D.7 of the Data Bank. A computer calculation gives a 95% confidence interval.

One-Sample Z: Fmarine

| Variable | N | Mean | StDev | 95.0% CI |
|---|---|---|---|---|
| Fmarine | 40 | 429.15 | 41.05 | ( 416.43, 441.87) |

(a) Does the 95% confidence interval cover the true mean growth of all female salmon in that marine environment?

(b) Why are you 95% confident that the interval ( 416.43, 441.87 ) covers the true mean?

**8.34** Refer to the data on the girth, in centimeters, of grizzly bears in Table D.6 of the Data Bank. A computer calculation gives

One-Sample Z: Girth

| Variable | N | Mean | StDev | 95.0% CI |
|---|---|---|---|---|
| Girth | 61 | 93.39 | 21.79 | ( 87.93, 98.86) |

(a) Does the 95% confidence interval cover the true mean girth of all grizzly bears in the area of the study? Explain.

(b) Why are you 95% confident that the interval ( 87.93, 98.86 ) covers the true mean?

**8.35** The amount of PCBs (polychlorinated biphenyls) was measured in 40 samples of soil that were treated with contaminated sludge. The following summary statistics were obtained.

$$\bar{x} = 3.56 \quad s = .5 \text{ ppm}$$

(a) Obtain a 95% confidence interval for the population mean $\mu$, amount of PCBs in the soil. Answer parts (b), (c), and (d) Yes, No, or Cannot tell. Explain your answer.

(b) Does the sample mean PCB content lie in your interval obtained in part (a)?

(c) Does the population mean PCB content lie in your interval obtained in part (a)?

(d) Is it likely that 95% of the data lie in your interval obtained in part (a)?

**8.36** A national fast food chain, with thousands of franchise locations, needed to audit the books at each location. They first selected a sample of 50 locations and performed the audit. They determined that a 95% confidence interval for the mean time to complete an audit is

( 28.4 hours, 52.7 hours )

Answer the following questions "Yes," "No," or "Cannot tell" and justify your answer.

(a) Does the population mean lie in the interval ( 28.4, 52.7 )?

(b) Does the sample mean lie in the interval ( 28.4, 52.7 )?

(c) For a future sample of 50 franchise locations, will the sample mean fall in the interval ( 28.4, 52.7 )?

(d) Does 95% of the sample data lie in the interval ( 28.4, 52.7 )?

# 4.   TESTING HYPOTHESES ABOUT A POPULATION MEAN

Broadly speaking, the goal of testing statistical hypotheses is to determine if a claim or conjecture about some feature of the population, a parameter, is strongly supported by the information obtained from the sample data. Here we illustrate the testing of hypotheses concerning a population mean $\mu$. The available data will be assumed to be a random sample of size $n$ from the population of interest. Further, the sample size $n$ will be large ($n > 30$ for a rule of thumb).

The formulation of a hypotheses testing problem and then the steps for solving it require a number of definitions and concepts. We will introduce these key statistical concepts

Null hypothesis and the alternative hypothesis

Type I and Type II errors

Level of significance

Rejection region

$P$–value

in the context of a specific problem to help integrate them with intuitive reasoning.

> **PROBLEM:   Can an upgrade reduce the mean transaction time at automated teller machines?**   At peak periods, customers are subject to unreasonably long waits before receiving cash. To help alleviate this difficulty, the bank wants to reduce the time it takes a customer to complete a transaction. From extensive records, it is found that the transaction times have a distribution with mean 270 and standard deviation 24 seconds. The teller machine vendor suggests that a new software and hardware upgrade will reduce the mean time for a customer to complete a transaction. For experimental verification, a random sample of 38 transaction times is taken at a machine with the upgrade and the sample mean $\overline{X}$ calculated. How should the result be used toward a statistical validation of the claim that the true (population) mean transaction time is less than 270 seconds?

Whenever we seek to establish a claim or conjecture on the basis of strong support from sample data, the problem is called one of hypothesis testing.

## 4.1  FORMULATING THE HYPOTHESES

In the language of statistics, the claim or the research hypothesis that we wish to establish is called the alternative hypothesis $H_1$. The opposite statement, one that nullifies the research hypothesis, is called the null hypothesis $H_0$. The word "null" in this context means that the assertion we are seeking to establish is actually void.

> ### Formulation of $H_0$ and $H_1$
>
> When our goal is to establish an assertion with substantive support obtained from the sample, the negation of the assertion is taken to be the null hypothesis $H_0$ and the assertion itself is taken to be the alternative hypothesis $H_1$.

Our initial question, "Is there strong evidence in support of the claim?" now translates to "Is there strong evidence for rejecting $H_0$?" The first version typically appears in the statement of a practical problem, whereas the second version is ingrained in the conduct of a statistical test. It is crucial to understand the correspondence between the two formulations of a question.

Before claiming that a statement is established statistically, adequate evidence from data must be produced to support it. A close analogy can be made to a criminal court trial where the jury clings to the null hypothesis of "not guilty" unless there is convincing evidence of guilt. The intent of the hearings is to establish the assertion that the accused is guilty, rather than to prove that he or she is innocent.

|  | Court Trial | Testing Statistical Hypothesis |
|---|---|---|
| **Requires strong evidence to establish:** | Guilt. | Conjecture (research hypothesis). |
| **Null hypothesis ($H_0$):** | Not guilty. | Conjecture is false. |
| **Alternative hypothesis ($H_1$):** | Guilty. | Conjecture is true. |
| **Attitude:** | Uphold "not guilty" unless there is a strong evidence of guilt. | Don't reject the null hypothesis unless it makes the sample data very unlikely to happen. |

False rejection of $H_0$ is a more serious error than failing to reject $H_0$ when $H_1$ is true.

Once $H_0$ and $H_1$ are formulated, our goal is to analyze the sample data in order to choose between them.

A **decision rule**, or a **test of the null hypothesis**, specifies a course of action by stating what sample information is to be used and how it is to be used in making a decision. Bear in mind that we are to make one of the following two decisions:

---

### Decisions

Either

Reject $H_0$ and conclude that $H_1$ is substantiated,

or

Retain $H_0$ and conclude that $H_1$ fails to be substantiated.

---

Rejection of $H_0$ amounts to saying that $H_1$ is substantiated, whereas nonrejection or retention of $H_0$ means that $H_1$ fails to be substantiated. A key point is that a decision to reject $H_0$ must be based on strong evidence. Otherwise, the claim $H_1$ could not be established beyond a reasonable doubt.

In our problem of evaluating the upgraded teller machine, let $\mu$ be the population mean transaction time. Because $\mu$ is claimed to be lower than 270 seconds, we formulate the alternative hypothesis as $H_1 : \mu < 270$. According to the description of the problem, the researcher does not care to distinguish between the situations that $\mu = 270$ and $\mu > 270$ for the claim is false in either case. For this reason, it is customary to write the null hypothesis simply as a statement of no difference. Accordingly, we formulate the

### Testing Problem

Test $H_0 : \mu = 270$     versus     $H_1 : \mu < 270$

## 4.2 TEST CRITERION AND REJECTION REGION

Naturally, the sample mean $\overline{X}$, calculated from the measurements of $n = 38$ randomly selected transaction times, ought to be the basis for rejecting $H_0$ or not. The question now is: For what sort of values of $\overline{X}$ should we reject $H_0$?

We first answer for the simplest setting where $\sigma$ is *known*. Because the claim states that $\mu$ is low (a left-sided alternative), only low values of $\overline{X}$ can contradict $H_0$ in favor of $H_1$. Therefore, a reasonable decision rule should be of the form

Reject $H_0$   if   $\overline{X} \leq c$

Retain $H_0$   if   $\overline{X} > c$

This decision rule is conveniently expressed as $R : \overline{X} \leq c$, where $R$ stands for the rejection of $H_0$. Also, in this context, the set of outcomes $[\overline{X} \leq c]$ is called the rejection region or critical region, and the cutoff point $c$ is called the critical value.

The cutoff point $c$ must be specified in order to fully describe a decision rule. To this end, we consider the case when $H_0$ holds, that is, $\mu = 270$. Rejection of $H_0$ would then be a wrong decision, amounting to a false acceptance of the claim—a serious error. For an adequate protection against this kind of error, we must ensure that $P[\overline{X} \le c]$ is very small when $\mu = 270$. For example, suppose that we wish to hold a low probability of $\alpha = .05$ for a wrong rejection of $H_0$. Then our task is to find the $c$ that makes

$$P[\overline{X} \le c] = .05 \quad \text{when} \quad \mu = 270$$

We know that, for large $n$, the distribution of $\overline{X}$ is approximately normal with mean $\mu$ and standard deviation $\sigma / \sqrt{n}$, whatever the form of the underlying population. Here $n = 38$ is large, and we initially assume that $\sigma$ is known. Specifically, we assume that $\sigma = 24$ seconds, the same standard deviation as with the original money machines. Then, when $\mu = 270$, the distribution of $\overline{X}$ is $N\left(270, 24 / \sqrt{38}\right)$ so

$$Z = \frac{\overline{X} - 270}{24 / \sqrt{38}}$$

has the $N(0, 1)$ distribution.

Because $P[Z \le -1.645] = .05$, the cutoff $c$ on the $\bar{x}$ scale must be 1.645 standard deviations below $\mu_0 = 270$ or

$$c = 270 - 1.645\left(\frac{24}{\sqrt{38}}\right) = 270 - 6.40 = 263.60$$

Our decision rule is now completely specified by the rejection region (see Figure 7)

$$R : \overline{X} \le 263.6$$

that has $\alpha = .05$ as the probability of wrongly rejecting $H_0$.

Instead of locating the rejection region on the scale of $\overline{X}$, we can cast the decision criterion on the standardized scale as well:

$$Z = \frac{\overline{X} - \mu_0}{\sigma / \sqrt{n}} = \frac{\overline{X} - 270}{24 / \sqrt{38}}$$

.05

$R : \overline{X} \le 263.6$     270.0     $\bar{x}$

$R : Z \le -1.645$     0     $z$

Figure 7   Rejection region with the cutoff $c = 263.6$.

and set the rejection region as $R : Z \leq -1.645$ (see Figure 7). This form is more convenient because the cutoff $-1.645$ is directly read off the normal table, whereas the determination of $c$ involves additional numerical work.

> The random variable $\overline{X}$ whose value serves to determine the action is called the **test statistic**.
>
> A **test of the null hypothesis** is a course of action specifying the set of values of a test statistic $\overline{X}$, for which $H_0$ is to be rejected.
>
> This set is called the **rejection region** of the test.

A test is completely specified by a test statistic and the rejection region.

## 4.3  TWO TYPES OF ERROR AND THEIR PROBABILITIES

Up to this point we only considered the probability of rejecting $H_0$ when, in fact, $H_0$ is true and illustrated how a decision rule is determined by setting this probability equal to .05. The following table shows all the consequences that might arise from the use of a decision rule.

| Decision Based on Sample | Unknown True Situation | |
|---|---|---|
| | $H_0$ True $\mu = 270$ | $H_1$ True $\mu < 270$ |
| Reject $H_0$ | Wrong rejection of $H_0$ (Type I error) | Correct decision |
| Retain $H_0$ | Correct decision | Wrong retention of $H_0$ (Type II error) |

In particular, when our sample-based decision is to reject $H_0$, either we have a correct decision (if $H_1$ is true) or we commit a Type I error (if $H_0$ is true). On the other hand, a decision to retain $H_0$ either constitutes a correct decision (if $H_0$ is true) or leads to a Type II error. To summarize:

> ### Two Types of Error
>
> Type I error: Rejection of $H_0$ when $H_0$ is true
>
> Type II error: Nonrejection of $H_0$ when $H_1$ is true
>
> $\alpha = $ Probability of making a Type I error
>
> (also called the level of significance)
>
> $\beta = $ Probability of making a Type II error

In our problem of evaluating the upgraded teller machine, the rejection region is of the form $R: \overline{X} \leq c$; so that,

$$\alpha = P[\overline{X} \leq c] \qquad \text{when} \qquad \mu = 270 \qquad (H_0 \text{ true})$$

$$\beta = P[\overline{X} > c] \qquad \text{when} \qquad \mu < 270 \qquad (H_1 \text{ true})$$

Of course, the probability $\beta$ depends on the numerical value of $\mu$ that prevails under $H_1$. Figure 8 shows the Type I error probability $\alpha$ as the shaded area under the normal curve that has $\mu = 270$ and the Type II error probability $\beta$ as the shaded area under the normal curve that has $\mu = 262$, a case of $H_1$ being true.

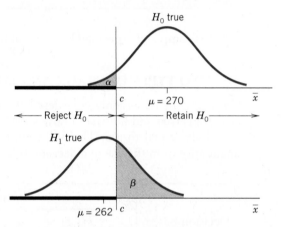

Figure 8   The error probabilities $\alpha$ and $\beta$.

From Figure 8, it is apparent that no choice of the cutoff $c$ can minimize both the error probabilities $\alpha$ and $\beta$. If $c$ is moved to the left, $\alpha$ gets smaller but $\beta$ gets larger, and if $c$ is moved to the right, just the opposite effects take place. In view of this dilemma and the fact that a wrong rejection of $H_0$ is the more serious error, we hold $\alpha$ at a predetermined low level such as .10, .05, or .01 when choosing a rejection region. We will not pursue the evaluation of $\beta$, but we do note that if the $\beta$ turns out to be uncomfortably large, the sample size must be increased.

## 4.4 PERFORMING A TEST

When determining the rejection region for this example, we assumed that $\sigma = 24$ seconds, the same standard deviation as with the original money machines. Then, when $\mu = 270$, the distribution of $\overline{X}$ is $N\left(270, 24 / \sqrt{38}\right)$ and the rejection region $R: \overline{X} \leq 263.6$ was arrived at by fixing $\alpha = .05$ and referring

$$Z = \frac{\overline{X} - 270}{24 / \sqrt{38}}$$

to the standard normal distribution.

In practice, we are usually not sure about the assumption that $\sigma = 24$, the standard deviation of the transaction times using the upgraded machine, is the same as with the original teller machine. But that does not cause any problem as long as the sample size is large. When $n$ is large ($n > 30$), the normal approximation for $\overline{X}$ remains valid even if $\sigma$ is estimated by the sample standard deviation $S$. Therefore, we employ the test statistic

$$Z = \frac{\overline{X} - \mu_0}{S / \sqrt{n}}$$

which has, approximately, a standard normal distribution. We use $Z$ to remind us of this fact. For testing $H_0 : \mu = \mu_0$ versus $H_1 : \mu < \mu_0$ with level of significance $\alpha$, we set the rejection region $R : Z \leq -z_\alpha$. This test is commonly called a **large sample normal test** or a **Z test**.

**Example 8**    A One-Sided Test of Hypotheses to Establish That $\mu < 270$

Referring to the automated teller machine transaction times, suppose that, from the measurements of a random sample of 38 transaction times, the sample mean and standard deviation are found to be 261 and 22 seconds, respectively. Test the null hypothesis $H_0 : \mu = 270$ versus $H_1 : \mu < 270$ using a 2.5% level of significance and state whether or not the claim $\mu < 270$ is substantiated.

SOLUTION    Because $n = 38$ and the null hypothesis specifies that $\mu$ has the value $\mu_0 = 270$, we employ the test statistic

$$Z = \frac{\overline{X} - 270}{S / \sqrt{38}}$$

The rejection region should consist of small values of $Z$ because $H_1$ is left-sided. For a 2.5% level of significance, we take $\alpha = .025$, and since $z_{.025} = 1.96$, the rejection region is (see Figure 9) $R : Z \leq -1.96$.
With the observed values $\bar{x} = 261$ and $s = 22$, we calculate the test statistic

$$z = \frac{261 - 270}{22 / \sqrt{38}} = -2.52$$

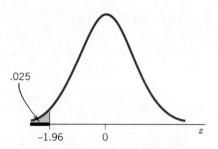

.025

−1.96        0

Figure 9    Rejection region for Z.

Because this observed $z$ is in $R$, the null hypothesis is rejected at the level of significance $\alpha = .025$. We conclude that the claim of a reduction in the mean transaction time is strongly supported by the data.

## 4.5  *P*-VALUE: HOW STRONG IS A REJECTION OF $H_0$?

Our test in Example 8 was based on the fixed level of significance $\alpha = .025$, and we rejected $H_0$ because the observed $z = -2.52$ fell in the rejection region $R: Z \leq -1.96$. A strong evidence against $H_0$ emerged due to the fact that a small $\alpha$ was used. The natural question at this point is: How small an $\alpha$ could we use and still arrive at the conclusion of rejecting $H_0$? To answer this question, we consider the observed $z = -2.52$ itself as the cutoff point (critical value) and calculate the rejection probability

$$P [ Z \leq -2.52 ] = .0059$$

The smallest possible $\alpha$ that would permit rejection of $H_0$, on the basis of the observed $z = -2.52$, is therefore .0059 (see Figure 10). It is called the significance probability or *P*-value of the observed $z$. This very small *P*-value, .0059, signifies a strong rejection of $H_0$ or that the result is highly statistically significant.

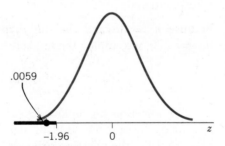

Figure 10   *P*-value with left-sided rejection region.

> The *P*-value is the probability, calculated under $H_0$, that the test statistic takes a value equal to or more extreme than the value actually observed.

The *P*-value serves as a measure of the strength of evidence against $H_0$. A small *P*-value means that the null hypothesis is strongly rejected or the result is highly statistically significant.

Our illustrations of the basic concepts of hypothesis tests thus far focused on a problem where the alternative hypothesis is of the form $H_1: \mu < \mu_0$, called a left-sided alternative. If the alternative hypothesis in a problem states that the true $\mu$ is larger than its null hypothesis value of $\mu_0$, we formulate the right-sided alternative $H_1 : \mu > \mu_0$ and use a right-sided rejection region $R : Z \geq z_\alpha$.

---

### The Steps for Testing Hypotheses

1. Formulate the null hypothesis $H_0$ and the alternative hypothesis $H_1$.
2. Test criterion: State the test statistic and the form of the rejection region.
3. With a specified $\alpha$, determine the rejection region.
4. Calculate the test statistic from the data.
5. Draw a conclusion: State whether or not $H_0$ is rejected at the specified $\alpha$ and interpret the conclusion in the context of the problem. Also, it is a good statistical practice to calculate the $P$–value and strengthen the conclusion.

---

We illustrate the right-sided case and the main steps for conducting a statistical test as summarized above.

**Example 9**    Evaluating a Weight Loss Diet—Calculation of a $P$–Value

A brochure inviting subscriptions for a new diet program states that the participants are expected to lose over 22 pounds in five weeks. Suppose that, from the data of the five-week weight losses of 56 participants, the sample mean and standard deviation are found to be 23.5 and 10.2 pounds, respectively. Could the statement in the brochure be substantiated on the basis of these findings? Test with $\alpha = .05$. Also calculate the $P$–value and interpret the result.

SOLUTION    Let $\mu$ denote the population mean weight loss from five weeks of participation in the program. Because our aim is to substantiate the assertion that $\mu > 22$ pounds, we formulate the hypotheses

$$H_0 : \mu = 22 \quad \text{versus} \quad H_1 : \mu > 22$$

The sample size is $n = 56$. Denoting the sample mean weight loss of the 56 participants by $\overline{X}$ and the standard deviation by $S$, our test statistic is

$$Z = \frac{\overline{X} - \mu_0}{S / \sqrt{n}} = \frac{\overline{X} - 22}{S / \sqrt{56}}$$

Since $H_1$ is right-sided, the rejection region should be right-sided $R : Z \geq c$. Because $z_{.05} = 1.645$, the test with level of significance .05 has the rejection region (see Figure 11) $R : Z \geq 1.645$.

Figure 11   Right-sided rejection region
with $\alpha = .05$.

With the observed values $\bar{x} = 23.5$ and $s = 10.2$, we calculate

$$z = \frac{23.5 - 22}{10.2 / \sqrt{56}} = 1.10$$

Because 1.10 is not in $R$, we do not reject the null hypothesis. We conclude that, with level of significance $\alpha = .05$, the stated claim that $\mu > 22$ is not substantiated.

Because our observed $z$ is 1.10 and larger values are more extreme, the significance probability of this result is

$$P\text{–value} = P\ [Z \geq 1.10] = .1357 \quad \text{(from the normal table)}$$

That is, .1357 is the smallest $\alpha$ at which $H_0$ could be rejected (Figure 12). This is not ordinarily considered a negligible chance so we conclude that the data do not provide a strong basis for rejection of $H_0$.

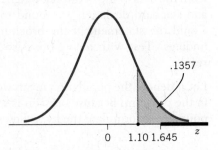

Figure 12   $P$–value with right-sided
rejection region.

The preceding hypotheses are called one-sided hypotheses, because the values of the parameter $\mu$ under the alternative hypothesis lie on one side of those under the null hypothesis. The corresponding tests are called one-sided tests or one-tailed tests. By contrast, we can have a problem of testing the null hypothesis

$$H_0 : \mu = \mu_0$$

versus the **two-sided alternative** or **two-sided hypothesis**

$$H_1: \mu \neq \mu_0$$

Here $H_0$ is to be rejected if $\overline{X}$ is too far away from $\mu_0$ in either direction, that is, if $Z$ is too small or too large. For a level $\alpha$ test, we divide the rejection probability $\alpha$ equally between the two tails and construct the rejection region

$$R: Z \leq -z_{\alpha/2} \quad \text{or} \quad Z \geq z_{\alpha/2}$$

which can be expressed in the more compact notation

$$R: |Z| \geq z_{\alpha/2}$$

**Example 10**    Testing Hypotheses about the Mean Time of Community Service

Consider the data of Example 1 concerning 40 observations on time devoted to community service in the past month. Do these data indicate that the population mean time is different from 2.6 hours?

SOLUTION    We are seeking evidence in support of $\mu \neq 2.6$ so the hypotheses should be formulated as

$$H_0: \mu = 2.6 \quad \text{versus} \quad H_1: \mu \neq 2.6$$

The sample size $n = 40$ being large, we will employ the test statistic

$$Z = \frac{\overline{X} - \mu_0}{S/\sqrt{n}} = \frac{\overline{X} - 2.6}{S/\sqrt{40}}$$

The two-sided form of $H_1$ dictates that the rejection region must also be two-sided.

Let us choose $\alpha = .05$, then $\alpha/2 = .025$ and $z_{.025} = 1.96$. Consequently, for $\alpha = .05$, the rejection region is (see Figure 13)

$$R: |Z| \geq 1.96$$

From Example 1, $\bar{x} = 4.55$ and $s = 5.17$, so the observed value of the test statistic is

$$z = \frac{\bar{x} - \mu_0}{s/\sqrt{n}} = \frac{4.55 - 2.6}{5.17/\sqrt{40}} = 2.39$$

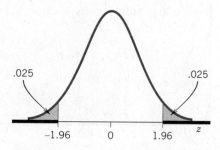

.025                                          .025

−1.96          0          1.96

Figure 13    Two-sided rejection region with $\alpha = .05$.

.0084                          .0084

-1.96      0      1.96    $z$

Figure 14   P–value with two-sided
rejection region.

Because $|z| = 2.39$ is larger than 1.96, we reject $H_0$ at $\alpha = .05$.

In fact, the large value $|z| = 2.39$ seems to indicate a much stronger rejection of $H_0$ than that arising from the choice of $\alpha = .05$. How small an $\alpha$ can we set and still reject $H_0$? This is precisely the idea underlying the significance probability or the $P$–value. We calculate (see Figure 14)

$$P\text{–value} = P\,[\,|Z| \geq 2.39\,]$$

$$= P\,[Z \leq -2.39\,] + P\,[Z \geq 2.39\,]$$

$$= 2 \times .0084 = .0168$$

With $\alpha$ as small as .0168, $H_0$ is still rejected. This very small $P$–value gives strong support for $H_1$.

In summary:

---

### Large Sample Tests for $\mu$

When the sample size is large, a $Z$ test concerning $\mu$ is based on the normal test statistic

$$Z = \frac{\overline{X} - \mu_0}{S / \sqrt{n}}$$

The rejection region is one- or two-sided depending on the alternative hypothesis. Specifically,

| | | |
|---|---|---|
| $H_1 : \mu > \mu_0$ | requires | $R : Z \geq z_\alpha$ |
| $H_1 : \mu < \mu_0$ | | $R : Z \leq -z_\alpha$ |
| $H_1 : \mu \neq \mu_0$ | | $R : |Z| \geq z_{\alpha/2}$ |

---

Because the central limit theorem prevails for large $n$, no assumption is required as to the shape of the population distribution.

# Exercises

**8.37** Stated here are some claims or research hypotheses that are to be substantiated by sample data. In each case, identify the null hypothesis $H_0$ and the alternative hypothesis $H_1$ in terms of the population mean $\mu$.

(a) The mean time a health insurance company takes to pay claims is less than 14 working days.

(b) The average person watching a movie at a local multiplex theater spends over $6.50 on refreshments.

(c) The mean hospital bill for a birth in the city is less than $7000.

(d) The mean time between purchases of a brand of mouthwash by loyal customers is different from 60 days.

**8.38** From an analysis of the sample data, suppose that the decision has been to reject the null hypothesis. In the context of each part ( a–d ) of Exercise 8.37, answer the following questions:

(a) In what circumstance is it a correct decision?

(b) When is it a wrong decision, and what type of error is then made?

**8.39** From an analysis of the sample data, suppose that the decision has been made to retain the null hypothesis. In the context of each part ( a–d ) of Exercise 8.37, answer the following questions.

(a) In what circumstance is it a correct decision?

(b) When is it a wrong decision, and what type of error is then made?

**8.40** For each situation ( a–d ) in Exercise 8.37, state which of the following three forms of the rejection region is appropriate when $\sigma$ is known.

$$R : \overline{X} \leq c \quad \text{( left-sided )}$$
$$R : \overline{X} \geq c \quad \text{( right-sided )}$$
$$R : \left| \overline{X} - \mu_0 \right| \geq c \quad \text{( two-sided )}$$

**8.41** Each part ( a–d ) of this problem gives the population standard deviation $\sigma$, the statement of a claim about $\mu$, the sample size $n$, and the desired level of significance $\alpha$. Formulate (i) the hypotheses, (ii) the test statistic $Z$, and (iii) the rejection region. (The answers to part (a) are provided.)

(a) $\sigma = 2$  claim: $\mu > 30$,  $n = 55$,  $\alpha = .05$

[Answers: (i) $H_0 : \mu = 30$,  $H_1 : \mu > 30$

(ii) $Z = \dfrac{\overline{X} - 30}{2 / \sqrt{55}}$  (iii) $R : Z \geq 1.645$ ]

(b) $\sigma = .085$  claim: $\mu < .15$
   $n = 125$   $\alpha = .025$

(c) $\sigma = 8.6$  claim: $\mu \neq 80$
   $n = 38$
   $\alpha = .01$

(d) $\sigma = 1.23$  claim: $\mu \neq 0$
   $n = 40$
   $\alpha = .06$

**8.42** Suppose that the observed values of the sample mean in the contexts of parts ( a–d ) of Exercise 8.41 are given as follows. Calculate the test statistic $Z$ and state the conclusion with the specified $\alpha$.

(a) $\bar{x} = 30.54$    (c) $\bar{x} = 77.35$

(b) $\bar{x} = .136$    (d) $\bar{x} = -.59$

**8.43** A market researcher wants to perform a test with the intent of establishing that his company's medium pump bottle of soap has a mean life greater than 40 days. The sample size is 70 and he knows that $\sigma = 5.6$.

(a) If you set the rejection region to be $R : \overline{X} \geq 41.31$, what is the level of significance of your test?

(b) Find the numerical value of $c$ so that the test $R : \overline{X} \geq c$ has a 5% level of significance.

**8.44** With reference to Exercise 8.15, perform a test with the intent of establishing that the mean number of items returned is greater than 2.0. Take $\alpha = .02$.

**8.45** With reference to Exercise 8.44,

(a) test
$$H_0 : \mu = 2.0 \qquad H_1 : \mu \neq 2.0$$
with $\alpha = .02$.

(b) Based on your decision in part (a), what error could you have possibly made? Explain in the context of the problem.

**8.46** An investigator at a large midwestern university wants to determine the typical weekly amount of time students work on part-time jobs. More particularly, he wants to test the null hypothesis that the mean time is 15 hours versus a two-sided alternative.

A sample of 39 students who hold part-time jobs is summarized by the computer output

**Descriptive Statistics: hours**

| Variable | N | Mean | Median | StDev |
|---|---|---|---|---|
| Hours | 39 | 16.69 | 15.00 | 7.61 |

(a) Perform the hypothesis test at the 1% level of significance.

(b) Calculate the significance probability and interpret the result.

**8.47** Refer to the data on the growth of female salmon in the marine environment in Table D.7 of the Data Bank. A computer calculation for a test of $H_0 : \mu = 411$ versus $H_1 : \mu \neq 411$ is given below.

**Test of mu = 411 vs mu not = 411**

| Variable | N | Mean | StDev |
|---|---|---|---|
| Fmarine | 40 | 429.15 | 41.05 |

| Variable | Z | P |
|---|---|---|
| Fmarine | 2.80 | 0.005 |

(a) What is the conclusion if you test with $\alpha = .01$?

(b) What mistake could you have made in part (a)?

(c) Before you collected the data, what was the probability of making the mistake in part (a)?

(d) Give a long-run relative frequency interpretation of the probability in part (c).

(e) Give the P–value of your test with $\alpha = .01$.

**8.48** Refer to the data on the girth, in centimeters, of grizzly bears in Table D.6 of the Data Bank. A computer calculation for a test of $H_0 : \mu = 100$ centimeters versus $H_1 : \mu \neq 100$ gives

**One-Sample Z: Girth**

**Test of mu = 100 vs mu not = 100**

| Variable | N | Mean | StDev |
|---|---|---|---|
| Girth | 61 | 93.39 | 21.79 |

| Variable | Z | P |
|---|---|---|
| Girth | -2.37 | 0.018 |

(a) What is the conclusion if you test with $\alpha = .02$?

(b) What mistake could you have made in part (a)?

(c) Before you collected the data, what was the probability of making the mistake in part (a)?

(d) Give a long-run relative frequency interpretation of the probability in part (c).

(e) Give the P–value of your test with $\alpha = .02$.

**8.49** A manager at a power company monitored the employee time required to process high-efficiency lamp bulb rebates. A random sample of 40 applications gave a sample mean time of 3.8 minutes and a standard deviation of 1.2 minutes.

(a) Is the claim that $\mu > 3.5$ minutes substantiated by these data? Test with $\alpha = .10$.

(b) Based on your decision in part (a), what error could you have possibly made? Explain in the context of the problem.

**8.50** Refer to Example 6 where, one day, the visits of $n = 48$ students to the social network site produced the summary statistics

$$\bar{x} = 35.96 \text{ minutes} \quad \text{and} \quad s = 29.11 \text{ minutes}$$

$$n = 48$$

(a) Conduct a test of hypotheses with the intent of establishing that mean connect time is greater than 30 minutes. Take $\alpha = .10$.

(b) Calculate the P–value.

(c) Would the test reject the null hypothesis with $\alpha = .05$? Explain your answer.

**8.51** A company wants to establish that the mean life of its batteries, when used in a wireless mouse, is over 183 days. The data consist of the life lengths of batteries in 64 different wireless mice.

(a) Formulate the null and alternative hypotheses.

(b) What is the conclusion to your test if $\bar{x} = 190.5$ and $s = 32$ days? Take $\alpha = .05$.

(c) Based on your decision in part (b), what error could you have possibly made?

**8.52** In a given situation, suppose $H_0$ was not rejected at $\alpha = .02$. Answer the following questions as "yes," "no," or "can't tell" as the case may be.

(a) Would $H_0$ also be retained at $\alpha = .01$?

(b) Would $H_0$ also be retained at $\alpha = .05$?

(c) Is the P–value smaller than .02?

**8.53** A company wishing to improve its customer service collected hold times from 75 randomly selected incoming calls to its hot line that were put on hold. These calls had sample mean hold time $\bar{x}$ = 3.4 minutes and $s$ = 2.4 minutes. Is the claim that $\mu$ > 3.0 minutes substantiated by these data? Test with $\alpha$ = .05.

**8.54** A company's mixed nuts are sold in cans and the label says that 25% of the contents is cashews. Suspecting that this might be an overstatement, an inspector takes a random sample of 35 cans and measures the percent weight of cashews [i.e., 100 ( weight of cashews / weight of all nuts )] in each can. The mean and standard deviation of these measurements are found to be 23.5 and 3.1, respectively. Do these results constitute strong evidence in support of the inspector's belief? (Answer by calculating and interpreting the $P$–value.)

**8.55** Biological oxygen demand (BOD) is an index of pollution that is monitored in the treated effluent of paper mills on a regular basis. From 43 determinations of BOD (in pounds per day) at a particular paper mill during the spring and summer months, the mean and standard deviation were found to be 3246 and 757, respectively. The company had set the target that the mean BOD should be 3000 pounds per day. Do the sample data indicate that the actual amount of BOD is significantly off the target? (Use $\alpha$ = .05.)

**8.56** Refer to Exercise 8.55. Along with the determinations of BOD, the discharge of suspended solids (SS) was also monitored at the same site. The mean and standard deviation of the 43 determinations of SS were found to be 5710 and 1720 pounds per day, respectively. Do these results strongly support the company's claim that the true mean SS is lower than 6000 pounds per day? (Answer by calculating and interpreting the $P$–value.)

# 5.  INFERENCES ABOUT A POPULATION PROPORTION

The reasoning leading to estimation of a mean also applies to the problem of estimation of a population proportion. Example 2 considers sampling $n$ = 200 student loan holders to infer about the proportion of the population that is very concerned about the time to repay. In terms of Bernoulli trials, success is having the characteristic concerned about the length of time. When $n$ elements are randomly sampled from the population, the data consists of the count $X$ of the number of sampled elements possessing the characteristic. Common sense suggests the sample proportion

$$\hat{p} = \frac{X}{n}$$

as an estimator of $p$. The hat notation reminds us that $\hat{p}$ is a statistic.

When the sample size $n$ is only a small fraction of the population size, the sample count $X$ has the binomial distribution with mean $np$ and standard deviation $\sqrt{npq}$, where $q = 1 - p$. Recall from Chapter 6 that, when $n$ is large, the binomial variable $X$ is well approximated by a normal with mean $np$ and standard deviation $\sqrt{npq}$, Example 8, Chapter 6, recasts this result as

$$Z = \frac{(X - np)/n}{(\sqrt{npq})/n} = \frac{\hat{p} - p}{\sqrt{pq/n}} \quad \text{is approximately } N(0, 1)$$

This last form, illustrated in Figure 15, is crucial to all inferences about a population proportion $p$. It shows that $\hat{p}$ is approximately normally distributed with mean $p$ and standard deviation $\sqrt{pq/n}$.

Figure 15    Approximate normal distribution of $\hat{p}$.

## 5.1  POINT ESTIMATION OF $p$

Intuitively, the sample proportion $\hat{p}$ is a reasonable estimator of the population proportion $p$. When the count $X$ has a binomial distribution,

$$E(X) = np \qquad sd(X) = \sqrt{npq}$$

Since $\hat{p} = X/n$, the properties of expectation give

$$E(\hat{p}) = p$$

$$sd(\hat{p}) = \sqrt{pq/n}$$

In other words, the sampling distribution of $\hat{p}$ has a mean equal to the population proportion. The second result shows that the standard error of the estimator $\hat{p}$ is

$$S.E.(\hat{p}) = \sqrt{\frac{pq}{n}}$$

The estimated standard error can be obtained by substituting the sample estimate $\hat{p}$ for $p$ and $\hat{q} = 1 - \hat{p}$ for $q$ in the formula, or

$$\text{Estimated S.E.}(\hat{p}) = \sqrt{\frac{\hat{p}\hat{q}}{n}}$$

When $n$ is large, prior to sampling, the probability is approximately .954 that the error of estimation $|\hat{p} - p|$ will be less than $2 \times ($ estimated S.E. $)$.

---

### Point Estimation of a Population Proportion

**Parameter:** Population proportion $p$

**Data:**      $X = $ Number having the characteristic in a random sample of size $n$

**Estimator:** $\hat{p} = \dfrac{X}{n}$

$$S.E.(\hat{p}) = \sqrt{\frac{pq}{n}} \qquad \text{and} \qquad \text{estimated S.E.}(\hat{p}) = \sqrt{\frac{\hat{p}\hat{q}}{n}}$$

For large $n$, an approximate $100(1 - \alpha)\%$ error margin is $z_{\alpha/2}\sqrt{\hat{p}\hat{q}/n}$.

**Example 11**    Estimating the Proportion of Purchasers

A large mail-order club that offers monthly specials wishes to try out a new item. A trial mailing is sent to a random sample of 250 members selected from the list of over 9000 subscribers. Based on this sample mailing, 70 of the members decide to purchase the item. Give a point estimate of the proportion of club members that could be expected to purchase the item and attach a 95.4% error margin.

SOLUTION    The number in the sample represents only a small fraction of the total membership, so the count can be treated as if it were a binomial variable.

Here $n = 250$ and $X = 70$, so the estimate of the population proportion is

$$\widehat{p} = \frac{70}{250} = .28$$

$$\text{Estimated S.E.}(\widehat{p}) = \sqrt{\frac{\widehat{p}\widehat{q}}{n}} = \sqrt{\frac{.28 \times .72}{250}} = .028$$

$$95.4\% \text{ error margin} = 2 \times .028 = .056$$

Therefore, the estimated proportion is $\widehat{p} = .28$, with a 95.4% error margin of .06 (rounded to two decimals).

---

## Novel Collection of Count Data

Producers of breakfast cereals continually experiment with new products. Each promising new cereal must be market-tested on a sample of potential purchasers. An added twist here is that youngsters are a major component of the market. In order to elicit accurate information from young people, one firm developed a smiling face scale.

After tasting a new product, respondents are asked to check one box to rate the taste. A good product should have most of the youngsters responding in the top two boxes. Grouping these into a single highest category and the lower three boxes into a lower category, we are in the situation of estimating the proportion of the market population that would rate taste in the highest category.

Out of a sample of 42 youngsters, 30 rated a new cereal in the top category.

## 5.2  CONFIDENCE INTERVAL FOR $p$

A large sample confidence interval for a population proportion can be obtained from the approximate normality of the sample proportion $\widehat{p}$. Since $\widehat{p}$ is nearly normal with mean $p$ and standard deviation $\sqrt{pq/n}$, the random interval $\widehat{p} \pm z_{\alpha/2}\sqrt{pq/n}$ is a candidate. However, the standard deviation involves the unknown parameter $p$, so we use the estimated standard deviation $\sqrt{\widehat{p}\widehat{q}/n}$ to set the endpoints of the confidence interval. Notice again that the common form of the confidence interval is

$$\text{Estimate} \ \pm \ (z \text{ value})(\text{estimated standard error})$$

---

### Large Sample Confidence Interval for $p$

For large $n$, a $100(1 - \alpha)\%$ confidence interval for $p$ is given by

$$\left( \widehat{p} - z_{\alpha/2}\sqrt{\frac{\widehat{p}\widehat{q}}{n}}, \qquad \widehat{p} + z_{\alpha/2}\sqrt{\frac{\widehat{p}\widehat{q}}{n}} \right)$$

---

**Example 12**    Confidence Interval for Proportion of Drivers Using Cell Phones

Research has shown it is equally dangerous to use hand-held or hands-free devices while driving. A study of about 5000 college students in California[3] revealed that 78 percent of them reported that they talked while driving.

   Find a 98% confidence interval for the proportion of all students who use a cell phone while driving.

SOLUTION    We will answer using sample size $n = 5000$. Let $p$ be the population proportion who use a cell phone while driving. Equivalently, it is the probability a randomly selected college student uses a cell phone while driving. The sample size $n = 5000$ is very large so a normal approximation to the distribution of $\widehat{p}$ is justified. Since $1 - \alpha = .98$ we have $\alpha/2 = .01$ and $z_{.01} = 2.33$. The observed $\widehat{p} = .78$ and $\widehat{q} = 1 - .78 = .22$. We calculate

$$z_{.01}\sqrt{\frac{\widehat{p}\widehat{q}}{n}} = 2.33\sqrt{\frac{.78 \times .22}{5000}} = 2.33 \times .0059 = .014$$

The 98% confidence interval for the population proportion of college students that use a cell phone while driving is $.78 \pm .014$, or $(.766, .794)$.

   Because our procedure produces true statements 98% of the time, we can be 98% confident that the proportion of college students who use cell phones while driving is between .766 and .794.

[3]University of California–San Diego Health Sciences news release April 24, 2012.

One word of caution. The sample was taken in southern California but, because of differences in regional driving habits, the extrapolation to the whole United States, or even all of California, may be misleading.

**Example 13**  Do Stradivarius Violins Sound Better than Other Violins?

Researcher Francis Schwarze and Swiss violin maker Michael Rhonheimer have discovered a method of treating wood with a special fungus. The fungus changes the cell structure of the wood by reducing its density and also making it more homogeneous. Four new violins were made of the same wood. Two were untreated, one was fungal treated for six months and another for nine months. The fifth was a legendary violin made in 1711 by Antonio Stradivarius and owned by a star violinist. In a blind test, held at a conference, he played the five instruments behind a curtain. Attendees were asked to choose which one was the Stradivarius. The results (courtesy of Francis Schwarze) for $n = 173$ respondents, are

| Violin | Un-treated 1 | Un-treated 2 | Treated 6 mo. | Treated 9 mo. | Stradivarius |
|--------|--------------|--------------|---------------|---------------|--------------|
| Frequency | 3 | 17 | 15 | 113 | 25 |

Find 99% confidence intervals for the probability that

(a)  A person will correctly choose the Stradivarius.

(b)  A person will choose the new violin with 9 months fungal treatment.

SOLUTION  The sample size $n = 173$ is large so a normal approximation is valid. Since $1 - \alpha = .99$, we have $\alpha / 2 = .005$ and $z_{.005} = 2.58$.

(a) Let $p$ be the probability that a person would correctly choose the Stradivarius. Because 25 persons, out of 173, correctly identified the Stradivarius, the observed

$$\hat{p} = \frac{25}{173} = .145 \quad \text{and} \quad \hat{q} = 1 - .145 = .855$$

We calculate

$$z_{.005}\sqrt{\frac{\hat{p}\hat{q}}{n}} = 2.58\sqrt{\frac{.145 \times .855}{173}} = 2.58 \times .0268 = .069$$

The 99% confidence interval for the population proportion that would correctly identify the Stradivarius is

$$.145 \pm .069 \quad \text{or} \quad (.076, .214)$$

The Stradivarius was second among the five but this choice just reached .20, the level of pure guessing.

(b) This time, let $p$ be the probability that a person incorrectly chooses the new violin having 9 months of fungal treatment. Because 113 out of 173 thought the new violin was the Stradivarius

$$\widehat{p} = \frac{113}{173} = .653 \quad \text{and} \quad \widehat{q} = 1 - .653 = .347$$

We calculate

$$z_{.005} \sqrt{\frac{\widehat{p}\widehat{q}}{n}} = 2.58 \sqrt{\frac{.653 \times .347}{173}} = 2.58 \times .0362 = .093$$

The 99% confidence interval for the population proportion that would incorrectly identify this violin as the Stradivarius is

$$.653 \pm .093 \quad \text{or} \quad (.560, .746)$$

Clearly over half of those in the sample chose the new violin with 9 months of fungal treatment. This result is stunning for those who believe the new violins are not as good as the old.

We have selected a small $\alpha = .01$ so, before sampling, each individual interval has probability .99 of covering the unknown proportion. This guarantees (see Chapter 14, Section 2) the probability that both will cover is at least .98. With at least 98% confidence, both intervals contain their respective proportions.

We do issue a caution because only one Stradivarius was included in the test although it presumably is one with very good sound qualities. Also, it is not certain all or even the majority of attendees at this conference were trained musicians.

## 5.3 DETERMINING THE SAMPLE SIZE

Note that, prior to sampling, the numerical estimate $\widehat{p}$ of $p$ is not available. Therefore, for a $100(1 - \alpha)\%$ error margin for the estimation of $p$, we use the expression $z_{\alpha/2}\sqrt{pq/n}$. The required sample size is obtained by equating $z_{\alpha/2}\sqrt{pq/n} = d$, where $d$ is the specified error margin. We then obtain

$$n = pq \left[ \frac{z_{\alpha/2}}{d} \right]^2$$

If the value of $p$ is known to be roughly in the neighborhood of a value $p^*$, then $n$ can be determined from

$$n = p^*(1 - p^*) \left[ \frac{z_{\alpha/2}}{d} \right]^2$$

Without prior knowledge of $p$, we can replace $p^*(1 - p^*)$ by its maximum possible value $\frac{1}{4}$ occuring at $p^* = \frac{1}{2}$, and determine $n$ from the relation

$$n = \frac{1}{4} \left[ \frac{z_{a/2}}{d} \right]^2$$

**Example 14**    Selecting a Sample Size for Estimating a Proportion

A public health survey is to be designed to estimate the proportion $p$ of a population having defective vision. How many persons should be examined if the public health commissioner wishes to be 98% certain that the error of estimation is below .05 when:

(a)   There is no knowledge about the value of $p$?

(b)   $p$ is known to be about .3?

SOLUTION    The tolerable error is $d = .05$. Also $1 - \alpha = .98$, so $\alpha / 2 = .01$. From the normal table, we know that $z_{.01} = 2.33$.

(a) Since $p$ is unknown, the conservative bound on $n$ yields

$$\frac{1}{4}\left[\frac{2.33}{.05}\right]^2 = 543$$

A sample of size 543 would suffice.

(b) If $p^* = .3$, the required sample size is

$$n = (.3 \times .7)\left[\frac{2.33}{.05}\right]^2 = 456$$

## 5.4  LARGE SAMPLE TESTS ABOUT $p$

We consider testing $H_0 : p = p_0$ versus $H_1 : p \neq p_0$. With a large number of trials $n$, the sample proportion

$$\hat{p} = \frac{X}{n}$$

is approximately normally distributed. Under the null hypothesis, $p$ has the specified value $p_0$ and the distribution of $\hat{p}$ is approximately $N(p_0, \sqrt{p_0 q_0 / n})$. Consequently, the standardized statistic

$$Z = \frac{\hat{p} - p_0}{\sqrt{p_0 q_0 / n}}$$

has approximately a $N(0, 1)$ distribution. Our rule of thumb is to treat $n > 30$ as large provided the condition on the expected number of successes and of failures is met. (See the discussion of the normal approximation to binomial on p. 255)

Since we are considering a two-sided alternative hypothesis, the rejection region of the level $\alpha$ test is given by

$$R: |Z| \geq z_{\alpha / 2}$$

For one-sided alternatives, the rejection region is one-sided.

---

**Large Sample Tests of $H_0: p = p_0$**

When the sample size $n$ is large, and neither $n\,p_0$ or $n\,(1 - p_0)$ are small, a $Z$ test is based on the normal statistic

$$Z = \frac{\widehat{p} - p_0}{\sqrt{p_0(1 - p_0)/n}} \quad \text{where} \quad \widehat{p} = \frac{X}{n}$$

The rejection region is one- or two-sided depending on the alternative hypothesis.

$$H_1: p > p_0 \quad \text{requires} \quad R: Z \geq z_\alpha$$

$$H_1: p < p_0 \quad \text{requires} \quad R: Z \leq -z_\alpha$$

$$H_1: p \neq p_0 \quad \text{requires} \quad R: |Z| \geq z_{\alpha/2}$$

---

**Example 15**    Testing for a Change in the Proportion below the Poverty Level

A five-year-old census recorded that 20% of the families in a large community lived below the poverty level. To determine if this percentage has changed, a random sample of 400 families is studied and 70 are found to be living below the poverty level. Does this finding indicate that the current percentage of families earning incomes below the poverty level has changed from what it was five years ago?

SOLUTION    Let $p$ denote the current population proportion of families living below the poverty level. Because we are seeking evidence to determine whether $p$ is *different* from .20, we wish to test

$$H_0: p = .20 \quad \text{versus} \quad H_1: p \neq .20$$

The sample size $n = 400$ being large, the $Z$ test is appropriate. The test statistic is

$$Z = \frac{\widehat{p} - .2}{\sqrt{.2 \times .8 / 400}}$$

If we set $\alpha = .05$, the rejection region is $R: |Z| \geq 1.96$. From the sample data the computed value of $Z$ is

$$z = \frac{(70/400) - .2}{\sqrt{.2 \times .8/400}} = \frac{.175 - .2}{.020} = -1.25$$

Because $|z| = 1.25$ is smaller than 1.96, the null hypothesis is not rejected at $\alpha = .05$. We conclude that the data do not provide strong evidence that a change in the percentage of families living below the poverty level has occurred.

The significance probability of the observed value of $Z$ is

$$P\text{–value} = P[\,|Z| \geq 1.25\,]$$
$$= P[\,Z \leq -1.25\,] + P[\,Z \geq 1.25\,]$$
$$= 2 \times .1056 = .2112$$

We would have to inflate $\alpha$ to more than .21 in order to reject the null hypothesis. Thus, the evidence against $H_0$ is really weak.

**Example 16**    A One-Sided Test Showing Smokers Have Fewer Males Babies

A study of 8960 births[4] revealed that, among the $n = 1103$ births to parents who both smoked, there were 480 males.

Construct a test of hypotheses, with $\alpha = .01$, with the intent of establishing that, when both parents smoke, the population proportion of male births is less than .5.

SOLUTION    The sample size is $n = 1103$ and the parameter $p$ is the population proportion of males born. We are trying to show that this proportion is less than .5 so we test $H_0: p = .5$ against a one-sided alternative. That is, we test

$$H_0 : p = .5 \qquad \text{versus} \qquad H_1 : p < .5$$

Because the sample size $n = 1103$ is very large, the $Z$ test is appropriate. The test statistic is

$$Z = \frac{\widehat{p} - .5}{\sqrt{.5 \times .5 / 1103}}$$

With $\alpha = .01$ the one-sided rejection region is $R: Z \leq -2.33$. See Figure 16.

.05

−2.33          0          z

Figure 16    Left-sided rejection region $Z \leq -2.33$.

From the sample data, $\widehat{p} = .435$ and the computed value of the test statistic $Z$ is

$$z = \frac{.435 - .5}{\sqrt{.5 \times .5 / 1103}} = \frac{-.065}{.01506} = -4.32$$

[4]The total sample size and percentages match G. Koshy et al., "Parental smoking and increased likelihood of female births," *Annals of Human Biology* **37**(6), (2010), pp. 789–800.

Because the observed value $z = -4.32$ is less than $-2.33$, we reject the null hypothesis in favor of the the alternative that $p$ is less than .5, at level of significance $\alpha = .01$.

The observed value $-4.32$ of $Z$ is to the left of the interval where the sampling distribution is exhibited in Figure 16. A computed calculation gives $P$–value $= P[Z < -4.32] = .0000$ indicating that the positive $P$–value rounds to zero to four decimal places. This extremely small $P$–value greatly strengthens the evidence against the null hypothesis that males and females are equally likely in the subpopulation that was sampled. When both parents smoke, males constitute a smaller proportion of all babies.

# Exercises

**8.57** Suppose that $n$ units are randomly sampled and $x$ number of the sampled units are found to have the characteristic of interest. In each case, (i) define $p$ in the context of the problem, (ii) provide a point estimate of $p$ and (iii) determine the 95% error margin.[5]

(a) A survey is conducted of $n = 986$ adults and $x = 295$ reported that reading is a favorite leisure time activity.

(b) A survey of $n = 440$ pet owners revealed that $x = 293$ buy their pets holiday presents.

**8.58** Software programs using artificial intelligence are now writing many stories. These so called robots have replaced humans as the authors of routine reports in sports, finance, and other areas. During the first seven days of September 2012, the football pages on the Big Ten network site contained 66 blogs of which 36 were written by robots.

(a) Find a 90% confidence interval for the population proportion written by robots.

(b) Does .5 appear to be a plausible value for the proportion given the 90% confidence interval.

(c) The data are likely representative of a week's activity during the football season. However, one day was a Saturday and Big Ten teams were involved in many games. The first task given to robot reporters was to convert box score data into stories and they keep up with this real time task by updating every quarter. On Saturdays, there is actually a group of stories

that go through the robot reporters. Comment on the Bernoulli assumptions of independence and constant probability of success.

**8.59** In a psychological experiment, individuals are permitted to react to a stimulus in one of two ways, say, $A$ or $B$. The experimenter wishes to estimate the proportion $p$ of persons exhibiting reaction $A$. How many persons should be included in the experiment to be 90% confident that the error of estimation is within .03 if the experimenter:

(a) Knows that $p$ is about .3?

(b) Has no idea about the value of $p$?

**8.60** A national safety council wishes to estimate the proportion of automobile accidents that involve pedestrians. How large a sample of accident records must be examined to be 98% certain that the estimate does not differ from the true proportion by more than .03? (The council believes that the true proportion is below .25.)

**8.61** An automobile club that pays for emergency road services (ERS) requested by its members wishes to estimate the proportions of the different types of ERS requests. Upon examining a sample of 2927 ERS calls, it finds that 1499 calls related to starting problems, 849 calls involved serious mechanical failures requiring towing, 498 calls involved flat tires or lockouts, and 81 calls were for other reasons.

(a) Estimate the true proportion of ERS calls that involved serious mechanical problems requiring towing and determine its 95% margin of error.

[5]These proportions of successes are close to those obtained in a recent Harris interactive poll.

(b) Calculate a 98% confidence interval for the true proportion of ERS calls that related to starting problems.

**8.62** Each year, an insurance company reviews its claim experience in order to set future rates. Regarding their damage-only automobile insurance policies, at least one claim was made on 2073 of the 12,299 policies in effect for the year. Treating these data as a random sample for the population of all possible damage-only policies that could be issued, find a 95% confidence interval for the population proportion of at least one claim.

**8.63** Out of a sample of 94 purchases at the drive-up window of a fast-food establishment, 27 were made with a major credit card.

(a) Estimate the proportion of sales made with a credit card.

(b) Obtain the estimated standard error.

(c) Find a 98% confidence interval for the population proportion of purchases paid with a major credit card.

**8.64** Refer to Example 16 and the babies born to parents that both smoke.

(a) Find a 90% confidence interval for the proportion of male babies born to parents that both smoke.

(b) Does $p$ lie in your interval obtained in part(a)?

(c) Why are you 95% confident about your interval in part(a)?

**8.65** Identify the null and the alternative hypotheses in the following situations.

(a) A university official believes that the proportion of students who currently hold part-time jobs has increased from the value .26 that prevailed four years ago.

(b) A cable company claims that, because of improved procedures, the proportion of its cable subscribers that have complaints against the cable company is now less than .13.

(c) Referring to part (b), suppose a consumer advocate feels the proportion of cable subscribers that have complaints against the cable company this year is greater than .13. She will conduct a survey to challenge the cable company's claim.

(d) An inspector wants to establish that 2 × 4 lumber at a mill does not meet a specification that requires at most 5% break under a standard load.

**8.66** Given here are the descriptive statements of some claims that one intends to establish on the basis of data. In each case, identify the null and the alternative hypotheses in terms of a population proportion $p$.

(a) Of smokers who eventually quit smoking, less than 40% are able to do so in just one attempt.

(b) On a particular freeway, over 25% of the cars that use a lane restricted exclusively to multi-passenger cars use the lane illegally.

(c) At a particular clinic, less than 20% of the patients wait over half an hour to see the doctor.

**8.67** Each part of this problem specifies a claim about a population proportion, the sample size $n$, and the desired level of significance $\alpha$. Formulate (i) the hypotheses, (ii) the test statistic, and (iii) the rejection region. (The answers to part (a) are provided for illustration.)

(a) Claim: $p < .32$  $n = 120$  $\alpha = .05$
[Answers:
(i) $H_0 : p = .32$  $H_1 : p < .32$
(ii) $Z = \dfrac{\hat{p} - .32}{\sqrt{.32 \times .68 / 120}} = \dfrac{\hat{p} - .32}{.0426}$
(iii) $R : Z \leq -1.645$

(b) Claim: $p < .75$  $n = 228$  $\alpha = .02$

(c) Claim: $p \neq .60$  $n = 77$  $\alpha = .02$

(d) Claim: $p < .56$  $n = .86$  $\alpha = .10$

**8.68** Given here are the observed sample proportions $\hat{p}$ in the contexts of parts (a–d) of Exercise 8.67. Calculate the test statistic and draw a conclusion of the test at the specified level of significance.

(a) $\hat{p} = .233$

(b) $\hat{p} = .818$

(c) $\hat{p} = .709$

(d) $\hat{p} = .387$

**8.69** A manager of a campus store that sells posters conjectures that more than 30% of all freshman dorm rooms have a poster of a rock group. From $n = 60$ rooms selected at random, an investigator records $X$ = number of rooms having a poster of a rock group.

(a) Formulate a null and alternative hypotheses for verifying the conjecture.

(b) Select a rejection region for which $\alpha = .05$.

(c) If $p = .4$, what error can be made? Explain in words in the context of the problem.

(d) What is the conclusion of your test if $X = 25$?

(e) Based on your conclusion in part (d), what error could you have made? Explain in the context of the problem.

**8.70** An educator wishes to test $H_0 : p = .3$ against $H_1 : p > .3$, where $p = $ proportion of college football players who graduate in four years.

(a) State the test statistic and the rejection region for a large sample test having $\alpha = .05$.

(b) If 19 out of a random sample of 48 players graduated in four years, what does the test conclude? Calculate the $P$-value and interpret the result.

**8.71** A concerned group of citizens wants to show that less than half the voters support the President's handling of a recent crisis. Let $p = $ proportion of voters who support the handling of the crisis.

(a) Determine $H_0$ and $H_1$.

(b) If a random sample of 500 voters gives 228 in support, what does the test conclude? Use $\alpha = .05$. Also evaluate the $P$-value.

**8.72** Refer to Exercise 8.61. Perform a test of hypotheses to determine whether the proportion of ERS calls involving flat tires or lockouts was significantly smaller than .19, the true proportion for previous years. (Use a 5% level of significance.)

**8.73** Refer to Example 12 where $n = 5000$ and proportion .78 of the students sometimes use cell phones while driving. Conduce a test of hypotheses, of size

.05, with the intent of establishing that the population proportion is greater than .75.

**8.74** An independent bank concerned about its customer base decided to conduct a survey of bank customers. Out of 505 customers who returned the survey form, 258 rated the overall bank services as excellent.

(a) Test, at level $\alpha = .10$, the null hypothesis that the proportion of customers who would rate the overall bank services as excellent is .46 versus a two-sided alternative.

(b) Calculate the $P$-value and comment on the strength of evidence.

**8.75** With reference to Exercise 8.74,

(a) Find a 90% confidence interval for the proportion of customers who would rate the overall bank services as excellent.

(b) The bank has 8200 customers. Convert your confidence interval for the proportion in part (a) into a 90% confidence interval for the total number of customers who would rate the overall bank services as excellent.

**8.76** From telephone interviews with 980 adults, it was found that 78% of those persons supported tougher legislation for antipollution measures. Does this poll substantiate the conjecture that more than 75% of the adult population are in favor of tougher legislation for antipollution measures? (Answer by calculating the $P$-value.)

**8.77** Refer to the box with the smiling face scale for rating cereals on page 339. Using the data that 30 out of 42 youngsters in a sample rated a cereal in the highest category, find an approximate 95% confidence interval for the corresponding population proportion.

## USING STATISTICS WISELY

1. Calculate the estimated standard error $s / \sqrt{n}$ to accompany the point estimate $\bar{x}$ of a population mean.

2. Understand the interpretation of a $100(1 - \alpha)\%$ confidence interval. When the sample size is large, before the data are collected

$$\left( \bar{X} - z_{\alpha/2} \frac{s}{\sqrt{n}}, \quad \bar{X} + z_{\alpha/2} \frac{s}{\sqrt{n}} \right)$$

is a random interval that will cover the fixed unknown mean $\mu$ with probability $1 - \alpha$. The long-run frequency interpretation of the probability $1 - \alpha$ says that, in many repeated applications of this method, about proportion $1 - \alpha$ of the times the interval will cover the respective population mean.

3. When conducting a test of hypothesis, formulate the alternative hypothesis as the assertion that the experiment seeks to confirm.

4. When the sample size is large, base a test of the null hypothesis $H_0 : \mu = \mu_0$ on the test statistic

$$\frac{\overline{X} - \mu_0}{S / \sqrt{n}}$$

which has, approximately a standard normal distribution. The rejection region is one-sided or two-sided corresponding to the alternative hypothesis.

5. Understand the interpretation of a level $\alpha$ test. If the null hypothesis is true, before the data are collected, the probability is $\alpha$ that the experiment will produce observations that lead to the rejection of the null hypothesis. Consequently, after many independent experiments, the proportion that lead to the incorrect rejection of the null hypothesis will be nearly $\alpha$.

6. To obtain a precise estimate of a proportion, it usually requires a sample size of a few hundred.

7. Remember that the statistical procedures presented in this chapter will not be valid if the large sample is not randomly selected but collected from convenient units.

## KEY IDEAS AND FORMULAS

Statistical concepts and methods provide the framework that allows us to learn about the population from a sample of observations. The process begins by modeling the population as a probability distribution, which has some numerical feature of interest called a **parameter**. Then, given a sample from this population distribution called the **data**, we make a generalization or **statistical inference** about the parameter.

Two basic forms of inference are (1) **estimation** of a population parameter and (2) **testing statistical hypotheses**.

A parameter can be estimated in two ways: by quoting (1) a single numerical value (**point estimation**) or (2) an interval of plausible values (**interval estimation**).

The statistic whose value gives a point estimate is called an **estimator**. The standard deviation of a point estimator is also called its **standard error**.

To be meaningful, a point estimate must be accompanied by an evaluation of its **error margin**.

A $100(1 - \alpha)\%$ **confidence interval** is an interval that, before sampling, will cover the true value of the parameter with probability $1 - \alpha$. The interval must be computable from the sample data.

If random samples are repeatedly drawn from a population and a $100(1 - \alpha)\%$ confidence interval is calculated from each, then about $100(1 - \alpha)\%$ of those intervals will include the true value of the parameter. We never know what happens in a single application. Our confidence draws from the success rate of $100(1 - \alpha)\%$ over many applications.

A statistical hypothesis is a statement about a population parameter.

A statement or claim, which is to be established with a strong support from the sample data, is formulated as the alternative hypothesis $(H_1)$. The null hypothesis $(H_0)$ says that the claim is void.

A test of the null hypothesis is a decision rule that tells us when to reject $H_0$ and when not to reject $H_0$. A test is specified by a test statistic and a rejection region (critical region).

A wrong decision may occur in one of the two ways:

A false rejection of $H_0$ (Type I error)

Failure to reject $H_0$ when $H_1$ is true (Type II error)

Errors cannot always be prevented when making a decision based on a sample. It is their probabilities that we attempt to keep small.

A Type I error is considered to be more serious. The maximum Type I error probability of a test is called its level of significance and is denoted by $\alpha$.

The significance probability or $P$–value of an observed test statistic is the smallest $\alpha$ for which this observation leads to the rejection of $H_0$.

**Main steps in testing statistical hypotheses**

1. Formulate the null hypothesis $H_0$ and the alternative hypothesis $H_1$.
2. Test criterion: State the test statistic and the form of the rejection region.
3. With a specified $\alpha$, determine the rejection region.
4. Calculate the test statistic from the data.
5. Draw a conclusion: State whether or not $H_0$ is rejected at the specified $\alpha$ and interpret the conclusion in the context of the problem. Also, it is a good statistical practice to calculate the $P$–value and strengthen the conclusion.

The Type II error probability is denoted by $\beta$.

### Inferences about a Population Mean When n Is Large

When $n$ is large, we need not be concerned about the shape of the population distribution. The central limit theorem tells us that the sample mean $\overline{X}$ is nearly normally distributed with mean $\mu$ and standard deviation $\sigma / \sqrt{n}$. Moreover, $\sigma / \sqrt{n}$ can be estimated by $S / \sqrt{n}$.

Parameter of interest is

$$\mu = \text{Population mean}$$

Inferences are based on

$$\overline{X} = \text{Sample mean}$$

1. A point estimator of $\mu$ is the sample mean $\overline{X}$.

$$\text{Estimated standard error} = \frac{S}{\sqrt{n}}$$

$$\text{Approximate } 100\,(1 - \alpha)\% \text{ error margin} = z_{\alpha/2}\,\frac{S}{\sqrt{n}}$$

2. A $100\,(1 - \alpha)\%$ confidence interval for $\mu$ is

$$\left( \overline{X} - z_{\alpha/2}\,\frac{S}{\sqrt{n}}, \quad \overline{X} + z_{\alpha/2}\,\frac{S}{\sqrt{n}} \right)$$

3. The test of the null hypothesis concerning $\mu$, called the **large sample normal test** or **$Z$ test**, uses the test statistic

$$Z = \frac{\overline{X} - \mu_0}{S/\sqrt{n}}$$

where $\mu_0$ is the value of $\mu$ that marks the boundary between $H_0$ and $H_1$. Given a level of significance $\alpha$,

Reject $H_0 : \mu = \mu_0$ in favor of $H_1 : \mu > \mu_0$ if $Z \geq z_\alpha$

Reject $H_0 : \mu = \mu_0$ in favor of $H_1 : \mu < \mu_0$ if $Z \leq -z_\alpha$

Reject $H_0 : \mu = \mu_0$ in favor of $H_1 : \mu \neq \mu_0$ if $|Z| \geq z_{\alpha/2}$

The first two alternative hypotheses, $H_1 : \mu > \mu_0$ and $H_1 : \mu < \mu_0$ are one-sided hypothesis and the third is a two-sided hypothesis. The rejection regions correspond to the alternative hypothesis so tests in the first two cases are called one-sided tests or one-tailed tests. Those in the third case are two-tailed tests.

### Inference about a Population Proportion When $n$ Is Large

Parameter of interest:

$p = $ Population proportion of individuals possessing stated characteristic
Inferences are based on

$$\widehat{p} = \frac{X}{n} \text{ the sample proportion.}$$

1. A point estimator of $p$ is $\widehat{p}$.

$$\text{Estimated standard error} = \sqrt{\frac{\widehat{p}\,\widehat{q}}{n}} \qquad \text{where } \widehat{q} = 1 - \widehat{p}$$

$$100\,(1 - \alpha)\% \text{ error margin} = z_{\alpha/2}\sqrt{\frac{\widehat{p}\,\widehat{q}}{n}}$$

2. A $100\,(1 - \alpha)\%$ confidence interval for $p$ is

$$\left( \widehat{p} - z_{\alpha/2}\sqrt{\frac{\widehat{p}\,\widehat{q}}{n}}, \quad \widehat{p} + z_{\alpha/2}\sqrt{\frac{\widehat{p}\,\widehat{q}}{n}} \right)$$

3. To test hypotheses about $p$, the test statistic is

$$Z = \frac{\hat{p} - p_0}{\sqrt{p_0 q_0 / n}}$$

where $p_0$ is the value of $p$ that marks the boundary between $H_0$ and $H_1$. The rejection region is right-sided, left-sided, or two-sided according to $H_1 : p > p_0$, $H_1 : p < p_0$, or $H_1 : p \neq p_0$, respectively.

# TECHNOLOGY

## Large sample confidence intervals and tests concerning a mean

The software programs use a known value for the population standard deviation $\sigma$. If this is not given in your application, you need to first obtain the sample standard deviation using the technology described in Chapter 2.

### MINITAB

### Confidence intervals for $\mu$

We illustrate the calculation of a 99% confidence interval for $\mu$ when we have determined that the sample standard deviation is 8.2 (or the known population $\sigma = 8.2$).

Data: *C1*

**Stat > Basic Statistics > 1-Sample Z.**

Type *C1* in **Samples in columns** and *8.2* in **Standard deviation**.

Click **Options** and type *99* in **Confidence level**.

Click **OK**. Click **OK**.

### Tests of hypotheses concerning $\mu$

We illustrate the calculation of an $\alpha = .01$ level test of $H_0 : \mu = 32$ versus a one-sided alternative $H_1 : \mu > 32$ when we have determined that the sample standard deviation is 8.2 (or the known population $\sigma = 8.2$).

Data: *C1*

**Stat > Basic Statistics > 1-Sample Z.**

Type *C1* in **Samples in Columns**.

Type *8.2* in **Sigma**. Click **Perform hypotheses test**. In **Hypothesized mean**, type *32*, the value of the mean under the null hypothesis.

Click **Options** and type *99* in **Confidence level**.

In the **Alternative** cell select **greater than**, the direction of the alternative hypothesis. Click **OK**. Click **OK**.

If the sample size and mean are available, instead of the second step, type these values in the corresponding cells.

## EXCEL

### Confidence intervals for $\mu$

We illustrate the calculation of a 99% confidence interval for $\mu$ when we have determined that the sample standard deviation is 8.2 (or the known population $\sigma = 8.2$) and the sample size is 100. Start with the data entered in column A.

Click the **Formulas** tab and select **More Functions**.

Choose **Statistical** and then **CONFIDENCE.NORM**.

Enter *1 − .99* or *.01* for **Alpha**, *8.2* for **Standard_dev**, and *100* for **size**. Click **OK**.

(Add and subtract this value to and from $\bar{x}$ to obtain the confidence interval.)

### Tests of hypotheses concerning $\mu$

We illustrate the calculation of a test of $H_0: \mu = 32$ versus a one-sided alternative $H_1: \mu > 32$ when we have determined that the sample standard deviation is 8.2 (or the known population $\sigma = 8.2$). Start with the data entered in column A.

Click the **Formulas** tab and then **More Functions**.

Choose **Statistical** and then **Z.TEST**.

Highlight the data in column A for **Array**. Enter *32* in **X** and *8.2* in **Sigma**.

(Leave **Sigma** blank and the sample standard deviation will be used.)

Click **OK**.

The program returns the one-sided *P*–value for right-sided alternatives. For two-sided alternatives, you need to double the *P*–value if $\bar{x}$ is above $\mu_0 = 32$.

## TI-84/-83 Plus

### Confidence intervals for $\mu$

We illustrate the calculation of a 99% confidence interval for $\mu$ when we have determined that the sample standard deviation is 8.2 (or the known population $\sigma = 8.2$). Start with the data entered in $L_1$.

Press **STAT** and select **TESTS** and then **7: Zinterval**.

Select **Data** for **Inpt**. With **List** set to $L_1$ and **FREQ** to 1,

Following $\sigma$: enter *8.2* and

enter *.99* following **C-Level**. Select **Calculate**.

Then press **ENTER**.

If instead the sample size and sample mean are available, the second step is Select **Stats** (instead of **Data**) and enter the sample size and mean for $n$ and $\bar{x}$.

### Tests of hypotheses concerning $\mu$

We illustrate the calculation of an $\alpha = .01$ level test of $H_0: \mu = 32$ versus a one-sided alternative $H_1: \mu > 32$ when we have determined that the sample standard deviation is 8.2 (or the known population $\sigma = 8.2$). Start with the data entered in column $\mathbf{L_1}$.

Press **STAT** and select **TESTS** and then **1: Z-Test**.

Select **Data** for **Inpt**. Enter 32 for $\mu_0$. Following $\sigma$: enter 8.2. With the **List** set to $\mathbf{L_1}$ and **Freq** to 1, select the direction of the alternate hypothesis. Select **Calculate**.

Press **ENTER**.

The calculator will return the $P$–value.

If instead the sample size and sample mean are available, the second step is Select **Stats** (instead of **Data**) and type in the sample size and mean.

## REVIEW EXERCISES

**8.78**    Refer to Exercise 2.4 where the number of automobile accidents reported per month were recorded for an intersection. The sample size is $n = 59$, $\bar{x} = 1.949$, and $s = 1.558$ accidents.

(a)  Give a point estimate of $\mu$, the mean number of accidents reported per month.

(b)  Determine the estimated standard error.

(c)  Calculate the 98% error margin.

**8.79**    A student in a large lecture section asked students how much they paid for a used copy of the text. The $n = 38$ responses yielded

$$\sum x_i = 3230.84 \text{ dollars} \quad \sum (x_i - \bar{x})^2 = 2028.35$$

(a)  Give a point estimate of $\mu$, the mean price paid.

(b)  Determine the estimated standard error.

(c)  Calculate the 95% error margin.

**8.80**    The time it takes for a taxi to drive from the office to the airport was recorded on 40 occasions. It was found that $\bar{x} = 47$ minutes and $s = 5$ minutes. Give

(a)  An estimate of $\mu$ = population mean time to drive.

(b)  An approximate 95.4% error margin.

**8.81**    By what factor should the sample size be increased to reduce the standard error of $\overline{X}$ to

(a)  One-half its original value?

(b)  One-fourth its original value?

**8.82**    A food service manager wants to be 95% certain that the error in the estimate of the mean number of sandwiches dispensed over the lunch hour is 10 or less. What sample size should be selected if a preliminary sample suggests

(a)  $\sigma = 40$?

(b)  $\sigma = 80$?

**8.83**    A zoologist wishes to estimate the mean blood sugar level of a species of animal when injected with a specified dosage of adrenaline. A sample of 55 animals of a common breed are injected with adrenaline, and their blood sugar measurements are recorded in units of milligrams per 100 milliliters of blood. The mean and standard deviation of these measurements are found to be 126.9 and 10.5, respectively.

(a)  Give a point estimate of the population mean and find a 95.4% error margin.

(b)  Determine a 90% confidence interval for the population mean.

**8.84**    Refer to Exercise 2.3 and the data on the number of extracurricular activities in which 30 students participated in the past week. These data have

$n = 30$ $\quad \bar{x} = 1.925 \quad$ $s = 1.607$ activities

(a)  Obtain a 98% confidence interval for $\mu$, the population mean number of activities.

(b) In a long series of experiments, each involving different students, what proportion of the intervals would cover $\mu$?

**8.85** After feeding a special diet to 80 mice, the scientist measures their weight in grams and obtains $\bar{x} = 35$ grams and $s = 4$ grams. He states that a 90% confidence interval for $\mu$ is given by

$$\left( 35 - 1.645 \frac{4}{\sqrt{80}}, \quad 35 + 1.645 \frac{4}{\sqrt{80}} \right)$$

or ( 34.26, 35.74 )

(a) Was the confidence interval calculated correctly? If not, provide the correct result.

(b) Does the interval ( 34.26, 35.74 ) cover the true mean? Explain your answer.

**8.86** The amount of PCBs was measured in 40 samples of soil that were treated with contaminated sludge. The following summary statistics were obtained.

$$\bar{x} = 3.56 \qquad s = .5 \text{ ppm}$$

(a) Perform a test of hypotheses with the intent of establishing that the mean PCB contamination is less than 3.7 p.p.m. Take $\alpha = .05$.

(b) What error could you have possibly made in part(a)?

**8.87** In each case, identify the null hypothesis ($H_0$) and the alternative hypothesis ($H_1$) using the appropriate symbol for the parameter of interest.

(a) A consumer group plans to test-drive several cars of a new model in order to document that its average highway mileage is less than 50 miles per gallon.

(b) Confirm the claim that the mean number of pages per transmission sent by a campus fax station is more than 3.4.

(c) A chiropractic method will be tried on a number of persons suffering from persistent backache in order to demonstrate the claim that its success rate is higher than 50%.

(d) The setting of an automatic dispenser needs adjustment when the mean fill differs from the intended amount of 16 ounces. Several fills will be accurately measured in order to decide if there is a need for resetting.

(e) The content of fat in a gourmet chocolate ice cream is more than the amount, 4%, that is printed on the label.

**8.88** A literary critic wants to establish that the mean number of words per sentence, appearing in a newly discovered short story, is different from 9.1 words. A sample of 36 sentences provided the data

$$\bar{x} = 8.6 \quad \text{and} \quad s = 1.2$$

(a) Formulate the null and alternative hypotheses. (Define any symbols you use.)

(b) Determine the test statistic.

(c) Give the form of the rejection region.

(d) What is the conclusion to your test? Take $\alpha = .10$.

(e) Calculate a P–value.

(f) Based on part (d), what error could you have possibly made?

**8.89** A test will be conducted to see how long a seven-ounce tube of toothpaste lasts. The researcher wants to establish that the mean time is greater than 30.5 days. From a random sample of size 75, an investigator obtains $\bar{x} = 32.3$ and $s = 6.2$ days.

(a) Formulate the null and alternative hypotheses. (Define any symbols you use.)

(b) Determine the test statistic.

(c) Give the form of the rejection region.

(d) What is the conclusion to your test? Take $\alpha = .10$.

(e) Calculate a P–value.

(f) Based on part (d), what error could you have possibly made?

**8.90** In a given situation, suppose $H_0$ was rejected at $\alpha = .05$. Answer the following questions as "yes," "no," or "can't tell" as the case may be.

(a) Would $H_0$ also be rejected at $\alpha = .03$?

(b) Would $H_0$ also be rejected at $\alpha = .10$?

(c) Is the P–value larger than .05?

**8.91** Refer to the data on the amount of reflected light from urban areas in Table D.3b of the Data Bank. A computer calculation for a test of $H_0: \mu = 84$ versus $H_1: \mu \neq 84$ has the output

```
Test  of  mu = 84  vs  mu  not  =  84

Variable    N      Mean      StDev
Lighturb    40    82.075     4.979


Variable        Z           P
Lighturb      -2.45       0.014
```

(a) What is the conclusion of the test when $\alpha = .03$?

(b) Use the value for $Z$ to test the null hypothesis $H_0: \mu = 84$ versus the one-sided alternative $H_1: \mu < 84$ at the $\alpha = .01$ level of significance.

**8.92**    Refer to Example 6 where, one day, the visits of $n = 48$ students to the social network site produced the summary statistics

$$\bar{x} = 35.96 \text{ minutes and } s = 29.11 \text{ minutes}$$

$$n = 48$$

(a) Calculate a 90% confidence interval for the mean daily time spent on the social network site.

(b) Compare your interval with the two confidence intervals obtained in Example 6. Comment.

**8.93**    The daily number of kayaks sold, $X$, at a water sports store has the probability distribution

| $x$ | 0 | 1 | 2 |
|---|---|---|---|
| $f(x)$ | .5 | .3 | .2 |

(a) Find the expected number of kayaks sold in a day.

(b) Find the standard deviation of the number of kayaks sold in a day.

(c) Suppose data from the next 64 different days give $\bar{x} = .84$ and standard deviation $s = .40$ number of kayaks sold. Can we conclude that the mean number of kayaks sold is greater than it used to be? Test with $\alpha = .05$.

**8.94**    In a large-scale, cost-of-living survey undertaken last January, weekly grocery expenses for families with one or two children were found to have a mean of $148 and a standard deviation of $25. To investigate the current situation, a random sample of families with one or two children is to be chosen and their last week's grocery expenses are to be recorded.

(a) How large a sample should be taken if one wants to be 95% sure that the error of estimation of the population mean grocery expenses per week for families with one or two children does not exceed $2? (Use the previous $s$ as an estimate of the current $\sigma$.)

(b) A random sample of 100 families is actually chosen, and from the data of their last week's grocery bills, the mean and the standard deviation are found to be $155 and $22, respectively.

Construct a 98% confidence interval for the current mean grocery expense per week for the population of families with one or two children.

**8.95**    In 2011, 67% of persons 22–33 years old held passports. From a random sample of size $n = 84$ collected today, 64 are passport holders.

(a) Estimate the proportion of persons 22–33 years old who hold passports today.

(b) Estimate a 95% error margin for your estimate.

**8.96**    With reference to Exercise 8.95,

(a) Conduct a test of size .05 with the intent of establishing that the proportion today is larger than .67.

(b) Find a 95% confidence interval for that proportion.

**8.97**    Out of a sample of $n = 625$ students interviewed, 139 had missed at least one class last week. Obtain a 95% confidence interval for $p = $ proportion of all students that missed at least one class last week.

**8.98**    With reference to Exercise 8.97, conduct a test with the intent of establishing that $p > .20$.

(a) Formulate the null and alternative hypotheses.

(b) Determine the test statistic.

(c) Give the form of the rejection region.

(d) What is the conclusion to your test? Take $\alpha = .05$.

(e) Calculate a $P$–value.

(f) Based on part (d), what error could you have possibly made?

**8.99**    From May 15 through June 30, 2012, American Airlines flew 152 flights from Chicago, Illinois, to Austin, Texas. Of these, 22 arrived late. Treat this as a random sample and conduct a test with the intent of establishing that the population proportion of late flights is greater than .10.

(a) Formulate the null and alternative hypotheses. Define any symbols.

(b) Determine the test statistic.

(c) Give the form of the rejection region.

(d) What is the conclusion to your test? Take $\alpha = .05$.

(e) Calculate a $P$–value.

(f) Based on part (d), what error could you have possibly made?

**8.100** Each year, an insurance company reviews its claim experience in order to set future rates. Regarding their damage-only automobile insurance policies, at least one claim was made on 2073 of the 12,299 policies in effect for the year. Treat these data as a random sample for the population of all possible damage-only policies that could be issued.

(a) Test, at level $\alpha = .05$, the null hypothesis that the probability of at least one claim is 0.16 versus a two-sided alternative.

(b) Calculate the $P$-value and comment on the strength of evidence.

**8.101** A genetic model suggests that 80% of the plants grown from a cross between two given strains of seeds will be of the dwarf variety. After breeding 200 of these plants, 136 were observed to be of the dwarf variety.

(a) Does this observation strongly contradict the genetic model?

(b) Construct a 95% confidence interval for the true proportion of dwarf plants obtained from the given cross.

**8.102** Internet prices for glasses are often considerably lower than those at the brick and mortar stores but quality is another question. To address that issue, some optometrists ordered 154 adult pairs of spectacles from various online stores.[6] Of the 154 pairs of spectacles, 44 pairs of eye wear contained at least one lens that did not pass optical analysis testing.

(a) Conduct a test of hypotheses with the intent of establishing that that the probablity of not passing the optical analysis testing is greater than .20. Use level of significance .025.

(b) Calculate the $P$-value and comment.

**8.103** A researcher asked participants to wear a simple counting device for a week and, for each day, to count the number of times they thought about food.[7]

The average of the seven daily counts, for each of $n = 65$ participants, are summarized by the statistics

$$n = 65 \qquad \bar{x} = 19.22 \qquad s = 17.66$$

Find a 95% confidence interval for the mean number of times a student thinks about food during one day.

**\*8.104** *Finding the power of a test.* Consider the problem of testing $H_0: \mu = 10$ versus $H_1: \mu > 10$ with $n = 64$, $\sigma = 2$ (known), and $\alpha = .025$. The rejection region of this test is given by

$$R: \frac{\bar{X} - 10}{2 / \sqrt{64}} \geq 1.96 \qquad \text{or}$$

$$R: \bar{X} \geq 10 + 1.96 \frac{2}{\sqrt{64}} = 10.49$$

Suppose we wish to calculate the power of this test at the alternative $\mu_1 = 11$. Power = the probability of rejecting the null hypothesis when the alternative is true. Since our test rejects the null hypothesis when $\bar{X} \geq 10.49$, its power at $\mu_1 = 11$ is the probability

$$P[\bar{X} \geq 10.49 \text{ when the true mean } \mu_1 = 11]$$

If the population mean is 11, we know that $\bar{X}$ has the normal distribution with mean 11 and sd = $\sigma / \sqrt{n} = 2 / \sqrt{64} = .25$. The standardized variable is

$$Z = \frac{\bar{X} - 11}{.25}$$

and we calculate

$$\text{Power} = P[\bar{X} \geq 10.49 \text{ when } \mu_1 = 11]$$
$$= P\left[Z \geq \frac{10.49 - 11}{.25}\right]$$
$$= P[Z \geq -2.04] = .9793$$
$$(\text{using normal table})$$

Following the above steps, calculate the power of this test at the alternative:

(a) $\mu_1 = 10.5$

(b) $\mu_1 = 10.8$

[6] K. Citek et al., "Safety and compliance of prescription spectacles ordered by the public via the internet," *Optometry* **82** (2011), pp. 549–555.

[7] T. Fisher, Z. Moore, and M. Pittenger, "Sex on the brain?: An examination of frequency of sexual cognitions as a function of gender, erotophilia, and social desirability," *Journal of Sex Research* **49**(1), (2012), pp. 69–77.

**8.105** Refer to the data on the computer attitude score (CAS) in Table D.4 of the Data Bank. A computer summary of a level $\alpha = .05$ test of $H_0 : \mu = 2.6$ versus a two-sided alternative and a 95% confidence interval is given below.

```
One-Sample Z: CAS


Test of mu = 2.6 vs mu not = 2.6


Variable    N     Mean     StDev
CAS        35    2.8157   0.4840


Variable      95.0% CI          Z       P
CAS        ( 2.6554, 2.9761)  2.64   0.008
```

(a) Will the 99% confidence interval for mean CAS be smaller or larger than the one in the printout? Verify your answer by determining the 99% confidence interval.

(b) Use the value for Z to test the null hypothesis $H_0 : \mu = 2.6$ versus the one-sided alternative $H_1 : \mu > 2.6$ with $\alpha = .05$.

**8.106** Refer to the data on percent malt extract in Table D.8 of the Data Bank. A computer summary of a level $\alpha = .05$ test of $H_0 : \mu = 77$ versus a two-sided alternative and a 95% confidence interval is given below.

```
One-Sample Z: malt extract (%)


Test of mu = 77 vs mu not = 77


Variable         N     Mean    StDev
malt extract    40    77.458   1.101


Variable            95.0% CI          Z       P
malt extract  ( 77.116, 77.799)  2.63  0.009
```

(a) Will the 98% confidence interval for mean malt extract be smaller or larger than the one in the printout? Verify your answer by determining the 98% confidence interval.

(b) Use the value for Z to test the null hypothesis $H_0 : \mu = 77.0$ versus the one-sided alternative $H_1 : \mu > 77.0$ with $\alpha = .05$ level.

## The Following Exercises Require a Computer

**8.107** Refer to the data on the heights of red pine seedlings in Exercise 8.4. Use MINITAB (or some other package program) to:

(a) Find a 97% percent confidence interval for the mean height.

(b) Test $H_0 : \mu = 1.9$ versus $H_1 : \mu \neq 1.9$ centimeters with $\alpha = .03$.

**8.108** Referring to speedy lizard data in Exercise 2.19, obtain a 95% confidence interval for the mean speed of that genus.

**8.109** Refer to the male salmon data given in Table D.7 of the Data Bank. Use MINITAB or some other package program to find a 90% large sample confidence interval for the mean freshwater growth.

**8.110** Refer to the physical fitness data given in Table D.5 of the Data Bank. Use MINITAB or some other package program to:

(a) Find a 97% large sample confidence interval for the pretest number of situps.

(b) Construct a histogram to determine if the underlying distribution is symmetric or has a long tail to one side.

**8.111** Refer to the sleep data given in Table D.10 of the Data Bank. Use MINITAB or some other package program to:

(a) Find a 95% large sample confidence interval for the mean number of breathing pauses per hour (BPH).

(b) Construct a histogram to determine if the underlying distribution is symmetric or has a long tail to one side.

**8.112** Refer to the grizzly bear data given in Table D.6 of the Data Bank. Use MINITAB or some other package program to:

(a) Find a 95% large sample confidence interval for the mean weight in pounds of all bears living in that area.

(b) Construct a histogram to determine if the underlying distribution is symmetric or has a long tail to one side.

# 9

# Small Sample Inferences for Normal Populations

Gary Braasch/The Image Bank/Getty Images

Both capturing and measuring make data collection difficult.

## Collecting a Sample of Lengths of Anacondas

Jesus Rivas, a herpetologist, is currently doing the definitive research on green anacondas. These snakes, some of the largest in the world, can grow to 25 feet in length. They have been known to swallow live goats and even people.

During the dry season Jesus Rivas and fellow researchers walk barefoot in shallow water in the Llanos grasslands shared by Venezuela and Colombia. When they feel a snake with their feet, they grab and hold it with the help of another person. After muzzling the snake with a sock and tape, they place a string along an imaginary centerline from head to tail. The measured length of string is the recorded length of the anaconda.

Females are typically larger than males. The lengths (feet) of 21 females are

| 10.2 | 11.4 | 13.6 | 17.1 | 16.5 | 11.8 | 15.6 |
| 11.8 | 11.3 | 11.9 | 9.6 | 14.4 | 13.2 | 13.5 |
| 12.4 | 12.1 | 11.6 | 8.6 | 13.0 | 16.3 | 14.4 |

If the captured snakes can be treated as a random sample, a 95% confidence interval for the mean length of a female anaconda in that area is given in the computer output

```
One-Sample T: Length(ft)

Variable       N     Mean   StDev       95.0% CI
Length(ft)    21   12.871   2.262   ( 11.842,  13.901)
```

In Chapter 8, we discussed inferences about a population mean when a large sample is available. Those methods are deeply rooted in the central limit theorem, which guarantees that the distribution of $\overline{X}$ is approximately normal. By the versatility of the central limit theorem, we did not need to know the specific form of the population distribution.

Many investigations, especially those involving costly experiments, require statistical inferences to be drawn from small samples ($n \leq 30$, as a rule of thumb). Since the sample mean $\overline{X}$ will still be used for inferences about $\mu$, we must address the question, "What is the sampling distribution of $\overline{X}$ when $n$ is not large?" Unlike the large sample situation, here we do not have an unqualified answer. In fact, when $n$ is small, the distribution of $\overline{X}$ does depend to a considerable extent on the form of the population distribution. With the central limit theorem no longer applicable, more information concerning the population is required for the development of statistical procedures. In other words, the appropriate methods of inference depend on the restrictions met by the population distribution.

In this chapter, we describe how to set confidence intervals and test hypotheses **when it is reasonable to assume that the population distribution is normal.**

We begin with inferences about the mean $\mu$ of a normal population. Guided by the development in Chapter 8, it is again natural to focus on the ratio

$$\frac{\overline{X} - \mu}{S / \sqrt{n}}$$

when $\sigma$ is also unknown. The sampling distribution of this ratio, called Student's *t* distribution, is introduced next.

# 1. STUDENT'S *t* DISTRIBUTION

When $\overline{X}$ is based on a random sample of size $n$ from a normal $N(\mu, \sigma)$ population, we know that $\overline{X}$ is exactly distributed as $N(\mu, \sigma/\sqrt{n})$. Consequently, the standardized variable

$$Z = \frac{\overline{X} - \mu}{\sigma/\sqrt{n}}$$

has the standard normal distribution.

Because $\sigma$ is typically unknown, an intuitive approach is to estimate $\sigma$ by the sample standard deviation $S$. Just as we did in the large sample situation, we consider the ratio

$$T = \frac{\overline{X} - \mu}{S/\sqrt{n}}$$

Its probability density function is still symmetric about zero. Although estimating $\sigma$ with $S$ does not appreciably alter the distribution in large samples, it does make a substantial difference if the sample is small. The new notation $T$ is required in order to distinguish it from the standard normal variable $Z$. In fact, this ratio is no longer standardized. Replacing $\sigma$ by the sample quantity $S$ introduces more variability in the ratio, making its standard deviation larger than 1.

The distribution of the ratio $T$ is known in statistical literature as "Student's *t* distribution." This distribution was first studied by a British chemist W. S. Gosset, who published his work in 1908 under the pseudonym "Student." The brewery for which he worked apparently did not want the competition to know that it was using statistical techniques to better understand and improve its fermentation process.

---

### Student's *t* Distribution

If $X_1, \ldots, X_n$ is a random sample from a normal population $N(\mu, \sigma)$ and

$$\overline{X} = \frac{1}{n} \sum X_i \quad \text{and} \quad S^2 = \frac{\sum (X_i - \overline{X})^2}{n - 1}$$

then the distribution of

$$T = \frac{\overline{X} - \mu}{S/\sqrt{n}}$$

is called Student's *t* distribution with $n - 1$ degrees of freedom

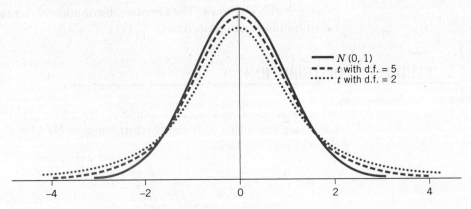

Figure 1    Comparison of $N(0, 1)$ and $t$ density curves.

The qualification "with $n - 1$ degrees of freedom" is necessary, because with each different sample size or value of $n - 1$, there is a different $t$ distribution. The choice $n - 1$ coincides with the divisor for the estimator $S^2$ that is based on $n - 1$ degrees of freedom.

The $t$ distributions are all symmetric around 0 but have tails that are more spread out than the $N(0, 1)$ distribution. However, with increasing degrees of freedom, the $t$ distributions tend to look more like the $N(0, 1)$ distribution. This agrees with our previous remark that for large $n$ the ratio

$$\frac{\overline{X} - \mu}{S / \sqrt{n}}$$

is approximately standard normal. The density curves for $t$ with 2 and 5 degrees of freedom are plotted in Figure 1 along with the $N(0, 1)$ curve.

Appendix B, Table 5, gives the upper $\alpha$ points $t_\alpha$ for some selected values of $\alpha$ and the degrees of freedom (abbreviated d.f.).

The curve is symmetric about zero so the lower $\alpha$ point is simply $-t_\alpha$. The entries in the last row marked "d.f. = infinity" in Appendix B, Table 5, are exactly the percentage points of the $N(0, 1)$ distribution.

**Example 1**    Obtaining Percentage Points of $t$ Distributions

Using Appendix B, Table 5, determine the upper .10 point of the $t$ distribution with 5 degrees of freedom. Also find the lower .10 point.

SOLUTION    With d.f. = 5, the upper .10 point of the $t$ distribution is found from Appendix B, Table 5, to be $t_{.10} = 1.476$. Since the curve is symmetric about 0, the lower .10 point is simply $-t_{.10} = -1.476$. See Figure 2.

Percentage Points of

*t* Distributions

| d.f. $\alpha$ | ... | .10 |
|---|---|---|
| . | . | |
| . | . | |
| . | . | |
| 5 | | → 1.476 |

Figure 2   The upper and lower .10
points of the *t* distribution with
d.f. = 5.

**Example 2**   Determining a Central Interval Having Probability .90

For the *t* distribution with d.f. = 9, find the number *b* such that
$P[-b < T < b] = .90$.

SOLUTION   In order for the probability in the interval $(-b, b)$ to be .90, we must have a
probability of .05 to the right of *b* and, correspondingly, a probability of .05 to
the left of $-b$ (see Figure 3). Thus, the number *b* is the upper $\alpha = .05$ point
of the *t* distribution. Reading Appendix B, Table 5, at $\alpha = .05$ and d.f. = 9,
we find $t_{.05} = 1.833$, so $b = 1.833$.

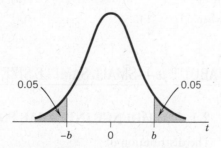

Figure 3   Finding the 5th percentile.

## Exercises

**9.1** Using the table for the $t$ distributions, find:
(a) The upper .05 point when d.f. = 6.
(b) The lower .025 point when d.f. = 10.
(c) The lower .01 point when d.f. = 9.
(d) The upper .10 point when d.f. = 13.

**9.2** Name the $t$ percentiles shown and find their values from Appendix B, Table 5.

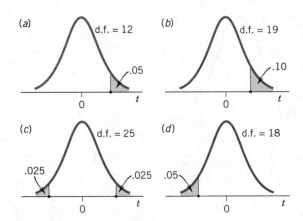

**9.3** Using the table for the $t$ distributions find:
(a) The 90th percentile of the $t$ distribution when d.f. = 8.
(b) The 99th percentile of the $t$ distribution when d.f. = 5.
(c) The 5th percentile of the $t$ distribution when d.f. = 22.
(d) The lower and upper quartiles of the $t$ distribution when d.f. = 17.

**9.4** Find the probability of
(a) $T < -1.761$ when d.f. = 14.

(b) $|T| > 2.306$ when d.f. = 8.
(c) $-1.734 < T < 1.734$ when d.f. = 18.
(d) $-1.812 < T < 2.764$ when d.f. = 10.

**9.5** In each case, find the number $b$ so that
(a) $P[T < b] = .95$ when d.f. = 7.
(b) $P[-b < T < b] = .95$ when d.f. = 16.
(c) $P[T > b] = .01$ when d.f. = 5.
(d) $P[T > b] = .99$ when d.f. = 11.

**9.6** Record the $t_{.05}$ values for d.f. of 5, 10, 15, 20, and 29. Does this percentile increase or decrease with increasing degrees of freedom?

**9.7** Using the table for the $t$ distributions, make an assessment for the probability of the stated event. (The answer to part (a) is provided.)
(a) $T > 2.6$ when d.f. = 7. (Answer: $P[T > 2.6]$ is between .01 and .025 because 2.6 lies between $t_{.05} = 2.365$ and $t_{.01} = 2.998$.)
(b) $T > 1.9$ when d.f. = 16.
(c) $T < -1.5$ when d.f. = 11.
(d) $|T| > 1.9$ when d.f. = 10.
(e) $|T| < 2.8$ when d.f. = 17.

**9.8** What can you say about the number $c$ in each case? Justify your answer. (The answer to part (a) is provided.)
(a) $P[T > c] = .03$ when d.f. = 5. (Answer: $c$ is between 2.015 and 2.571 because $t_{.05} = 2.015$ and $t_{.025} = 2.571$.)
(b) $P[T > c] = .016$ when d.f. = 11.
(c) $P[T < -c] = .004$ when d.f. = 13.
(d) $P[|T| > c] = .03$ when d.f. = 6.
(e) $P[|T| < c] = .96$ when d.f. = 27.

## 2. INFERENCES ABOUT $\mu$—SMALL SAMPLE SIZE

### 2.1 CONFIDENCE INTERVAL FOR $\mu$

The distribution of

$$T = \frac{\overline{X} - \mu}{S/\sqrt{n}}$$

provides the key for determining a **confidence interval for** $\mu$, the mean of a normal population. For a $100(1 - \alpha)\%$ confidence interval, we consult the $t$ table (Appendix B, Table 5) and find $t_{\alpha/2}$, the upper $\alpha/2$ point of the $t$ distribution with $n - 1$ degrees of freedom (see Figure 4).

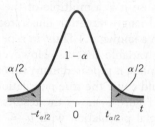

Figure 4    $t_{\alpha/2}$ and the probabilities.

Since $\dfrac{\overline{X} - \mu}{S/\sqrt{n}}$ has the $t$ distribution with d.f. $= n - 1$, we have

$$1 - \alpha = P\left[ -t_{\alpha/2} < \frac{\overline{X} - \mu}{S/\sqrt{n}} < t_{\alpha/2} \right]$$

$$= P\left[ -t_{\alpha/2}\frac{S}{\sqrt{n}} < \overline{X} - \mu < t_{\alpha/2}\frac{S}{\sqrt{n}} \right]$$

In order to obtain a confidence interval, let us rearrange the terms inside the second set of brackets so that only the parameter $\mu$ remains in the center. The above probability statement then becomes

$$P\left[ \overline{X} - t_{\alpha/2}\frac{S}{\sqrt{n}} < \mu < \overline{X} + t_{\alpha/2}\frac{S}{\sqrt{n}} \right] = 1 - \alpha$$

which is precisely in the form required for a confidence statement about $\mu$. The probability is $1 - \alpha$ that the random interval $\overline{X} - t_{\alpha/2}S/\sqrt{n}$ to $\overline{X} + t_{\alpha/2}S/\sqrt{n}$ will cover the true population mean $\mu$. This argument is virtually the same as in Section 3 of Chapter 8. Only now the unknown $\sigma$ is replaced by the sample standard deviation $S$, and the $t$ percentage point is used instead of the standard normal percentage point.

---

**A $100(1 - \alpha)\%$ Confidence Interval for a Normal Population Mean**

$$\left( \overline{X} - t_{\alpha/2}\frac{S}{\sqrt{n}}, \quad \overline{X} + t_{\alpha/2}\frac{S}{\sqrt{n}} \right)$$

where $t_{\alpha/2}$ is the upper $\alpha/2$ point of the $t$ distribution with d.f. $= n - 1$.

Let us review the meaning of a confidence interval in the present context. Imagine that random samples of size $n$ are repeatedly drawn from a normal population and the interval $(\bar{x} - t_{\alpha/2}s/\sqrt{n}, \quad \bar{x} + t_{\alpha/2}s/\sqrt{n})$ calculated in each case. The interval is centered at $\bar{x}$ so the center varies from sample to sample. The length of an interval, $2t_{\alpha/2}s/\sqrt{n}$, also varies from sample to sample because it is a multiple of the sample standard deviation $s$. (This is unlike the fixed-length situation illustrated in Figure 4 of Chapter 8, which was concerned with a known $\sigma$.) Thus, in repeated sampling, the intervals have variable centers and variable lengths. However, our confidence statement means that if the sampling is repeated many times, about $100(1 - \alpha)\%$ of the resulting intervals would cover the true population mean $\mu$.

Figure 5 shows the results of drawing 10 samples of size $n = 7$ from the normal population with $\mu = 100$ and $\sigma = 10$. Selecting $\alpha = .05$, we find that the value of $t_{.025}$ with 6 d.f. is 2.447, so the 95% confidence interval is $\bar{X} \pm 2.447\ S/\sqrt{7}$. In the first sample, $\bar{x} = 103.88$ and $s = 7.96$, so the interval is $(96.52, 111.24)$. The 95% confidence intervals are shown by the vertical line segments. Nine of 10 cover $\mu$.

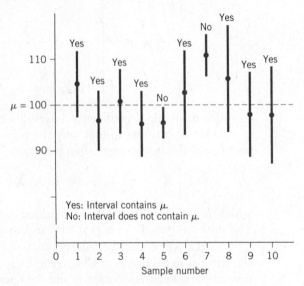

Figure 5    Behavior of confidence intervals based on the $t$ distribution.

**Example 3**    Interpreting a Confidence Interval

The weights (pounds) of $n = 8$ female wolves captured in the Yukon-Charley Rivers National Reserve (see Table D.9 of the Data Bank) are

| 57 | 84 | 90 | 71 | 71 | 77 | 68 | 73 |
|----|----|----|----|----|----|----|----|

Treating these weights as a random sample from a normal distribution:

    (a) Find a 90% confidence interval for the population mean weight of all female wolves living on the Yukon-Charley Rivers National Reserve.

    (b) Is $\mu$ included in this interval?

SOLUTION

    (a) If we assume that the weights are normally distributed, a 90% confidence interval for the mean weight $\mu$ is given by

$$\left( \overline{X} - t_{.05} \frac{S}{\sqrt{n}}, \quad \overline{X} - t_{.05} \frac{S}{\sqrt{n}} \right)$$

where $n = 8$. The $t$ statistic is based on $n - 1 = 8 - 1 = 7$ degrees of freedom so, consulting the $t$ table, we find $t_{.05} = 1.895$. Beforehand, this is a random interval.

To determine the confidence interval from the given sample, we first obtain the summary statistics

$$\overline{x} = \frac{591.000}{8} = 73.875 \qquad s^2 = \frac{708.875}{7} = 101.268 \qquad \text{so}$$

$$s = \sqrt{101.268} = 10.063$$

The 90% confidence interval for $\mu$ is then

$$73.875 \pm 1.895 \times \frac{10.063}{\sqrt{8}} = 73.875 \pm 6.742 \quad \text{or} \ (\,67.13,\ 80.62\,)$$

We are 90% confident that the mean weight of all female wolves is between 67.13 and 80.62 pounds. We have this confidence because, over many occasions of sampling, approximately 90% of the intervals calculated using this procedure will contain the true mean.

    (b) We will never know if a single realization of the confidence interval, such as $(\,67.13, 80.62\,)$, covers the unknown $\mu$. It is unknown because it is based on every female wolf in the very large reserve. Our confidence in the method is based on the high percentage of times that $\mu$ is covered by intervals in repeated samplings.

**Remark:** When repeated independent measurements are made on the same material and any variation in the measurements is basically due to experimental error (possibly compounded by nonhomogeneous materials), the normal model is often found to be appropriate. It is still necessary to graph the individual data points (too few here) in a dot diagram and normal-scores plot to reveal any wild observations or serious departures from normality. In all small sample situations, it is important to remember that the validity of a confidence interval rests on the reasonableness of the model assumed for the population.

Recall from the previous distribution chapter that the length of a $100(1 - \alpha)\%$ confidence interval for the mean $\mu$ of a normal $\mu$ is $2 z_{\alpha/2} \sigma / \sqrt{n}$ when $\sigma$ is known, whereas it is $2 t_{\alpha/2} S / \sqrt{n}$ when $\sigma$ is unknown. Given a small sample size $n$ and consequently a small number of degrees of freedom $(n - 1)$, the extra variability caused by estimating $\sigma$ with $S$ makes the $t$ percentage point $t_{\alpha/2}$ much larger than the normal percentage point $z_{\alpha/2}$. For instance, with d.f. $= 4$, we have $t_{.025} = 2.776$, which is considerably larger than $z_{.025} = 1.96$. Thus, when $\sigma$ is unknown, the confidence estimation of $\mu$ based on a very small sample size is expected to produce a much less precise inference (namely, a wide confidence interval) compared to the situation when $\sigma$ is known. With increasing $n$, $\sigma$ can be more closely estimated by $S$ and the difference between $t_{\alpha/2}$ and $z_{\alpha/2}$ tends to diminish.

## 2.2 HYPOTHESES TESTS FOR $\mu$

The steps for conducting a test of hypotheses concerning a population mean were presented in the previous chapter. If the sample size is small, basically the same procedure can be followed provided it is reasonable to assume that the population distribution is normal. However, in the small sample situation, our test statistic

$$T = \frac{\overline{X} - \mu_0}{S / \sqrt{n}}$$

has Student's $t$ distribution with $n - 1$ degrees of freedom.

The $t$ table (Appendix B, Table 5) is used to determine the rejection region to test hypotheses about $\mu$.

---

### Hypotheses Tests for $\mu$—Small Samples

To test $H_0 : \mu = \mu_0$ concerning the mean of a **normal population**, the test statistic is

$$T = \frac{\overline{X} - \mu_0}{S / \sqrt{n}}$$

which has Student's $t$ distribution with $n - 1$ degrees of freedom:

| | |
|---|---|
| $H_1 : \mu > \mu_0$ | R: $T \geq t_\alpha$ |
| $H_1 : \mu < \mu_0$ | R: $T \leq -t_\alpha$ |
| $H_1 : \mu \neq \mu_0$ | R: $T \geq t_{\alpha/2}$ |

The test is called a **Student's $t$ test** or simply a $t$ test.

---

**Example 4**    A Student's $t$ Test to Confirm the Water Is Safe

A city health department wishes to determine if the mean bacteria count per unit volume of water at a lake beach is within the safety level of 200. A researcher collected 10 water samples of unit volume and found the bacteria counts to be

| | | | | |
|---|---|---|---|---|
| 175 | 190 | 205 | 193 | 184 |
| 207 | 204 | 193 | 196 | 180 |

Do the data strongly indicate that there is no cause for concern? Test with $\alpha = .05$.

SOLUTION    Let $\mu$ denote the current (population) mean bacteria count per unit volume of water. Then, the statement "no cause for concern" translates to $\mu < 200$, and the researcher is seeking strong evidence in support of this hypothesis. So the formulation of the null and alternative hypotheses should be

$$H_0 : \mu = 200 \quad \text{versus} \quad H_1 : \mu < 200$$

Since the counts are spread over a wide range, an approximation by a continuous distribution is not unrealistic for inference about the mean. Assuming further that the measurements constitute a sample from a normal population, we employ the $t$ test with

$$T = \frac{\overline{X} - 200}{S / \sqrt{10}} \quad \text{d.f.} = 9$$

We test at the level of significance $\alpha = .05$. Since $H_1$ is left-sided, we set the rejection region $T \leq -t_{.05}$. From the $t$ table we find that $t_{.05}$ with d.f. $= 9$ is 1.833, so our rejection region is $R: T \leq -1.833$ as in Figure 6. Computations from the sample data yield

$$\overline{x} = 192.7$$

$$s = 10.81$$

$$t = \frac{192.7 - 200}{10.81 / \sqrt{10}} = \frac{-7.3}{3.418} = -2.14$$

Because the observed value $t = -2.14$ is smaller than $-1.833$, the null hypothesis is rejected at $\alpha = .05$. On the basis of the data obtained from these 10 measurements, there does seem to be strong evidence that the true mean is within the safety level.

Values of $T$ smaller than $-2.14$ are more extreme evidence in favor of the alternative hypothesis. Since, with 9 degrees of freedom, $t_{.05} = 1.833$

and $t_{.025} = 2.262$, the $P$-value is between .05 and .025. The accompanying computer output gives

$$P\text{-value} = P[\, T \le -2.14 \,] = .031 \quad (\text{see Figure 7})$$

The moderately small $P$-value strengthens the evidence that the mean bacteria count is within the safety level.

Figure 6 Left-sided rejection region $T \le -1.833$.

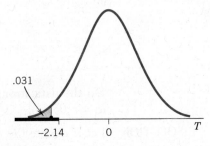

Figure 7 $P$-value for left-sided rejection region, $T$ statistic.

---

### Computer Solution to Example 4

(see the Techology section for the MINITAB COMMANDS)

**T Test of the Mean**

Test of mu = 200.00 vs mu < 200.00

| Variable | N | Mean | StDev | SE Mean | T | P |
|----------|-----|--------|-------|---------|-------|-------|
| Bac Coun | 10 | 192.70 | 10.81 | 3.42 | −2.14 | 0.031 |

---

## Exercises

**9.9** The weights from a random sample of 20 golden retriever dogs have mean 76.1 pounds and standard deviation 5.9 pounds. Assume that the weights of the dogs have a normal distribution.

(a) Construct a 98% confidence interval for the population mean.

(b) What is the length of this confidence interval? What is its center?

(c) If a 98% confidence interval were calculated from another random sample of size $n = 20$, would it have the same length as that found in part (b)? Why or why not?

**9.10** Recorded here are the germination times (number of days) for seven seeds of a new strain of snap bean.

$$12 \quad 16 \quad 15 \quad 20 \quad 17 \quad 11 \quad 18$$

Stating any assumptions that you make, determine a 95% confidence interval for the true mean germination time for this strain.

**9.11** A zoologist collected 20 wild lizards in the southwestern United States. The total length (mm) of each was measured.

179 157 169 146 143 131 159 142 141 130
142 116 130 140 138 137 134 114 90 114

Obtain a 95% confidence interval for the mean length.

**9.12** One approach to investigating the consequences of a major oil spill in the Gulf of Mexico is to look at the amount of heavy metals that have been incorporated in the shells of oysters. Oysters continually build their shells, and if vanadium, cobalt, or chromium from the oil are incorporated in the shells, this might seriously effect the food chain. Suppose that, from a sample of size $n = 12$ shells, the scientists obtain the summary statistics $\bar{x} = 45$ ppm and $s = 5.9$ ppm for the amount of heavy metals in the oyster shells. Assuming that the population distribution is normal, obtain a 95% confidence interval for the mean amount of heavy metals in oyster shells.

**9.13** Eighteen samples of seaweed, each weighing 50-kilograms, are collected to study the feasibility of extracting protein for use in animal feed. The 18 determinations of protein yield have sample mean 3.6 kilograms and standard deviation .8 kilogram. Determine a 95% confidence interval for the mean protein yield per 50 kilograms of seaweed. What did you assume about the population?

**9.14** The monthly rent (dollars) for a two-bedroom apartment on the west side of town was recorded for a sample of ten apartments.

950 745 875 827 1030 920 840 1085 650 895

Obtain a 95% confidence interval for the mean monthly rent for two-bedroom apartments.

**9.15** From a random sample of size 12, one has calculated the 95% confidence interval for $\mu$ and obtained the result ( 38.6, 46.2 ).

(a) What are the $\bar{x}$ and $s$ for that sample?

(b) Calculate a 98% confidence interval for $\mu$.

**9.16** From a random sample of size 18, a researcher states that ( 11.1, 15.7 ) inches is a 90% confidence interval for $\mu$, the mean length of bass caught in a small lake. A normal distribution was assumed. Using the 90% confidence interval, obtain:

(a) A point estimate of $\mu$ and its 90% margin of error.

(b) A 95% confidence interval for $\mu$.

**9.17** Henry Cavendish (1731–1810) provided direct experimental evidence of Newton's law of universal gravitation, which specifies the force of attraction between two masses. In an experiment with known masses determined by weighing, the measured force can also be used to calculate a value for the density of the earth. The values of the earth's density from Cavendish's renowned experiment in time order by column are

| | | | |
|------|------|------|------|
| 5.36 | 5.62 | 5.27 | 5.46 |
| 5.29 | 5.29 | 5.39 | 5.30 |
| 5.58 | 5.44 | 5.42 | 5.75 |
| 5.65 | 5.34 | 5.47 | 5.68 |
| 5.57 | 5.79 | 5.63 | 5.85 |
| 5.53 | 5.10 | 5.34 | |

[These data were published in *Philosophical Transactions*, **17** (1798), p. 469.] Find a 99% confidence interval for the density of the earth.

**9.18** One of the first investigations of the amount of persistent organic pollutants present in soil within an urban setting was motivated by a major flood seventy days earlier.[1] One series of measurements of the total amount of PCB congeners consisted of 14 observations measured in the units ng g$^{-1}$ d.w. It is usual to express the amounts in terms of their natural logarithm

3.30 3.85 4.08 2.56 2.94 3.18 2.12
1.92 3.83 3.40 5.48 3.69 3.33 3.40

Find a 95% confidence interval for the mean of the logarithm of the amounts of PCB congeners.

**9.19** Refer to Exercise 9.18. Do these data support the claim that the mean logarithm of the amount of PCB congeners is less than 3.8? Take $\alpha = .05$.

**9.20** Refer to Exercise 9.14. Do these data support the claim that the mean monthly rent for a two-bedroom apartment differs from 775 dollars? Take $\alpha = .05$.

**9.21** The data on the lengths of anacondas on the front piece of the chapter yield a 95% confidence interval for the population mean length of all anaconda snakes in the area of the study.

```
Variable      N Mean     StDev 95.0% CI
Length(ft) 21 12.871 2.262 ( 11.842,13.901)
```

[1]A. Martinex et al., Spatial distribution of chlordanes and PCB congeners in soil in Cedar Rapids, Iowa, USA, "*Environmental Pollution*" **101**, (2012), pp. 222–228.

(a) Is the population mean length of all female anacondas living in the study area contained in this interval?

(b) Explain why you are 95% confident that it is contained in the interval.

**9.22** Refer to the data on the weight (pounds) of male wolves given in Table D.9 of the Data Bank. A computer calculation gives a 95% confidence interval.

```
Variable N    Mean   StDev   95.0% CI
Malewt   11   91.91  12.38  ( 83.59,100.23)
```

(a) Is the population mean weight for all male wolves in the Yukon-Charley Rivers National Reserve contained in this interval?

(b) Explain why you are 95% confident that it is contained in the interval.

**9.23** The following measurements of the diameters (in feet) of Indian mounds in southern Wisconsin were gathered by examining reports in *Wisconsin Archeologist* (courtesy of J. Williams).

$$22 \quad 24 \quad 24 \quad 30 \quad 22 \quad 20 \quad 28$$
$$30 \quad 24 \quad 34 \quad 36 \quad 15 \quad 37$$

(a) Do these data substantiate the conjecture that the population mean diameter is larger than 21 feet? Test at $\alpha = .01$.

(b) Determine a 90% confidence interval for the population mean diameter of Indian mounds.

**9.24** Measurements of the acidity (pH) of rain samples were recorded at 13 sites in an industrial region.

$$3.5 \quad 5.1 \quad 5.0 \quad 3.6 \quad 4.8 \quad 3.6 \quad 4.7$$
$$4.3 \quad 4.2 \quad 4.5 \quad 4.9 \quad 4.7 \quad 4.8$$

Determine a 95% confidence interval for the mean acidity of rain in that region.

**9.25** Refer to Exercise 9.11, where a zoologist collected 20 wild lizards in the southwestern United States. Do these data substantiate a claim that the mean length is greater than 128 mm? Test with $\alpha = .05$.

**9.26** Geologists dating rock, using a strontium-isotope technique, provided the ages 5.2 and 4.4 million years for two specimens. Treating these as a random sample of size 2 from a normal distribution

(a) Obtain a 90% confidence interval for $\mu$, the true age of the rock formation.

(b) Geologists do get an estimate of error from the strontium-isotope analysis so they usually do not take duplicate readings. From a statistical point of view, two observations are much better than one. What happens to the length of the confidence interval if the sample size were increased from 2 to 4? Answer by considering the ratio of the lengths.

**9.27** The data on the weight (lb) of female wolves, from Table D.9 of the Data Bank, are

$$57 \quad 84 \quad 90 \quad 71 \quad 71 \quad 77 \quad 68 \quad 73$$

Test the null hypothesis that the mean weight of females is 83 pounds versus a two-sided alternative. Take $\alpha = .05$.

**9.28** The ability of a grocery store scanner to read accurately is measured in terms of maximum attenuation (db). In one test with 20 different products, the values of this measurement had a mean 10.7 and standard deviation 2.4. The normal assumption is reasonable.

(a) Is there strong evidence that $\mu$, the population mean maximum attenuation for all possible products, is greater than 9.25 db?

(b) Give a 98% confidence interval for the mean maximum attenuation.

**9.29** The mean drying time of a brand of spray paint is known to be 90 seconds. The research division of the company that produces this paint contemplates that adding a new chemical ingredient to the paint will accelerate the drying process. To investigate this conjecture, the paint with the chemical additive is sprayed on 15 surfaces and the drying times are recorded. The mean and the standard deviation computed from these measurements are 86 and 4.5 seconds, respectively.

(a) Do these data provide strong evidence that the mean drying time is reduced by the addition of the new chemical?

(b) Construct a 98% confidence interval for the mean drying time of the paint with the chemical additive.

(c) What did you assume about the population distribution?

**9.30** A few years ago, noon bicycle traffic past a busy section of campus had a mean of $\mu = 300$. To see if any change in traffic has occurred, counts were taken for a sample of 15 weekdays. It was found that $\bar{x} = 340$ and $s = 30$.

(a) Construct an $\alpha = .05$ test of $H_0: \mu = 300$ against the alternative that some change has occurred.

(b) Obtain a 95% confidence interval for $\mu$.

**9.31** Refer to the computer anxiety scores for female accounting students in Table D.4 of the Data Bank. A computer calculation for a test of $H_0: \mu = 2$ versus $H_1: \mu \neq 2$ is given here.

```
Test of mu = 2 vs mu not = 2

Variable      N    Mean    StDev
FCARS        15   2.514   0.773

Variable    95.0% CI          T     P
FCARS     ( 2.086, 2.942)   2.58  0.022
```

(a) What is the conclusion if you test with $\alpha = .05$?

(b) What mistake could you have made in part (a)?

(c) Before you collected the data, what was the probability of making the mistake in part (a)?

(d) Give a long-run relative frequency interpretation of the probability in part (c).

# 3. RELATIONSHIP BETWEEN TESTS AND CONFIDENCE INTERVALS

By now the careful reader should have observed a similarity between the formulas we use in testing hypotheses and in estimation by a confidence interval. To clarify the link between these two concepts, let us consider again the inferences about the mean $\mu$ of a normal population.

A $100(1 - \alpha)$% confidence interval for $\mu$ is

$$\left( \overline{X} - t_{\alpha/2} \frac{S}{\sqrt{n}}, \quad \overline{X} + t_{\alpha/2} \frac{S}{\sqrt{n}} \right)$$

because before the sample is taken, the probability that

$$\overline{X} - t_{\alpha/2} \frac{S}{\sqrt{n}} < \mu < \overline{X} + t_{\alpha/2} \frac{S}{\sqrt{n}}$$

is $1 - \alpha$. On the other hand, the rejection region of a level $\alpha$ test for $H_0: \mu = \mu_0$ versus the two-sided alternative $H_1: \mu \neq \mu_0$ is

$$R: \quad \left| \frac{\overline{X} - \mu_0}{S/\sqrt{n}} \right| \geq t_{\alpha/2}$$

Let us use the name "acceptance region" to mean the opposite (or complement) of the rejection region. Reversing the inequality in $R$, we obtain

Acceptance region

$$-t_{\alpha/2} < \frac{\overline{X} - \mu_0}{S/\sqrt{n}} < t_{\alpha/2}$$

which can also be written as

Acceptance region

$$\overline{X} - t_{\alpha/2} \frac{S}{\sqrt{n}} < \mu_0 < \overline{X} + t_{\alpha/2} \frac{S}{\sqrt{n}}$$

The latter expression shows that any given null hypothesis $\mu_0$ will be accepted (more precisely, will not be rejected) at level $\alpha$ if $\mu_0$ lies within the $100(1 - \alpha)\%$ confidence interval.

Having established a $100(1 - \alpha)\%$ confidence interval for $\mu$, we know at once that all possible null hypotheses values $\mu_0$ lying outside this interval will be rejected by a two-sided test with level of significance $\alpha$ and all those lying inside will not be rejected.

**Example 5**   Relation between a 95% Confidence Interval and Two-Sided $\alpha = .05$ Test

A random sample of size $n = 9$ from a normal population produced the mean $\bar{x} = 8.3$ and the standard deviation $s = 1.2$. Obtain a 95% confidence interval for $\mu$ and also test $H_0: \mu = 8.5$ versus $H_1: \mu \neq 8.5$ with $\alpha = .05$.

SOLUTION   A 95% confidence interval has the form

$$\left( \overline{X} - t_{.025} \frac{S}{\sqrt{n}}, \quad \overline{X} + t_{.025} \frac{S}{\sqrt{n}} \right)$$

where $t_{.025} = 2.306$ corresponds to $n - 1 = 8$ degrees of freedom. Here $\bar{x} = 8.3$ and $s = 1.2$, so the interval becomes

$$\left( 8.3 - 2.306 \frac{1.2}{\sqrt{9}}, \quad 8.3 + 2.306 \frac{1.2}{\sqrt{9}} \right) = (7.4, 9.2)$$

Turning now to the problem of testing $H_0: \mu = 8.5$, we observe that the value 8.5 lies in the 95% confidence interval we have just calculated. Using the correspondence between confidence interval and acceptance region, we can at once conclude that $H_0: \mu = 8.5$ should not be rejected at $\alpha = .05$. Alternatively, a formal step-by-step solution can be based on the test statistic

$$T = \frac{\overline{X} - 8.5}{S / \sqrt{n}}$$

The rejection region consists of both large and small values.

Rejection region

$$\left| \frac{\overline{X} - 8.5}{S / \sqrt{n}} \right| \geq t_{.025} = 2.306$$

Now the observed value $|t| = \sqrt{9} \, |8.3 - 8.5| / 1.2 = .5$ does not fall in the rejection region, so the null hypothesis $H_0: \mu = 8.5$ is not rejected at $\alpha = .05$. This conclusion agrees with the one we arrived at from the confidence interval.

This relationship indicates how confidence estimation and tests of hypotheses with two-sided alternatives are really integrated in a common framework. A confidence interval statement is regarded as a more comprehensive inference procedure than testing a single null hypothesis, because a confidence interval statement in effect tests many null hypotheses at the same time.

## Exercises

**9.32** Based on a random sample of tail lengths for 15 male kites, an investigator calculates the 95% confidence interval ( 183.0, 195.0 ) mm based on the $t$ distribution. The normal assumption is reasonable.

(a) What is the conclusion of the $t$ test for $H_0 : \mu = 190.5$ versus $H_1 : \mu \neq 190.5$?

(b) What is the conclusion if $H_0 : \mu = 182.2$?

**9.33** In Example 3, the 90% confidence interval for the mean weight of female wolves was found to be ( 67.13, 80.62 ) pounds.

(a) What is the conclusion of testing $H_0 : \mu = 81$ versus $H_1 : \mu \neq 81$ at level $\alpha = .10$?

(b) What is the conclusion if $H_0 : \mu = 69$?

**9.34** The petal width (mm) of one kind of iris has a normal distribution. Suppose that, from a random sample of widths, the $t$ based 90% confidence interval for the population mean width is ( 16.8, 19.6 ) mm. Answer each question "yes," "no," or "can't tell," and justify your answer. On the basis of the same sample:

(a) Would $H_0 : \mu = 20$ be rejected in favor of $H_1 : \mu \neq 20$ at $\alpha = .10$?

(b) Would $H_0 : \mu = 18$ be rejected in favor of $H_1 : \mu \neq 18$ at $\alpha = .10$?

*(c) Would $H_0 : \mu = 17$ be rejected in favor of $H_1 : \mu \neq 17$ at $\alpha = .05$?

*(d) Would $H_0 : \mu = 22$ be rejected in favor of $H_1 : \mu \neq 22$ at $\alpha = .01$?

**9.35** Recorded here are the amounts of decrease in percent of body fat for eight participants in an exercise program over three weeks.

| 1.8 | 10.6 | −1.2 | 12.9 | 15.1 | −2.0 | 6.2 | 10.8 |

(a) Construct a 95% confidence interval for the population mean amount $\mu$ of decrease in percent body fat over the three-week program.

(b) If you were to test $H_0 : \mu = 15$ versus $H_1 : \mu \neq 15$ at $\alpha = 0.5$, what would you conclude from your result in part (a)?

(c) Perform the hypothesis test indicated in part (b) and confirm your conclusion.

**9.36** Refer to the data in Exercise 9.35.

(a) Construct a 90% confidence interval for $\mu$.

(b) If you were to test $H_0 : \mu = 10$ versus $H_1 : \mu \neq 10$ at $\alpha = .10$, what would you conclude from your result in part (a)? Why?

(c) Perform the hypothesis test indicated in part (b) and confirm your conclusion.

**9.37** Establish the connection between the large sample $Z$ test, which rejects $H_0 : \mu = \mu_0$ in favor of $H_1 : \mu \neq \mu_0$, at $\alpha = .05$, if

$$Z = \frac{\overline{X} - \mu_0}{S / \sqrt{n}} \geq 1.96 \quad \text{or}$$

$$\frac{\overline{X} - \mu_0}{S / \sqrt{n}} \leq -1.96$$

and the 95% confidence interval

$$\overline{X} - 1.96 \frac{S}{\sqrt{n}} \quad \text{to} \quad \overline{X} + 1.96 \frac{S}{\sqrt{n}}$$

# 4. INFERENCES ABOUT THE STANDARD DEVIATION $\sigma$ (THE CHI-SQUARE DISTRIBUTION)

Aside from inferences about the population mean, the population variability may also be of interest. Apart from the record of a baseball player's batting average, information on the variability of the player's performance from one game to the next may be an indicator of reliability. Uniformity is often a criterion of production quality for a manufacturing process. The quality engineer must ensure that the variability of the measurements does not exceed a specified limit. It may also be important to ensure sufficient uniformity of the inputted raw material for trouble-free operation of the machines. In this section, we consider inferences for the standard deviation $\sigma$ of a population under the assumption that the population distribution is normal. In contrast to the inference procedures concerning the population mean $\mu$, the methods presented here can frequently produce misleading inferences when the normal assumption is violated.

To make inferences about $\sigma^2$, the natural choice of a statistic is its sample analog, which is the sample variance,

$$S^2 = \frac{\sum\limits_{i=1}^{n} (X_i - \overline{X})^2}{n - 1}$$

We take $S^2$ as the point estimator of $\sigma^2$ and its square root $S$ as the **point estimator** of $\sigma$. To estimate by confidence intervals and test hypotheses, we must consider the sampling distribution of $S^2$. To do this, we introduce a new distribution, called the $\chi^2$ distribution (read "chi-square distribution"), whose form depends on $n - 1$.

---

## $\chi^2$ **Distribution**

Let $X_1, \ldots, X_n$ be a random sample from a normal population $N(\mu, \sigma)$. Then the distribution of

$$\chi^2 = \frac{\sum\limits_{i=1}^{n} (X_i - \overline{X})^2}{\sigma^2} = \frac{(n-1)S^2}{\sigma^2}$$

is called the $\chi^2$ distribution with $n - 1$ degrees of freedom.

---

Unlike the normal or $t$ distribution, the probability density curve of a $\chi^2$ distribution is an asymmetric curve stretching over the positive side of the line and having a long right tail. The form of the curve depends on the value of the degrees of freedom. A typical $\chi^2$ curve is illustrated in Figure 8.

Appendix B, Table 6, provides the upper $\alpha$ points of $\chi^2$ distributions for various values of $\alpha$ and the degrees of freedom. As in both the cases of the $t$ and the normal distributions, the upper $\alpha$ point $\chi_\alpha^2$ denotes the $\chi^2$ value such that

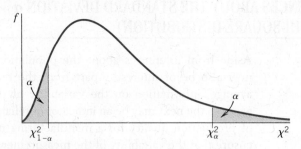

Figure 8   Probability density curve of a $\chi^2$ distribution.

the area to the right is $\alpha$. The lower $\alpha$ point or $100\,\alpha$ th percentile, read from the column $\chi^2_{1-\alpha}$ in the table, has an area $1 - \alpha$ to the right. For example, the lower .05 point is obtained from the table by reading the $\chi^2_{.95}$ column, whereas the upper .05 point is obtained by reading the column $\chi^2_{.05}$.

**Example 6**   Finding Percentage Points of the $\chi^2$ Distribution

Find the upper .05 point of the $\chi^2$ distribution with 17 degrees of freedom. Also find the lower .05 point.

SOLUTION

Percentage Points of the $\chi^2$ Distributions
(Appendix B, Table 6)

| d.f. \ $\alpha$ | . . . | .95 | . . . | .05 |
|---|---|---|---|---|
| 17 | - - -▶ | 8.67 | - - -▶ | 27.59 |

The upper .05 point is read from the column labeled $\alpha = .05$. We find $\chi^2_{.05} = 27.59$ for 17 d.f. The lower .05 point is read from the column $\alpha = .95$. We find $\chi^2_{.95} = 8.67$, as the lower .05 point.

The $\chi^2$ is the basic distribution for constructing confidence intervals for $\sigma^2$ or $\sigma$. We outline the steps in terms of a 95% confidence interval for $\sigma^2$. Dividing the probability $\alpha = .05$ equally between the two tails of the $\chi^2$ distribution and using the notation just explained, we have

$$P\left[ \chi^2_{.975} < \frac{(n-1)S^2}{\sigma^2} < \chi^2_{.025} \right] = .95$$

where the percentage points are read from the $\chi^2$ table at d.f. $= n - 1$. Because

$$\frac{(n-1)S^2}{\sigma^2} < \chi^2_{.025} \quad \text{is equivalent to} \quad \frac{(n-1)S^2}{\chi^2_{.025}} < \sigma^2$$

and

$$\chi^2_{.975} < \frac{(n-1)S^2}{\sigma^2} \quad \text{is equivalent to} \quad \sigma^2 < \frac{(n-1)S^2}{\chi^2_{.975}}$$

we have

$$P\left[ \frac{(n-1)S^2}{\chi^2_{.025}} < \sigma^2 < \frac{(n-1)S^2}{\chi^2_{.975}} \right] = .95$$

This last statement, concerning a random interval covering $\sigma^2$, provides a 95% confidence interval for $\sigma^2$.

---

A 95% confidence interval for $\sigma^2$ is

$$\left( \frac{(n-1)S^2}{\chi^2_{.025}}, \quad \frac{(n-1)S^2}{\chi^2_{.975}} \right)$$

---

A confidence interval for $\sigma$ can be obtained by taking the square root of the endpoints of the interval for $\sigma$.

---

A 95% confidence interval for $\sigma$ is

$$\left( S\sqrt{\frac{n-1}{\chi^2_{.025}}}, \quad S\sqrt{\frac{n-1}{\chi^2_{.975}}} \right)$$

---

**Example 7**    A Confidence Interval for Standard Deviation of Hours of Sleep

Refer to the data on hours of sleep in Example 6, Table 4, of Chapter 2. Fifty-nine students reported the hours of sleep the previous night. Obtain 90% confidence intervals for $\sigma^2$ and $\sigma$, using the summary statistics

$$n = 59 \qquad \bar{x} = 7.181 \qquad s = 1.282$$

SOLUTION    Here $n = 59$, so d.f. $= n - 1 = 58$. A computer calculation, or interpolation in the $\chi^2$ table, gives $\chi^2_{.95} = 41.49$ and $\chi^2_{.05} = 76.78$. Using the preceding probability statements, we determine that a 90% confidence interval for the variance $\sigma^2$ is

$$\left( \frac{58 \times (1.282)^2}{76.78}, \quad \frac{58 \times (1.282)^2}{41.49} \right) = (1.242, 2.298)$$

and the corresponding interval for $\sigma$ is $(\sqrt{1.242}, \sqrt{2.297}) = (1.11, 1.52)$ hours. We are 90% confident that the standard deviation of number of hours slept is between 1.11 and 1.52 hours. We have this confidence because 90% of the intervals calculated by this procedure in repeated samples will cover the true $\sigma$.

You may verify that the pattern of a normal-scores plot does not contradict the assumption of normality.

---

It is instructive to note that the midpoint of the confidence interval for $\sigma^2$ in Example 7 is not $s^2 = 1.644 = (1.282)^2$ which is the best point estimate.

This is in sharp contrast to the confidence intervals for $\mu$, and it serves to accent the difference in logic between interval and point estimation.

For a test of the null hypothesis $H_0 : \sigma^2 = \sigma_0^2$ it is natural to employ the statistic $S^2$. If the alternative hypothesis is one-sided, say $H_1 : \sigma^2 > \sigma_0^2$, then the rejection region should consist of large values of $S^2$ or alternatively large values of the

$$\text{Test statistic} \quad \chi^2 = \frac{(n-1)S^2}{\sigma_0^2} \quad \text{d.f.} = n-1$$

To **test hypothesis about** $\sigma$, a level $\alpha$ rejection region for the one-sided alternative $H_1 : \sigma^2 > \sigma_0^2$ is

$$R : \frac{(n-1)S^2}{\sigma_0^2} \geq \chi_\alpha^2 \quad \text{d.f.} = n-1$$

For a two-sided alternative $H_1 : \sigma^2 \neq \sigma_0^2$, a level $\alpha$ rejection region is

$$R : \frac{(n-1)S^2}{\sigma_0^2} \leq \chi_{1-\alpha/2}^2 \quad \text{or} \quad \frac{(n-1)S^2}{\sigma_0^2} \geq \chi_{\alpha/2}^2$$

Once again, we remind the reader that the inference procedures for $\sigma$ presented in this section are extremely sensitive to departures from a normal population.

**Example 8**   Testing the Standard Deviation of Green Gasoline

One company, actively pursuing the making of green gasoline, starts with biomass in the form of sucrose and converts it into gasoline using catalytic reactions. At one step in a pilot plant process, the output includes carbon chains of length 3. Fifteen runs with same catalyst produced the product volumes (liter)

| | | | | | | | |
|---|---|---|---|---|---|---|---|
| 2.79 | 2.88 | 2.09 | 2.32 | 3.51 | 3.31 | 3.17 | 3.62 |
| 2.79 | 3.94 | 2.34 | 3.62 | 3.22 | 2.80 | 2.70 | |

While mean product volume is the prime parameter, it is also important to control variation. Conduct a test with the intent of showing that the population standard deviation $\sigma$ is less than .8 liter. Use $\alpha = .05$.

SOLUTION   We first calculate the summary statistics $\bar{x} = 3.007$ and $s^2 = .2889$ so $s = .5374$. The claim becomes the alternative hypothesis. Since $\sigma < .8$ is equivalent to $\sigma^2 < .64 = .8^2$, we actually test

$$H_0 : \sigma^2 = .64 \quad \text{versus} \quad H_1 : \sigma^2 < .64$$

with $\alpha = .05$.

Because $n = 15$, the d.f. $= 14$. Also, $1 - \alpha = 1 - .05 = .95$. Consulting the $\chi^2$ table with 14 d.f., we find $\chi_{.95}^2 = 6.57$ so the rejection region is

$$R : \frac{(n-1)S^2}{\sigma_0^2} \leq \chi_{.95}^2 = 6.57$$

The observed value of the test statistic is

$$\frac{(n-1)S^2}{\sigma_0^2} = \frac{14 \times .2889}{.64} = 6.320$$

Consequently, at significance level .05, we reject the null hypothesis and conclude that the population variance is less than .64. Equivalently, the population standard deviation is less than .80 liter.

The left-sided rejection region is illustrated in Figure 9 and the observed value $\chi^2 = 6.32$ is close to the boundary. A computer calculation gives $P$-value $= .042$ which only slightly strengthens the evidence against the null hypothesis $\sigma = .80$ liter.

Figure 9   Left-sided rejection region $\chi^2 \leq 6.57$.

## Exercises

**9.38** Using the table for the $\chi^2$ distributions, find:

(a) The upper 5% point when d.f. $= 9$.

(b) The upper 1% point when d.f. $= 14$.

(c) The lower 2.5% point when d.f. $= 7$.

(d) The lower 1% point when d.f. $= 22$.

**9.39** Name the $\chi^2$ percentiles shown and find their values from Appendix B, Table 6.

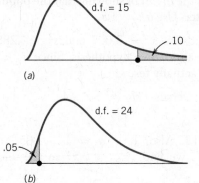

(a)

(b)

(c) Find the percentile in part (a) if d.f. $= 40$.

(d) Find the percentile in part (b) if d.f. $= 8$.

**9.40** Find the probability of

(a) $\chi^2 > 31.33$ when d.f. $= 18$.

(b) $\chi^2 < 1.15$ when d.f. $= 5$.

(c) $3.24 < \chi^2 < 18.31$ when d.f. $= 10$.

(d) $3.49 < \chi^2 < 20.09$ when d.f. $= 8$.

**9.41** Beginning students in accounting took a test and a computer anxiety score (CARS) was assigned to each student on the basis of their answers to nineteen questions on the test (see Table D.4 of the Data Bank). A small population standard deviation would indicate that computer anxiety is nearly the same for all female beginning accounting students.

The scores for 15 female students are

2.90  1.00  1.90  2.37  3.32  3.79  3.26  1.90
1.84  2.58  1.58  2.90  2.42  3.42  2.53

Assuming that the distribution of scores of all college students can be modeled as a normal distribution

(a) Obtain a point estimate of the population standard deviation $\sigma$.

(b) Construct a 95% confidence interval for $\sigma$.

(c) Examine whether or not your point estimate is located at the center of the confidence interval.

**9.42** Find a 90% confidence interval for $\sigma$ based on the $n = 40$ measurements of heights of red pine seedlings given in Exercise 8.4. State any assumption you make about the population. (Note: $s = .475$ for this data set.)

**9.43** Refer to Exercise 9.42. A related species has population standard deviation $\sigma = .6$. Do the data provide strong evidence that the red pine population standard deviation is smaller than .6? Test with $\alpha = .05$.

**9.44** Plastic sheets produced by a machine are periodically monitored for possible fluctuations in thickness. Uncontrollable heterogeneity in the viscosity of the liquid mold makes some variation in thickness measurements unavoidable. However, if the true standard deviation of thickness exceeds 1.5 millimeters, there is cause to be concerned about the product quality. Thickness measurements (in millimeters) of 10 specimens produced on a particular shift resulted in the following data.

| 226 | 228 | 226 | 225 | 232 |
| 228 | 227 | 229 | 225 | 230 |

Do the data substantiate the suspicion that the process variability exceeded the stated level on this particular shift? (Test at $\alpha = .05$.) State the assumption you make about the population distribution.

**9.45** Refer to Exercise 9.44. Construct a 95% confidence interval for the true standard deviation of the thickness of sheets produced on this shift.

**9.46** During manufacture, the thickness of laser printer paper is monitored. Data from several random samples each day during the year suggest that thickness follows a normal distribution. A sample of $n = 10$ thickness measurements (ten-thousandths inch) yields the 95% confidence interval $(4.33, 11.50)$ for $\sigma^2$.

(a) What was the standard deviation $s$ for the sample? (*Hint*: Examine how $s$ enters the formula of a confidence interval.)

(b) Calculate a 90% confidence interval for $\sigma$.

**9.47** Refer to the data of lizard lengths in Exercise 9.11.

(a) Determine a 90% confidence interval for the population standard deviation $\sigma$.

(b) Should $H_0 : \sigma = 9$ be rejected in favor of $H_1 : \sigma \neq 9$ at $\alpha = .10$? [Answer by using your result in part (a).]

**9.48** Referring to the data in Exercise 9.17, determine a 99% confidence interval for the population standard deviation of the density measurements.

**9.49** Referring to Exercise 9.23, construct a 95% confidence interval for the population standard deviation of the diameters of Indian mounds.

**9.50** Do the data in Exercise 9.24 substantiate the conjecture that the true standard deviation of the acidity measurements is larger than 0.4? Test at $\alpha = .05$.

# 5. ROBUSTNESS OF INFERENCE PROCEDURES

The small sample methods for both confidence interval estimation and hypothesis testing presuppose that the sample is obtained from a normal population. Users of these methods would naturally ask:

1. What method can be used to determine if the population distribution is nearly normal?

2. What can go wrong if the population distribution is nonnormal?

3. What procedures should be used if it is nonnormal?

4. If the observations are not independent, is this serious?

A1. To answer the first question, we could construct the dot diagram or normal-scores plot. These may indicate a wild observation or a striking departure from normality. If none of these is visible, the investigator would feel more secure using the preceding inference procedures. However, any plot based on a small sample cannot provide convincing justification for normality. Lacking sufficient observations to justify or refute the normal assumption, we are led to a consideration of the second question.

A2. Confidence intervals and tests of hypotheses concerning $\mu$ are based on Student's $t$ distribution. If the population is nonnormal, the actual percentage points may differ substantially from the tabulated values. When we say that $\overline{X} \pm t_{.025} S / \sqrt{n}$ is a 95% confidence interval for $\mu$, the true probability that this random interval will contain $\mu$ may be, say, 85% or 99%. Fortunately, the effects on inferences about $\mu$ using the $t$ statistic are not too serious if the sample size is at least moderately large (say, 15). In larger samples, such disturbances tend to disappear due to the central limit theorem. We express this fact by saying that **inferences about $\mu$ using the $t$ statistic are reasonably** "robust." However, this qualitative discussion should not be considered a blanket endorsement for $t$. When the sample size is small, a wild observation or a distribution with long tails can produce misleading results.

Unfortunately, inferences about $\sigma$ using the $\chi^2$ distribution may be seriously affected by nonnormality even with large samples. We express this by saying that **inferences about $\sigma$ using the $\chi^2$ distribution are not "robust" against departures of the population distribution from normality**.

A3. We cannot give a specific answer to the third question without knowing something about the nature of nonnormality. Dot diagrams or histograms of the original data may suggest some transformations that will bring the shape of the distribution closer to normality. If it is possible to obtain a transformation that leads to reasonably normal data plots, the problem can then be recast in terms of the transformed data. Otherwise, users can benefit from consulting with a statistician.

A4. A basic assumption throughout Chapters 8 and 9 is that the sample is drawn at random, so the observations are independent of one another. If the sampling is made in such a manner that the observations are dependent, however, all the inferential procedures we discussed here for small as well as large samples may be seriously in error. This applies to both the level of significance of a test and a stated confidence level. Concerned with the possible effect of a drug on blood pressure, suppose that an investigator includes 5 patients in an experiment and makes 4 successive measurements on each. This does *not* yield a random sample of size 5 $\times$ 4 $=$ 20, because the 4 measurements made on each person are likely to be dependent. This type of sampling requires a more sophisticated method of analysis. An investigator who is sampling opinions about a political issue may choose 100 families at random and record the opinions of both the husband and wife in each family. This also does *not* provide a random sample of size 100 $\times$ 2 $=$ 200, although it may be a convenient sampling method. When measurements are made closely together in time or distance, there is a danger of losing independence because adjacent observations are more likely to be similar than observations that are made farther apart. Because independence is the most crucial assumption, we must be constantly alert to detect such violations. Prior to a formal analysis of the data, a close scrutiny of the sampling process is imperative.

## USING STATISTICS WISELY

1. The inferences in this chapter, based on the $t$ distribution, require that the distribution of the individual observations be normal. With very small sample sizes, say 10 or smaller, it is not possible to check this assumption. The best we can do is make dot diagrams, or normal-scores plots, to make sure there are no obvious outliers. With somewhat larger sample sizes serious asymmetry can be recognized. Recall, however, the central limit result that the distribution of $\overline{X}$ becomes more nearly normal with increasing sample size whatever the population. This tends to make inferences based on the $t$ statistic relatively insensitive to small or moderate departures from a normal population as long as sample size is greater than about 15.

2. Understand the interpretation of a $100(1 - \alpha)\%$ confidence interval. For normal populations, before the data are collected,

$$\left( \overline{X} - t_{\alpha/2} \frac{S}{\sqrt{n}}, \qquad \overline{X} + t_{\alpha/2} \frac{S}{\sqrt{n}} \right)$$

is a random interval that will cover the fixed unknown mean $\mu$ with probability $1 - \alpha$. Once a numerical value for the interval is obtained, the interval is fixed and we say we have $100(1 - \alpha)\%$ confidence that the mean is contained in the interval. After many applications of this procedure, to different samples from independent experiments, approximately proportion $1 - \alpha$ of the intervals will cover the respective population mean.

3. Formulate, as the alternative hypothesis the assertion that the experiment seeks to confirm. Then, base a test of the null hypothesis $H_0 : \mu = \mu_0$ on the test statistic

$$T = \frac{\overline{X} - \mu_0}{S/\sqrt{n}}$$

which has a $t$ distribution with $n - 1$ degrees of freedom. The rejection region is one-sided or two-sided corresponding to the alternative hypothesis.

## KEY IDEAS AND FORMULAS

When the sample size is small, additional conditions need to be imposed on the population. In this chapter, we **assume that the population distribution is normal**.

Inferences about the mean of a normal population are based on

$$T = \frac{\overline{X} - \mu}{S/\sqrt{n}}$$

which has Student's $t$ distribution with $n - 1$ degrees of freedom.

Inferences about the standard deviation of a normal population are based on $(n - 1)S^2/\sigma^2$, which has a $\chi^2$ distribution with $n - 1$ degrees of freedom.

Moderate departures from a normal population distribution do not seriously affect inferences based on Student's $t$. These procedures are "robust."

Nonnormality can seriously affect inferences about $\sigma$.

### Inferences about a Normal Population Mean

When $n$ is small, we assume that the population is approximately normal. Inference procedures are derived from Student's $t$ sampling distribution of

$$T = \frac{\overline{X} - \mu}{S / \sqrt{n}}$$

1. A $100(1 - \alpha)\%$ confidence interval for $\mu$ is

$$\left( \overline{X} - t_{\alpha/2} \frac{S}{\sqrt{n}}, \quad \overline{X} + t_{\alpha/2} \frac{S}{\sqrt{n}} \right)$$

2. To test hypotheses about $\mu$, using Student's $t$ test, the test statistic is

$$T = \frac{\overline{X} - \mu_0}{S / \sqrt{n}}$$

Given a level of significance $\alpha$, the $t$ test will:

Reject $H_0 : \mu = \mu_0$ in favor of $H_1 : \mu > \mu_0$ if $T \geq t_\alpha$

Reject $H_0 : \mu = \mu_0$ in favor of $H_1 : \mu < \mu_0$ if $T \leq -t_\alpha$

Reject $H_0 : \mu = \mu_0$ in favor of $H_1 : \mu \neq \mu_0$ if $|T| \geq t_{\alpha/2}$

### Inferences about a Normal Population Standard Deviation

Inferences are derived from the $\chi^2$ distribution for $(n - 1) S^2 / \sigma^2$.

1. A point estimator of $\sigma$ is the sample standard deviation $S$.

2. A 95% confidence interval **for** $\sigma$ is

$$\left( S \sqrt{\frac{n - 1}{\chi^2_{.025}}}, \quad S \sqrt{\frac{n - 1}{\chi^2_{.975}}} \right)$$

3. To test hypotheses about $\sigma$, the test statistic is

$$\frac{(n - 1) S^2}{\sigma_0^2}$$

Given a level of significance $\alpha$,

$$\left. \begin{array}{l} \text{Reject } H_0 : \sigma = \sigma_0 \\ \text{in favor of } H_1 : \sigma < \sigma_0 \end{array} \right\} \text{ if } \quad \frac{(n - 1) S^2}{\sigma_0^2} \leq \chi^2_{1 - \alpha}$$

$$\left. \begin{array}{l} \text{Reject } H_0 : \sigma = \sigma_0 \\ \text{in favor of } H_1 : \sigma > \sigma_0 \end{array} \right\} \text{ if } \quad \frac{(n - 1) S^2}{\sigma_0^2} \geq \chi^2_{\alpha}$$

$$\left. \begin{array}{l} \text{Reject } H_0 : \sigma = \sigma_0 \\ \text{in favor of } H_1 : \sigma \neq \sigma_0 \end{array} \right\} \text{ if } \quad \frac{(n - 1) S^2}{\sigma_0^2} \leq \chi^2_{1 - \alpha/2}$$

$$\text{or} \quad \frac{(n - 1) S^2}{\sigma_0^2} \geq \chi^2_{\alpha/2}$$

## TECHNOLOGY

### *Confidence intervals and tests concerning a normal mean*

#### MINITAB

#### *Confidence intervals for* $\mu$

We illustrate the calculation of a 99% confidence interval for $\mu$ based on the $t$ distribution.

> Data: C1
>
> **Stat** > **Basic Statistics** > **1-Sample t.**
>
> Type *C1* in **Samples in Columns.**
>
> Click **Options** and type *99* in **Confidence level.**
>
> Click **OK**. Click **OK**.

#### *Tests of hypotheses concerning* $\mu$

We illustrate the calculation of an $\alpha = .01$ level test of $H_0 : \mu = 32$ versus a one-sided alternative, $H_1 : \mu > 32$.

> Data: *C1*
>
> **Stat** > **Basic Statistics** > **1-Sample t.**
>
> Type *C1* in **Samples in.** Click **Perform hypotheses test** and following **Hypothesized mean**, type *32*, the value of the mean under the null hypothesis. Click **Options** and type *99* in **Confidence level.**
>
> In the **Alternative** cell select *greater than*, the direction of the alternative hypothesis. Click **OK**. Click **OK**.

If the sample size, mean, and standard deviation are available, instead of the second step, type these values in the corresponding cells.

#### EXCEL

#### *Confidence intervals for* $\mu$

We illustrate the calculation of a 99% confidence interval for $\mu$ based on the $t$ distribution. First obtain the standard deviation $s$ as described in Chapter 2, Technology.

> Click the **Formulas** tab and then **More Functions**. Choose **Statistical** and the **CONFIDENCE.T**. Enter .01 in **Alpha**, the value of $s$ in **standard_dev**, and the number of observations in **size**. Click **OK**.

#### *Tests of hypotheses concerning* $\mu$

First obtain $\bar{x}$ and $s$ as described in Chapter 2, Technology. Then calculate the value of the $t$ statistic. The command **DIST.T** evaluates the cumulative distribution at the observed value of $t$ and this can be converted to the $P$–value.

### TI-84/-83 PLUS

#### Confidence intervals for $\mu$

We illustrate the calculation of a 99% confidence interval for $\mu$. Start with the data entered in $\mathsf{L_1}$.

> Press **STAT** and select **TESTS** and then 8: **Tinterval**. Press **ENTER**.
>
> Select **Data** for Inpt: Set List to $\mathsf{L_1}$, Freq to 1, and C-Level: to .99.
> Select **Calculate**.
>
> Then press **ENTER**.

If, instead, the sample size, mean, and standard deviation are available, the second step is:

> Select **Stats** (instead of **Data**) and enter the sample size, mean, and standard deviation.

#### Tests of hypotheses concerning $\mu$

We illustrate the calculation of an $\alpha = .01$ level test of $H_0: \mu = 32$ versus a one-sided alternative $H_1: \mu > 32$. Start with the data entered in $\mathsf{L_1}$.

> Press **STAT** and select **TESTS** and then 2: **T-Test**. Press **ENTER**.
>
> Select **Data** for Inpt: Enter 32 for $\mu_0$. Set List to $\mathsf{L_1}$ and Freq to 1.
>
> Select the direction of the alternative hypothesis.
>
> Select **Calculate**. Press **ENTER**.
>
> The calculator will return the $P$–value.

If, instead, the sample size, mean, and standard deviation are available, the second step is:

> Select **Stats** (instead of **Data**) and enter the sample size, mean, and standard deviation.

---

## REVIEW EXERCISES

**9.51**  Using the table of percentage points for the $t$ distributions, find

(a) $t_{.05}$ when d.f. $= 7$.

(b) $t_{.025}$ when d.f. $= 11$.

(c) The lower .05 point when d.f. $= 7$.

(d) The lower .05 point when d.f. $= 11$.

**9.52**  Using the table for the $t$ distributions, find the probability of

(a) $T > 2.720$ when d.f. $= 22$.

(b) $T < 3.250$ when d.f. $= 9$.

(c) $|T| < 2.567$ when d.f. $= 17$.

(d) $-1.383 < T < 2.262$ when d.f. $= 9$.

**9.53** A $t$ distribution assigns more probability to large values than the standard normal.

(a) Find $t_{.05}$ for d.f. $= 12$ and then evaluate $P[\,Z > t_{.05}\,]$. Verify that $P[\,T > t_{.05}\,]$ is greater than $P[\,Z > t_{.05}\,]$.

(b) Examine the relation for d.f. of 5 and 20, and comment.

**9.54** Measurements of the amount of suspended solids in river water on 14 Monday mornings yield $\bar{x} = 47$ and $s = 9.4$ mg/l. Obtain a 95% confidence interval for the mean amount of suspended solids. State any assumption you make about the population.

**9.55** Determine a 99% confidence interval for $\mu$ using the data in Exercise 9.54.

**9.56** The time to blossom of 21 plants has $\bar{x} = 38.4$ days and $s = 5.1$ days. Give a 95% confidence interval for the mean time to blossom.

**9.57** Refer to Example 8 concerning the product volume for green gasoline. Obtain

(a) A point estimate of $\mu$ and its 95% error margin.

(b) A 90% confidence interval for the mean.

(c) Explain why you are 90% confident that the interval in part (b) covers the true unknown mean.

**9.58** Refer to Exercise 9.54. The water quality is acceptable if the mean amount of suspended solids is less than 49 mg/l. Construct an $\alpha = .05$ test to establish that the quality is acceptable.

(a) Specify $H_0$ and $H_1$.

(b) State the test statistic.

(c) What does the test conclude?

**9.59** Refer to Exercise 9.56. Do these data provide strong evidence that the mean time to blossom is less than 42 days? Test with $\alpha = .01$.

(a) Formulate the null and alternative hypotheses.

(b) Determine the test statistic.

(c) Give the form of the rejection region.

(d) What is the conclusion to your test?

(e) What error could you have made in light of your decision in part (d)?

(f) What can you say about the $P$–value?

**9.60** In a study of head injuries to ice hockey players the number of hits were monitored electronically by equipping approved helmets with accelerometers that measured head acceleration $(\text{m/s}^2)$ when a player was hit. The researches counted the number of serious hits, those greater than 30 m/s$^2$, accumulated by each of the 13 players on the team. The counts covered a wide range and tended to have a long right-hand tail so we took the square root and obtained the measurements[2]

| | | | | | | |
|------|------|------|------|------|------|------|
| 3.74 | 1.73 | 3.16 | 4.00 | 3.61 | 5.10 | 6.40 |
| 4.58 | 2.65 | 2.83 | 3.61 | 3.00 | 3.61 | |

where the largest values indicate larger chances of injury.

Conduct a test of hypotheses with the intent of establishing that the mean of the square root, $\mu$, is greater than 2.8. Take $\alpha = .05$.

**9.61** Referring to Exercise 9.60, test $H_0 : \mu = 2.8$ versus $H_1 : \mu \neq 2.8$ using $\alpha = .02$.

**9.62** Refer to Example 8 concerning the yield of green gasoline. Conduct a test of hypothesis which is intended to show that the mean product volume is greater than 2.75 liters.

(a) Formulate the null and alternative hypotheses.

(b) Determine the test statistic.

(c) Give the form of the rejection region.

(d) What is the conclusion to your test? Take $\alpha = .05$.

(e) What error could you have made in light of your decision in part (d)?

(f) What can you say about the $P$–value?

**9.63** The supplier of a particular brand of vitamin pills claims that the average potency of these pills after a certain exposure to heat and humidity is at

---

[2]J. Cubos et al., "Relationships among risk factors for concussion in minor ice hockey," *ASTM International*, Selected Technical Papers STP 1615, eds. R. Greenwald and A. Ashare, *Safety in Ice Hockey* **5**, (2009), 260–276.

least 65. Before buying these pills, a distributor wants to verify the supplier's claim is valid. To this end, the distributor chooses a random sample of 9 pills from a batch and measures their potency after the specified exposure.

(a) Formulate the hypotheses about the mean potency $\mu$.

(b) Determine the rejection region of the test with $\alpha = .05$. State any assumption you make about the population.

(c) The data are 63, 72, 64, 69, 59, 65, 66, 64, 65. Apply the test and state your conclusion.

**9.64**    A weight loss program advertises "LOSE 40 POUNDS IN 4 MONTHS." A random sample of $n = 25$ customers has $\bar{x} = 32$ pounds lost and $s = 12$. To contradict this claim test $H_0 : \mu = 40$ against $H_1 : \mu < 40$ with $\alpha = .05$.

**9.65**    A car advertisement asserts that with the new collapsible bumper system, the average body repair cost for the damages sustained in a collision impact of 10 miles per hour does not exceed $1500. To test the validity of this claim, 5 cars are crashed into a stone barrier at an impact force of 10 miles per hour and their subsequent body repair costs are recorded. The mean and the standard deviation are found to be $1620 and $90, respectively. Do these data strongly contradict the advertiser's claim?

**9.66**    Combustion efficiency measurements were recorded for 10 home heating furnaces of a new model. The sample mean and standard deviation were found to be 73.2 and 2.74, respectively. Do these results provide strong evidence that the average efficiency of the new model is higher than 70? (Test at $\alpha = .05$. Comment also on the $P$–value.)

**9.67**    A person with asthma took measurements by blowing into a peak-flow meter on seven consecutive days.

| | | | | | | |
|---|---|---|---|---|---|---|
| 429 | 425 | 471 | 422 | 432 | 444 | 454 |

(a) Obtain a 95% confidence interval for the population mean peak-flow.

(b) Conduct an $\alpha = .10$ level test of $H_0 : \mu = 453$ versus $H_1 : \mu \neq 453$.

**9.68**    Times to finish a sixteen ounce bottle of mayonnaise, were recorded by a sample of 11 purchasers. It is determined that $\sum x_i = 645.7$ days and $\sum (x_i - \bar{x})^2 = 198.41$.

(a) Obtain a point estimate of the population mean life $\mu$ and its 90% error margin.

(b) Obtain a 95% confidence interval for the mean.

(c) Obtain a 95% confidence interval for $\sigma$.

(d) What did you assume about the population in your answers to part (a) and part (b)?

**9.69**    Using the table of percentage points of the $\chi^2$ distributions, find:

(a) $\chi^2_{.05}$ with d.f. $= 6$.

(b) $\chi^2_{.025}$ with d.f. $= 22$.

(c) The lower .05 point with d.f. $= 6$.

(d) The lower .025 point with d.f. $= 22$.

**9.70**    Using the table for the $\chi^2$ distributions, find:

(a) The 90th percentile of $\chi^2$ when d.f. $= 11$.

(b) The 10th percentile of $\chi^2$ when d.f. $= 8$.

(c) The median of $\chi^2$ when d.f. $= 23$.

(d) The 1st percentile of $\chi^2$ when d.f. $= 40$.

**9.71**    Refer to Exercise 9.68. Conduct a test of hypothesis with the intent of establishing that the mean bottle life is different from 55.0 days.

(a) Formulate the null and alternative hypotheses.

(b) Determine the test statistic.

(c) Give the form of the rejection region.

(d) What is the conclusion to your test? Take $\alpha = .05$.

(e) What error could you have made in light of your decision in part (d)?

(f) What can you say about the $P$–value?

**9.72**    Refer to the data of Exercise 9.66. Is there strong evidence that the standard deviation for the efficiency of the new model is below .30?

**9.73**    Refer to the water quality data in Example 4. Perform a test of hypotheses with the intent of showing that the population standard deviation is less than 18.0. Take $\alpha = .05$.

**9.74** Refer to the data on the head length (cm) of male grizzly bears given in Table D.6 of the Data Bank. A computer calculation for a test of $H_0 : \mu = 21$ versus $H_1 : \mu \neq 21$ is given below.

```
Test of mu = 21 vs mu not = 21
```

| Variable | N | Mean | StDev |
|---|---|---|---|
| Mhdln | 25 | 18.636 | 3.697 |

| Variable | 95.0% CI | T | P |
|---|---|---|---|
| Mhdln | ( 17.110, 20.162) | -3.20 | 0.004 |

(a) What is the conclusion if you test with $\alpha = .01$?

(b) What mistake could you have made in part (a)?

(c) Before you collected the data, what was the probability of making the mistake in part (a)?

(d) Give a long-run relative frequency interpretation of the probability in part (c).

**9.75** Refer to the computer output concerning the head length (cm) of male grizzly bears in Exercise 9.74.

(a) Is the population mean head length for all male bears in the study area contained in this interval?

(b) Explain why you are 95% confident that it is contained in the interval.

## The Following Exercises Require a Computer

**9.76** Refer to the data on the length (cm) of male grizzly bears given in Table D.6 of the Data Bank.

(a) Find a 99% confidence interval for the population mean.

(b) Is the population mean length for all male grizzly bears in Alaska contained in this interval?

(c) Explain why you are 99% confident that it is contained in the interval.

**9.77** Refer to the computer anxiety scores (CARS) for males in Table D.4 of the Data Bank. Conduct an $\alpha = .025$ level of $H_0 : \mu = 2.7$ versus $H_1 : \mu > 2.7$.

**9.78** Researchers[3] asked participants to carry a counting device for one week and to record the number of times they thought about food each day. The average of the seven counts is intended to represent a typical day. For 19 males, we have

| | | | | | | |
|---|---|---|---|---|---|---|
| 15.9 | 31.9 | 12.6 | 5.4 | 22.6 | 3.1 | 18.9 |
| 12.0 | 11.0 | 21.9 | 4.6 | 57.7 | 50.7 | 12.1 |
| 8.7 | 3.3 | 30.0 | 21.4 | 85.4 | | |

times per day.

(a) Make a graphical summary, for instance a dot diagram or normal-scores plot, that confirms the long right-hand tail.

(b) Take the one-fourth power, or square root twice, and make the same graphical summary. Comment.

(c) Obtain a 95% confidence for the mean of the fourth root of values.

(d) If you take the fourth power of the endpoints of the interval in Part (c), you will not have a 95% confidence interval for the mean on the original scale. Explain why you could call this a 95% confidence interval for the median on the original scale.

[3] T. Fisher, A. Moore, and M. Pittenger. "Sex on the brain?: An examination of frequency of sexual cognitions as a function of gender, erotophilia, and social desirability," *Journal of Sex Research*, **21**(1), (2012), pp. 69–77.

# 10

# Comparing Two Treatments

©Wavebreak Media Ltd. / Corbis

Are video game playing skills transferable?

## Does Playing Action Video Games Modify Ability to See Objects?

Video game playing is now a widespread activity in our society. Two scientists[1] wondered if the game playing itself could alter players' visual abilities and improve performance on related tasks. As one part of their experiment, they briefly flashed different numbers of target squares on the screen. The count of the number of target squares flashed that could be unerringly apprehended was recorded for a sample of 13 regular game players and 13 nonplayers. (courtesy of G. Green).

Number of Items Apprehended

|  | Sample Size | Mean | Standard Deviation |
|---|---|---|---|
| Game players | $n_1 = 13$ | $\bar{x}_1 = 4.89$ | $s_1 = 1.58$ |
| Nongame players | $n_2 = 13$ | $\bar{x}_2 = 3.27$ | $s_2 = 1.15$ |

The difference in sample means, $4.89 - 3.27 = 1.62$ items apprehended, is substantial. A 95% confidence interval for the difference of population means is

$$( 0.50, \ 2.74 ) \qquad \text{items unerringly apprehended}$$

With 95% confidence we can assert that mean difference is greater than .5 but not greater than 2.74 items unerringly apprehended. These data provide strong evidence that regular game players can unerringly apprehend more items in this new task.

---

In virtually every area of human activity, new procedures are invented and existing techniques revised. Advances occur whenever a new technique proves to be better than the old. To compare them, we conduct experiments, collect data about their performance, and then draw conclusions from statistical analyses. The manner in which sample data are collected, called an **experimental design** or **sampling design**, is crucial to an investigation. In this chapter, we introduce two experimental designs that are most fundamental to a comparative study: (1) independent samples and (2) a matched pairs sample.

As we shall see, the methods of analyzing the data are quite different for these two processes of sampling. First, we outline a few illustrative situations where a comparison of two methods requires statistical analysis of data.

**Example 1**   Agricultural Field Trials for Pesticide Residual in Vegetables

The amount of pesticides in the food supply is a major health problem. To ascertain whether a new pesticide will result in less residue in vegetables, field

[1]G. S. Green and D. Bavelier, "Action video game modifies visual selective attention," *Nature* **423** (May 29, 2003), pp. 534–537.

trials must be performed by applying the new and current pesticides to nearly identical farm plots. For instance, the new pesticide could be applied to about one-half of the plot and the current pesticide to the rest but leaving a sufficient buffer zone in between.

Several different fields with different climate and soil conditions and several different crops should be included in the study. A conclusion about pesticide residue in lettuce grown in sandy soil in southern Wisconsin, while of some interest, is not general enough.

**Example 2** Drug Evaluation

Pharmaceutical researchers strive to synthesize chemicals to improve their efficiency in curing diseases. New chemicals may result from educated guesses concerning potential biological reactions, but evaluations must be based on their effects on diseased animals or human beings. To compare the effectiveness of two drugs in controlling tumors in mice, several mice of an identical breed are taken as experimental subjects. After infecting them with cancer cells, some are subsequently treated with drug 1 and others with drug 2. The data of tumor sizes for the two groups then provides a basis for comparing the drugs. When testing the drugs on human subjects, the experiment takes a different form. Artificially infecting them with cancer cells is absurd! In fact, it would be criminal. Instead, the drugs are administered to cancer patients who are available for the study. In contrast, with a pool of mice of an "identical breed," the available subjects may be of varying conditions of general health, prognosis of the disease, and other factors.

# 1. TWO DESIGNS: INDEPENDENT SAMPLES AND MATCHED PAIRS SAMPLE

When discussing a comparative study, the common statistical term **treatment** is used to refer to the things that are being compared. The basic unit that is exposed to one treatment or another is called an **experimental unit** or **experimental subject**, and the characteristic that is recorded after the application of a treatment to a subject is called the **response**. For instance, the two treatments in Example 1 are the current and new pesticides, the experimental subjects are the agricultural plots, and the response is the amount of residue in vegetables.

The term **experimental design** refers to the manner in which subjects are chosen and assigned to treatments. For comparing two treatments, the two basic types of design are:

1. Independent samples (complete randomization).
2. Matched pairs sample (randomization within each matched pair).

The case of independent samples arises when the subjects are randomly divided into two groups, one group is assigned to treatment 1 and the other to treatment 2. The response measurements for the two treatments are then unrelated because they arise from separate and unrelated groups of subjects. Consequently, each set of response measurements can be considered a sample from a population, and we can speak in terms of a comparison between two population distributions.

With the matched pairs design, the experimental subjects are chosen in pairs so that the members in each pair are alike, whereas those in different pairs may be substantially dissimilar. One member of each pair receives treatment 1 and the other treatment 2. Example 3 illustrates these ideas.

**Example 3**  Independent Samples versus Matched Pairs Design

To compare the effectiveness of two drugs in curing a disease, suppose that 8 patients are included in a clinical study. Here, the time to cure is the response of interest. Figure 1a portrays a design of independent samples where the 8 patients are randomly split into groups of 4, one group is treated with drug 1, and the other with drug 2. The observations for drug 1 will have no relation to those for drug 2 because the selection of patients in the two groups is left completely to chance.

To conduct a matched pairs design, one would first select the patients in pairs. The two patients in each pair should be as alike as possible in regard to their physiological conditions; for instance, they should be of the same gender and age group and have about the same severity of the disease. These preexisting conditions may be different from one pair to another. Having paired the subjects, we randomly select one member from each pair to be treated with drug 1 and the other with drug 2. Figure 1b shows this matched pairs design.

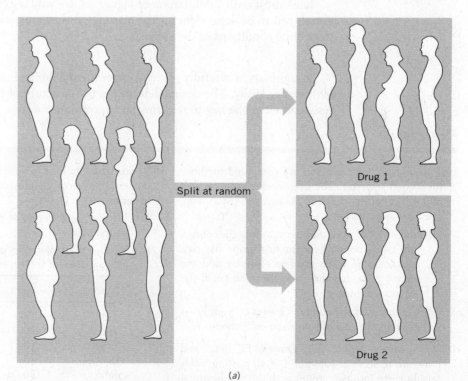

Split at random

Drug 1

Drug 2

(a)

Figure 1a   Independent samples, each of size 4.

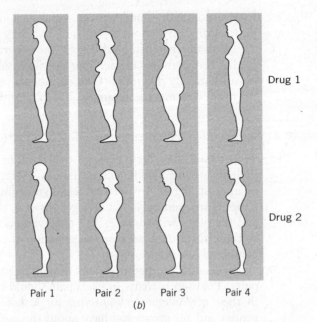

Pair 1    Pair 2    Pair 3    Pair 4
(b)

Figure 1b   Matched pairs design with four pairs of subjects. Separate random assignment of Drug 1 each pair.

In contrast with the situation of Figure 1a, we would expect the responses
of each pair to be dependent for the reason that they are governed by the same
preexisting conditions of the subjects.

In summary, a carefully planned experimental design is crucial to a successful
comparative study. The design determines the structure of the data. In turn, the
design provides the key to selecting an appropriate analysis.

## Exercises

**10.1** Grades for first semester are compared to those
for second semester. The five one-semester courses
biology, chemistry, English, history, and psychology
must be taken next year.

(a) Make a list of all possible ways to split the courses
into two groups where the first group has two
courses to be taken the first semester and the
second group has three courses to be taken the
second semester.

(b) Confirm that the number of ways on your list is
given by the rule of combinations 5 choose 2.

**10.2** Six mice—Alpha, Tau, Omega, Pi, Beta, and
Phi—are to serve as subjects. List all possible ways
to split them into two groups with the first having 4
mice and the second 2 mice.

**10.3** Six students in a psychology course have volunteered to serve as subjects in a matched pairs
experiment.

(a) List all possible sets of pairings if subjects are
paired by age.

(b) If subjects are paired by gender, how many pairs
are available for the experiment?

| Name | Age | Gender |
|------|-----|--------|
| Tom | 18 | M |
| Sue | 20 | F |
| Erik | 18 | M |
| Grace | 20 | F |
| John | 20 | M |
| Roger | 18 | M |

**10.4** Identify the following as either matched pair or independent samples. Also identify the experimental units, treatments, and response in each case.

(a) Twelve persons are given a high-potency vitamin C capsule once a day. Another twelve do not take extra vitamin C. Investigators record the number of colds in 5 winter months.

(b) One self-fertilized plant and one cross-fertilized plant are grown in each of 7 pots. Their heights are measured after 3 months.

(c) Ten newly married couples are interviewed. Both the husband and wife respond to the question, "How many children would you like to have?"

(d) Learning times are recorded for 5 dogs trained by a reward method and 3 dogs trained by a reward-punishment method.

## 2.  INFERENCES ABOUT THE DIFFERENCE OF MEANS—INDEPENDENT LARGE SAMPLES

Here we discuss the methods of statistical inference for comparing two treatments or two populations on the basis of independent samples. Recall that with the **independent samples design**, a collection of $n_1 + n_2$ subjects is randomly divided into two groups and the responses are recorded. We conceptualize population 1 as the collection of responses that would result if a vast number of subjects were given treatment 1. Similarly, population 2 refers to the population of responses under treatment 2. The design of independent samples can then be viewed as one that produces unrelated random samples from two populations (see Figure 2). In other situations, the populations to be compared may be quite real entities. For instance, one may wish to compare the residential property values in the east suburb of a city to those in the west suburb. Here the issue of assigning experimental subjects to treatments does not arise. The collection of all residential properties in each suburb constitutes a population from which a sample will be drawn at random.

Figure 2   Independent random samples.

With the design of independent samples, we obtain

| Sample | Summary Statistics | | |
|---|---|---|---|
| $X_1, X_2, \ldots, X_{n_1}$ from population 1 | $\overline{X} = \dfrac{1}{n_1} \sum\limits_{i=1}^{n_1} X_i$ | | $S_1^2 = \dfrac{\sum\limits_{i=1}^{n_1} (X_i - \overline{X})^2}{n_1 - 1}$ |
| $Y_1, Y_2, \ldots, Y_{n_2}$ from population 2 | $\overline{Y} = \dfrac{1}{n_2} \sum\limits_{i=1}^{n_2} Y_i$ | | $S_2^2 = \dfrac{\sum\limits_{i=1}^{n_2} (Y_i - \overline{Y})^2}{n_2 - 1}$ |

To make confidence statements or to test hypotheses, we specify a statistical model for the data.

---

### Statistical Model: Independent Random Samples

1. $X_1, X_2, \ldots, X_{n_1}$ is a random sample of size $n_1$ from population 1 whose mean is denoted by $\mu_1$ and standard deviation by $\sigma_1$.
2. $Y_1, Y_2, \ldots, Y_{n_2}$ is a random sample of size $n_2$ from population 2 whose mean is denoted by $\mu_2$ and standard deviation by $\sigma_2$.
3. The samples are independent. In other words, the response measurements under one treatment are unrelated to the response measurements under the other treatment.

---

We now set our goal toward drawing a comparison between the mean responses of the two treatments or populations. In statistical language, we are interested in making inferences about the parameter

$$\mu_1 - \mu_2 = (\text{Mean of population 1}) - (\text{Mean of population 2})$$

When the sample sizes are large, no additional assumptions are required.

## 2.1 ESTIMATION

Inferences about the difference $\mu_1 - \mu_2$ are naturally based on its estimate $\overline{X} - \overline{Y}$, the difference between the sample means. When both sample sizes $n_1$ and $n_2$ are large (say, greater than 30), $\overline{X}$ and $\overline{Y}$ are each approximately normal and their difference $\overline{X} - \overline{Y}$ is approximately normal with

Mean

$$E(\overline{X} - \overline{Y}) = \mu_1 - \mu_2$$

Variance

$$\text{Var}(\overline{X} - \overline{Y}) = \frac{\sigma_1^2}{n_1} + \frac{\sigma_2^2}{n_2}$$

Standard error

$$\text{S.E.}(\overline{X} - \overline{Y}) = \sqrt{\frac{\sigma_1^2}{n_1} + \frac{\sigma_2^2}{n_2}}$$

***Note:*** Because the entities $\overline{X}$ and $\overline{Y}$ vary in repeated sampling and independently of each other, the distance between them becomes more variable than the individual members. This explains the mathematical fact that the variance of the difference $\overline{X} - \overline{Y}$ equals the *sum* of the variances of $\overline{X}$ and $\overline{Y}$.

When $n_1$ and $n_2$ are both large, the normal approximation remains valid if $\sigma_1^2$ and $\sigma_2^2$ are replaced by their estimators

$$S_1^2 = \frac{\sum_{i=1}^{n_1} (X_i - \overline{X})^2}{n_1 - 1} \quad \text{and} \quad S_2^2 = \frac{\sum_{i=1}^{n_2} (Y_i - \overline{Y})^2}{n_2 - 1}$$

We conclude that, when the sample sizes $n_1$ and $n_2$ are large,

$$Z = \frac{(\overline{X} - \overline{Y}) - (\mu_1 - \mu_2)}{\sqrt{\dfrac{S_1^2}{n_1} + \dfrac{S_2^2}{n_2}}} \quad \text{is approximately } N(0, 1)$$

A confidence interval for $\mu_1 - \mu_2$ is constructed from this sampling distribution. Similar to the single sample confidence interval, our confidence interval here is again of the form

Estimate of parameter $\pm$ ( $z$ value ) ( estimated standard error )

---

### Large Samples Confidence Interval for $\mu_1 - \mu_2$

When $n_1$ and $n_2$ are greater than 30, an approximate $100(1 - \alpha)\%$ confidence interval for $\mu_1 - \mu_2$ is given by

$$\left( \overline{X} - \overline{Y} - z_{\alpha/2} \sqrt{\frac{S_1^2}{n_1} + \frac{S_2^2}{n_2}}, \quad \overline{X} - \overline{Y} + z_{\alpha/2} \sqrt{\frac{S_1^2}{n_1} + \frac{S_2^2}{n_2}} \right)$$

where $z_{\alpha/2}$ is the upper $\alpha/2$ point of $N(0, 1)$.

---

**Example 4**  Large Samples Confidence Interval for Difference in Mean Job Satisfaction

A considerable proportion of a person's time is spent working, and satisfaction with the job and satisfaction with life tend to be related. Job satisfaction is typically measured on a four point scale

| *Very Dissatisfied* | *A Little Dissatisfied* | *Moderately Satisfied* | *Very Satisfied* |
|---|---|---|---|
| | | | |

A numerical scale is created by assigning 1 to very dissatisfied, 2 to a little dissatisfied, 3 to moderately satisfied, and 4 to very satisfied.

The responses of 226 firefighters and 247 office supervisors, presented in Exercise 10.12, yielded the summary statistics

|  | Fire-fighter | Office Supervisor |
|---|---|---|
| Mean | 3.673 | 3.547 |
| sd | .7235 | .6089 |

Construct a 95% confidence interval for difference in mean job satisfaction scores.

SOLUTION   Let $\mu_1$ be the mean job satisfaction for firefighters and $\mu_2$ the mean for office supervisors. We have

$$n_1 = 226 \qquad \bar{x} = 3.673 \qquad s_1 = .7235$$
$$n_2 = 247 \qquad \bar{y} = 3.547 \qquad s_2 = .6089$$

$$\sqrt{\frac{s_1^2}{n_1} + \frac{s_2^2}{n_2}} = \sqrt{\frac{(.7235)^2}{226} + \frac{(.6089)^2}{247}} = .06018$$

For a 95% confidence interval, we use $z_{.025} = 1.96$ and calculate

$$\bar{x} - \bar{y} = 3.673 - 3.547 = .126$$

$$z_{.025}\sqrt{\frac{s_1^2}{n_1} + \frac{s_2^2}{n_2}} = 1.96 \times .06018 = .121$$

Therefore, a 95% confidence interval for $\mu_1 - \mu_2$ is given by

$$.126 \pm .121 \qquad \text{or} \qquad (.005, .247)$$

We are 95% confident that the mean score for firefighters is .005 units to .247 units higher than the mean score of office supervisors. This interval contains inconsequential positive values as well as positive values that are possibly important differences on the satisfaction scale.

## 2.2 HYPOTHESES TESTING

Let us turn our attention to testing hypotheses concerning $\mu_1 - \mu_2$. A test of the null hypothesis that the two population means are the same, $H_0 : \mu_1 - \mu_2 = 0$, employs the test statistic

$$Z = \frac{\bar{X} - \bar{Y}}{\sqrt{\dfrac{S_1^2}{n_1} + \dfrac{S_2^2}{n_2}}}$$

which is approximately $N(0, 1)$ when $\mu_1 - \mu_2 = 0$.

**Example 5**   Testing Equality of Mean Job Satisfaction

Do the data in Example 4 provide strong evidence that the mean job satisfaction of firefighters is different from the mean job satisfaction of office supervisors? Test at $\alpha = .02$.

SOLUTION   Because we are asked to show that the two means are different, we formulate the problem as testing

$$H_0: \mu_1 - \mu_2 = 0 \quad \text{versus} \quad H_1: \mu_1 - \mu_2 \neq 0$$

We use the test statistic

$$Z = \frac{\overline{X} - \overline{Y}}{\sqrt{\dfrac{S_1^2}{n_1} + \dfrac{S_2^2}{n_2}}}$$

and set a two-sided rejection region. Specifically, with $\alpha = .02$, we have $\alpha/2 = .01$ and $z_{\alpha/2} = 2.33$, so the rejection region is $R: |Z| \geq 2.33$.

Using the sample data given in Example 4, we calculate

$$z = \frac{3.673 - 3.547}{\sqrt{\dfrac{(.7235)^2}{226} + \dfrac{(.6089)^2}{247}}} = \frac{.126}{.0618} = 2.04$$

Because the observed value $z = 2.04$ does not lie in the rejection region, we fail to reject the null hyphothesis at level $\alpha = .02$. The evidence against equal means is only moderately strong since the $P$–value is

$$P[Z < -2.04] + P[Z > 2.04] = .0207 + .0207 = .0414 \text{ (see Figure 3)}$$

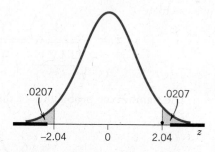

Figure 3   $P$–value with two-sided rejection region.

**Example 6**   Large Samples Test with a One-Sided Alternative

In June two years ago, chemical analyses were made of 85 water samples (each of unit volume) taken from various parts of a city lake, and the measurements of chlorine content were recorded. During the next two winters, the use of road salt was substantially reduced in the catchment areas of the lake. This June, 110 water samples were analyzed and their chlorine contents recorded. Calculations of the mean and the standard deviation for the two sets of data give

|  | Chlorine Content | |
| --- | --- | --- |
|  | Two Years Ago | Current Year |
| Mean | 18.3 | 17.8 |
| Standard deviation | 1.2 | 1.8 |

Test the claim that lower salt usage has reduced the amount of chlorine in the lake. Base your decision on the $P$–value.

SOLUTION    Let $\mu_1$ be the population mean two years ago and $\mu_2$ the population mean in the current year. Because the claim is that $\mu_2$ is less than $\mu_1$, we formulate the hypotheses

$$H_0 : \mu_1 - \mu_2 = 0 \quad \text{versus} \quad H_1 : \mu_1 - \mu_2 > 0$$

With the test statistic

$$Z = \frac{\overline{X} - \overline{Y}}{\sqrt{\dfrac{S_1^2}{n_1} + \dfrac{S_2^2}{n_2}}}$$

the rejection region should be of the form $R : Z \geq c$ because $H_1$ is right-sided. Using the data

$$n_1 = 85 \qquad \overline{x} = 18.3 \qquad s_1 = 1.2$$
$$n_2 = 110 \qquad \overline{y} = 17.8 \qquad s_2 = 1.8$$

we calculate

$$z = \frac{18.3 - 17.8}{\sqrt{\dfrac{(1.2)^2}{85} + \dfrac{(1.8)^2}{110}}} = \frac{.5}{.2154} = 2.32$$

The significance probability of this observed value is (see Figure 4)

Figure 4   $P$–value $= .0102.$

$$P\text{–value} = P[\, Z \geq 2.32 \,] = .0102$$

Because the $P$–value is very small, we conclude that there is strong evidence to reject $H_0$ and support $H_1$.

We summarize the procedure for testing $\mu_1 - \mu_2 = \delta_0$ where $\delta_0$ is specified under the null hypothesis. The case $\mu_1 = \mu_2$ corresponds to $\delta_0 = 0$.

---

### Testing $H_0: \mu_1 - \mu_2 = \delta_0$ with Large Samples

Test statistic:

$$Z = \frac{\overline{X} - \overline{Y} - \delta_0}{\sqrt{\dfrac{S_1^2}{n_1} + \dfrac{S_2^2}{n_2}}}$$

| Alternative Hypothesis | Level $\alpha$ Rejection Region |
|---|---|
| $H_1: \mu_1 - \mu_2 > \delta_0$ | $R: Z \geq z_\alpha$ |
| $H_1: \mu_1 - \mu_2 < \delta_0$ | $R: Z \leq -z_\alpha$ |
| $H_1: \mu_1 - \mu_2 \neq \delta_0$ | $R: |Z| \geq z_{\alpha/2}$ |

---

Finally, we would like to emphasize that with large samples we can also learn about other differences between the two populations.

**Example 7**  Large Samples Reveal Additional Differences between Populations

Natural resource managers have attempted to use the Satellite Landsat Multispectral Scanner data for improved landcover classification. When the satellite was flying over country known to consist of forest, the following intensities were recorded on the near-infrared band of a thermatic mapper. The sample has already been ordered.

```
77   77   78   78   81   81   82   82   82   82   82   83   83   84  84  84
84   85   86   86   86   86   86   87   87   87   87   87   87   87  89  89
89   89   89   89   89   90   90   90   91   91   91   91   91   91  91  91
91   91   93   93   93   93   93   93   94   94   94   94   94   94  94  94
94   94   94   94   95   95   95   95   95   96   96   96   96   96  96  97
97   97   97   97   97   97   97   97   98   99  100  100  100  100
100 100  100  100  100  101  101  101  101  101  101  102
102 102  102  102  102  103  103  104  104  104 .105  107
```

When the satellite was flying over urban areas, the intensities of reflected light on the same near-infrared band were

71  72  73  74  75  77  78  79  79  79  79  80  80  80  81  81  81
82  82  82  82  84  84  84  84  84  84  85  85  85  85  85  85  86
86  87  88  90  91  94

If the means are different, the readings could be used to tell urban from forest area. Obtain a 95% confidence interval for the difference in mean radiance levels.

SOLUTION    Computer calculations give

|                      | Forest  | Urban   |
|----------------------|---------|---------|
| Number               | 118     | 40      |
| Mean                 | 92.932  | 82.075  |
| Standard deviation   | 6.9328  | 4.9789  |

and, for large sample sizes, the approximate 95% confidence interval for $\mu_1 - \mu_2$ is given by

$$\left( \overline{X} - \overline{Y} - z_{.025} \sqrt{\frac{S_1^2}{n_1} + \frac{S_2^2}{n_2}} , \quad \overline{X} - \overline{Y} + z_{.025} \sqrt{\frac{S_1^2}{n_1} + \frac{S_2^2}{n_2}} \right)$$

Since $z_{.025} = 1.96$, the 95% confidence interval is calculated as

$$92.932 - 82.075 \pm 1.96 \sqrt{\frac{(6.9328)^2}{118} + \frac{(4.9789)^2}{40}} \quad \text{or} \quad (8.87, 12.84)$$

The mean for the forest is 8.87 to 12.84 levels of radiance higher than the mean for the urban areas.

Because the sample sizes are large, we can also learn about other differences between the two populations. The stem-and-leaf displays and boxplots in Figure 5 reveal that there is some difference in the standard deviation as well as

Figure 5   Stem-and-leaf displays and boxplots give additional information about population differences.

the means. The graphs further indicate a range of high readings that are more likely to come from forests than urban areas. This feature has proven helpful in discriminating between forest and urban areas on the basis of near-infrared readings.

## Exercises

**10.5** According to the assumptions underlying the two independent samples design,

(a) Are two observations from the same sample, say $X_2$ and $X_4$,
   (i) Independent?
   (ii) Do they always have the same distribution?

(b) Are two observations from different samples, say $X_2$ and $Y_2$,
   (i) Independent?
   (ii) Do they always have the same distribution?

**10.6** According to the assumptions underlying the two independent samples design, let the first population be $N(8, 2)$ and the second be $N(10, 4)$. Identify the values of the parameters $\mu_1$, $\sigma_1$, $\mu_2$, $\sigma_2$.

**10.7** One semester, an instructor taught the same computer course at two universities located in different cities. He was able to give the same final at both locations. The student's scores provided the summary statistics.

| Sample 1 | Sample 2 |
|---|---|
| $n_1 = 52$ | $n_2 = 44$ |
| $\bar{x} = 73$ | $\bar{y} = 66$ |
| $s_1^2 = 151$ | $s_2^2 = 142$ |

(a) Obtain a point estimate of $\mu_1 - \mu_2$.
(b) Estimate the S.E.$(\bar{X} - \bar{Y})$.
(c) Construct a 95% confidence interval for $\mu_1 - \mu_2$.

**10.8** Rural and urban students are to be compared on the basis of their scores on a nationwide musical aptitude test. Two random samples of sizes 90 and 100 are selected from rural and urban seventh-grade students. The summary statistics from the test scores are

|  | Rural | Urban |
|---|---|---|
| Sample size | 90 | 100 |
| Mean | 76.4 | 81.2 |
| Standard deviation | 8.2 | 7.6 |

Construct a 98% confidence interval for the difference in population mean scores between urban and rural students.

**10.9** Perform a test to determine if there is a significant difference between the population mean scores in Exercise 10.8. Use $\alpha = .05$.

**10.10** A linguist wants to compare the writing styles in two magazines and one measure is the number of words per sentence. On the basis of 50 randomly selected sentences from each source, she finds

Magazine 1  $n_1 = 50$  $\bar{x} = 12.6$  $s_1 = 4.2$

Magazine 2  $n_2 = 50$  $\bar{y} = 9.5$  $s_2 = 1.9$

Determine a 98% confidence interval for the difference in mean number of words per sentence.

**10.11** Refer to Exercise 10.10. Perform a test of hypothesis that is intended to show that the mean for magazine 1 is more than 2 words larger than the mean for magazine 2.

(a) Formulate the null and alternative hypotheses. (Define any symbols you use.)

(b) State the test statistic and the rejection region with $\alpha = .05$.

(c) Perform the test at $\alpha = .05$. Also, find the P–value and comment.

**10.12** Refer to Example 4. Workers in three occupations were questioned about satisfaction with their jobs. Suppose the number of responses in each of the four categories are

| Occupation | Very Dis- satisfied | A Little Dis- satisfied | Moder- ately Satisfied | Very Satis- fied | Sample Size |
|---|---|---|---|---|---|
| Firefighters | 6 | 16 | 24 | 180 | 226 |
| Office Supervisors | 1 | 12 | 85 | 149 | 247 |
| Clergy | 1 | 5 | 6 | 78 | 90 |

Assign 1 to very dissatisfied, 2 to a little dissatisfied, 3 to moderately satisfied, and 4 to very satisfied. (These data have means and proportions *very satisfied* similar to those reported in a recent large scale survey.)

(a) Calculate the sample mean and variance separately for firefighters and office supervisors. The scores for clergy have mean 3.79 and standard deviation .590.

(b) Construct a 95% confidence interval for the difference in mean satisfaction score for clergy and office supervisors.

**10.13** Refer to the confidence interval obtained in Exercise 10.12(b). If you were to test the null hypothesis that the mean satisfaction scores are equal versus the two-sided alternatives, what would be the conclusion of your test with $\alpha = .05$? (See Section 3 of Chapter 9.)

**10.14** In a study of interspousal aggression and its possible effect on child behavior, the behavior problem checklist (BPC) scores were recorded for 47 children whose parents were classified as aggressive. The sample mean and standard deviation were 7.92 and 3.45, respectively. For a sample of 38 children whose parents were classified as nonaggressive, the mean and standard deviation of the BPC scores were 5.80 and 2.87, respectively. Do these observations substantiate the conjecture that the children of aggressive families have a higher mean BPC than those of nonaggressive families? (Answer by calculating the $P$–value.)

**10.15** An experiment was conducted to study the effect of eating walnuts on the quality of sperm.[2] The subjects were healthy men 21–35 years old. The treatment group received 75 gm of whole-shelled walnuts a day for 12 weeks. This was a single blind trial where the assessors of sperm quality were not

told the group membership. Measurements were made at the start of study and again at the 12-th week visit. The statistical analysis is based on the change in vitality, *final − initial*, where vitality is the percent of sperm alive. The summary statistics are

Vitality (% alive)

| | Walnut Group | Control Group |
|---|---|---|
| Number | 55 | 52 |
| Mean | 5.5 | .51 |
| Standard deviation | 10.0 | 7.4 |

Is the mean change for the walnut eating group significantly higher than for the control group? Answer by calculating the $P$–value.

**10.16** Referring to the data in Exercise 10.15, determine a 99% confidence interval for the difference of mean change.

**10.17** Psychologists have made extensive studies on the relationship between child abuse and later criminal behavior. Consider a study that consisted of the follow-ups of 52 boys who were abused in their preschool years and 67 boys who were not abused. The data of the number of criminal offenses of those boys in their teens yielded the following summary statistics.

| | Abused | Nonabused |
|---|---|---|
| Mean | 2.48 | 1.57 |
| Standard deviation | 1.94 | 1.31 |

Is the mean number of criminal offenses significantly higher for the abused group than that for the nonabused group? Answer by calculating the $P$–value.

**10.18** The generic output below summarizes the data, given in Table D.5 of the Data Bank, on the pretest percent body fat in male and female students.

| Gender | N | Mean | StDev |
|---|---|---|---|
| M | 40 | 14.38 | 7.34 |
| F | 43 | 23.72 | 5.78 |

Find a 99% confidence interval for the difference of mean percent body fat.

[2]W. Robbins et al., "Walnuts improve semen quality in men consuming a Western-style diet: Randomized control dietary intervention trial," *Biology of Reproduction* 87(4), (2012): 101, pp 1–8.

# 3. INFERENCES ABOUT THE DIFFERENCE OF MEANS—INDEPENDENT SMALL SAMPLES FROM NORMAL POPULATIONS

Not surprisingly, more distributional structure must be imposed before inference procedures can be developed for small samples. Here, we require that both population distributions be normal. A restriction to normal populations is not new and was already required for small sample inferences about a single mean. We treat the equal variance case first. It is the simplest.

## 3.1 INFERENCES: NORMAL POPULATIONS WITH EQUAL VARIANCES

The additional assumptions specify that the normal populations are normal and they also require the two standard deviations to be equal.

---

**Additional Assumptions When the Sample Sizes Are Small**

1. Both populations are normal.
2. The population standard deviations $\sigma_1$ and $\sigma_2$ are equal.

---

The assumption requiring equal variability is somewhat artificial but we reserve comment until later. Letting $\sigma$ denote the common standard deviation, we summarize.

---

**Small Samples Assumptions**

1. $X_1, X_2, \ldots, X_{n_1}$ is a random sample from $N(\mu_1, \sigma)$.
2. $Y_1, Y_2, \ldots, Y_{n_2}$ is a random sample from $N(\mu_2, \sigma)$.
   (*Note*: $\sigma$ is the same for both distributions.)
3. $X_1, X_2, \ldots, X_{n_1}$ and $Y_1, Y_2, \ldots, Y_{n_2}$ are independent.

---

### Estimation

Again, $\overline{X} - \overline{Y}$ is our choice for a statistic.

$$\text{Mean of } (\overline{X} - \overline{Y}) = E(\overline{X} - \overline{Y}) = \mu_1 - \mu_2$$

$$\text{Var}(\overline{X} - \overline{Y}) = \frac{\sigma^2}{n_1} + \frac{\sigma^2}{n_2} = \sigma^2 \left( \frac{1}{n_1} + \frac{1}{n_2} \right)$$

The common variance $\sigma^2$ can be estimated by combining information provided by both samples. Specifically, the sum $\sum_{i=1}^{n_1} (X_i - \overline{X})^2$ incorporates $n_1 - 1$

pieces of information about $\sigma^2$, in view of the constraint that the deviations $X_i - \overline{X}$ sum to zero. Independently of this, $\sum_{i=1}^{n_2} (Y_i - \overline{Y})^2$ contains $n_2 - 1$ pieces of information about $\sigma^2$. These two quantities can then be combined,

$$\sum (X_i - \overline{X})^2 + \sum (Y_i - \overline{Y})^2$$

to obtain a pooled estimate of the common $\sigma^2$. The proper divisor is the sum of the component degrees of freedom, or $(n_1 - 1) + (n_2 - 1) = n_1 + n_2 - 2$.

---

### Pooled Estimator of the Common $\sigma^2$

$$S_{\text{pooled}}^2 = \frac{\sum_{i=1}^{n_1} (X_i - \overline{X})^2 + \sum_{i=1}^{n_2} (Y_i - \overline{Y})^2}{n_1 + n_2 - 2}$$

$$= \frac{(n_1 - 1) S_1^2 + (n_2 - 1) S_2^2}{n_1 + n_2 - 2}$$

---

**Example 8**   Calculating the Pooled Estimate of Variance

Calculate the observed value $s_{\text{pooled}}^2$ from these two samples.

| | | | | | |
|---|---|---|---|---|---|
| Sample from population 1: | 8 | 5 | 7 | 6 | 9 | 7 |
| Sample from population 2: | 2 | 6 | 4 | 7 | 6 | |

SOLUTION   The sample means are

$$\overline{x} = \frac{\sum x_i}{6} = \frac{42}{6} = 7 \qquad \overline{y} = \frac{\sum y_i}{5} = \frac{25}{5} = 5$$

Furthermore,

$$(6 - 1) s_1^2 = \sum (x_i - \overline{x})^2$$
$$= (8 - 7)^2 + (5 - 7)^2 + (7 - 7)^2 + (6 - 7)^2$$
$$+ (9 - 7)^2 + (7 - 7)^2 = 10$$
$$(5 - 1) s_2^2 = \sum (y_i - \overline{y})^2$$
$$= (2 - 5)^2 + (6 - 5)^2 + (4 - 5)^2$$
$$+ (7 - 5)^2 + (6 - 5)^2 = 16$$

Thus, $s_1^2 = 2$, $s_2^2 = 4$, and the pooled variance is

$$s_{pooled}^2 = \frac{\sum (x_i - \bar{x})^2 + \sum (y_i - \bar{y})^2}{n_1 + n_2 - 2} = \frac{10 + 16}{6 + 5 - 2} = 2.89$$

The pooled variance is closer to 2 than 4 because the first sample size is larger.

These arithmetic details serve to demonstrate the concept of pooling. Using a calculator with a "standard deviation" key, one can directly get the sample standard deviations $s_1 = 1.414$ and $s_2 = 2.000$. Noting that $n_1 = 6$ and $n_2 = 5$, we can then calculate the pooled variance as

$$s_{pooled}^2 = \frac{5 (1.414)^2 + 4 (2.000)^2}{9} = 2.89$$

Employing the pooled estimator $\sqrt{S_{pooled}^2}$ for the common $\sigma$, we obtain a Student's $t$ variable that is basic to inferences about $\mu_1 - \mu_2$.

$$T = \frac{(\bar{X} - \bar{Y}) - (\mu_1 - \mu_2)}{S_{pooled} \sqrt{\dfrac{1}{n_1} + \dfrac{1}{n_2}}}$$

has Student's $t$ distribution with $n_1 + n_2 - 2$ degrees of freedom.

We can now obtain confidence intervals for $\mu_1 - \mu_2$, which are again of the form

Estimate of parameter $\pm$ ($t$ value) $\times$ (Estimated standard error)

**Confidence Interval for $\mu_1 - \mu_2$ Small Samples and $\sigma = \sigma_2$**

A $100(1 - \alpha)\%$ confidence interval for $\mu_1 - \mu_2$ is given by

$$\bar{X} - \bar{Y} \pm t_{\alpha/2} S_{pooled} \sqrt{\frac{1}{n_1} + \frac{1}{n_2}}$$

where

$$S_{pooled}^2 = \frac{(n_1 - 1) S_1^2 + (n_2 - 1) S_2^2}{n_1 + n_2 - 2}$$

and $t_{\alpha/2}$ is the upper $\alpha/2$ point of the $t$ distribution with d.f. $= n_1 + n_2 - 2$.

**Example 9**    Calculating a Small Samples Confidence Interval

Beginning male and female accounting students were given a test and, on the basis of their answers, were assigned a computer anxiety score (CARS). Using the data given in Table D.4 of the Data Bank, obtain a 95% confidence interval for the difference in mean computer anxiety score between beginning male and female accounting students.

SOLUTION    The dot diagrams of these data, plotted in Figure 6, give the appearance of approximately equal amounts of variation.

Figure 6    Dot diagrams of the computer anxiety data in Example 9.

We assume that the CARS data for both females and males are random samples from normal populations with means $\mu_1$ and $\mu_2$, respectively, and a common standard deviation $\sigma$. Computations from the data provide the summary statistics:

$$
\begin{array}{ccc}
 & \text{Female CARS} & \\
n_1 = 15 & \bar{x} = 2.514 & s_1 = .773 \\
 & \text{Male CARS} & \\
n_2 = 20 & \bar{y} = 2.963 & s_2 = .525
\end{array}
$$

We calculate

$$
S_{pooled} = \sqrt{\frac{14(.773)^2 + 19(.525)^2}{15 + 20 - 2}} = .642
$$

With a 95% confidence interval $\alpha / 2 = .025$ and consulting the $t$ table, we find (interpolating) that $t_{.025} = 2.035$ for d.f. $= n_1 + n_2 - 2 = 33$. Thus a 95% confidence interval for $\mu_1 - \mu_2$ becomes

$$
\bar{x} - \bar{y} \pm t_{0.25}\, s_{pooled} \sqrt{\frac{1}{n_1} + \frac{1}{n_2}}
$$

$$
= 2.514 - 2.963 \pm 2.035 \times .642 \sqrt{\frac{1}{15} + \frac{1}{20}}
$$

$$
= -.449 \pm .446 \quad \text{or} \quad (-.895, -.003)
$$

We can be 95% confident that the mean computer anxiety score for female beginning accounting students can be .003 to .895 units lower than the mean score for males.

This interval is quite wide. Certainly the values near $-.003$ represent technically insignificant differences.

## Tests of Hypotheses

Tests of hypotheses concerning the difference in means are based on a statistic having student's $t$ distribution.

---

**Testing $H_0 : \mu_1 - \mu_2 = \delta_0$ with Small Samples and $\sigma_1 = \sigma_2$**

Test statistic:

$$T = \frac{(\bar{X} - \bar{Y}) - \delta_0}{S_{\text{pooled}} \sqrt{\dfrac{1}{n_1} + \dfrac{1}{n_2}}} \qquad \text{d.f.} = n_1 + n_2 - 2$$

| **Alternative Hypothesis** | **Level $\alpha$ Rejection Region** |
|---|---|
| $H_1 : \mu_1 - \mu_2 > \delta_0$ | $R : T \geq t_\alpha$ |
| $H_1 : \mu_1 - \mu_2 < \delta_0$ | $R : T \leq -t_\alpha$ |
| $H_1 : \mu_1 - \mu_2 \neq \delta_0$ | $R : |T| \geq t_{\alpha/2}$ |

---

**Example 10**   Testing the Equality of Mean Computer Anxiety Scores

Refer to the computer anxiety scores (CARS) described in Example 9 and the summary statistics

Female CARS

$n_1 = 15 \qquad \bar{x} = 2.514 \qquad s_1 = .773$

Male CARS

$n_2 = 20 \qquad \bar{y} = 2.963 \qquad s_2 = .525$

Do these data strongly indicate that the mean score for females is lower than that for males? Test at level $\alpha = .05$.

SOLUTION    We are seeking strong evidence in support of the hypothesis that the mean computer anxiety score for females ($\mu_1$) is less than the mean score for males. Therefore the alternative hypothesis should be taken as $H_1 : \mu_1 < \mu_2$ or $H_1 : \mu_1 - \mu_2 < 0$, and our problem can be stated as testing

$$H_0 : \mu_1 - \mu_2 = 0 \qquad \text{versus} \qquad H_1 : \mu_1 - \mu_2 < 0$$

We employ the test statistic

$$T = \frac{\overline{X} - \overline{Y}}{S_{pooled}\sqrt{\dfrac{1}{n_1} + \dfrac{1}{n_2}}} \qquad \text{d.f.} = n_1 + n_2 - 2$$

and set the left-sided rejection region $R: T \leq -t_{.05}$. For d.f. $= n_1 + n_2 - 2 = 33$, we approximate the tabled value as $t_{.05} = 1.692$. The rejection region is $R: T \leq -1.692$ as illustrated in Figure 7.

.05

−1.692          0                    $t$

Figure 7    Rejection region for Example 10.

With $S_{pooled} = .642$ already calculated in Example 9, the observed value of the test statistic $T$ is

$$t = \frac{2.514 - 2.963}{.642\sqrt{\dfrac{1}{15} + \dfrac{1}{20}}} = \frac{-.449}{.2193} = -2.05$$

This value lies in the rejection region $R$. Consequently, at the .05 level of significance, we reject the null hypothesis in favor of the alternative hypothesis that males have a higher mean computer anxiety score.

A computer calculation gives a $P$-value of about .025 so the evidence of $H_1$ is moderately strong.

**Example 11**    A Confidence Interval Shows Mothers of Males Have Milk with Higher Energy Content

Researchers, interested in the energy content of mothers' milk, conducted a study with 25 well-nourished, healthy mothers exclusively breast feeding their 2- to 5-month old babies. All of the participants lived in Massachusetts.[3] Thirteen of the mothers had sons and twelve had daughters. The summary statistics

---

[3]C. E. Powe, C . D. Knott, and N. Conklin-Brittain, "Infant sex predicts breast milk energy content," *American Journal of Human Biology* **22** (2010), pp. 50–54.

Energy ( kcal/100 ml )

|  | Males | Females |
|---|---|---|
| Number | 13 | 12 |
| Mean | 75.56 | 60.81 |
| Standard deviation | 19.37 | 15.64 |

are obtained in the study.

Does the gender of baby effect the amount of energy available in the mothers' milk? If so, what can you say about the mean difference?

Base your answers on a 95% confidence interval for the difference of mean energy density between the mothers' milk of male babies and and the mothers' milk of female babies.

SOLUTION Since the two standard deviations are not very different, we use the pooled estimate of variance

$$s_{pooled} = \sqrt{\frac{13(19.37)^2 + 12(15.64)^2}{13 + 12 - 2}} = 17.684$$

For a 95% confidence interval, $\alpha / 2 = .025$, and we find $t_{.025} = 2.069$ for d.f. $= 23$. The resulting 95% confidence interval is

$$\bar{x} - \bar{y} \pm t_{.025} \, s_{pooled} \sqrt{\frac{1}{n_1} + \frac{1}{n_2}}$$

$$= 75.56 - 60.81 \pm 2.069 \times 17.684 \sqrt{\frac{1}{13} + \frac{1}{12}}$$

$$= 14.75 \pm 14.647$$

or $.10 < \mu_1 - \mu_2 < 29.40$

The 95% confidence interval includes only positive values for $\mu_1 - \mu_2$. With 95% confidence, we conclude that the mean energy density is larger when the baby is male than when it is female. However, the mean difference could be almost as small as .10 kcal/100ml which is likely not scientifically significant. It also could be almost as large as 29.4 kcal/100ml which is an important difference. A larger study is required to more accurately assess the mean difference.

This study involved well nourished healthy mothers in Massachusetts but they were not a random sample of new mothers. Generalizations are difficult, especially extrapolations to under-nourished mothers.

## Deciding Whether or not to Pool

Our preceding discussion of large and small sample inferences raises a few questions:

For small sample inference, why do we assume the population standard deviations to be equal when no such assumption was needed in the large samples case?

When should we be wary about this assumption, and what procedure should we use when the assumption is not reasonable?

Learning statistics would be a step simpler if the ratio

$$\frac{(\bar{X} - \bar{Y}) - (\mu_1 - \mu_2)}{\sqrt{\dfrac{S_1^2}{n_1} + \dfrac{S_2^2}{n_2}}}$$

had a $t$ distribution for small samples from normal populations. Unfortunately, statistical theory proves it otherwise. The distribution of this ratio is *not* a $t$ and, worse yet, it depends on the unknown quantity $\sigma_1 / \sigma_2$. The assumption $\sigma_1 = \sigma_2$ and the change of the denominator to $S_{\text{pooled}} \sqrt{1/n_1 + 1/n_2}$ allow the $t$-based inferences to be valid. However, the $\sigma_1 = \sigma_2$ restriction and accompanying pooling are not needed in large samples where a normal approximation holds.

With regard to the second question, the relative magnitude of the two sample standard deviations $s_1$ and $s_2$ should be the prime consideration. The assumption $\sigma_1 = \sigma_2$ is reasonable if $s_1 / s_2$ is not very much different from 1. As a working rule, the range of values $\frac{1}{2} \leq s_1 / s_2 \leq 2$ may be taken as reasonable cases for making the assumption $\sigma_1 = \sigma_2$ and hence for pooling. If $s_1 / s_2$ is seen to be smaller than $\frac{1}{2}$ or larger than 2, the assumption $\sigma_1 = \sigma_2$ would be suspect. In that case, some approximate methods of inference about $\mu_1 - \mu_2$ are available.

## 3.2 INFERENCES: NORMAL POPULATIONS WITH UNEQUAL VARIANCES

When the sample sizes are not large and the observed standard deviations quite different, we can still proceed to test hypotheses and obtain confidence intervals provided that the populations are normal. The procedure we introduce is somewhat complex but is widely used in most statistical software programs. Consequently, using software is the preferred approach.

Inference procedures are developed based on a statistic that is the same as the large sample statistic for comparing two means. However, we use the notation $T^*$, where

$$T^* = \frac{(\bar{X} - \bar{Y}) - (\mu_1 - \mu_2)}{\sqrt{\dfrac{S_1^2}{n_1} + \dfrac{S_2^2}{n_2}}}$$

Because the sample sizes are small, we chose to approximate the distribution of $T^*$ by a $t$ distribution. But, its degrees of freedom are estimated using the sample standard deviations. As a consequence, the distribution of $T^*$ is only approximately a $t$ distribution.

This approach of approximating the distribution of $T^*$ by a $t$, is sometimes called the Satterthwaite correction after the person who derived the approximation.

---

**Normal Populations with $\sigma_1 \neq \sigma_2$**

Let $\delta = \mu_1 - \mu_2$. When the sample sizes $n_1$ and $n_2$ are not large and $\sigma_1 \neq \sigma_2$

$$T^* = \frac{\overline{X} - \overline{Y} - \delta}{\sqrt{\dfrac{S_1^2}{n_1} + \dfrac{S_2^2}{n_2}}}$$

is approximately distributed as a $t$ with estimated degrees of freedom.
  The estimated degrees of freedom depend on the observed values of the sample variances $s_1^2$ and $s_2^2$.

estimated degrees of freedom =

$$\nu = \frac{\left(\dfrac{s_1^2}{n_1} + \dfrac{s_2^2}{n_2}\right)^2}{\dfrac{1}{n_1 - 1}\left(\dfrac{s_1^2}{n_1}\right)^2 + \dfrac{1}{n_2 - 1}\left(\dfrac{s_2^2}{n_2}\right)^2}$$

---

The estimated degrees of freedom $\nu$ are often rounded down to an integer so a $t$ table can be consulted.

**Example 12** Testing Equality and Confidence Intervals Concerning Mean of Green Gas

One process of making green gasoline, not just a gasoline additive, takes biomass in the form of sucrose and converts it into gasoline using catalytic reactions. This research is still at the pilot plant stage. At one step in a pilot plant process, the product volume (liters) consists of carbon chains of length 3. Nine runs were made with each of two catalysts and the product volumes measured.

| catalyst 1 | 1.86 | 2.05 | 2.06 | 1.88 | 1.75 | 1.64 | 1.86 | 1.75 | 2.13 |
|---|---|---|---|---|---|---|---|---|---|
| catalyst 2 | .32 | 1.32 | .93 | .84 | .55 | .84 | .37 | .52 | .34 |

The sample sizes $n_1 = n_2 = 9$ and the summary statistics are

$$\overline{x} = 1.887, \quad s_1^2 = .0269 \quad \overline{y} = .670 \quad s_2^2 = .1133$$

(a)  Is the mean yield with catalyst 1 more than .80 liters higher than the yield with catalyst 2? Test with $\alpha = 0.05$.

(b)  Find a 98% confidence interval for the difference of means.

SOLUTION     (a)  The test concerns the $\delta = \mu_1 - \mu_2$ and we wish to show that $\delta$ is greater than .80. Therefore we formulate the testing problem as

$$H_0: \mu_1 - \mu_2 = .80 \quad \text{versus} \quad H_1: \mu_1 - \mu_2 > .80$$

The sample sizes $n_1 = n_2 = 9$ are small and there are no outliers or obvious departures from normality. However, $s_2^2 = .1133$ is more than four times $s_1^2 = .0269$. We should not pool.

To apply the approximate $t$, we first estimate the degrees of freedom

$$\frac{\left(\dfrac{S_1^2}{n_1} + \dfrac{S_2^2}{n_2}\right)^2}{\dfrac{1}{n_1-1}\left(\dfrac{S_1^2}{n_1}\right)^2 + \dfrac{1}{n_2-1}\left(\dfrac{S_2^2}{n_2}\right)^2} = \frac{\left(\dfrac{.0269}{9} + \dfrac{.1133}{9}\right)^2}{\dfrac{1}{9-1}\left(\dfrac{.0269}{9}\right)^2 + \dfrac{1}{9-1}\left(\dfrac{.1133}{9}\right)^2} = 11.60$$

To use the $t$ table, we round down to 11 and then obtain $t_{0.025} = 2.201$ for 11 degrees of freedom.

The rejection region becomes

$$T^* = \frac{\overline{X} - \overline{Y} - \delta_0}{\sqrt{\dfrac{S_1^2}{n_1} + \dfrac{S_2^2}{n_2}}} \geq 2.201$$

The calculated value of the test statistic is

$$t^* = \frac{1.887 - .670 - .80}{\sqrt{\dfrac{.0269}{9} + \dfrac{.1133}{9}}} = \frac{.417}{.1248} = 3.34$$

which falls in the rejection region. So, we reject the null hypothesis at $\alpha = .05$ and conclude that the mean product volume from catalyst 1 is more than .80 liters higher.

(b) Since $t_{.01} = 2.718$, the approximate 98% confidence interval is

$$\overline{x} - \overline{y} \pm 2.718 \sqrt{\frac{s_1^2}{n_1} + \frac{s_2^2}{n_2}} = 1.887 - .670 \pm 2.718(.1248) \text{ or } (.88, 1.56)$$

## Exercises

**10.19** Suppose that measurements of the size of butterfly wings (cm) for two related species yielded the data

| Species 1 | 6 | 4 | 7 | 3 |
|---|---|---|---|---|
| Species 2 | 6 | 9 | 6 | |

(a) Calculate $s_{pooled}^2$.

(b) Give an estimate of the common standard deviation for the wing size for the two species.

(c) Evaluate the $t$ statistic for testing equality of the two population mean wing sizes.

**10.20** Three male and three female students recorded the number of times they used a credit card in one week.

| Male | 8 | 4 | 6 |
|---|---|---|---|
| Female | 1 | 5 | 3 |

(a) Calculate $s_{pooled}^2$.

(b) Give an estimate of the common standard deviation for the number of uses.

(c) Evaluate the $t$ statistic for testing equality of the two population mean number of uses.

**10.21** Two gel pens, Gel-1 and Gel-2, are compared on the basis of the number of weeks before they stop writing. Out of 27 persons available, 13 are randomly chosen to receive Gel-1 and the other 14 receive Gel-2. These are the only pens they use for writing.

$$n_1 = 13, \bar{x} = 19, \Sigma (x_i - \bar{x})^2 = 28$$

$$n_2 = 14, \bar{y} = 17, \Sigma (y_i - \bar{y})^2 = 32$$

(a) Obtain $s^2_{pooled}$.

(b) Test $H_0: \mu_1 = \mu_2$ against $H_1: \mu_1 > \mu_2$ with $\alpha = .05$.

(c) Determine a 95% confidence interval for $\mu_1 - \mu_2$.

**10.22** The data on the weight (lb) of male and female wolves, from Table D.9 of the Data Bank, are

| Female | 57 84 90 71 71 77 68 73 |
|---|---|
| Male | 71 93 101 84 88 117 86 86 93 86 106 |

(a) Test the null hypothesis that the mean weights of males and females are equal versus a two-sided alternative. Take $\alpha = .05$.

(b) Obtain a 95% confidence interval for the difference of population mean weights.

(c) State any assumptions you make about the populations.

**10.23** To compare two programs for training industrial workers to perform a skilled job, 20 workers are included in an experiment. Of these, 10 are selected at random and trained by method 1; the remaining 10 workers are trained by method 2. After completion of training, all the workers are subjected to a time-and-motion test that records the speed of performance of a skilled job. The following data are obtained.

| | Time (minutes) |
|---|---|
| Method 1 | 15 20 11 23 16 21 18 16 27 24 |
| Method 2 | 23 31 13 19 23 17 28 26 25 28 |

(a) Can you conclude from the data that the mean job time is significantly less after training with method 1 than after training with method 2? (Test with $\alpha = .05$.)

(b) State the assumptions you make for the population distributions.

(c) Construct a 95% confidence interval for the population mean difference in job times between the two methods.

**10.24** Does the shape of the glass influence how fast a person drinks? As part of a larger study, some participants drank lager from a 12-ounce glass that was either curved or had straight sides. To distract them, they were told their task was to find words in a puzzle but they were videoed and the times they took to finish the drink were recorded.[4] The summary statistics are

| | Time (min) | |
|---|---|---|
| | Straight | Curved |
| Number | 20 | 19 |
| Mean | 11.30 | 7.12 |
| Standard Deviation | 4.43 | 4.78 |

(a) Calculate $s^2_{pooled}$. Is it reasonable to pool the variances in this case?

(b) Conduct a test of hypotheses with the intent of establishing that the mean time to finish drinking lager from the curved glass is the smallest. Take $\alpha = .025$.

**10.25** Given here are the sample sizes and the sample standard deviations for independent random samples from two populations. For each case, state which of the three tests you would use in testing hypotheses about $\mu_1 - \mu_2$: (1) $Z$ test, (2) $t$ test with pooling, (3) approximate $t$ test without pooling. Also, state any assumptions you would make about the population distributions.

(a) $n_1 = 35, s_1 = 12.2$
$n_2 = 50, s_2 = 8.6$

(b) $n_1 = 8, s_1 = 0.86$
$n_2 = 7, s_2 = 1.12$

[4]A Attwood et al., "Glass shape influences consumption rate for alcoholic beverages," *PLoS ONE* 7(8), (2012), e43007. Summary statistics from calculations based on their Figure 2.

(c)    $n_1 = 8$,  $s_1 = 1.54$

   $n_2 = 15$,  $s_2 = 5.36$

(d)    $n_1 = 70$,  $s_1 = 6.2$

   $n_2 = 60$,  $s_2 = 2.1$

**10.26**  To compare the effectiveness of isometric and isotonic exercise methods, 20 obese male students are included in an experiment: 10 are selected at random and assigned to one exercise method; the remaining 10 are assigned to the other exercise method. After five weeks, the reductions in abdomen measurements are recorded in centimeters, and the following results are obtained.

|  | Isometric Method A | Isotonic Method B |
|---|---|---|
| Mean | 2.4 | 3.2 |
| Standard deviation | 0.8 | 1.0 |

(a) Do these data support the claim that the isotonic method is more effective?

(b) Construct a 95% confidence interval for $\mu_B - \mu_A$.

**10.27**  Refer to Exercise 10.26.

(a) Aside from the type of exercise method, identify a few other factors that are likely to have an important effect on the amount of reduction accomplished in a five-week period.

(b) If you were to design this experiment, describe how you would divide the 20 students into two groups.

**10.28**  Refer to Exercise 10.25 (c). In addition, $\bar{x} = 21.1$ and $\bar{y} = 15.4$. Test $H_0 : \mu_1 - \mu_2 = 0$ versus $H_1 : \mu_1 - \mu_2 \leq 0$ at level $\alpha = .05$.

**10.29**  Refer to Example 12. Using the approximate $t$ distribution, obtain a 95% confidence interval for the difference of means.

# 4.  RANDOMIZATION AND ITS ROLE IN INFERENCE

We have presented the methods of drawing inferences about the difference between two population means. Let us now turn to some important questions regarding the design of the experiment or data collection procedure. The manner in which experimental subjects are chosen for the two treatment groups can be crucial. For example, suppose that a remedial-reading instructor has developed a new teaching technique and is permitted to use the new method to instruct half the pupils in the class. The instructor might choose the most alert or the students who are more promising in some other way, leaving the weaker students to be taught in the conventional manner. Clearly, a comparison between the reading achievements of these two groups would not just be a comparison of two teaching methods. A similar fallacy can result in comparing the nutritional quality of a new lunch package if the new diet is given to a group of children suffering from malnutrition and the conventional diet is given to a group of children who are already in good health.

When the assignment of treatments to experimental units is under our control, steps can be taken to ensure a valid comparison between the two treatments. At the core lies the principle of impartial selection, or **randomization**. The choice of the experimental units for one treatment or the other must be made by a chance mechanism that does not favor one particular selection over any other. It must not be left to the discretion of the experimenters because, even unconsciously, they may be partial to one treatment.

Suppose that a comparative experiment is to be run with $N$ experimental units, of which $n_1$ units are to be assigned to treatment 1 and the remaining $n_2 = N - n_1$ units are to be assigned to treatment 2. The principle of randomization tells us that the $n_1$ units for treatment 1 must be chosen at random from the available collection of $N$ units—that is, in a manner such that all $\binom{N}{n_1}$ possible choices are equally likely to be selected.

---

### Randomization Procedure for Comparing Two Treatments

From the available $N = n_1 + n_2$ experimental units, choose $n_1$ units at random to receive treatment 1 and assign the remaining $n_2$ units to treatment 2. The random choice entails that all $\binom{N}{n_1}$ possible selections are equally likely to be chosen.

---

As a practical method of random selection, we can label the available units from 1 to $N$. Then read random digits from Table 1, Appendix B, until $n_1$ different numbers between 1 and $N$ are obtained. These $n_1$ experimental units receive treatment 1 and the remaining units receive treatment 2. For a quicker and more efficient means of random sampling, one can use the computer (see, for instance, the Technology section of Chapter 4).

Although randomization is not a difficult concept, it is one of the most fundamental principles of a good experimental design. It guarantees that uncontrolled sources of variation have the same chance of favoring the response of treatment 1 as they do of favoring the response of treatment 2. Any systematic effects of uncontrolled variables, such as age, strength, resistance, or intelligence, are chopped up or confused in their attempt to influence the treatment responses.

---

Randomization prevents uncontrolled sources of variation from influencing the responses in a systematic manner.

---

Of course, in many cases, the investigator does not have the luxury of randomization. Consider comparing crime rates of cities before and after a new law. Aside from a package of criminal laws, other factors such as poverty, inflation, and unemployment play a significant role in the prevalence of crime. As long as these contingent factors cannot be regulated during the observation period, caution should be exercised in crediting the new law if a decline in the crime rate is observed or discrediting the new law if an increase in the crime rate is observed. When randomization cannot be performed, extreme caution must be exercised in crediting an apparent difference in means to a difference in treatments. The differences may well be due to another factor.

# *Exercises*

**10.30** Randomly allocate 2 subjects from among

   Al, Bob, Carol, Ellen, John

to be in the control group. The others will receive a treatment.

**10.31** Randomly allocate three subjects from among 6 mice,

   Alpha, Tau, Omega, Pi, Beta, Phi

to group 1.

**10.32** Early studies showed a disproportionate number of heavy smokers among lung cancer patients. One scientist theorized that the presence of a particular gene could tend to make a person want to smoke and be susceptible to lung cancer.

(a) How would randomization settle this question?

(b) Would such a randomization be ethical with human subjects?

**10.33** Observations on 10 mothers who nursed their babies and 8 who did not revealed that nursing mothers felt warmer toward their babies. Can we conclude that nursing affects a mother's feelings toward her child?

**10.34** Suppose that you are asked to design an experiment to study the effect of a hormone injection on the weight gain of pregnant rats during gestation. You have decided to inject 6 of the 12 rats available for the experiment and retain the other 6 as controls.

(a) Briefly explain why it is important to randomly divide the rats into the two groups. What might be wrong with the experimental results if you choose to give the hormone treatment to 6 rats that are easy to grab from their cages?

(b) Suppose that the 12 rats are tagged with serial numbers from 1 through 12 and 12 marbles identical in appearance are also numbered from 1 through 12. How can you use these marbles to randomly select the rats in the treatment and control groups?

# 5. MATCHED PAIRS COMPARISONS

In comparing two treatments, it is desirable that the experimental units or subjects be as alike as possible, so that a difference in responses between the two groups can be attributed to differences in treatments. If some identifiable conditions vary over the units in an uncontrolled manner, they could introduce a large variability in the measurements. In turn, this could obscure a real difference in treatment effects. On the other hand, the requirement that all subjects be alike may impose a severe limitation on the number of subjects available for a comparative experiment. To compare two analgesics, for example, it would be impractical to look for a sizable number of patients who are of the same sex, age, and general health condition and who have the same severity of pain. Aside from the question of practicality, we would rarely want to confine a comparison to such a narrow group. A broader scope of inference can be attained by applying the treatments on a variety of patients of both sexes and different age groups and health conditions.

The concept of **matching** or **blocking** is fundamental to providing a compromise between the two conflicting requirements that the experimental units be alike and also of different kinds. The procedure consists of choosing units in pairs or blocks so that the units in each block are similar and those in different blocks are dissimilar. One of the units in each block is assigned to treatment 1, the other to treatment 2. This process preserves the effectiveness of a comparison within

---

**Matched Pairs Design**

| Matched pair | Experimental units | |
|:---:|:---:|:---:|
| 1 | ② | ① |
| 2 | ① | ② |
| 3 | ① | ② |
| ⋮ | ⋮ | ⋮ |
| $n$ | ② | ① |

Units in each pair are alike, whereas units in different pairs may be dissimilar. In each pair, a unit is chosen at random to receive treatment 1, the other unit receives treatment 2.

---

each block and permits a diversity of conditions to exist in different blocks. Of course, the treatments must be allotted to each pair randomly to avoid selection bias. This design is called a **matched pairs design** or sampling. For example, in studying how two different environments influence the learning capacities of preschoolers, it is desirable to remove the effect of heredity: Ideally, this is accomplished by working with twins.

In a matched pairs design, the response of an experimental unit is influenced by:

1.  The conditions prevailing in the block (pair).
2.  A treatment effect.

By taking the difference between the two observations in a block, we can filter out the common block effect. These differences then permit us to focus on the effects of treatments that are freed from undesirable sources of variation.

---

**Pairing (or Blocking)**

Pairing similar experimental units according to some identifiable characteristic(s) serves to remove this source of variation from the experiment.

---

The structure of the observations in a paired comparison is given below, where $X$ and $Y$ denote the responses to treatments 1 and 2, respectively. The difference between the responses in each pair is recorded in the last column, and the summary statistics are also presented.

**Structure of Data for a Matched Pair Comparison**

| Pair | Treatment 1 | Treatment 2 | Difference |
|------|-------------|-------------|------------|
| 1 | $X_1$ | $Y_1$ | $D_1 = X_1 - Y_1$ |
| 2 | $X_2$ | $Y_2$ | $D_2 = X_2 - Y_2$ |
| . | . | . | . |
| . | . | . | . |
| . | . | . | . |
| $n$ | $X_n$ | $Y_n$ | $D_n = X_n = Y_n$ |

The differences $D_1, D_2, \ldots, D_n$ are a random sample.
Summary statistics:

$$\bar{D} = \frac{1}{n} \sum_{i=1}^{n} D_i \qquad S_D^2 = \frac{\sum_{i=1}^{n} (D_i - \bar{D})^2}{n - 1}$$

Although the pairs $(X_i, Y_i)$ are independent of one another, $X_i$ and $Y_i$ within the $i$th pair will usually be dependent. In fact, if the pairing of experimental units is effective, we would expect $X_i$ and $Y_i$ to be relatively large or small together. Expressed in another way, we would expect $(X_i, Y_i)$ to have a high positive correlation. Because the differences $D_i = X_i - Y_i, i = 1, 2, \ldots, n$ are freed from the block effects, it is reasonable to assume that they constitute a random sample from a population with mean $\mu_D$ and variance $\sigma_D^2$, where $\mu_D$ represents the mean difference of the treatment effects. In other words,

$$E(D_i) = \mu_D \qquad \text{and} \qquad \text{Var}(D_i) = \sigma_D^2 \quad \text{for} \quad i = 1, \ldots, n$$

If the mean difference $\mu_D$ is zero, then the two treatments can be considered equivalent. A positive $\mu_D$ signifies that treatment 1 has a higher mean response than treatment 2. Considering $D_1, \ldots, D_n$ to be a single random sample from a population, we can immediately apply the techniques discussed in Chapters 8 and 9 to learn about the population mean $\mu_D$.

## 5.1 INFERENCES FROM A LARGE NUMBER OF MATCHED PAIRS

As we learned in Chapter 8, the assumption of an underlying normal distribution can be relaxed when the sample size is large. The central limit theorem applied to the differences $D_1, \ldots, D_n$ suggests that when $n$ is large, say, greater than 30,

$$\frac{\bar{D} - \mu_D}{S_D / \sqrt{n}} \qquad \text{is approximately } N(0, 1)$$

Inferences can then be based on the percentage points of the $N(0, 1)$ distribution or, equivalently, those of the $t$ distribution, with the degrees of freedom marked "infinity."

**Matched Pairs**

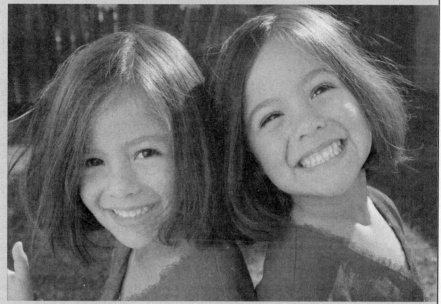

digitalskillet / iStockphoto

Identical twins are the epitome of matched pair experimental subjects. They are matched with respect to not only age but also a multitude of genetic factors. Social scientists, trying to determine the influence of environment and heredity, have been especially interested in studying identical twins who were raised apart. Observed differences in IQ and behavior are then supposedly due to environmental factors.

When the subjects are animals like mice, two from the same litter can be paired. Going one step further, genetic engineers can now provide two identical plants or small animals by cloning these subjects.

**Example 13**    Does Conditioning Reduce Percent Body Fat?

A conditioning class is designed to introduce students to a variety of training techniques to improve fitness and flexibility. The percent body fat was measured at the start of the class and at the end of the semester. The data for 81 students are given in Table D.5 of the Data Bank.

(a)  Obtain a 98% confidence interval for the mean reduction in percent body fat.

(b)  Test, at $\alpha = .01$, to establish the claim that the conditioning class reduces the mean percent body fat.

SOLUTION

(a) Each subject represents a block that produces one measurement of percent body fat at the start of the semester ($x$) and one at the end ($y$). The 81 paired differences $d_i = x_i - y_i$ are summarized using a computer.

| | N | Mean | StDev | SE Mean |
|---|---|---|---|---|
| Difference | 81 | 3.322 | 2.728 | 0.303 |

That is, $\bar{d} = 3.322$ and $s_D = 2.728$. The sample size 81 is large so there is no need to assume that the population is normal. Since, from the normal table, $z_{.01} = 2.33$, the 98% confidence interval becomes

$$\left( \bar{d} - 2.33 \frac{s_D}{\sqrt{81}}, \quad \bar{d} + 2.33 \frac{s_D}{\sqrt{81}} \right)$$

$$\left( 3.322 - 2.33 \times \frac{2.728}{\sqrt{81}}, \quad 3.322 + 2.33 \times \frac{2.728}{\sqrt{81}} \right) =$$

$$( 3.322 - .706, \quad 3.322 + .706 )$$

or $( 2.62, 4.03 )$ percent. We are 98% confident that the mean reduction in body fat is 2.62 to 4.03 percent.

(b) Because the claim is that $\mu_D > 0$, the initial reading tends to be higher than at the end of class, we formulate:

$$H_0 : \mu_D = 0 \quad \text{versus} \quad H_1 : \mu_D > 0$$

The test statistic

$$Z = \frac{\bar{D}}{S_D / \sqrt{n}}$$

is approximately normally distributed so the rejection region is $R : Z \geq z_{.01} = 2.33$. The observed value of the test statistic

$$z = \frac{\bar{d}}{S_D / \sqrt{81}} = \frac{3.322}{2.728 / \sqrt{81}} = \frac{3.322}{.303} = 10.96$$

falls in the rejection region. Consequently $H_0$ is rejected in favor of $H_1$ at level $\alpha = .01$. We conclude that the conditioning class does reduce the mean percent body fat. The value of the test statistic is so far in the rejection region that the $P$-value is .0000 to at least four places. The evidence in support of $H_1$ is very very strong.

## 5.2 INFERENCES FROM A SMALL NUMBER OF MATCHED PAIRS

When the sample size is not large, we make the additional assumption that the distribution of the differences is normal.

In summary,

---

**Small Samples Inferences about the Mean Difference $\mu_D$**

Assume that the differences $D_i = X_i - Y_i$ are a random sample from an $N(\mu_D, \sigma_D)$ distribution. Let

$$\bar{D} = \frac{\sum\limits_{i=1}^{n} D_i}{n} \quad \text{and} \quad S_D = \sqrt{\frac{\sum\limits_{i=1}^{n} (D_i - \bar{D})^2}{n-1}}$$

Then:

1. A $100(1 - \alpha)\%$ confidence interval for $\mu_D$ is given by

$$\left( \bar{D} - t_{\alpha/2} \frac{S_D}{\sqrt{n}}, \quad \bar{D} + t_{\alpha/2} \frac{S_D}{\sqrt{n}} \right)$$

where $t_{\alpha/2}$ is based on $n - 1$ degrees of freedom.

2. A test of $H_0: \mu_D = \mu_{D0}$ is based on the test statistic

$$T = \frac{\bar{D} - \mu_{D0}}{S_D / \sqrt{n}} \quad \text{d.f.} = n - 1$$

---

**Example 14**  Does a Pill Incidentally Reduce Blood Pressure?

A medical researcher wishes to determine if a pill has the undesirable side effect of reducing the blood pressure of the user. The study involves recording the initial blood pressures of 15 college-age women. After they use the pill regularly for six months, their blood pressures are again recorded. The researcher wishes to draw inferences about the effect of the pill on blood pressure from the observations given in Table 1.

(a)  Calculate a 95% confidence interval for the mean reduction in blood pressure.

(b)  Do the data substantiate the claim that use of the pill reduces blood pressure? Test at $\alpha = .01$.

**TABLE 1**  Blood-Pressure Measurements before and after Use of Pill

| | Subject | | | | | | | | | | | | | | |
|---|---|---|---|---|---|---|---|---|---|---|---|---|---|---|---|
| | 1 | 2 | 3 | 4 | 5 | 6 | 7 | 8 | 9 | 10 | 11 | 12 | 13 | 14 | 15 |
| Before ($x$) | 70 | 80 | 72 | 76 | 76 | 76 | 72 | 78 | 82 | 64 | 74 | 92 | 74 | 68 | 84 |
| After ($y$) | 68 | 72 | 62 | 70 | 58 | 66 | 68 | 52 | 64 | 72 | 74 | 60 | 74 | 72 | 74 |
| $d = x - y$ | 2 | 8 | 10 | 6 | 18 | 10 | 4 | 26 | 18 | −8 | 0 | 32 | 0 | −4 | 10 |

Courtesy of a family planning clinic.

SOLUTION

(a) Here each subject represents a block generating a pair of measurements: one before using the pill and the other after using the pill. The paired differences $d_i = x_i - y_i$ are computed in the last row of Table 1, and we calculate the summary statistics

$$\bar{d} = \frac{\sum d_i}{15} = 8.80 \quad s_D = \sqrt{\frac{\sum (d_i - \bar{d})^2}{14}} = 10.98$$

If we assume that the paired differences constitute a random sample from a normal population $N(\mu_D, \sigma_D)$, a 95% confidence interval for the mean difference $\mu_D$ is given by

$$\bar{D} \pm t_{.025} \frac{s_D}{\sqrt{15}}$$

where $t_{.025}$ is based on d.f. $= 14$. From the $t$ table, we find $t_{.025} = 2.145$. The 95% confidence interval is then computed as

$$8.80 \pm 2.145 \times \frac{10.98}{\sqrt{15}} = 8.80 \pm 6.08 \text{ or } (2.72, 14.88)$$

This means that we are 95% confident the mean reduction of blood pressure is between 2.72 and 14.88.

(b) Because the claim is that $\mu_D > 0$, we formulate

$$H_0: \mu_D = 0 \quad \text{versus} \quad H_1: \mu_D > 0$$

We employ the test statistic $T = \dfrac{\bar{D}}{s_D / \sqrt{n}}$, d.f. $= 14$ and set a right-sided rejection region. With d.f. $= 14$, we find $t_{.01} = 2.624$, so the rejection region is $R: T \geq 2.624$.

The observed value of the test statistic

$$t = \frac{\bar{d}}{s_D / \sqrt{n}} = \frac{8.80}{10.98 / \sqrt{15}} = \frac{8.80}{2.84} = 3.10$$

falls in the rejection region. Consequently, $H_0$ is rejected in favor of $H_1$ at $\alpha = .01$. We conclude that a reduction in blood pressure following use of the pill is strongly supported by the data.

Further, as apparent from Figure 8, the observed value $t = 3.10$ has a small $P$-value. A software calculation gives

$$P\text{-value} = P[T \geq 3.10] = .004$$

*Note:* To be more convinced that the pill causes the reduction in blood pressure, it is advisable to measure the blood pressures of the same subjects once again after they have stopped using the pill for a period of time. This amounts to performing the experiment in reverse order to check the findings of the first stage.

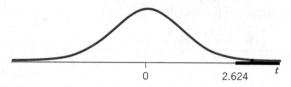

Figure 8   Rejection region $R: T \geq 2.624$ and the observed value $t = 3.10$.

## 5.3 RANDOMIZATION WITH MATCHED PAIRS

Example 14 is a typical before-after situation. Data gathered to determine the effectiveness of a safety program or an exercise program would have the same structure. In such cases, there is really no way to choose how to order the experiments within a pair. The before situation must precede the after situation. If something other than the institution of the program causes performance to improve, the improvement will be incorrectly credited to the program. However, when the order of the application of treatments can be determined by the investigator, something can be done about such systematic influences. Suppose that a coin is flipped to select the treatment for the first unit in each pair. Then the other treatment is applied to the second unit. Because the coin is flipped again for each new pair, any uncontrolled variable has an equal chance of helping the performance of either treatment 1 or treatment 2. After eliminating an identified source of variation by pairing, we return to randomization in an attempt to reduce the systematic effects of any uncontrolled sources of variation.

---

### Randomization with Pairing

After pairing, the assignment of treatments should be randomized for each pair.

---

Randomization within each pair chops up or diffuses any systematic influences that we are unable to control.

## Exercises

**10.35** Given the following matched pairs sample,

| x | y |
|---|---|
| 6 | 3 |
| 4 | 1 |
| 8 | 3 |
| 6 | 5 |
| 9 | 7 |
| 6 | 8 |

(a) Evaluate the $t$ statistic

$$t = \frac{\bar{d}}{S_D / \sqrt{n}}.$$

(b) How many degrees of freedom does this $t$ have?

**10.36** Two sun blockers are to be compared. One blocker is rubbed on one side of a subject's back and the other blocker is rubbed on the other side. Each subject then lies in the sun for two hours. After waiting an additional hour, each side is rated according to redness.

(a) Evaluate the $t$ statistic

$$t = \frac{\bar{d}}{S_D / \sqrt{n}}.$$

(b) How many degrees of freedom does this $t$ have?

| Subject No. | Blocker 1 | Blocker 2 |
|---|---|---|
| 1 | 2 | 2 |
| 2 | 7 | 5 |
| 3 | 8 | 4 |
| 4 | 3 | 1 |
| 5 | 5 | 3 |

**10.37** It is claimed that an industrial safety program is effective in reducing the loss of working hours due to factory accidents. The following data are collected concerning the weekly loss of working hours due to accidents in six plants both before and after the safety program is instituted.

| | Plant | | | | | |
|---|---|---|---|---|---|---|
| | 1 | 2 | 3 | 4 | 5 | 6 |
| Before | 12 | 30 | 15 | 37 | 29 | 15 |
| After | 10 | 29 | 16 | 35 | 26 | 16 |

Do the data substantiate the claim? Use $\alpha = .05$.

**10.38** A manufacturer claims his boot waterproofing is better than the major brand. Five pairs of shoes are available for a test.

(a) Explain how you would conduct a paired sample test.

(b) Write down your assignment of waterproofing to each shoe. How did you randomize?

**10.39** A food scientist wants to study whether quality differences exist between yogurt made from skim milk with and without the preculture of a particular type of bacteria, called Psychrotrops (PC). Samples of skim milk are procured from seven dairy farms. One-half of the milk sampled from each farm is inoculated with PC, and the other half is not. After yogurt is made with these milk samples, the firmness of the curd is measured, and those measurements are given below.

| | Dairy Farm | | | | | | |
|---|---|---|---|---|---|---|---|
| Curd Firmness | A | B | C | D | E | F | G |
| With PC | 68 | 75 | 62 | 86 | 52 | 46 | 72 |
| Without PC | 61 | 69 | 64 | 76 | 52 | 38 | 68 |

(a) Do these data substantiate the conjecture that the treatment of PC results in a higher degree of curd firmness? Test at $\alpha = .05$.

(b) Determine a 90% confidence interval for the mean increase of curd firmness due to the PC treatment.

**10.40** A study is to be made of the relative effectiveness of two kinds of cough medicines in increasing sleep. Six people with colds are given medicine $A$ the first night and medicine $B$ the second night. Their hours of sleep each night are recorded.

| | Subject | | | | | |
|---|---|---|---|---|---|---|
| | 1 | 2 | 3 | 4 | 5 | 6 |
| Medicine $A$ | 4.8 | 4.1 | 5.8 | 4.9 | 5.1 | 7.4 |
| Medicine $B$ | 3.9 | 4.2 | 5.0 | 4.9 | 5.2 | 7.1 |

(a) Establish a 95% confidence interval for the mean change in hours of sleep when switching from medicine $A$ to medicine $B$.

(b) How and what would you randomize in this study? Briefly explain your reason for randomization.

**10.41** Two methods of memorizing difficult material are being tested to determine if one produces better retention. Nine pairs of students are included in the study. The students in each pair are matched according to IQ and academic background and then assigned to the two methods at random. A memorization test is given to all the students, and the following scores are obtained:

| | Pair | | | | | | | | |
|---|---|---|---|---|---|---|---|---|---|
| | 1 | 2 | 3 | 4 | 5 | 6 | 7 | 8 | 9 |
| Method $A$ | 90 | 86 | 72 | 65 | 44 | 52 | 66 | 38 | 83 |
| Method $B$ | 85 | 87 | 70 | 62 | 44 | 53 | 62 | 35 | 86 |

At $\alpha = .05$, test to determine if there is a significant difference in the effectiveness of the two methods.

**10.42** In an experiment conducted to see if electrical pricing policies can affect consumer behavior, 10 homeowners in Wisconsin had to pay a premium for power use during the peak hours. They were offered lower off-peak rates. For each home, the July on-peak usage (kilowatt hours) under the pricing experiment was compared to the previous July usage.

| Year | |
| --- | --- |
| Previous | Experimental |
| 200 | 160 |
| 180 | 175 |
| 240 | 210 |
| 425 | 370 |
| 120 | 110 |
| 333 | 298 |
| 418 | 368 |
| 380 | 250 |
| 340 | 305 |
| 516 | 477 |

(a) Find a 95% confidence interval for the mean decrease.

(b) Test $H_0 : \mu_D = 0$ against $H_1 : \mu_D \neq 0$ at level $\alpha = .05$.

(c) Comment on the feasibility of randomization of treatments.

(d) Without randomization, in what way could the results in parts (a) and (b) be misleading?

(*Hint*: What if air conditioner use is a prime factor, and the July with experimental pricing was cooler than the previous July?)

**10.43** Laser guns for detecting speeders require an accurate measurement of distance. In one daily calibration test, police measure the known distances 164 feet and 104 feet. The difference in readings should be 60 but there is some variation. The differences (courtesy of Madison Police Department) for ten guns are

60.6  60.4  60.4  60.5  60.3  60.4  60.2  60.1  60.4  60.4

(a) Do the data provide strong evidence that the mean difference is not equal to 60 feet?

(b) Construct a 95% confidence interval for the mean difference.

(c) Check that all of the values for mean distance in the confidence interval, as well as all ten observations, are less than 1 percent of the exact 60 feet. Laser guns are much more accurate than the old radar guns.

**10.44** A study about the effects of marijuana use concludes that persons who became persistent users before they reached the age of 18, and who remained persistent users until age 38, will lose several IQ points over that time span.[5] The summary data, for the change in IQ from before marijuana use to age 38, are

$$n = 23 \qquad \bar{d} = -8.0 \qquad s_D = 14.5$$

(a) Do the data provide strong evidence that the mean difference is not equal to 0?

(b) Construct a 95% confidence interval for the mean difference.

# 6. CHOOSING BETWEEN INDEPENDENT SAMPLES AND A MATCHED PAIRS SAMPLE

When planning an experiment to compare two treatments, we often have the option of either designing two independent samples or designing a sample with paired observations. Therefore, some comments about the pros and cons of these two sampling methods are in order here. Because a paired sample with $n$ pairs of observations contains $2n$ measurements, a comparable situation would be two independent samples with $n$ observations in each. First, note that the sample mean difference is the same whether or not the samples are paired. This is

[5]Data consistent with graphical summary in M. Meier, et al., "Persistent cannabis users show neuropsychological decline from childhood to midlife," *Proceedings of the National Academy of Science Plus* **109** (2012), pp. 2657–2664.

because

$$\bar{D} = \frac{1}{n}\sum (X_i - Y_i) = \bar{X} - \bar{Y}$$

Therefore, using either sampling design, the confidence intervals for the difference between treatment effects have the common form

$$(\bar{X} - \bar{Y}) \pm t_{\alpha/2} (\text{estimated standard error})$$

However, the estimated standard error as well as the degrees of freedom for $t$ are different between the two situations.

|  | Independent Samples $(n_1 = n_2 = n)$ | Paired Sample $(n$ Pairs$)$ |
|---|---|---|
| Estimated standard error | $S_{\text{pooled}} \sqrt{\dfrac{1}{n} + \dfrac{1}{n}}$ | $\dfrac{S_D}{\sqrt{n}}$ |
| d.f. of $t$ | $2n - 2$ | $n - 1$ |

Because the length of a confidence interval is determined by these two components, we now examine their behavior under the two competing sampling schemes.

Paired sampling results in a loss of degrees of freedom and, consequently, a larger value of $t_{\alpha/2}$. For instance, with a paired sample of $n = 10$, we have $t_{.05} = 1.833$ with d.f. $= 9$. But the $t$ value associated with independent samples, each of size 10, is $t_{.05} = 1.734$ with d.f. $= 18$. Thus, if the estimated standard errors are equal, then a loss of degrees of freedom tends to make confidence intervals larger for paired samples. Likewise, in testing hypotheses, a loss of degrees of freedom for the $t$ test results in a loss of power to detect real differences in the population means.

The merit of paired sampling emerges when we turn our attention to the other component. If experimental units are paired so that an interfering factor is held nearly constant between members of each pair, the treatment responses $X$ and $Y$ within each pair will be equally affected by this factor. If the prevailing condition in a pair causes the $X$ measurement to be large, it will also cause the corresponding $Y$ measurement to be large and vice versa. As a result, the variance of the difference $X - Y$ will be smaller in the case of an effective pairing than it will be in the case of independent random variables. The estimated standard deviation will be typically smaller as well. With an effective pairing, the reduction in the standard deviation usually more than compensates for the loss of degrees of freedom.

In Example 14, concerning the effect of a pill in reducing blood pressure, we note that a number of important factors (age, weight, height, general health, etc.) affect a person's blood pressure. By measuring the blood pressure of the same person before and after use of the pill, these influencing factors can be held nearly constant for each pair of measurements. On the other hand, independent samples of one group of persons using the pill and a separate control group of persons not using the pill are apt to produce a greater variability in blood-pressure measurements if all the persons selected are not similar in age, weight, height, and general health.

In summary, paired sampling is preferable to independent sampling when an appreciable reduction in variability can be anticipated by means of pairing. When the experimental units are already alike or their dissimilarities cannot be linked to identifiable factors, an arbitrary pairing may fail to achieve a reduction in variance. The loss of degrees of freedom will then make a paired comparison less precise.

# 7. COMPARING TWO POPULATION PROPORTIONS

We are often interested in comparing two populations with regard to the rate of incidence of a particular characteristic. Comparing the jobless rates in two cities, the percentages of female employees in two categories of jobs, and infant mortality in two ethnic groups are just a few examples. The parameters of interest are the proportion of members in each population that possess the characteristic. Let

$p_1$ = proportion possessing the characteristic in Population 1

$p_2$ = proportion possessing the characteristic in Population 2

Our goals in this section are to construct confidence intervals for $p_1 - p_2$ and test the null hypothesis, $H_0 : p_1 = p_2$, that the rates are the same for two populations. The methods would also apply to the problems of comparison between two treatments, where the response of a subject falls into one of two possible categories that we may technically call "success" and "failure." The success rates for the two treatments can then be identified as the two population proportions $p_1$ and $p_2$.

The form of the data is displayed in Table 2, where $X$ and $Y$ denote the numbers of successes in independent random samples of sizes $n_1$ and $n_2$ taken from Population 1 and Population 2, respectively.

**TABLE 2**   Independent Samples from Two Dichotomous Populations

|  | No. of Successes | No. of Failures | Sample Size |
|---|---|---|---|
| Population 1 | $X$ | $n_1 - X$ | $n_1$ |
| Population 2 | $Y$ | $n_2 - Y$ | $n_2$ |

## 7.1 ESTIMATION

The population proportions of successes $p_1$ and $p_2$ are estimated by the corresponding sample proportions

$$\widehat{p}_1 = \frac{X}{n_1} \quad \text{and} \quad \widehat{p}_2 = \frac{Y}{n_2}$$

Naturally, $\widehat{p}_1 - \widehat{p}_2$ serves to estimate the difference $p_1 - p_2$. Its standard error is given by

$$\text{S.E.}(\widehat{p}_1 - \widehat{p}_2) = \sqrt{\frac{p_1 q_1}{n_1} + \frac{p_2 q_2}{n_2}}$$

where $q_1 = 1 - p_1$ and $q_2 = 1 - p_2$. This formula of the standard error stems from the fact that because $\widehat{p}_1$ and $\widehat{p}_2$ are based on independent samples, the variance of their difference equals the sum of their individual variances.

We can calculate the estimated standard error of $\widehat{p}_1 - \widehat{p}_2$ by using the above expression with the population proportions replaced by the corresponding sample proportions. Moreover, when $n_1$ and $n_2$ are large, the estimator $\widehat{p}_1 - \widehat{p}_2$ is approximately normally distributed. Specifically,

$$Z = \frac{(\widehat{p}_1 - \widehat{p}_2) - (p_1 - p_2)}{\text{Estimated standard error}} \quad \text{is approximately } N(0, 1)$$

and this can be the basis for constructing confidence intervals for $p_1 - p_2$.

---

**Large Samples Confidence Interval for $p_1 - p_2$**

An approximate $100(1 - \alpha)\%$ confidence interval for $p_1 - p_2$ is

$$(\widehat{p}_1 - \widehat{p}_2) \pm z_{\alpha/2}\sqrt{\frac{\widehat{p}_1(1 - \widehat{p}_1)}{n_1} + \frac{\widehat{p}_2(1 - \widehat{p}_2)}{n_2}}$$

provided the sample sizes $n_1$ and $n_2$ are large.

---

**Example 15**  A Confidence Interval for a Difference in Success Rates

An investigation comparing a medicated patch with the unmedicated control patch for helping smokers quit the habit was discussed on page 91. At the end of the study, the number of persons in each group who were abstinent and who were smoking are repeated in Table 3.

Determine a 95% confidence interval for the difference in success probabilities.

**TABLE 3**  Quitting Smoking

|                    | Abstinent | Smoking | Total |
|--------------------|-----------|---------|-------|
| Medicated patch    | 21        | 36      | 57    |
| Unmedicated patch  | 11        | 44      | 55    |
|                    | 32        | 80      | 112   |

SOLUTION    Let $p_1$ and $p_2$ denote the probabilities of quitting smoking with the medicated and unmedicated patches, respectively. We calculate

$$\hat{p}_1 = \frac{21}{57} = .3684 \qquad \hat{p}_2 = \frac{11}{55} = .2000$$

$$\hat{p}_1 - \hat{p}_2 = .1684$$

$$\sqrt{\frac{\hat{p}_1 \hat{q}_1}{n_1} + \frac{\hat{p}_2 \hat{q}_2}{n_2}} = \sqrt{\frac{.3684 \times .6316}{57} + \frac{.2000 \times .8000}{55}} = .0836$$

A 95% confidence interval for $p_1 - p_2$ is

$$(.1684 - 1.96 \times .0836, .1684 + 1.96 \times .0836) \quad \text{or} \quad (.005, .332)$$

The confidence interval only covers positive values. With 95% confidence the success rate with the medicated patch is .005 to .332 higher than for the control group that received the untreated patches. The lower value is so close to 0 that it is still plausible that the medicated patch is not very effective.

**Note:**    A confidence interval for each of the population proportions $p_1$ and $p_2$ can be determined by using the method described in Section 5 of Chapter 8. For instance, with the data of Example 15, a 90% confidence interval for $p_1$ is calculated as

$$\hat{p}_1 \pm 1.645 \sqrt{\frac{\hat{p}_1 \hat{q}_1}{n_1}} = .3684 \pm 1.645 \sqrt{\frac{.3684 \times .6316}{57}}$$

$$= .3684 \pm .1051 \quad \text{or} \quad (.263, .474)$$

## 7.2 TESTING STATISTICAL HYPOTHESES

In order to formulate a test of $H_0 : p_1 = p_2$ when the sample sizes $n_1$ and $n_2$ are large, we again turn to the fact that $\hat{p}_1 - \hat{p}_2$ is approximately normally distributed. But now we note that under $H_0$ the mean of this normal distribution is $p_1 - p_2 = 0$ and the standard deviation is

$$\sqrt{pq} \sqrt{\frac{1}{n_1} + \frac{1}{n_2}}$$

where $p$ stands for the common probability of success $p_1 = p_2$ and $q = 1 - p$. The unknown common $p$ is estimated by pooling information from the two samples. The proportion of successes in the combined sample provides

$$\text{Pooled estimate} \quad \hat{p} = \frac{X + Y}{n_1 + n_2}$$

$$\text{or, alternatively,} \quad \hat{p} = \frac{n_1 \hat{p}_1 + n_2 \hat{p}_2}{n_1 + n_2}$$

$$\text{Also, estimated S.E.} (\hat{p}_1 - \hat{p}_2) = \sqrt{\hat{p}\hat{q}} \sqrt{\frac{1}{n_1} + \frac{1}{n_2}}$$

In summary,

---

### Testing $H_0 : p_1 = p_2$ with Large Samples

Test statistic:

$$Z = \frac{\widehat{p}_1 - \widehat{p}_2}{\sqrt{\widehat{p}\,(1 - \widehat{p})}\,\sqrt{\dfrac{1}{n_1} + \dfrac{1}{n_2}}} \qquad \text{where } \widehat{p} = \frac{X + Y}{n_1 + n_2}$$

| Alternative Hypothesis | Level $\alpha$ Rejection Region |
|---|---|
| $H_1 : p_1 > p_2$ | $R : Z \geq z_\alpha$ |
| $H_1 : p_1 < p_2$ | $R : Z \leq -z_\alpha$ |
| $H_1 : p_1 \neq p_2$ | $R : |Z| \geq z_\alpha$ |

---

**Example 16**   Using A/B Testing to Establish a More Effective Web Site

Many e-commerce companies are currently implementing a strategy called A/B testing to improve their Internet sales. One particular product is selected and then one element of its Web page is changed. For instance, a picture may be added or an existing picture changed. Next, a fraction of the incoming traffic to the product page is directed to the modified page. Each day, for a week or so, the fraction of visitors to the product page who purchase is recorded both for the original page and for the modified page. The two cases give rise to the terminology A and B.

Now that we described the comparison of proportions from two independent samples, it is clear that A/B testing, originally described on page 3, is really this well known statistical problem.

The Web site of one major online retailer gets about 2.4 million hits each day. If a change of the main page is to be compared with the existing page, there will be millions of hits for each in a day. Further downstream, the object may be to improve sales performance from anyone reaching a designated product page.

In one instance, where a picture was added and the goal was to increase the number of purchasers, the total weekly counts are

|  | Original Page | Modified Page |
|---|---|---|
| Number of visitors | 3050 | 3307 |
| Number that purchase | 311 | 417 |

Is there strong evidence the modified page is more effective?

SOLUTION   The samples sizes are very large and because many such changes will ultimately be tested we take $\alpha$ quite small. (The reason for small $\alpha$ is explained in Chapter 14). We test with $\alpha = .005$. We want to establish that $p_1 < p_2$ so that is the alternative hypothesis and we test

$$H_0 : p_1 = p_2 \qquad \text{versus} \qquad H_1 : p_1 < p_2$$

or equivalently as

$$H_0 : p_1 - p_2 = 0 \qquad \text{versus} \qquad H_1 : p_1 - p_2 < 0$$

The sample sizes $n_1 = 3050$ and $n_2 = 3307$ being very large, we will employ the test statistic

$$Z = \frac{\hat{p}_1 - \hat{p}_2}{\sqrt{\hat{p}\hat{q}}\sqrt{\dfrac{1}{n_1} + \dfrac{1}{n_2}}}$$

and set the left-sided rejection region in view of the fact that $H_1$ is left-sided. With $\alpha = .005$, the rejection region is $R : Z \leq -2.576$ as shown in Figure 9.

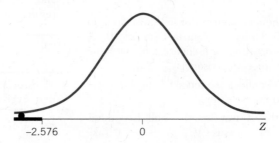

$$-2.576 \qquad \qquad 0 \qquad \qquad \qquad Z$$

Figure 9    Rejection region $R : Z \leq -2.576$ and the observed value $z = -3.105$.

We calculate

$$\hat{p}_1 = \frac{311}{3050} = .1020 \qquad \qquad \hat{p}_2 = \frac{417}{3307} = .1261$$

$$\text{pooled estimate} \quad \hat{p} = \frac{311 + 417}{3050 + 3307} = .1145$$

The observed value of the test statistic

$$z = \frac{.1020 - .1261}{\sqrt{.1145 \times .8855}\sqrt{\dfrac{1}{3050} + \dfrac{1}{3307}}} = -3.105$$

lies in the rejection region $R$ so, at level $\alpha = .005$, we reject the null hypothesis of equal probabilities in favor of the alternative that the modified page leads to more purchasers. Because $\alpha$ is so small, the data give very strong evidence against the null hypothesis.

The significance probability

$$P\text{–value} = P[Z \leq -3.105] = .001$$

further strengthens the conclusion.

Because the modified page is more successful than the original, it becomes the new standard and an additional change is tested for improvement. Because of high traffic at the site, even seemingly marginal improvements in purchase or drop-off rates will create substantial additional sales.

# Exercises

**10.45** Refer to the measurement of job satisfaction in Example 4. Rather than average score, researchers often prefer the proportion who respond "very satisfied." Using the data in Exercise 10.12, compare the proportion for clergy, $p_1$, with the proportion for office supervisors, $p_2$.

(a) Find a 95% confidence interval for the difference of proportions $p_1 - p_2$.

(b) Perform the $Z$ test for the null hypothesis $H_0: p_1 = p_2$ versus $H_1: p_1 > p_2$. Take $\alpha = .05$.

**10.46** Refer to the measurement of job satisfaction in Exercise 10.45. Using the data in Exercise 10.12, compare the proportion for firefighters, $p_1$, with the proportion for office supervisors, $p_2$.

(a) Find a 95% confidence interval for the difference of proportions $p_1 - p_2$.

(b) Perform the $Z$ test for the null hypothesis $H_0: p_1 = p_2$ versus $H_1: p_1 > p_2$. Take $\alpha = .05$.

**10.47** In a study of the relationship between temperament and personality, 49 female high school students who had a high level of reactivity (HRL) and 54 students who had a low level of reactivity (LRL) were classified according to their attitude to group pressure with the following results.

| | Attitude | | |
|---|---|---|---|
| Reactivity | Submissive | Resistant | Total |
| HRL | 34 | 15 | 49 |
| LRL | 12 | 42 | 54 |

Is resistance to group pressure significantly lower in the HRL group than the LRL group? Answer by calculating the $P$–value.

**10.48** Refer to Exercise 10.17 concerning a study on the relationship between child abuse and later criminal behavior. Suppose that from follow-ups of 85 boys who were abused in their preschool years and 120 boys who were not abused, it was found that 21 boys in the abused group and 11 boys in the nonabused group were chronic offenders in their teens. Do these data substantiate the conjecture that abused boys are more prone to be chronic offenders than nonabused boys? Test at $\alpha = .01$.

**10.49** The popular disinfectant Listerine is named after Joseph Lister, a British physician who pioneered the use of antiseptics. Lister conjectured that human infections might have an organic origin and thus could be prevented by using a disinfectant. Over a period of several years, he performed 75 amputations: 40 using carbolic acid as a disinfectant and 35 without any disinfectant. The following results were obtained.

| | Patient Survived | Patient Died | Total |
|---|---|---|---|
| With carbolic acid | 34 | 6 | 40 |
| Without carbolic acid | 19 | 16 | 35 |

Are the survival rates significantly different between the two groups? Test at $\alpha = .05$ and calculate the $P$–value.

**10.50** Referring to the data of Exercise 10.49, calculate a 95% confidence interval for the difference between the survival rates for the two groups.

**10.51** Random samples of 250 persons in the 30- to 40-year age group and 250 persons in the 60- to 70-year age group are asked about the average number of hours they sleep per night, and the following summary data are recorded.

| | Hours of Sleep | | |
|---|---|---|---|
| Age | $\leq 8$ | $> 8$ | Total |
| 30–40 | 173 | 77 | 250 |
| 60–70 | 120 | 130 | 250 |
| Total | 293 | 207 | 500 |

Do these data demonstrate that the proportion of persons who have $\leq 8$ hours of sleep per night is significantly higher for the age group 30 to 40 than that for the age group 60 to 70? Answer by calculating the $P$–value.

**10.52** Referring to Exercise 10.51, denote by $p_1$ and $p_2$ the population proportions in the two groups who have $\leq$ 8 hours of sleep per night. Construct a 95% confidence interval for $p_1 - p_2$.

**10.53** A medical researcher conjectures that smoking can result in wrinkled skin around the eyes. By observing 150 smokers and 250 nonsmokers, the researcher finds that 95 of the smokers and 103 of the nonsmokers have prominent wrinkles around their eyes.

(a) Do these data substantiate the belief that prominent wrinkles around eyes are more prevalent among smokers than nonsmokers? Answer by calculating the $P$-value.

(b) If the results are statistically significant, can the researcher readily conclude that smoking causes wrinkles around the eyes? Why or why not?

**10.54** In the survey on which Table D.12 of the Data Bank is based, a larger number of persons was asked to respond to the statement "I would characterize my political beliefs as liberal" on a seven point Likert scale from strongly disagree (1) to strongly agree (7). A count of those who strongly disagree, for each of three age groups, yielded the summary statistics.

| Age | $x_i$ | $n_i$ | $p_i$ |
|-----|-------|-------|-------|
| 19–24 | 3 | 47 | .0638 |
| 25–34 | 13 | 64 | .2031 |
| 35–44 | 14 | 93 | .1505 |

(a) Test equality of the proportions for the last two age groups versus a two-sided alternative. Take $\alpha = .05$.

(b) Test equality of the proportions for the first two age groups versus a two-sided alternative. Take $\alpha = .05$.

(c) Explain why the large samples test in part (b) is not valid.

**10.55** A major clinical trial of a new vaccine for type-B hepatitis was conducted with a high-risk group of 1083 male volunteers. From this group, 549 men were given the vaccine and the other 534 a placebo. A follow-up of all these individuals yielded the data:

| | Follow-up | | |
|---|---|---|---|
| | Got Hepatitis | Did Not Get Hepatitis | Total |
| Vaccine | 11 | 538 | 549 |
| Placebo | 70 | 464 | 534 |

(a) Do these observations testify that the vaccine is effective? Use $\alpha = .01$.

(b) Construct a 95% confidence interval for the difference between the incidence rates of hepatitis among the vaccinated and nonvaccinated individuals in the high-risk group.

**10.56** Records of drivers with a major medical condition (diabetes, heart condition, or epilepsy) and also a group of drivers with no known health conditions were retrieved from a motor vehicle department. Drivers in each group were classified according to their driving record in the last year.

| | Traffic Violations | | |
|---|---|---|---|
| Medical Condition | None | One or More | Total |
| Diabetes | 119 | 41 | 160 |
| Heart condition | 121 | 39 | 160 |
| Epilepsy | 72 | 78 | 150 |
| None (control) | 157 | 43 | 200 |

Let $p_D$, $p_H$, $p_E$, and $p_C$ denote the population proportions of drivers having one or more traffic violations in the last year for the four groups "diabetes," "heart condition," "epilepsy," and "control," respectively.

(a) Test $H_0 : p_D = p_C$ versus $H_1 : p_D > p_C$ at $\alpha = .10$.

(b) Is there strong evidence that $p_E$ is higher than $p_C$? Answer by calculating the $P$-value.

**10.57** Refer to Exercise 10.56.

(a) Construct a 95% confidence interval for $p_E - p_H$.

(b) Construct a 90% confidence interval for $p_H - p_C$.

(c) Construct 95% confidence intervals for $p_D$, $p_H$, $p_E$, and $p_C$, individually.

## USING STATISTICS WISELY

1. In all cases where a sample size is small and normality is assumed, each sample should be graphed in a dot plot to reveal any obvious outliers which could invalidate the inferences based on the normal theory.

2. When comparing two treatments using the independent samples design, randomly assign the treatments to groups whenever possible.

3. When comparing two treatments using the independent samples design, but the sample sizes are not large, you must decide whether or not to pool the estimates of variance.

$$s^2_{pooled} = \frac{(n_1 - 1)s_1^2 + (n_2 - 1)s_2^2}{(n_1 - 1) + (n_2 - 1)}$$

Pooling the estimates efficiently uses the information in samples provided that the variances are the same.

4. However, do not pool the two sample variances $s_1^2$ and $s_2^2$ if they are very different. We suggest a factor of 4 as being too different. In that case, we recommend using the approximate $t$ statistic with the Satterthwaite correction.

5. When comparing two treatments under a matched pairs design, whenever possible, assign the treatments within each pair at random.

6. Under a matched pairs design, calculate the difference of the two responses for each pair. Then, analyze the differences using the results for the one sample problem. Also, the differences should be plotted to reveal any obvious outliers and, for moderate sample sizes, departures from normality.

## KEY IDEAS AND FORMULAS

In any comparative study of treatments, products, methods, and so on, the term treatment refers to the things being compared. The basic unit or object, to which one of the treatments is applied, is called an **experimental unit** or an **experimental subject**. The **response variable** is the characteristic that is recorded on each unit.

The specification of which treatment to compare and method of assigning experimental units is called the **experimental (or sampling) design**. The choice of appropriate statistical methods for making inferences depends heavily on the experimental design chosen for data collection.

A carefully designed experiment is fundamental to the success of a comparative study.

The most basic experimental designs to compare two treatments are **independent samples** and **matched pairs sample**.

The **independent samples design** requires the subjects to be randomly selected for assignment to each treatment. Randomization prevents uncontrolled factors from systematically favoring one treatment over the other.

With a matched pairs design, subjects in each pair are alike, while those in different pairs may be dissimilar. For each pair, the two treatments should be randomly allocated to the members.

The idea of matching or blocking experimental units is to remove a known source of variation from comparisons. Pairing subjects according to some feature prevents that source of variation from interfering with treatment comparisons. By contrast, random allocation of subjects according to the independent random sampling design spreads these variations between the two treatments.

### Inferences with Two Independent Random Samples

1. *Large samples.* When $n_1$ and $n_2$ are both greater than 30, inferences about $\mu_1 - \mu_2$ are based on the fact that

$$\frac{(\bar{X} - \bar{Y}) - (\mu_1 - \mu_2)}{\sqrt{\dfrac{S_1^2}{n_1} + \dfrac{S_2^2}{n_2}}} \quad \text{is approximately } N(0, 1)$$

A $100(1 - \alpha)\%$ confidence interval for $\mu_1 - \mu_2$ is

$$\bar{X} - \bar{Y} \pm z_{\alpha/2} \sqrt{\frac{S_1^2}{n_1} + \frac{S_2^2}{n_2}}$$

To test $H_0 : \mu_1 - \mu_2 = \delta_0$, we use the normal test statistic

$$Z = \frac{(\bar{X} - \bar{Y}) - \delta_0}{\sqrt{\dfrac{S_1^2}{n_1} + \dfrac{S_2^2}{n_2}}}$$

No assumptions are needed in regard to the shape of the population distributions.

2. *Small samples.* When $n_1$ and $n_2$ are small, inferences using the $t$ distribution require the assumptions:

(a) Both populations are normal.

(b) $\sigma_1 = \sigma_2$.

The common $\sigma^2$ is estimated by

$$S_{\text{pooled}}^2 = \frac{(n_1 - 1)S_1^2 + (n_2 - 1)S_2^2}{n_1 + n_2 - 2}$$

Inferences about $\mu_1 - \mu_2$ are based on

$$T = \frac{(\bar{X} - \bar{Y}) - (\mu_1 - \mu_2)}{S_{\text{pooled}} \sqrt{\dfrac{1}{n_1} + \dfrac{1}{n_2}}} \qquad \text{d.f.} = n_1 + n_2 - 2$$

A $100(1 - \alpha)\%$ confidence interval for $\mu_1 - \mu_2$ is

$$(\bar{X} - \bar{Y}) \pm t_{\alpha/2} \, S_{\text{pooled}} \sqrt{\frac{1}{n_1} + \frac{1}{n_2}}$$

To test $H_0: \mu_1 - \mu_2 = \delta_0$, the test statistic is

$$T = \frac{(\bar{X} - \bar{Y}) - \delta_0}{S_{\text{pooled}} \sqrt{\dfrac{1}{n_1} + \dfrac{1}{n_2}}} \qquad \text{d.f.} = n_1 + n_2 - 2$$

### Inferences with a Matched Pair Sample

With a paired sample $(X_1, Y_1), \ldots, (X_n, Y_n)$, the first step is to calculate the differences $D_i = X_i - Y_i$, their mean $\bar{D}$, and standard deviation $S_D$.

If $n$ is small, we assume that the $D_i$'s are normally distributed $N(\mu_D, \sigma_D)$. Inferences about $\mu_D$ are based on

$$T = \frac{\bar{D} - \mu_D}{S_D / \sqrt{n}} \qquad \text{d.f.} = n - 1$$

A $100(1 - \alpha)\%$ confidence interval for $\mu_D$ is

$$\bar{D} \pm t_{\alpha/2} \, S_D / \sqrt{n}$$

The test of $H_0: \mu_D = \mu_{D0}$ is performed with the test statistic:

$$T = \frac{\bar{D} - \mu_{D0}}{S_D / \sqrt{n}} \qquad \text{d.f.} = n - 1$$

If $n$ is large, the assumption of normal distribution for the $D_i$'s is not needed. Inferences are based on the fact that

$$Z = \frac{\bar{D} - \mu_{D0}}{S_D / \sqrt{n}} \qquad \text{is approximately } N(0, 1)$$

### Summary of Inferences about Means

Table 4 summarizes all of the statistical procedures we have considered for making inferences about (1) a single mean, (2) the difference of two means, or (3) a mean difference for a pair of observations.

### Comparing Two Binomial Proportions—Large Samples

Data:

$X = $ No. of successes in $n_1$ trials with success probability $P(S) = p_1$

$Y = $ No. of successes in $n_2$ trials with success probability $P(S) = p_2$

To test $H_0: p_1 = p_2$ versus $H_1: p_1 \neq p_2$, use the $Z$ test:

$$Z = \frac{\hat{p}_1 - \hat{p}_2}{\sqrt{\hat{p}(1 - \hat{p})} \sqrt{\dfrac{1}{n_1} + \dfrac{1}{n_2}}} \qquad \text{with} \quad R: |Z| \geq z_{\alpha/2}$$

**TABLE 4**  General Formulas for Inferences about a Mean $(\mu)$, Difference of Two Means $(\mu_1 - \mu_2)$

$$\text{Confidence interval} = \text{Point estimator} \pm (\text{Tabled value})(\text{Estimated or true std. dev.})$$

$$\text{Test statistic} = \frac{\text{Point estimator} - \text{Parameter value at } H_0 \text{ (null hypothesis)}}{(\text{Estimated or true}) \text{ std. dev. of point estimator}}$$

| | Ch. 8 | Ch. 9 | Ch. 10 | Independent Samples | | Matched Samples |
|---|---|---|---|---|---|---|
| | General | Normal with unknown $\sigma$ | Normal $N(\mu_1, \sigma_1), N(\mu_2, \sigma_2)$ $\sigma_1 = \sigma_2 = \sigma$ | Normal $N(\mu_1, \sigma_1), N(\mu_2, \sigma_2)$ $\sigma_1 \neq \sigma_2$ | General | Normal for the difference $D_i = X_i - Y_i$ |
| Population(s) | Mean $\mu$ | Mean $\mu$ | $\mu_1 - \mu_2$ | $\mu_1 - \mu_2$ | $\mu_1 - \mu_2$ | $\mu_D = \mu_1 - \mu_2$ |
| Inference on | | | | | | |
| Sample(s) | $X_1, \ldots, X_n$ | $X_1, \ldots, X_n$ | $X_1, \ldots, X_{n_1}$ $Y_1, \ldots, Y_{n_2}$ | $X_1, \ldots, X_{n_1}$ $Y_1, \ldots, Y_{n_2}$ | $X_1, \ldots, X_{n_1}$ $Y_1, \ldots, Y_{n_2}$ | $X_1 \; Y_1 \quad D_1 = X_1 - Y_1$ $X_2 \; Y_2 \quad D_2 = X_2 - Y_2$ $\cdots \qquad \cdots$ $X_n \; Y_n \quad D_n = X_n - Y_n$ |
| Sample size $n$ | Large $n > 30$ | $n \geq 2$ | $n_1 \geq 2$ $n_2 \geq 2$ | $n_1 \geq 2$ $n_2 \geq 2$ | $n_1 > 30$ $n_2 > 30$ | $n \geq 2$ |
| Point estimator | $\bar{X}$ | $\bar{X}$ | $\bar{X} - \bar{Y}$ | $\bar{X} - \bar{Y}$ | $\bar{X} - \bar{Y}$ | $\bar{D} = \bar{X} - \bar{Y}$ |
| Variance of point estimator | $\dfrac{\sigma^2}{n}$ | $\dfrac{\sigma^2}{n}$ | $\sigma^2 \left( \dfrac{1}{n_1} + \dfrac{1}{n_2} \right)$ | $\dfrac{\sigma_1^2}{n_1} + \dfrac{\sigma_2^2}{n_2}$ | $\dfrac{\sigma_1^2}{n_1} + \dfrac{\sigma_2^2}{n_2}$ | $\dfrac{\sigma_D^2}{n}$ |
| Std. dev. of point estimator | $\dfrac{\sigma}{\sqrt{n}}$ | $\dfrac{\sigma}{\sqrt{n}}$ | $\sigma \sqrt{\dfrac{1}{n_1} + \dfrac{1}{n_2}}$ | $\sqrt{\dfrac{\sigma_1^2}{n_1} + \dfrac{\sigma_2^2}{n_2}}$ | $\sqrt{\dfrac{\sigma_1^2}{n_1} + \dfrac{\sigma_2^2}{n_2}}$ | $\dfrac{\sigma_D}{\sqrt{n}}$ |
| Estimated std. dev. | $\dfrac{S}{\sqrt{n}}$ | $\dfrac{S}{\sqrt{n}}$ | $S_{pooled} \sqrt{\dfrac{1}{n_1} + \dfrac{1}{n_2}}$ | $\sqrt{\dfrac{S_1^2}{n_1} + \dfrac{S_2^2}{n_2}}$ | $\sqrt{\dfrac{S_1^2}{n_1} + \dfrac{S_2^2}{n_2}}$ | $\dfrac{S_D}{\sqrt{n}}$ |
| Distribution | Normal | $t$ with d.f. $= n - 1$ | $t$ with d.f. $= n_1 + n_2 - 2$ | $t$ with estimated d.f.[a] | Normal | $t$ with d.f. $= n - 1$ |
| Test statistic | $\dfrac{\bar{X} - \mu_0}{S/\sqrt{n}}$ | $\dfrac{\bar{X} - \mu_0}{S/\sqrt{n}}$ | $\dfrac{(\bar{X} - \bar{Y}) - \delta_0}{S_{pooled} \sqrt{\dfrac{1}{n_1} + \dfrac{1}{n_2}}}$ $S^2_{pooled} = \dfrac{(n_1 - 1)S_1^2 + (n_2 - 1)S_2^2}{n_1 + n_2 - 2}$ | $\dfrac{(\bar{X} - \bar{Y}) - \delta_0}{\sqrt{\dfrac{S_1^2}{n_1} + \dfrac{S_2^2}{n_2}}}$ | $\dfrac{(\bar{X} - \bar{Y}) - \delta_0}{\sqrt{\dfrac{S_1^2}{n_1} + \dfrac{S_2^2}{n_2}}}$ | $\dfrac{\bar{D} - \mu_{D0}}{S_D / \sqrt{n}}$ $S_D = $ sample std. dev. of the $D_i$'s |

[a] estimated by $\left( \dfrac{s_1^2}{n_1} + \dfrac{s_2^2}{n_2} \right)^2 \Bigg/ \left[ \dfrac{(s_1^2/n_1)^2}{(n_1 - 1)} + \dfrac{(s_2^2/n_2)^2}{(n_2 - 1)} \right]^2$

where

$$\widehat{p}_1 = \frac{X}{n_1} \qquad \widehat{p}_2 = \frac{Y}{n_2} \qquad \widehat{p} = \frac{X + Y}{n_1 + n_2}$$

To test $H_0 : p_1 = p_2$ versus $H_1 : p_1 > p_2$, use the $Z$ test with $R : Z \geq z_\alpha$. A $100(1 - \alpha)\%$ confidence interval for $p_1 - p_2$ is

$$(\widehat{p}_1 - \widehat{p}_2) \pm z_{\alpha/2} \sqrt{\frac{\widehat{p}_1(1 - \widehat{p}_1)}{n_1} + \frac{\widehat{p}_2(1 - \widehat{p}_2)}{n_2}}$$

## TECHNOLOGY

### Confidence intervals and tests for comparing means

### MINITAB

### Matched pair samples

We illustrate with the calculation of a 98% confidence interval and .02 level test.

Enter the first sample in *C1* and second sample in *C2*.

**Dialog box:**

**Stat > Basic Statistics > Paired t.**

Type *C1* in **First Sample** and *C2* in **Second Sample**.

Click **Options**. Type *98* in **Confidence level**.

Enter the value of the mean under the null hypothesis 0 in **Test mean** and choose *not equal* under the **Alternative**. Click **OK**. Click **OK**.

### Two-sample t tests and confidence intervals

We illustrate with the calculation of a 98% confidence interval for $\mu_1 - \mu_2$ and .02 level test of the null hypothesis of no difference in means.

**Data:**

*C1* First sample

*C2* Second sample

**Dialog box:**

**Stat > Basic Statistics > 2-Sample t.**

Select **Samples in different columns**. Type *C1* in **First** and *C2* in **Second**.

To pool the estimates of variance, click the box **Assume equal variances**. Click **Options**. Type *98* in **Confidence level**, the null hypothesis value of the mean 0 in **Test mean**, and select *not equal* under **Alternative**. Click **OK**. Click **OK**.

The approximate *t* procedure is used if the **assume equal variances** box is not clicked.

## EXCEL

### *Matched pairs t tests*

We illustrate with the calculation of a test of the null hypothesis $H_0: \mu_D = 0$ versus a two-sided alternative with $\alpha = .02$. Begin with values for the first variable in column A and the second in column B.

Click the **Data** tab and select **Data Analysis**.

Select **t-Test: Paired Two-Sample for Means**. Click **OK**.

With the cursor in **Variable 1 Range**, highlight the data in column A.

With the cursor in **Variable 2 Range**, highlight the data in column B.

Type the hypothesized value *0* after **Hypothesized Mean Difference**. Enter *.02* for **Alpha**. Click **OK**.

The program returns a summary that includes the value of the $t$ statistic, the $P$–value for a one-sided test (actually the smallest tail probability), and the $P$–value for a two-sided test.

### *Two-sample t tests*

We illustrate with a test of the null hypothesis of no difference between the two means. Begin with the first sample in column A and the second in column B.

Click the **Data** tab and select **Data Analysis**.

Select **t-Test: Two-Sample Assuming Unequal Variances**. Click **OK** or, to pool, Select **t-Test: Two-Sample Assuming Equal Variances** and click **OK**.

With the cursor in **Variable 1 Range**, highlight the data in column A.

With the cursor in **Variable 2 Range**, highlight the data in column B.

Type the hypothesized value *0* after **Hypothesized mean difference**.

Enter *.02* for **Alpha**. Click **OK**.

The program returns the one-sided and two-sided $P$–values.

## TI-84/-83 PLUS

### *Matched Pairs Samples*

### *Confidence intervals*

We illustrate the calculation of a 98% confidence interval. Start with the values for the first variable entered in $L_1$ and the second in $L_2$. Let $L_3 = L_2 - L_1$

Press **STAT** and select **TESTS** and then **8: Tinterval**. Press **ENTER**.

Select **Data** for **Inpt**. Set **List** to $L_3$ and **Freq** to 1.

Enter *.98* following **C–Level**: Select **Calculate**.

Then press **ENTER**.

### *Tests*

We illustrate the calculation of a test of the null hypothesis of 0 mean difference. Start with the first sample entered in $L_1$ and the second in $L_2$. Then let $L_3 = L_2 - L_1$ or $L_3 = L_1 - L_2$ depending on how the alternative is defined.

> Press **STAT** and select **TESTS** and then **2: T-Test**.
>
> Select **Data** for **Inpt**. Enter $0$ for $\mu_0$. Set **List** to $L_3$ and **Freq** to **1**.
>
> Select the direction of the alternative hypothesis for the mean difference. Select **Calculate**.
>
> Press **ENTER**.

If, instead, the sample size, mean, and standard deviation are available, the second step is

> Select **Stats** and type in the sample sizes, means, and standard deviations.

### Two-sample t tests

We illustrate with the calculation of a test of the null hypothesis of no difference between the two means. Start with the data entered in $L_1$ and $L_2$.

> Press **STAT** and select **TESTS** and then **4: 2-SampT-Test**.
>
> Select **Data** for **Inpt** and set **List1** to $L_1$, **List2** to $L_2$, **Freq1** to **1**, and **Freq2** to **1**.
>
> Select the direction of the alternative hypothesis. Set **Pooled** to *NO* if you do not wish to pool.
>
> Select **Calculate** or **Draw** and press **ENTER**.

The calculator will return the $P$–value. **Draw** will draw the $t$ distribution and shade the area of the $P$–value. If, instead, the sample sizes, means, and standard deviations are available, the second step is

> Select **Stats** and enter the sample sizes, means, and standard deviations.

### Two-sample confidence intervals

The command **0:2-SampT-Int** replaces **4:2-SampT-test**.

## REVIEW EXERCISES

**10.58** Refer to Table D.12 of the Data Bank. The responses, of persons in the first and second age groups, on the frequency of credit card use question, have the summary statistics

| Age | $n_i$ | $\bar{x}_i$ | $s_i$ |
|---|---|---|---|
| 19–24 | 18 | 3.000 | 1.609 |
| 25–34 | 28 | 4.429 | 1.643 |

(a) Obtain a 98% confidence interval for the difference in the two population mean frequency of credit card use scores.

(b) Test for equality of the two means versus a two-sided alternative. Use $\alpha = .02$.

**10.59** Two versions of a new simplified tax form, A1 and A2, need to be evaluated with respect to the time, in hours, it takes to complete the form. Forty persons were selected to fill out Form A1 and 40 were selected to fill out Form A2. The summary statistics are:

$$\text{Form A1:} \quad \bar{x} = 12.2, \quad s_1 = 1.1$$

$$\text{Form A2:} \quad \bar{y} = 7.2, \quad s_2 = 3.4$$

(a) Find a 95% confidence interval for the difference in means.

(b) If only 80 persons are available for this study, how would you choose the 40 to use form A1?

**10.60** Refer to Table D.12 of the Data Bank. The responses, of persons in the first and third age groups to the average monthly amount charged, have the summary statistics

| Age | $n_i$ | $\bar{x}_i$ | $s_i$ |
|---|---|---|---|
| 19–24 | 18 | 177.2 | 235.0 |
| 35–44 | 48 | 563.9 | 831.3 |

Determine a 95% confidence interval for the difference of means.

**10.61** Refer to the data in Exercise 10.60. Suppose you wish to establish that the mean response of 35–44 year olds is larger than that of 19–24 year olds by more than 150 dollars.

(a) Formulate the null hypothesis and the alternative hypothesis.

(b) State the test statistic and the rejection region with $\alpha = .05$.

(c) Perform the test at $\alpha = .05$. Also, find the P–value and comment.

**10.62** A study of postoperative pain relief is conducted to determine if drug $A$ has a significantly longer duration of pain relief than drug $B$. Observations of the hours of pain relief are recorded for 55 patients given drug $A$ and 58 patients given drug $B$. The summary statistics are

| | $A$ | $B$ |
|---|---|---|
| Mean | 4.64 | 4.03 |
| Standard deviation | 1.25 | 1.82 |

(a) Formulate $H_0$ and $H_1$.

(b) State the test statistic and the rejection region with $\alpha = .10$.

(c) State the conclusion of your test with $\alpha = .10$. Also, find the P–value and comment.

**10.63** Consider the data of Exercise 10.62.

(a) Construct a 90% confidence interval for $\mu_A - \mu_B$.

(b) Give a 95% confidence interval for $\mu_A$ using the data of drug $A$ alone. (*Note*: Refer to Chapter 8 page 317.)

**10.64** Obtain $s^2_{pooled}$ for the gaming data in the chapter opening.

**10.65** Given the following two samples,

$$8 \quad 11 \quad 5 \quad 9 \quad 7 \quad \text{and} \quad 5 \quad 3 \quad 4 \quad 8$$

obtain (a) $s^2_{pooled}$ and (b) the value of the $t$ statistic for testing $H_0 : \mu_1 - \mu_2 = 2$. State the d.f. of the $t$.

**10.66** A fruit grower wishes to evaluate a new spray that is claimed to reduce the loss due to damage by insects. To this end, he performs an experiment with 27 trees in his orchard by treating 12 of those trees with the new spray and the other 15 trees with the standard spray. From the data of fruit yield (in pounds) of those trees, the following summary statistics were found.

| | New Spray | Standard Spray |
|---|---|---|
| Mean yield | 249 | 233 |
| Standard deviation | 19 | 45 |

Do these data substantiate the claim that a higher yield should result from the use of the new spray? State the assumptions you make and test at $\alpha = .05$.

**10.67** Referring to Exercise 10.66, construct an approximate $t$ 95% confidence interval for the difference in mean yields between the new spray and the standard spray.

**10.68** Refer to Exercise 10.59 but suppose, instead, that the sample sizes are 8 and 7 with the same summary statistics:

$$\text{Form A1:} \quad \bar{x} = 12.2, \quad s_1 = 1.1$$

$$\text{Form A2:} \quad \bar{y} = 7.2, \quad s_2 = 3.4$$

(a) Test, with the approximate $t$ procedure, that the population means are different. Use $\alpha = .05$.

(b) What assumptions did you make for your answer?

**10.69** An investigation is conducted to determine if the mean age of welfare recipients differs between two cities $A$ and $B$. Random samples of 75 and 100 welfare recipients are selected from city $A$ and city $B$, respectively, and the following computations are made.

|                    | City $A$ | City $B$ |
|--------------------|----------|----------|
| Mean               | 37.8     | 43.2     |
| Standard deviation | 6.8      | 7.5      |

(a) Do the data provide strong evidence that the mean ages are different in city $A$ and city $B$? (Test at $\alpha = .02$.)

(b) Construct a 98% confidence interval for the difference in mean ages between $A$ and $B$.

(c) Construct a 98% confidence interval for the mean age for city $A$ and city $B$ individually. (*Note*: Refer to Chapter 8 page 317.)

**10.70** Refer to the data on the weight of wolves in Table D.9 of the Data Bank. A computer analysis produces the output

```
Two-sample T for Male wt vs Female wt
              N      Mean      StDev
Male wt      11      91.9      12.4
Female wt     8      73.9      10.1
T-Test of difference = 0 (vs not =):
T-Value = 3.38 P-Value = 0.004 DF = 17
```

(a) What is the conclusion to testing the equality of mean weights at level $\alpha = .05$?

(b) Test the null hypothesis that males weigh an average of 5 pounds more than females against a two-sided alternative. Take $\alpha = .05$.

**10.71** Refer to the computer attitude scores (CAS) of students given in Table D.4 of the Data Bank. A computer analysis produces the output

```
Two-sample T for CAS
sex     N      Mean      StDev
F      15      2.643     0.554
M      20      2.945     0.390
```

```
Difference = mu M — mu F
Estimate for difference: 0.302
90\% CI for the difference
(0.032, 0.571)
T-Test of difference = 0
(vs not  =):
T-Value  = 1.89 P-Value  = 0.067
DF  = 33
Both use Pooled StDev  = 0.467
```

(a) What is the conclusion to testing the equality of mean computer attitude scores at level $\alpha = .05$?

(b) Find a 95% confidence interval for the mean attitude score for males minus the mean attitude score for females.

(c) Test the null hypothesis that, on average, males score .1 lower than females against a two-sided alternative. Take $\alpha = .05$.

**10.72** To compare the crop yields from two strains of wheat, $A$ and $B$, an experiment was conducted at eight farms located in different parts of a state. At each farm, strain $A$ was grown on one plot and strain $B$ on another; all 16 plots were of equal sizes. Given below are data of yield in pounds per plot.

|          | Farm |    |    |    |    |    |    |    |
|----------|------|----|----|----|----|----|----|----|
|          | 1    | 2  | 3  | 4  | 5  | 6  | 7  | 8  |
| Strain $A$ | 23 | 39 | 19 | 43 | 33 | 29 | 28 | 42 |
| Strain $B$ | 18 | 33 | 21 | 34 | 33 | 20 | 21 | 40 |

(a) Is there strong evidence that strain $A$ has a higher mean yield than strain $B$? Test at $\alpha = .05$.

(b) What should be randomized in this experiment and how?

**10.73** Refer to the problem stated in Exercise 10.72, but now suppose that the study was conducted at 16 farms, of which 8 were selected for planting strain $A$ and the other 8 for strain $B$. Here also the plots used were all of equal sizes. Recorded below are the data of yields in pounds per plot.

| Strain $A$ | 23 | 39 | 19 | 28 | 42 | 43 | 33 | 29 |
|----------|----|----|----|----|----|----|----|----|
| Strain $B$ | 20 | 21 | 40 | 34 | 33 | 18 | 33 | 21 |

(a) Is there strong evidence that strain $A$ has a higher mean yield than strain $B$? State the assumptions you make and use $\alpha = .05$.

(b) What should be randomized in this experiment and how?

(c) Check to see that each data set here is just a scrambled form of the data set in Exercise 10.72. Briefly explain why the conclusion of your test is different in the two situations.

**10.74** In each of the following cases, how would you select the experimental units and conduct the experiment—matched pairs or independent samples?

(a) Compare the mileage obtained from two gasolines. Twelve SUVs of various sizes are available.

(b) Compare the drying times of two latex-based interior paints. Ten walls are available.

(c) Compare two methods of teaching swimming. Twenty five-year-old girls are available.

**10.75** A sample of river water is divided into two specimens. One is randomly selected to be sent to Lab $A$ and the other is sent to Lab $B$. This is repeated for a total of nine times. The measurement of suspended solids at Lab $B$ is subtracted from that of Lab $A$ to obtain the differences

12    10    15    42    11    −4    −2    10    −7

(a) Is there strong evidence that the mean difference is not zero? Test with $\alpha = .02$.

(b) Construct a 90% confidence interval for the mean difference of the suspended solids measurements.

**10.76** Two scales are available at a campus athletic facility. A student wonders if, on average, they give the same reading for weight. She and four others weigh themselves on both scales. The readings are

| Person | Scale 1 | Scale 2 |
|--------|---------|---------|
| 1 | 113.8 | 114.1 |
| 2 | 218.7 | 217.2 |
| 3 | 149.2 | 147.3 |
| 4 | 104.9 | 103.5 |
| 5 | 166.6 | 165.7 |

(a) Find a 95% confidence interval for the mean difference in scale readings.

(b) Based on your answer to part (a), what is the conclusion to testing, with $\alpha = .05$, that the mean difference is 0 versus a two-sided alternative hypothesis? Explain your reasoning.

(c) Explain how you would randomize in this experiment.

**10.77** An experiment is conducted to determine if the use of a special chemical additive with a standard fertilizer accelerates plant growth. Ten locations are included in the study. At each location, two plants growing in close proximity are treated. One is given the standard fertilizer, the other the standard fertilizer with the chemical additive. Plant growth after four weeks is measured in centimeters. Do the following data substantiate the claim that use of the chemical additive accelerates plant growth? State the assumptions that you make and devise an appropriate test of the hypothesis. Take $\alpha = .05$.

| | Location | | | | | | | | | |
|---|---|---|---|---|---|---|---|---|---|---|
| | 1 | 2 | 3 | 4 | 5 | 6 | 7 | 8 | 9 | 10 |
| Without additive | 20 | 31 | 17 | 22 | 19 | 32 | 25 | 18 | 21 | 19 |
| With additive | 23 | 34 | 16 | 21 | 22 | 31 | 29 | 20 | 25 | 23 |

**10.78** Referring to Exercise 10.77, suppose that the two plants at each location are situated in the east-west direction. In designing this experiment, you must decide which of the two plants at each location—the one in the east or the one in the west—is to be given the chemical additive.

(a) Explain how, by repeatedly tossing a coin, you can randomly allocate the treatments to the plants at the 10 locations.

(b) Perform the randomization by actually tossing a coin 10 times.

**10.79** Students can bike to a park on the other side of a lake by going around one side of the lake or the other. After much discussion about which was faster, they decided to perform an experiment. Among the

12 students available, 6 were randomly selected to follow Path *A* on one side of the lake and the rest followed Path *B* on the other side. They all went on different days so the conclusion would apply to a variety of conditions.

| | Travel Time (minutes) | | | | | |
|---|---|---|---|---|---|---|
| Path *A* | 10 | 12 | 15 | 11 | 16 | 11 |
| Path *B* | 12 | 15 | 17 | 13 | 18 | 16 |

(a) Is there a significant difference between the mean travel times between the two paths? State the assumptions you have made in performing the test.

(b) Suggest an alternative design for this study that would make the comparison more effective.

**10.80** Five pairs of tests are conducted to compare two methods of making rope. Each sample batch contains enough hemp to make two ropes. The tensile strength measurements are

| | Test | | | | |
|---|---|---|---|---|---|
| | 1 | 2 | 3 | 4 | 5 |
| Method 1 | 14 | 12 | 18 | 16 | 15 |
| Method 2 | 16 | 15 | 17 | 16 | 14 |

(a) Treat the data as 5 paired observations and calculate a 95% confidence interval for the mean difference in tensile strengths between ropes made by the two methods.

(b) Repeat the calculation of a 95% confidence interval treating the data as independent random samples.

(c) Briefly discuss the conditions under which each type of analysis would be appropriate.

**10.81** An experiment was conducted to study whether cloud seeding reduces the occurrences of hail. At a hail-prone geographical area, seeding was done on 50 stormy days and another 165 stormy days were also observed without seeding. The following counts were obtained.

| | Days | |
|---|---|---|
| | Seeded | Not Seeded |
| Hail | 7 | 43 |
| No hail | 43 | 122 |
| Total | 50 | 165 |

Do these data substantiate the conjecture that seeding reduces the chance of hail? (Answer by determining the $P$-value.)

**10.82** Example 16, Chapter 8, presents the result that 158 male babies were born to 363 parents both of whom smoke. The same study found that 2685 male babies were born to 5045 parents, neither of which smoked. Obtain a 95% confidence interval of the increase in the proportion of male births when nonsmokers are compared with smokers.

**10.83** A study (courtesy of R. Golubjatnikov) is undertaken to compare the rates of prevalence of CF antibody to parainfluenza I virus among boys and girls in the age group 5 to 9 years. Among 113 boys tested, 34 are found to have the antibody; among 139 girls tested, 54 have the antibody. Do the data provide strong evidence that the rate of prevalence of the antibody is significantly higher in girls than boys? Use $\alpha = .05$. Also, find the $P$-value.

**10.84** In the survey on which Table D.12 of the Data Bank is based, a large number of persons was each asked "Is it preferable to have 1% of your monthly charges donated to a charity of your choice rather than credited back to your account?" The full sample, larger than that in Table D.12, yielded the summary statistics for the number of yes responses.

| Gender | $x_i$ | $n_i$ | $\hat{p}_i$ |
|---|---|---|---|
| Female | 48 | 195 | .2462 |
| Male | 52 | 294 | .1769 |

(a) Do the data provide evidence that females are more willing than males to donate to charity, in terms of having a larger proportion that would say "yes" to the question? Test with $\alpha = .05$.

(b) Obtain a 95% confidence interval for the difference in population proportions.

## The Following Exercises Require a Computer

**10.85** We illustrate the MINITAB commands and output for the two-sample $t$ test. The approximate $t$ test is performed if you don't click **Assume equal variances**.

---

**Data**

C1: 5 2 8 3

C2: 8 9 6

**Dialog box:**

**Stat > Basic Statistics > 2-Sample t**

Click **Samples in different columns.**

Type C1 in **First**. C2 in **Second**.

Click **OK**. Click **Assume equal variances**.

Click **OK**.

---

```
Two-Sample T-Test and CI: C1, C2
Two-sample T for C1 vs C2

       N    Mean    StDev   SE Mean
C1     4    4.50    2.65       1.3
C2     3    7.67    1.53      0.88

Difference = mu (C1) — mu (C2)
Estimate for difference:  -3.17
95% CI for difference:
(-7.61, 1.28)
T-Test of difference = 0 (vs
not  =): T-Value =   -1.83
P-value =  0.127 DF =  5
Both use Pooled StDev = 2.2657
```

(a) From the output, what is the conclusion to testing $H_0: \mu_1 - \mu_2 = 0$ versus a two-sided alternative at level $\alpha = .05$?

(b) Refer to Table 12 of the Data Bank. Find a 97% confidence interval for the difference of mean monthly amount charged by the first and second age groups.

**10.86** Refer to the alligator data in Table D.11 of the Data Bank. Using the data on testosterone $x_4$ from the Lake Apopka alligators, find a 95% confidence interval for the difference of means between males and females. There should be a large difference for healthy alligators. Comment on your conclusion.

**10.87** Refer to the alligator data in Table D.11 of the Data Bank. Using the data on testosterone $x_4$ for male alligators, compare the means for the two lake regions.

(a) Should you pool the variances with these data?

(b) Find a 90% confidence interval for the difference of means between the two lakes. Use the approximate $t$ procedure on page 413.

(c) Which population has the highest mean and how much higher is it?

**10.88** Refer to the bear data in Table D.6 of the Data Bank. Compare the head widths of males and females by obtaining a 95% confidence interval for the difference of means and also test equality versus a two-sided alternative with $\alpha = .05$.

**10.89** Refer to the marine growth of salmon data in Table D.7 of the Data Bank. Compare the mean growth of males and females by obtaining a 95% confidence interval for the difference of means and also test equality versus a two-sided alternative with $\alpha = .05$.

**10.90** Refer to the physical fitness data in Table D.5 of the Data Bank. Find a 95% confidence interval for the mean difference of the pretest minus posttest number of situps. Also test that the mean difference is zero versus a two-sided alternative with $\alpha = .05$.

**10.91** Refer to the physical fitness data in Table D.5 of the Data Bank. Find a 95% confidence interval for the mean difference of the pretest minus posttest time to complete the rowing test. Also test that the mean difference is zero versus a two-sided alternative with $\alpha = .05$.

# 11

# Regression Analysis I
# Simple Linear Regression

© Rafael Macia/Science Source

## The Highest Roller Coasters Are Fastest

Some roller coasters are designed to twist riders and turn them upside down.
Others are designed to provide fast rides over large drops. Among the 12 tallest
roller coasters in the world, the maximum height (feet) is related to top speed

(miles per hour). Each data point, consisting of the pair of values (height, speed), represents one roller coaster. The fitted line predicts an increase in top speed of .19 miles per hour for each foot of height, or 19 miles per hour for each 100 feet in height.

This linear relation can be used to predict the top speed of future roller coasters. For instance, that of the next 410 foot roller coaster.

Speed = 33.45 + 0.1929 Height

Except for the brief treatment in Sections 4 and 5 of Chapter 3, we have only discussed statistical inferences based on the sample measurements of a single variable. In many investigations, two or more variables are observed for each experimental unit in order to determine:

1. Whether the variables are related.
2. How strong the relationships appear to be.
3. Whether one variable of primary interest can be predicted from observations on the others.

To address these issues, we review and then expand the treatment in Chapter 3, which is restricted to a descriptive viewpoint.

In this chapter, we take the additional step of including the omnipresent random variation as an error term in the model. Then, on the basis of the model, we can test whether one variable actually influences the other. Further, we produce confidence interval answers when using the estimated straight line for prediction. The correlation coefficient is shown to measure the strength of the linear relationship.

# 1. REGRESSION WITH A SINGLE PREDICTOR

**Regression analysis** concerns the study of relationships between quantitative variables with the object of identifying, estimating, and validating the relationship. The estimated relationship can then be used to predict one variable from the value of the other variables(s).

In this chapter, we confine our attention to study of the relation between two variables. A regression problem involving a single predictor is also called simple regression.

One may be curious about why the study of relationships of variables has been given the rather unusual name "regression." Historically, the word regression was first used in its present technical context by a British scientist, Sir Francis Galton, who analyzed the heights of sons and the average heights of their parents. From his observations, Galton concluded that sons of very tall (short) parents were generally taller (shorter) than the average but not as tall (short) as their parents. This result was published in 1885 under the title "Regression Toward Mediocrity in Hereditary Stature." In this context, "regression toward mediocrity" meant that the sons' heights tended to revert toward the average rather than progress to more extremes. However, in the course of time, the word regression became synonymous with the statistical study of relation among variables.

## 1.1 HOW DOES RANDOMNESS IMPACT THE STUDY OF A RELATION?

Studies of relation among variables abound in virtually all disciplines of science and the humanities. We outline just a few illustrative situations in order to bring the object of regression analysis into sharp focus. The examples progress from a case where beforehand there is an underlying straight line model that is masked by random disturbances to a case where the data may or may not reveal some relationship along a line or curve.

**Example 1** A Straight Line Model Masked by Random Disturbances

A factory manufactures items in batches and the production manager wishes to relate the production cost $y$ of a batch to the batch size $x$. Certain costs are practically constant, regardless of the batch size $x$. Building costs and administrative and supervisory salaries are some examples. Let us denote the fixed costs collectively by $F$. Certain other costs may be directly proportional to the number of units produced. For example, both the raw materials and labor required to produce the product are included in this category. Let C denote the cost of producing one item. In the absence of any other factors, we can then expect to have the relation

$$y = F + Cx$$

In reality, other factors also affect the production cost, often in unpredictable ways. Machines occasionally break down and result in lost time and added expenses for repair. Variation of the quality of the raw materials may also cause occasional slowdown of the production process. Thus, an ideal relation can be masked by random disturbances. Consequently, the relationship between $y$ and $x$ must be investigated by a statistical analysis of the cost and batch-size data.

**Example 2**   Expect an Increasing Relation but Not Necessarily a Straight Line

Suppose that the yield $y$ of tomato plants in an agricultural experiment is to be studied in relation to the dosage $x$ of a certain fertilizer, while other contributing factors such as irrigation and soil dressing are to remain as constant as possible. The experiment consists of applying different dosages of the fertilizer, over the range of interest, in different plots and then recording the tomato yield from these plots. Different dosages of the fertilizer will typically produce different yields, but the relationship is not expected to follow a precise mathematical formula. Aside from unpredictable chance variations, the underlying form of the relation is not known.

**Example 3**   A Scatter Diagram May Reveal an Empirical Relation

The aptitude of a newly trained operator for performing a skilled job depends on both the duration of the training period and the nature of the training program. To evaluate the effectiveness of the training program, we must conduct an experimental study of the relation between growth in skill or learning $y$ and duration $x$ of the training. It is too much to expect a precise mathematical relation simply because no two human beings are exactly alike. However, an analysis of the data of the two variables could help us to assess the nature of the relation and utilize it in evaluating a training program.

In each of these three cases, for any fixed value $x$, randomness causes the observed values of the response to spread over a range of values. Consequently, any relation with $x$ must actually be formulated in terms of the expected value of the response variable and how it varies as $x$ changes.

These examples illustrate the simplest settings for regression analysis where one wishes to determine how one variable is related to one other variable. In more complex situations several variables may be interrelated, or one variable of major interest may depend on several influencing variables. Regression analysis extends to these multivariate problems. (See Section 2, Chapter 12.)

RAJ

Fine but you are an exception. Statistics,[1] based on extensive data, confirm that earnings typically increase with each additional step in education.

Even though randomness is omnipresent, regression analysis allows us to identify it and estimate relationships.

## 1.2  THE FIRST STEP IN REGRESSION ANALYSIS — PLOT THE DATA

Our aim is to study the relation between two variables $x$ and $y$ and use it to predict $y$ from $x$. The variable $x$ acts as an independent variable whose values are controlled by the experimenter. The variable $y$ depends on $x$ and is also subjected to unaccountable variations or errors.

---

### Notation

$x$ =  independent variable, also called predictor variable, explanatory variable, causal variable, or input variable
$y$ =  dependent or response variable

---

For clarity, we introduce the main ideas of regression in the context of a specific experiment. This experiment, described in Example 4, and the data set of Table 1 will be referred to throughout this chapter. By so doing, we provide a flavor of the subject matter interpretation of the various inferences associated with a regression analysis.

**Example 4**  Relief from Symptoms of Allergy Related to Dosage

In one stage of the development of a new drug for an allergy, an experiment is conducted to study how different dosages of the drug affect the duration of relief from the allergic symptoms. Ten patients are included in the experiment. Each patient receives a specified dosage of the drug and is asked to report back as soon as the protection of the drug seems to wear off. The observations are recorded in Table 1, which shows the dosage $x$ and duration of relief $y$ for the 10 patients.

[1]Median weekly earnings in 2012, Bureau of Labor Statistics, Current Population Survey.

**TABLE 1**  Dosage $x$ (in milligrams) and the
Number of Hours of Relief $y$ from
Allergy for Ten Patients

| Dosage $x$ | Duration of Relief $y$ |
|---|---|
| 3 | 9 |
| 3 | 5 |
| 4 | 12 |
| 5 | 9 |
| 6 | 14 |
| 6 | 16 |
| 7 | 22 |
| 8 | 18 |
| 8 | 24 |
| 9 | 22 |

Seven different dosages are used in the experiment, and some of these are repeated for more than one patient. A glance at the table shows that $y$ generally increases with $x$, but it is difficult to say much more about the form of the relation simply by looking at this tabular data.

For a generic experiment, we use $n$ to denote the sample size or the number of runs of the experiment. Each run gives a pair of observations $(x, y)$ in which $x$ is the fixed setting of the independent variable and $y$ denotes the corresponding response. See Table 2.

**TABLE 2**  Data Structure for a Simple Regression

| Setting of the Independent Variable | Response |
|---|---|
| $x_1$ | $y_1$ |
| $x_2$ | $y_2$ |
| $x_3$ | $y_3$ |
| . | . |
| . | . |
| . | . |
| $x_n$ | $y_n$ |

We always begin our analysis by plotting the data because the eye can easily detect patterns along a line or curve.

### First Step in the Analysis

Plotting a scatter diagram is an important preliminary step prior to undertaking a formal statistical analysis of the relationship between two variables.

The existence of any increasing, or decreasing, relationship is readily apparent and preliminary judgments can be reached whether or not it is a straight line relation.

The scatter diagram of the observations in Table 1 appears in Figure 1. This scatter diagram reveals that the relationship is approximately linear in nature; that is, the points seem to cluster around a straight line. Because a linear relation is the simplest relationship to handle mathematically, we present the details of the statistical regression analysis for this case. Other situations can often be reduced to this case by applying a suitable transformation to one or both variables.

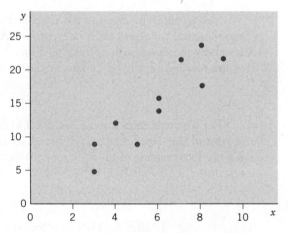

Figure 1    Scatter diagram of the data of Table 1.

## 2.  A STRAIGHT LINE REGRESSION MODEL

Recall that if the relation between $y$ and $x$ is exactly a straight line, then the variables are connected by the formula

$$y = \beta_0 + \beta_1 x$$

where $\beta_0$ indicates the intercept of the line with the $y$ axis, and $\beta_1$ represents the slope of the line, or the change in $y$ per unit change in $x$ (see Figure 2).

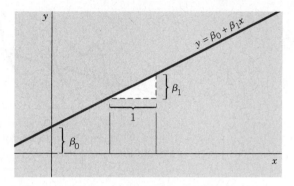

Figure 2   Graph of straight line $y = \beta_0 + \beta_1 x$.

Statistical ideas must be introduced into the study of relation when the points in a scatter diagram do not lie perfectly on a line, as in Figure 1. We think of these data as observations on an underlying linear relation that is being masked by random disturbances or experimental errors due in part to differences in severity of allergy, physical condition of subjects, their environment, and so on. All of the variables that influence the response, days of relief, are not even known, yet alone measured. The effects of all these variables are modeled as unobservable random variables. Given this viewpoint, we formulate the following linear regression model as a tentative representation of the mode of relationship between $y$ and $x$.

---

### Statistical Model for a Straight Line Regression

We assume that the response $Y$ is a random variable that is related to the input variable $x$ by

$$Y_i = \beta_0 + \beta_1 x_i + e_i \qquad i = 1, \ldots, n$$

where:
1. $Y_i$ denotes the response corresponding to the $i$th experimental run in which the input variable $x$ is set at the value $x_i$.
2. $e_1, \ldots, e_n$ are the unknown error components that are superimposed on the true linear relation. These are **unobservable random variables**, which we assume are independently and normally distributed with mean zero and an unknown standard deviation $\sigma$.
3. The parameters $\beta_0$ and $\beta_1$, which together locate the straight line, are unknown.

---

The mean of the response $Y_i$, corresponding to the level $x_i$ of the controlled variable, is $\beta_0 + \beta_1 x_i$.

Figure 3   Normal distributions of $Y$ with means on a straight line.

Analogous to using the Greek letter $\mu$ to denote a single population mean, we use the Greek letters $\beta_0$ and $\beta_1$ to denote the slope and intercept of the unknown line on which the means lie as $x$ varies.

Further, according to this model, the observation $Y_i$ is one observation from the normal distribution with mean $\beta_0 + \beta_1 x_i$ and standard deviation $\sigma$. One interpretation of this is that as we attempt to observe the true value on the line, nature adds the random error $e$ to this quantity. This statistical model is illustrated in Figure 3, which shows a few normal distributions for the response variable $Y$ for different values of the input variable $x$. All these distributions have the same standard deviation and their means lie on the unknown true straight line $\beta_0 + \beta_1 x$. Aside from the fact that $\sigma$ is unknown, the line on which the means of these normal distributions are located is also unknown. In fact, an important objective of the statistical analysis is to estimate this line.

## Exercises

**11.1**  Plot the line $y = 2 + 3x$ by locating the points for $x = 1$ and $x = 4$. What is its intercept? What is its slope?

**11.2**  A store manager has determined that the monthly profit $y$ realized from selling a particular brand of car battery is given by

$$y = 12x - 75$$

where $x$ denotes the number of these batteries sold in a month.

(a)  If 41 batteries were sold in a month, what was the profit?

(b)  At least how many batteries must be sold in a month in order to make a profit?

**11.3**  Identify the predictor variable $x$ and the response variable $y$ in each of the following situations.

(a)  A training director wishes to study the relationship between the duration of training for new recruits and their performance in a skilled job.

(b)  The aim of a study is to relate the carbon monoxide level in blood samples from smokers with the average number of cigarettes they smoke per day.

(c)  An agronomist wishes to investigate the growth rate of a fungus in relation to the level of humidity in the environment.

(d)  A market analyst wishes to relate the expenditures incurred in promoting a product in test markets and the subsequent amount of product sales.

**11.4**  Identify the values of the parameters $\beta_0$, $\beta_1$, and $\sigma$ in the statistical model

$$Y = 4 + 2x + e$$

where $e$ is a normal random variable with mean 0 and standard deviation 5.

**11.5** Identify the values of the parameters $\beta_0$, $\beta_1$, and $\sigma$ in the statistical model

$$Y = 6 - 3x + e$$

where $e$ is a normal random variable with mean 0 and standard deviation 3.

**11.6** Under the linear regression model:

(a) Determine the mean and standard deviation of $Y$, for $x = 4$, when $\beta_0 = 1$, $\beta_1 = 3$, and $\sigma = 2$.

(b) Repeat part (a) with $x = 2$.

**11.7** Under the linear regression model:

(a) Determine the mean and standard deviation of $Y$, for $x = 1$, when $\beta_0 = 3$, $\beta_1 = -2$, and $\sigma = 3$.

(b) Repeat part (a) with $x = 2$.

**11.8** Graph the straight line for the means of the linear regression model

$$Y = \beta_0 + \beta_1 x + e$$

having $\beta_0 = -3$, $\beta_1 = 4$.

**11.9** Graph the straight line for the means of a linear regression model $Y = \beta_0 + \beta_1 x + e$ having $\beta_0 = 7$ and $\beta_1 = 2$.

**11.10** Consider the linear regression model

$$Y = \beta_0 + \beta_1 x + e$$

where $\beta_0 = -2$, $\beta_1 = -1$, and the normal random variable $e$ has standard deviation 3.

(a) What is the mean of the response $Y$ when $x = 3$? When $x = 5$?

(b) Will the response at $x = 3$ always be larger than that at $x = 6$? Explain.

**11.11** Consider the following linear regression model

$$Y = \beta_0 + \beta_1 x + e$$

where $\beta_0 = 4$, $\beta_1 = 3$, and the normal random variable $e$ has the standard deviation 4.

(a) What is the mean of the response $Y$ when $x = 4$? When $x = 5$?

(b) Will the response at $x = 5$ always be larger than that at $x = 4$? Explain.

## 3. THE METHOD OF LEAST SQUARES

Let us tentatively assume that the preceding formulation of the straight line model is correct. We can then proceed to estimate the regression line and solve a few related inference problems. The problem of estimating the regression parameters $\beta_0$ and $\beta_1$ can be viewed as fitting the best straight line to the $y$ to $x$ relationship exhibited in the scatter diagram.

One can draw a line by eyeballing the scatter diagram, but such a judgment may be open to dispute. Moreover, statistical inferences cannot be based on a line that is estimated subjectively. On the other hand, the **method of least squares** is an objective and efficient method of determining the best fitting straight line. Moreover, this method is quite versatile because its application extends beyond the simple **straight line regression model**.

Suppose that an arbitrary line $y = b_0 + b_1 x$ is drawn on the scatter diagram as it is in Figure 4. In contrast to the Greek letters that locate the true line, we use the Latin letters $b_0$ and $b_1$ to denote the intercept and slope of an arbitrary, or trial, line. At the value $x_i$ of the independent variable, the $y$ value predicted by this line is $b_0 + b_1 x_i$ whereas the observed value is $y_i$. The discrepancy between the observed and predicted $y$ values is then $y_i - b_0 - b_1 x_i = d_i$, which is the **vertical** distance of the point from the line.

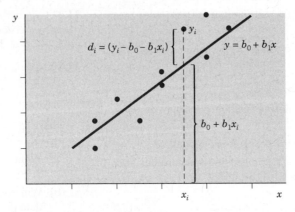

Figure 4 Deviations of the observations from a line $y = b_0 + b_1 x$.

We square each discrepancy and then take their sum

$$D = \sum_{i=1}^{n} d_i^2 = \sum_{i=1}^{n} (y_i - b_0 - b_1 x_i)^2$$

as an overall measure of the discrepancy of the observed points from the trial line $y = b_0 + b_1 x$. The magnitude of $D$ obviously depends on the line that is drawn. In other words, it depends on the parameters $b_0$ and $b_1$, the two quantities that determine the trial line. A good fit will make $D$ as small as possible. We now state the principle of least squares in general terms to indicate its usefulness to fitting many other models.

---

### The Principle of Least Squares

Determine the values for the parameters so that the overall discrepancy

$$D = \sum (\text{Observed response} - \text{Predicted response})^2$$

is minimized.
    The parameter values thus determined are called the **least squares estimates**.

---

For the straight line model, the least squares principle involves the determination of $b_0$ and $b_1$ to minimize.

$$D = \sum_{i=1}^{n} (y_i - b_0 - b_1 x_i)^2$$

The particular values $b_0$ and $b_1$ that minimize the sum of squares are denoted by $\widehat{\beta}_0$ and $\widehat{\beta}_1$, respectively. The hat notation, or $\widehat{\phantom{x}}$, over a parameter indicates that this random variable is an estimate of the corresponding parameter. They are called the least squares estimates of the regression parameters $\beta_0$ and $\beta_1$. The best fitting straight line or best fitting regression line is then given by the equation

$$\widehat{y} = \widehat{\beta}_0 + \widehat{\beta}_1 x$$

where the hat over $y$ indicates that it is an estimated quantity obtained by the least squares method.

　　To describe the formulas for the least squares estimators, we first introduce some basic notation.

---

**Basic Notation**

$$\bar{x} = \frac{1}{n}\sum x \qquad \bar{y} = \frac{1}{n}\sum y$$

$$S_{xx} = \sum (x - \bar{x})^2 = \sum x^2 - \frac{\left(\sum x\right)^2}{n}$$

$$S_{yy} = \sum (y - \bar{y})^2 = \sum y^2 - \frac{\left(\sum y\right)^2}{n}$$

$$S_{xy} = \sum (x - \bar{x})(y - \bar{y}) = \sum xy - \frac{\left(\sum x\right)\left(\sum y\right)}{n}$$

---

The quantities $\bar{x}$ and $\bar{y}$ are the sample means of the $x$ and $y$ values, $S_{xx}$ and $S_{yy}$ are the sums of squared deviations from the means, and $S_{xy}$ is the sum of the cross products of deviations. These five summary statistics are the key ingredients for calculating the least squares estimates and handling the inference problems associated with the linear regression model. (The reader may review Sections 4 and 5 of Chapter 3 where calculations of these statistics were illustrated.)

　　The formulas for the least squares estimators are

---

**Least squares estimator of $\beta_0$**

$$\widehat{\beta}_0 = \bar{y} - \widehat{\beta}_1 \bar{x}$$

**Least squares estimator of $\beta_1$**

$$\widehat{\beta}_1 = \frac{S_{xy}}{S_{xx}}$$

---

The estimates $\widehat{\beta}_0$ and $\widehat{\beta}_1$ can then be used to locate the best fitting line:

---

### Fitted (or estimated) regression line

$$\widehat{y} = \widehat{\beta}_0 + \widehat{\beta}_1 x$$

---

As we have already explained, this line provides the best fit to the data in the sense that the sum of squares of the deviations, or

$$\sum_{i=1}^{n} (y_i - \widehat{\beta}_0 - \widehat{\beta}_1 x_i)^2$$

is the smallest.

The individual deviations of the observations $y_i$ from the fitted values $\widehat{y}_i = \widehat{\beta}_0 + \widehat{\beta}_1 x_i$ are called the residuals, and we denote these by $\widehat{e}_i$.

---

### Residuals

$$\widehat{e}_i = y_i - \widehat{\beta}_0 - \widehat{\beta}_1 x_i \qquad i = 1, \ldots, n$$

---

Some residuals are positive and some negative, and a property of the least squares fit is that the **sum of the residuals is always zero**.

In Chapter 12, we discuss how the residuals can be used to check the assumptions of a regression model. For now, the sum of squares of the residuals is a quantity of interest because it leads to an estimate of the variance $\sigma^2$ of the error distributions illustrated in Figure 3. The residual sum of squares is also called the sum of squares due to error and is abbreviated as SSE.

---

The residual sum of squares or the sum of squares due to error is

$$\text{SSE} = \sum_{i=1}^{n} \widehat{e}_i^2 = S_{yy} - \frac{S_{xy}^2}{S_{xx}}$$

---

The second expression for SSE, which follows after some algebraic manipulations (see Exercise 11.24), is handy for directly calculating SSE. However, we stress the importance of determining the individual residuals for their role in model checking (see Section 3, Chapter 12).

An estimate of variance $\sigma^2$ is obtained by dividing SSE by $n - 2$. The reduction by 2 is because two degrees of freedom are lost from estimating the two parameters $\beta_0$ and $\beta_1$.

### Estimate of Variance

The estimator of the error variance $\sigma^2$ is

$$S^2 = \frac{\text{SSE}}{n - 2}$$

In applying the least squares method to a given data set, we first compute the basic quantities $\bar{x}$, $\bar{y}$, $S_{xx}$, $S_{yy}$, and $S_{xy}$. Then the preceding formulas can be used to obtain the least squares regression line, the residuals, and the value of SSE. Computations for the data given in Table 1 are illustrated in Table 3.

**TABLE 3**   Computations for the Least Squares Line, SSE, and Residuals Using the Data of Table 1

| $x$ | $y$ | $x^2$ | $y^2$ | $xy$ | $\widehat{\beta}_0 + \widehat{\beta}_1 x$ | Residual $\widehat{e}$ |
|---|---|---|---|---|---|---|
| 3 | 9 | 9 | 81 | 27 | 7.15 | 1.85 |
| 3 | 5 | 9 | 25 | 15 | 7.15 | −2.15 |
| 4 | 12 | 16 | 144 | 48 | 9.89 | 2.11 |
| 5 | 9 | 25 | 81 | 45 | 12.63 | −3.63 |
| 6 | 14 | 36 | 196 | 84 | 15.37 | −1.37 |
| 6 | 16 | 36 | 256 | 96 | 15.37 | .63 |
| 7 | 22 | 49 | 484 | 154 | 18.11 | 3.89 |
| 8 | 18 | 64 | 324 | 144 | 20.85 | −2.85 |
| 8 | 24 | 64 | 576 | 192 | 20.85 | 3.15 |
| 9 | 22 | 81 | 484 | 198 | 23.59 | −1.59 |
| Total    59 | 151 | 389 | 2651 | 1003 | | .04    (rounding error) |

$\bar{x} = 5.9, \quad \bar{y} = 15.1$

$\widehat{\beta}_1 = \dfrac{112.1}{40.9} = 2.74$

$S_{xx} = 389 - \dfrac{(59)^2}{10} = 40.9$

$\widehat{\beta}_0 = 15.1 - 2.74 \times 5.9 = -1.07$

$S_{yy} = 2651 - \dfrac{(151)^2}{10} = 370.9$

$\text{SSE} = 370.9 - \dfrac{(112.1)^2}{40.9} = 63.6528$

$S_{xy} = 1003 - \dfrac{59 \times 151}{10} = 112.1$

The data in the first two columns yield the next three columns. Then, the sum of entries in a column are obtained so $\bar{x}$, $\bar{y}$, $S_{xx}$, $S_{yy}$, and $S_{xy}$ can be calculated. From these, $\widehat{\beta}_0$, $\widehat{\beta}_1$, and SSE are obtained.

The equation of the line fitted by the least squares method is then

$$\widehat{y} = -1.07 + 2.74x$$

Figure 5 shows a plot of the data along with the fitted regression line.

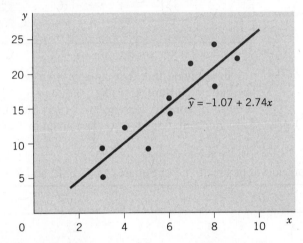

Figure 5    The least squares regression line for the data given in Table 1.

The residuals $\widehat{e}_i = y_i - \widehat{y}_i = y_i + 1.07 - 2.74x_i$ are computed in the last column of Table 3. The sum of squares of the residuals is

$$\sum_{i=1}^{n} \widehat{e}_i^{\,2} = (1.85)^2 + (-2.15)^2 + (2.11)^2 + \cdots + (-1.59)^2$$

$$= 63.653$$

which agrees with our previous calculations of SSE, except for the error due to rounding. Theoretically, the sum of the residuals should be zero, and the difference between the sum .04 and zero is also due to rounding.

The estimate of the variance $\sigma^2$ is

$$s^2 = \frac{\text{SSE}}{n-2} = \frac{63.6528}{8} = 7.96$$

The calculations involved in a regression analysis become increasingly tedious with larger data sets. Access to a computer proves to be a considerable advantage. Table 4 illustrates a part of the computer-based analysis of linear regression using the data of Example 4 and the MINITAB package. For a more complete regression analysis, see Table 5 on page 476.

**TABLE 4**  Regression Analysis of the Data in
Table 1, Example 4, Using MINITAB

**Data: C11T3**
C1: 3 3 4 5 6 6 7 8 8 9
C2: 9 5 12 9 14 16 22 18 24 22
**Dialog box:**
**Stat > Regression > Regression**
Type C2 in **Response**
Type C1 in **Predictors**. Click **OK**.
**Output:**
**Regression Analysis**
The regression equation is
$y = -1.07 + 2.74x$

---

**Example 5**  Predicting the Weight of Polar Bears from Body Measurements

Polar bears are among the world's largest land carnivores and males are typically larger than females. The U.S. Geological Survey monitors the health of the population. Monitoring procedures include making measurements on weight and girth at the heart, when the polar bear is sedated.

The measurements on $x = $ girth ( cm ) and $y = $ weight ( lb ) for $n = 30$ adult male bears, given in Table D.13 of the Data Bank, are summarized by

$$n = 30 \qquad \bar{x} = 149.737 \qquad \bar{y} = 837.533$$
$$\sum (x - \bar{x})^2 = 4367.99 \quad \sum (x - \bar{x})(y - \bar{y}) = 41915.5 \quad \sum (y - \bar{y})^2 = 492089$$

(a)  Obtain the equation of the fitted regression line.

(b)  Use the least squares line to obtain a point estimate of the weight of a bear which has girth 165 cm.

(c)  Specify two variables that likely contribute the error term $e$ in the model.

(d)  Calculate the residual sum of squares and estimate $\sigma^2$.

SOLUTION

(a)  We have

$$\widehat{\beta_1} = \frac{S_{xy}}{S_{xx}} = \frac{41915.5}{4367.99} = 9.596$$

$$\widehat{\beta_0} = \bar{y} - \widehat{\beta_1}\bar{x} = 837.533 - 9.596 \times 149.737 = -599.3$$

So the fitted regression line is $-599.3 + 9.596x$.

(b)  According to the straight line regression model, the weight of the new polar bear with girth 165 cm is given by $\beta_0 + \beta_1( 165 ) + e^*$

where the error $e^*$ has mean 0. Consequently, the expected value of the weight is $\beta_0 + \beta_1(165)$ and its least squares estimate

$$\widehat{\beta}_0 + \widehat{\beta}_1(165) = -599.3 + 9.596(165) = 984.0 \quad \text{pounds}$$

is the predicted weight.

   A good prediction equation permits the investigator to sometimes forgo weighing the polar bear in difficult circumstances when a large scale is not available.

(c) Any variable not specifically in the regression equation must be, by default, included in the error term. For instance, age would likely influence the weight as would the abundance of food in the previous six months.

(d)

$$\text{SSE} = S_{yy} - \frac{S_{xy}^2}{S_{xx}} = 492089 - \frac{41915.5^2}{4367.99} = 89865$$

and

$$s^2 = \frac{\text{SSE}}{n-2} = \frac{89865}{28} = 3209$$

Starting in Section 5, we use the estimate of $\sigma^2$ to set confidence intervals and conduct tests of hypotheses concerning $\beta_0$, $\beta_1$ and predicted values.

## Exercises

**11.12** A student collected data on the number of large pizzas consumed, $y$, while $x$ students are watching a professional football game on TV. Suppose that the data from five games are:

| $x$ | 2 | 5 | 6 | 3 | 4 |
|-----|---|---|---|---|---|
| $y$ | 1 | 6 | 10 | 3 | 5 |

(a) Construct a scatter diagram.

(b) Calculate $\bar{x}$, $\bar{y}$, $S_{xx}$, $S_{xy}$, and $S_{yy}$.

(c) Calculate the least squares estimates $\widehat{\beta}_0$ and $\widehat{\beta}_1$.

(d) Determine the fitted line and draw the line on the scatter diagram.

**11.13** The office manager at a real estate firm makes a pot of coffee every morning. The time before it runs out, $y$, in hours depends on the number of persons working inside that day, $x$. Suppose that the pairs of $(x, y)$ values from six days are:

| $x$ | 1 | 2 | 3 | 3 | 4 | 5 |
|-----|---|---|---|---|---|---|
| $y$ | 8 | 4 | 5 | 3 | 3 | 1 |

(a) Plot the scatter diagram.

(b) Calculate $\bar{x}$, $\bar{y}$, $S_{xx}$, $S_{xy}$, and $S_{yy}$.

(c) Calculate the least squares estimates $\widehat{\beta}_0$ and $\widehat{\beta}_1$.

(d) Determine the fitted line and draw the line on the scatter diagram.

**11.14** Refer to Exercise 11.12.

(a) Find the residuals and verify that they sum to zero.

(b) Calculate the residual sum of squares SSE by
  (i) Adding the squares of the residuals.
  (ii) Using the formula $\text{SSE} = S_{yy} - S_{xy}^2 / S_{xx}$

(c) Obtain the estimate of $\sigma^2$.

**11.15** Refer to Exercise 11.13.

(a) Find the residuals and verify that they sum to zero.

(b) Calculate the residual sums of squares SSE by
  (i) Adding the squares of the residuals.
  (ii) Using the formula SSE $= S_{yy} - S_{xy}^2 / S_{xx}$

(c) Obtain the estimate of $\sigma^2$.

**11.16** A help desk devoted to student software problems also receives phone calls. The number of persons that can be served in person, within one hour, is the response y. The predictor variable, x, is the number of phone calls answered.

| x | 0 | 1 | 2 | 3 | 4 |
|---|---|---|---|---|---|
| y | 7 | 8 | 5 | 4 | 1 |

(a) Calculate $\bar{x}$, $\bar{y}$, $S_{xx}$, $S_{xy}$, and $S_{yy}$.

(b) Calculate the least squares estimates $\widehat{\beta}_0$ and $\widehat{\beta}_1$.

(c) Determine the fitted line.

(d) Use the fitted line to give a point estimate of the predicted number of persons served when 3 calls are answered.

**11.17** A student hourly employee does small secretarial projects. The number of projects she completes in a day is the response variable y. The number of hours she works in a day is the predictor variable x.

| x | 1 | 2 | 4 | 6 | 7 |
|---|---|---|---|---|---|
| y | 4 | 3 | 6 | 8 | 9 |

(a) Calculate $\bar{x}$, $\bar{y}$, $S_{xx}$, $S_{xy}$, and $S_{yy}$.

(b) Calculate the least squares estimates $\widehat{\beta}_0$ and $\widehat{\beta}_1$.

(c) Determine the fitted line.

(d) Use the fitted line to give a point estimate of the predicted number of projects completed when she works 6.5 hours.

**11.18** To be able to better predict the output of an oil field, researchers[2] collected data that included y = the yearly output (100,000 barrels) and the number

of new wells drilled (1000) in the previous year. The summary statistics are

$$n = 18 \qquad \bar{x} = 14.85 \qquad \bar{y} = 4.685$$

$$\sum (x - \bar{x})^2 = 1032.87$$

$$\sum (x - \bar{x})(y - \bar{y}) = 311.164$$

$$\sum (y - \bar{y})^2 = 96.1544$$

(a) Obtain the equation of the best fitting straight line.

(b) Calculate the residual sum of squares.

(c) Estimate $\sigma^2$.

**11.19** To extend known results on the social hierarchy of monkeys, researchers[3] scanned n = 14 healthy persons using positron emission tomography to image dopamine type 2/3 in the brain. The social status of each person was assessed using the Bartlett Simplified Measure of Social Status. Letting x = dopamine binding potential and y = social status, the data can be summarized by

$$n = 14 \qquad \bar{x} = 2.526 \qquad \bar{y} = 33.416$$

$$\sum (x - \bar{x})^2 = 1.607$$

$$\sum (x - \bar{x})(y - \bar{y}) = 19.305$$

$$\sum (y - \bar{y})^2 = 448.839$$

Using these summary statistics

(a) Obtain the equation of the best fitting straight line.

(b) Calculate the residual sum of squares.

(c) Estimate $\sigma^2$.

**11.20** The data on female wolves in Table D.9 of the Data Bank concerning body weight (lb) and body length (cm) are

| Weight | 57 | 84 | 90 | 71 | 77 | 68 | 73 |
|---|---|---|---|---|---|---|---|
| Body length | 123 | 129 | 143 | 125 | 122 | 125 | 122 |

(a) Obtain the least squares fit of body weight to the predictor body length.

[2] I. Mustafar and R. Razali, "A study of prediction of output in oil field using multiple linear regression," *International Journal of Applied Science and Technology* 1 (4), (2011), pp. 107–113.
[3] Data from Figure 1, D. Martinez et al., "Dopamine type 2/3 receptor availability in the striatum and social status in humans," *Biological Psychiatry* 67 (1), (2010), pp. 275–278.

(b) Calculate the residual sum of squares.

(c) Estimate $\sigma^2$.

**11.21** Refer to the data on female wolves in Exercise 11.20.

(a) Obtain the least squares fit of body length to the predictor body weight.

(b) Calculate the residual sum of squares.

(c) Estimate $\sigma^2$.

(d) Compare your answer in part (a) with your answer to part (a) of Exercise 11.20. Should the two answers be the same? Why or why not?

**11.22** Using the formulas of $\widehat{\beta}_1$ and SSE, show that SSE can also be expressed as

(a) SSE $= S_{yy} - \widehat{\beta}_1 S_{xy}$

(b) SSE $= S_{yy} - \widehat{\beta}_1^2 S_{xx}$

**11.23** Referring to the formulas of $\widehat{\beta}_0$ and $\widehat{\beta}_1$, show that the point $(\bar{x}, \bar{y})$ lies on the fitted regression line.

**11.24** To see why the residuals always sum to zero, refer to the formulas of $\widehat{\beta}_0$ and $\widehat{\beta}_1$ and verify that

(a) The predicted values are

$$\widehat{y}_i = \bar{y} + \widehat{\beta}_1 (x_i - \bar{x})$$

(b) The residuals are

$$\widehat{e}_i = y_i - \widehat{y}_i = (y_i - \bar{y}) - \widehat{\beta}_1 (x_i - \bar{x})$$

Then show that $\sum\limits_{i=1}^{n} \widehat{e}_i = 0$

(c) Verify that $\sum\limits_{i=1}^{n} \widehat{e}_i^2 = S_{yy} + \widehat{\beta}_1^2 S_{xx} - 2\widehat{\beta}_1 S_{xy} = S_{yy} - S_{xy}^2 / S_{xx}$

# 4. THE SAMPLING VARIABILITY OF THE LEAST SQUARES ESTIMATORS—TOOLS FOR INFERENCE

It is important to remember that the line $\widehat{y} = \widehat{\beta}_0 + \widehat{\beta}_1 x$ obtained by the principle of least squares is an **estimate** of the unknown true regression line $y = \beta_0 + \beta_1 x$. In our drug evaluation problem (Example 4), the estimated line is

$$\widehat{y} = -1.07 + 2.74x$$

Its slope $\widehat{\beta}_1 = 2.74$ suggests that the mean duration of relief increases by 2.74 hours for each unit dosage of the drug. Also, if we were to estimate the expected duration of relief for a specified dosage $x^* = 4.5$ milligrams, we would naturally use the fitted regression line to calculate the estimate $-1.07 + 2.74 \times 4.5 = 11.26$ hours. A few questions concerning these estimates naturally arise at this point.

1. In light of the value 2.74 for $\widehat{\beta}_1$, could the slope $\beta_1$ of the true regression line be as much as 4? Could it be zero so that the true regression line is $y = \beta_0$, which does not depend on $x$? What are the plausible values for $\beta_1$?

2. How much uncertainty should be attached to the estimated duration of 11.26 hours corresponding to the given dosage $x^* = 4.5$?

To answer these and related questions, we must know something about the sampling distributions of the least squares estimators. These sampling distributions will enable us to test hypotheses and set confidence intervals for the parameters $\beta_0$ and $\beta_1$ that determine the straight line and for the straight line itself. Again, the $t$ distribution is relevant.

1. The standard deviations (also called standard errors) of the least squares estimators are

$$\text{S.E.}(\widehat{\beta}_1) = \frac{\sigma}{\sqrt{S_{xx}}} \qquad \text{S.E.}(\widehat{\beta}_0) = \sigma\sqrt{\frac{1}{n} + \frac{\bar{x}^2}{S_{xx}}}$$

To estimate the standard error, use

$$S = \sqrt{\frac{\text{SSE}}{n-2}} \quad \text{in place of } \sigma$$

2. Inferences about the slope $\beta_1$ are based on the $t$ distribution

$$T = \frac{\widehat{\beta}_1 - \beta_1}{S/\sqrt{S_{xx}}} \qquad \text{d.f.} = n - 2$$

Inferences about the intercept $\beta_0$ are based on the $t$ distribution

$$T = \frac{\widehat{\beta}_0 - \beta_0}{S\sqrt{\dfrac{1}{n} + \dfrac{\bar{x}^2}{S_{xx}}}} \qquad \text{d.f.} = n - 2$$

3. At a specified value $x = x^*$, the expected response is $\beta_0 + \beta_1 x^*$. This is estimated by $\widehat{\beta}_0 + \widehat{\beta}_1 x^*$ with

Estimated standard error

$$S\sqrt{\frac{1}{n} + \frac{(x^* - \bar{x})^2}{S_{xx}}}$$

Inferences about $\beta_0 + \beta_1 x^*$ are based on the $t$ distribution

$$T = \frac{(\widehat{\beta}_0 + \widehat{\beta}_1 x^*) - (\beta_0 + \beta_1 x^*)}{S\sqrt{\dfrac{1}{n} + \dfrac{(x^* - \bar{x})^2}{S_{xx}}}} \qquad \text{d.f.} = n - 2$$

# 5. IMPORTANT INFERENCE PROBLEMS

We are now prepared to test hypotheses, construct confidence intervals, and make predictions in the context of straight line regression.

## 5.1 INFERENCE CONCERNING THE SLOPE $\beta_1$

In a regression analysis problem, it is of special interest to determine whether the expected response does or does not vary with the magnitude of the input variable $x$. According to the linear regression model,

$$\text{Expected response} = \beta_0 + \beta_1 x$$

This does not change with a change in $x$ if and only if $\beta_1 = 0$. We can therefore test the null hypothesis $H_0 : \beta_1 = 0$ against a one- or a two-sided alternative, depending on the nature of the relation that is anticipated. If we refer to the boxed statement (2) of Section 4, the null hypothesis $H_0 : \beta_1 = 0$ is to be tested using the test statistic

$$T = \frac{\widehat{\beta}_1}{S / \sqrt{S_{xx}}} \qquad \text{d.f.} = n - 2$$

**Example 6**   A Test to Establish that Duration of Relief Increases with Dosage

Do the data given in Table 1 constitute strong evidence that the mean duration of relief increases with higher dosages of the drug?

SOLUTION   For an increasing relation, we must have $\beta_1 > 0$. Therefore, we are to test the null hypothesis $H_0 : \beta_1 = 0$ versus the one-sided alternative $H_1 : \beta_1 > 0$.
    We select $\alpha = .05$. Since $t_{.05} = 1.860$, with d.f. $= 8$ we set the rejection region $R : T \geq 1.860$ as shown in Figure 6.

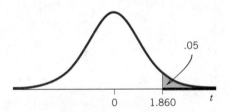

Figure 6   The rejection region
$R : T \geq 1.860$.

Using the calculations that follow Table 3, we have

$$\widehat{\beta}_1 = 2.74$$

$$s^2 = \frac{\text{SSE}}{n - 2} = \frac{63.6528}{8} = 7.9566, \quad s = 2.8207$$

$$\text{Estimated S.E.}(\widehat{\beta}_1) = \frac{s}{\sqrt{S_{xx}}} = \frac{2.8207}{\sqrt{40.90}} = .441$$

$$\text{Test statistic} \quad t = \frac{2.74}{.441} = 6.213$$

The observed $t$ value is in the rejection region, so $H_0$ is rejected. Moreover, 6.213 is much larger than $t_{.005} = 3.355$, so the $P$–value is much smaller than .005.

A computer calculation gives $P[T > 6.213] = .0001$. There is strong evidence that larger dosages of the drug tend to increase the duration of relief over the range covered in the study.

A warning is in order here concerning the interpretation of the test of $H_0 : \beta_1 = 0$. If $H_0$ is not rejected, we may be tempted to conclude that $y$ does not depend on $x$. Such an unqualified statement may be erroneous. First, the absence of a linear relation has only been established over the range of the $x$ values in the experiment. It may be that $x$ was just not varied enough to influence $y$. Second, the interpretation of lack of dependence on $x$ is valid only if our model formulation is correct. If the scatter diagram depicts a relation on a curve but we inadvertently formulate a straight line model and test $H_0 : \beta_1 = 0$, the conclusion that $H_0$ is not rejected should be interpreted to mean "no linear relation," rather than "no relation." We elaborate on this point further in Section 6. Our present viewpoint is to assume that the model is correctly formulated and discuss the various inference problems associated with it.

More generally, we may test whether or not $\beta_1$ is equal to some specified value $\beta_{10}$, not necessarily zero.

The **test of the null hypothesis**

$$H_0 : \beta_1 = \beta_{10}$$

is based on

$$T = \frac{\widehat{\beta}_1 - \beta_{10}}{S/\sqrt{S_{xx}}} \quad \text{d.f.} = n - 2$$

In addition to testing hypotheses, we can provide a confidence interval for the parameter $\beta_1$ using the $t$ distribution.

A $100(1 - \alpha)\%$ **confidence interval** for $\beta_1$ is

$$\left( \widehat{\beta}_1 - t_{\alpha/2} \frac{S}{\sqrt{S_{xx}}}, \quad \widehat{\beta}_1 + t_{\alpha/2} \frac{S}{\sqrt{S_{xx}}} \right)$$

where $t_{\alpha/2}$ is the upper $\alpha/2$ point of the $t$ distribution with d.f. $= n - 2$.

**Example 7**    A Confidence Interval for $\beta_1$

Construct a 95% confidence interval for the slope of the regression line in reference to the drug trial data of Table 1.

SOLUTION    In Example 6, we found that $\widehat{\beta}_1 = 2.74$ and $s / \sqrt{S_{xx}} = .441$. The required confidence interval is given by

$$2.74 \pm 2.306 \times .441 = 2.74 \pm 1.02 \quad \text{or} \quad (1.72, 3.76)$$

We are 95% confident that by adding one extra milligram to the dosage, the mean duration of relief would increase somewhere between 1.72 and 3.76 hours.

## 5.2 INFERENCE ABOUT THE INTERCEPT $\beta_0$

Although somewhat less important in practice, inferences similar to those outlined in Section 5.1 can be provided for the parameter $\beta_0$. The procedures are again based on the $t$ distribution with d.f. $= n - 2$, stated for $\widehat{\beta}_0$ in Section 4. In particular,

> A $100(1 - \alpha)\%$ **confidence interval** for $\beta_0$ is
>
> $$\left( \widehat{\beta}_0 - t_{\alpha/2} S \sqrt{\frac{1}{n} + \frac{\bar{x}^2}{S_{xx}}}, \quad \widehat{\beta}_0 + t_{\alpha/2} S \sqrt{\frac{1}{n} + \frac{\bar{x}^2}{S_{xx}}} \right)$$

To illustrate this formula, let us consider the data of Table 1. In Table 3, we find $\widehat{\beta}_0 = -1.07$, $\bar{x} = 5.9$, and $S_{xx} = 40.9$. Also, $s = 2.8207$. Therefore, a 95% confidence interval for $\beta_0$ is calculated as

$$-1.07 \pm 2.306 \times 2.8207 \sqrt{\frac{1}{10} + \frac{(5.9)^2}{40.9}}$$

$$= -1.07 \pm 6.34 \quad \text{or} \quad (-7.41, 5.27)$$

Note that $\beta_0$ represents the mean response corresponding to the value 0 for the input variable $x$. In the drug evaluation problem the parameter $\beta_0$ is of little practical interest because the range of $x$ values covered in the experiment was 3 to 9 and it would be unrealistic to extend the line to $x = 0$. In fact, the estimate $\widehat{\beta}_0 = -1.07$ does not have an interpretation as a (time) duration of relief.

## 5.3 ESTIMATION OF THE MEAN RESPONSE FOR A SPECIFIED X VALUE

Often, the objective in a regression study is to employ the fitted regression in estimating the expected response corresponding to a specified level of the input variable. For example, we may want to estimate the expected duration of relief for a specified dosage $x^*$ of the drug. According to the linear model described in Section 2, the expected response at a value $x^*$ of the input variable $x$ is given by $\beta_0 + \beta_1 x^*$. The expected response is estimated by $\widehat{\beta}_0 + \widehat{\beta}_1 x^*$ which is the ordinate of the fitted regression line at $x = x^*$. Referring to statement (3) of Section 4, we determine that the $t$ distribution can be used to construct confidence intervals or test hypotheses.

A $100(1 - \alpha)\%$ confidence interval for the expected response $\beta_0 + \beta_1 x^*$ is

$$\widehat{\beta}_0 + \widehat{\beta}_1 x^* \pm t_{\alpha/2} S \sqrt{\frac{1}{n} + \frac{(x^* - \bar{x})^2}{S_{xx}}}$$

To **test the hypothesis** that $\beta_0 + \beta_1 x^* = \mu_0$, some specified value, we use

$$T = \frac{\widehat{\beta}_0 + \widehat{\beta}_1 x^* - \mu_0}{S \sqrt{\frac{1}{n} + \frac{(x^* - \bar{x})^2}{S_{xx}}}} \qquad \text{d.f.} = n - 2$$

**Example 8**    A Confidence Interval for the Expected Duration of Relief

Again consider the data given in Table 1 and the calculations for the regression analysis given in Table 3. Obtain a 95% confidence interval for the expected duration of relief when the dosage is (a) $x^* = 6$ and (b) $x^* = 9.5$.

SOLUTION

(a)  The fitted regression line is

$$\widehat{y} = -1.07 + 2.74x$$

The expected duration of relief corresponding to the dosage $x^* = 6$ milligrams of the drug is estimated as

$$\widehat{\beta}_0 + \widehat{\beta}_1 x^* = -1.07 + 2.74 \times 6 = 15.37 \text{ hours}$$

$$\text{Estimated standard error} = s \sqrt{\frac{1}{10} + \frac{(6 - 5.9)^2}{40.9}}$$

$$= 2.8207 \times .3166 = .893$$

A 95% confidence interval for the mean duration of relief with the dosage $x^* = 6$ is therefore

$$15.37 \pm t_{.025} \times .893 = 15.37 \pm 2.306 \times .893$$
$$= 15.37 \pm 2.06 \quad \text{or} \quad (13.31, \ 17.43)$$

We are 95% confident that 6 milligrams of the drug produces an average duration of relief that is between about 13.31 and 17.43 hours.

(b)  Suppose that we also wish to estimate the mean duration of relief under the dosage $x^* = 9.5$. We follow the same steps to calculate the point estimate.

$$\widehat{\beta}_0 + \widehat{\beta}_1 x^* = -1.07 + 2.74 \times 9.5 = 24.96 \text{ hours}$$

$$\text{Estimated standard error} = 2.8207 \sqrt{\frac{1}{10} + \frac{(9.5 - 5.9)^2}{40.9}}$$

$$= 1.821$$

A 95% confidence interval is

$$24.96 \pm 2.306 \times 1.821 = 24.96 \pm 4.20 \quad \text{or} \quad (20.76, \ 29.16)$$

We are 95% confident that 9.5 milligrams of the drug produces an average of 20.76 to 29.16 hours of relief. Note that the interval is much larger than the one for 6 milligrams.

The formula for the standard error shows that when $x^*$ is close to $\bar{x}$, the standard error is smaller than it is when $x^*$ is far removed from $\bar{x}$. This is confirmed by Example 8, where the standard error at $x^* = 9.5$ can be seen to be more than twice as large as the value at $x^* = 6$. Consequently, the confidence interval for the former is also wider. In general, estimation is more precise near the mean $\bar{x}$ than it is for values of the $x$ variable that lie far from the mean.

***Caution concerning extrapolation:*** Extreme caution should be exercised in extending a fitted regression line to make long-range predictions far away from the range of $x$ values covered in the experiment. Not only does the confidence interval become so wide that predictions based on it can be extremely unreliable, but an even greater danger exists. If the pattern of the relationship between the variables changes drastically at a distant value of $x$, the data provide no information with which to detect such a change. Figure 7 illustrates this situation. We would observe a good linear relationship if we experimented with $x$ values in the 5 to 10 range, but if the fitted line were extended to estimate the response at $x^* = 20$, then our estimate would drastically miss the mark.

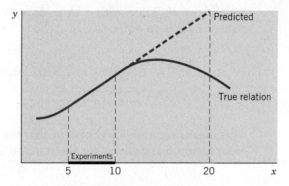

Figure 7   Danger in long-range prediction.

## 5.4 PREDICTION OF A SINGLE RESPONSE FOR A SPECIFIED X VALUE

Suppose that we give a specified dosage $x^*$ of the drug to a **single** patient and we want to predict the duration of relief from the symptoms of allergy. This problem is different from the one considered in Section 5.3, where we were interested in estimating the mean duration of relief for the population of **all** patients given the dosage $x^*$. The prediction is still determined from the fitted line; that is, the predicted value of the response is $\widehat{\beta}_0 + \widehat{\beta}_1 x^*$ as it was in the preceding case. However, the standard error of the prediction here is larger, because a single observation is more uncertain than the mean of the population distribution. We now give the formula of the estimated standard error for this case.

> The **estimated standard error when predicting a single observation** $y$ at a given $x^*$ is
>
> $$S\sqrt{1 + \frac{1}{n} + \frac{(x^* - \bar{x})^2}{S_{xx}}}$$

The formula for the confidence interval must be modified accordingly. We call the resulting interval a prediction interval because it pertains to a future observation.

**Example 9**   Calculating a Prediction Interval for a Future Trial

Once again, consider the drug trial data given in Table 1. A new trial is to be made on a single patient with the dosage $x^* = 6.5$ milligrams. Predict the duration of relief and give a 95% prediction interval for the duration of relief.

SOLUTION   The predicted duration of relief is

$$\widehat{\beta}_0 + \widehat{\beta}_1 x^* = -1.07 + 2.74 \times 6.5 = 16.74 \text{ hours}$$

Since $t_{.025} = 2.306$ with d.f. $= 8$, a 95% prediction interval for the new patient's duration of relief is

$$16.74 \pm 2.306 \times 2.8207 \sqrt{1 + \frac{1}{10} + \frac{(6.5 - 5.9)^2}{40.9}}$$

$$= 16.74 \pm 6.85 \quad \text{or} \quad (9.89, 23.59)$$

This means we are 95% confident that this particular patient will have relief from symptoms of allergy for about 9.9 to 23.6 hours.

In the preceding discussion, we have used the data of Example 4 to illustrate the various inferences associated with a straight line regression model. Example 10 gives applications to a different data set.

**Example 10**    Prediction after Fitting a Straight Line Relation of a Human Development Index to Internet Usage

One measure of the development of a country is the Human Development Index (HDI) which combines life expectancy, literacy, educational attainment, and gross domestic product per capita into an index whose values lie between 0 and 1, inclusive.

We randomly selected fifteen countries, of the 152 countries, below the top twenty-five most developed countries on the list. HDI is the response variable $y$, and Internet usage per 100 persons, $x$, is the predictor variable. The data, given in Exercise 11.31, have the summary statistics

$$n = 15 \qquad \bar{x} = 30.133 \qquad \bar{y} = .6100$$
$$S_{xx} = 8062.39 \quad S_{yy} = .35350 \quad S_{xy} = 49.729$$

(a)  Determine the equation of the best fitting straight line.

(b)  Do the data substantiate the claim that Internet usage per 100 persons is a good predictor of HDI and that large values of both variables tend to occur together?

(c)  Estimate the mean value of HDI for 25 Internet users per 100 persons and construct a 95% confidence interval.

(d)  Find the predicted $y$ for $x = 95$ Internet users per 100 persons.

SOLUTION

(a)    $\hat{\beta}_1 = \dfrac{S_{xy}}{S_{xx}} = \dfrac{49.729}{8062.39} = .00617$

$\hat{\beta}_0 = \bar{y} - \hat{\beta}_1 \bar{x} = .6100 - .00617 \times 30.193 = .4241$

So, the equation of the fitted line is

$$\hat{y} = .4241 + .00617x$$

(b) To answer this question, we decide to test $H_0: \beta_1 = 0$ versus $H_1: \beta_1 > 0$. The test statistic is

$$T = \frac{\widehat{\beta}_1}{S / \sqrt{S_{xx}}}$$

We select $\alpha = .01$. Since $t_{.01} = 2.650$ with d.f. $= 13$, we set the right-sided rejection region $R: T \geq 2.650$ as shown in Figure 8.

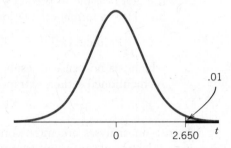

Figure 8   The rejection region $R: T \geq 2.650$.

We calculate

$$\text{SSE} = S_{yy} - \frac{S_{xy}^2}{S_{xx}} = .35350 - \frac{(49.729)^2}{8062.39} = .04678$$

$$s = \sqrt{\frac{\text{SSE}}{n - 2}} = \sqrt{\frac{.04678}{13}} = .0600$$

$$\text{Estimated S.E.}(\widehat{\beta}_1) = \frac{s}{\sqrt{S_{xx}}} = \frac{.0600}{\sqrt{8062.39}} = .000668$$

The $t$ statistic has the value

$$t = \frac{.00617}{.000668} = 9.24$$

Since the observed $t = 9.24$ is greater than 2.650, $H_0$ is rejected with $\alpha = .01$. The $P$-value is much less than .0001.
   We conclude that larger values of Internet users per 100 persons significantly increases the expected HDI, within the range of values of $x$ included in the data.

(c) The expected value of the HDI corresponding to $x^* = 25$ Internet users per 100 is estimated as

$$\widehat{\beta}_0 + \widehat{\beta}_1 x^* = .4241 + .00617(25) = .5784$$

and its

$$\text{Estimated S.E.} = s \sqrt{\frac{1}{15} + \frac{(25 - 30.133)^2}{8062.39}} = .0159$$

Since $t_{.025} = 2.160$ for d.f. $= 13$, the required confidence interval is

$$.5734 \pm .0159 \times .0238 = .5734 \pm .0343 \quad \text{or} \quad (.514, .613)$$

We are 95% confident that the expected value of HDI, for $x^* = 18$, is between .514 and .613.

(d) Since $x = 95$ is far above the largest value of 77.0 users per 100, it is not sensible to predict $y$ at $x = 95$ using the fitted regression line. Here a formal calculation gives

$$\text{Predicted HDI} = .4241 + .00617(95) = 1.010$$

which is a nonsensical result for an index that should not exceed 1. As mentioned earlier, extrapolation typically gives unreliable results.

Regression analyses are most conveniently done on a computer. A more complete selection of the output from the computer software package MINITAB, for the data in Example 4, is given in Table 5.

The output of the computer software package SAS for the data in Example 4 is given in Table 6. Notice the similarity of information in Tables 5 and 6. Both include the least squares estimates of the coefficients, their estimated standard deviations, and the $t$ test for testing that the coefficient is zero. The estimate of $\sigma^2$ is presented as the mean square error in the analysis of variance table.

**TABLE 5**  MINITAB Computer Output for the Data in Example 4

```
The regression equation is
Relief = -1.07 + 2.74 Dosage
```

| Predictor | Coef | SE Coef | T | P |
|---|---|---|---|---|
| Constant | -1.071 | 2.751 | -0.39 | 0.707 |
| Dosage | 2.7408 | 0.4411 | 6.21 | 0.000 |

```
S = 2.82074      R-Sq = 82.8%
```

Analysis of Variance

| Source | DF | SS | MS | F | P |
|---|---|---|---|---|---|
| Regression | 1 | 307.25 | 307.25 | 38.62 | 0.000 |
| Residual Error | 8 | 63.65 | 7.96 | | |
| Total | 9 | 370.90 | | | |

**TABLE 6**    SAS Computer Output for the Data in Example 4

Model: MODEL 1
Dependent Variable:  y

Analysis of Variance

| Source | DF | Sum of Squares | Mean Square | F Value | Pr > F |
|---|---|---|---|---|---|
| Model | 1 | 307.24719 | 307.24719 | 38.62 | 0.0003 |
| Error | 8 | 63.65281 | 7.95660 | | |
| Corrected Total | 9 | 370.90000 | | | |

Root MSE    2.82074    R-Square    0.8284

Parameter Estimates

| Variable | DF | Parameter Estimate | Standard Error | t Value | Pr>-t- |
|---|---|---|---|---|---|
| Intercep | 1 | -1.07090 | 2.75091 | -0.39 | 0.7072 |
| x | 1 | 2.74083 | 0.44106 | 6.21 | 0.0003 |

**Example 11**    Predicting the Number of Situps after a Semester of Conditioning

University students taking a physical fitness class were asked to count the number of situps they could do at the start of the class and again at the end of the semester.

Refer to the physical fitness data, on numbers of situps, in Table D.5 of the Data Bank.

(a)  Find the least squares fitted line to predict the posttest number of situps from the pretest number at the start of the conditioning class.

(b)  Find a 95% confidence interval for the mean number of posttest situps for persons who can perform 35 situps in the pretest. Also find a 95% prediction interval for the number of posttest situps that will be performed by a new person this semester who does 35 situps in the pretest.

(c)  Repeat part (b), but replace the number of pretest situps with 20.

SOLUTION    The scatter plot in Figure 9, on page 478, suggests that a straight line may model the expected value of posttest situps given the number of pretest situps. Here $x$ is the number of pretest situps and $y$ is the number of posttest situps. We use MINITAB statistical software to obtain the output.

Figure 9  Scatter plot of number of situps.

**Regression Analysis: Post Situps versus Pre Situps**

**The regression equation is**
**Post Situps = 10.3 + 0.899 Pre Situps**

| Predictor | Coef | SE Coef | T | P |
|---|---|---|---|---|
| Constant | 10.331 | 2.533 | 4.08 | 0.000 |
| Pre Situps | 0.89904 | 0.06388 | 14.07 | 0.000 |

**S = 5.17893  R-Sq = 71.5%**

**Analysis of Variance**

| Source | DF | SS | MS | F | P |
|---|---|---|---|---|---|
| Regression | 1 | 5312.9 | 5312.9 | 198.09 | 0.000 |
| Residual Error | 79 | 2118.9 | 26.8 | | |
| Total | 80 | 7431.8 | | | |

**Predicted Values for New Observations**

**New**
| Obs | Pre Sit | Fit | SE Fit | 95% CI | 95% PI |
|---|---|---|---|---|---|
| 1 | 35.0 | 41.797 | 0.620 | (40.563, 43.032) | (31.415, 52.179) |
| 2 | 20.0 | 28.312 | 1.321 | (25.682, 30.941) | (17.673, 38.950) |

From the output $\widehat{y} = \widehat{\beta}_0 + \widehat{\beta}_1 x = 10.3 + 0.899\ x$ and $s^2 = (5.1789)^2 = 26.8$ is the estimate of $\sigma^2$.

We have selected the option in MINITAB to obtain the two confidence intervals and prediction intervals given in the output. The prediction intervals pertain to the posttest number of situps performed by a specific new person. The first is for a person who performed 35 situps in the pretest. The prediction intervals are wider than the corresponding confidence intervals for the expected number of posttest situps for the population of all students who would do 35 situps in the pretest. The same relation holds, as it must, for 20 pretest situps.

## Exercises

**11.25** We all typically go to the shortest line in the grocery store. Data were collected on the number of carts ahead in line and the total time to check out (minutes), including time in line, on five occasions.

| Number of Carts | Time to Check Out |
|---|---|
| 1 | 5 |
| 2 | 11 |
| 3 | 9 |
| 4 | 14 |
| 5 | 16 |

(a) Calculate the least squares estimates $\widehat{\beta}_0$ and $\widehat{\beta}_1$. Also estimate the error variance $\sigma^2$.

(b) Test $H_0 : \beta_1 = 0$ versus $H_1 : \beta_1 \neq 0$ with $\alpha = .05$.

(c) Estimate the expected $y$ value corresponding to $x = 3$ carts and give a 90% confidence interval.

**11.26** Refer to Exercise 11.25. Construct a 90% confidence interval for the intercept $\beta_0$. Interpret.

**11.27** Refer to Exercise 11.25. Obtain a 95% confidence interval for $\beta_1$. Interpret.

**11.28** An engineer found that by adding small amounts of a compound to rechargeable batteries during manufacture, she could extend their lifetimes. She experimented with different amounts of the additive ($g$) and measured the hours they lasted in a laptop.

| Amount of Additive | Life (hours) |
|---|---|
| 0 | 3.9 |
| 1 | 4.0 |
| 2 | 4.5 |
| 3 | 4.6 |
| 4 | 5.0 |

(a) Calculate the least squares estimates $\widehat{\beta}_0$ and $\widehat{\beta}_1$. Also estimate the error variance $\sigma^2$.

(b) Test $H_0 : \beta_1 = 1$ versus $H_1 : \beta_1 \neq 1$ with $\alpha = .05$.

(c) Estimate the expected $y$ value corresponding to $x = 3.5$ and give a 95% confidence interval. Interpret.

(d) Construct a 90% confidence interval for the intercept $\beta_0$. Interpret.

**11.29** For a random sample of seven homes that were recently sold in a city suburb, the assessed values $x$ and the selling prices $y$ are

| ($1000) | | ($1000) | |
|---|---|---|---|
| x | y | x | y |
| 283.5 | 288.0 | 310.2 | 311.0 |
| 290.0 | 291.2 | 294.6 | 299.0 |
| 270.5 | 276.2 | 320.0 | 318.0 |
| 300.8 | 307.0 | | |

(a) Plot the scatter diagram.

(b) Determine the equation of the least squares regression line and draw this line on the scatter diagram.

(c) Construct a 95% confidence interval for the slope of the regression line. Interpret.

**11.30** Refer to the data in Exercise 11.29.

(a) Estimate the expected selling price of homes that were assessed at $290,000 and construct a 95% confidence interval. Interpret.

(b) For a single home that was assessed at $290,000, give a 95% prediction interval for the selling price. Interpret.

**11.31** One measure of the development of a country is the Human Development Index (HDI). Life expectancy, literacy, educational attainment, and gross domestic product per capita are combined into an index between 0 and 1, inclusive with 1 being the highest development. The United Nations Development Program reports values for 177 countries. We

randomly selected fifteen countries, below the top twenty-five. Both HDI and the predictor variable $x =$ Internet usage per 100 persons are obtained from their reports.

$$n = 15 \qquad \bar{x} = 30.133 \qquad \bar{y} = .6100$$

$$S_{xx} = 8062.39 \quad S_{xy} = 49.729 \quad S_{yy} = .35350$$

(a) Obtain a 95% confidence interval for the mean HDI when Internet usage per 100 persons is 22. Compare the width of the interval with that of the interval in Example 10.

(b) Obtain a 95% prediction interval for a single country with Internet usage 22 per one hundred persons. Interpret.

(c) Does your analysis show that Internet usage causes HDI to increase?

**11.32** Refer to Exercise 11.31.

(a) Obtain the least squares estimates by fitting a straight line to the response Internet usage using the predictor variable HDI.

(b) Test, with $\alpha = .05$, $H_0$: $\beta_1 = 0$ versus a two-sided alternative.

(c) Obtain a 95% confidence interval for the mean Internet usage per 100 persons, when the HDI is .650. Interpret.

(d) Obtain a 95% prediction interval for Internet usage in a single country with HDI .650. Interpret.

**11.33** According to the computer output in Table 8:

(a) What model is fitted?

(b) Test, with $\alpha = .05$, if the $x$ term is needed in the model.

**11.34** According to the computer output in Table 8:

(a) Predict the mean response when $x = 5000$.

(b) Find a 90% confidence interval for the mean response when $x = 5000$. You will need the additional information $n = 30$, $\bar{x} = 8354$, and $\Sigma (x_i - \bar{x})^2 = 97,599,296$.

**11.35** According to the computer output in Table 9:

(a) What model is fitted?

(b) Test, with $\alpha = .05$, if the $x$ term is needed in the model.

**TABLE 7**   Human Development Index

| Country | Internet / 100 | HDL |
|---|---|---|
| Poland | 64.9 | 0.813 |
| Bahrain | 77.0 | 0.806 |
| Uruguay | 51.4 | 0.783 |
| Bulgaria | 51.0 | 0.771 |
| Ukraine | 30.6 | 0.729 |
| Brazil | 45.0 | 0.718 |
| Dominican Republic | 35.5 | 0.689 |
| Moldova | 38.0 | 0.649 |
| India | 10.1 | 0.547 |
| Madagascar | 1.9 | 0.480 |
| Tanzania | 12.0 | 0.468 |
| Nepal | 9.0 | 0.458 |
| Uganda | 13.0 | 0.446 |
| Zambia | 11.5 | 0.430 |
| Ethiopia | 1.1 | 0.363 |

*Source*: Human Development 2011 reports at UNDP web site http://hdr.undp.org/en/statistics/hdi/

**TABLE 8**  Computer Output for Exercises 11.33 and 11.34

```
THE REGRESSION EQUATION IS
Y = 994 + 0.104x

PREDICTOR      COEF       STDEV      T-RATIO        P
CONSTANT       994.0      254.7        3.90      0.001
X              0.10373    0.02978      3.48      0.002

S = 299.4     R-SQ = 30.2%

ANALYSIS OF VARIANCE

SOURCE         DF         SS          MS        F         P
REGRESSION      1      1087765     1087765    12.14     0.002
ERROR          28      2509820       89636
TOTAL          29      3597585
```

**TABLE 9**  Computer Output for Exercises 11.35 and 11.36

```
THE REGRESSION EQUATION IS
Y = 0.338 + 0.831x

PREDICTOR      COEF       STDEV      T-RATIO        P
CONSTANT       0.3381     0.1579       2.14      0.043
X              0.83099    0.08702      9.55      0.000

S = 0.1208    R-SQ = 79.9%

ANALYSIS OF VARIANCE

SOURCE         DF         SS          MS        F         P
REGRESSION      1       1.3318      1.3318    91.20     0.000
ERROR          23       0.3359      0.0146
TOTAL          24       1.6676
```

**11.36** According to the computer output in Table 9:

(a) Predict the mean response when $x = 3$.

(b) Find a 90% confidence interval for the mean response when $x = 3$. You will need the additional information $n = 25$, $\bar{x} = 1.793$, and $\Sigma (x_i - \bar{x})^2 = 1.848$.

(c) Find a 90% confidence interval for the mean response when $x = 2$. Interpret.

**11.37** Consider the data on male wolves in Table D.9 of the Data Bank concerning age (years) and canine length (mm).

(a) Obtain the least squares fit of canine length to the predictor age.

(b) Test $H_0 : \beta_1 = 0$ versus $H_1 : \beta_1 \neq 0$ with $\alpha = .05$.

(c) Obtain a 90% confidence interval for the canine length when the age is $x = 4$.

(d) Obtain a 90% prediction interval for the canine length of an individual wolf when the age is $x = 4$.

# 6. THE STRENGTH OF A LINEAR RELATION

To arrive at a measure of adequacy of the straight line model, we examine how much of the variation in the response variable is explained by the fitted regression line. To this end, we view an observed $y_i$ as consisting of two components.

$$\underset{\substack{\text{Observed} \\ y \text{ value}}}{y_i} = \underset{\substack{\text{Explained by} \\ \text{linear relation}}}{(\widehat{\beta}_0 + \widehat{\beta}_1 x_i)} + \underset{\substack{\text{Residual or} \\ \text{deviation from} \\ \text{linear relation}}}{(y_i - \widehat{\beta}_0 - \widehat{\beta}_1 x_i)}$$

In an ideal situation where all the points lie exactly on the line, the residuals are all zero, and the $y$ values are completely accounted for or **explained** by the linear dependence on $x$.

We can consider the sum of squares of the residuals

$$\text{SSE} = \sum_{i=1}^{n} (y_i - \widehat{\beta}_0 - \widehat{\beta}_1 x_i)^2 = S_{yy} - \frac{S_{xy}^2}{S_{xx}}$$

to be an overall measure of the discrepancy or departure from linearity. The total variability of the $y$ values is reflected in the **total sum of squares**

$$S_{yy} = \sum_{i=1}^{n} (y_i - \bar{y})^2$$

of which SSE forms a part. The difference

$$S_{yy} - \text{SSE} = S_{yy} - \left( S_{yy} - \frac{S_{xy}^2}{S_{xx}} \right)$$

$$= \frac{S_{xy}^2}{S_{xx}}$$

forms the other part. Paralleling the decomposition of the observation $y_i$, as a residual plus a part due to regression, we consider a decomposition of the variability of the $y$ values.

$$S_{yy} = (S_{yy} - SSE) + SSE$$

---

### Decomposition of Variability

$$S_{yy} \quad = \quad \frac{S_{xy}^2}{S_{xx}} \quad + \quad SSE$$

| Total variability of $Y$ | Variability explained by the linear relation | Residual or unexplained variability |

---

The first term on the right-hand side of this equality is called the **sum of squares (SS) due to regression**. Likewise, the total variability $S_{yy}$ is also called the **total SS of $y$**. In order for the straight line model to be considered as providing a good fit to the data, the SS due to the linear regression should comprise a major portion of $S_{yy}$. In an ideal situation in which all points lie on the line, SSE is zero, so $S_{yy}$ is completely explained by the fact that the $x$ values vary in the experiment. That is, the linear relationship between $y$ and $x$ is solely responsible for the variability in the $y$ values.

As an index of how well the straight line model fits, it is then reasonable to consider the proportion of the $y$ variability explained by the linear relation

$$R^2 = \frac{\text{SS due to linear regression}}{\text{Total SS of } y} = \frac{S_{xy}^2 / S_{xx}}{S_{yy}} = \frac{S_{xy}^2}{S_{xx}S_{yy}}$$

From Section 4 of Chapter 3, recall that the quantity

$$r = \frac{S_{xy}}{\sqrt{S_{xx} S_{yy}}}$$

is named the **sample correlation coefficient**. Thus, the square of the sample correlation coefficient represents the proportion of the $y$ variability explained by the linear relation.

---

The **strength of a linear relation** is measured by

$$R^2 = r^2 = \frac{S_{xy}^2}{S_{xx}S_{yy}}$$

which is the square of the sample correlation coefficient $r$.

The value of $r$ is always between $-1$ and $1$, inclusive whereas $r^2$ is always between $0$ and $1$. Also, the estimated slope and the correlation coefficient are related (see Exercise 11.47).

$$r = \widehat{\beta}_1 \frac{\sqrt{S_{xx}}}{\sqrt{S_{yy}}}$$

so that the estimated slope $\widehat{\beta}_1$ and $r$ have the same sign.

**Example 12**  The Proportion of Variability in Duration Explained by Dosage

Let us consider the drug trial data in Table 1. From the calculations provided in Table 3,

$$S_{xx} = 40.9 \qquad S_{yy} = 370.9 \qquad S_{xy} = 112.1$$

Fitted regression line

$$\widehat{y} = -1.07 + 2.74x$$

How much of the variability in $y$ is explained by the linear regression model?

SOLUTION   To answer this question, we calculate

$$r^2 = \frac{S_{xy}^2}{S_{xx}S_{yy}} = \frac{(112.1)^2}{40.9 \times 370.9} = .83$$

This means that 83% of the variability in the duration of relief $y$ is explained by the linear regression on dosage. The linear model seems quite satisfactory in this respect.

**Example 13**  Proportion of Variation Explained in Number of Situps

Refer to physical fitness data in Table D.5 of the Data Bank. Using the data on numbers of situps, find the proportion of variation in the posttest number of situps explained by the pretest number that was obtained at the beginning of the conditioning class.

SOLUTION   Repeating the relevant part of the computer output from Example 11,

```
The regression equation is
Post Situps = 10.3 + 0.899 Pre Situps

Predictor        Coef      SE Coef        T          P
Constant       10.331        2.533     4.08      0.000
Pre Situps    0.89904      0.06388    14.07      0.000

S = 5.17893     R-Sq = 71.5%
```

Analysis of Variance

| Source | DF | SS | MS | F | P |
|--------|-----|---------|--------|--------|-------|
| Regression | 1 | 5312.9 | 5312.9 | 198.09 | 0.000 |
| Residual Error | 79 | 2118.9 | 26.8 | | |
| Total | 80 | 7431.8 | | | |

we find R-Sq = 71.5%, or proportion .715. From the analysis-of-variance table we could also have calculated

$$\frac{\text{Sum of squares regression}}{\text{Total sum of squares}} = \frac{5312.9}{7431.8} = .715$$

Using a person's pretest number of situps to predict their posttest number of situps explains 71.5% of the variation in the posttest number.

When the value of $r^2$ is small, we can only conclude that a straight line relation does not give a good fit to the data. Such a case may arise due to the following reasons:

1.  There is little relation between the variables in the sense that the scatter diagram fails to exhibit any pattern, as illustrated in Figure 10a. In this case, the use of a different regression model is not likely to reduce the SSE or explain a substantial part of $S_{yy}$.

(a)

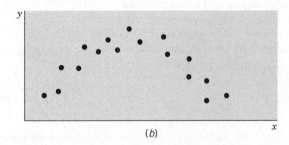

(b)

Figure 10   Scatter diagram patterns: (a) No relation. (b) A nonlinear relation.

2.  There is a prominent relation but it is nonlinear in nature; that is, the scatter is banded around a curve rather than a line. The part of $S_{yy}$ that is explained by straight line regression is small because the model is inappropriate. Some other relationship may improve the fit substantially. Figure 10b illustrates such a case, where the SSE can be reduced by fitting a suitable curve to the data.

## Exercises

**11.38** Based on a study of over 143,000 students, from 110 schools, investigators[4] estimate the correlation between SAT scores and first year college grade point ratios by $r = .35$.

If you predict first year college grades from the SAT score, what percentage of the variation in grades would be explained by the least squares straight line based on the SAT Score?

**11.39** To extend known results on the social hierarchy in monkeys, researchers[5] scanned $n = 14$ healthy persons using positron emission tomography to image dopamine type 2/3 in the brain. The social status of each person was assessed using the Bartlett Simplified Measure of Social Status. Letting $x = $ dopamine binding potential and $y = $ social status, from Exercise 11.19 the summary statistics are

$$\bar{x} = 2.526 \quad n = 14 \quad \bar{y} = 33.416$$

$$\sum (x - \bar{x})^2 = 1.607$$

$$\sum (x - \bar{x})(y - \bar{y}) = 19.305$$

$$\sum (y - \bar{y})^2 = 448.839$$

(a)  If you predict social status from the dopamine binding potential, what percentage of the variation in social status will be explained by the least squares straight line based on dopamine?

(b)  If you predict dopamine binding potential from the social status, what percentage of the variation in dopamine will be explained by the least squares straight line based on social status?

(c)  Comment on your two answers.

**11.40** Refer to Example 10 and Exercise 11.31, concerning the prediction of a human development index by Internet usage, where

$$\bar{x} = 30.133 \quad n = 15 \quad \bar{y} = .6100$$

$$\sum (x - \bar{x})^2 = 8062.39$$

$$\sum (x - \bar{x})(y - \bar{y}) = 49.729$$

$$\sum (y - \bar{y})^2 = .35350$$

Determine the proportion of variation in $y$ that is explained by linear regression.

**11.41** Refer to Exercise 11.40 but consider the prediction of Internet usage when the human development index is the predictor variable.

(a)  Determine the proportion of variation in Internet usage that is explained by linear regression.

(b)  Compare your answer in part (a) with that of Exercise 11.40. Comment.

**11.42** Refer to Exercise 11.25.

(a)  What proportion of the $y$ variability is explained by the linear regression on $x$?

(b)  Find the sample correlation coefficient.

**11.43** Refer to Exercise 11.28.

(a)  What proportion of $y$ variability is explained by the linear regression on $x$?

(b)  Find the sample correlation coefficient.

[4]P. Sackett et al., "The role of socioeconomic status in SAT–grade relationships and in college admissions decisions," *Psychological Science*, **23** (9), (2012) pp. 1000–1007.

[5]D. Martinez et al., "Dopamine type 2/3 receptor availability in the striatum and social status in humans," *Biological Psychiatry* **67** (1), (2010), pp. 275–278.

**11.44** Refer to Exercise 11.33. According to the computer output in Table 8, find the proportion of $y$ variability explained by $x$. Also, calculate $R^2$ from the analysis of variance table.

**11.45** Refer to Exercise 11.35. According to the computer output in Table 9, find the proportion of $y$ variability explained by $x$.

**11.46** Scientists,[6] looking for ways to reduce the amount of organic aerosols (OA) in the atmosphere, measured the nighttime rate of production of $NO_3$ and also the rate of production of particular organic nitrates which comprise about 30% of the organic aerosol present. The summary statistics for the production of $x =$ low reactivity $NO_3$ ( ppbv / hr ) and $y =$ production of particular organic nitrates (ppbv/hr) are

$$\bar{x} = .8523 \qquad n = 13 \qquad \bar{y} = .0208$$

$$\sum (x - \bar{x})^2 = 1.949$$

$$\sum (x - \bar{x})(y - \bar{y}) = .02738$$

$$\sum (y - \bar{y})^2 = .000930$$

(a) Calculate the correlation coefficient $r$.

(b) What proportion of the variation in OA is explained by the least squares straight line fit to the amount of low reactivity $NO_3$?

*__11.47__ (a) Show that the sample correlation coefficient $r$ and the slope $\widehat{\beta}_1$ of the fitted regression line are related as

$$r = \frac{\widehat{\beta}_1 \sqrt{S_{xx}}}{\sqrt{S_{yy}}}$$

(b) Show that SSE $= (1 - r^2) S_{yy}$.

*__11.48__ Show that the SS due to regression, $S_{xy}^2 / S_{xx}$, can also be expressed as $\widehat{\beta}_1^2 S_{xx}$.

# 7. REMARKS ABOUT THE STRAIGHT LINE MODEL ASSUMPTIONS

A regression study is not completed by performing a few routine hypothesis tests and constructing confidence intervals for parameters on the basis of the formulas given in Section 4. Such conclusions can be seriously misleading if the assumptions made in the model formulations are grossly incompatible with the data. It is therefore essential to check the data carefully for indications of any violation of the assumptions. To review, the assumptions involved in the formulation of our straight line model are briefly stated again.

1. The underlying relation is linear.
2. Independence of errors.
3. Constant variance.
4. Normal distribution.

Of course, when the general nature of the relationship between $y$ and $x$ forms a curve rather than a straight line, the prediction obtained from fitting a straight line model to the data may produce nonsensical results. Often, a suitable transformation of the data reduces a nonlinear relation to one that is approximately linear in form. A few simple transformations are discussed in Chapter 12. Violating the assumption of independence is perhaps the most serious matter, because this can drastically distort the conclusions drawn from the $t$ tests and the confidence statements associated with interval estimation. The implications of assumptions 3 and 4 were illustrated earlier in Figure 3. If

[6]Data from Figure 2, A. Rollins et al., "Evidence for $NO_x$ control over nighttime SOA formation," *Science* **133** (September 7, 2012), pp. 1210–1212.

the scatter diagram shows different amounts of variability in the $y$ values for different levels of $x$, then the assumption of constant variance may have been violated. Here, again, an appropriate transformation of the data often helps to stabilize the variance. Finally, using the $t$ distribution in hypothesis testing and confidence interval estimation is valid as long as the errors are approximately normally distributed. A moderate departure from normality does not impair the conclusions, especially when the data set is large. In other words, a violation of assumption 4 alone is not as serious as a violation of any of the other assumptions. Methods of checking the residuals to detect any serious violation of the model assumptions are discussed in Chapter 12.

## USING STATISTICS WISELY

1. As a first step, plot the response variable versus the predictor variable. Examine the plot to see if a linear or some other relationship exists.

2. Apply the principal of least squares to obtain estimates of the coefficients when fitting a straight line model.

3. Determine the $100(1 - \alpha)\%$ confidence intervals for the slope and intercept parameters. You can perform a test of hypotheses and look at $P$-values to decide whether or not the parameters are zero. If not, you can use the fitted line for prediction.

4. Don't use the fitted line to make predictions beyond the range of the data. The model may be different over that range.

5. Don't assume a large $r^2$, or correlation, implies a causal relationship.

## KEY IDEAS AND FORMULAS

In its simplest form, regression analysis deals with studying the manner in which the response variable $y$ depends on a predictor variable $x$. Sometimes, the response variable is called the dependent variable and predictor variable is called the independent or input variable.

The first important step in studying the relation between the variables $y$ and $x$ is to plot the scatter diagram of the data $(x_i, y_i)$, $i = 1, \ldots, n$. If this plot indicates an approximate linear relation, a straight line regression model is formulated:

$$\text{Response} = \text{A straight line in } x + \text{Random error}$$
$$Y_i = \beta_0 + \beta_1 x_i + e_i$$

The random errors are assumed to be independent, normally distributed, and have mean 0 and equal standard deviations $\sigma$.

The least squares estimate $\widehat{\beta}_0$ and least squares estimate $\widehat{\beta}_1$ are obtained by the method of least squares, which minimizes the sum of squared deviations

$\sum (y_i - b_0 - b_1 x_i)^2$. The least squares estimates $\widehat{\beta}_0$ and $\widehat{\beta}_1$ determine the best fitting regression line $\widehat{y} = \widehat{\beta}_0 + \widehat{\beta}_1 x$, which serves to predict $y$ from $x$.

The differences $y_i - \widehat{y}_i = $ Observed response $-$ Predicted response are called the residuals.

The adequacy of a straight line fit is measured by $r^2$, which represents the proportion of $y$ variability that is explained by the linear relation between $y$ and $x$. A low value of $r^2$ only indicates that a linear relation is not appropriate—there may still be a relation on a curve.

Least squares estimators

$$\widehat{\beta}_1 = \frac{S_{xy}}{S_{xx}} \qquad \widehat{\beta}_0 = \bar{y} - \widehat{\beta}_1 \bar{x}$$

Best fitting straight line

$$\widehat{y} = \widehat{\beta}_0 + \widehat{\beta}_1 x$$

Residuals

$$\widehat{e}_i = y_i - \widehat{y}_i = y_i - \widehat{\beta}_0 - \widehat{\beta}_1 x_i$$

Residual sum of squares

$$\text{SSE} = \sum_{i=1}^{n} \widehat{e}_i^2 = S_{yy} - \frac{S_{xy}^2}{S_{xx}}$$

Estimate of variance $\sigma^2$

$$S^2 = \frac{\text{SSE}}{n-2}$$

Inferences

1. Inferences concerning the slope $\beta_1$ are based on the

Estimator $\widehat{\beta}_1$

$$\text{Estimated S.E.} = \frac{S}{\sqrt{S_{xx}}}$$

and the sampling distribution

$$T = \frac{\widehat{\beta}_1 - \beta_1}{S / \sqrt{S_{xx}}} \qquad \text{d.f.} = n - 2$$

A $100(1 - \alpha)\%$ confidence interval for $\beta_1$ is

$$\widehat{\beta}_1 \pm t_{\alpha/2} \frac{S}{\sqrt{S_{xx}}}$$

The test statistic for testing $H_0 : \beta_1 = \beta_{10}$ is

$$T = \frac{\widehat{\beta}_1 - \beta_{10}}{S / \sqrt{S_{xx}}} \qquad \text{d.f.} = n - 2$$

2. Inferences concerning the **intercept** $\beta_0$ are based on the

$$\text{Estimator } \widehat{\beta}_0$$

$$\text{Estimated S.E.} = S\sqrt{\frac{1}{n} + \frac{\bar{x}^2}{S_{xx}}}$$

and the sampling distribution

$$T = \frac{\widehat{\beta}_0 - \beta_0}{S\sqrt{\dfrac{1}{n} + \dfrac{\bar{x}^2}{S_{xx}}}} \qquad \text{d.f.} = n - 2$$

A $100(1 - \alpha)\%$ confidence interval for $\beta_0$ is

$$\widehat{\beta}_0 \pm t_{\alpha/2} S\sqrt{\frac{1}{n} + \frac{\bar{x}^2}{S_{xx}}}$$

3. At a specified $x = x^*$, the expected response is $\beta_0 + \beta_1 x^*$. Inferences about the **expected response** are based on the

$$\text{Estimator } \widehat{\beta}_0 + \widehat{\beta}_1 x^*$$

$$\text{Estimated S.E.} = S\sqrt{\frac{1}{n} + \frac{(x^* - \bar{x})^2}{S_{xx}}}$$

A $100(1 - \alpha)\%$ confidence interval for the expected response at $x^*$ is given by

$$\widehat{\beta}_0 + \widehat{\beta}_1 x^* \pm t_{\alpha/2} S\sqrt{\frac{1}{n} + \frac{(x^* - \bar{x})^2}{S_{xx}}}$$

4. A **single response** at a specified $x = x^*$ is predicted by $\widehat{\beta}_0 + \widehat{\beta}_1 x^*$ with

$$\text{Estimated S.E.} = S\sqrt{1 + \frac{1}{n} + \frac{(x^* - \bar{x})^2}{S_{xx}}}$$

A $100(1 - \alpha)\%$ **prediction interval** for a single response is

$$\widehat{\beta}_0 + \widehat{\beta}_1 x^* \pm t_{\alpha/2} S\sqrt{1 + \frac{1}{n} + \frac{(x^* - \bar{x})^2}{S_{xx}}}$$

### Decomposition of Variability

The total sum of squares $S_{yy}$ is the sum of two components, the sum of squares due to regression $S_{xy}^2 / S_{xx}$ and the sum of squares due to error

$$S_{yy} = \frac{S_{xy}^2}{S_{xx}} + \text{SSE}$$

$$\text{Variability explained by the linear relation} = \frac{S_{xy}^2}{S_{xx}} = \widehat{\beta}_1^2 \, S_{xx}$$

$$\text{Unexplained variability due to error} = \text{SSE}$$

$$\text{Total } y \text{ variability} = S_{yy}$$

The strength of a linear relation, or proportion of $y$ variability explained by linear regression

$$r^2 = \frac{S_{xy}^2}{S_{xx} \, S_{yy}}$$

Sample correlation coefficient

$$r = \frac{S_{xy}}{\sqrt{S_{xx} \, S_{yy}}}$$

# TECHNOLOGY

### *Fitting a straight line and calculating the correlation coefficient*

## MINITAB

### *Fitting a straight line—regression analysis*

Begin with the values for the predictor variable $x$ in C*1* and the response variable $y$ in C*2*.

> Stat > Regression > Regression.
> Type C*2* in **Response**. Type C*1* in **Predictors**.
> Click **OK**.

To calculate the correlation coefficient, start as above with data in C*1* and C*2*.

> Stat > Basic Statistics > Correlation.
> Type C*1* C*2* in **Variables**. Click **OK**.

## EXCEL

### *Fitting a straight line—regression analysis*

The commands for creating a scatter plot and adding a fitted line are given in Chapter 3.

To obtain a more complete statistical analysis and diagnostic plots, begin with the values of the predictor variable in column A and the values of the response variable in column B.

Click on the **Data** tab and then **Data Analysis**.

Select **Regression**. Click **OK**.

With the cursor in the **Y Range**, highlight the data in column B.

With the cursor in the **X Range**, highlight the data in column A.

Check boxes for **Residuals, Residual Plots,** and **Line Fit Plot.** Click **OK**.

To calculate the correlation coefficient, begin with the first variable in column A and the second in column B.

Select the **FORMULAS** tab and then **More Functions**.

Choose **Statistical** and then **CORREL**.

Highlight the data in column A for **Array1** and highlight the data in column B for **Array2**. Click **OK**.

## TI-84/-83 PLUS

### *Fitting a straight line—regression analysis*

Enter the values of the predictor variable in $L_1$ and those of the response variable in $L_2$.

Select **STAT**, then **CALC**, and then **8;LinReg (ax + b)**.

With **LinReg** on the Home screen, press **Enter**.

The calculator will return the intercept $a$, slope $b$, and correlation coefficient $r$. If $r$ is not shown, go to the **2nd 0** to get step. **CATALOG** and select **DiagnosticON**. Press **ENTER** twice. Then, repeat through the step. **LinReg**.

## REVIEW EXERCISES

**11.49** Concerns that were raised for the environment near a government facility led to a study of plants. Since leaf area is difficult to measure, the leaf area ( $cm^2$ ) was fit to

$$x = \text{Leaf length} \times \text{Leaf width}$$

using a least squares approach. For data collected one year, the fitted regression line is

$$\hat{y} = .2 + 0.5\, x$$

and $s^2 = (0.3)^2$. Comment on the size of the slope. Should it be positive or negative, less than one, equal to one, or greater than one?

**11.50** Last week's total number of hours worked by a student, $y$, depends on the number of days, $x$, he reported to work last week. Suppose the data from nine students provided

| $x$ | 1 | 1 | 1 | 2 | 3 | 3 | 3 | 4 | 5 |
|-----|---|---|---|---|---|---|---|---|---|
| $y$ | 8 | 6 | 7 | 10 | 15 | 12 | 13 | 19 | 18 |

(a) Plot the scatter diagram.

(b) Calculate $\bar{x}$, $\bar{y}$, $S_{xx}$, $S_{yy}$, and $S_{xy}$.

(c) Determine the equation of the least squares fitted line and draw the line on the scatter diagram.

(d) Find the predicted number of hours $y$ corresponding to $x = 3$ days.

**11.51** Refer to Exercise 11.50.

(a) Find the residuals.

(b) Calculate the SSE by (i) summing the squares of the residuals and also (ii) using the formula $SSE = S_{yy} - S_{xy}^2 / S_{xx}$.

(c) Estimate the error variance.

**11.52** Refer to Exercise 11.50.

(a) Construct a 95% confidence interval for the slope of the regression line.

(b) Obtain a 90% confidence interval for the expected $y$ value corresponding to $x = 4$ days.

**11.53** An experiment is conducted to determine how the strength $y$ of plastic fiber depends on the size $x$ of the droplets of a mixing polymer in suspension. Data of $(x, y)$ values, obtained from 15 runs of the experiment, have yielded the following summary statistics.

$$\bar{x} = 8.3 \qquad \bar{y} = 54.8$$

$$S_{xx} = 5.6 \qquad S_{xy} = -12.4 \qquad S_{yy} = 38.7$$

(a) Obtain the equation of the least squares regression line.

(b) Test the null hypothesis $H_0 : \beta_1 = -2$ against the alternative $H_1 : \beta_1 < -2$, with $\alpha = .05$.

(c) Estimate the expected fiber strength for droplet size $x = 10$ and set a 95% confidence interval.

**11.54** Refer to Exercise 11.53.

(a) Obtain the decomposition of the total $y$ variability into two parts: one explained by linear relation and one not explained.

(b) What proportion of the $y$ variability is explained by the straight line regression?

(c) Calculate the sample correlation coefficient between $x$ and $y$.

**11.55** A recent graduate moving to a new job collected a sample of monthly rent (dollars) and size (square feet) of 2-bedroom apartments in one area of a midwest city.

| Size | Rent | Size | Rent |
|------|------|------|------|
| 1077 | 1075 | 1204 | 1125 |
| 1288 | 1270 | 820  | 845  |
| 842  | 760  | 945  | 885  |
| 1210 | 1175 | 1192 | 1035 |

(a) Plot the scatter diagram and find the least squares fit of a straight line.

(b) Do these data substantiate the claim that the monthly rent increases with the size of the apartment? (Test with $\alpha = .05$).

(c) Give a 95% confidence interval for the expected increase in rent for one additional square foot.

(d) Give a 95% prediction interval for the monthly rent of a specific apartment having 1025 square feet.

**11.56** Refer to Exercise 11.55.

(a) Calculate the sample correlation coefficient.

(b) What proportion of the $y$ variability is explained by the fitted regression line?

**11.57** A top Internet site features used cars for sale. The age, $x$, measured in years and the price, $y$, expressed in thousands of dollars was recorded from a sample of listings for a popular midsize foreign car.

| Age   | 1    | 2    | 3    | 3    | 4    |
|-------|------|------|------|------|------|
| Price | 27.9 | 23.0 | 21.6 | 20.9 | 20.0 |

| Age   | 5    | 5    | 6    | 6    | 7   |
|-------|------|------|------|------|-----|
| Price | 18.5 | 17.6 | 17.0 | 15.5 | 9.5 |

(a) Plot the scatter diagram.

(b) Determine the equation of the least squares regression line and draw this line on the scatter diagram.

(c) Construct a 95% confidence interval for the slope of the regression line.

**11.58** Refer to Exercise 11.57.

(a) From the fitted regression line, determine the predicted value for the average selling price of a 5-year-old car and construct a 95% confidence interval.

(b) Determine the predicted value for a 5-year-old car to be listed on the site next week. Construct a 90% prediction interval.

(c) Is it justifiable to predict the selling price of a 15-year-old car from the fitted regression line? Give reasons for your answer.

**11.59** Again referring to Exercise 11.57, find the sample correlation coefficient between age and selling price. What proportion of the variability in price is explained by the fitted straight line? Comment on the adequacy of the straight line fit.

## The Following Exercises Require a Computer

**11.60** *Using the computer.* The calculations involved in a regression analysis become increasingly tedious with larger data sets. Access to a computer proves to be of considerable advantage. We repeat here a computer-based analysis of linear regression using the data of Example 4 and the MINITAB package.

The sequence of steps in MINITAB:

---

**Data: C11T3**

*C1:* 3 3  4 5  6  6 7  8  8  9

*C2:* 9 5 12 9 14 16 22 18 24 22

**Dialog box:**

**Stat > Regression > Regression**

Type *C2* in **Response**

Type *C1* in **Predictors**. Click **OK**.

---

produces all the results that are basic to a linear regression analysis. The important pieces in the output are shown in Table 10.

Compare Table 10 with the calculations illustrated in Sections 3 to 6. In particular, identify:

(a) The least squares estimates.

(b) The SSE.

(c) The estimated standard errors of $\widehat{\beta}_0$ and $\widehat{\beta}_1$.

(d) The $t$ statistics for testing $H_0 : \beta_0 = 0$ and $H_0 : \beta_1 = 0$.

(e) $r^2$.

(f) The decomposition of the total sum of squares into the sum of squares explained by the linear regression and the residual sum of squares.

**11.61** Consider the data on wolves in Table D.9 of the Data Bank concerning body length (cm) and weight (lb). Calculate the correlation coefficient $r$ and $r^2$ for

(a) All wolves.

(b) Male wolves.

(c) Female wolves.

(d) Comment on the differences in your answers. Make a multiple scatter diagram (see Chapter 3) to clarify the situation.

**11.62** Consider the data on all of the wolves in Table D.9 of the Data Bank concerning body length (cm)

---

**TABLE 10    MINITAB Regression Analysis of the Data in Example 4**

The regression equation is
$Y = -1.07 + 2.74x$

| Predictor | Coef | SE Coef | T | P |
|---|---|---|---|---|
| Constant | −1.071 | 2.751 | −0.39 | 0.707 |
| X | 2.7408 | 0.4411 | 6.21 | 0.000 |

S = 2.821     R-Sq = 82.8%

Analysis of Variance

| Source | DF | SS | MS | F | P |
|---|---|---|---|---|---|
| Regression | 1 | 307.25 | 307.25 | 38.62 | 0.000 |
| Error | 8 | 63.65 | 7.96 | | |
| Total | 9 | 370.90 | | | |

and weight (lb). Using MINITAB or some other software program:

(a) Plot weight versus body length.

(b) Obtain the least squares fit of weight to the predictor variable body length.

(c) Test $H_0: \beta_1 = 0$ versus $H_1: \beta_1 > 0$ with $\alpha = .05$.

**11.63** Refer to Exercise 11.62 and a least squares fit using the data on all of the wolves in Table D.9 of the Data Bank concerning body length (cm) and weight (lb). There is one obvious outlier, row 18 with body length 123 and weight 106, indicated in the MINITAB output. Drop this observation.

(a) Obtain the least squares fit of weight to the predictor variable body length.

(b) Test $H_0: \beta_1 = 0$ versus $H_1: \beta_1 > 0$ with $\alpha = .05$.

(c) Comment on any important differences between your answers to parts (a) and (b) and the answer to Exercise 11.62.

**11.64** Many college students obtain college degree credits by demonstrating their proficiency on exams developed as part of the College Level Examination Program (CLEP). Based on their scores on the College Qualification Test (CQT), it would be helpful if students could predict their scores on a corresponding portion of the CLEP exam. The following data (courtesy of R. W. Johnson) are for $x$ = Total CQT score and $y$ = Mathematical CLEP score.

(a) Find the least squares fit of a straight line.

(b) Construct a 95% confidence interval for the slope.

| $x$ | $y$ | $x$ | $y$ |
|-----|-----|-----|-----|
| 170 | 698 | 174 | 645 |
| 147 | 518 | 128 | 578 |
| 166 | 725 | 152 | 625 |
| 125 | 485 | 157 | 558 |
| 182 | 745 | 174 | 698 |
| 133 | 538 | 185 | 745 |
| 146 | 485 | 171 | 611 |
| 125 | 625 | 102 | 458 |
| 136 | 471 | 150 | 538 |
| 179 | 798 | 192 | 778 |

(c) Construct a 95% prediction interval for the CLEP score of a student who obtains a CQT score of 150.

(d) Repeat part (c) with $x = 175$ and $x = 195$.

**11.65** Use MINITAB or some other software program to regress the marine growth on freshwater growth for the fish growth data in Table D.7 of the Data Bank. Do separate regression analyses for:

(a) All fish.

(b) Males.

(c) Females.

Your analysis should include (i) a scatter diagram, (ii) a fitted line, (iii) a determination if $\beta_1$ differs from zero. Also (iv) find a 95% confidence interval for the population mean when the freshwater growth is 100.

**11.66** Use MINITAB or some other software to obtain the scatter diagram, correlation coefficient, and the regression line of the final time to run 1.5 miles on the initial times given in Table D.5 of the Data Bank.

**11.67** The data on the maximum height (feet) and top speed (mph) of the 12 highest roller coasters, displayed in the chapter opener, are

| Height | Speed |
|--------|-------|
| 456 | 128 |
| 420 | 120 |
| 415 | 100 |
| 377 | 107 |
| 318 | 95 |
| 310 | 93 |
| 263 | 81 |
| 259 | 81 |
| 249 | 78 |
| 245 | 85 |
| 240 | 79 |
| 235 | 85 |

(a) Use MINITAB or some other software program to determine the proportion of variation in speed due to regression on height.

(b) What top speed is predicted for a new roller coaster of height 425 feet?

(c) What top speed is predicted for a new roller coaster of height 490 feet? What additional danger is there in this prediction?

# 12

# Regression Analysis II
## Multiple Linear Regression and Other Topics

1   Nonlinear Relations and Linearizing Transformations
2   Multiple Linear Regression
3   Residual Plots to Check the Adequacy of a Statistical Model

© David Hall/Science Source

## Micronutrients and Kelp Cultures: Evidence for Cobalt and Manganese Deficiency in Southern California Deep Seawater

Scientists have quantified the manner in which naturally occurring copper and zinc concentrations are toxic to marine organisms. They studied the effects of micronutrients on the growth of commercially significant giant kelp.

A regression analysis based on the least squares method can lead to models that relate growth to specific toxic substances.

The results are that toxic copper (Cu) and Zinc (Zn) ion concentrations may partially control growth. Also, cobalt (Co) and manganese (Mn) deficiencies exhibit some control in the growth.[1]

A least squares fit of gametophytic growth data in the defined medium generated the expression

$$Y = 136 + 8x_{Mn} - 5x_{Cu} + 7x_{Co}$$
$$-7x_{Zn}x_{Cu} - 15x_{Zn}^2 - 27x_{Mn}^2 - 12x_{Cu}^2$$
$$-18x_{Co}^2 - 6x_{Cu}x_{Zn}^2 - 6x_{Cu}x_{Mn}^2$$

where $Y$ is mean gametophytic length in micrometers. The authors consider the fit of the experimental data to this equation as excellent.

**Here, several variables are important for predicting growth.**

The basic ideas of regression analysis have a much broader scope of application than the straight line model of Chapter 11. In this chapter, our goal is to extend the ideas of regression analysis in two important directions.

1. To handle nonlinear relations by means of appropriate transformations applied to one or both variables.

2. To accommodate several predictor variables into a regression model.

These extensions enable the reader to appreciate the breadth of regression techniques that are applicable to real-life problems. We then discuss some graphical procedures that are helpful in detecting any serious violation of the assumptions that underlie a regression analysis.

# 1. NONLINEAR RELATIONS AND LINEARIZING TRANSFORMATIONS

When studying the relation between two variables $y$ and $x$, a scatter plot of the data often indicates that a relationship, although present, is far from linear. This can be established on a statistical basis by checking that the value of $r^2$ is small so a straight line fit is not adequate.

Statistical procedures for handling nonlinear relationships are more complicated than those for handling linear relationships, with the exception of a specific type of model called the **polynomial regression model**, which is discussed in

[1] J. S. Kuwabara, "Micronutrients and kelp cultures: evidence for cobalt and manganese deficiency in southern California deep sea waters," *Science* **216** (June 11, 1982), pp. 1219–1221. Copyright © 1982 by AAAS.

Section 2. In some situations, however, it may be possible to transform the variables $x$ and/or $y$ in such a way that the new relationship is close to being linear. A linear regression model can then be formulated in terms of the transformed variables, and the appropriate analysis can be based on the transformed data.

Transformations are often motivated by the pattern of data. Sometimes, when the scatter diagram exhibits a relationship on a curve in which the $y$ values increase too fast in comparison with the $x$ values, a plot of $\sqrt{y}$ or some other fractional power of $y$ can help to linearize the relation. This situation is illustrated in Example 1.

**Example 1**    Transforming the Response to Approximate a Linear Relation

To determine the maximum stopping ability of cars when their brakes are fully applied, 10 cars are driven each at a specified speed and the distance each requires to come to a complete stop is measured. The various initial speeds selected for each of the 10 cars and the stopping distances recorded are given in Table 1. Can the data be transformed to a nearly straight line relationship?

**TABLE 1**   Data on Speed and Stopping Distance

| Initial speed $x$ (mph) | 20 | 20 | 30 | 30 | 30 | 40 | 40 | 50 | 50 | 60 |
|---|---|---|---|---|---|---|---|---|---|---|
| Stopping distance $y$ (ft) | 16.3 | 26.7 | 39.2 | 63.5 | 51.3 | 98.4 | 65.7 | 104.1 | 155.6 | 217.2 |

SOLUTION   The scatter diagram for the data appears in Figure 1. The relation deviates from a straight line most markedly in that $y$ increases at a much faster rate at large $x$ than at small $x$. This suggests that we can try to linearize the relation by plotting $\sqrt{y}$ or some other fractional power of $y$ with $x$.

Figure 1   Scatter diagram of the data given in Table 1.

We try the transformed data $\sqrt{y}$ given in Table 2. The scatter diagram for these data, which exhibits an approximate linear relation, appears in Figure 2.

**TABLE 2**   Data on Speed and Square Root of Stopping Distance

| $x$ | 20 | 20 | 30 | 30 | 30 | 40 | 40 | 50 | 50 | 60 |
|---|---|---|---|---|---|---|---|---|---|---|
| $y' = \sqrt{y}$ | 4.037 | 5.167 | 6.261 | 7.969 | 7.162 | 9.920 | 8.106 | 10.203 | 12.474 | 14.738 |

Figure 2    Scatter diagram of the transformed data given in Table 2.

With the aid of a standard computer program for regression analysis (see Exercise 12.27), the following results are obtained by transforming the original data.

$$\bar{x} = 37 \qquad \bar{y'} = 8.604$$

$$S_{xx} = 1610 \qquad S_{y'y'} = 97.773 \qquad S_{xy'} = 381.621$$

$$\widehat{\beta}_0 = -.167 \qquad \widehat{\beta}_1 = .237$$

Thus, the equation of the fitted line is

$$\widehat{y'} = -.167 + .237x$$

The proportion of the $y'$ variation that is explained by the straight line model is

$$r^2 = \frac{(381.621)^2}{(1610)(97.773)} = .925$$

A few common nonlinear models and their corresponding linearizing transformations are given in Table 3.

**TABLE 3**   Some Nonlinear Models and Their Linearizing Transformations

| Nonlinear Model | Transformation | Transformed Model $y' = \beta_0 + \beta_1 x'$ | | |
|---|---|---|---|---|
| (a) $y = ae^{bx}$ | $y' = \log_e y \quad x' = x$ | $\beta_0 = \log_e a$ | $\beta_1 = b$ | |
| (b) $y = ax^b$ | $y' = \log y \quad x' = \log x$ | $\beta_0 = \log a$ | $\beta_1 = b$ | |
| (c) $y = \dfrac{1}{a + bx}$ | $y' = \dfrac{1}{y} \quad x' = x$ | $\beta_0 = a$ | $\beta_1 = b$ | |
| (d) $y = a + b\sqrt{x}$ | $y' = y \quad x' = \sqrt{x}$ | $\beta_0 = a$ | $\beta_1 = b$ | |

In some situations, a specific nonlinear relation is strongly suggested by either the data or a theoretical consideration. Even when initial information about the form is lacking, a study of the scatter diagram often indicates the appropriate linearizing transformation.

**Example 2**   How Far Do the Elephants Go from the River?

The Masai Mara National Reserve is a main gathering place for large animals during the wet season. An aerial survey provided information that included a count of the number of elephants and their distance to a major river.[2]

**TABLE 4**   Data on the Number of Elephants and Distance to the River

| Distance(km) | 3 | 6 | 10.5 | 12 | 15 | 18 | 21 | 24 | 27 |
|---|---|---|---|---|---|---|---|---|---|
| Count | 1845 | 561 | 145 | 127 | 109 | 181 | 72 | 18 | 13 |

(a)   Make a scatter diagram of count versus distance.

(b)   Take the logarithm, base 10, of both variables and make a scatter diagram.

(c)   Fit a straight line by least squares to the appropriate data.

SOLUTION   The scatter diagrams are shown in Figure 3a and 3b. Although the pattern is somewhat ragged because of the hump near the right-hand end, the log-scale is reasonable for a first straight line fit.

---

[2]Data read from Figure 3(a). Kiambi et al., "Wet season aerial count of large herbivores in Masai Mara National Reserve and adjacent community areas," Kenya Wildlife Service, Mara Research Station, 2012.

**Figure 3**    Scatter diagrams. (*a*) Count versus distance. (*b*) Log (count) versus log (distance).

(c) MINITAB software produces the calculations

**The regression equation is**
**log (count = 4.33 - 2.04 log(distance)**

| Predictor | Coef | SE Coef | T | P |
|---|---|---|---|---|
| Constant | 4.3310 | 0.3541 | 12.23 | 0.000 |
| log (distance) | -2.0372 | 0.3103 | -6.56 | 0.000 |

```
S = 0.270752    R-Sq = 86.0%
```

**Analysis of Variance**

| Source | DF | SS | MS | F | P |
|---|---|---|---|---|---|
| Regression | 1 | 3.1591 | 3.1591 | 43.09 | 0.000 |
| Residual Error | 7 | 0.5131 | 0.0733 | | |
| Total | 8 | 3.6722 | | | |

The negative estimate of slope, $-2.037$, means that the estimated number decreases as the distance increases. Also, the $P$-value for testing $\widehat{\beta}_1 = 0$ is less than .001.

Note that $r^2 = .86$ is quite large despite the raggedness of pattern in the scatter diagram.

Returning to the original scale, the regression equation becomes

$$\text{count} = 21429\,(\text{distance})^{-2.037}$$

which is close to equation given in the reference.

Once the data are entered on a computer, it is easy to obtain the transformed data $1/y$, $\log_e y$, $y^{1/2}$, and $y^{1/4}$. Note $y^{1/4}$ is obtained by taking the square root of $y^{1/2}$. A scatter plot of $\sqrt{y}$ versus $\log_e x$ or any number of others can then be constructed and examined for a linear relation. Under relation (a) in Table 3, the graph of $\log_e y$ versus $x$ would be linear.

We must remember that all inferences about the transformed model are based on the assumptions of a linear relation and independent normal errors with constant variance. Before we can trust these inferences, this transformed model must be scrutinized to determine whether any serious violation of these assumptions may have occurred (see Section 3).

## Exercises

**12.1**  Developers have built a small robotic vehicle that can travel over rough terrain. They recorded the time $y$, in minutes, that it takes to travel a fixed distance over various but similar terrains. For a fixed run, the robot's motor is set at a nominal speed $x$, in feet per second, but this varied from run to run.

| $x$ | .5 | 1 | 2 | 4 | 5 | 6 | 7 |
|---|---|---|---|---|---|---|---|
| $y$ | 4.6 | 3.2 | 2.1 | 1.7 | .9 | .7 | .8 |

(a)  Plot the scatter diagram.

(b)  Obtain the best fitting straight line and draw it on the scatter diagram.

(c)  What proportion of the $y$ variability is explained by the fitted line?

**12.2**  Refer to the data of Exercise 12.1.

(a)  Consider the reciprocal transformation $y' = 1/y$ and plot the scatter diagram of $y'$ versus $x$.

(b)  Fit a straight line regression to the transformed data.

(c)  Calculate $r^2$ and comment on the adequacy of the fit.

**12.3** Find a linearizing transformation in each case.

(a) $y = \dfrac{1}{(a + bx)^3}$

(b) $\dfrac{1}{y} = a + \dfrac{b}{1 + x}$

**12.4** An experiment was conducted for the purpose of studying the effect of temperature on the life-length of an electrical insulation. Specimens of the insulation were tested under fixed temperatures, and their times to failure recorded.

| Temperature $x$ ($^\circ$C) | Failure Time $y$ (thousand hours) |
|---|---|
| 180 | 7.3, 7.9, 8.5, 9.6, 10.3 |
| 210 | 1.7, 2.5, 2.6, 3.1 |
| 230 | 1.2, 1.4, 1.6, 1.9 |
| 250 | .6, .7, 1.0, 1.1, 1.2 |

(a) Fit a straight line regression to the transformed data

$$x' = \frac{1}{x} \quad \text{and} \quad y' = \log y$$

(b) Is there strong evidence that an increase in temperature reduces the life of the insulation?

(c) Comment on the adequacy of the fitted line.

**12.5** In an experiment (courtesy of W. Burkholder) involving stored-product beetles (*Trogoderma glabrum*) and their sex-attractant pheromone, the pheromone is placed in a pit-trap in the centers of identical square arenas. Marked beetles are then released along the diagonals of each square at various distances from the pheromone source. After 48 hours, the pit-traps are inspected. Control pit-traps containing no pheromone captured no beetles.

| Release Distance (centimeters) | No of Beetles Captured out of 8 |
|---|---|
| 6.25 | 5, 3, 4, 6 |
| 12.5 | 5, 2, 5, 4 |
| 24 | 4, 5, 3, 0 |
| 50 | 3, 4, 2, 2 |
| 100 | 1, 2, 2, 3 |

(a) Plot the original data with $y$ = number of beetles captured. Also plot $y$ with $x = \log_e$ (distance).

(b) Fit a straight line by least squares to the appropriate graph in part (a).

(c) Construct a 95% confidence interval for $\beta_1$.

(d) Establish a 95% confidence interval for the mean at a release distance of 18 cm.

# 2. MULTIPLE LINEAR REGRESSION

A response variable $y$ may depend on a predictor variable $x$ but, after a straight line fit, it may turn out that the unexplained variation is large, so $r^2$ is small and a poor fit is indicated. At the same time, an attempt to transform one or both of the variables may fail to dramatically improve the value of $r^2$. This difficulty may well be due to the fact that the response depends on not just $x$ but other factors as well. When used alone, $x$ fails to be a good predictor of $y$ because of the effects of those other influencing variables. For instance, the yield of a crop depends on not only the amount of fertilizer but also on the rainfall and average temperature during the growing season. Cool weather and no rain could completely cancel the choice of a correct fertilizer.

To obtain a useful prediction model, one should record the observations of all variables that may significantly affect the response. These other variables may then be incorporated explicitly into the regression analysis. The name multiple regression refers to a model of relationship where the response depends on two or more predictor variables. Here, we discuss the main ideas of a multiple regression analysis in the setting of two predictor variables.

Suppose that the response variable $y$ in an experiment is expected to be influenced by two input variables $x_1$ and $x_2$, and the data relevant to these input variables are recorded along with the measurements of $y$. With $n$ runs of an experiment, we have a data set of the form shown in Table 5.

**TABLE 5**   Data Structure for Multiple Regression with Two Input Variables

| Experimental | Input Variables | | Response |
|:---:|:---:|:---:|:---:|
| Run | $x_1$ | $x_2$ | $y$ |
| 1 | $x_{11}$ | $x_{12}$ | $y_1$ |
| 2 | $x_{21}$ | $x_{22}$ | $y_2$ |
| . | . | . | . |
| . | . | . | . |
| . | . | . | . |
| $i$ | $x_{i1}$ | $x_{i2}$ | $y_i$ |
| . | . | . | . |
| . | . | . | . |
| . | . | . | . |
| $n$ | $x_{n1}$ | $x_{n2}$ | $y_n$ |

By analogy with the simple linear regression model, we can then tentatively formulate:

---

### A Multiple Regression Model

$$Y_i = \beta_0 + \beta_1 x_{i1} + \beta_2 x_{i2} + e_i \qquad i = 1, \ldots, n$$

where $x_{i1}$ and $x_{i2}$ are the values of the two input variables for the $i$th experimental run and $Y_i$ is the corresponding response.

The error components $e_i$ are assumed to be independent normal variables with mean 0 and variance $\sigma^2$.

The regression parameters $\beta_0$, $\beta_1$, and $\beta_2$ are unknown and so is $\sigma^2$.

---

This model suggests that aside from the random error, the response varies linearly with each of the independent variables when the other remains fixed.

The principle of least squares is again useful in estimating the regression parameters. For this model, we are required to vary $b_0$, $b_1$, and $b_2$ simultaneously to minimize the sum of squared deviations

$$\sum_{i=1}^{n} (y_i - b_0 - b_1 x_{i1} - b_2 x_{i2})^2$$

The least squares estimates $\widehat{\beta}_0$, $\widehat{\beta}_1$, and $\widehat{\beta}_2$ are the solutions to the following equations, which are extensions of the corresponding equations for fitting the straight line model (see Section 3 of Chapter 11.)

$$\widehat{\beta}_1 S_{11} + \widehat{\beta}_2 S_{12} = S_{1y}$$
$$\widehat{\beta}_1 S_{12} + \widehat{\beta}_2 S_{22} = S_{2y}$$
$$\widehat{\beta}_0 = \bar{y} - \widehat{\beta}_1 \bar{x}_1 - \widehat{\beta}_2 \bar{x}_2$$

where $S_{11}$, $S_{12}$, and so on, are the sums of squares and cross products of deviations of the variables indicated in the suffix. They are computed just as in a straight line regression model. Methods are available for interval estimation, hypothesis testing, and examining the adequacy of fit. In principle, these methods are similar to those used in the simple regression model, but the algebraic formulas are more complex and hand computations become more tedious. However, a multiple regression analysis is easily performed on a computer with the aid of the standard packages such as MINITAB, SAS, or SPSS. We illustrate the various aspects of a multiple regression analysis with the data of Example 3 and computer-based calculations.

**Example 3**    Interpreting the Regression of Blood Pressure on Weight and Age

We are interested in studying the systolic blood pressure $y$ in relation to weight $x_1$ and age $x_2$ in a class of males of approximately the same height. From 13 subjects preselected according to weight and age, the data set listed in Table 6 was obtained.

**TABLE 6**    The Data of $x_1$ = Weight in Pounds, $x_2$ = Age, and $y$ = Blood Pressure of 13 Males

| $x_1$ | $x_2$ | $y$ |
|-------|-------|-----|
| 152 | 50 | 120 |
| 183 | 20 | 141 |
| 171 | 20 | 124 |
| 165 | 30 | 126 |
| 158 | 30 | 117 |
| 161 | 50 | 129 |
| 149 | 60 | 123 |
| 158 | 50 | 125 |
| 170 | 40 | 132 |
| 153 | 55 | 123 |
| 164 | 40 | 132 |
| 190 | 40 | 155 |
| 185 | 20 | 147 |

Use a computer package to perform a regression analysis using the model

$$Y_i = \beta_0 + \beta_1 x_{i1} + \beta_2 x_{i2} + e_i$$

SOLUTION    To use MINITAB, we first enter the data of $x_1$, $x_2$, and $y$ in three different columns and then use the regression command,

> **Data**: C12T5
>
> *C1*: 152 183 171 · · ·185
>
> C2:  50  20  20 · · · 20
>
> C3: 120 141 124 · · ·147
>
> **Dialog box**:
>
> **Stat > Regression > Regression**
>
> Type C3 in **Response**.
>
> Type *C1* and C2 in **Predictors**.
>
> Click **OK**.

With the last command, the computer executes a multiple regression analysis. We focus our attention on the principal aspects of the output, as shown in Table 7.

**TABLE 7**    Regression Analysis of the Data in Table 6: Selected MINITAB Output

①
The regression equation is
Y = −65.1 + 1.08 Weight + 0.425 Age

| Predictor | Coef | SE Coef | T | P |
|---|---|---|---|---|
| Constant | −65.10 | 14.94 | −4.36 | 0.001 |
| Weight | ② 1.07710 | 0.07707 | 13.98 | 0.000 |
| Age | 0.42541 | 0.07315 | ④ 5.82 | 0.000 |

③ S = 2.50861    ⑤ R−SQ = 95.8%

Analysis of Variance

| Source | DF | SS | MS | F | P |
|---|---|---|---|---|---|
| Regression | 2 | 1423.84 | 711.92 | 113.13 | 0.000 |
| Residual Error | 10 | ⑥ 62.93 | 6.29 | | |
| Total | 12 | 1486.77 | | | |

We now proceed to interpret the results in Table 7 and use them to make further statistical inferences.

(i)   The equation of the fitted linear regression is

$$① \quad \hat{y} = -65.1 + 1.08\,x_1 + .425\,x_2$$

This means that the mean blood pressure increases by 1.08 if weight $x_1$ increases by one pound and age $x_2$ remains fixed. Similarly, a 1-year increase in age with the weight held fixed will only increase the mean blood pressure by .425.

(ii) The estimated regression coefficient and the corresponding estimated standard errors are

$$\begin{array}{lll} \hat{\beta}_0 = -65.10 & \text{Estimated S.E. } (\hat{\beta}_0) = 14.94 \\ ② \quad \hat{\beta}_1 = 1.07710 & \text{Estimated S.E. } (\hat{\beta}_1) = .07707 \\ \hat{\beta}_2 = .42541 & \text{Estimated S.E. } (\hat{\beta}_2) = .07315 \end{array}$$

③ Further, the error standard deviation $\sigma$ is estimated by $s = 2.50861$ with

$$\begin{aligned} \text{Degrees of freedom} &= n - (\text{No. of input variables}) - 1 \\ &= 13 - 2 - 1 \\ &= 10 \end{aligned}$$

These results are useful in interval estimation and hypothesis tests about the regression coefficients. In particular, a $100(1 - \alpha)\%$ confidence interval for a coefficient $\beta$ is given by

$$\text{Estimated coefficient} \pm t_{\alpha/2} (\text{Estimated S.E.})$$

where $t_{\alpha/2}$ is the upper $\alpha/2$ point of the distribution with d.f. $= 10$. For instance, a 95% confidence interval for $\beta_1$ is

$$1.07710 \pm 2.228 \times .07707$$

$$= 1.07710 \pm .17171 \quad \text{or} \quad (.905, 1.249)$$

To test the null hypothesis that a particular coefficient $\beta$ is zero, we employ the test statistic

$$t = \frac{\text{Estimated coefficient} - 0}{\text{Estimated S.E.}} \qquad \text{d.f.} = 10$$

These $t$-ratios appear in Table 7. Suppose that we wish to examine whether the mean blood pressure significantly increases with age. In the language of hypothesis testing, this problem translates to one of testing $H_0 : \beta_2 = 0$ versus $H_1 : \beta_2 > 0$. The observed value of the ④ test statistic is $t = 5.82$ with d.f. $= 10$. Since this is larger than the tabulated value $t_{.01} = 2.764$, the null hypothesis is rejected in favor of $H_1$, with $\alpha = .01$. In fact, it is rejected even with $\alpha = .005$.

(iii) In Table 7, the result "R-SQ $= 95.8\%$" or

$$⑤ \quad R^2 = .958$$

tells us that 95.8% of the variability of $y$ is explained by the fitted multiple regression of $y$ on $x_1$ and $x_2$. The "analysis of variance" shows the decomposition of the total variability $\Sigma(y_i - \bar{y})^2 = 1486.77$ into the two components.

$$⑥ \quad 1486.77 \quad = \quad 1423.84 \quad + \quad 62.93$$

| Total variability of $y$ | Variability explained by the regression of $y$ on $x_1$ and $x_2$ | Residual or unexplained variability |

Thus,

$$R^2 = \frac{1423.84}{1486.77} = .958$$

and $\sigma^2$ is estimated by $s^2 = 62.93 / 10 = 6.293$, so $s = 2.509$ [which checks with $s$ from (ii)].

---

**TABLE 8**   A Regression Analysis of the Data in Example 3 Using SAS

Model: MODEL1
Dependent Variable: y

### Analysis of Variance

| Source | DF | | Sum of Squares | Mean Square | F Value | Pr > F |
|--------|-----|----|----------------|-------------|---------|--------|
| Model | 2 | | 1423.83797 | 711.91898 | 113.13 | <.0001 |
| Error | 10 | ⑥ | 62.93126 | 6.29313 | | |
| Corrected Total | 12 | | 1486.76923 | | | |

③ Root MSE 2.50861     ⑤ R-Square 0.9577

### Parameter Estimates

| Variable | DF | | Parameter Estimate | Standard Error | | t Value | Pr > |t| |
|----------|-----|----|--------------------|----------------|----|---------|----------|
| Intercept | 1 | | −65.09968 | 14.94458 | | −4.36 | 0.0014 |
| x1 | 1 | ② | 1.07710 | 0.07707 | ④ | 13.98 | <.0001 |
| x2 | 1 | | 0.42541 | 0.07315 | | 5.82 | 0.0002 |

The **square of the multiple correlation coefficient** $R^2$ gives the proportion of variability in $y$ explained by the fitted multiple regression.

The output from the SAS package is given in Table 8. The quantities needed in our analysis have been labeled with the same circled numbers as in the MINITAB output.

**Example 4**    Computer-Aided Regression Analysis—Two Predictors

The times for 81 students to complete a rowing test both before and after completing a one-semester conditioning course are given in Table D.5 in the Data Bank. It may be that not only the pretest rowing time but also gender would be useful for predicting the posttest rowing time. Perform a regression analysis.

SOLUTION    We use MINITAB to obtain the output

```
Regression Analysis: Post row versus Pre row, Gender

The regression equation is
Post row = 97.3 + 0.726 Pre row + 32.1 Gender

Predictor      Coef     SE Coef        T        P
Constant      97.33       31.68     3.07    0.003
Pre row     0.72573     0.05487    13.23    0.000
Gender       32.083       9.756     3.29    0.002


S = 31.8137     R-Sq = 85.7%

Analysis of Variance

Source           DF        SS         MS        F        P
Regression        2     471547     235774   232.95    0.000
Residual Error   78      78945       1012
Total            80     550492
```

Which variables should be used to predict posttest rowing time? Reading from the column of $P$–values for the individual coefficients, the largest is only .003. The constant term and the coefficients of pretest rowing time and gender are significantly different from 0. All three terms are needed in the model.

The plot of residuals versus fit in Figure 4 reveals a constant width band so there is no evidence against the assumption of constant variance. The one large negative residual is case 17 and the two large positive residuals are cases 29 and 70.

Figure 4   Residuals of posttest row times versus fits.

## 2.1 POLYNOMIAL REGRESSION

A scatter diagram may exhibit a relationship on a curve for which a suitable linearizing transformation cannot be constructed. Another method of handling such a nonlinear relation is to include terms with higher powers of $x$ in the model $Y = \beta_0 + \beta_1 x + e$. In this instance, by including the second power of $x$, we obtain the model

$$Y_i = \beta_0 + \beta_1 x_i + \beta_2 x_i^2 + e_i \qquad i = 1, \ldots, n$$

which states that aside from the error components $e_i$, the response $y$ is a **quadratic function** (or a **second-degree polynomial**) of the independent variable $x$. Such a model is called a **polynomial regression** model of $y$ with $x$, and the highest power of $x$ that occurs in the model is called the **degree** or the **order** of the polynomial regression. It is interesting to note that the analysis of a polynomial regression model does not require any special techniques other than those used in multiple regression analysis. By identifying $x$ and $x^2$ as the two variables $x_1$ and $x_2$, respectively, this second-degree polynomial model reduces to the form of a multiple regression model:

$$Y_i = \beta_0 + \beta_1 x_{i1} + \beta_2 x_{i2} + e_i \qquad i = 1, \ldots, n$$

where $x_{i1} = x_i$ and $x_{i2} = x_i^2$. In fact, both these types of models and many more types are special cases of a class called **general linear models** [1, 2]. The term linear refers to the property that the mean, $E(Y_i)$, is linear in the parameters. There are no such terms as $\beta_1^2$ or $\beta_1 \beta_2$.

Before we talk about the general linear model, let's look at an example which analyzes a second-degree polynomial model.

**Example 5**    Fitting a Quadratic Relation of a Human Development Index to
Internet Usage

Refer to Chapter 11, Example 10, concerning the development of a country measured by the Human Development Index (HDI) and the predictor variable Internet usage per 100 persons. Although we randomly selected only fifteen countries, of the 152 countries, below the twenty-five most developed there is still an indication that the relation is increasing less rapidly for high Internet usage. Fit a quadratic and test whether or not a squared term is required.

SOLUTION    We fit the quadratic model $Y_i = \beta_0 + \beta_1 x_i + \beta_2 x_i^2 + e_i$ using MINITAB and obtain the output.

```
The regression equation is
HDI = 0.377 + 0.0110 Internet - 0.000068 Internet Sq


Predictor              Coef        SE Coef         T        P
Constant            0.37731       0.02837      13.30    0.000
Internet           0.010974      0.001993       5.51    0.000
Internet Sq      -0.00006779    0.00002697     -2.51    0.027


S = 0.0505342      R-Sq = 91.3%


Analysis of Variance


Source             DF        SS           MS         F        P
Regression          2     0.32286      0.16143     63.21    0.000
Residual Error     12     0.03064      0.00255
Total              14     0.35350
```

According to the output, the estimated coefficient of $x^2$ is $\widehat{\beta}_2 = -.0000678$ and the $t$ test for testing $H_0 : \beta_2 = 0$ versus $H_1 : \beta_2 \neq 0$ has P–value

$$P[\, T < -2.51\,] + P[\, T > 2.51\,] = .027$$

as indicated under P in the MINITAB output. This gives moderate evidence that a quadratic term is needed. Note that the proportion of variation in HDI explained by Internet usage has increased to .913 as indicated in the output R-Sq $= 91.3\%$.

The data and fitted curve, shown in Figure 5, illustrate the bend in the straight line relation at the few higher values of Internet usage. The quadratic fit is still not ideal because it starts to turn down over the range of the experiment, whereas the underlying relation is likely to always increase. This reminds us that there is no "true model," but proposed models are only approximations. However, some models are useful.

Figure 5   Quadratic fit to HDI by Internet usage.

## 2.2  GENERAL LINEAR MODEL

By virtue of its wide applicability, the multiple linear regression model plays a prominent role in the portfolio of a statistician. Although a complete analysis cannot be given here, the general structure of a multiple regression model merits further attention. We have already mentioned that most least squares analyses of multiple linear regression models are carried out with the aid of a computer. All the programs for implementing the analysis require the investigator to provide the values of the response $y_i$ and the $p$ input variables $x_{i1}, \ldots, x_{ip}$ for each run $i = 1, 2, \ldots, n$. In writing $1 \cdot \beta_0$, where 1 is the known value of an extra "dummy" input variable corresponding to $\beta_0$, the model is

$$\overset{\text{Observation}}{Y_i} \quad = \overset{\text{Input variables}}{1 \cdot \beta_0 + x_{i1}\,\beta_1 + x_{i2}\,\beta_2 + \cdots + x_{ip}\,\beta_p} + \overset{\text{Error}}{e_i}$$

This is called a **linear model** because it is linear in the $\beta_i$'s. That is, there are no terms such as $\beta_1 \beta_2$ or $\beta_1^2$.

The basic quantities can be arranged in the form of these arrays, which are denoted by boldface letters.

$$\overset{\text{Observation}}{\mathbf{y} = \begin{bmatrix} y_1 \\ y_2 \\ \cdot \\ \cdot \\ \cdot \\ y_i \\ \cdot \\ \cdot \\ \cdot \\ y_n \end{bmatrix}} \qquad \overset{\text{Input variables}}{\mathbf{X} = \begin{bmatrix} 1 & x_{11} & \cdots & x_{1p} \\ 1 & x_{21} & \cdots & x_{2p} \\ \cdot & \cdot & \cdot & \cdot \\ \cdot & \cdot & \cdot & \cdot \\ \cdot & \cdot & \cdot & \cdot \\ 1 & x_{i1} & \cdots & x_{ip} \\ \cdot & \cdot & \cdot & \cdot \\ \cdot & \cdot & \cdot & \cdot \\ \cdot & \cdot & \cdot & \cdot \\ 1 & x_{n1} & \cdots & x_{np} \end{bmatrix}}$$

Only the arrays **y** and **X** are required to obtain the least squares estimates of $\beta_0, \beta_1, \ldots, \beta_p$ that minimize

$$\sum_{i=1}^{n} (y_i - b_0 - x_{i1}b_1 - \cdots - x_{ip}b_p)^2$$

The input array **X** is called the **design matrix**.
In the same vein, setting

$$\mathbf{e} = \begin{bmatrix} e_1 \\ e_2 \\ \cdot \\ \cdot \\ \cdot \\ e_n \end{bmatrix} \quad \text{and} \quad \boldsymbol{\beta} = \begin{bmatrix} \beta_0 \\ \beta_1 \\ \cdot \\ \cdot \\ \cdot \\ \beta_p \end{bmatrix}$$

we can write the model in the suggested form

|  | Observation | | Design matrix | Parameter | | Error |
|---|---|---|---|---|---|---|
|  | **y** | = | **X** | $\boldsymbol{\beta}$ | + | **e** |

which forms the basis for a thorough but more advanced treatment of regression.

## Exercises

**12.6** A student fit the regression model $Y = \beta_0 + \beta_1 x_1 + \beta_2 x_2 + e$ to data from the fifty states and Washington, D.C., so $n = 51$. The response $y =$ median income in thousands of dollars and the two predictor variables are $x_1 =$ median monthly housing costs for home owners and $x_2 =$ percentage of persons below the poverty level in the last twelvemonths. The least squares estimates are

$$\hat{\beta}_0 = 66.168 \qquad \hat{\beta}_1 = .0156 \qquad \hat{\beta}_2 = -1.560$$

Predict the response for

(a) $x_1 = 1100, x_2 = 13$

(b) $x_1 = 1100, x_2 = 15$

(c) $x_1 = 1500, x_2 = 15$

**12.7** Consider the multiple linear regression model

$$Y = \beta_0 + \beta_1 x_1 + \beta_2 x_2 + e$$

where $\beta_0 = -1, \beta_1 = -2, \beta_2 = 3$, and the normal random variable $e$ has standard deviation 3. What is the mean of the response $Y$ when $x_1 = 3$ and $x_2 = -2$?

**12.8** In Exercise 12.6, the residual sum of squares (SSE) is 531.6 and the SS due to regression is 4337.7.

(a) Estimate the error standard deviation $\sigma$. State the degrees of freedom.

(b) Find $R^2$ and interpret the result.

**12.9** Refer again to Exercise 12.6 and assume that the assumptions about the model prevail. The estimated standard errors of $\hat{\beta}_1$, and $\hat{\beta}_2$ are .00161, and .1862, respectively.

(a) Determine a 95% confidence interval for $\beta_2$.

(b) Test $H_0 : \beta_1 = .0125$ versus $H_1 : \beta_1 > .0125$, with $\alpha = .05$.

**12.10** Consider the data on all of the wolves in Table D.9 of the Data Bank concerning age (years) and canine length (mm).

(a) Obtain the least squares fit of the straight line regression model $Y = \beta_0 + \beta_1 x + e$ to predict canine length from age.

(b) Obtain the least squares fit of the multiple regression model $Y = \beta_0 + \beta_1 x_1 + \beta_2 x_2 + e$ to predict canine length using age $x_1$ and body length $x_2$.

(c) What is the predicted canine length for a wolf of age 2.5 and body length 127?

(d) What proportion of the $y$ variability is explained by the fitted model in part (b)?

(e) Obtain 95% confidence intervals for $\beta_0$, $\beta_1$, and $\beta_2$.

**12.11** Consider the response variable miles per gallon on highways and the two predictor variables $x_1$ = engine volume (liter) and $x_2$ = size of battery (volt). Using a recent government Fuel Economy Guide, and the data on hybrid-electric cars and SUVs, we obtain the regression analysis given in Table 9.

(a) Identify the least squares estimates $\widehat{\beta}_0$, $\widehat{\beta}_1$, and $\widehat{\beta}_2$.

(b) What model is suggested from this analysis?

(c) What is the proportion of $y$ variability explained by the regression on $x_1$ and $x_2$?

(d) Estimate $\sigma^2$.

**12.12** Referring to Example 5, Chapter 11, and Table D.13 of the Data Bank, we now use two predictors, $x_1$ = girth (cm) and $x_2$ = length to predict $y$ =

weight (lb) of a polar bear. The output from a regression analysis is given in Table 10.

(a) How many polar bears were included in the analysis?

(b) Identify the least squares estimates $\widehat{\beta}_0$, $\widehat{\beta}_1$, and $\widehat{\beta}_2$.

(c) What model is suggested from this analysis?

(d) What is the proportion of $y$ variability explained by the regression on $x_1$ and $x_2$?

**12.13** With reference to Exercise 12.11:

(a) Test $H_0 : \beta_1 = 0$ versus $H_1 : \beta_1 \neq 0$ with $\alpha = .05$.

(b) Test $H_0 : \beta_2 = 0$ versus $H_1 : \beta_2 \neq 0$ with $\alpha = .05$.

(c) Estimate the expected $y$ value corresponding to $x_1 = 3.2$ and $x_2 = 200$.

(d) Construct a 90% confidence interval for the intercept $\beta_0$.

**12.14** Referring to Exercise 12.12, we have added one more predictor $x_3$ = skull width (cm). The output for a regression analysis is shown in Table 11.

---

**TABLE 9**    Computer Output of a Regression Analysis to Be Used for Exercise 12.11

```
Regression Analysis: y versus x1, x2

The Regression equation is
y = 45.3 - 3.22 x1 - 0.0207 X2
```

| Predictor | Coef | SE Coef | T | P |
|-----------|------|---------|---|---|
| Constant | 45.299 | 2.345 | 19.32 | 0.000 |
| x1 | -3.2243 | 0.4562 | -7.07 | 0.000 |
| x2 | -0.020661 | 0.008574 | -2.41 | 0.026 |

```
S = 3.40356    R-SQ = 79.5%
```

Analysis of Variance

| Source | DF | SS | MS | F | P |
|--------|----|----|----|----|----|
| Regression | 2 | 897.53 | 448.77 | 38.74 | 0.000 |
| Error | 20 | 231.68 | 11.58 | | |
| Total | 22 | 1129.22 | | | |

**TABLE 10**    Computer Output of a Regression Analysis of Polar Bear Data to Be Used for Exercise 12.12

```
The regression equation is
y = - 1162 +  8.70 x1 + 2.71 x2

Predictor       Coef     SE Coef         T        P
Constant     -1162.3       244.3     -4.76    0.000
x1            8.7022      0.8502     10.24    0.000
x2             2.706       1.030      2.63    0.014


S = 51.4915      R-Sq = 85.5%


Analysis of Variance


Source           DF          SS         MS        F        P
Regression        2      420502     210251    79.30    0.000
Residual Error   27       71587       2651
Total            29      492089
```

**TABLE 11**    Computer Output of a Regression Analysis of Polar Bear Data to Be Used for Exercise 12.14

```
The regression equation is
y = - 1581 + 5.89 x1 + 3.19 x2 + 28.2 x3

Predictor       Coef     SE Coef         T         P
Constant     -1580.7       275.7     -5.73     0.000
x1            5.886       1.343      4.38     0.000
x2           3.1903      0.9568      3.33     0.003
x3            28.24       11.00      2.57     0.016


S = 46.8673      R-Sq = 88.4%


Analysis of Variance


Source           DF          SS         MS        F        P
Regression        3      434979     144993    66.01    0.000
Residual Error   26       57110       2197
Total            29      492089
```

(a) Test $H_0 : \beta_3 = 0$ versus $H_0 : \beta_3 \neq 0$ with $\alpha = .05$.

(b) Test $H_0 : \beta_2 = 0$ versus $H_0 : \beta_2 \neq 0$ with $\alpha = .05$.

(c) Estimate the expected value of weight $(Y)$ corresponding to $x_1 = 150$, $x_2 = 250$, and $x_3 = 25$.

(d) What is the proportion of $y$ variability explained by the regression on $x_1$, $x_2$, and $x_3$?

## 3. RESIDUAL PLOTS TO CHECK THE ADEQUACY OF A STATISTICAL MODEL

> **General Attitude Toward a Statistical Model**
>
> A regression analysis is not completed by fitting a model by least squares, providing confidence intervals, and testing various hypotheses. These steps tell only half the story: the statistical inferences that can be made when the postulated model is adequate. In most studies, we cannot be sure that a particular model is nearly correct. Therefore, we should adopt the following strategy.
>
> 1. Tentatively entertain a model.
>
> 2. Obtain least squares estimates and compute the residuals.
>
> 3. Review the model by examining the residuals.

Step 3 often suggests methods of appropriately modifying the model. We then return to step 1, where the modified model is entertained, and this **iteration** is continued until a model is obtained for which the data do not seem to contradict the assumptions made about the model.

Once a model is fitted by least squares, all the information on variation that cannot be explained by the model is contained in the residuals

$$\widehat{e}_i = y_i - \widehat{y}_i \qquad i = 1, 2, \ldots, n$$

where $y_i$ is the observed value and $\widehat{y}_i$ denotes the corresponding value predicted by the fitted model. For example, in the case of a simple linear regression model, $\widehat{y}_i = \widehat{\beta}_0 + \widehat{\beta}_1 x_i$.

Recall from our discussion of the straight line model in Chapter 11 that we have made the assumptions of independence, constant variance, and a normal distribution for the error components $e_i$. The inference procedures are based on these assumptions. When the model is correct, the residuals can be considered as estimates of the errors $e_i$ that are distributed as $N(0, \sigma)$.

To determine the merits of the tentatively entertained model, we can examine the residuals by plotting them in various ways. Then if we recognize any systematic pattern formed by the plotted residuals, we would suspect that some assumptions regarding the model are invalid. There are many ways to plot the residuals, depending on what aspect is to be examined. We mention a few of these here to illustrate the techniques.

### 3.1 HISTOGRAM OR DOT DIAGRAM OF RESIDUALS

To picture the overall behavior of the residuals, we can plot a **histogram** for a large number of observations or a **dot diagram** for fewer observations. For example, in a dot diagram like the one in Figure 6a, the residuals seem to behave like a sample from a normal population and there do not appear to be any "wild"

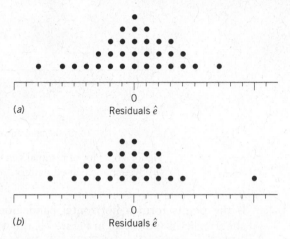

Figure 6   Dot diagram of residuals. (*a*) Normal pattern. (*b*) One large residual.

observations. In contrast, Figure 6*b* illustrates a situation in which the distribution appears to be quite normal except for a single residual that lies far to the right of the others. The circumstances that produced the associated observation demand a close scrutiny.

## 3.2  PLOT OF RESIDUAL VERSUS PREDICTED VALUE

A plot of the residuals $\hat{e}_i$ versus the **predicted value** $\hat{y}_i$ often helps to detect the inadequacies of an assumed relation or a violation of the assumption of constant error variance. Figure 7 illustrates some typical phenomena.

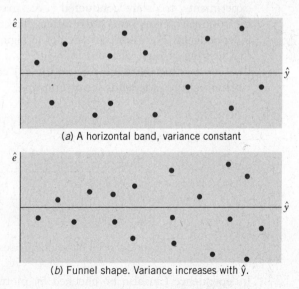

(*a*) A horizontal band, variance constant

(*b*) Funnel shape. Variance increases with $\hat{y}$.

Figure 7a   Plot of residual versus predicted value. (*a*) Constant spread. (*b*) Increasing spread.

(c) Systematic curved pattern.

Figure 7b    Plot of residual versus predicted value.
(c) Systematic curved pattern.

If the points form a horizontal band around zero, as in Figure 7a, then no abnormality is indicated. In Figure 7b, the width of the band increases noticeably with increasing values of $\hat{y}$. This indicates that the error variance $\sigma^2$ tends to increase with an increasing level of response. We would then suspect the validity of the assumption of constant variance in the model. Figure 7c shows residuals that form a systematic pattern. Instead of being randomly distributed around the $\hat{y}$ axis, they tend first to increase steadily and then decrease. This would lead us to suspect that the model is inadequate and a squared term or some other nonlinear $x$ term should be considered.

### 3.3  PLOT OF RESIDUAL VERSUS TIME ORDER

The most crucial assumption in a regression analysis is that the errors $e_i$ are independent. Lack of independence frequently occurs in business and economic applications, where the observations are collected in a time sequence with the intention of using regression techniques to predict future trends. In many other experiments, trials are conducted successively in time. In any event, a plot of the residuals versus **time order** often detects a violation of the assumption of independence. For example, the plot in Figure 8 exhibits a systematic pattern in that a string of high values is followed by a string of low values. This indicates that consecutive residuals are (positively) correlated, and we would suspect a violation of the independence assumption.

Figure 8    Plot of residual versus time order.

Independence can also be checked by plotting the successive pairs $(\hat{e}_i, \hat{e}_{i-1})$, where $\hat{e}_1$ indicates the residual from the first $y$ value observed, $\hat{e}_2$ indicates the second, and so on. Independence is suggested if the scatter diagram is a patternless cluster, whereas points clustered along a line suggest a lack of independence between adjacent observations.

**Example 6**    Checking the Model for Rowing Time

In Example 4, we obtained the regression equation

$$\text{Post row} = 97.3 + .726 \text{ Pre row} + 32.1 \text{ Gender}$$

and $s = 31.8137$ Before making any inferences, the model assumptions need to be checked. Construct the three graphs

(a)   Histogram of residuals.

(b)   Residuals versus predicted values.

(c)   Residuals versus observation order.

Comment on any patterns that reveal violations of the assumptions concerning the model.

SOLUTION    The MINITAB regression *four-in-one* graphics option creates four graphs including the three requested.

Figure 9    Three residual plots for posttest rowing time using MINITAB.

There are no obvious patterns that suggest a systematic departure from the model or time dependence in the order recorded. A normal-scores plot would better assess the normal assumption but some lack of normality is not a problem when the sample size is this large.

It is important to remember that our confidence in statistical inference procedures is related to the validity of the assumptions about them. A mechanically made inference may be misleading if some model assumption is grossly violated. An **examination of the residuals** is an important part of regression analysis, because it helps to detect any inconsistency between the data and the postulated model.

If no serious violation of the assumption is exposed in the process of examining residuals, we consider the model adequate and proceed with the relevant inferences. Otherwise, we must search for a more appropriate model.

### References

1. S. Chatterjee and A. Hadi, *Regression Analysis by Example*, 5th ed., John Wiley & Sons: New York, 2012.

2. D. Montgomery, E. Peck, and G. Vinning, *Introduction to Linear Regression Analysis*, 5th. ed., John Wiley & Sons: New York, 2012.

## USING STATISTICS WISELY

1. Always, as a first step, plot the response variable versus the predictor variable. If there is more than one predictor variable, make separate plots for each. Examine the plot to see if a linear or other relationship exists.

2. Do not routinely accept the regression analysis presented in computer output. Instead, criticize the model by inspecting the residuals for outliers and moderate to severe lack of normality. A normal-scores plot is useful if there are more than 20 or so residuals. That plot may suggest a transformation.

3. Plot the residuals versus predicted value to check the assumption of constant variance. Plot the residuals in time order if that is appropriate. A trend over time would cast doubt on the assumption of independent errors.

## KEY IDEAS AND FORMULAS

When a scatter diagram shows relationship on a curve, it may be possible to choose a **transformation** of one or both variables such that the transformed data exhibit a linear relation. A simple linear regression analysis can then be performed on the transformed data.

**Multiple regression analysis** is a versatile technique of building a prediction model with several input variables. In addition to obtaining the least squares fit, we can construct confidence intervals and test hypotheses about the influence of each input variable.

A polynomial regression model is a special case of multiple regression where the powers $x$, $x^2$, $x^3$, and so on, of a single predictor $x$ play the role of the individual predictors.

The highest power of $x$ that occurs in the model is called the **degree** or **order** of the regression model. A **quadratic function,** or **second-degree polynomial,** is commonly fit as an alternative to a straight line.

Both the polynomial regression model and the multiple regression with several predictors are special cases of **general linear models**.

The measure $R^2$, called the **square of the multiple correlation coefficient,** represents the proportion of $y$ variability that is explained by the fitted multiple regression model.

$$R^2 = \frac{\text{Regression SS}}{\text{Total SS}}$$

To safeguard against a misuse of regression analysis, we must scrutinize the data for agreement with the model assumptions. An examination of the residuals, especially by graphical plots, including a **dot diagram** or **histogram**, a plot versus **predicted value**, and a plot versus **time order**, is essential for detecting possible violations of the assumptions and also identifying the appropriate modifications of an initial model.

# TECHNOLOGY

### *Regression with two or more predictors and quadratic regression*

## MINITAB

### *Regression with two or more predictors*

Begin with the values for the two predictor variables in C2 and C3 and the response variable $y$ in C1.

> Stat > Regression > Regression.
> Type C1 in **Response**. Type C2 and C3 in **Predictors**.
> Select **Graphs**. Click **Four in one**. Click **OK**.
> Click **OK**.

The graphics step produces four residual plots: including histogram, residual versus fit, and residual versus order.

### *Transforming data*

We illustrate with the predictor variable $x$ in C2 and transforming to $\log(x)$, where the logarithm is base 10 in C3.

> Calc > Calculator.
> Type C3 in **Store result in variable** and $LOGT(C2)$ in **Expression**.
> Click **OK**.

### *Fitting a quadratic regression model*

With the values of $x$ in *C1* and the $y$ values in *C2*, you must select:

Stat > Regression > Fitted Line Plot.

Enter *C2* in **Response (Y)** and enter *C1* in **Predictor (X)**.

Under **Type of Regression Model** choose **Quadratic**. Click **OK**.

## TI-84/-83 PLUS

### *Fitting a quadratic regression model*

Enter the values of the predictor variable in $\mathbf{L_1}$ and those of the response variable in $\mathbf{L_2}$.

Select **STAT**, then **CALC**, and then **5: QuadReg**.

With **QuadReg** on the Home screen press **Enter**.

The calculator will give $a$, $b$, and $c$ in the equation

$$y = ax^2 + bx + c$$

## REVIEW EXERCISES

**12.15**    An environmental scientist identified a point source for E. Coli at the edge of a stream. She then measured $y$ = E. Coli, in colony forming units per 100 ml of water, at different distances, in feet, downstream from the point source. Suppose she obtains the following pairs of $(x, y)$.

| $x$ | 100 | 150 | 250 | 250 | 400 | 650 | 1000 | 1600 |
|-----|-----|-----|-----|-----|-----|-----|------|------|
| $y$ | 21  | 20  | 24  | 17  | 18  | 10  | 11   | 9    |

(a) Transform the $x$ values to $x' = \log_{10} x$ and plot the scatter diagram of $y$ versus $x'$.

(b) Fit a straight line regression to the transformed data.

(c) Obtain a 90% confidence interval for the slope of the regression line.

(d) Estimate the expected $y$ value corresponding to $x = 300$ and give a 95% confidence interval.

*12.16    Obtain a linearizing transformation in each case.

(a) $y = \dfrac{1}{(1 + ae^{bx})^2}$

(b) $y = e^{ax^b}$

**12.17**    A genetic experiment is undertaken to study the competition between two types of female flies (*Drosophila melanogaster*) in cages with one male genotype acting as a substrate. The independent variable $x$ is the time spent in cages, and the dependent variable $y$ is the ratio of the numbers of type 1 to type 2 females. The following data (courtesy of C. Denniston) are recorded.

(a) Plot the scatter diagram of $y$ versus $x$ and determine if a linear model of relation is appropriate.

(b) Determine if a linear relation is plausible for the transformed data $y' = \log_{10} y$.

(c) Fit a straight line regression to the transformed data.

| Time $x$ (days) | No. Type 1 | No. Type 2 | $y = \dfrac{\text{No.Type 1}}{\text{No.Type 2}}$ |
|---|---|---|---|
| 17 | 137 | 586 | .23 |
| 31 | 278 | 479 | .58 |
| 45 | 331 | 167 | 1.98 |
| 59 | 769 | 227 | 3.39 |
| 73 | 976 | 75 | 13.01 |

**12.18** Some researchers[3] illustrate a multiple linear regression equation to predict the yearly output of an oil field. We use their latest 20 years of data and change the units of three of their variables to

$y = \log_e$ (yearly output) where output is measured in 100,000 barrels

$x_1 = \log_e$ (number of startup wells) where the number of new startup wells is measured in 1000.

$x_2 = \log_e$ (water volume) where last year's injected water volume is measured in 100,000 gallons

The least squared fit, to $n = 20$ years of data, gives the results

$$\widehat{\beta}_0 = -.6415 \quad \widehat{\beta}_1 = .2869 \quad \widehat{\beta}_2 = .6532$$

$$\text{SS due to regression} = 6.2696$$

$$\text{SSE} = .0431$$

(a) Predict the response for
   (i) $x_1 = 1.5$ and $x_2 = 2.5$
   (ii) $x_1 = 2.5$ and $x_2 = 3.5$

(b) Estimate the error standard deviation $\sigma$ and state the degrees of freedom.

(c) What proportion of the $y$ variability is explained by the fitted regression?

**12.19** Refer to Exercise 12.18. The estimated standard errors of $\widehat{\beta}_1$ and $\widehat{\beta}_2$ are .0606 and .0632, respectively.

(a) Obtain a 90% confidence interval for $\beta_1$.

(b) Test $H_0: \beta_2 = .5$ versus $H_1: \beta_2 \neq .5$ with $\alpha = .05$.

**12.20** A second-degree polynomial $\widehat{y} = \widehat{\beta}_0 + \widehat{\beta}_1 x + \widehat{\beta}_2 x^2$ is fitted to a response $y$, and the

following predicted values and residuals are obtained.

| $\widehat{y}$ | Residual |
|---|---|
| 4.01 | .28 |
| 5.53 | −.33 |
| 6.21 | −.21 |
| 6.85 | .24 |
| 8.17 | −.97 |
| 8.34 | .46 |
| 8.81 | .79 |
| 9.62 | −1.02 |
| 10.05 | 1.35 |
| 10.55 | −1.55 |
| 10.77 | .63 |
| 10.77 | 1.73 |
| 10.94 | −2.14 |
| 10.98 | 1.92 |
| 10.98 | −1.18 |

Do the assumptions appear to be violated?

**12.21** The following predicted values and residuals are obtained in an experiment conducted to determine the degree to which the yield of an important chemical in the manufacture of penicillin is dependent on sugar concentration (the time order of the experiments is given in parentheses).

| Predicted | Residual |
|---|---|
| 2.2 (9) | −1 |
| 3.1 (6) | −2 |
| 2.5 (13) | 3 |
| 3.3 (1) | −3 |
| 2.3 (7) | −1 |
| 3.6 (14) | 5 |
| 2.6 (8) | 0 |
| 2.5 (3) | 0 |
| 3.0 (12) | 3 |
| 3.2 (4) | −2 |
| 2.9 (11) | 2 |
| 3.3 (2) | −5 |
| 2.7 (10) | 0 |
| 3.2 (5) | 1 |

---

[3]I. Mustafar and R. Razali, "A study of prediction of output in oil fields using multiple linear regression," *International Journal of Applied Science and Technology* **1** (4), (2011), pp. 107–113.

(a) Plot the residuals against the predicted values and also against the time order.

(b) Do the basic assumptions appear to be violated?

**12.22**   An experimenter obtains the following residuals after fitting a quadratic expression in $x$.

| $x = 11$ | $x = 2$ | $x = 3$ | $x = 4$ | $x = 5$ |
|---|---|---|---|---|
| −.1 | 1.3 | −.1 | 0 | −.2 |
| 0 | −.2 | −.3 | .2 | 0 |
| −.2 | −.1 | .1 | −.1 | −.2 |
| .6 | −.3 | .4 | 0 | −.2 |
| −.1 | .1 | −.1 | −.2 | −.3 |
|  |  | .1 |  | −.1 |

Do the basic assumptions appear to be violated?

**12.23**   An interested student used the method of least squares to fit the straight line $\hat{y} = 850.7 + 320.3x$ to gross domestic product, $y$, in real dollars. The results for 26 recent years, $x = 1, 2, \ldots, 26$, appear here. Which assumption(s) for a linear regression model appear to be seriously violated by the data? (*Note:* Regression methods are usually not appropriate for this type of data.)

| Year | $y$ | $\hat{y}$ | Residual |
|---|---|---|---|
| 1 | 1637.7 | 1171.0 | 466.7 |
| 2 | 1824.6 | 1491.4 | 333.2 |
| 3 | 2030.1 | 1811.7 | 218.4 |
| 4 | 2293.8 | 2132.0 | 161.8 |
| 5 | 2562.2 | 2452.3 | 109.9 |
| 6 | 2788.1 | 2772.6 | 15.5 |
| 7 | 3126.8 | 3092.9 | 33.9 |
| 8 | 3253.2 | 3413.3 | −160.1 |
| 9 | 3534.6 | 3733.6 | −199.0 |
| 10 | 3930.9 | 4053.9 | −123.0 |
| 11 | 4217.5 | 4374.2 | −156.7 |
| 12 | 4460.1 | 4694.5 | −234.4 |
| 13 | 4736.4 | 5014.8 | −278.4 |
| 14 | 5100.4 | 5335.2 | −234.8 |
| 15 | 5482.1 | 5655.5 | −173.4 |
| 16 | 5800.5 | 5975.8 | −175.3 |
| 17 | 5992.1 | 6296.1 | −304.0 |
| 18 | 6342.3 | 6616.4 | −274.1 |
| 19 | 6667.4 | 6936.8 | −269.4 |
| 20 | 7085.2 | 7257.1 | −171.9 |
| 21 | 7414.7 | 7577.4 | −162.7 |
| 22 | 7838.5 | 7897.7 | −59.2 |
| 23 | 8332.4 | 8218.0 | 114.4 |
| 24 | 8793.5 | 8538.3 | 255.2 |
| 25 | 9353.5 | 8858.7 | 494.8 |
| 26 | 9951.5 | 9179.0 | 772.5 |

## The Following Exercises Require a Computer

**12.24**   Consider the data on male wolves in Table D.9 of the Data Bank concerning age (years) and canine length (mm).

(a) Obtain the least squares fit of canine length to the predictor age.

(b) Obtain the least squares fit of canine length to a quadratic function of the predictor age. The MINITAB commands are

```
Data: DBT9
C2:  4       2      4 . . .  0
C6: 28.7   27.0   27.2 . . . 24.5
Dialog box:
Stat > Regression > Fitted line plot
Type C6 in Response.
Type C2 in Predictor.
Click Quadratic. Click OK.
```

(c) What proportion of the $y$ variability is explained by the quadratic regression model?

(d) Compare the estimated standard deviations, $s$, of the random error term in parts (a) and (b).

**12.25**   The resident population of the United States grew from 1910 to 2010 but the growth was not linear. The response variable is $y$ = population in millions and, to simplify the calculations, the predictor variable is $x$ = year − 1900.

| $x$ | 10 | 20 | 30 | 40 | 50 | 60 |
|---|---|---|---|---|---|---|
| $y$ | 92.2 | 106.0 | 123.2 | 132.2 | 151.3 | 179.3 |

| x | 70 | 80 | 90 | 100 | 110 |
|---|----|----|----|-----|-----|
| y | 203.3 | 226.5 | 248.7 | 281.4 | 310.2 |

(a) Fit a quadratic regression model
$Y = \beta_0 + \beta_1 x + \beta_2 x^2 + e$ to these data.

(b) What proportion of the $y$ variability is explained by the quadratic regression model?

(c) Test $H_0: \beta_1 = 0$ versus $H_1: \beta_1 > 0$ with $\alpha = .05$.

**12.26** Listed below are the price quotations on an Internet site for a midsize foreign used car along with their age and odometer mileage.

| Age (years) $x_1$ | Mileage $x_2$ (thousand miles) | Price $y$ (thousand of dollars) |
|---|---|---|
| 1 | 9 | 27.9 |
| 2 | 8 | 23.0 |
| 3 | 27 | 21.6 |
| 3 | 32 | 20.9 |
| 4 | 34 | 20.0 |
| 5 | 48 | 18.5 |
| 5 | 54 | 17.6 |
| 6 | 44 | 17.0 |
| 6 | 80 | 15.5 |
| 7 | 122 | 9.5 |

Perform a multiple regression analysis of these data. In particular

(a) Determine the equation for predicting the price from age and mileage. Interpret the meaning of the coefficients $\widehat{\beta}_1$ and $\widehat{\beta}_2$.

(b) Give 95% confidence intervals for $\beta_1$ and $\beta_2$.

(c) Obtain $R^2$ and interpret the result.

**12.27** Refer to the data of speed $x$ and stopping distance $y$ given in Table 1. The MINITAB commands for fitting a straight line regression to $y' = \sqrt{y}$ and $x$ are

Data: C12T1

C1: 20 20 30 30 30 40 40 50 50 60

C2: 16.3 26.7 39.2 63.5 51.3 98.4 65.7
104.1 155.6 217.2

Dialog box:

Calc > Calculator

Type SQRT( C2 ) in the **Expression** box.

Type C3 in **Store** box. Click **OK**.

Stat > Regression > Regression

Type C3 in **Response**.

Type C1 in **Predictors**. Click **OK**.

(a) Obtain the computer output and identify the equation of the fitted line and the value of $r^2$ (see Example 1).

(b) Give a 95% confidence interval for the slope.

(c) Obtain a 95% confidence interval for the expected $y'$ value at $x = 45$.

**12.28** A forester seeking information on basic tree dimensions obtains the following measurements of the diameters 4.5 feet above the ground and the heights of 12 sugar maple trees (courtesy of A. Ek). The forester wishes to determine if the diameter measurements can be used to predict the tree height.

(a) Plot the scatter diagram and determine if a straight line relation is appropriate.

(b) Determine an appropriate linearizing transformation. In particular, try $x' = \log x$, $y' = \log y$.

(c) Fit a straight line regression to the transformed data.

| Diameter $x$ (inches) | Height $y$ (feet) |
|---|---|
| .9 | 18 |
| 1.2 | 26 |
| 2.9 | 32 |
| 3.1 | 36 |
| 3.3 | 44.5 |
| 3.9 | 35.6 |
| 4.3 | 40.5 |
| 6.2 | 57.5 |
| 9.6 | 67.3 |
| 12.6 | 84 |
| 16.1 | 67 |
| 25.8 | 87.5 |

(d) What proportion of variability is explained by the fitted model?

**12.29**    Recorded here are the scores $x_1$ and $x_2$ in two midterm examinations, the GPA $x_3$, and the final examination score $y$ for 20 students in a statistics class.

| $x_1$ | $x_2$ | $x_3$ | $y$ | $x_1$ | $x_2$ | $x_3$ | $y$ |
|------|------|------|----|------|------|------|----|
| 87  | 25 | 2.9 | 60 | 93  | 60 | 3.2 | 44 |
| 100 | 84 | 3.3 | 80 | 92  | 69 | 3.1 | 53 |
| 91  | 52 | 3.5 | 73 | 100 | 86 | 3.6 | 86 |
| 85  | 60 | 3.7 | 83 | 80  | 67 | 3.5 | 59 |
| 56  | 76 | 2.8 | 33 | 100 | 96 | 3.8 | 81 |
| 81  | 28 | 3.1 | 65 | 69  | 51 | 2.8 | 20 |
| 85  | 67 | 3.1 | 53 | 80  | 75 | 3.6 | 64 |
| 96  | 83 | 3.0 | 68 | 74  | 70 | 3.1 | 38 |
| 79  | 60 | 3.7 | 88 | 79  | 66 | 2.9 | 77 |
| 96  | 69 | 3.7 | 89 | 95  | 83 | 3.3 | 47 |

(a) Ignoring the data of GPA and the first midterm score, fit a simple linear regression of $y$ on $x_2$. Compute $r^2$.

(b) Fit a multiple linear regression to predict the final examination score from the GPA and the scores in the midterms. Compute $R^2$.

(c) Interpret the values of $r^2$ and $R^2$ obtained in parts (a) and (b).

**12.30**    Refer to Exercise 11.64.

(a) Fit a quadratic model $Y = \beta_0 + \beta_1 x + \beta_2 x^2 + e$ to the data for CLEP scores $y$ and CQT scores $x$.

(b) Use the fitted regression to predict the expected CLEP score when $x = 160$.

(c) Compute $r^2$ for fitting a line and $R^2$ for fitting a quadratic expression. Interpret these values and comment on the improvement of fit.

**\*12.31**    Write the design matrix $X$ for fitting a multiple regression model to the data of Exercise 12.26.

**\*12.32**    Write the design matrix $X$ for fitting a quadratic regression model using the data of Exercise 12.25.

**12.33**    Refer to the physical conditioning data given in Table D.5 of the Data Bank. Use MINITAB or some other package to fit a regression of the final number of situps on the initial number of situps and the gender of the student.

**12.34**    Refer to the physical fitness data in Table D.5 of the Data Bank. Use both the data on the pretest run time and gender for predicting the posttest run time. Obtain the least squares fit and plot the residuals versus fitted value.

# 13

# Analysis of Categorical Data

Courtesy of Richard Johnson

## Are Firm Mattresses Really Best?

Although it is commonly believed that firm mattresses can help persons with low-back pain, experimental evidence is lacking. Medical researchers conducted a carefully designed experiment.[1] Of the patients admitted to the study, 158 were randomly selected to sleep on a new firm mattress and the others were given new medium-firm mattresses. The patients were not told about the firmness of their new mattresses although most perceived the correct firmness. The persons who installed the new beds and the research assistants who collected the data from patients were also kept unaware of the firmness of the mattresses.

This design includes meritable features: (1) the random assignment of treatments (mattress firmness) and (2) the *double blind* feature where patient and researcher are not told which treatment has been assigned.

The patients were asked to report whether or not they had low-back pain upon arising during the 90-day follow-up period.

| Firmness | Low Back Pain | | Total |
|---|---|---|---|
| | No | Yes | |
| Firm | 36 | 122 | 158 |
| Medium firm | 55 | 95 | 150 |

The proportion with no low-back pain upon arising is $.228 = 36/158$ for those who slept on firm mattresses and the proportion is $.367 = 55/150$ for those who slept on medium-firm mattresses. These data provide convincing evidence that medium-firm mattresses reduce low-back pain upon arising. The $P$-value for the test of equality of the proportions of persons in each population who would answer *No* versus a two-sided alternative is $.008$. In less than 1 in 100 times would chance variation give an observed difference this large or larger.

The term categorical data, introduced in Section 1 of Chapter 3, refers to observations that are only classified into categories. Such data are abundant in most fields of quantitative study, particularly in the social sciences. In a study of a persons current hair color, an observation can be classified as blond, brown, black, red, or other. In a survey of job compatibility, employed persons may be classified as being satisfied, neutral, or dissatisfied with their jobs. Testing problems that arise from categorical data are the subject of this chapter.

# 1. FORMULATING TESTING PROBLEMS CONCERNING CATEGORICAL DATA

The expression **categorical data** refers to observations that are classified in categories so the data set consists of the frequency, or count, for each category. In plant breeding, each offspring of a cross-fertilization may be classified into one of several genotypes; manufactured items may be sorted into categories called "free of defects," "slightly blemished," and "rejects"; and tablet owners may rate

---

[1] F. Kovacs et al., "Effect of firmness of mattress on chronic non-specific low-back pain: Randomized, double-blind, controlled, multicentre trial," *The Lancet* **362** (November 15, 2003), pp. 1599–1604.

their Internet provider's service as very poor, poor, average, good, or very good. Usually, a category is defined by a qualitative trait. However, the categories can also arise by grouping observations on an original numerical measurement scale. Based on specific ranges of values, income can be categorized as high, medium, or low, and rainfall can be categorized as heavy, moderate, or light.

The next three examples present the testing problems addressed in this chapter under the umbrella term of chi-square, or $\chi^2$, tests.

**Example 1**  One Sample Classified in Several Categories

The offspring produced by a cross between two given types of plants can be any of the three genotypes denoted by $A$, $B$, and $C$. A theoretical model of gene inheritance suggests that the offspring of types $A$, $B$, and $C$ should be in the ratio $1:2:1$. For experimental verification, 100 plants are bred by crossing the two given types. Their genetic classifications are recorded in Table 1. Do these data contradict the genetic model?

**TABLE 1**  Classification of Crossbred Plants

| Genotype | $A$ | $B$ | $C$ | Total |
|---|---|---|---|---|
| Observed frequency | 18 | 55 | 27 | 100 |

Let us denote the population proportions or the probabilities of the genotypes $A$, $B$, and $C$ by $p_A$, $p_B$, and $p_C$, respectively. Since the genetic model states that these probabilities are in the ratio $1:2:1$, our object is to test the null hypothesis

$$H_0: p_A = \frac{1}{4} \quad p_B = \frac{2}{4} \quad p_C = \frac{1}{4}$$

Here the data consist of frequency counts of a random sample classified in three categories or cells, the null hypothesis specifies the numerical values of the cell probabilities, and we wish to examine if the observed frequencies contradict the null hypothesis.

**Example 2**  Independent Samples Classified in Several Categories

To compare the effectiveness of two diets, $A$ and $B$, 150 infants were included in a study. Diet $A$ was given to 80 randomly selected infants and diet $B$ to the other 70 infants. At a later time, the health of each infant was observed and classified into one of the three categories "excellent," "average," and "poor." From the frequency counts recorded in Table 2, on page 330, we wish to test the null hypothesis that there is no difference between the quality of the two diets.

The results for the sample of infants receiving diet $A$ appear in the first row and those for the sample receiving diet $B$ in the second row. These two samples are independent. For a descriptive summary of these data, it is proper to compute the relative frequencies for each row. These are given in Table 2(a).

The (unknown) population proportions or probabilities are entered in Table 2(b). They allow us to describe the null hypothesis more clearly.

**TABLE 2**    Health under Two Different Diets

|  | Excellent | Average | Poor | Sample Size |
|---|---|---|---|---|
| Diet A | 37 | 24 | 19 | 80 |
| Diet B | 17 | 33 | 20 | 70 |
| Total | 54 | 57 | 39 | 150 |

**Table 2(a)**    Relative Frequencies (from Table 2)

|  | Excellent | Average | Poor | Total |
|---|---|---|---|---|
| Diet A | .46 | .30 | .24 | 1 |
| Diet B | .24 | .47 | .29 | 1 |

**Table 2(b)**    Population Proportions or Probabilities

|  | Excellent | Average | Poor | Total |
|---|---|---|---|---|
| Diet A | $p_{A1}$ | $p_{A2}$ | $p_{A3}$ | 1 |
| Diet B | $p_{B1}$ | $p_{B2}$ | $p_{B3}$ | 1 |

The null hypothesis of "no difference" is equivalent to the statement that, for each response category, the probability is the same for diet $A$ and diet $B$. Consequently, we formulate

$$H_0 : p_{A1} = p_{B1} \qquad p_{A2} = p_{B2} \qquad p_{A3} = p_{B3}$$

Note that although $H_0$ specifies a structure for the cell probabilities, it does not give the numerical value of the common probability in each column.

**Example 3**    One Sample Simultaneously Classified According to Two Characteristics

Improperly disclosed personal information has become a major source leading to credit fraud. In a survey conducted to help investigate the pervasiveness of the problem, 395 respondents reported receiving notification that their personal data had been compromised. Each of these cases were then classified according to type of institution sending the notification and also whether or not harm was caused to the person's credit. Harm to credit primarily includes instances of charging merchandise or taking money from bank accounts. Attention was restricted to institutions categorized as

Government (federal, state, and local)

Financial (bank, investment organization, credit firm, or insurance)

Other Commercial (Internet, retail, or communications company).

The cell frequencies for the two-way classification[2] are given in Table 3.

**TABLE 3**   Type of Organization and Fraudulent Use of Personal Information

|  | Government | Financial | Other Commercial | Total |
|---|---|---|---|---|
| Harm | 30 | 24 | 21 | 75 |
| No Harm | 182 | 104 | 34 | 320 |
| Total | 212 | 128 | 55 | 395 |

Do the data indicate that the pattern of cases of harmed credit, caused by the release of personal data, differs among the three types of organizations?

Unlike Example 2, here we have a single random sample, but each sampled individual elicits two types of response: source of notification and whether or not harm was done to credit. In the present context, the null hypothesis of "no difference" amounts to saying the outcome on credit and type of institution giving notification are independent. In other words, whether or not harm was done is unrelated to, or independent of, the type of institution that improperly disclosed personal information. A formal specification of this null hypothesis, in terms of the cell probabilities, is deferred until Section 4.

Frequency count data that arise from a classification of the sample observations according to two or more characteristics are called **cross-tabulated data** or a **contingency table**. If only two characteristics are observed and the contingency table has $r$ rows and $c$ columns, it is designated as an $r \times c$ table. Thus, Tables 2 and 3 are both $2 \times 3$ contingency tables.

Although Tables 2 and 3 have the same appearance, there is a fundamental difference in regard to the method of sampling. The row totals 80 and 70 in Table 2 are the predetermined sample sizes; these are not outcomes of random sampling, as are the column totals. By contrast, both sets of marginal totals in Table 3 are outcomes of random sampling—none were fixed beforehand. To draw the distinction, one often refers to Table 2 as a $2 \times 3$ contingency table with fixed row totals. In Sections 3 and 4, we will see that the formulation of the null hypothesis is different for the two situations.

## 2. PEARSON'S $\chi^2$ TEST FOR GOODNESS OF FIT

We first consider the type of problem illustrated in Example 1, where the data consist of frequency counts observed from a random sample and the null hypothesis specifies the unknown cell probabilities. Our primary goal is to test if the model given by the null hypothesis fits the data, and this is appropriately called **testing for goodness of fit**.

[2]The resulting proportions are close to those reported in a recent survey by Harris Interactive.

For general discussion, suppose a random sample of size $n$ is classified into $k$ categories or cells labeled $1, 2, \ldots, k$ and let $n_1, n_2, \ldots, n_k$ denote the respective cell frequencies. If we denote the cell probabilities by $p_1, p_2, \ldots, p_k$, a null hypothesis that completely specifies the cell probabilities is of the form

$$H_0: \quad p_1 = p_{10}, \ldots, p_k = p_{k0}$$

where $p_{10}, \ldots, p_{k0}$ are given numerical values that satisfy $p_{10} + \cdots + p_{k0} = 1$.

From Chapter 5 recall that if the probability of an event is $p$, then the expected number of occurrences of the event in $n$ trials is $np$. Therefore, once the cell probabilities are specified, the expected cell frequencies can be readily computed by **multiplying** these probabilities by the sample size $n$. A goodness of fit test attempts to determine if a conspicuous discrepancy exists between the observed cell frequencies and those expected under $H_0$. (See Table 4.)

**TABLE 4**   The Basis of a Goodness of Fit Test

| Cells | 1 | 2 | $\cdots$ | $k$ | Total |
|---|---|---|---|---|---|
| Observed frequency O | $n_1$ | $n_2$ | $\cdots$ | $n_k$ | $n$ |
| Probability under $H_0$ | $p_{10}$ | $p_{20}$ | $\cdots$ | $p_{k0}$ | 1 |
| Expected frequency $E$ under $H_0$ | $np_{10}$ | $np_{20}$ | $\cdots$ | $np_{k0}$ | $n$ |

A useful measure for the overall discrepancy between the observed and expected frequencies is given by the **chi-square** or $\chi^2$ statistic.

$$\chi^2 = \sum_{i=1}^{k} \frac{(n_i - np_{i0})^2}{np_{i0}} = \sum_{\text{cells}} \frac{(O - E)^2}{E}$$

where O and $E$ symbolize an observed frequency and the corresponding expected frequency. The discrepancy in each cell is measured by the squared difference between the observed and the expected frequencies divided by the expected frequency. The $\chi^2$ measure is the sum of these quantities for all cells.

Notice that the term $(O - E)^2 / E$ depends on more than just the difference $O - E$ between the observed and expected value. For instance $(95 - 100)^2 / 100 = .25$ is much smaller than $(5 - 10)^2 / 10 = 2.5$ although the difference is 5 in both cases. It is the relative difference between O and $E$, not just the actual difference $O - E$, that determines the contribution of $(O - E)^2 / E$ to the value of $\chi^2$.

The $\chi^2$ statistic was originally proposed by Karl Pearson (1857–1936), who found the distribution for large $n$ to be approximately a $\chi^2$ distribution with d.f. $= k - 1$. Due to this distribution, the statistic is denoted by $\chi^2$ and called **Pearson's $\chi^2$ statistic for goodness of fit**. Because a large value of the overall discrepancy indicates a disagreement between the data and the null hypothesis, the upper tail of the $\chi^2$ distribution constitutes the rejection region.

A description of the $\chi^2$ table (Appendix B, Table 6) and how to obtain an upper $\alpha$ point, $\chi^2_\alpha$, is given in Section 4 of Chapter 9.

---

### Pearson's $\chi^2$ Test for Goodness of Fit (Based on Large $n$)

**Null hypothesis**

$$H_0: p_1 = p_{10}, \ldots, p_k = p_{k0}$$

**Test statistic**

$$\chi^2 = \sum_{i=1}^{k} \frac{(n_i - np_{i0})^2}{np_{i0}} = \sum_{\text{cells}} \frac{(O - E)^2}{E}$$

**Rejection region**

$$\chi^2 \geq \chi^2_\alpha$$

where $\chi^2_\alpha$ is the upper $\alpha$ point of the $\chi^2$ distribution with

$$\text{d.f.} = k - 1 = (\text{Number of cells}) - 1$$

---

It should be remembered that Pearson's $\chi^2$ test is an approximate test that is valid only for large samples. As a rule of thumb, $n$ should be large enough so that the expected frequency of each cell is at least 5.

**Example 4**   $\chi^2$ Goodness of Fit for a Genetic Model

Referring to Example 1, test the goodness of fit of the genetic model to the data in Table 1. Take $\alpha = .05$.

SOLUTION   Following the structure of Table 4, the computations for the $\chi^2$ statistic are exhibited in Table 5 where the last line gives the calculation

$$\sum_{\text{cells}} \frac{(O - E)^2}{E} = \frac{(18 - 25)^2}{25} + \frac{(55 - 50)^2}{50} + \frac{(27 - 25)^2}{25}$$

$$= 1.96 \qquad + .50 \qquad + .16 \qquad = 2.62$$

**TABLE 5**   The $\chi^2$ Goodness of Fit Test for the Data in Table 1

| Cell | A | B | C | Total |
|---|---|---|---|---|
| Observed frequency O | 18 | 55 | 27 | 100 |
| Probability under $H_0$ | .25 | .50 | .25 | 1.0 |
| Expected frequency E | 25 | 50 | 25 | 100 |
| $\dfrac{(O - E)^2}{E}$ | 1.96 | .50 | .16 | $2.62 = \chi^2$<br>d.f. $= 2$ |

We use the $\chi^2$ statistic with rejection region $R$: $\chi^2 \geq 5.99$ since $\chi^2_{.05} = 5.99$ with d.f. $= 2$ (Appendix B, Table 6). Because the observed $\chi^2 = 2.62$ is smaller than this value, the null hypothesis is not rejected at $\alpha = .05$. We conclude that the data in Table 1 do not contradict the genetic model.

The $P$–value $= .270$, shown in Figure 1, confirms the conclusion.

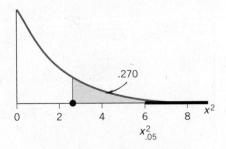

Figure 1    The $P$–value for Example 4 based on the $\chi^2$ distribution with 2 d.f.

The $\chi^2$ statistic measures the overall discrepancy between the observed frequencies and those expected under a given null hypothesis. Example 4 demonstrates its application when the frequency counts arise from a single random sample and the categories refer to only one characteristic—namely, the genotype of the offspring. Basically, the same principle extends to testing hypotheses with more complex types of categorical data such as the contingency tables illustrated in Examples 2 and 3. In preparation for these developments, we state two fundamental properties of the $\chi^2$ statistic:

---

### Properties of Pearson's $\chi^2$ Statistic

1. **Additivity:** If $\chi^2$ statistics are computed from independent samples, then their sum is also a $\chi^2$ statistic whose d.f. equals the sum of the d.f.'s of the components.

2. **Loss of d.f. due to estimation of parameters:** If $H_0$ does not completely specify the cell probabilities, then some parameters have to be estimated in order to obtain the expected cell frequencies. In that case, the d.f. of $\chi^2$ is reduced by the number of parameters estimated.

   d.f. of $\chi^2 = (\text{No. of cells}) - 1 - (\text{No. of parameters estimated})$

# Exercises

**13.1** Given below are the frequencies observed from 320 tosses of a die. Do these data cast doubt on the fairness of the die?

| Face No. | 1 | 2 | 3 | 4 | 5 | 6 | Total |
|----------|----|----|----|----|----|----|-------|
| Frequency | 39 | 63 | 56 | 67 | 57 | 38 | 320 |

**13.2** Recorded here is the frequency distribution of the blood types of 100 persons who have volunteered to donate blood at a plasma center.

| Blood type | O | A | B | AB | Total |
|------------|----|----|----|----|-------|
| Frequency | 40 | 44 | 10 | 6 | 100 |

Test the goodness of fit of the model, which assumes that the four blood types are equally likely in the population of plasma donors. Use $\alpha = .05$.

**13.3** Referring to the data in Exercise 13.2, test the null hypothesis that the probability of the blood types O, A, B, and AB is in the ratios $4:4:1:1$. Use $\alpha = .05$.

**13.4** A market researcher wishes to assess consumers' preference among three different colors available on name-brand dishwashers. The following frequencies were observed from a random sample of 150 recent sales.

| Color | White | Stainless | Black | Total |
|-------|-------|-----------|-------|-------|
| Frequency | 63 | 56 | 31 | 150 |

Test the null hypothesis, at $\alpha = .05$, that all three colors are equally popular.

**13.5** A shipment of assorted nuts is labeled as having 45% walnuts, 20% hazelnuts, 20% almonds, and 15% pistachios. By randomly picking several scoops of nuts from this shipment, an inspector finds the counts shown.

|  | Counts |
|-----------|--------|
| Walnuts | 95 |
| Hazelnuts | 70 |
| Almonds | 33 |
| Pistachios | 42 |
| Total | 240 |

Could these findings be a strong basis for an accusation of mislabeling? Test at $\alpha = .025$.

**13.6** Cross-fertilizing a pure strain of red flowers with a pure strain of white flowers produces pink hybrids that have one gene of each type. Crossing these hybrids can lead to any one of four possible gene pairs. Under Mendel's theory, these four are equally likely, so

$$P(\text{white}) = \frac{1}{4} \quad P(\text{pink}) = \frac{1}{2} \quad P(\text{red}) = \frac{1}{4}$$

An experiment carried out by Correns, one of Mendel's followers, resulted in the frequencies 141, 291, and 132 for the white, pink, and red flowers, respectively. (*Source*: W. Johannsen, 1909, *Elements of the Precise Theory of Heredity*, G. Fischer, Jena.)

Do these observations appear to contradict the probabilities suggested by Mendel's theory?

**13.7** According to the records of the National Safety Council, accidental deaths in the United States during a recent year had the following distribution according to the principal types of accidents.

| Motor Vehicle | Poison | Falls | Choking | Drowning | Other |
|---------|--------|-------|---------|----------|-------|
| 40.5% | 18.5% | 15.7% | 4.1% | 3.2% | 18.0% |

In a particular geographical region, the accidental deaths, classified according to the principal types of accidents, yielded the following frequency distribution.

| Motor Vehicle | Poison | Falls | Choking | Drowning | Other | Total |
|---------|--------|-------|---------|----------|-------|-------|
| 356 | 207 | 125 | 33 | 26 | 161 | 908 |

Do these data show a significantly different distribution of accidental deaths as compared to that for the entire United States? Test at $\alpha = .05$.

**13.8** The following table, based on government data, shows the frequency distribution of births by day of the week for all registered births in the United States in a recent year. Test the null hypothesis, at $\alpha = .01$, that all seven days of the week are equally likely for childbirth.

Number of Births (in 10,000) by Day of the Week, United States

| | Number of births |
|---|---|
| Mon. | 62.9 |
| Tues. | 69.3 |
| Wed. | 67.8 |
| Thurs. | 67.7 |
| Fri. | 64.3 |
| Sat. | 43.2 |
| Sun. | 37.9 |
| All Days | 413.1 |

*Source*: National Vital Statistics *Reports*, **60** (1) (November 2, 2011).

**13.9** Observations of 80 litters, each containing 3 rabbits, reveal the following frequency distribution of the number of male rabbits per litter.

| Number of males in litter | 0 | 1 | 2 | 3 | Total |
|---|---|---|---|---|---|
| Number of litters | | 19 | 32 | 22 | 7 | 80 |

Under the model of Bernoulli trials for the sex of rabbits, the probability distribution of the number of males per litter should be binomial with 3 trials and $p$ = probability of a male birth. From these data, the parameter $p$ is estimated as

$$\hat{p} = \frac{\text{Total number of males in 80 litters}}{\text{Total number of rabbits in 80 litters}}$$

$$= \frac{97}{240} \approx .4$$

(a) Using the binomial table for three trials and $p$ = .4, determine the cell probabilities.

(b) Perform the $\chi^2$ test for goodness of fit. (In determining the d.f., note that one parameter has been estimated from the data.)

**\*13.10** *Alternative expressions for Pearson's $\chi^2$.*

(a) By expanding the square on the right-hand side of

$$\chi^2 = \sum_{\text{cells}} \frac{(n_i - np_{i0})^2}{np_{i0}}$$

show that the $\chi^2$ statistic can also be expressed as

$$\chi^2 = \sum_{\text{cells}} \frac{n_i^2}{np_{i0}} - n \qquad \text{that is,}$$

$$\chi^2 = \sum_{\text{cells}} \frac{O^2}{E} - n$$

(b) Express the $\chi^2$ statistic as

$$\chi^2 = n \sum (\hat{p}_i - p_{i0})^2 / p_{i0}$$

where $\hat{p}_i = n_i / n$ is the observed relative frequency. Note that doubling the sample size, while maintaining the same observed and hypothesized proportions, doubles the value of $\chi^2$. This results in a smaller $P$-value.

# 3. CONTINGENCY TABLE WITH ONE MARGIN FIXED (TEST OF HOMOGENEITY)

From each population, we draw a random sample of a predetermined sample size and classify each response in categories. These data form a two-way contingency table where one classification refers to the populations and the other to the response under study. Our objective is to test whether the populations are alike, or **homogeneous**, with respect to cell probabilities. To do so, we will determine if the observed proportions in each response category are nearly the same for all populations.

Let us pursue our development of the $\chi^2$ test of homogenity with the data of Table 2.

**Example 5**    Developing a $\chi^2$ Test to Compare Two Diets

Referring to Example 2, test the null hypothesis that there is no difference between the quality of the two diets.

SOLUTION    For ease of reference, the data in Table 2 are reproduced in Table 6. Here the populations correspond to the two diets and the response is recorded in three categories. The row totals 80 and 70 are the fixed sample sizes.

**TABLE 6**    A 2 × 3 Contingency Table with Fixed Row Totals

|  | Excellent | Average | Poor | Total |
|---|---|---|---|---|
| Diet $A$ | 37 | 24 | 19 | 80 |
| Diet $B$ | 17 | 33 | 20 | 70 |
| Total | 54 | 57 | 39 | 150 |

We have already formulated the null hypothesis of "homogeneity" or "no difference between the diets" as [see Table 2(b)]

$$H_0: \quad p_{A1} = p_{B1} \qquad p_{A2} = p_{B2} \qquad p_{A3} = p_{B3}$$

If we denote these common probabilities under $H_0$ by $p_1$, $p_2$, and $p_3$, respectively, the expected cell frequencies in each row would be obtained by multiplying these probabilities by the sample size. In particular, the expected frequencies in the first row are $80\,p_1$, $80\,p_2$, and $80\,p_3$, and those in the second row are $70\,p_1$, $70\,p_2$, and $70\,p_3$. However, the $p_i$'s are not specified by $H_0$. Therefore, we have to estimate these parameters in order to obtain the numerical values of the expected frequencies.

The column totals 54, 57, and 39 in Table 6 are the frequency counts of the three response categories in the combined sample of size 150. Under $H_0$, the estimated probabilities are

$$\widehat{p}_1 = \frac{54}{150} \qquad \widehat{p}_2 = \frac{57}{150} \qquad \widehat{p}_3 = \frac{39}{150}$$

We use these estimates to calculate the expected frequencies in the first row as

$$80 \times \frac{54}{150} = \frac{80 \times 54}{150} = 28.8 \qquad \frac{80 \times 57}{150} = 30.4 \qquad \frac{80 \times 39}{150} = 20.8$$

and similarly for the second row. Referring to Table 6, notice the interesting pattern in these calculations:

$$\text{Expected cell frequency} = \frac{\text{Row total} \times \text{Column total}}{\text{Grand total}}$$

Table 7(a) presents the observed frequencies $O$ along with the expected frequencies $E$. The latter are given in parentheses. Table 7(b) computes the

**Table 7(a)**    The Observed and Expected Frequencies of the Data in Table 6

|        | Excellent | Average | Poor |
|--------|-----------|---------|------|
| Diet $A$ | 37 (28.8) | 24 (30.4) | 19 (20.8) |
| Diet $B$ | 17 (25.2) | 33 (26.6) | 37 (18.2) |

**Table 7(b)**    The Values of $(O - E)^2 / E$

|        | Excellent | Average | Poor |       |
|--------|-----------|---------|------|-------|
| Diet $A$ | 2.335 | 1.347 | .156 |       |
| Diet $B$ | 2.668 | 1.540 | .178 |       |
|        |           |         |      | $8.224 = \chi^2$ |

discrepancy measure $(O - E)^2 / E$ for the individual cells. Adding these over all the cells, we obtain the value of the $\chi^2$ statistic.

$$\chi^2 = 2.335 + 1.347 + .156 +$$
$$2.668 + 1.540 + .178 = 8.224$$

In order to determine the degrees of freedom, we employ the properties of the $\chi^2$ statistic stated in Section 2. Our $\chi^2$ statistic has been computed from two independent samples; each contributes $3 - 1 = 2$ d.f. because there are three categories. The added d.f. $= 2 + 2 = 4$ must now be reduced by the number of parameters we have estimated. Since $p_1$, $p_2$, and $p_3$ satisfy the relation $p_1 + p_2 + p_3 = 1$, there are really two undetermined parameters among them. Therefore, our $\chi^2$ statistic has d.f. $= 4 - 2 = 2$.

With d.f. $= 2$, the tabulated upper 5% point of $\chi^2$ is 5.99 (Appendix B, Table 6). Since the observed $\chi^2 = 8.224$ is larger, the null hypothesis is rejected at $\alpha = .05$. A computer calculation gives the $P$-value .016. Therefore, a significant difference between the quality of the two diets is indicated by the data.

Having obtained a significant $\chi^2$, we should now examine Tables 7(a) and 7(b) and try to locate the source of the significance. We find that large contributions to $\chi^2$ come from the "excellent" category, where the relative frequency is 37 / 80, or 46%, for diet $A$ and 17 / 70, or 24%, for diet $B$. These data indicate that diet $A$ is superior.

Motivated by Example 5, we are now ready to describe the $\chi^2$ test procedure for an $r \times c$ contingency table that has independent samples from $r$ populations that are classified in $c$ response categories. As we have seen before, the expected frequency of a cell is given by (row total $\times$ column total) / grand total. With regard to the d.f. of the $\chi^2$ for an $r \times c$ table, we note that each of the $r$ rows contributes $c - 1$ d.f.'s so the total contribution is $r(c - 1)$. Since $p_1 + p_2 + \cdots + p_c = 1$, only $c - 1$ of the parameters have to be estimated,

$$
\begin{aligned}
\text{d.f. of } \chi^2 &= r(c - 1) - (c - 1) \\
&= (r - 1)(c - 1) \\
&= (\text{No. of rows} - 1) \times (\text{No. of columns} - 1)
\end{aligned}
$$

---

### The $\chi^2$ Test of Homogeneity in a Contingency Table

#### Null hypothesis

In each response category, the probabilities are equal for all the populations.

#### Test statistic

$$
\chi^2 = \sum_{\text{cells}} \frac{(O - E)^2}{E} \qquad
\begin{cases}
O = \text{Observed cell frequency} \\
E = \dfrac{\text{Row total} \times \text{Column total}}{\text{Grand total}}
\end{cases}
$$

$$
\text{d.f.} = (\text{No. of rows} - 1) \times (\text{No. of columns} - 1)
$$

#### Rejection region

$$
\chi^2 \geq \chi^2_\alpha
$$

---

**Example 6**   A Designed Experiment to Curb Cheating on Tax Forms

Governments require many written forms that rely on the honesty of the person reporting. A designed experiment investigates a possible way to curb cheating on tax forms. A total of $n = 101$ participants were randomly assigned to each of three experimental conditions: write signature at the top of the tax form before filling out form, sign at the bottom after filling out form, and no signature required.

Participants took a timed 5 minute test using a worksheet containing twenty simple math puzzles. They had to identify, among 12 three-digit numbers, which two added up to ten and each correct answer was worth $1. Subjects were told this payment was unusually large because they would be taxed at 20% of their reported earnings. The participants self-reported their results on a one page tax form and signed it as assigned. Clearly, they could make more money by cheating and reporting an inflated number of correct answers.

Unknown to the participants, an imperceptible code enabled the investigators to identify both the actual and reported number of correct solutions.

**TABLE 8**   Reporting of Number Correct by the Participants

|  |  | Over Reported | Not Over Reported | Sample Size |
|---|---|---|---|---|
| Signature | Top | 13 ( 20.79 ) | 22 ( 14.21 ) | 35 |
|  | Bottom | 26 ( 19.60 ) | 7 ( 13.40 ) | 33 |
|  | None | 21 ( 19.60 ) | 12 ( 13.40 ) | 33 |

The observed frequency counts are given in Table 8 where the expected cell frequencies are shown in parentheses.[3]

Are any of three signing conditions best at reducing the probability that a person will report more than the actual number of problems solved?

SOLUTION   Let $p_T$, $p_B$, and $p_N$ be the probabilities of over reporting for the groups where the signature is on top, bottom, and not required. Based on independent random samples from three binomial populations, we want to test

$$H_0 : p_T = p_B = p_N$$

The expected cell frequencies, shown in parentheses in Table 8, are computed by multiplying the row and column totals and then dividing by the total sample size 101. The $\chi^2$ statistic is computed in Table 9.

**TABLE 9**   The Values of $( O - E )^2 / E$ for the Data in Table 8

|  |  | Over Reported | Not Over Reported |  |
|---|---|---|---|---|
| Signature | Top | 2.919 | 4.271 |  |
|  | Bottom | 2.090 | 3.057 |  |
|  | None | .100 | .146 |  |

$$12.58 = \chi^2$$
$$(3 - 1)(2 - 1) = 2 = \text{d.f. of } \chi^2$$

With d.f. = 2, the tabulated upper 5% point of the $\chi^2$ is 5.99. Consequently, the null hypothesis is rejected at $\alpha = .05$. Computer software produces $P$–value = .002 as shown in Figure 2.

The two largest contributions to the $\chi^2$ statistic are produced by (i) the large relative frequency of persons who do not over report when they sign at the top before filling out the form and (ii) the small relative frequency of persons who do not over report when their signature is at the bottom. Also, quite large contributions come from the small relative frequency of over reporting when

[3]L. Shu, et al., "Signing at the beginning makes ethics salient and decreases dishonest self-reports in comparison to signing at the end," *Proceedings of the National Academy of Sciences* 2012, **109** (38), pp. 15197–15200.

Figure 2   The $P$-value for Example 6
based on the $\chi^2$ distribution with 2 d.f.

the signature is at the top and the large relative frequency when the signature is at the bottom.

To further analyze the differences, we first calculate

$$\widehat{p}_T = \frac{13}{35} = .3714 \qquad \widehat{p}_B = \frac{26}{33} = .7879 \qquad \widehat{p}_N = \frac{21}{33} = .6364$$

The 95% large sample confidence interval for the difference in probabilities for bottom and top signing is

$$p_B - p_T : .7879 - .3714 \pm 1.96 \sqrt{\frac{.7879\,(.2121)}{33} + \frac{.3714\,(.7286)}{35}}$$

$$\text{or} \quad (.204, .629)$$

The probability of over reporting, for the group where subjects signed ahead of time on the top of the page, is .204 to .629 lower than for the group where no signature was required. Perhaps the simple task of having people sign a tax form, before they fill it out, would reduce cheating.

As is the case with the comparison of two populations, whenever possible it is desirable to take the active step of randomly assigning subjects to treatment groups. This key step for a designed experiment prevents uncontrolled sources of variation from systematically influencing the results. For instance, in the example above, an honest person has approximately equal chances of being assigned to any of the three groups.

**Example 7**   The $\chi^2$ Test for $2 \times 2$ Contingency Table

To determine the possible effect of a chemical treatment on the rate of seed germination, 100 chemically treated seeds and 150 untreated seeds are sown. The numbers of seeds that germinate are recorded in Table 10. Do the data provide strong evidence that the rate of germination is different for the treated and untreated seeds?

SOLUTION   Let $p_1$ and $p_2$ denote the probabilities of germination for the chemically treated seeds and the untreated seeds, respectively. We wish to test the null hypothesis

**TABLE 10**    Germination of Seeds

|          | Germinated    | Not Germinated | Total |
|----------|---------------|----------------|-------|
| Treated  | 84 ( 86.40 )  | 16 ( 13.60 )   | 100   |
| Untreated| 132 ( 129.60 )| 18 ( 20.40 )   | 150   |
| Total    | 216           | 34             | 250   |

$H_0 : p_1 = p_2$ versus $H_1 : p_1 \neq p_2$. For the $\chi^2$ test, we calculate the expected frequencies in the usual way. These are given in parentheses in Table 10. The computed value of $\chi^2$ is

$$\chi^2 = .067 + .424 + .044 + .282$$

$$= .817$$

$$\text{d.f.} = ( 2 - 1 )( 2 - 1 ) = 1$$

The tabulated 5% value of $\chi^2$ with d.f. $= 1$ is 3.84. Because the observed $\chi^2 = .817$ is smaller, the null hypothesis is not rejected at $\alpha = .05$. The $P$–value is .366. The rate of germination is not significantly different between the treated and untreated seeds.

## 3.1 ANOTHER METHOD OF ANALYZING A 2 × 2 CONTINGENCY TABLE

In light of Example 7, we note that a 2 × 2 contingency table, with one margin fixed, is essentially a display of independent random samples from two dichotomous (i.e., two-category) populations. This structure is shown in Table 11, where we have labeled the two categories "success" and "failure." Here $X$ and $Y$ denote the numbers of successes in independent random samples of sizes $n_1$ and $n_2$ taken from population 1 and population 2, respectively.

**TABLE 11**    Independent Samples from Two Dichotomous Populations

|              | No. of Successes | No. of Failures | Sample Size |
|--------------|------------------|-----------------|-------------|
| Population 1 | $X$              | $n_1 - X$       | $n_1$       |
| Population 2 | $Y$              | $n_2 - Y$       | $n_2$       |

If we let $p_1$ and $p_2$ denote the probabilities of success for populations 1 and 2, respectively, our objective is to test the null hypothesis $H_0 : p_1 = p_2$. The sample proportions

$$\hat{p}_1 = \frac{X}{n_1} \quad \text{and} \quad \hat{p}_2 = \frac{Y}{n_2}$$

provide estimates of $p_1$ and $p_2$. When the sample sizes are large, a test of $H_0 : p_1 = p_2$ can be based on (see Section 7 of Chapter 10)

**Test statistic**

$$Z = \frac{\widehat{p}_1 - \widehat{p}_2}{\sqrt{\widehat{p}(1 - \widehat{p})}\sqrt{\dfrac{1}{n_1} + \dfrac{1}{n_2}}} \qquad \text{where } \widehat{p} = \frac{X + Y}{n_1 + n_2}$$

The level $\alpha$ rejection region is $|Z| \geq z_{\alpha/2}$, $Z \leq -z_\alpha$, or $Z \geq z_\alpha$ according to whether the alternative hypothesis is $p_1 \neq p_2$, $p_1 < p_2$, or $p_1 > p_2$. Here $z_\alpha$ denotes the upper $\alpha$ point of the $N(0, 1)$ distribution.

Although the test statistics $Z$ and

$$\chi^2 = \sum_{\text{cells}} \frac{(O - E)^2}{E}$$

appear to have quite different forms, there is an exact relation between them—namely,

$$Z^2 = \chi^2 \quad (\text{for a } 2 \times 2 \text{ contingency table})$$

Also, $z^2_{\alpha/2}$ is the same as the upper $\alpha$ point of $\chi^2$ with d.f. $= 1$. For instance, with $\alpha = .05$, $z^2_{.025} = (1.96)^2 = 3.8416$, which is also the upper 5% point of $\chi^2$ with d.f. $= 1$ (see Appendix B, Table 6). Thus, the two test procedures are equivalent when the alternative hypothesis is two-sided. However, if the alternative hypothesis is one-sided, such as $H_1 : p_1 > p_2$, only the $Z$ test is appropriate.

**Example 8**   Conducting a $Z$ test to Compare Two Proportions

Use the $Z$ test with the data of Example 7. Also, determine a 95% confidence interval for $p_1 - p_2$.

SOLUTION   We calculate

$$\widehat{p}_1 = \frac{84}{100} = .84 \qquad \widehat{p}_2 = \frac{132}{150} = .88$$

$$\text{Pooled estimate } \widehat{p} = \frac{84 + 132}{100 + 150} = .864$$

$$Z = \frac{\widehat{p}_1 - \widehat{p}_2}{\sqrt{\widehat{p}(1 - \widehat{p})}\sqrt{\dfrac{1}{n_1} + \dfrac{1}{n_2}}}$$

$$= \frac{-.04}{\sqrt{.864 \times .136}\sqrt{\dfrac{1}{100} + \dfrac{1}{150}}} = -.904$$

Because the observed $|Z|$ is smaller than $z_{.025} = 1.96$, the null hypothesis is not rejected at $\alpha = .05$. Note that $Z^2 = (-.904)^2 = .817$ agrees with the result $\chi^2 = .817$ found in Example 7.

For the confidence interval, we calculate

$$\hat{p}_1 - \hat{p}_2 = -.04$$

$$\sqrt{\frac{\hat{p}_1(1-\hat{p}_1)}{n_1} + \frac{\hat{p}_2(1-\hat{p}_2)}{n_2}} = \sqrt{\frac{.84 \times .16}{100} + \frac{.88 \times .12}{150}}$$

$$= .045$$

A 95% confidence interval for $p_1 - p_2$ is

$$-.04 \pm 1.96 \times .045 = -.04 \pm .09 \quad \text{or} \quad (-.13, .05)$$

## Exercises

**13.11** Among a sample of 800 adult males, 414 said they usually open all of their mail. Among 900 adult females, 532 said they usually open all of their mail.[4]

(a) Construct a two-way table based on these frequencies.

(b) Formulate the null hypothesis.

(c) Conduct a $\chi^2$ test of your null hypothesis. Use $\alpha = .05$.

(d) Comment on the form of any departure from the null hypothesis.

**13.12** Chapter 3, Table 1, gives the counts of underclassmen and upperclassmen according to hours worked.

|  | No Job | Work 10 Hours or Less | Work More Than 10 Hours | Total |
|---|---|---|---|---|
| Underclassmen | 132 | 28 | 20 | 180 |
| Upperclassmen | 124 | 44 | 52 | 220 |
| Total | 256 | 72 | 72 | 400 |

(a) Formulate the null hypothesis of no difference in working hours.

(b) Conduct a $\chi^2$ test of your null hypothesis. Use $\alpha = .05$.

(c) Comment on the form of any departure from the null hypothesis.

**13.13** Nausea from air sickness affects some travelers. In a comparative study of the effectiveness of two brands of motion sickness pills, brand $A$ pills were given to 45 persons randomly selected from a group of 90 air travelers, while the other 45 persons were given brand $B$ pills. The following results were obtained.

|  | \multicolumn{4}{c}{Degree of Nausea} | |
|---|---|---|---|---|---|
|  | None | Slight | Moderate | Severe | Total |
| Brand $A$ | 18 | 17 | 6 | 4 | 45 |
| Brand $B$ | 11 | 14 | 14 | 6 | 45 |

Do these observations demonstrate that the two brands of pills are significantly different in quality? Test at $\alpha = .05$.

**13.14** Refer to the hardness of mattresses data in the beginning of the chapter. Confirm that these data establish a difference in the proportions who did not have lower back pain using:

(a) The $\chi^2$ test with level $\alpha = .01$.

(b) The $Z$ test with level $\alpha = .01$. Calculate the $P$-value.

**13.15** Referring to Example 6, but using only the two conditions "signature at bottom" after filling out the form and "no signature required."

(a) Conduct a $\chi^2$ test. Take $\alpha = .05$.

(b) Does the conclusion of your test prove that the two probabilities $p_N$ and $p_B$ are equal? Clearly explain the proper conclusion of your test.

---

[4]These proportions are close to those obtained in a recent Gallup survey.

**13.16** Referring to Example 6, but using only the two conditions "signature at the top" and "no signature required,"

(a) Make a 2 × 2 table and test $H_0 : p_T = p_N$ versus $H_1 : p_T < p_N$ at $\alpha = .02$.

(b) Obtain a 95% confidence interval for $p_N - p_T$.

**13.17** Concern over the quality of eyeglasses purchased over the Internet led to a study where many glasses were individually ordered from several different online vendors. Using only those lenses with antireflection coating, 61 had a thickness at the center greater than 1.9 mm and 31 were thinner. Of the 61 thicker lenses, 12 failed impact testing while 18 of the 31 thinner lens failed.[5]

(a) Perform a $\chi^2$ test of homogeneity. Take $\alpha = .05$.

(b) Comment on any pattern in the relative frequencies and what it means to this problem.

**13.18** Osteoporosis, or a loss of bone minerals, is a common cause of broken bones in the elderly. A researcher on aging conjectures that bone mineral loss can be reduced by regular physical therapy or certain kinds of physical activity. A study is conducted on 200 elderly subjects of approximately the same age divided into control, physical therapy, and physical activity groups. After a suitable period of time, the nature of change in bone mineral content is observed.

|  | Change in Bone Mineral | | | |
|---|---|---|---|---|
|  | Appreciable Loss | Little Change | Appreciable Increase | Total |
| Control | 38 | 15 | 7 | 60 |
| Therapy | 22 | 32 | 16 | 70 |
| Activity | 15 | 30 | 25 | 70 |
| Total | 75 | 77 | 48 | 200 |

Do these data indicate that the change in bone mineral varies for the different groups?

## The Following Exercises Require a Computer

**13.19** *Using the computer.* The analysis of a contingency table can be conveniently done on a computer. For an illustration, we present here a MINITAB analysis of the data in Table 6, Example 5.

---

Data: C13T6

C1: 37 17

C2: 24 33

C3: 19 20

Dialog box:

**Stat > Tables > Chisquare Test**
Type *C1-C3* in **Columns containing the table**. Click **OK**.

---

The **output** is as follows:

```
Chi-Square Test: C1, C2, C3

Expected counts are printed below
observed counts
Chi-Square contributions are printed
below expected counts
              C1      C2      C3   Total
   1          37      24      19      80
           28.80   30.40   20.80
           2.335   1.347   0.156

   2          17      33      20      70
           25.20   26.60   18.20
           2.668   1.540   0.178

Total         54      57      39     150

Chi-Sq = 8.224, DF = 2, P-Value =
0.016
```

(a) Compare this output with the calculations presented in Example 5.

(b) Do Exercise 13.11 on a computer.

(c) Do Exercise 13.18 on a computer.

**13.20** With reference to the sleep data in Table D.10 of the Data Bank, make two categories of snorers on the basis of the last variable: those who responded three or more times a week so their score is coded 4 or 5 and those who responded less than three times a week. Test the equality of proportions for males and females using the $\chi^2$ test with $\alpha = .05$.

---

[5]K. Citek et al., "Safety and compliance of prescription spectacles ordered by the public via the Internet," *Optometry* **82** (2011), pp. 549–555.

# 4. CONTINGENCY TABLE WITH NEITHER MARGIN FIXED (TEST OF INDEPENDENCE)

When two traits are observed for each element of a random sample, the data can be simultaneously classified with respect to these traits. We then obtain a two-way contingency table in which both sets of marginal totals are random. An illustration was already provided in Example 3. To cite a few other examples: A random sample of employed persons may be classified according to educational attainment and type of occupation; college students may be classified according to the year in college and attitude toward a dormitory regulation; flowering plants may be classified with respect to type of foliage and size of flower.

A typical inferential aspect of cross-tabulation is the study of whether the two characteristics appear to be manifested independently or certain levels of one characteristic tend to be associated or contingent with some levels of another.

**Example 9**    Developing a $\chi^2$ Test of Independence

Analyze the data in Example 3 concerning credit fraud and release of personal information.

SOLUTION    The $2 \times 3$ contingency table of Example 3 is given in Table 12. Here a single random sample of 395 persons, who were notified that personal credit information was released by the institution, were classified according to type of institution and whether or not personal credit was harmed.

**TABLE 12**    Type of Organization and Fraudulent Use of Personal Information

|  | Government | Financial | Other Commercial | Total |
|---|---|---|---|---|
| Harm | 30 | 24 | 21 | 75 |
| No Harm | 182 | 104 | 34 | 320 |
| Total | 212 | 128 | 55 | 395 |

Dividing the cell frequencies by the sample size 395, we obtain the relative frequencies shown in Table 13(a). Its row marginal totals .190 and .810

**Table 13(a)**    Proportion of Observations in Each Cell

|  | Government | Financial | Other Commercial | Total |
|---|---|---|---|---|
| Harm | .076 | .061 | .053 | .190 |
| No Harm | .461 | .263 | .086 | .810 |
| Total | .537 | .324 | .139 | 1.000 |

represent the sample proportions of cases where credit harm was inflicted and where it was not, respectively. Likewise, the column marginal totals show the sample proportions for the three types of institutions. For instance, Government was the source of proportion .537, or over half of the releases.

Imagine a classification of the entire population. The unknown population proportions (i.e., the probabilities of the cells) are represented by the entries in Table 13(b), where the suffixes H and N stand for Harm and No Harm, and 1, 2, and 3 refer to "Government," "Financial," and "Other Commercial" institutions. Table 13(b) is the population analogue of Table 13(a), which shows the sample proportions. For instance

$$\text{Cell probability } p_{H1} = P(\text{Harm and Government})$$
$$\text{Row marginal probability } p_H = P(\text{Harm})$$
$$\text{Column marginal probability } p_1 = P(\text{Government})$$

**Table 13(b)**   Probability of Each Cell

|  | Government | Financial | Other Commercial | Row Marginal Probability |
|---|---|---|---|---|
| Harm | $p_{H1}$ | $p_{H2}$ | $p_{H3}$ | $p_H$ |
| No Harm | $p_{N1}$ | $p_{N2}$ | $p_{N3}$ | $p_N$ |
| Column marginal probability | $p_1$ | $p_2$ | $p_3$ | 1 |

We are concerned with testing the null hypothesis that the two classifications are independent. Recall from Chapter 4 that the probability of the intersection of independent events is the product of their probabilities. Thus, the independence of the two classifications means that $p_{H1} = p_H p_1$, $p_{H2} = p_H p_2$, and so on.

Therefore, the null hypothesis of independence can be formalized as

$H_0$:   Each cell probability is the product of the
corresponding pair of marginal probabilities.

To construct a $\chi^2$ test, we need to determine the expected frequencies. Under $H_0$, the expected cell frequencies are

$$395\, p_H p_1 \quad 395\, p_H p_2 \quad 395\, p_H p_3$$
$$395\, p_N p_1 \quad 395\, p_N p_2 \quad 395\, p_N p_3$$

These involve the unknown marginal probabilities that must be estimated from the data. Referring to Table 12, we calculate the estimates as

$$\widehat{p}_H = \frac{75}{395} \quad \widehat{p}_N = \frac{320}{395}$$

$$\widehat{p}_1 = \frac{212}{395} \quad \widehat{p}_2 = \frac{128}{395} \quad \widehat{p}_3 = \frac{55}{395}$$

Then, the expected frequency in the first cell is estimated as

$$395 \, \widehat{p}_H \, \widehat{p}_1 = 395 \times \frac{75}{395} \times \frac{212}{395}$$

$$= \frac{75 \times 212}{395} = 40.25$$

Notice that the expected frequency for each cell of Table 12 is of the form

$$\frac{\text{Row total} \times \text{Column total}}{\text{Grand total}}$$

Table 14(a) presents the observed cell frequencies along with the expected frequencies shown in parentheses. The quantities $(O - E)^2 / E$ and the $\chi^2$ statistic are then calculated in Table 14(b).

**Table 14(a)**    The Observed and Expected Cell Frequencies for the Credit Fraud Data in Table 12

|          | Government | Financial | Other Commercial |
|----------|------------|-----------|------------------|
| Harm     | 30 (40.25) | 24 (24.30) | 21 (10.44) |
| No Harm  | 182 (171.75) | 104 (103.70) | 34 (44.56) |

**Table 14(b)**    The Values of $(O - E)^2 / E$

|          | Government | Financial | Other Commercial | Total |
|----------|------------|-----------|------------------|-------|
| Harm     | 2.612 | .004 | 10.672 | |
| No Harm  | .612 | .001 | 2.501 | |
|          |       |      |        | $16.402 = \chi^2$ |

Having calculated the $\chi^2$ statistic, it now remains to determine its d.f. by invoking the properties we stated in Section 2. Because we have a single random sample, only the property (2) is relevant to this problem. Since $p_H + p_N = 1$ and $p_1 + p_2 + p_3 = 1$, we have really estimated $1 + 2 = 3$ parameters. Hence,

$$\text{d.f. of } \chi^2 = (\text{No. of cells}) - 1 - (\text{No. of parameters estimated})$$

$$= 6 - 1 - 3$$

$$= 2$$

We choose level of significance $\alpha = .05$ and the tabulated upper 5% point of $\chi^2$ with d.f. $= 2$ is 5.99. Because the observed $\chi^2$ is larger than the

Figure 3    The $P$–value for Example 9
based on the $\chi^2$ distribution with 2 d.f.

tabulated value, the null hypothesis of independence is rejected at $\alpha = .05$. The very small $P$–value of .0003, shown in Figure 3, provides strong evidence against independence.

Continuing our analysis, we estimate the three conditional probabilities

$$\widehat{P}(\text{Harm} \mid \text{Government}) = \frac{30}{212} = .142$$

$$\widehat{P}(\text{Harm} \mid \text{Financial}) = \frac{24}{128} = .188$$

$$\widehat{P}(\text{Harm} \mid \text{Other}) = \frac{21}{55} = .382$$

The probability of harm is clearly highest when data are released from other financial institutions.

Inspection of Table 14(b) reveals that moderately large contributions from (i) the cell where Government released personal data has a smaller than expected number of cases where credit is harmed and (ii) the cell where Other Commercial institutions have smaller than expected number of cases where credit is not harmed. However, a gigantic contribution to $\chi^2$ comes from the cell where personal data released by Other Commercial institutions has much larger than expected number of cases where credit is harmed.

In all cases where you are notified that your personal information has been compromised, you need to check your credit reports and carefully read credit card statements. This is particularly true if other commercial institutions are involved.

From our analysis of the contingency table in Example 9, the $\chi^2$ test of independence in a general $r \times c$ contingency table is readily apparent. In fact, it is much the same as the test of homogeneity described in Section 3. The expected cell frequencies are determined in the same way—namely,

$$\text{Expected cell frequency} = \frac{\text{Row total} \times \text{Column total}}{\text{Grand total}}$$

and the test statistic is again

$$\chi^2 = \sum_{\text{cells}} \frac{(O - E)^2}{E}$$

With regard to the d.f. of $\chi^2$ in the present case, we initially have $rc - 1$ d.f. because there are $rc$ cells into which a single random sample is classified. From this, we must subtract the number of estimated parameters. This number is $(r - 1) + (c - 1)$ because there are $r - 1$ parameters among the row marginal probabilities and $c - 1$ parameters among the column marginal probabilities. Therefore,

$$
\begin{aligned}
\text{d.f. of } \chi^2 &= rc - 1 - (r - 1) - (c - 1) \\
&= rc - r - c + 1 \\
&= (r - 1)(c - 1) \\
&= (\text{No. of rows} - 1) \times (\text{No. of columns} - 1)
\end{aligned}
$$

which is identical to the d.f. of $\chi^2$ for testing homogeneity. In summary, the $\chi^2$ test statistic, its d.f., and the rejection region for testing independence are the same as when testing homogeneity. It is only the statement of the null hypothesis that is different between the two situations.

---

### The Null Hypothesis of Independence

$H_0$ : Each cell probability equals the product of the corresponding row and column marginal probabilities.

---

## 4.1 SPURIOUS DEPENDENCE

When the $\chi^2$ test leads to a rejection of the null hypothesis of independence, we conclude that the data provide evidence of a statistical association between the two characteristics. However, we must refrain from making the hasty interpretation that these characteristics are directly related. A claim of casual relationship must draw from common sense, which statistical evidence must not be allowed to supersede.

Two characteristics may appear to be strongly related due to the common influence of a third factor that is not included in the study. In such cases, the dependence is called a spurious dependence. For instance, if a sample of individuals is classified in a 2 × 2 contingency table according to whether or not they are heavy drinkers and whether or not they suffer from respiratory trouble, we would probably find a moderately high value for $\chi^2$ and conclude that a moderately strong statistical association exists between drinking habit and lung condition. But the reason for the association may be that most heavy drinkers are also heavy smokers and the smoking habit is a direct cause of respiratory trouble. This discussion should remind the reader of a similar warning given in Chapter 3 regarding the interpretation of a correlation coefficient between two sets of measurements.

# Exercises

**13.21** Applicants for public assistance are allowed an appeals process when they feel unfairly treated. At such a hearing, the applicant may choose self-representation or representation by an attorney. The appeal may result in an increase, decrease, or no change of the aid recommendation. Court records of 320 appeals cases provided the following data.

| Type of Representation | Amount of Aid | | |
| --- | --- | --- | --- |
| | Increase | Unchanged | Decrease |
| Self | 59 | 108 | 17 |
| Attorney | 70 | 63 | 3 |

Test the null hypothesis that the appeals decision and the type of representation are independent. Test at $\alpha = .05$.

**13.22** A consultant to all kinds of retailers suggests that they have plenty of baskets available for customers. He bases his suggestion on data collected by watching videos from hidden cameras. Suppose that out of 200 customers, 80 picked up a basket when they entered the store. Among those who picked up baskets 60 persons made purchases while only 41 of those without baskets made a purchase.

(a) Conduct a $\chi^2$ test of independence between purchasing and the decision to pick up a basket. Use $\alpha = .05$.

(b) The consultant may be confusing association with cause and effect. For instance, do you think the decision to pick up a basket has anything to do with the intent to purchase when a person enters the store? Comment.

**13.23** A survey was conducted by sampling 400 persons who were questioned regarding union membership and attitude toward decreased national spending on social welfare programs. The cross-tabulated frequency counts are presented.

| | Support | Indifferent | Opposed | Total |
| --- | --- | --- | --- | --- |
| Union | 112 | 36 | 28 | 176 |
| Nonunion | 84 | 68 | 72 | 224 |
| Total | 196 | 104 | 100 | 400 |

Can these observed differences be explained by chance or are attitude and membership status associated?

**13.24** Lying can have major economic and social implications. A double blind trial was conducted to study the effect of testosterone on lying. Among 91 healthy men, $n_1 = 46$ received 50 mg of testosterone on a skin patch while the control group, $n_2 = 45$, received a placebo look alike patch. Subsequently, subjects performed a task where their payoff depended on the self-reported outcome of a roll of a die. Subjects received the number of Euro that they reported for their roll of the die unless a 6 appeared and then the payoff was 0. Clearly, they could increase their payoff by lying without the danger of being caught.

The data on the number of times a subject reported a 5, the most lucrative outcome, are

| Placebo: 28 out of 45 times | Testosterone: 16 out of 46 times |
| --- | --- |

*Source:* Wibral M, et. al., "Testosterone administration reduces lying in men," PLoS ONE 7 (10): (2012) e46774, doi:10.1371/journal.pone.0046774.

(a) Conduct a $\chi^2$ test of homogeneity at level $\alpha = .05$.

(b) Has the testosterone increased or decreased lying? Base your answer on a large sample confidence interval for the difference of two proportions for reporting 5.

**13.25** In a study of factors that regulate behavior, three kinds of subjects are identified: overcontrollers, average controllers, and undercontrollers, with the first group being most inhibited. Each subject is given the routine task of filling a box with buttons and all subjects are told they can stop whenever they wish. Whenever a subject indicates he or she wishes to stop, the experimenter asks, "Don't you really want to continue?" The number of subjects in each group who stop and the number who continue are given in the following table.

| Controller | Continue | Stop | Total |
|------------|----------|------|-------|
| Over       | 9        | 9    | 18    |
| Average    | 8        | 12   | 20    |
| Under      | 3        | 14   | 17    |
| Total      | 20       | 35   | 55    |

Are controller group and continue/stop decision associated?

**13.26** Out of 120 members of a club who responded to a survey, 80 said that the golfing facilities were influential, 53 said the dining facilities were influential, and 25 said both were influential in their decision to join the club.

(a) Construct a two-way table of frequencies with two categories influential/not influential for each facility, golf and dining.

(b) Formulate the null hypothesis.

(c) Conduct a $\chi^2$ test of your null hypothesis. Use $\alpha = .05$.

(d) Comment on the form of any departure from the null hypothesis.

### The Following Exercises Require a Computer

**13.27** *Using the computer.* The analysis of a contingency table becomes tedious especially when the size of the table is large. Using a computer makes the task quite easy. We illustrate this by using MINITAB to analyze the data in Table 12, Example 9.

Data: C13T12

C*1*: 30 182

C2: 24 104

C3: 21 34

Dialog box:

**Stat > Tables > Chisquare Test**

Type C*1*-C3 in **Columns containing the table**. Click **OK**.

The **output** is as follows:

```
Chi-Square Test: C1, C2, C3
Expected counts are printed below
     observed counts
Chi-Square contributions are printed
   below expected counts
             C1        C2        C3   Total
1            30        24        21      75
          40.25     24.30     10.44
          2.612     0.004    10.672

2           182       104        34     320
         171.75    103.70     44.56
          0.612     0.001     2.501

Total       212       128        55     395

Chi-Sq = 16.402, DF = 2,
P-Value = 0.000
```

(a) Compare this output with the calculations presented in Example 9.

(b) Do Exercise 13.24 on a computer.

**13.28** Do Exercise 13.26 on a computer.

---

## USING STATISTICS WISELY

1. Moderately large sample sizes are required to detect differences in proportions. Usually 50 to 100 observations from each population are needed.

2. Although the $\chi^2$ test statistic is the same for testing independence and for testing equality of proportions, you should be clear which null hypothesis you are testing. When you sample from separate populations, the $\chi^2$ test concerns the equality of proportions. If a single sample is cross-classified according to two characteristics, the $\chi^2$ test concerns independence.

3. Don't routinely apply the inference procedures for comparing proportions when it is obvious that the outcomes have a time order dependence.

## KEY IDEAS AND FORMULAS

The term categorical data refers to observations that are only classified into categories so the data consist of frequency counts for the categories. When the frequency counts arise because observations are classified according to two or more characteristics, they are called cross-tabulated data or a contingency table. A chi-square ($\chi^2$) statistic compares the observed frequencies with those expected under a null hypothesis.

When a chi-square test of independence leads to the rejection of the null hypothesis, we say we have established a statistical association. There is no assertion of a causal relation as it may be a spurious dependence caused by a third variable.

### Pearson's $\chi^2$ Test for Goodness of Fit

Data: Observed cell frequencies $n_1, \ldots, n_k$ from a random sample of size $n$ classified into $k$ cells.

The null hypothesis specifies the cell probabilities

$$H_0: \quad p_1 = p_{10}, \ldots, p_k = p_{k0}$$

**Test statistic**

$$\chi^2 = \sum_{i=1}^{k} \frac{(n_i - np_{i0})^2}{np_{i0}} \qquad \text{d.f.} = k - 1$$

**Rejection region** $\chi^2 \geq \chi_\alpha^2$

### $\chi^2$ Test of Homogeneity in an $r \times c$ Contingency Table

Data: Independent random samples from $r$ populations, each sample classified in $c$ response categories.

Null hypothesis: In each response category, the probabilities are equal for all the populations.

**Test statistic**

$$\chi^2 = \sum_{\text{cells}} \frac{(O - E)^2}{E} \qquad \text{d.f.} = (r - 1)(c - 1)$$

where for each cell

$$O = \text{Observed cell frequency}$$

$$E = \frac{\text{Row total} \times \text{Column total}}{\text{Grand total}}$$

**Rejection region** $\chi^2 \geq \chi_\alpha^2$

### $\chi^2$ Test of Independence in an $r \times c$ Contingency Table

Data: A random sample of size $n$ is simultaneously classified with respect to two characteristics, one has $r$ categories and the other $c$ categories.

Null hypothesis: The two classifications are independent; that is, each cell probability is the product of the row and column marginal probabilities.

Test statistic and rejection region: Same as when testing homogeneity.

### Limitation

All inference procedures of this chapter require large samples. The $\chi^2$ tests are appropriate if no expected cell frequency is too small ($\geq 5$ is normally required).

## TECHNOLOGY

### Conducting a $\chi^2$ test

### MINITAB

### Conducting a $\chi^2$ test

We illustrate the calculation of the $\chi^2$ statistic with an example. Enter the counts

```
30  42  28
 9  10  31
```

in columns 1 to 3. Select the MINITAB commands

**Stat > Tables > Chisquare Test.**

Type $C1 - C3$ in **columns containing the table.**

Click **OK.**

### EXCEL

### Calculating a $\chi^2$ statistic

Enter observed values in a rectangular range of cells.
Enter the expected values in another rectangular range of cells.

Click on the **Formulas** tab and then **More Functions.** Select **Statistical,** and then **CHISQ.TEST.**

With the cursor in the textbox for **Actual_range,** highlight the observed values.

With the cursor in the textbox for **Expected_range,** highlight the expected values.

Click **OK.**

The software will return the $P$-value.

**TI-84/-83 PLUS**

### Calculating a $\chi^2$ statistic

We first enter the observed counts and expected values as matrices.

Press **2nd**, then **MATRIX** and select **EDIT**.

Press **ENTER** and type 2 × 3 which is the number or rows and number of columns.

Type each entry for **[ A ]**, an observed value, and press **ENTER** each time.

Press **2nd** then **QUIT**. Press **MATRIX** and select **EDIT** again.

Select **MATRIX [ B ]**, press **ENTER**, and type the number of rows, number of columns, and enter the expected values. Press **2nd** then **QUIT**.

Select **STAT** then **TESTS** and then **C: $X^2$-Test**. Press **ENTER**.

Arrow down to select **Calculate**. Enter the appropriate matrix names (matrices [A] and [B] are the defaults) and then press **ENTER**.

The software returns the value of the $\chi^2$ statistic, the $P$–value, and the degrees of freedom.

## REVIEW EXERCISES

**13.29** To examine the quality of a random number generator, frequency counts of the individual integers are recorded from an output of 500 integers. The concept of randomness implies that the integers 0, 1, . . . , 9 are equally likely. Based on the observed frequency counts, would you suspect any bias of the random number generator? Answer by performing the $\chi^2$ test.

| Integer | 0 | 1 | 2 | 3 | 4 | 5 | 6 | 7 | 8 | 9 | Total |
|---|---|---|---|---|---|---|---|---|---|---|---|
| Frequency | 41 | 58 | 51 | 61 | 39 | 56 | 45 | 35 | 62 | 52 | 500 |

**13.30** The following table records the observed number of births at a hospital in four consecutive quarterly periods.

| Quarters | Jan.–March | April–June | July–Sept. | Oct.–Dec. |
|---|---|---|---|---|
| Number of births | 55 | 29 | 26 | 41 |

It is conjectured that twice as many babies are born during the Jan.–March quarter than any of the other three quarters. At $\alpha = .10$, test if these data strongly contradict the stated conjecture.

**13.31** A community paper in the Spanish language was delivered to many sites in the San Francisco Bay area. As a check on the circulation numbers that are important to advertisers, a survey was conducted at four drop sites. The number of papers delivered to each site and the number remaining after 3 days was recorded.

| | Papers Delivered | Papers Remaining |
|---|---|---|
| Site 1 | 50 | 17 |
| Site 2 | 47 | 12 |
| Site 3 | 48 | 7 |
| Site 4 | 50 | 21 |

(a) Formulate a null hypothesis of no difference in the proportions of papers taken from the sites.

(b) Conduct a $\chi^2$ test of your null hypothesis. Use $\alpha = .05$.

(c) Let $p_1$, $p_2$, $p_3$, and $p_4$ denote the probabilities that a paper will be taken from drop sites 1, 2, 3, and 4, respectively. Construct the four individual 95% confidence intervals and plot these intervals.

**13.32** Referring to the data for site 3 and site 4 in Exercise 13.31, make a 2 × 2 table and test $H_0: p_3 = p_4$ versus $H_1: p_3 > p_4$ at $\alpha = .05$.

(a) Is there strong evidence that the probability a paper will be taken from the drop site is higher for site 3 than for site 4? Answer by calculating the $P$–value.

(b) Construct a 95% confidence interval for $P_3 - p_4$.

**13.33** Refer to Exercise 3.48 and the data concerning a vaccine for type B hepatitis.

|  | Hepatitis | No Hepatitis | Total |
|---|---|---|---|
| Vaccinated | 11 | 538 | 549 |
| Not vaccinated | 70 | 464 | 534 |
| Total | 81 | 1002 | 1083 |

Do these data indicate that there is a different rate of incidence of hepatitis between the vaccinated and nonvaccinated participants? Use the $\chi^2$ test for homogeneity in a contingency table.

**13.34** Refer to the data in Exercise 13.33.

(a) Use the $Z$ test for testing the equality of two population proportions with a two-sided alternative. Verify the relation $\chi^2 = Z^2$ by comparing their numerical values.

(b) If the alternative is that the incidence rate is lower for the vaccinated group, which of the two tests should be used?

**13.35** To compare the effectiveness of four drugs in relieving postoperative pain, an experiment was done by randomly assigning 195 surgical patients to the drugs under study. Recorded here are the number of patients assigned to each drug and the number of patients who were free of pain for a period of five hours.

|  | Free of Pain | No. of Patients Assigned |
|---|---|---|
| Drug 1 | 23 | 53 |
| Drug 2 | 30 | 47 |
| Drug 3 | 19 | 51 |
| Drug 4 | 29 | 44 |

(a) Make a 4 × 2 contingency table showing the counts of patients who were free of pain and those who had pain, and test the null hypothesis that all four drugs are equally effective. (Use $\alpha = .05$.)

(b) Let $p_1$, $p_2$, $p_3$, and $p_4$ denote the population proportions of patients who would be free of pain under the use of drugs 1, 2, 3, and 4, respectively. Calculate a 90% confidence interval for each of these probabilities individually.

**13.36** Using the data for drugs 1 and 3 in Exercise 13.35, make a 2 × 2 contingency table and test $H_0: p_1 = p_3$ versus $H_1: p_1 \neq p_3$ at $\alpha = .05$ employing:

(a) The $\chi^2$ test.

(b) The $Z$ test.

**13.37** In a study conducted to monitor the quality of eyeglasses purchased over the Internet, 154 pairs of glasses were received from individual orders placed with several different online vendors. The lenses were subject to both an optical analysis and a standard impact test. A pair of glasses passes a test only if both lenses pass. The outcomes for these two tests are summarized as

Number of Pairs of Glasses

|  | Passed Optical | Failed Optical | Total |
|---|---|---|---|
| Passed Impact | 85 | 34 | 119 |
| Failed Impact | 25 | 10 | 35 |
| Total | 110 | 44 | 154 |

*Source*: K. Citek et al., "Safety and compliance of prescription spectacles ordered by the public via the Internet," *Optometry* 82 (2011), pp. 549–555.

(a) Perform a $\chi^2$ test of dependence. Take $\alpha = .01$.

(b) This is a very unusual case. Does it prove independence? Explain.

**13.38** In a study on the effect of diet and lifestyle on heart disease, 96 patients with severe coronary blockage were randomly assigned, 49 to an experimental group and 47 to a control group. The patients in the experimental group had a low-fat vegetarian diet, regular exercise, and stress-management training, whereas those in the control group had a low-fat diet and moderate exercise. The condition of their coronary blockage was monitored throughout the study period, and the following results were noted.

| Group | Coronary Blockage Worsened | No Change | Improved | Total |
|---|---|---|---|---|
| Experimental | 4 | 8 | 37 | 49 |
| Control | 8 | 25 | 14 | 47 |
| | | | | 96 |

Analyze the data to determine if the changes in coronary blockage were significantly different between the two groups of patients.

**13.39** Based on interviews of couples seeking divorces, a social worker compiles the following data related to the period of acquaintanceship before marriage and the duration of marriage.

| Acquaintanceship Before Marriage | Duration of Marriage ≤ 4 years | > 4 years | Total |
|---|---|---|---|
| Under $\frac{1}{2}$ year | 11 | 8 | 19 |
| $\frac{1}{2}$ – $1\frac{1}{2}$ years | 28 | 24 | 52 |
| Over $1\frac{1}{2}$ years | 21 | 19 | 40 |
| Total | 60 | 51 | 111 |

Perform a test to determine if the data substantiate an association between the stability of a marriage and the period of acquaintanceship prior to marriage.

**13.40** By polling a random sample of 350 undergraduate students, a campus press obtains the following frequency counts regarding student attitude toward a proposed change in dormitory regulations.

| | Favor | Indifferent | Oppose | Total |
|---|---|---|---|---|
| Male | 95 | 72 | 19 | 186 |
| Female | 53 | 79 | 32 | 164 |
| Total | 148 | 151 | 51 | 350 |

Are attitude toward the proposal and gender associated?

**13.41** In a genetic study of chromosome structures, 132 individuals are classified according to the type of structural chromosome aberration and carriers in their parents. The following counts are obtained.

| Type of Aberration | Carrier One Parent | Neither Parent | Total |
|---|---|---|---|
| Presumably Innocuous | 27 | 20 | 47 |
| Substantially Unbalanced | 36 | 49 | 85 |
| Total | 63 | 69 | 132 |

Test the null hypothesis that type of aberration is independent of parental carrier.

**13.42** A random sample of 130 business executives was classified according to age and the degree of risk aversion as measured by a psychological test.

| Age | Degree of Risk Aversion Low | Medium | High | Total |
|---|---|---|---|---|
| Below 45 | 14 | 22 | 7 | 43 |
| 45–55 | 16 | 33 | 12 | 61 |
| Over 55 | 4 | 15 | 7 | 26 |
| | | | | 130 |

Do these data demonstrate an association between risk aversion and age?

**13.43** *Pooling contingency tables can produce spurious association.* A large organization is being investigated to determine if its recruitment is sex-biased. Tables 15 and 16, respectively, show the classification of applicants for secretarial and for sales positions according to gender and result of interview.

Table 17 is an aggregation of the corresponding entries of Table 15 and Table 16.

**TABLE 15**   Secretarial Positions

|        | Offered | Denied | Total |
|--------|---------|--------|-------|
| Male   | 25      | 50     | 75    |
| Female | 75      | 150    | 225   |
| Total  | 100     | 200    | 300   |

**TABLE 16**   Sales Positions

|        | Offered | Denied | Total |
|--------|---------|--------|-------|
| Male   | 150     | 50     | 200   |
| Female | 75      | 25     | 100   |
| Total  | 225     | 75     | 300   |

**TABLE 17**   Secretarial and Sales Positions

|        | Offered | Denied | Total |
|--------|---------|--------|-------|
| Male   | 175     | 100    | 275   |
| Female | 150     | 175    | 325   |
| Total  | 325     | 275    | 600   |

(a) Verify that the $\chi^2$ statistic for testing independence is zero for each of the data sets given in Tables 15 and 16.

(b) For the pooled data given in Table 17, compute the value of the $\chi^2$ statistic and test the null hypothesis of independence.

(c) Explain the paradoxical result that there is no sex bias in any job category, but the combined data indicate sex discrimination.

# 14

# Analysis of Variance (ANOVA)

©Richard Levine/Alamy Limited

## Which Brand HDTV Has the Clearest Picture?

A good experimental design is to collect samples of several HDTVs of each brand and measure their picture clarity. The statistical technique called "analysis of variance" enables us to verify differences among the brands.

In Chapter 10, we introduced methods for comparing two population means. When several means must be compared, more general methods are required. We now become acquainted with the powerful technique called **analysis of variance (ANOVA)** that allows us to analyze and interpret observations from several populations. This versatile statistical tool partitions the total variation in a data set according to the sources of variation that are present. In the context of comparing $k$ population means, the two sources of variation are (1) differences **between** means or **treatments** and (2) **within** population variation **(error)**. We restrict our discussion to this case, although ANOVA techniques apply to much more complex situations.

In this chapter, you will learn how to test for differences among several means and to make confidence statements about pairs of means.

# 1. COMPARISON OF SEVERAL TREATMENTS—ONE-WAY ANALYSIS OF VARIANCE

It is usually more expedient in terms of both time and expense to simultaneously compare several treatments than it is to conduct several comparative trials of two treatments at a time. The term **one-way analysis of variance** is synonymous with independent random sampling from several populations when each population is identified as the population of responses under a particular treatment. Let treatment 1 be applied to $n_1$ experimental units, treatment 2 to $n_2$ units, . . . , treatment $k$ to $n_k$ units. In a completely randomized design, $n_1$ experimental units selected at random from the available collection of $n = n_1 + n_2 + \cdots + n_k$ units are to receive treatment 1, $n_2$ units randomly selected from the remaining units are to receive treatment 2, and proceeding in this manner, treatment $k$ is to be applied to the remaining $n_k$ units. The special case of this design for a comparison of $k = 2$ treatments has already been discussed in Section 1 of Chapter 10. The data structure for the response measurements can be represented by the format shown in Table 1, where $y_{ij}$ is the $j$th observation on treatment $i$. The summary statistics appear in the last two columns.

Before proceeding with the general case of $k$ treatments, it would be instructive to explain the reasoning behind the analysis of variance and the associated calculations in terms of a numerical example.

**Example 1**    The Structure of Data from an Experiment for Comparing Four Means

Four models of high-end earbuds are compared for sound reproduction. There are 5 sets of earbud A, 4 of earbud B, 7 of earbud C, and 6 of earbud D available for testing. The quality of sound reproduction can be determined objectively by measuring audio signals received by a robot head wearing earbuds and then comparing them with the known signal wave that was sent. Quantitatively, the measure of sound distortion called total harmonic distortion is an overall

**TABLE 1** Data Structure for the Completely Randomized Design with $k$ Treatments

| | Observations | Mean | Sum of Squares |
|---|---|---|---|
| Treatment 1 | $y_{11}, y_{12}, \ldots, y_{1n_1}$ | $\bar{y}_1$ | $\sum_{j=1}^{n_1} (y_{1j} - \bar{y}_1)^2$ |
| Treatment 2 | $y_{21}, y_{22}, \ldots, y_{2n_2}$ | $\bar{y}_2$ | $\sum_{j=1}^{n_2} (y_{2j} - \bar{y}_2)^2$ |
| $\vdots$ | $\vdots$ | $\vdots$ | $\vdots$ |
| Treatment $k$ | $y_{k1}, y_{k2}, \ldots, y_{kn_k}$ | $\bar{y}_k$ | $\sum_{j=1}^{n_k} (y_{kj} - \bar{y}_k)^2$ |

$$\text{Grand mean } \bar{y} = \frac{\text{Sum of all observations}}{n_1 + n_2 + \cdots + n_k} = \frac{n_1 \bar{y}_1 + \cdots + n_k \bar{y}_k}{n_1 + \cdots + n_k}$$

measure of the discrepancy, in percent. Because the values are substantially below 1% for high quality earbuds, we give the values for distortion in hundredths of a percent so 10 is .1% and so on. Suppose the test results for sound distortion produce the data in Table 2.

**TABLE 2** Sound Distortion Obtained with Four Brands of Earbuds

| Earbud | Observations | Mean | Sum of Squares |
|---|---|---|---|
| $A$ | 10 15 8 12 15 | $\bar{y}_1 = 12$ | $\sum_{j=1}^{5} (y_{1j} - \bar{y}_1)^2 = 38$ |
| $B$ | 14 18 21 15 | $\bar{y}_2 = 17$ | $\sum_{j=1}^{4} (y_{2j} - \bar{y}_2)^2 = 30$ |
| $C$ | 17 16 14 15 17 15 18 | $\bar{y}_3 = 16$ | $\sum_{j=1}^{7} (y_{3j} - \bar{y}_3)^2 = 12$ |
| $D$ | 12 15 17 15 16 15 | $\bar{y}_4 = 15$ | $\sum_{j=1}^{6} (y_{4j} - \bar{y}_4)^2 = 14$ |
| | Grand mean $\bar{y} = 15$ | | |

Two questions immediately come to mind. Does any significant difference exist among the mean distortions obtained using the four brands of earbuds? Can we establish confidence intervals for the mean differences between brands?

An analysis of the results essentially consists of decomposing the observations into contributions from different sources. We reason that the deviation of an individual observation from the grand mean, $y_{ij} - \bar{y}$, is partly due to differences among the mean qualities of the brands and partly due to random variation in measurements within the same group. This suggests the following decomposition.

$$\text{Observation} = \binom{\text{Grand}}{\text{mean}} + \binom{\text{Deviation due}}{\text{to treatment}} + (\text{Residual})$$

$$y_{ij} = \bar{y} + (\bar{y}_i - \bar{y}) + (y_{ij} - \bar{y}_i)$$

For the data given in Table 2, the decomposition of all the observations can be presented in the form of the following arrays:

**Observations**

$$y_{ij}$$

$$\begin{bmatrix} 10 & 15 & 8 & 12 & 15 \\ 14 & 18 & 21 & 15 \\ 17 & 16 & 14 & 15 & 17 & 15 & 18 \\ 12 & 15 & 17 & 15 & 16 & 15 \end{bmatrix}$$

**Grand mean**

$$\bar{y}$$

**Treatment effects**

$$(\bar{y}_i - \bar{y})$$

$$= \begin{bmatrix} 15 & 15 & 15 & 15 & 15 \\ 15 & 15 & 15 & 15 \\ 15 & 15 & 15 & 15 & 15 & 15 & 15 \\ 15 & 15 & 15 & 15 & 15 & 15 \end{bmatrix} + \begin{bmatrix} -3 & -3 & -3 & -3 & -3 \\ 2 & 2 & 2 & 2 \\ 1 & 1 & 1 & 1 & 1 & 1 & 1 \\ 0 & 0 & 0 & 0 & 0 & 0 \end{bmatrix}$$

**Residuals**

$$(y_{ij} - \bar{y}_i)$$

$$+ \begin{bmatrix} -2 & 3 & -4 & 0 & 3 \\ -3 & 1 & 4 & -2 \\ 1 & 0 & -2 & -1 & 1 & -1 & 2 \\ -3 & 0 & 2 & 0 & 1 & 0 \end{bmatrix}$$

For instance, the upper left-hand entries of the arrays show that

$$10 = 15 + (-3) + (-2)$$

$$y_{11} = \bar{y} + (\bar{y}_1 - \bar{y}) + (y_{11} - \bar{y}_1)$$

If there is really no difference in the mean distortions obtained using the four brands of earbuds, we can expect the entries of the second array on the right-hand side of the equation, whose terms are $\bar{y}_i - \bar{y}$, to be close to zero. As an overall measure of the amount of variation due to differences in the treatment means, we calculate the sum of squares of all the entries in this array, or

$$\underbrace{(-3)^2 + \cdots + (-3)^2}_{n_1 = 5} + \underbrace{2^2 + \cdots + 2^2}_{n_2 = 4} + \underbrace{1^2 + \cdots + 1^2}_{n_3 = 7} + \underbrace{0^2 + \cdots + 0^2}_{n_4 = 6}$$

$$= 5(-3)^2 + 4(2)^2 + 7(1)^2 + 6(0)^2$$

$$= 68$$

Thus, the sum of squares due to differences in the treatment means, also called the treatment sum of squares, is given by

$$\text{Treatment sum of squares} = \sum_{i=1}^{4} n_i (\bar{y}_i - \bar{y})^2 = 68$$

The last array consists of the entries $y_{ij} - \bar{y}_i$ that are the deviations of individual observations from the corresponding treatment mean. These deviations reflect inherent variabilities in the material, fabrication, and the measuring device and are called the **residuals**. The overall variation due to random errors is measured by the sum of squares of all these residuals

$$(-2)^2 + 3^2 + (-4)^2 + \cdots + 1^2 + 0^2 = 94$$

Thus, we obtain

$$\text{Error sum of squares} = \sum_{i=1}^{4} \sum_{j=1}^{n_i} (y_{ij} - \bar{y}_i)^2 = 94$$

The double summation indicates that the elements are summed within each row and then over different rows. Alternatively, referring to the last column in Table 2, we obtain

$$\text{Error sum of squares} = \sum_{j=1}^{5} (y_{1j} - \bar{y}_1)^2 + \sum_{j=1}^{4} (y_{2j} - \bar{y}_2)^2$$

$$+ \sum_{j=1}^{7} (y_{3j} - \bar{y}_3)^2 + \sum_{j=1}^{6} (y_{4j} - \bar{y}_4)^2$$

$$= 38 + 30 + 12 + 14 = 94$$

Finally, the deviations of individual observations from the grand mean $y_{ij} - \bar{y}$ are given by the array

$$\text{Deviations} = \begin{bmatrix} -5 & 0 & -7 & -3 & 0 & & & \\ -1 & 3 & 6 & 0 & & & & \\ 2 & 1 & -1 & 0 & 2 & 0 & 3 \\ -3 & 0 & 2 & 0 & 1 & 0 & \end{bmatrix}$$

The total variation present in the data is measured by the sum of squares of all these deviations.

$$\text{Total sum of squares} = \sum_{i=1}^{4} \sum_{j=1}^{n_i} (y_{ij} - \bar{y})^2$$

$$= (-5)^2 + 0^2 + (-7)^2 + \cdots + 0^2$$

$$= 162$$

Note that the total sum of squares is the sum of the treatment sum of squares and the error sum of squares, or,

$$162 = 68 + 94$$

It is time to turn our attention to another property of this decomposition, the degrees of freedom associated with the sums of squares. In general terms:

$$\begin{pmatrix} \text{Degrees of} \\ \text{freedom} \\ \text{associated with a} \\ \text{sum of squares} \end{pmatrix} = \begin{pmatrix} \text{Number of} \\ \text{elements} \\ \text{whose squares} \\ \text{are summed} \end{pmatrix} - \begin{pmatrix} \text{Number of linear} \\ \text{constraints} \\ \text{satisfied by the} \\ \text{elements} \end{pmatrix}$$

In our present example, the treatment sum of squares is the sum of four terms $n_1(\bar{y}_1 - \bar{y})^2 + n_2(\bar{y}_2 - \bar{y})^2 + n_3(\bar{y}_3 - \bar{y})^2 + n_4(\bar{y}_4 - \bar{y})^2$, where the elements satisfy the single constraint

$$n_1(\bar{y}_1 - \bar{y}) + n_2(\bar{y}_2 - \bar{y}) + n_3(\bar{y}_3 - \bar{y}) + n_4(\bar{y}_4 - \bar{y}) = 0$$

This equality holds because the grand mean $\bar{y}$ is a weighted average of the treatment means, or

$$\bar{y} = \frac{n_1\bar{y}_1 + n_2\bar{y}_2 + n_3\bar{y}_3 + n_4\bar{y}_4}{n_1 + n_2 + n_3 + n_4}$$

Consequently, the number of degrees of freedom associated with the treatment sum of squares is $4 - 1 = 3$. To determine the degrees of freedom for the error sum of squares, we note that the entries $y_{ij} - \bar{y}_i$ in each row of the residual array sum to zero and there are 4 rows. The number of degrees of freedom for the error sum of squares is then $(n_1 + n_2 + n_3 + n_4) - 4 = 22 - 4 = 18$. Finally, the number of degrees of freedom for the total sum of squares is $(n_1 + n_2 + n_3 + n_4) - 1 = 22 - 1 = 21$, because the 22 entries $(y_{ij} - \bar{y})$ whose squares are summed satisfy the single constraint that their total is zero. Note that the degrees of freedom for the total sum of squares is the sum of the degrees of freedom for treatment and error, or,

$$21 = 3 + 18.$$

We summarize the calculations thus far in Table 3.

**TABLE 3**  ANOVA Table for Distortion Data

| Source | Sum of Squares | d.f. |
|--------|----------------|------|
| Treatment | 68 | 3 |
| Error | 94 | 18 |
| Total | 162 | 21 |

Guided by this numerical example, we now present the general formulas for the analysis of variance for a comparison of $k$ treatments using the data structure given in Table 1. Beginning with the basic decomposition

$$( y_{ij} - \bar{y} ) = ( \bar{y}_i - \bar{y} ) + ( y_{ij} - \bar{y}_i )$$

and squaring each side of the equation, we obtain

$$( y_{ij} - \bar{y} )^2 = ( \bar{y}_i - \bar{y} )^2 + ( y_{ij} - \bar{y}_i )^2 + 2( \bar{y}_i - \bar{y} )( y_{ij} - \bar{y}_i )$$

When summed over $j = 1, \ldots, n_i$ the last term on the right-hand side of this equation reduces to zero due to the relation $\sum_{j=1}^{n_i} ( y_{ij} - \bar{y}_i ) = 0$. Therefore, summing each side of the preceding relation over $j = 1, \ldots, n_i$ and $i = 1, \ldots, k$ provides the decomposition

$$\underset{\substack{\uparrow \\ \text{Total SS}}}{\sum_{i=1}^{k} \sum_{j=1}^{n_i} ( y_{ij} - \bar{y} )^2} = \underset{\substack{\uparrow \\ \text{Treatment SS}}}{\sum_{i=1}^{k} n_i ( \bar{y}_i - \bar{y} )^2} + \underset{\substack{\uparrow \\ \text{Residual SS} \\ \text{or error SS}}}{\sum_{i=1}^{k} \sum_{j=1}^{n_i} ( y_{ij} - \bar{y}_i )^2}$$

$$\text{d.f.} = \sum_{i=1}^{k} n_i - 1 \qquad \text{d.f.} = k - 1 \qquad \text{d.f.} = \sum_{i=1}^{k} n_i - k$$

We summarize the two key results.

$$\text{Total SS} = \text{Treatment SS} + \text{Error SS}$$

$$\text{d.f.}_{\text{Total}} = \text{d.f.}_{\text{Treatment}} + \text{d.f.}_{\text{Error}}$$

It is customary to present the decomposition of the sum of squares and the degrees of freedom in a tabular form called the **analysis of variance table**, abbreviated as **ANOVA table**. Table 4 contains the additional column for the **mean square** associated with a component, which is defined as

$$\text{Mean square} = \frac{\text{Sum of squares}}{\text{d.f.}}$$

In the ANOVA Table for comparing $k$ treatments, we set $n = \sum_{i=1}^{k} n_i$.

**TABLE 4**   ANOVA Table for Comparing $k$ Treatments

| Source | Sum of Squares | d.f. | Mean Square |
|---|---|---|---|
| Treatment | $SS_T = \sum_{i=1}^{k} n_i ( \bar{y}_i - \bar{y} )^2$ | $k - 1$ | $MS_T = \dfrac{SS_T}{k - 1}$ |
| Error | $SSE = \sum_{i=1}^{k} \sum_{j=1}^{n_i} ( y_{ij} - \bar{y}_i )^2$ | $n - k$ | $MSE = \dfrac{SSE}{n - k}$ |
| Total | $\sum_{i=1}^{k} \sum_{j=1}^{n_i} ( y_{ij} - \bar{y} )^2$ | $n - 1$ | |

## 1.1  GUIDE TO HAND CALCULATION

When performing an ANOVA on a calculator, it is convenient to express the sums of squares in an alternative form. These employ the treatment totals

$$T_i = \sum_{j=1}^{n_i} y_{ij} = \text{Sum of all responses under treatment } i$$

$$T = \sum_{i=1}^{k} T_i = \sum_{i=1}^{k} \sum_{j=1}^{n_i} y_{ij} = \text{Sum of all observations}$$

to calculate the sums of squares:

$$\text{Total SS} = \sum_{i=1}^{k} \sum_{j=1}^{n_i} y_{ij}^2 - \frac{T^2}{n} \qquad \text{where} \qquad n = \sum_{i=1}^{k} n_i$$

$$SS_T = \sum_{i=1}^{k} \frac{T_i^2}{n_i} - \frac{T^2}{n}$$

$$SSE = \text{Total SS} - SS_T$$

Notice that the SSE can be obtained by subtraction.

**Example 2**  Calculating Sums of Squares Using the Alternative Formulas

Obtain the Total SS, $SS_T$, and SSE for the data in Example 1 using the alternative form of calculation.

SOLUTION

$$T_1 = 10 + 15 + 8 + 12 + 15 = 60 \qquad\qquad n_1 = 5$$
$$T_2 = 14 + 18 + 21 + 15 = 68 \qquad\qquad n_2 = 4$$
$$T_3 = 17 + 16 + 14 + 15 + 17 + 15 + 18 = 112 \quad n_3 = 7$$
$$T_4 = 12 + 15 + 17 + 15 + 16 + 15 = 90 \qquad\quad n_4 = 6$$

and

$$T = T_1 + T_2 + T_3 + T_4 \qquad\qquad n = n_1 + n_2 + n_3 + n_4$$
$$= 60 + 68 + 112 + 90 = 330 \qquad = 5 + 4 + 7 + 6 = 22$$

Since

$$\sum_{i=1}^{4} \sum_{j=1}^{n_i} y_{ij}^2 = (10)^2 + (15)^2 + \cdots + (16)^2 + (15)^2 = 5112$$

$$\text{Total SS} = 5112 - \frac{(330)^2}{22} = 162$$

$$\text{SS}_T = \frac{(60)^2}{5} + \frac{(68)^2}{4} + \frac{(112)^2}{7} + \frac{(90)^2}{6} - \frac{(330)^2}{22} = 68$$

$$\text{SSE} = \text{Total SS} - \text{SS}_T = 162 - 68 = 94$$

## Exercises

**14.1** Water collected in a single bottle from a river is divided into eight specimens. Two specimens are randomly selected and sent to Lab A, two to Lab B, two to Lab C, and two to Lab D. The amount of heavy metals (ppm) is measured for each specimen. Calling each lab a treatment, suppose the data are

| Treatment | Observations |
|-----------|--------------|
| A | 5  9 |
| B | 8  4 |
| C | 4  2 |
| D | 7  9 |

(a) Obtain the arrays that show a decomposition for the observations.

(b) Find the sum of squares for each array.

(c) Determine the degrees of freedom for each sum of squares.

(d) Summarize by an ANOVA table.

**14.2** Subjects must press a button when they hear a signal. The three treatments are three different intensities of the signal. The time elapsed between presentation of the signal and when the button is pushed, are recorded in hundredths of a second. Suppose the data are

| Treatment | Observations |
|-----------|--------------|
| 1 | 35  24  28  21 |
| 2 | 19  14  14  13 |
| 3 | 21  16  21  14 |

(a) Obtain the arrays that show a decomposition for the observations.

(b) Find the sum of squares for each array.

(c) Determine the degrees of freedom for each sum of squares.

(d) Summarize by an ANOVA table.

**14.3** It was decided to vary the experiment in Exercise 14.1 by dropping Lab D. Water from a single collection bottle was divided into eleven specimens. The amount of heavy metals (ppm) is measured for each specimen. Suppose the data are as follows and repeat Exercise 14.1.

| Treatment | Observations |
|-----------|--------------|
| A | 7  5  4  4 |
| B | 6  1  2 |
| C | 2  1  0  1 |

**14.4** The abilities of six different brand athletic socks to wick moisture are rated. The rating is based on the time to reach 1% moisture content after being subjected to a gentle mist. Use the relations for sums of squares and d.f. to complete the following ANOVA table:

| Source | Sum of Squares | d.f. |
|--------|----------------|------|
| Treatment | 34 | 5 |
| Error |  |  |
| Total | 92 | 25 |

**14.5** Spots cannot always be removed by dry cleaning. Suppose the records from four different dry cleaning establishments yield the following data on number of unremovable spots per day.

| Treatment | Observations |
|-----------|--------------|
| 1 | 2 1 3 |
| 2 | 1 5 |
| 3 | 9 5 6 4 |
| 4 | 3 4 5 |

Provide a decomposition of the observations and obtain the ANOVA table.

**14.6** Referring to the lying experiment in Chapter 13. Example 6, the subjects were also able to cheat on the tax form by claiming excessive travel expenses for participating in the experiment. The subjects were randomly assigned to the three groups so the average amounts claimed should be similar unless the position of the signature makes a difference. The summary is

Dollar Amount Claimed for Travel (dollars)

| | Signature | |
|---|---|---|
| Top | Bottom | None |
| $n_1 = 35$ | $n_2 = 33$ | $n_3 = 33$ |
| $\bar{y}_1 = 5.27$ | $\bar{y}_2 = 9.62$ | $\bar{y}_3 = 8.45$ |
| $s_1 = 4.43$ | $s_2 = 6.20$ | $s_3 = 5.92$ |

(a) Recall that $(n_i - 1)s_i^2 = \sum_{j=1}^{n_i}(y_{ij} - \bar{y}_i)^2$
so that

$$\text{SSE} = \sum_{i=1}^{k}\sum_{j=1}^{n_i}(y_{ij} - \bar{y}_i)^2 = \sum_{i=1}^{k}(n_i - 1)s_i^2$$

Calculate the SSE and enter the value in the ANOVA table.

| Source | Sum of Squares | d.f. | Mean Square |
|--------|----------------|------|-------------|
| Treatment | 346.79 | 2 | |
| Error | | | |
| Total | | 100 | |

(b) Complete the ANOVA table.

**14.7** Reading levels vary between different magazines. To avoid difficulties caused by different typefaces and sizes, an investigator just counted the number of letters and punctuation signs. Random samples of 20 sentences were selected from *The New Yorker*, *Sports Illustrated*, and *National Geographic*. The resulting summary statistics are:

| *The New Yorker* | *Sports Illustrated* | *National Geographic* |
|------------------|----------------------|------------------------|
| $\bar{y}_1 = 94.4$ | $\bar{y}_2 = 92.9$ | $\bar{y}_3 = 75.5$ |
| $s_1 = 58.4$ | $s_2 = 54.2$ | $s_3 = 38.1$ |
| $n_1 = 20$ | $n_2 = 20$ | $n_3 = 20$ |

where $s_i^2 = \sum_{j=1}^{n_i}(y_{ij} - \bar{y}_i)^2 / (n_i - 1)$

Present the ANOVA table for these data.

# 2. POPULATION MODEL AND INFERENCES FOR A ONE-WAY ANALYSIS OF VARIANCE

To implement a formal statistical test for no difference among treatment effects, we need to have a population model for the experiment. To this end, we assume that the response measurements with the $i$th treatment constitute a random sample from a normal population with a mean of $\mu_i$ and a **common variance** of $\sigma^2$. The samples are assumed to be mutually independent.

**Population Model for Comparing $k$ Treatments**

$$Y_{ij} = \mu_i + e_{ij} \quad j = 1, \ldots, n_i \quad \text{and} \quad i = 1, \ldots, k$$

where $\mu_i = $ $i$th treatment mean. The errors $e_{ij}$ are all independently distributed as $N(0, \sigma)$.

Before presenting the test for equality of means based on this model, we give a second parametrization that previews the formulation of statistical models for more complicated designs. For each $i$, the mean $\mu_i$ is considered to be the sum of an overall effect $\mu$, common to all treatments, and an effect due only to the $i$th treatment.

$$\mu_i = \mu + \alpha_i \qquad \text{subject to the constraint } \sum_{i=1}^{k} n_i \alpha_i = 0$$

The right-hand side, $\mu + \alpha_i$, is estimated by $\bar{y} + (\bar{y}_i - \bar{y})$ in the decomposition on page 562 and the estimated treatment effects satisfy the same constraint as the $\alpha_i$'s.

## 2.1  *F* DISTRIBUTION

The $F$ test will determine if significant differences exist between the $k$ sample means. The null hypothesis that no difference exists among the $k$ population means can now be phrased as follows:

$$H_0: \mu_1 = \mu_2 = \cdots = \mu_k$$

The alternative hypothesis is that not all the $\mu_i$'s are equal. Seeking a criterion to test the null hypothesis, we observe that when the population means are all equal, $\bar{y}_i - \bar{y}$ is expected to be small, and consequently, the treatment mean square $\sum n_i (\bar{y}_i - \bar{y})^2 / (k - 1)$ is expected to be small. On the other hand, it is likely to be large when the means differ markedly. The error mean square, which provides an estimate of $\sigma^2$, can be used as a yardstick for determining how large a treatment mean square should be before it indicates significant differences. Its value is not influenced by differences in the treatment means.

Statistical distribution theory tells us that under $H_0$ the ratio

$$F = \frac{\text{Treatment mean square}}{\text{Error mean square}} = \frac{\text{Treatment SS} / (k - 1)}{\text{Error SS} / (n - k)}$$

has an $F$ distribution with d.f. $= (k - 1, n - k)$, where $n = \sum n_i$.

Notice that an $F$ distribution is specified in terms of its numerator degrees of freedom $v_1 = k - 1$ and denominator degrees of freedom $v_2 = n - k$. We denote

$$F_\alpha(v_1, v_2) = \text{Upper } \alpha \text{ point of the } F \text{ distribution with } (v_1, v_2) \text{ d.f.}$$

which is also called the upper $100\alpha$-th percentage point.

The upper $\alpha = .05$ and $\alpha = .10$ points are given in Appendix B, Table 7, for several pairs of d.f. With $v_1 = 7$ and $v_2 = 15$, for $\alpha = .05$, we read from column $v_1 = 7$ and row $v_2 = 15$ to obtain $F_{.05}(7, 15) = 2.71$ (see Table 5).

**TABLE 5** Percentage Points of
$F(v_1, v_2)$ Distributions
$\alpha = .05$.

| $v_2$ \ $v_1$ | $\cdots$ | 7 | $\cdots$ |
|---|---|---|---|
| . | | . | |
| . | | . | |
| . | | . | |
| 15 | $\cdots$ | 2.71 | |
| . | | | |
| . | | | |
| . | | | |

We summarize the $F$ test introduced above.

---

### $F$ Test for Equality of Means

Reject $H_0 : \mu_1 = \mu_2 = \cdots = \mu_k$ if

$$F = \frac{\text{Treatment SS}/(k-1)}{\text{Error SS}/(n-k)} \geq F_\alpha(k-1, n-k)$$

where $n = \sum_{i=1}^{k} n_i$ and $F_\alpha(k-1, n-k)$ is the upper $\alpha$ point of the $F$ distribution with d.f. $= (k-1, n-k)$.

---

The computed value of the $F$-ratio is usually presented in the last column of the ANOVA table.

---

**Example 3** The $F$ Test for Testing the Null Hypothesis of No Difference in Sound Distortion Means

Construct the ANOVA table for the data given in Example 1 concerning a comparison of four brands of earbuds. Test the null hypothesis that the means are equal. Use $\alpha = .05$.

SOLUTION Using our earlier calculations for the component sums of squares, we construct the ANOVA table that appears in Table 6.

A test of the null hypothesis $H_0 : \mu_1 = \mu_2 = \mu_3 = \mu_4$ is performed by comparing the observed $F$ value 4.34 with the tabulated value of $F$ with d.f. $= (3, 18)$. At a .05 level of significance, the tabulated value is found to be 3.16. Because this is exceeded by the observed value, we conclude that there

**TABLE 6**   ANOVA Table for the Data Given in Example 1

| Source | Sum of Squares | d.f. | Mean Square | F-ratio |
|--------|----------------|------|-------------|---------|
| Treatment | 68 | 3 | 22.67 | $\dfrac{22.67}{5.22} = 4.34$ |
| Error | 94 | 18 | 5.22 | |
| Total | 162 | 21 | | |

is a significant difference among the four mean sound distortions. A computer calculation gives $P$–value $= .018$, as illustrated in Figure 1, which further strengthens the conclusion of unequal mean distortions for the earbuds.

Figure 1   The rejection region and
$P$–value for the $F$ test in Example 3.

Creating the ANOVA table via computer software is the best method and the $P$–value is included.

Table 7 gives some typical output from a computer program where the term factor is used instead of treatment. The MINITAB commands for obtaining these results are given in Exercise 14.37.

**TABLE 7**   Computer Output: One-Way Analysis of Variance for Sound
Distortion Data

```
One-way Analysis of Variance

Analysis of Variance
Source    DF        SS        MS        F         P
Factor     3     68.00     22.67      4.34     0.018
Error     18     94.00      5.22
Total     21    162.00
```

# Exercises

**14.8** Using the table of percentage points for the $F$ distribution, find

(a) The upper 5% point when $v_1 = 6$ and $v_2 = 11$.

(b) The upper 5% point when $v_1 = 11$ and $v_2 = 6$.

**14.9** Using Appendix B, Table 7, find the upper 10% point of $F$ for

(a) d.f. $= (3, 5)$   (b) d.f. $= (3, 10)$

(c) d.f. $= (3, 15)$   (d) d.f. $= (3, 30)$

(e) What effect does increasing the denominator d.f. have?

**14.10** A psychologist investigating the connection between music and memory, randomly assigns subjects to one of three treatment groups. The first group hears white noise, the second Mozart, and the third heavy metal. With the appropriate background sound turned on, subjects visually study a picture. The picture is removed, the sound turned off, and then the subjects are asked to answer a questionnaire concerning the image. Suppose the number of wrong answers result in the following ANOVA table.

| Source | Sum of Squares | d.f. |
|---|---|---|
| Treatment | 12 | 2 |
| Error | 104 | 41 |
| Total | 116 | 43 |

Carry out the $F$ test for equality of means taking $\alpha = .10$.

**14.11** Based on the current General Social Survey, an index is created from the few questions asking about the degree of confidence in government. Six age groups are compared. Suppose the values of the index result in the following ANOVA table.

| Source | Sum of Squares | d.f. |
|---|---|---|
| Treatment | 23 | 5 |
| Error | 56 | 30 |
| Total | 79 | 35 |

Carry out the $F$ test for equality of means taking $\alpha = .05$.

**14.12** Using the data from Exercise 14.1, test for equality of means using $\alpha = .05$.

**14.13** Test for equality of means based on the data in Exercise 14.2. Take $\alpha = .05$.

**14.14** Three bread recipes are to be compared with respect to density of the loaf. Five loaves will be baked using each recipe.

(a) If one loaf is made and baked at a time, how would you select the order?

(b) Given the following data, conduct an $F$ test for equality of means. Take $\alpha = .05$.

| Recipe | Observation |
|---|---|
| 1 | .95 .86 .71 .72 .74 |
| 2 | .71 .85 .62 .72 .64 |
| 3 | .69 .68 .51 .73 .44 |

**14.15** Test for equality of means based on the travel expense data in Exercise 14.6. Take $\alpha = .05$.

**14.16** Refer to the data on reading levels in Exercise 14.7. Test for equality of means. Take $\alpha = .05$.

# 3. SIMULTANEOUS CONFIDENCE INTERVALS

The ANOVA $F$ test is only the initial step in our analysis. It determines if significant differences exist among the treatment means. Our goal should be more than to merely conclude that treatment differences are indicated by the data. Rather, we must detect likenesses and differences among the treatments.

The problem of estimating differences in treatment means is of even greater importance than the overall $F$ test.

Referring to the comparison of $k$ treatments using the data structure given in Table 1, let us examine how a confidence interval can be established for $\mu_1 - \mu_2$, the mean difference between treatment 1 and treatment 2. The statistic

$$T = \frac{(\bar{Y}_1 - \bar{Y}_2) - (\mu_1 - \mu_2)}{\sqrt{\dfrac{\text{SSE}}{n - k}} \sqrt{\dfrac{1}{n_1} + \dfrac{1}{n_2}}}$$

has a $t$ distribution with d.f. $= n - k$, and this can be employed to construct a confidence interval for $\mu_1 - \mu_2$. More generally:

---

### Confidence Interval for a Single Difference

A $100(1 - \alpha)\%$ confidence interval for $\mu_i - \mu_{i'}$, the difference of means for treatment $i$ and treatment $i'$ is given by

$$(\bar{Y}_i - \bar{Y}_{i'}) \pm t_{\alpha/2} S \sqrt{\frac{1}{n_i} + \frac{1}{n_{i'}}}$$

where

$$S = \sqrt{\text{MSE}} = \sqrt{\frac{\text{SSE}}{n - k}}$$

and $t_{\alpha/2}$ is the upper $\alpha/2$ point of $t$ with d.f. $= n - k$.

---

Even though the confidence interval is for the difference of two particular means, the mean square error is computed using the data from all $k$ treatments. This is because we assume that the variance is the same for all treatments.

If the $F$ test first shows a significant difference in means, then some statisticians feel that it is reasonable to compare means pairwise according to the preceding intervals. However, many statisticians prefer a more conservative procedure based on the following reasoning.

Without the provision that the $F$ test is significant, the preceding method provides **individual** confidence intervals for pairwise differences. However, with $k = 4$ treatments, there are $\binom{4}{2} = 6$ pairwise differences $\mu_i - \mu_{i'}$, and this procedure applied to all pairs yields six confidence statements, each having a $100(1 - \alpha)\%$ level of confidence. It is difficult to determine what level of confidence will be achieved for claiming that **all** six of these statements are correct. To overcome this dilemma, procedures have been developed for several confidence intervals to be constructed in such a manner that the joint probability that all the statements are true is guaranteed not to fall below a predetermined level. Such intervals are called multiple confidence intervals or simultaneous confidence intervals. Numerous methods proposed in the statistical literature have achieved varying degrees of success. We present one that can be used simply and conveniently in general applications.

The procedure, called the multiple-$t$ confidence intervals, consists of setting confidence intervals for the differences $\mu_i - \mu_{i'}$ in much the same way we just did for the individual differences, except that a different percentage point is read from the $t$ table.

Operationally, the construction of these confidence intervals does not require any new concepts or calculations, but it usually involves some nonstandard percentage point of $t$. For example, with $k = 3$ and $1 - \alpha = .95$, if we want to set simultaneous intervals for all $m = \binom{k}{2} = 3$ pairwise differences, we require that the upper $\alpha / (2m) = .05 / 6 = .00833$ point of a $t$ distribution.

---

### Multiple-$t$ Confidence Intervals

A set of $100(1 - \alpha)\%$ simultaneous confidence intervals for $m$ number of pairwise differences $\mu_i - \mu_{i'}$ is given by

$$(\bar{Y}_i - \bar{Y}_{i'}) \pm t_{\alpha / 2m} S \sqrt{\frac{1}{n_i} + \frac{1}{n_{i'}}}$$

where $S = \sqrt{\text{MSE}}$, $m$ = the number of confidence statements, and $t_{\alpha / 2m}$ = the upper $\alpha / (2m)$ point of $t$ with d.f. $= n - k$.

Prior to sampling, the probability of all the $m$ statements being correct is at least $1 - \alpha$.

---

**Example 4**   Calculating Multiple-$t$ Confidence Intervals to Reveal Which means Differ

An experiment is conducted to determine the soil moisture deficit resulting from varying amounts of residual timber left after cutting trees in a forest. The three treatments are treatment 1: no timber left; treatment 2: 2000 bd ft left; treatment 3: 8000 bd ft left. (Board feet is a particular unit of measurement of timber volume.) The measurements of moisture deficit are given in Table 8. Perform the ANOVA test and construct 95% multiple-$t$ confidence intervals for the treatment differences.

SOLUTION   Our analysis employs convenient alternative forms of the expressions for sums of squares involving totals.

**TABLE 8**  Moisture Deficit in Soil

| Treatment | Observations | Total | Mean |
|---|---|---|---|
| 1 | 1.52 1.38 1.29 1.48 1.63 | $T_1 = 7.30$ | $\bar{y}_1 = 1.460$ |
| 2 | 1.63 1.82 1.35 1.03 2.30 1.45 | $T_2 = 9.58$ | $\bar{y}_2 = 1.597$ |
| 3 | 2.56 3.32 2.76 2.63 2.12 2.78 | $T_3 = 16.17$ | $\bar{y}_3 = 2.695$ |
| | | Grand total | Grand mean |
| | | $T = 33.05$ | $\bar{y} = 1.944$ |

The total number of observations $n = 5 + 6 + 6 = 17$.

$$\text{Total SS} = \sum_{i=1}^{3} \sum_{j=1}^{n_i} (y_{ij} - \bar{y})^2 = \sum_{i=1}^{3} \sum_{j=1}^{n_i} y_{ij}^2 - \frac{T^2}{n}$$

$$= 71.3047 - 64.2531 = 7.0516$$

$$\text{Treatment SS} = \sum_{i=1}^{3} n_i (\bar{y}_i - \bar{y})^2 = \sum_{i=1}^{3} \frac{T_i^2}{n_i} - \frac{T^2}{n}$$

$$= 69.5322 - 64.2531 = 5.2791$$

$$\text{Error SS} = \text{Total SS} - \text{Treatment SS} = 1.7725$$

The ANOVA table appears in Table 9.

**TABLE 9**  ANOVA Table for Comparison of
Moisture Deficit

| Source | Sum of Squares | d.f. | Mean Square | F-ratio |
|---|---|---|---|---|
| Treatment | 5.2791 | 2 | 2.640 | 20.8 |
| Error | 1.7725 | 14 | .127 | |
| Total | 7.0516 | 16 | | |

Because the observed value of $F$ is larger than the tabulated value $F_{.05}(2, 14) = 3.74$, the null hypothesis of no difference in the treatment effects is rejected at $\alpha = .05$. The $P$–value is less than .0001 which provides very strong evidence against the null hypothesis.

In constructing a set of 95% multiple-$t$ confidence intervals for pairwise differences, note that there are $\binom{3}{2} = 3$ pairs, so

$$\frac{\alpha}{2m} = \frac{.05}{(2 \times 3)} = .00833$$

From Appendix B, Table 5, the upper .00833 point of $t$ with d.f. $= 14$ is 2.718. The simultaneous confidence intervals are calculated as follows:

$$\mu_2 - \mu_1: \quad (1.597 - 1.460) \pm 2.718 \times .356 \times \sqrt{\frac{1}{6} + \frac{1}{5}}$$

$$= (-.45, .72)$$

$$\mu_3 - \mu_2: \quad (2.695 - 1.597) \pm 2.718 \times .356 \times \sqrt{\frac{1}{6} + \frac{1}{6}}$$

$$= (.54, 1.66)$$

$$\mu_3 - \mu_1: \quad (2.695 - 1.460) \pm 2.718 \times .356 \times \sqrt{\frac{1}{6} + \frac{1}{5}}$$

$$= (.65, 1.82)$$

These confidence intervals indicate that treatments 1 and 2 do not differ appreciably, but the mean for treatment 3 is considerably higher than the means for treatments 1 and 2.

## Exercises

**14.17** Taking $\alpha = .05$ and $n - k = 26$, determine the appropriate percentile of the $t$ distribution when calculating the multiple-$t$ confidence intervals with (a) $m = 3$ and (b) $m = 5$.

**14.18** Construct the 90% multiple-$t$ confidence intervals using the sound distortion data in Example 1.

**14.19** As suggested by the photo in the beginning of this chapter, consumers can rate new HDTVs. Many Web sites use a five point scale. Although the individual responses are not normal, the central limit result applied to each of the four samples does justify treating the treatment means as if the normal assumption prevailed. The summary statistics of ratings for four of the most popular sets at one Web site are

$$n_1 = 14 \quad \bar{y}_1 = 4.50$$

$$n_2 = 20 \quad \bar{y}_2 = 3.60 \quad s = 1.56$$

$$n_3 = 30 \quad \bar{y}_3 = 4.13$$

$$n_4 = 15 \quad \bar{y}_4 = 4.56$$

Use $\alpha = .10$ and determine:

(a) $t$ intervals for each of the six differences of means.

(b) The six multiple-$t$ intervals.

**14.20** A study was conducted to examine the degree that the length of gestation influences school achievement for children born in the normal range of 37 to 41 weeks gestation. Academic achievement was measured by a third grade reading test that we summarize for the gestation periods 37, 39, and 41 weeks.

| Third Grade Reading Scores Gestation Weeks | | |
|---|---|---|
| 37 | 39 | 41 |
| $n_1 = 12184$ | $n_2 = 35197$ | $n_3 = 22091$ |
| $\bar{x}_1 = 50.335$ | $\bar{x}_2 = 50.971$ | $\bar{x}_3 = 51.155$ |
| $s_1 = 9.603$ | $s_2 = 9.756$ | $s_3 = 9.364$ |

*Source*: Means and standard deviation from Figure 1, K. Noble et al., "Academic achievement varies with gestational age among children born at term," *Pediatrics* **130** (2), (2012), pp. 257–264.

(a) Complete the ANOVA table.

| Source | Sum of Squares | d.f. | Mean Square | F-ratio |
|---|---|---|---|---|
| Treatment | | | | |
| Error | 6410377 | 69469 | | |
| Total | | 69471 | | |

(b) State the conclusion to the $F$ test for equality of means. Take $\alpha = .05$.

(c) Calculate 95% simultaneous confidence intervals for the differences in means.

(d) Specify which population means are different and which are not, as well as their orderings. What is your overall conclusion regarding the reading scores and gestation length.

**14.21** Determine the expression for the length of the $t$ interval for $\mu_1 - \mu_2$ and the multiple-$t$ interval for $\mu_1 - \mu_2$ when $m = 10$. The ratio of lengths does not depend on the data. Evaluate this ratio for $\alpha = .10$ and $n - k = 15$.

## 4. GRAPHICAL DIAGNOSTICS AND DISPLAYS TO SUPPLEMENT ANOVA

In addition to testing hypotheses and setting confidence intervals, an analysis of data must include a critical examination of the assumptions involved in a model. As in regression analysis, of which analysis of variance is a special case, the residuals must be examined for evidence of serious violations of the assumptions. This aspect of the analysis is not addressed in the ANOVA table summary.

**Example 5** Plotting Residuals

Determine the residuals for the moisture data given in Example 4 (see Table 10) and graphically examine them for possible violations of the assumptions.

SOLUTION

**TABLE 10** Residuals $y_{ij} - \bar{y}_i$ for the Data Given in Table 8

| Treatment | Residuals | | | | | |
|---|---|---|---|---|---|---|
| 1 | .06 | −.08 | −.17 | .02 | .17 | |
| 2 | .03 | .22 | −.25 | −.57 | .70 | −.15 |
| 3 | −.14 | .63 | .07 | −.07 | −.58 | .09 |

The residual plots of these data are shown in Figure 2, where the combined dot diagram is presented in (a) and the dot diagrams of residuals corresponding to individual treatments appear in (b).

From an examination of the dot diagrams, the variability in the points for treatment 1 appears to be somewhat smaller than the variabilities in the points for treatments 2 and 3. However, given so few observations, it is difficult to determine if this has occurred by chance or if treatment 1 actually has a smaller variance. A few more observations are usually necessary to obtain a meaningful pattern for the individual treatment plots.

Fortunately, the ANOVA testing procedure is **robust** in the sense that small or moderate departures from normality and constant variance do not seriously affect its performance.

(a) Combined residual plot

(b) Residuals with individual treatment

Figure 2    Residual plots for the data given in Example 4.

In addition to the ANOVA a graphical portrayal of the data, as a box plot for each treatment, conveys important information available for making comparisons of populations.

**Example 6**    Box Plots Reveal Differences between Populations

The sepal width was measured on 50 iris flowers for each of three varieties, *Iris setosa*, *Iris versicolor*, and *Iris virginica*. A computer calculation produced the summary shown in the following ANOVA table.

| Source | SS | d.f. | F |
|---|---|---|---|
| Treatment | 11.345 | 2 | 49.16 |
| Error | 16.962 | 147 | |
| Total | 28.307 | 149 | |

Since $F_{.05}(2, 147) \approx 3.05$, we reject the null hypothesis of equal sepal width means at the 5% level of significance.

| Treatment | Sample Mean |
|---|---|
| *Iris setosa* | 3.428 |
| *Iris versicolor* | 2.770 |
| *Iris virginica* | 2.974 |

A calculation of multiple-*t* confidence intervals shows that all population means differ from one another (see Exercise 14.38). Display the data on the three varieties in boxplots.

SOLUTION    The data are given in Exercise 14.38. Boxplots graphically display the variation in the sepal width measurements. From Figure 3, we see that the *Iris setosa* typically has larger sepal width. R. A. Fisher, who developed analysis of variance, used these data along with other lengths and widths to introduce a statistical technique for identifying varieties of plants.

Figure 3    Boxplots for the three iris samples.

# Exercises

**14.22** Referring to the sound distortion data in Example 1, graph the residuals in

(a) A dot plot using all of the residuals.

(b) Individual dot plots for each model of earbuds.

**14.23** Referring to the density of bread data in Exercise 14.14, plot the residuals

(a) In a dot plot using all of the residuals.

(b) Individual dot plots for each recipe.

**14.24** Refer to the MINITAB commands in Exercise 14.37 but also check the box **Store residuals**. Next, referring to Exercise 6.55, graph the residuals from an ANOVA of the iris data in Exercise 14.38

(a) In a normal-scores plot using all of the residuals.

(b) Individual normal-scores plots for each type of iris.

# 5.  RANDOMIZED BLOCK EXPERIMENTS FOR COMPARING *k* TREATMENTS

Just as we can pair like subjects or experimental units to improve upon the procedure of taking two independent samples, we can also arrange, or **block** subjects into homogeneous groups of size *k* when comparing *k* treatments. Then if each treatment is applied to exactly one unit in the block and comparisons are only drawn between treatment responses from the same block, extraneous variability should be greatly reduced. It is this concept of **blocking** that underlies the **randomized block design**. When effective blocking variables are chosen, we can more efficiently test for differences in treatments.

The term "block design" originated from the design of agricultural field trials, where "block" refers to a group of adjacent plots. A few typical examples for which the block design may be appropriate are clinical trials to compare several competing drugs, where the experimental subjects are grouped in blocks according to age group and severity of symptoms; psychological experiments comparing several stimuli, where subjects may be blocked according to socioeconomic background; and comparison of several techniques for storing fruit or vegetables, where each incoming shipment is regarded as a block.

As its name implies, **randomization** is a basic part of the block design. This time, once the grouping of like experimental subjects in blocks is accomplished, we randomly select one subject from the first block to receive treatment 1, one of the remaining subjects to receive treatment 2, and so on. The same procedure is repeated with a new randomization for each of the remaining blocks.

Once the data are obtained, they can be arranged in rows according to the treatments and in columns according to the blocks. If we designate the measurement corresponding to treatment $i$ and block $j$ by $y_{ij}$, the data structure of a randomized block design with $b$ blocks and $k$ treatments is shown in Table 11.

The row and column means are denoted by

$$i\text{th treatment (row) mean } \bar{y}_{i\cdot} = \frac{1}{b} \sum_{j=1}^{b} y_{ij}$$

$$j\text{th block (column) mean } \bar{y}_{\cdot j} = \frac{1}{k} \sum_{i=1}^{k} y_{ij}$$

and the overall mean, or

$$\text{grand mean } \bar{y}_{\cdot\cdot} = \frac{1}{bk} \sum_{i=1}^{k} \sum_{j=1}^{b} y_{ij}$$

These means are shown in the margins of the table. Here an overbar on $y$ indicates an average and a dot in the subscript denotes that the average is taken over the subscript appearing in that place.

**TABLE 11**  Data Structure of a Randomized Block Design with $b$ Blocks and $k$ Treatments

|  | Block 1 | Block 2 | $\cdots$ | Block $b$ | Treatment Means |
|---|---|---|---|---|---|
| Treatment 1 | $y_{11}$ | $y_{12}$ | $\cdots$ | $y_{1b}$ | $\bar{y}_{1\cdot}$ |
| Treatment 2 | $y_{21}$ | $y_{22}$ | $\cdots$ | $y_{2b}$ | $\bar{y}_{2\cdot}$ |
| $\cdot$ | $\cdot$ | $\cdot$ | $\cdots$ | $\cdot$ | $\cdot$ |
| $\cdot$ | $\cdot$ | $\cdot$ | $\cdots$ | $\cdot$ | $\cdot$ |
| $\cdot$ | $\cdot$ | $\cdot$ | $\cdots$ | $\cdot$ | $\cdot$ |
| Treatment $k$ | $y_{k1}$ | $y_{k2}$ | $\cdots$ | $y_{kb}$ | $\bar{y}_{k\cdot}$ |
| Block means | $\bar{y}_{\cdot 1}$ | $\bar{y}_{\cdot 2}$ | $\cdots$ | $\bar{y}_{\cdot b}$ | $\bar{y}_{\cdot\cdot}$ |

We now discuss the analysis of variance for a randomized block design with illustrative calculations based on the data given in Example 7.

**Example 7**  The Structure of Data from a Randomized Block Experiment

The cutting speeds of four types of tools are being compared in an experiment. Five materials of varying degrees of hardness are to be used as experimental blocks. The data pertaining to measurements of cutting time in seconds appear in Table 12.

**TABLE 12**  Measurements of Cutting Time According to Types of Tool (Treatments) and Hardness of Material (Blocks)

| Treatment | Block 1 | 2 | 3 | 4 | 5 | Treatment Means |
|---|---|---|---|---|---|---|
| 1 | 12 | 2 | 8 | 1 | 7 | $\bar{y}_{1\cdot} = 6$ |
| 2 | 20 | 14 | 17 | 12 | 17 | $\bar{y}_{2\cdot} = 16$ |
| 3 | 13 | 7 | 13 | 8 | 14 | $\bar{y}_{3\cdot} = 11$ |
| 4 | 11 | 5 | 10 | 3 | 6 | $\bar{y}_{4\cdot} = 7$ |
| Block means | $\bar{y}_{\cdot 1} = 14$ | $\bar{y}_{\cdot 2} = 7$ | $\bar{y}_{\cdot 3} = 12$ | $\bar{y}_{\cdot 4} = 6$ | $\bar{y}_{\cdot 5} = 11$ | $\bar{y}_{\cdot\cdot} = \dfrac{200}{20} = 10$ |

The observations form a two-way table, and their decomposition indicates both a term for row (treatment) deviations and column (block) deviations.

> **Decomposition of Observations**
>
> $$\begin{array}{ccccccc} \text{Obser-} & = & \text{Grand} & + & \text{Deviation} & + & \text{Deviation} & + & \text{Residual} \\ \text{vation} & & \text{mean} & & \text{due to} & & \text{due to} & & \\ & & & & \text{treatment} & & \text{block} & & \end{array}$$
>
> $$y_{ij} = \bar{y}_{\cdot\cdot} + (\bar{y}_{i\cdot} - \bar{y}_{\cdot\cdot}) + (\bar{y}_{\cdot j} - \bar{y}_{\cdot\cdot}) + (y_{ij} - \bar{y}_{i\cdot} - \bar{y}_{\cdot j} + \bar{y}_{\cdot\cdot})$$

Taking the observation $y_{11} = 12$ in Example 7, we obtain

$$12 = 10 + (6 - 10) + (14 - 10) + (12 - 14 - 6 + 10)$$
$$= 10 + (-4) + (4) + (2)$$

Table 13 contains the results of the decomposition of all the observations. Table 14 gives an alternative format.

**TABLE 13**    Decomposition of Observations for the Randomized Block
Experiment in Table 12

$$
\begin{bmatrix}
12 & 2 & 8 & 1 & 7 \\
20 & 14 & 17 & 12 & 17 \\
13 & 7 & 13 & 8 & 14 \\
11 & 5 & 10 & 3 & 6
\end{bmatrix}
=
\begin{bmatrix}
10 & 10 & 10 & 10 & 10 \\
10 & 10 & 10 & 10 & 10 \\
10 & 10 & 10 & 10 & 10 \\
10 & 10 & 10 & 10 & 10
\end{bmatrix}
+
\begin{bmatrix}
-4 & -4 & -4 & -4 & -4 \\
6 & 6 & 6 & 6 & 6 \\
1 & 1 & 1 & 1 & 1 \\
-3 & -3 & -3 & -3 & -3
\end{bmatrix}
$$

Observation $y_{ij}$ = Grand Mean $\bar{y}_{..}$ + Treatment Effect $\bar{y}_{.j} - \bar{y}_{..}$

$$
+
\begin{bmatrix}
4 & -3 & 2 & -4 & 1 \\
4 & -3 & 2 & -4 & 1 \\
4 & -3 & 2 & -4 & 1 \\
4 & -3 & 2 & -4 & 1
\end{bmatrix}
+
\begin{bmatrix}
2 & -1 & 0 & -1 & 0 \\
0 & 1 & -1 & 0 & 0 \\
-2 & -1 & 0 & 1 & 2 \\
0 & 1 & 1 & 0 & -2
\end{bmatrix}
$$

+ Block Effect $\bar{y}_{i.} - \bar{y}_{..}$ + Residual $y_{ij} - \bar{y}_{i.} - \bar{y}_{.j} + \bar{y}_{..}$

**TABLE 14**    An Alternative Format of the Decomposition Table
for a Randomized Block Experiment

| Treatment | Block 1 | ... | Block $j$ | ... $b$ | Deviation of Treatment Mean from Grand Mean |
|---|---|---|---|---|---|
| 1 | | | | | $(\bar{y}_1. - \bar{y}_{..})$ |
| . | | | | | . |
| . | | | | | . |
| $i$ | ... | $(y_{ij} - \bar{y}_{i.} - \bar{y}_{.j} + \bar{y}_{..})$ | ... | | $(\bar{y}_{i.} - \bar{y}_{..})$ |
| . | | | | | . |
| . | | | | | . |
| $k$ | | | | | $(\bar{y}_{k.} - \bar{y}_{..})$ |
| Deviation of block mean from grand mean | $(\bar{y}_{.1} - \bar{y}_{..})$ ... | | $(\bar{y}_{.j} - \bar{y}_{..})$ ... | $(\bar{y}_{.b} - \bar{y}_{..})$ | Grand mean $\bar{y}_{..}$ |

The number of distinct entries $\bar{y}_{i.} - \bar{y}_{..}$ in the treatment effects array is $k$, and the single constraint is that they must sum to zero. Thus, $k - 1$ degrees of freedom are associated with the treatment sum of squares.

### Sum of Squares Due to Treatment

$$SS_T = \sum_{i=1}^{k} \sum_{j=1}^{b} (\bar{y}_{i.} - \bar{y}_{..})^2$$

$$= b \sum_{i=1}^{k} (\bar{y}_{i.} - \bar{y}_{..})^2$$

with d.f. $= k - 1$.

In the array of treatment effects in Table 13, each entry appears $b = 5$ times, once in each block. The treatment sum of squares for this example is then

$$SS_T = 5(-4)^2 + 5(6)^2 + 5(1)^2 + 5(-3)^2 = 310$$

with d.f. $= 4 - 1 = 3$.

In a similar manner, we obtain the block sum of squares.

### Sum of Squares Due to Block

$$SS_B = \sum_{i=1}^{k} \sum_{j=1}^{b} (\bar{y}_{.j} - \bar{y}_{..})^2$$

$$= k \sum_{j=1}^{b} (\bar{y}_{.j} - \bar{y}_{..})^2$$

with d.f. $= b - 1$.

Referring to the array of block effects in Table 13, we find the block sum of squares for our example to be

$$SS_B = 4(4)^2 + 4(-3)^2 + 4(2)^2 + 4(-4)^2 + 4(1)^2 = 184$$

with d.f. $= 5 - 1 = 4$.

The number of degrees of freedom associated with the residual array is $(b - 1)(k - 1)$. To understand why, note that among the $b \times k$ residuals, the following constraints are satisfied. One constraint is that the sum of all entries must be zero. The fact that all row sums are zero introduces $k - 1$ additional constraints. This is so because having fixed any $k - 1$ row totals and the grand total, the remaining row total is automatically fixed. By the same reasoning, $b - 1$ additional constraints arise from the fact that all column totals are zero. Consequently, the number of degrees of freedom for the residual is

$$bk - 1 - (k - 1) - (b - 1) = (b - 1)(k - 1)$$

---

### Residual Sum of Squares or Error Sum of Squares

$$SSE = \sum_{i=1}^{k} \sum_{j=1}^{b} (y_{ij} - \bar{y}_{i.} - \bar{y}_{.j} + \bar{y}_{..})^2$$

with d.f. $= (b - 1)(k - 1)$.

---

In our example,

$$SSE = 2^2 + 0^2 + (-2)^2 + 0^2 + \cdots + (-2)^2 = 24$$

with d.f. $= (5 - 1)(4 - 1) = 12$.

Finally, the total sum of squares equals the sum of squares of each observation about the grand mean, or

$$\text{Total sum of squares} = \sum_{i=1}^{k} \sum_{j=1}^{b} (y_{ij} - \bar{y}_{..})^2 = \sum_{i=1}^{k} \sum_{j=1}^{b} y_{ij}^2 - bk\bar{y}_{..}^2.$$

with d.f. $= bk - 1$. In our example, we sum the square of each entry in the array for $y_{ij}$ and subtract the sum of squares of entries in the $\bar{y}_{..}$ array.

$$\text{Total sum of squares} = (12)^2 + (2)^2 + \cdots + (6)^2$$
$$- [(10)^2 + (10)^2 + \cdots + (10)^2]$$
$$= 518$$

with d.f. $= (5)(4) - 1 = 19$. This provides a check on our previous calculations, because those sums of squares and degrees of freedom must sum to these totals.

$$SS: 518 = 310 + 184 + 24$$
$$d.f:\ 19 = \quad 3 + \quad 4 + 12$$

These calculations are conveniently summarized in the ANOVA tables shown in Table 15 for the data in our numerical example and for the general

**TABLE 15**   ANOVA Table for the Data Given in Example 7

| Source | Sum of Squares | d.f. | Mean Square | F-Ratio |
|--------|----------------|------|-------------|---------|
| Treatments | 310 | 3 | 103.3 | 51.7 |
| Blocks | 184 | 4 | 46 | 23 |
| Residual | 24 | 12 | 2 | |
| Total | 518 | 19 | | |

case in Table 16. The last column of $F$-ratios will be explained after we discuss the population model.

**TABLE 16** ANOVA Table for a Randomized Block Design

| Source | Sum of Squares | d.f. | Mean Square | F-Ratio |
|---|---|---|---|---|
| Treatments | $SS_T = b\sum_{i=1}^{k}(\bar{y}_{i\cdot} - \bar{y}_{\cdot\cdot})^2$ | $k-1$ | $MS_T = \dfrac{SS_T}{k-1}$ | $\dfrac{MS_T}{MSE}$ |
| Blocks | $SS_B = k\sum_{j=1}^{b}(\bar{y}_{\cdot j} - \bar{y}_{\cdot\cdot})^2$ | $b-1$ | $MS_B = \dfrac{SS_B}{b-1}$ | $\dfrac{MS_B}{MSE}$ |
| Residual | $SSE = \sum_{i=1}^{k}\sum_{j=1}^{b}(y_{ij} - \bar{y}_{i\cdot} - \bar{y}_{\cdot j} + \bar{y}_{\cdot\cdot})^2$ | $(b-1)(k-1)$ | $MSE = \dfrac{SSE}{(b-1)(k-1)}$ | |
| Total | $\sum_{i=1}^{k}\sum_{j=1}^{b}(y_{ij} - \bar{y}_{\cdot\cdot})^2$ | $bk-1$ | | |

Again, a statistical test of treatment differences is based on an underlying population model.

---

**Population Model for a Randomized Block Experiment**

$$\text{Observation} = \begin{matrix}\text{Overall}\\\text{mean}\end{matrix} + \begin{matrix}\text{Treatment}\\\text{effect}\end{matrix} + \begin{matrix}\text{Block}\\\text{effect}\end{matrix} + \text{Error}$$

$$Y_{ij} = \mu + \alpha_i + \beta_j + e_{ij}$$

for $i = 1, \ldots, k$ and $j = 1, \ldots, b$, where the parameters satisfy

$$\sum_{i=1}^{k} \alpha_i = 0 \qquad \sum_{j=1}^{b} \beta_j = 0$$

and the $e_{ij}$ are random errors independently distributed as $N(0, \sigma)$.

---

Tests for the absence of treatment differences or differences in block effects can now be performed by comparing the corresponding mean square with the yardstick of the error mean square by using an $F$ test.

Reject $H_0: \alpha_1 = \cdots = \alpha_k = 0$ (no treatment differences) if

$$\frac{MS_T}{MSE} > F_\alpha(k-1, (b-1)(k-1))$$

Reject $H_0: \beta_1 = \cdots = \beta_b = 0$ (no block differences) if

$$\frac{MS_B}{MSE} > F_\alpha(b-1, (b-1)(k-1))$$

To test the hypothesis of no treatment differences for the analysis of variance in Table 16, we find that the tabulated .05 point of $F(3, 12)$ is 3.49, a value far exceeded by the observed $F$-ratio for treatment effect. We therefore conclude that a highly significant treatment difference is indicated by the data. A computer calculation gives a $P$–value that is much smaller than .0001. This very strongly supports our conclusion.

The block effects are also highly significant, because the observed $F$ value of 23 is much larger than the tabulated value $F_{.05}(4, 12) = 3.26$. The $P$–value is less than .0001 which strongly supports the decision to conduct a randomized block experiment. Significant blocking effects are expected since the blocking variable, hardness, is chosen because it is known to be a good predictor of tool life.

Again, we stress that a serious violation of the model assumptions is likely to jeopardize the conclusions drawn from the preceding analyses and a careful examination of the residuals should be an integral part of the analysis. In addition to plotting the whole set of residuals in a graph, separate plots for individual treatments and individual blocks should also be studied. When observations are collected over time, a plot of the residuals versus the time order is also important.

## 5.1 CONFIDENCE INTERVALS FOR TREATMENT DIFFERENCES

In addition to performing the overall $F$ test for detecting treatment differences, the experimenter typically establishes confidence intervals to compare specific pairs of treatments. This is particularly important when the $F$ test leads to a rejection of the null hypothesis, thus signifying the presence of treatment differences.

We next estimate the difference $\alpha_i - \alpha_{i'}$ of the mean responses of treatments $i$ and $i'$. Because $\bar{Y}_{i\bullet} - \bar{Y}_{i'\bullet}$ is normally distributed with a mean of $\alpha_i - \alpha_{i'}$ and a variance of

$$\sigma^2\left(\frac{1}{b} + \frac{1}{b}\right) = \sigma^2\left(\frac{2}{b}\right)$$

the ratio

$$T = \frac{(\bar{Y}_{i\bullet} - \bar{Y}_{i'\bullet}) - (\alpha_i - \alpha_{i'})}{\sqrt{MSE}\ \sqrt{2/b}}$$

has a $t$ distribution with d.f. $= (b - 1)(k - 1)$. This result can be used to construct a 95% confidence interval

$$( \bar{Y}_{i\bullet} - \bar{Y}_{i'\bullet} - t_{.025} \sqrt{\text{MSE}} \sqrt{\frac{2}{b}}, \ \bar{Y}_{i\bullet} - \bar{Y}_{i'\bullet} + t_{.025} \sqrt{\text{MSE}} \sqrt{\frac{2}{b}} )$$

for an individual difference $\alpha_i - \alpha_{i'}$.

When several such pairwise comparisons are to be integrated into a combined confidence statement, the concept of simultaneous confidence intervals, discussed in Section 3, is again applied.

## Exercises

**14.25** The hours of relief are measured under a placebo, Brand A, and Brand B sinus medicines. Call these treatment 1, treatment 2, and treatment 3, respectively. They are each given in succession, in random order, to each subject with two days between each treatment. Suppose the data are

| Treatment | Block | | | |
|---|---|---|---|---|
| | 1 | 2 | 3 | 4 |
| 1 | 11 | 10 | 7 | 0 |
| 2 | 7 | 8 | 7 | 2 |
| 3 | 15 | 6 | 13 | 10 |

(a) Provide a decomposition for the observations from this randomized block experiment.

(b) Find the sum of squares for each array.

(c) Determine the degrees of freedom by checking the constraints for each array.

**14.26** Suppose you wish to compare three different brands of tick collars for dogs. You have available three of each of the breeds Poodle, Lab, Collie, and Dachshund. Explain how you would assign a brand of tick collar to each of the 12 dogs in order to conduct a randomized block experiment.

**14.27** Refer to Exercise 14.25. Present the ANOVA table. What conclusions can you draw from the two $F$ tests? Take $\alpha = .05$.

**14.28** The yield, in pounds, of three types of heritage tomatoes (treatments) is obtained after planting them in three equal sized plots within a site (block). Types are randomly assigned to plots. A total of four different sites are used. Suppose the data are

| Treatment | Block | | | |
|---|---|---|---|---|
| | 1 | 2 | 3 | 4 |
| 1 | 35 | 24 | 28 | 21 |
| 2 | 19 | 14 | 14 | 13 |
| 3 | 21 | 16 | 21 | 14 |

(a) Provide a decomposition for the observations from this randomized block experiment.

(b) Find the sum of squares for each array.

(c) Determine the degrees of freedom by checking the constraints for each array.

**14.29** Three loaves of bread, each made according to a different recipe, are baked in one oven at the same time. Because of possible uncontrolled variations in oven performance, each baking is treated as a block. This procedure is repeated five times, and the following measurements of density are obtained.

| Recipe | Block | | | | |
|---|---|---|---|---|---|
| | 1 | 2 | 3 | 4 | 5 |
| 1 | .95 | .86 | .71 | .72 | .74 |
| 2 | .71 | .85 | .62 | .72 | .64 |
| 3 | .69 | .88 | .51 | .73 | .44 |

(a) How should the three oven positions of the three loaves be selected for each trial?

(b) Perform an analysis of variance for these data.

**14.30** Refer to Exercise 14.28. Present the ANOVA table. What conclusions can you draw from the two $F$ tests? Take $\alpha = .05$.

**14.31** As part of a cooperative study on the nutritional quality of oats, 6 varieties of oat kernels with their hulls removed are subjected to a mineral analysis. The plants are grown according to a randomized block design, and the measurements of protein by percent of dry weight are recorded in Table 17.

(a) Perform an analysis of variance for these data.

(b) Calculate and plot the residuals. Does the model appear to be adequate?

**14.32** Referring to Exercise 14.31, suppose that variety 6 is of special interest. Construct simultaneous 90% confidence intervals for the differences between the mean of variety 6 and each of the other means.

**TABLE 17**  Protein (prct. dry wt.)

| Treatment | Block 1 | 2 | 3 | 4 | 5 | 6 |
|-----------|------|------|------|------|------|------|
| 1 | 19.09 | 20.29 | 20.31 | 19.60 | 18.62 | 20.10 |
| 2 | 16.28 | 17.88 | 16.88 | 17.57 | 16.72 | 17.32 |
| 3 | 16.31 | 18.17 | 17.38 | 17.53 | 16.34 | 17.88 |
| 4 | 17.50 | 18.05 | 17.59 | 17.64 | 17.38 | 18.04 |
| 5 | 16.25 | 16.92 | 15.88 | 14.78 | 15.97 | 16.66 |
| 6 | 21.09 | 21.37 | 21.38 | 20.52 | 21.09 | 21.58 |

Data courtesy of D. Peterson, L. Schrader, and V. Youngs.

## USING STATISTICS WISELY

1. When collecting data according to a one-way ANOVA design, conduct the trials in random order if at all possible.

2. Do not routinely accept the analysis of variance generated by statistical software. Instead, inspect the residuals for outliers or patterns indicating that the variance is not constant across treatments.

## KEY IDEAS AND FORMULAS

Several populations can be compared using Analysis of variance (ANOVA).

A One-way analysis of variance design specifies taking independent random samples from each population.

All of the observations are assumed to have a common variance. Then the ANOVA consists of a separation of the total sum of squares into components due to different sources of variation.

When samples are taken from $k$ populations the two sources of variation are the **within** population variation or **error** and the differences **between** population means or **treatments**. The $j$th observation on the $i$th treatment is $y_{ij}$. The error variation is estimated using the **residuals** or $y_{ij} - \bar{y}_i$ which are the deviations of the observations from their respective sample means.

The total variation in the observations $y_{ij}$ is expressed as

$$\text{Total sum of squares} = \sum_{i=1}^{k} \sum_{j=1}^{n_i} (y_{ij} - \bar{y})^2$$

$\sum_{i=1}^{k} \sum_{j=1}^{n_i} (y_{ij} - \bar{y})^2$ is partitioned into the two components:

1. **Treatment sum of squares:**

$$SS_T = \sum_{i=1}^{k} n_i (\bar{y}_i - \bar{y})^2$$

2. **Error sum of squares:**

$$SSE = \sum_{i=1}^{k} \sum_{j=1}^{n_i} (y_{ij} - \bar{y}_i)^2$$

This decomposition of sums of squares is summarized in an **analysis of variance (ANOVA) table.** This table includes the mean square for each sum of squares.

$$\text{Mean square} = \frac{\text{Sum of squares}}{\text{Degrees of freedom}}$$

The **$F$ statistic** = (Treatment mean square)/(Error mean square) which has an **$F$ distribution** with $k - 1$ and $n - k$ degrees of freedom where $n = \sum n_i$.

To ensure an overall confidence level for all confidence intervals, it is desirable to use a **simultaneous (multiple) confidence intervals** procedure when calculating confidence intervals for the many mean differences. The **multiple-$t$ confidence intervals** for all $m = k(k - 1)/2$ statements about the mean differences $\mu_i - \mu_{i'}$ are

$$\left( \bar{Y}_i - \bar{Y}_{i'} - t_{\alpha/2m} \sqrt{\frac{MSE}{n - k}} , \quad \bar{Y}_i - \bar{Y}_{i'} + t_{\alpha/2m} \sqrt{\frac{MSE}{n - k}} \right)$$

Fortunately, the one-way analysis of variance is **robust** with respect to small or moderate departures from the assumptions of normal errors with common variance.

The technique of **blocking**, or grouping experimental units into homogeneous sets, can reduce extraneous variation and sharpen comparisons among treatments.

The analysis of variance of a **randomized block experiment** is based on the partition of the total sum of squares $\sum_{i=1}^{k} \sum_{j=1}^{b} (y_{ij} - \bar{y}_{..})^2$ into three components:

1. Treatment sum of squares:

$$SS_T = b \sum_{i=1}^{k} (\bar{y}_{i.} - \bar{y}_{..})^2$$

2. Block sum of squares:

$$SS_B = k \sum_{j=1}^{b} (\bar{y}_{.j} - \bar{y}_{..})^2$$

3. Error sum of squares:

$$SSE = \sum_{i=1}^{k} \sum_{j=1}^{b} (y_{ij} - \bar{y}_{i.} - \bar{y}_{.j} + \bar{y}_{..})^2$$

# TECHNOLOGY

## *One-way analysis of variance (ANOVA)*

### MINITAB

We illustrate with the data from different populations in separate columns. For example, with data from 3 populations in columns *C1*, *C2*, and *C3*:

Data

*C1*:  6  10   8

*C2*: 13  14  15  14

*C3*:  3   7   5

**Stat > ANOVA > One-way (Unstacked)**

Type *C1–C3* in **Responses (in separate columns:)**

Click on **Graphs**. Select **Three in one**. Click **OK**.

Click **OK**.

In addition to the analysis, this produces a histogram of the residuals as well as a plot of residuals versus fitted values.

### EXCEL

Enter the data from the different populations in separate columns, for instance A, B, C, etc., with a label in the first cell of each column.

Click the **DATA** tab and then **Data Analysis**. Select **Anova:Single Factor** and click **OK**. Highlight the data to fill in **Input Range** and click columns after **Grouped By**. Check **Labels in First Row**.

Enter **Alpha** and click **OK**.

The output includes the ANOVA table and $P$-value for test of equality of the means.

## TI-84/-83 PLUS

Store the data in lists, one for each population. With 3 populations enter data in $L_1$, $L_2$ and $L_3$.

Select **STAT** > **TESTS** > **ANOVA**. Press **ENTER**.

Type $(L_1, L_2, L_3)$ to obtain $ANOVA(L_1, L_2, L_3)$.

Press **Enter**.

The software returns the value of the $F$ statistic and the $P$-value, along with other values from the ANOVA table.

## REVIEW EXERCISES

**14.33** Food scientists investigated whether making a cheese sauce by a continuous process or batch process made any difference to taste. They also included a third treatment, a carefully selected ideal product, that served as a control. Treatment 1 is the ideal, treatment 2 is the continuous process, and treatment 3 is the batch process. Suppose the sensory ratings by trained testers are

| Treatment 1 | Treatment 2 | Treatment 3 |
|---|---|---|
| 19 | 16 | 13 |
| 18 | 11 | 16 |
| 21 | 13 | 18 |
| 18 | 14 | 11 |
|  | 11 | 15 |
|  |  | 11 |

Provide a decomposition for the observations one-way analysis of variance design with three treatments.

**14.34** Compute the sums of squares and construct the ANOVA table for the data given in Exercise 14.33.

**14.35** Refer to the data on travel expenses in Exercise 14.6 and the $F$-test in Exercise 14.15. It is the ordering of the population means, and the sizes of their differences, that is the key finding of study.

(a) Calculate 95% simultaneous confidence intervals for the differences in means.

(b) Are these conclusions, concerning the position of the signature and dishonest self-reporting, consistent with those in Chapter 13, Example 6? Comment.

**14.36** As part of the multilab study, four fabrics are tested for flammability at the National Bureau of Standards. The following burn times in seconds are recorded after a paper tab is ignited on the hem of a dress made of each fabric.

| Fabric 1 | Fabric 2 | Fabric 3 | Fabric 4 |
|----------|----------|----------|----------|
| 17.8 | 11.2 | 11.8 | 14.9 |
| 16.2 | 11.4 | 11.0 | 10.8 |
| 17.5 | 15.8 | 10.0 | 12.8 |
| 17.4 | 10.0 | 9.2 | 10.7 |
| 15.0 | 10.4 | 9.2 | 10.7 |

(a) State the statistical model and present the ANOVA table. With $\alpha = .05$, test the null hypothesis of no difference in the degree of flammability for the four fabrics.

(b) If the null hypothesis is rejected, construct simultaneous confidence intervals to determine the fabric(s) with the lowest mean burn time.

(c) Plot the residuals and comment on the plausibility of the assumptions.

(d) If the tests had been conducted one at a time on a single mannequin, how would you have randomized the fabrics tested in this experiment?

## The Following Exercises Require a Computer

**14.37** *Using the computer.* MINITAB can be used for ANOVA. Start with the data on each treatment, from Example 1, set in separate columns. The sequence of commands and output is:

Data

C1: 10 15  8 12 15

C2: 14 18 21 15

C3: 17 16 14 15 17 15 18

C4: 12 15 17 15 16 15

Dialog box:

Stat > ANOVA > One-way (Unstacked)

Type *C1–C4* in Responses. Click **OK**.

The **Output** is as follows:

One-way ANOVA: C1, C2, C3, C4

| Source | DF | SS | MS | F | P |
|--------|----|----|----|----|----|
| Factor | 3 | 68.00 | 22.67 | 4.34 | 0.018 |
| Error | 18 | 94.00 | 5.22 | | |
| Total | 21 | 162.00 | | | |

S = 2.285   R-Sq = 41.98%

| Level | N | Mean | StDev |
|-------|---|------|-------|
| C1 | 5 | 12.000 | 3.082 |
| C2 | 4 | 17.000 | 3.162 |
| C3 | 7 | 16.000 | 1.414 |
| C4 | 6 | 15.000 | 1.673 |

Individual 95% CIs For Mean Based on Pooled StDev

```
-+-------+-------+-------+-----
(-----*------)
               (------*------)
             (------*------)
           (------*------)
-+-------+-------+----+------
10.0    12.5    15.0    17.5
```

Pooled StDev = 2.285

Use computer software to analyze the moisture data in Table 8.

**14.38** The iris data described in Example 6 are given in the stem-and-leaf diagrams below.

```
LEAF UNIT  =  0.10
SETOSA              VERSICOLOR        VIRGINICA
                    2 0
2 3                 2 22333           2 2
2                   2 4445555         2 5555
2                   2 66677777        2 667777
2 9                 2 8888889999999   2 8888888899
3 0000001111        3 00000000111     3 0000000000001111
3 2222233           3 2223            3 22222333
3 444444444555555   3 4               3 44
3 666777                              3 6
3 888899                              3 88
4 01                                  3 44
4 2                                   3 6
4 4                                   3 88
```

The MINITAB output for the analysis of the iris data is given below.

```
One-way ANOVA:

Source   DF      SS      MS       F      P
Iris      2   11.345   5.672   49.16  0.000
Error   147   16.962   0.115
Total   149   28.307
S =  0.3397 R-Sq =  40.08%
```

```
Level   N        Mean        Stdev
1      50       3.4280      0.3791
2      50       2.7700      0.3138
3      50       2.9740      0.3225
```

```
Pooled StDev = 0.3397
```

```
Individual 95% CIs For Mean Based on
Pooled StDev
---+---------+---------+---------+------
                       (---*---)
(---*---)
        (---*---)
---+---------+---------+---------+------
 2.75      3.00      3.25      3.50
```

(a) Identify the SSE and its degrees of freedom. Also locate $s$.

(b) Check the calculation of $F$ from the given sums of squares and d.f.

(c) Is there one population with highest mean or are two or more alike? Use multiple-$t$ confidence intervals with $\alpha = .05$.

**14.39** Three different chemicals are compared on their ability to make fabric stain-resistent. Four bolts of cloth, manufactured weeks apart, are used. Because the results are expected to vary from bolt of cloth to bolt, three small samples are cut from each bolt and a different chemical is applied to each. The assignment of chemicals is random. This process is repeated for each of the four bolts. Suppose the measurements of stain-resistance are

| Treatment | Block | | | |
|---|---|---|---|---|
|  | 1 | 2 | 3 | 4 |
| 1 | 8 | 9 | 1 | 6 |
| 2 | 5 | 12 | 0 | 11 |
| 3 | 8 | 15 | 8 | 13 |

(a) Provide a decomposition for the observations from this randomized block experiment.

(b) Find the sum of squares for each array.

(c) Determine the degrees of freedom by checking the constraints for each array.

**14.40** Refer to the output here concerning the time (min) it took four different persons (blocks) to complete three different tasks.

(a) Identify the SSE and its degrees of freedom.

(b) Are the block means different? Check the calculation of $F$ for blocks from the given sums of squares and degrees of freedom.

(c) Are the mean task times different? Check the calculation of $F$ for treatments (tasks) from the given sums of squares and degrees of freedom.

(d) Use multiple-$t$ 95% confidence intervals to investigate differences between mean task times.

```
Two-way Analysis of Variance
Analysis of Variance for Time
Source       DF      SS      MS      F      P
Block         3   90.92   30.31   7.96  0.016
Treatment     2   71.17   35.58   9.35  0.014
Error         6   22.83    3.81
Total        11  184.92
```

| Block | Mean | Treatment | Mean |
|---|---|---|---|
| 1 | 9.0 | 1 | 8.00 |
| 2 | 13.7 | 2 | 9.50 |
| 3 | 6.7 | 3 | 13.75 |
| 4 | 12.3 |  |  |

# 15

# Nonparametric Inference

Alija/iStockphoto

Wines can be ranked without reference to a quantitative scale of measurement. Individuals use non-quantitative characteristics to help select their favorite wines.

## Selecting the Best Vintage

Using their professional judgment, wine critics do give a numeral rating as a guide to overall qualitative placement and wine quality. However, ratings among critics differ on the same bottle of wine and even the meaning of 90 points on one scale is different from that of 90 points on another critic's scale. Ratings can never substitute for your own palate or your own wine tasting rankings.

The term nonparametric pertains to inference procedures that apply to any population distribution, not just a normal distribution. Nonparametric procedures typically utilize simple aspects of the sample data such as order relationships. Stretching or compressing the scale of measurement does not change them. Consequently, the null distribution of a nonparametric test statistic can be determined without regard to the underlying distribution.

What type of observations are especially suited to a nonparametric analysis? Characteristics like degree of apathy, taste preference, and surface gloss cannot be evaluated on an objective numerical scale, and an assignment of numbers is, therefore, bound to be arbitrary. Also, when people are asked to express their views by marking on a five-point rating scale,

the numbers have little physical meaning beyond the fact that higher scores indicate greater agreement. Data of this type are called **ordinal data**, because only the order of the numbers is meaningful and the distance between the two numbers does not lend itself to practical interpretation. Nonparametric procedures that utilize information only on order or rank are particularly suited to measurements on an ordinal scale.

# 1. THE WILCOXON RANK-SUM TEST FOR COMPARING TWO TREATMENTS

**Nonparametric** refers to inference procedures that do not require the population distribution to be normal or some other specific form. Because they rely only on ordering, they remain unchanged when the scale of measurement is stretched or compressed. The null distribution of test statistics remains the same whatever the form of the population distribution. For this reason, the tests are also called **distribution-free tests** and it is their strongest advantage.

The problem of comparing two populations based on independent random samples has already been discussed in Section 1 of Chapter 10. Under the assumption of normality and equal standard deviations, the parametric inference procedures were based on Student's $t$ statistic. Here we describe a useful nonparametric procedure named after its proposer F. Wilcoxon (1945). An equivalent alternative version was independently proposed by H. Mann and D. Whitney (1947).

For a comparative study of two treatments $A$ and $B$, a set of $n = n_A + n_B$ experimental units is randomly divided into two groups of sizes $n_A$ and $n_B$, respectively. Treatment $A$ is applied to the $n_A$ units, and treatment $B$ to the other $n_B$ units. The response measurements, recorded in a slightly different notation than before, are

| | | | |
|---|---|---|---|
| Treatment $A$ | $X_1$, | $X_2$, | $\cdots$ , $X_{n_A}$ |
| Treatment $B$ | $Y_1$, | $Y_2$, | $\cdots$ , $Y_{n_B}$ |

These data constitute independent random samples from two populations. Assuming that larger responses indicate a better treatment, we wish to test the null hypothesis that there is no difference between the two treatment effects versus the one-sided alternative that treatment $A$ is more effective than treatment $B$. In the present nonparametric setting, we only assume that the distributions are continuous.

---

### Model: Both Population Distributions Are Continuous

**Hypotheses**

$H_0$ : The two population distributions are identical.
$H_1$ : The distribution of population $A$ is **shifted** to the right of the distribution of population $B$.

---

Note that no assumption is made regarding the *shape* of the population distribution. This is in sharp contrast to our $t$ test in Chapter 10, where we assumed that the population distributions were normal with equal standard deviations. Figure 1 illustrates the hypotheses $H_0$ and $H_1$.

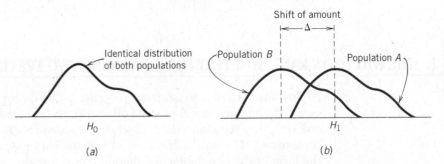

Figure 1   The two hypotheses $H_0$ and $H_1$ (a) Null distribution. (b) A shift alternative.

The basic concept underlying the rank-sum test can now be explained by the following intuitive line of reasoning. Suppose that the two sets of observations are plotted on the same diagram using different markings $A$ and $B$ to identify their sources. Under $H_0$, the samples come from the same population, so that the two sets of points should be well mixed. However, if the larger observations are more often associated with the first sample, for example, we can infer that

Figure 2   Combined plot of the two samples and the combined sample ranks. (*a*) Mixed ranks. (*b*) Higher ranks are mostly *A*.

population *A* is possibly shifted to the right of population *B*. These two situations are diagrammed in Figure 2, where the combined set of points in each case is serially numbered from left to right. These numbers are called the **combined sample ranks**. In Figure 2*a*, large as well as small ranks are associated with each sample, whereas in Figure 2*b*, most of the larger ranks are associated with the first sample. Therefore, if we consider the sum of the ranks associated with the first sample as a test statistic, a large value of this statistic should reflect that the first population is located to the right of the second.

To establish a rejection region with a specified level of significance, we must consider the distribution of the rank-sum statistic under the null hypothesis. This concept is explored in Example 1, where small sample sizes are investigated for easy enumeration.

**Example 1**   Determining the Null Distribution of the Rank-Sum Statistic

To determine if a new hybrid seedling produces a bushier flowering plant than a currently popular variety, a horticulturist plants 2 new hybrid seedlings and 3 currently popular seedlings in a garden plot. After the plants mature, the following measurements of shrub girth in inches are recorded.

|  | Shrub Girth (in inches) | | |
|---|---|---|---|
| Treatment *A* (new hybrid) | 31.8 | 39.1 | |
| Treatment *B* (current variety) | 35.5 | 27.6 | 21.3 |

Do these data indicate that the new hybrid produces larger shrubs than the current variety?

SOLUTION   We wish to test the null hypothesis

$$H_0 : A \text{ and } B \text{ populations are identical}$$

versus the alternative hypothesis

$$H_1 : \text{Population } A \text{ is shifted from } B \text{ toward larger values}$$

For the rank-sum test, the two samples are placed together and ranked from smallest to largest:

| Combined sample ordered observations | 21.3 | 27.6 | 31.8 | 35.5 | 39.1 |
|---|---|---|---|---|---|
| Ranks | 1 | 2 | 3 | 4 | 5 |
| Treatment | B | B | A | B | A |

$$\text{Rank sum for } A \quad W_A = 3 + 5 = 8$$
$$\text{Rank sum for } B \quad W_B = 1 + 2 + 4 = 7$$

Because larger measurements and therefore higher ranks for treatment $A$ tend to support $H_1$, the rejection region of our test should consist of large values for $W_A$:

$$\text{Reject } H_0 \qquad \text{if } W_A \geq c$$

To determine the critical value $c$ so that the Type I error probability is controlled at a specified level $\alpha$, we evaluate the probability distribution of $W_A$ under $H_0$. When the two samples come from the same population, every pair of integers out of $\{1, 2, 3, 4, 5\}$ is equally likely to be the ranks for the two $A$ measurements. There are $\binom{5}{2} = 10$ potential pairs, so that each collection of possible ranks has a probability of $\frac{1}{10} = .1$ under $H_0$. These rank collections are listed in Table 1 with their corresponding $W_A$ values. The null distribution of $W_A$ can be obtained immediately from Table 1 by collecting the probabilities of identical values (see Table 2).

**TABLE 1**   Rank Collections for Treatment $A$ with Sample Sizes $n_A = 2$, $n_B = 3$

| Ranks of $A$ | Rank Sum $W_A$ | Probability |
|---|---|---|
| 1, 2 | 3 | .1 |
| 1, 3 | 4 | .1 |
| 1, 4 | 5 | .1 |
| 1, 5 | 6 | .1 |
| 2, 3 | 5 | .1 |
| 2, 4 | 6 | .1 |
| 2, 5 | 7 | .1 |
| 3, 4 | 7 | .1 |
| 3, 5 | 8 | .1 |
| 4, 5 | 9 | .1 |
| | Total | 1.0 |

**TABLE 2**  Distribution of the Rank Sum $W_A$ for Sample Sizes $n_A = 2, n_B = 3$

| Value of $W_A$ | 3 | 4 | 5 | 6 | 7 | 8 | 9 |
|---|---|---|---|---|---|---|---|
| Probability | .1 | .1 | .2 | .2 | .2 | .1 | .1 |

The observed value $W_A = 8$ has the significance probability $P_{H_0}(W_A \geq 8) = .1 + .1 = .2$. In other words, we must tolerate a Type I error probability of .2 in order to reject $H_0$. The rank-sum test leads us to conclude that the evidence is not sufficiently strong to reject $H_0$. Note that even if the $A$ measurements did receive the highest ranks of 4 and 5, a significance level of $\alpha = .1$ would be required to reject $H_0$.

Guided by Example 1, we now state the rank-sum test procedure in a general setting.

---

### Wilcoxon Rank-Sum Test

Let $X_1, \ldots, X_{n_A}$ and $Y_1, \ldots, Y_{n_B}$ be independent random samples from continuous populations $A$ and $B$, respectively. To test

$$H_0 : \text{The populations are identical:}$$

1. Rank the combined sample of $n = n_A + n_B$ observations in increasing order of magnitude.
2. Find the rank sum $W_A$ of the first sample.
3. (a) For $H_1$: Population $A$ is shifted to the right of population $B$; set the rejection region at the upper tail of $W_A$.
   (b) For $H_1$: Population $A$ is shifted to the left of population $B$; set the rejection region at the lower tail of $W_A$.
   (c) For $H_1$: Populations are different; set the rejection region at both tails of $W_A$ having equal probabilities.

---

Determining the null distribution of the rank-sum statistic by direct enumeration becomes more tedious as the sample sizes increase. However, tables for the null distribution of this statistic have been prepared for small samples, and an approximation is available for large samples. To explain the use of Appendix B, Table 8, first we note some features of the rank sums $W_A$ and $W_B$.

The total of the two rank sums $W_A + W_B$ is a constant, which is the sum of the integers $1, 2, \ldots, n$, where $n$ is the combined sample size. For instance, in Example 1,

$$W_A + W_B = (3 + 5) + (1 + 2 + 4)$$
$$= 1 + 2 + 3 + 4 + 5 = 15$$

Therefore, a test that rejects $H_0$ for large values of $W_A$ is equivalent to a test that rejects $H_0$ for small values of $W_B$. We can just as easily designate $W_B$ the test statistic and set the rejection region at the lower tail. Consequently, we can always concentrate on the rank sum of the smaller sample and set the rejection region at the lower (or upper) tail, depending on whether the alternative hypothesis states that the corresponding population distribution is shifted to the left (or right).

Second, the distribution of each of the rank-sum statistics $W_A$ and $W_B$ is symmetric. In fact, $W_A$ is symmetric about $n_A(n_A + n_B + 1)/2$ and $W_B$ is symmetric about $n_B(n_A + n_B + 1)/2$. Table 2 illustrates the symmetry of the $W_A$ distribution for the case $n_A = 2$, $n_B = 3$. This symmetry also holds for the other test statistic calculated from the larger sample size.

## 1.1 THE USE OF APPENDIX B, TABLE 8

The Wilcoxon rank-sum test statistic is taken as

$W_S = $ sum of ranks of the smaller sample in the combined sample ranking

When the sample sizes are equal, take the sum of ranks for either of the samples. Appendix B, Table 8 gives the upper- as well as the lower-tail probabilities:

$$\text{Upper-tail probability} \quad P[W_S \geq x]$$

$$\text{Lower-tail probability} \quad P[W_S \leq x^*]$$

By the symmetry of the distribution, these probabilities are equal when $x$ and $x^*$ are at equal distances from the center. The table includes the $x^*$ values corresponding to the $x$'s at the upper tail.

**Example 2**   Using Table 8 in Appendix B to Set the Rejection Region
Find $P[W_S \geq 25]$ and $P[W_S \leq 8]$ when

$$\text{Smaller sample size} = 3$$

$$\text{Larger sample size} = 7$$

SOLUTION   From Table 8, we read $P = P[W_S \geq x]$ opposite the entry $x = 25$, so $.033 = P[W_S \geq 25]$.

The lower-tail entry $P[W_S \leq 8]$ is obtained by reading $P[W_S \leq x^*]$ opposite $x^* = 8$. We find $P[W_S \leq 8] = .033$ illustrating the symmetry of $W_S$.

$$P = P[W_S \geq x] = P[W_S \leq x^*]$$

Smaller Sample Size $=$ 3, Larger
Sample Size $=$ 7

|   | $x$ | $P$ | $x^*$ |
|---|-----|-----|-------|
|   | 22 | — | 11 |
|   | 23 | — | 10 |
|   | 24 | — | 9 |
| → | 25 | .033 | 8 |
|   | 26 | — | 7 |
|   | 27 | — | 6 |

The steps to follow when using Appendix B, Table 8 in performing a rank-sum test are:

Use the rank sum $W_S$ of the smaller sample as the test statistic. (If the sample sizes are equal, take either rank sum as $W_S$.)

1. If $H_1$ states that the population corresponding to $W_S$ is shifted to the right of the other population, set a rejection region of the form $W_S \geq c$ and take $c$ as the smallest $x$ value for which $P \leq \alpha$.

2. If $H_1$ states that the population corresponding to $W_S$ is shifted to the left, set a rejection region of the form $W_S \leq c$ and take $c$ as the largest $x^*$ value for which $P \leq \alpha$.

3. If $H_1$ states that the population corresponding to $W_S$ is shifted in either direction, set a rejection region of the form $W_S \leq c_1$ or $W_S \geq c_2$ and read $c_1$ from the $x^*$ column and $c_2$ from the $x$ column, so that $P \leq \alpha / 2$.

**Example 3**   Apply the Rank-Sum Test to Compare Two Geological Formations

Two geological formations are compared with respect to richness of mineral content. The mineral contents of 7 specimens of ore collected from formation 1 and 5 specimens collected from formation 2 are measured by chemical analysis. The following data are obtained:

| | Mineral Content | | | | | | |
|---|---|---|---|---|---|---|---|
| Formation 1 | 7.6 | 11.1 | 6.8 | 9.8 | 4.9 | 6.1 | 15.1 |
| Formation 2 | 4.7 | 6.4 | 4.1 | 3.7 | 3.9 | | |

Do the data provide strong evidence that formation 1 has a higher mineral content than formation 2? Test with $\alpha$ near .05.

SOLUTION   To use the rank-sum test, first we rank the combined sample and determine the sum of ranks for the second sample, which has the smaller size. The observations from the second sample and their ranks are underlined here for quick identification:

| Combined ordered values | 3.7 | 3.9 | 4.1 | 4.7 | 4.9 | 6.1 | 6.4 | 6.8 | 7.6 | 9.8 | 11.1 | 15.1 |
|---|---|---|---|---|---|---|---|---|---|---|---|---|
| Ranks | 1 | 2 | 3 | 4 | 5 | 6 | 7 | 8 | 9 | 10 | 11 | 12 |

The observed value of the rank-sum statistic is

$$W_S = 1 + 2 + 3 + 4 + 7 = 17$$

We wish to test the null hypothesis that the two population distributions are identical versus the alternative hypothesis that the second population, corresponding to $W_S$, lies to the left of the first. The rejection region is therefore at the lower tail of $W_S$.

Reading Appendix B, Table 8 with smaller sample size = 5 and larger sample size = 7, we find $P[W_S \leq 21] = .037$ and $P[W_S \leq 22] = .053$. Hence, the rejection region with $\alpha = .053$ is established as $W_S \leq 22$. Because the observed value falls in this region, the null hypothesis is rejected at $\alpha = .053$. In fact, it would be rejected if $\alpha$ were as low as $P[W_S \leq 17] = .009$.

**Example 4**   Comparing Two Flame-Retardant Materials

Flame-retardant materials are tested by igniting a paper tab on the hem of a dress worn by a mannequin. One response is the vertical length of damage to the fabric measured in inches. The following data (courtesy of B. Joiner) for 5 samples, each taken from two fabrics, were obtained by researchers at the National Bureau of Standards as part of a larger cooperative study.

| Fabric A | 5.7 | 7.3 | 7.6 | 6.0 | 6.5 |
|---|---|---|---|---|---|
| Fabric B | 4.9 | 7.4 | 5.3 | 4.6 | 6.2 |

Do the data provide strong evidence that a difference in flammability exists between the two fabrics? Test with $\alpha$ near .05.

SOLUTION   The sample sizes are equal, so that we can take the rank sum of either sample as the test statistic. We compute the rank sum for the second sample.

| Ordered values | 4.6 | 4.9 | 5.3 | 5.7 | 6.0 | 6.2 | 6.5 | 7.3 | 7.4 | 7.6 |
|---|---|---|---|---|---|---|---|---|---|---|
| Ranks | 1 | 2 | 3 | 4 | 5 | 6 | 7 | 8 | 9 | 10 |

$$W_S = 1 + 2 + 3 + 6 + 9 = 21$$

Because the alternative hypothesis is two-sided, the rejection region includes both tails of $W_S$. From Appendix B, Table 8, we find that

$$P[W_S \geq 37] = .028 = P[W_S \leq 18]$$

Thus with $\alpha = .056$, the rejection region is $W_S \geq 37$ or $W_S \leq 18$. The observed value does not fall in the rejection region so the null hypothesis is not rejected at $\alpha = .056$.

## 1.2 LARGE SAMPLE APPROXIMATION

When the sample sizes are large, the null distribution of the rank-sum statistic is approximately normal and the test can therefore be performed using the normal table. Specifically, with $W_A$ denoting the rank sum of the sample of size $n_A$, suppose that both $n_A$ and $n_B$ are large. Then $W_A$ is approximately normally distributed. Under $H_0$, the distribution of $W_A$ has

$$\text{Mean} = \frac{n_A(n_A + n_B + 1)}{2}$$

$$\text{Variance} = \frac{n_A n_B(n_A + n_B + 1)}{12}$$

---

**Large Sample Approximation to the Rank-Sum Statistic**

$$Z = \frac{W_A - n_A(n_A + n_B + 1)/2}{\sqrt{n_A n_B(n_A + n_B + 1)/12}}$$

is approximately $N(0, 1)$ when $H_0$ is true.

---

The rejection region for the $Z$ statistic can be determined by using the standard normal table.

**Example 5**  The Error When Using the Large Sample Approximation

Investigate the amount of error involved in the large sample approximation to the distribution of the rank-sum statistic when $n_A = 9$, $n_B = 10$, and $\alpha = .05$.

SOLUTION  The approximate one-sided rejection region is

$$R: \frac{W_A - 9(20)/2}{\sqrt{9 \times 10 \times 20/12}} = \frac{W_A - 90}{12.247} \geq 1.645$$

which simplifies to $R: W_A \geq 110.1$. From Appendix B, Table 8, we find $P[W_S \geq 110] = .056$ and $P[W_S \geq 111] = .047$, which are quite close to $\alpha = .05$. The error decreases with increasing sample sizes.

## 1.3 HANDLING TIED OBSERVATIONS

In the preceding examples, observations in the combined sample are all distinct and therefore the ranks are determined without any ambiguity. Often, however, due to imprecision in the measuring scale or a basic discreteness of the scale, such as a five-point preference rating scale, observed values may be repeated in one or both samples. For example, consider the two samples

| Sample 1 | 20 | 24 | 22 | 24 | 26 |
|----------|----|----|----|----|----|
| Sample 2 | 26 | 28 | 26 | 30 | 18 |

The ordered combined sample is

$$18 \quad 20 \quad 22 \quad \underbrace{24 \quad 24}_{\text{Tie}} \quad \underbrace{26 \quad 26 \quad 26}_{\text{Tie}} \quad 28 \quad 30$$

Here two ties are present; the first has 2 elements, and the second 3. The two positions occupied by 24 are eligible for the ranks 4 and 5, and we assign the average rank $(4 + 5)/2 = 4.5$ to each of these observations. Similarly, the three tied observations 26, eligible for the ranks 6, 7, and 8, are each assigned the average rank $(6 + 7 + 8)/3 = 7$. After assigning average ranks to the tied observations and usual ranks to the other observations, the rank-sum statistic can then be calculated.

When ties are present in small samples, the distribution in Appendix B, Table 8 no longer holds exactly.

## Exercises

**15.1** Independent random samples of sizes $n_A = 4$ and $n_B = 2$ are taken from two continuous populations.

(a) Enumerate all possible collections of ranks associated with the smaller sample in the combined sample ranking. Attach probabilities to these rank collections under the null hypothesis that the populations are identical.

(b) Obtain the null distribution of $W_S = $ sum of ranks of the smaller sample. Verify that the tail probabilities agree with the tabulated values.

**15.2** Independent samples of sizes $n_A = 2$ and $n_B = 2$ are taken from two continuous populations.

(a) Enumerate all possible collections of ranks associated with population A. Also attach probabilities to these rank collections assuming that the populations are identical.

(b) Obtain the null distribution of $W_A$.

**15.3** Using Appendix B, Table 8, find:

(a) $P[W_S \geq 39]$ when $n_A = 5$, $n_B = 6$.

(b) $P[W_S \leq 15]$ when $n_A = 6$, $n_B = 4$.

(c) The point $c$ such that $P[W_S \geq c]$ is close to .05 when $n_A = 7$, $n_B = 7$.

**15.4** Using Appendix B, Table 8, find:

(a) $P[W_S \geq 57]$ when $n_A = 6$, $n_B = 8$.

(b) $P[W_S \leq 31]$ when $n_A = 8$, $n_B = 6$.

(c) $P[W_S \geq 38$ or $W_S \leq 22]$ when $n_A = 5$ and $n_B = 6$.

(d) The point $c$ such that $P[W_S \leq c]$ is close to .05 when $n_A = 4$, $n_B = 7$.

(e) The points $c_1$ and $c_2$ such that $P[W_S \leq c_1] = P[W_S \geq c_2]$ is about .025 when $n_A = 7$, $n_B = 9$.

**15.5** See Table D.10 in the Data Bank. The number of breathing pauses per hour (BPH) helps determine a sleeping disorder. We took a random sample of 3 males, population A, and a random sample of 2 females and obtained the values of BPH

| Males | 10.39 | 7.61 | 2.42 |
|---|---|---|---|
| Females | 2.58 | .41 | |

(a) Evaluate $W_A$.

(b) Evaluate $W_S$.

**15.6** The following data pertain to the serum calcium measurements in units of IU/L and the serum alkaline phosphate measurements in units of $\mu g/ml$ for two breeds of pigs, Chester White and Hampshire:

**Chester White**

| Calcium | 116 | 112 | 82 | 63 | 117 | 69 | 79 | 87 |
|---|---|---|---|---|---|---|---|---|
| Phosphate | 47 | 48 | 57 | 75 | 65 | 99 | 97 | 110 |

**Hampshire**

| Calcium | 62 | 59 | 80 | 105 | 60 | 71 | 103 | 100 |
|---|---|---|---|---|---|---|---|---|
| Phosphate | 230 | 182 | 162 | 78 | 220 | 172 | 79 | 58 |

Using the Wilcoxon rank-sum procedure, test if the serum calcium level is different for the two breeds.

**15.7** Referring to the data in Exercise 15.6, is there strong evidence of a difference in the serum phosphate level between the two breeds?

**15.8** A project (courtesy of Howard Garber) is constructed to prevent the decline of intellectual performance in children who have a high risk of the most common type of mental retardation, called cultural-familial. It is believed that this can be accomplished by a comprehensive family intervention program. Seventeen children in the high-risk category are chosen in early childhood and given special schooling until the age of $4\frac{1}{2}$. Another 17 children in the same high-risk category form the control group. Measurements of the psycholinguistic quotient (PLQ) are recorded for the control and the experimental groups at the age of $4\frac{1}{2}$ years.

Do the data at the bottom of the page strongly indicate improved PLQs for the children who received special schooling? Use the Wilcoxon rank-sum test with a large sample approximation: Use $\alpha = .05$.

**15.9** The possible synergetic effect of insecticides and herbicides is a matter of concern to many environmentalists. It is feared that farmers who apply both herbicides and insecticides to a crop may enhance the toxicity of the insecticide beyond the desired level. An experiment is conducted with a particular insecticide and herbicide to determine the toxicity of the treatments.

*Treatment 1:* A concentration of .25 $\mu g$ per gram of soil of insecticide with no herbicide.

*Treatment 2:* Same dosage of insecticide used in treatment 1 plus 100 $\mu g$ of herbicide per gram of soil.

**PLQ at Age $4\frac{1}{2}$ Years**

| Experimental group | 105.4 | 118.1 | 127.2 | 110.9 | 109.3 | 121.8 | 112.7 | 120.3 |
|---|---|---|---|---|---|---|---|---|
| Control group | 79.6 | 87.3 | 79.6 | 76.8 | 79.6 | 98.2 | 88.9 | 70.9 |

**PLQ at Age $4\frac{1}{2}$ Years**

| Experimental group | 110.9 | 120.0 | 100.0 | 122.8 | 121.8 | 112.9 | 107.0 | 113.7 | 103.6 |
|---|---|---|---|---|---|---|---|---|---|
| Control group | 87.0 | 77.0 | 96.4 | 100.0 | 103.7 | 61.2 | 91.1 | 87.0 | 76.4 |

Several batches of fruit flies are exposed to each treatment, and the mortality percent is recorded as a measure of toxicity. The following data are obtained:

| Treatment 1 | Treatment 2 |
|:---:|:---:|
| 40 | 36 |
| 28 | 49 |
| 31 | 56 |
| 38 | 25 |
| 43 | 37 |
| 46 | 30 |
| 29 | 41 |
| 18 | |

Determine if the data strongly indicate different toxicity levels among the treatments.

**15.10** Morphologic measurements of a particular type of fossil excavated from two geological sites provided the following data:

| Site $A$ | Site $B$ |
|:---:|:---:|
| 1.49 | 1.31 |
| 1.32 | 1.46 |
| 2.01 | 1.86 |
| 1.59 | 1.58 |
| 1.76 | 1.64 |

Do the data strongly indicate that fossils at the sites differ with respect to the particular morphology measured?

**15.11** If $n_A = 1$ and $n_B = 9$, find

(a) The rank configuration that most strongly supports $H_1$: Population $A$ is shifted to the right of population $B$.

(b) The null probability of $W_A = 10$.

(c) Is it possible to have $\alpha = .05$ with these sample sizes?

**15.12** One aspect of a study of gender differences involves the play behavior of monkeys during the first year of life (courtesy of H. Harlow, U. W. Primate Laboratory). Six male and six female monkeys are observed in groups of four families during several ten-minute test sessions. The mean total number of times each monkey initiates play with another age mate is recorded.

| Males | 3.64 | 3.11 | 3.80 | 3.58 | 4.55 | 3.92 |
|:---|:---:|:---:|:---:|:---:|:---:|:---:|
| Females | 1.91 | 2.06 | 1.78 | 2.00 | 1.30 | 2.32 |

(a) Plot the observations.

(b) Test for equality using the Wilcoxon rank-sum test with $\alpha$ approximately .05.

(c) Determine the significance probability.

# 2.  MATCHED PAIRS COMPARISONS

In the presence of extensive dissimilarity in the experimental units, two treatments can be compared more efficiently if alike units are paired and the two treatments applied one to each member of the pair. In this section, we discuss two nonparametric tests, the **sign test** and the **Wilcoxon signed-rank test**, that can be safely applied to paired differences when the assumption of normality is suspect. The data structure of a matched pairs experiment is given in Table 3, where the observations on the $i$th pair are denoted by $(X_i, Y_i)$. The null hypothesis of primary interest is that there is no difference, or

$$H_0: \text{No difference in the treatment effects}$$

**TABLE 3**    Data Structure of Matched Pairs Sampling

| Pair | Treatment $A$ | Treatment $B$ | Difference $A - B$ |
|------|---------------|---------------|--------------------|
| 1 | $X_1$ | $Y_1$ | $D_1$ |
| 2 | $X_2$ | $Y_2$ | $D_2$ |
| . | . | . | . |
| . | . | . | . |
| . | . | . | . |
| $n$ | $X_n$ | $Y_n$ | $D_n$ |

## 2.1 THE SIGN TEST

This nonparametric test is notable for its intuitive appeal and ease of application. As its name suggests, the sign test is based on the signs of the response differences $D_i$. The test statistic is

$S$ = number of pairs in which treatment $A$ has a higher response than treatment $B$

= number of positive signs among the differences $D_1, \ldots, D_n$

When the two treatment effects are actually alike, the response difference $D_i$ in each pair is as likely to be positive as it is to be negative. Moreover, if measurements are made on a continuous scale, the possibility of identical responses in a pair can be neglected. The null hypothesis is then formulated as

$$H_0 : P\,[+] \ = \ .5 \ = \ P\,[-]$$

If we identify a plus sign as a success, the test statistic $S$ is simply the number of successes in $n$ trials and therefore has a binomial distribution with $p = .5$ under $H_0$. If the alternative hypothesis states that treatment $A$ has higher responses than treatment $B$, which is translated $P\,[+] > .5$, then large values of $S$ should be in the rejection region. For two-sided alternatives $H_1 : P\,[+] \neq .5$, a two-tailed test should be employed.

**Example 6**    Applying the Sign Test to Compare Two Types of Spark Plugs

Mileage tests are conducted to compare a new versus a conventional spark plug. A sample of 12 cars ranging from subcompacts to full-sized sedans are included in the study. The gasoline mileage for each car is recorded, once with the conventional plug and once with the new plug. The results are given in Table 4. Test the null hypothesis of no difference versus the one-sided alternative that the new plug is better. Use the sign test and take $\alpha \leq .05$.

SOLUTION    We are to test

$$H_0 : \text{No difference between } A \text{ and } B, \text{ or } P\,[+] \ = \ .5$$

**TABLE 4**    Mileage Data

| Car Number | New A | Conventional B | Difference A − B |
|:---:|:---:|:---:|:---:|
| 1 | 26.4 | 24.3 | +2.1 |
| 2 | 20.3 | 19.8 | + .5 |
| 3 | 25.8 | 26.9 | −1.1 |
| 4 | 26.5 | 27.2 | − .7 |
| 5 | 32.5 | 30.5 | +2.0 |
| 6 | 38.3 | 37.9 | + .4 |
| 7 | 22.1 | 22.4 | − .3 |
| 8 | 30.1 | 28.6 | +1.5 |
| 9 | 22.9 | 23.1 | − .2 |
| 10 | 32.6 | 31.6 | +1.0 |
| 11 | 27.3 | 25.5 | +1.8 |
| 12 | 29.4 | 28.6 | + .8 |

versus the one-sided alternative

$H_1$: The new plug $A$ is better than the conventional plug $B$, or $P[+] > .5$

Looking at the differences $A - B$, we can see that there are 8 plus signs in the sample of size $n = 12$. Thus, the observed value of the sign test statistic is $S = 8$. We will reject $H_0$ for large values of $S$. Consulting the binomial table for $n = 12$ and $p = .5$, we find $P[S \geq 9] = .073$ and $P[S \geq 10] = .019$. If we wish to control $\alpha$ below .05, the rejection region should be established at $S \geq 10$. The observed value $S = 8$ is too low to be in the rejection region, so that at the level of significance $\alpha = .019$, the data do not sustain the claim of mileage improvement.

The significance probability of the observed value is $P[S \geq 8] = .194$.

An application of the sign test does not require the numerical values of the differences to be calculated. The number of positive signs can be obtained by glancing at the data. Even when a response cannot be measured on a well-defined numerical scale, we can often determine which of the two responses in a pair is better. This is the only information that is required to conduct a sign test.

For large samples, the sign test can be performed by using the normal approximation to the binomial distribution. With large $n$, the binomial distribution with $p = .5$ is close to the normal distribution with mean $n/2$ and standard deviation $\sqrt{n/4}$.

> **Large Sample Approximation to the Sign Test Statistic**
>
> Under $H_0$,
>
> $$Z = \frac{S - n/2}{\sqrt{n/4}}$$
>
> is approximately distributed as $N(0, 1)$.

**Example 7**  Applying the Sign Test to a Large Sample of Beer Preferences

In a TV commercial filmed live, 100 persons tasted two beers $A$ and $B$ and each selected their favorite. A total of $S = 57$ preferred beer $A$. Does this provide strong evidence that $A$ is more popular?

SOLUTION  According to the large sample approximation, to the binomial

$$Z = \frac{S - n/2}{\sqrt{n/4}} = \frac{57 - 50}{\sqrt{25}} = 1.4$$

The significance probability $P[Z > 1.4] = .0808$ is not small enough to provide strong support to the claim that beer $A$ is more popular.

## 2.2  HANDLING TIES

When the two responses in a pair are exactly equal, we say that there is a **tie**. Because a tied pair has zero difference, it does not have a positive or a negative sign. In the presence of ties, the sign test is performed by discarding the tied pairs, thereby reducing the sample size. For instance, when a sample of $n = 20$ pairs has 10 plus signs, 6 minus signs, and 4 ties, the sign test is performed with the effective sample size $n = 20 - 4 = 16$ and $S = 10$.

## 2.3  THE WILCOXON SIGNED-RANK TEST

We have already noted that the sign test extends to ordinal data for which the responses in a pair can be compared without being measured on a numerical scale. However, when numerical measurements are available, the sign test may result in a considerable loss of information because it includes only the signs of the differences and disregards their magnitudes. Compare the two sets of paired differences plotted in the dot diagrams in Figure 3. In both cases, there are $n = 6$ data points with 4 positive signs, so that the sign test will lead to identical conclusions. However, the plot in Figure 3b exhibits more of a shift toward the positive side, because the positive differences are farther away from zero than the negative differences. Instead of attaching equal weights to all the positive signs, as is done in the sign test, we should attach larger weights to the plus signs

Figure 3   Two plots of paired differences with the same number of + signs but with different locations for the distributions.

that are farther away from zero. This is precisely the concept underlying the signed-rank test.

In the signed-rank test, the paired differences are ordered according to their numerical values without regard to signs, and then the ranks associated with the positive observations are added to form the test statistic. To illustrate, we refer to the mileage data given in Example 6 where the paired differences appear in the last column of Table 4. We attach ranks by arranging these differences in increasing order of their **absolute** values and record the corresponding signs.

| Paired differences | 2.1 | .5 | −1.1 | −.7 | 2.0 | .4 | −.3 | 1.5 | −.2 | 1.0 | 1.8 | .8 |
|---|---|---|---|---|---|---|---|---|---|---|---|---|
| Ordered absolute values | .2 | .3 | .4 | .5 | .7 | .8 | 1.0 | 1.1 | 1.5 | 1.8 | 2.0 | 2.1 |
| Ranks | 1 | 2 | 3 | 4 | 5 | 6 | 7 | 8 | 9 | 10 | 11 | 12 |
| Signs | − | − | + | + | − | + | + | − | + | + | + | + |

The signed-rank statistic $T^+$ is then calculated as

$$T^+ = \text{sum of the ranks associated with positive observations}$$
$$= 3 + 4 + 6 + 7 + 9 + 10 + 11 + 12$$
$$= 62$$

If the null hypothesis of no difference in treatment effects is true, then the paired differences $D_1, D_2, \ldots, D_n$ constitute a random sample from a population that is symmetric about zero. On the other hand, the alternate hypothesis that treatment $A$ is better asserts that the distribution is shifted from zero toward positive values. Under $H_1$, not only are more plus signs anticipated, but the positive signs are also likely to be associated with larger ranks. Consequently, $T^+$ is expected to be large under the one-sided alternative, and we select a rejection region in the upper tail of $T^+$.

---

**Steps in the Signed-Rank Test**

1.  Calculate the differences $D_i = X_i - Y_i$,      $i = 1, \ldots, n$.
2.  Assign ranks by arranging the absolute values of the $D_i$ in increasing order; also record the corresponding signs.
3.  Calculate the signed-rank statistic $T^+$ = sum of ranks of positive differences $D_i$.
4.  Set the rejection region at the upper tail, lower tail, or at both tails of $T^+$, according to whether treatment $A$ is stated to have a higher, lower, or different response than treatment $B$ under the alternative hypothesis.

---

Selected tail probabilities of the null distribution of $T^+$ are given in Appendix B, Table 9 for $n = 3$ to $n = 15$.

## 2.4  USING APPENDIX B, TABLE 9

By symmetry of the distribution around $n(n + 1)/4$, we obtain

$$P[T^+ \geq x] = P[T^+ \leq x^*]$$

when $x^* = n(n + 1)/2 - x$. The $x$ and $x^*$ values in Appendix B, Table 9 satisfy this relation. To illustrate the use of this table, we refer once again to the mileage data given in Example 6. There, $n = 12$ and the observed value of $T^+$ is found to be 62. From the table, we find $P[T^+ \geq 61] = .046$. Thus, the null hypothesis is rejected at the level of significance $\alpha = .046$, and a significant mileage improvement using the new type of spark plug is indicated.

$$P = P[T^+ \geq x] = P[T^+ \leq x^*]$$

$$n = 12$$

| $x$ | $P$ | $x^*$ |
|---|---|---|
| 56 | . | 22 |
| . | . | . |
| . | . | . |
| . | . | . |
| → 61 | .046 | 17 |
| 62 | | 16 |
| . | . | . |
| . | . | . |
| . | . | . |
| 68 | . | 10 |

With increasing sample size $n$, the null distribution of $T^+$ is closely approximated by a normal curve, with mean $n(n + 1)/4$ and variance $n(n + 1)(2n + 1)/24$.

---

### Large Sample Approximation to Signed-Rank Statistic

$$Z = \frac{T^+ - n(n + 1)/4}{\sqrt{n(n + 1)(2n + 1)/24}}$$

is approximately distributed as $N(0, 1)$.

---

This result can be used to perform the signed-rank test with large samples.

**Example 8** Applying the Signed-Rank Test to Compare Spark Plugs

Refer to the mileage data in Example 6. Obtain the significance probability for the signed-rank test using (a) the exact distribution in Appendix B, Table 9 and (b) the large sample approximation.

SOLUTION

(a) For the mileage data, $T^+ = 62$ and $n = 12$. From Appendix B, Table 9, the exact significance probability is $P[T^+ \geq 62] = .039$.

(b) The normal approximation to this probability uses

$$z = \frac{62 - 12(13)/4}{\sqrt{12(13)(25)/24}} = \frac{23}{12.75} = 1.804$$

From the normal table, we approximate $P[T^+ \geq 62]$ by $P[Z \geq 1.804] = .036$.

The normal approximation improves with increasing sample size.

### 2.5  *HANDLING TIES

In computing the signed-rank statistic, ties may occur in two ways: Some of the differences $D_i$ may be zero or some nonzero differences $D_i$ may have the same absolute value. The first type of tie is handled by discarding the zero values and then reducing the sample size. The second type of tie is handled by ranking the remaining absolute values and then assigning the average rank to each absolute difference in a group of tied values.

# Exercises

**15.13** In a taste test of two chocolate chip cookie recipes, 13 out of 18 subjects favored recipe $A$. Using the sign test, find the significance probability when $H_1$ states that recipe $A$ is preferable.

**15.14** Two critics rate the service at six award-winning restaurants on a continuous 0 to 10 scale. Is there a difference between the critics' ratings?

(a) Use the sign test with $\alpha$ below .05.

(b) Find the significance probability.

| Restaurant | Service Rating | |
|---|---|---|
| | Critic 1 | Critic 2 |
| 1 | 6.1 | 7.3 |
| 2 | 5.2 | 5.5 |
| 3 | 8.9 | 9.1 |
| 4 | 7.4 | 7.0 |
| 5 | 4.3 | 5.1 |
| 6 | 9.7 | 9.8 |

**15.15** A social researcher interviews 25 newly married couples. Each husband and wife are independently asked the question: "How many children would you like to have?" The following data are obtained.

| Couple | Answer of | |
|---|---|---|
| | Husband | Wife |
| 1 | 3 | 2 |
| 2 | 1 | 1 |
| 3 | 2 | 1 |
| 4 | 2 | 3 |
| 5 | 5 | 1 |
| 6 | 0 | 1 |
| 7 | 0 | 2 |
| 8 | 1 | 3 |
| 9 | 2 | 2 |
| 10 | 3 | 1 |
| 11 | 4 | 2 |
| 12 | 1 | 2 |
| 13 | 3 | 3 |
| 14 | 2 | 1 |
| 15 | 3 | 2 |
| 16 | 2 | 2 |

| Couple | Answer of | |
|---|---|---|
| | Husband | Wife |
| 17 | 0 | 0 |
| 18 | 1 | 2 |
| 19 | 2 | 1 |
| 20 | 3 | 2 |
| 21 | 4 | 3 |
| 22 | 3 | 1 |
| 23 | 0 | 0 |
| 24 | 1 | 2 |
| 25 | 1 | 1 |

Do the data show a significant difference of opinion between husbands and wives regarding an ideal family size? Use the sign test with $\alpha$ close to .05.

**15.16** Use Appendix B, Table 9, to find:

(a) $P[T^+ \geq 54]$ when $n = 11$.

(b) $P[T^+ \leq 32]$ when $n = 15$.

(c) The value of $c$ so that $P[T^+ \geq c]$ is nearly .05 when $n = 14$.

**15.17** Use Appendix B, Table 9, to find:

(a) $P[T^+ \geq 65]$ when $n = 12$.

(b) $P[T^+ \leq 10]$ when $n = 10$.

(c) The value $c$ such that $P[T^+ \geq c] = .039$ when $n = 8$.

(d) The values $c_1$ and $c_2$ such that $P[T^+ \leq c_1] = P[T^+ \geq c_2] = .027$ when $n = 11$.

**15.18** Referring to Exercise 15.14, apply the Wilcoxon signed-rank test with $\alpha$ near .05.

**15.19** The null distribution of the Wilcoxon signed-rank statistic $T^+$ is determined from the fact that under the null hypothesis of a symmetric distribution about zero, each of the ranks $1, 2, \ldots, n$ is equally likely to be associated with a positive sign or a negative sign. Moreover, the signs are independent of the ranks.

(a) Considering the case $n = 3$, identify all $2^3 = 8$ possible associations of signs with the ranks 1, 2, and 3, and determine the value of $T^+$ for each association.

(b) Assigning the equal probability of $\frac{1}{8}$ to each case, obtain the distribution of $T^+$ and verify that the tail probabilities agree with the tabulated values.

**15.20** A married couple's monthly credit charges are divided into his and hers and the difference, husband's minus wife's, is calculated. A random sample of 30 married couples yielded the Wilcoxon signed-rank statistic $T^+ = 325$. What is the significance probability if the alternative is two-sided?

**15.21** In Example 14 of Chapter 10, we presented data on the blood pressure of 15 persons before and after they took a pill.

| Before | After | Difference |
|--------|-------|------------|
| 70 | 68 | 2 |
| 80 | 72 | 8 |
| 72 | 62 | 10 |
| 76 | 70 | 6 |
| 76 | 58 | 18 |
| 76 | 66 | 10 |
| 72 | 68 | 4 |
| 78 | 52 | 26 |
| 82 | 64 | 18 |
| 64 | 72 | -8 |
| 74 | 74 | 0 |
| 92 | 60 | 32 |
| 74 | 74 | 0 |
| 68 | 72 | -4 |
| 84 | 74 | 10 |

(a) Perform a sign test, with $\alpha$ near .05, to determine if blood pressure has decreased after taking the pill.

(b) Perform a Wilcoxon signed-rank test to determine if blood pressure has decreased after taking the pill.

**15.22** Charles Darwin performed an experiment to determine if self-fertilized and cross-fertilized plants have different growth rates. Pairs of *Zea mays* plants, one self- and the other cross-fertilized, were planted in pots, and their heights were measured after a specified period of time. The data Darwin obtained were

| Pair | Cross- | Self- |
|------|--------|-------|
| | Plant height (in $\frac{1}{8}$ inches) | |
| 1 | 188 | 139 |
| 2 | 96 | 163 |
| 3 | 168 | 160 |
| 4 | 176 | 160 |
| 5 | 153 | 147 |
| 6 | 172 | 149 |
| 7 | 177 | 149 |
| 8 | 163 | 122 |
| 9 | 146 | 132 |
| 10 | 173 | 144 |
| 11 | 186 | 130 |
| 12 | 168 | 144 |
| 13 | 177 | 102 |
| 14 | 184 | 124 |
| 15 | 96 | 144 |

*Source*: C. Darwin, *The effects of cross- and self-fertilization in the vegetable kingdom*, D. Appleton and Co., New York, 1902.

(a) Calculate the paired differences and plot a dot diagram for the data. Does the assumption of normality seem plausible?

(b) Perform the Wilcoxon signed-rank test to determine if cross-fertilized plants have a higher growth rate than self-fertilized plants.

# 3.  MEASURE OF CORRELATION BASED ON RANKS

Ranks may also be employed to determine the degree of association between two random variables. These two variables could be mathematical ability and musical aptitude or the aggressiveness scores of first- and second-born sons on a psychological test. We encountered this same general problem in Chapter 3, where we introduced Pearson's product moment correlation coefficient

$$r = \frac{\sum_{i=1}^{n} (X_i - \bar{X})(Y_i - \bar{Y})}{\sqrt{\sum_{i=1}^{n} (X_i - \bar{X})^2} \sqrt{\sum_{i=1}^{n} (Y_i - \bar{Y})^2}}$$

as a measure of association between $X$ and $Y$. Serving as a descriptive statistic, $r$ provides a numerical value for the amount of linear dependence between $X$ and $Y$.

---

### Structure of the Observations

The $n$ pairs $(X_1, Y_1)$, $(X_2, Y_2)$, ..., $(X_n, Y_n)$ are independent, and each pair has the same continuous bivariate distribution. The $X_1, \ldots, X_n$ are then ranked among *themselves*, and the $Y_1, \ldots, Y_n$ are ranked among *themselves*:

| Pair no. | 1 | 2 | $\cdots$ | $n$ |
|---|---|---|---|---|
| Ranks of $X_i$ | $R_1$ | $R_2$ | $\cdots$ | $R_n$ |
| Ranks of $Y_i$ | $S_1$ | $S_2$ | $\cdots$ | $S_n$ |

---

Before we present a measure of association, we note a few simplifying properties. Because each of the ranks, 1, 2, ..., $n$ must occur exactly once in the set $R_1, R_2, \ldots, R_n$, it can be shown that

$$\bar{R} = \frac{1 + 2 + \cdots + n}{n} = \frac{n+1}{2}$$

$$\sum_{i=1}^{n} (R_i - \bar{R})^2 = \frac{n(n^2 - 1)}{12}$$

for all possible outcomes. Similarly,

$$\bar{S} = \frac{n+1}{2} \quad \text{and} \quad \sum_{i=1}^{n} (S_i - \bar{S})^2 = \frac{n(n^2 - 1)}{12}$$

A measure of correlation is defined by C. Spearman that is analogous to Pearson's correlation $r$, except that Spearman replaces the observations with their ranks. Spearman's rank correlation $r_{Sp}$ is defined by

$$r_{Sp} = \frac{\sum_{i=1}^{n}(R_i - \bar{R})(S_i - \bar{S})}{\sqrt{\sum_{i=1}^{n}(R_i - \bar{R})^2}\sqrt{\sum_{i=1}^{n}(S_i - \bar{S})^2}}$$

$$= \frac{\sum_{i=1}^{n}\left(R_i - \frac{n+1}{2}\right)\left(S_i - \frac{n+1}{2}\right)}{n(n^2 - 1)/12}$$

This rank correlation shares the properties of $r$ that $-1 \leq r_{Sp} \leq 1$ and that values near $+1$ indicate a tendency for the larger values of $X$ to be paired with the larger values of $Y$. However, the rank correlation captures wider types of dependence, because its interpretation does not require the relationship to be linear.

---

**Spearman's Rank Correlation**

$$r_{Sp} = \frac{\sum_{i=1}^{n}\left(R_i - \frac{n+1}{2}\right)\left(S_i - \frac{n+1}{2}\right)}{n(n^2 - 1)/12}$$

1. $-1 \leq r_{Sp} \leq 1$.
2. $r_{Sp}$ near $+1$ indicates a tendency for the larger values of $X$ to be associated with the larger values of $Y$. Values near $-1$ indicate the opposite relationship.
3. The association need not be linear; only an increasing/decreasing relationship is required to produce values of $r_{Sp}$ near 1 or $-1$.

---

**Example 9**   Calculating Spearman's Rank Correlation

An interviewer, in charge of hiring several persons for a call center, wishes to determine the strength of the relationship between ranks given on the basis of an interview and scores on an aptitude test. The data for six applicants are

| Interview rank | 5 | 2 | 3 | 1 | 6 | 4 |
|---|---|---|---|---|---|---|
| Aptitude score | 47 | 32 | 29 | 28 | 56 | 38 |

Calculate $r_{Sp}$.

SOLUTION   There are 6 ranks, so that $\bar{R} = (n + 1)/2 = 7/2 = 3.5$ and $n(n^2 - 1)/12 = 35/2 = 17.5$. Ranking the aptitude scores, we obtain

| Interview $R_i$ | 5 | 2 | 3 | 1 | 6 | 4 |
|---|---|---|---|---|---|---|
| Aptitude $S_i$ | 5 | 3 | 2 | 1 | 6 | 4 |

Thus,

$$\sum_{i=1}^{n} \left( R_i - \frac{n+1}{2} \right) \left( S_i - \frac{n+1}{2} \right)$$

$$= (5 - 3.5)(5 - 3.5) + (2 - 3.5)(3 - 3.5) + \cdots$$
$$+ (4 - 3.5)(4 - 3.5)$$

$$= 1.5(1.5) + (-1.5)(-.5) + \cdots + (.5)(.5)$$

$$= 16.5$$

and

$$r_{Sp} = \frac{16.5}{17.5} = .943$$

The relationship between interview rank and aptitude score appears to be quite strong.

Figure 4 helps to stress the point that $r_{Sp}$ is a measure of any monotone relationship, not merely a linear relation.

$r_{Sp} = .99$

$r$ is inappropriate

Figure 4   $r_{Sp}$ is a measure of any monotone relationship.

A large sample approximation to the distribution of $r_{Sp}$ is available.

If $X$ and $Y$ are independent,

$$\sqrt{n - 1}\ r_{Sp} \text{ is approximately distributed as } N(0, 1)$$

provided that the sample size is large.

This approximation leads to a convenient form of a test for independence. Reject

$$H_0 : X \text{ and } Y \text{ are independent}$$

in favor of

$$H_1 : \text{Large values of } X \text{ and } Y \text{ tend to occur together}$$
$$\text{and small values tend to occur together}$$

if

$$\sqrt{n - 1} \; r_{Sp} \geq z_a$$

Recall that $z_\alpha$ is the upper $\alpha$ point of a standard normal distribution. Two-tailed tests can also be conducted.

**Example 10**   Establishing Dependence When Large $X$ and $Y$ Tend to Occur Together and so Do Small $X$ and $Y$

The grade point average (GPA) and Scholastic Achievement Test (SAT) scores for 40 applicants yielded $r_{Sp}$ = .4. Do large values of GPA and SAT tend to occur together? That is, test for lack of independence using $\alpha$ = .05.

SOLUTION   For $\alpha$ = .05, the rejection region is $\sqrt{n - 1} \; r_{Sp} \geq z_{.05}$ = 1.96. Since

$$\sqrt{n - 1} \; r_{Sp} = \sqrt{39}(.4) = 2.498$$

we reject $H_0$: $X$ and $Y$ are independent at level $\alpha$ = .05. Large values of GPA and SAT tend to occur together and so do small values.

## Exercises

**15.23**   Refer to Exercise 11.31 and the first four countries in Table 7. The number of Internet users per one hundred residents and the human development index (HDI) are

| Internet/100 | 64.9 | 77.0 | 51.4 | 51.0 |
|---|---|---|---|---|
| HDI | .813 | .806 | .783 | .771 |

Calculate Spearman's rank correlation.

**15.24**   Refer to Exercise 11.67 and the height and speed of the four tallest roller coasters. Given the data

| Height | 456 | 420 | 415 | 377 |
|---|---|---|---|---|
| Speed | 128 | 120 | 100 | 107 |

Evaluate $r_{Sp}$.

**15.25**   The following scores are obtained on a test of dexterity and aggression administered to a random sample of 10 high school seniors.

| Student | 1 | 2 | 3 | 4 | 5 | 6 | 7 | 8 | 9 | 10 |
|---|---|---|---|---|---|---|---|---|---|---|
| Dexterity | 23 | 29 | 45 | 36 | 49 | 41 | 30 | 15 | 42 | 38 |
| Aggression | 45 | 48 | 16 | 28 | 38 | 21 | 36 | 18 | 31 | 37 |

Evaluate Spearman's statistic.

**15.26**   Referring to Example 10, determine the significance probability of $r_{Sp}$ = .4, using the one-sided test, when $n$ = 40.

# 4. CONCLUDING REMARKS

In contrast to nonparametric procedures, Student's $t$ and the chi-square statistic $(n - 1)S^2 / \sigma^2$ were developed to make inferences about the parameters $\mu$ and $\sigma$ of a normal population. These *normal-theory parametric* procedures can be seriously affected when the sample size is small and the underlying distribution is not normal. Drastic departures from normality can occur in the forms of conspicuous asymmetry, sharp peaks, or heavy tails. For instance, a $t$ test with an intended level of significance of $\alpha = .05$ may have an actual Type I error probability far in excess of this value. These effects are most pronounced for small or moderate sample sizes precisely when it is most difficult to assess the shape of the population. The selection of a parametric procedure leaves the data analyst with the question: Does my normality assumption make sense in the present situation? To avoid this risk, a nonparametric method could be used in which inferences rest on the safer ground of distribution-free properties.

When the data constitute measurements on a meaningful numerical scale and the assumption of normality holds, parametric procedures are certainly more efficient in the sense that tests have higher power than their nonparametric counterparts. This brings to mind the old adage, "You get what you pay for." A willingness to assume more about the population form leads to improved inference procedures. However, trying to get too much for your money by assuming more about the population than is reasonable can lead to the "purchase" of invalid conclusions. A choice between the parametric and nonparametric approach should be guided by a consideration of loss of efficiency and the degree of protection desired against possible violations of the assumptions.

Tests are judged by two criteria: control of the Type I error probability and the power to detect alternatives. Nonparametric tests guarantee the desired control of the Type I error probability $\alpha$, whatever the form of the distribution. However, a parametric test established at $\alpha = .05$ for a normal distribution may suffer a much larger $\alpha$ when a departure from normality occurs. This is particularly true with small sample sizes. To achieve universal protection, nonparametric tests, quite expectedly, must forfeit some power to detect alternatives when normality actually prevails. As plausible as this argument sounds, it is rather surprising that the loss in power is often marginal with such simple procedures as the Wilcoxon rank-sum test and the signed-rank test.

Finally, the presence of dependence among the observations affects the usefulness of nonparametric and parametric methods in much the same manner. When either method is used, the level of significance of the test may seriously differ from the nominal value selected by the analyst.

> ***Caution:*** When successive observations are dependent, nonparametric test procedures lose their distribution-free property, and conclusions drawn from them can be seriously misleading.

## USING STATISTICS WISELY

1. A one-sample nonparametric test will provide valid inferences with a small sample size where it may not be possible to check the assumption of normality. Of course, the power of the rank test will generally be less than the normal theory paired $t$ test when normality holds.

2. When the two sample sizes are small, it is a good idea to conduct a two-sample Wilcoxon rank-sum test. If software is available, also obtain the corresponding confidence interval for the difference in location. These provide a baseline comparison for the result based on the $t$ distribution.

3. If the sample sizes are large enough so the dot diagrams reveal a difference in both location and spread, the Wilcoxon rank-sum test is not appropriate.

4. Remember that nonparametric tests can produce invalid inferences if there is time dependence between the observations.

## KEY IDEAS AND FORMULAS

Nonparametric tests obtain their **distribution-free** character because rank orders of the observations do not depend on the shape of the population distribution.

The Wilcoxon rank-sum test, based on the test statistic

$$W_A = \text{sum of ranks of the } n_A \text{ observations}$$
$$\text{from population } A, \text{ among all } n_A + n_B \text{ observations}$$

applies to the comparison of two populations. It uses the **combined sample ranks**.

In the paired-sample situation, equality of treatments can be tested using either the sign test based on the statistic

$$S = \text{No. of positive differences}$$

or the Wilcoxon signed-rank based on the statistic

$$T^+ = \text{sum, over positive differences,}$$
$$\text{of the ranks of their absolute values}$$

The level of a nonparametric test holds whatever the form of the (continuous) population distribution.

Any tie in the observations requires specific handling.

# TECHNOLOGY

*Nonparametric tests and confidence intervals*

**MINITAB**

*One sample—inference about median*

Start with the data in *C1*. To find a 95% confidence interval for the median using the sign test:

> Stat > Nonparametrics > 1-Sample Sign.
> Type *C1* in **Variables**. Click **Confidence interval** and type *0.95* in **Level**. Click **OK**.

To test a hypothesis concerning the median, instead of **Confidence interval**:

> Click **Test Median** and choose the form of the **Alternative** hypothesis. You cannot set the level.

To find a 95% confidence interval for the median using the Wilcoxon signed-rank test:

> Stat > Nonparametrics > 1-Sample Wilcoxon.
> Type *C1* in **Variables**. Click **Confidence interval** and type *0.95* in **Level**. Click **OK**.

To test a hypothesis concerning the median, instead of **Confidence interval**:

> Click **Test Median** and choose the form of the **Alternative** hypothesis.

*Two-sample Wilcoxon test for equality of populations*

Start with the data from the first population in *C1* and the data from the second in C2. To test at level $\alpha = .05$:

> Stat > Nonparametrics > Mann-Whitney.
> Type *C1* in **First Sample** and C2 in **Second Sample**. Type *0.95* in **Confidence level** and select the form of the **Alternative**. Click **OK**.

The output includes a confidence interval for the difference in locations.

# REVIEW EXERCISES

**15.27** Three male and three female students recorded the number of times they used a credit card in one week.

| Male   | 8 | 4 | 6 |
|--------|---|---|---|
| Female | 1 | 5 | 3 |

Designate males as population A. Evaluate $W_A$.

**15.28** Using Appendix B, Table 8, find:

(a) $P[W_S \geq 42]$ when $n_1 = 5, n_2 = 7$.

(b) $P[W_S \leq 25]$ when $n_1 = 6, n_2 = 6$.

(c) $P [ W_S \geq 81$ or $W_S \leq 45 ]$ when $n_1 = 10$, $n_2 = 7$.

(d) The point $c$ such that $P [ W_S \leq c ] = .036$ when $n_1 = 8, n_2 = 4$.

(e) The points $c_1$ and $c_2$ such that $P [ W_S \geq c_2 ] = P [ W_S \leq c_1 ] = .05$ when $n_1 = 3, n_2 = 9$.

**15.29** (a) Evaluate all possible rank configurations associated with treatment $A$ when $n_A = 3$ and $n_B = 2$.

(b) Determine the null distribution of $W_A$.

**15.30** Five finalists in a figure-skating contest are rated by two judges on a 10-point scale as follows:

| Contestants | A | B | C | D | E |
|---|---|---|---|---|---|
| Judge 1 | 6 | 9 | 2 | 8 | 5 |
| Judge 2 | 8 | 10 | 4 | 7 | 3 |

Calculate the Spearman's rank correlation $r_{Sp}$ between the two ratings.

**15.31** Using Appendix B, Table 9, find:

(a) $P [ T^+ \geq 28 ]$ when $n = 8$.

(b) $P [ T^+ \leq 5 ]$ when $n = 7$.

(c) The point $c$ such that $P [ T^+ \leq c ]$ is approximately .05 when $n = 13$.

**15.32** Referring to Exercise 15.30, calculate:

(a) The sign test statistic.

(b) The significance probability when the alternative is that Judge 2 gives higher scores than Judge 1.

**15.33** In a study of the cognitive capacities of non-human primates, 19 monkeys of the same age are randomly divided into two groups of 10 and 9. The groups are trained by two different teaching methods to recollect an acoustic stimulus. The monkeys' scores on a subsequent test are seen below.

Do the data strongly indicate a difference in the recollection abilities of monkeys trained by the two methods? Use the Wilcoxon rank-sum test with $\alpha$ close to .10.

**15.34** A mixture of compounds called phenolics occurs in wood waste products. It has been found that when phenolics are present in large quantities, the waste becomes unsuitable for use as a livestock feed. To compare two species of wood, a dairy scientist measures the percentage content of phenolics from 6 batches of waste of species $A$ and 7 batches of waste of species $B$. The following data are obtained.

Percentage of Phenolics

| Species $A$ | 2.38 | 4.19 | 1.39 | 3.73 | 2.86 | 1.21 | |
|---|---|---|---|---|---|---|---|
| Species $B$ | 4.67 | 5.38 | 3.89 | 4.67 | 3.58 | 4.96 | 3.98 |

Use the Wilcoxon rank-sum test to determine if the phenolics content of species $B$ is significantly higher than that of species $A$. Use $\alpha$ close to .05.

**15.35** (a) Calculate Spearman's rank correlation for the data on Chester Whites in Exercise 15.6.

(b) Test for independence of calcium and phosphate levels using the rejection region
$$\sqrt{n - 1} \; r_{Sp} \geq 1.96 \quad \text{or} \quad \leq -1.96$$

(c) What is the approximate level of significance?

**15.36** In the study described in Exercise 1.23, golfers were asked to estimate the size of the hole (cm) in the green by selecting among nine in a board. The size selected and their score for the day for three golfers are

| Hole size | 11.5 | 10.0 | 10.5 |
|---|---|---|---|
| Score | 84 | 104 | 94 |

Evaluate $r_{Sp}$.

Memory Scores

| Method 1 | 167 | 149 | 137 | 178 | 179 | 155 | 164 | 104 | 151 | 150 |
|---|---|---|---|---|---|---|---|---|---|---|
| Method 2 | 98 | 127 | 140 | 103 | 116 | 105 | 100 | 95 | 131 | |

**15.37**    Refer to Exercise 10.76. Evaluate:

(a)  Sign test statistic.

(b)  Signed-rank statistic.

*  **15.38**    *Confidence interval for median using the sign test.*    Let $X_1, \ldots, X_n$ be a random sample from a continuous population whose median is denoted by $M$. For testing $H_0 : M = M_0$, we can use the sign test statistic $S = $ No. of $X_i > M_0, i = 1, \ldots, n$. $H_0$ is rejected at level $\alpha$ in favor of $H_1 : M \neq M_0$ if $S \leq r$ or $S \geq n - r + 1$, where $r$ is the largest integer satisfying

$$\sum_{x=0}^{r} b(x; n, .5) \leq \alpha / 2$$

If we repeat this test procedure for all possible values of $M_0$, a $100(1 - \alpha)\%$ confidence interval for $M$ is then the range of values $M_0$ so that $S$ is in the acceptance region. Ordering the observations from smallest to largest, verify that this confidence interval becomes

$(r + 1)$st smallest to $(r + 1)$st largest observation

(a)  Refer to Example 6. Using the sign test, construct a confidence interval for the population median of the differences $A - B$, with a level of confidence close to 95%.

(b)  Repeat part (a) using Darwin's data given in Exercise 15.22.

# Summation Notation

## A1.1  SUMMATION AND ITS PROPERTIES

The addition of numbers is basic to our study of statistics. To avoid a detailed and repeated writing of this operation, the symbol $\Sigma$ (the Greek capital letter *sigma*) is used as mathematical shorthand for the operation of addition.

---

**Summation Notation $\Sigma$**

The notation $\sum_{i=1}^{n} x_i$ represents the sum of $n$ numbers $x_1, x_2, \ldots, x_n$ and is read as the sum of all $x_i$ with $i$ ranging from 1 to $n$.

$$\sum_{i=1}^{n} x_i = x_1 + x_2 + \cdots + x_n$$

---

The term following the sign $\Sigma$ indicates the quantities that are being summed, and the notations on the bottom and the top of the $\Sigma$ specify the range of the terms being added. For instance,

$$\sum_{i=1}^{3} x_i = x_1 + x_2 + x_3$$

$$\sum_{i=1}^{4} (x_i - 3) = (x_1 - 3) + (x_2 - 3) + (x_3 - 3) + (x_4 - 3)$$

**Example** Suppose that the four measurements in a data set are given as $x_1 = 2$, $x_2 = 5, x_3 = 3, x_4 = 4$. Compute the numerical values of

(a) $\sum_{i=1}^{4} x_i$        (b) $\sum_{i=1}^{4} 6$        (c) $\sum_{i=1}^{4} 2x_i$

(d) $\sum_{i=1}^{4} (x_i - 3)$        (e) $\sum_{i=1}^{4} x_i^2$        (f) $\sum_{i=1}^{4} (x_i - 3)^2$

SOLUTION

(a) $\sum_{i=1}^{4} x_i = x_1 + x_2 + x_3 + x_4 = 2 + 5 + 3 + 4 = 14$

(b) $\sum_{i=1}^{4} 6 = 6 + 6 + 6 + 6 = 4(6) = 24$

(c) $\sum_{i=1}^{4} 2x_i = 2x_1 + 2x_2 + 2x_3 + 2x_4 = 2 \left( \sum_{i=1}^{4} x_i \right)$

$$= 2 \times 14 = 28$$

(d) $\sum_{i=1}^{4} (x_i - 3) = (x_1 - 3) + (x_2 - 3)$

$$+ (x_3 - 3) + (x_4 - 3)$$

$$= \sum_{i=1}^{4} x_i - 4(3) = 14 - 12 = 2$$

(e) $\sum_{i=1}^{4} x_i^2 = x_1^2 + x_2^2 + x_3^2 + x_4^2 = 2^2 + 5^2 + 3^2 + 4^2 = 54$

(f) $\sum_{i=1}^{4} (x_i - 3)^2 = (x_1 - 3)^2 + (x_2 - 3)^2$

$$+ (x_3 - 3)^2 + (x_4 - 3)^2$$

$$= (2 - 3)^2 + (5 - 3)^2$$

$$+ (3 - 3)^2 + (4 - 3)^2$$

$$= 1 + 4 + 0 + 1 = 6$$

Alternatively, noting that $(x_i - 3)^2 = x_i^2 - 6x_i + 9$, we can write

$$\sum_{i=1}^{4} (x_i - 3)^2 = \sum_{i=1}^{4} (x_i^2 - 6x_i + 9)$$

$$= (x_1^2 - 6x_1 + 9) + (x_2^2 - 6x_2 + 9)$$

$$+ (x_3^2 - 6x_3 + 9) + (x_4^2 - 6x_4 + 9)$$

$$= \sum_{i=1}^{4} x_i^2 - 6 \left( \sum_{i=1}^{4} x_i \right) + 4(9)$$

$$= 54 - 6(14) + 36 = 6$$

A few basic properties of the summation operation are apparent from the numerical demonstration in the example.

---

**Some Basic Properties of Summation**

If $a$ and $b$ are fixed numbers,

$$\sum_{i=1}^{n} bx_i = b \sum_{i=1}^{n} x_i$$

$$\sum_{i=1}^{n} (bx_i + a) = b \sum_{i=1}^{n} x_i + na$$

$$\sum_{i=1}^{n} (x_i - a)^2 = \sum_{i=1}^{n} x_i^2 - 2a \sum_{i=1}^{n} x_i + na^2$$

---

## Exercises

1. Demonstrate your familiarity with the summation notation by evaluating the following expressions when $x_1 = 4$, $x_2 = -2$, $x_3 = 1$.

   (a) $\sum_{i=1}^{3} x_i$

   (b) $\sum_{i=1}^{3} 7$

   (c) $\sum_{i=1}^{3} 5x_i$

   (d) $\sum_{i=1}^{3} (x_i - 2)$

   (e) $\sum_{i=1}^{3} (x_i - 3)$

   (f) $\sum_{i=1}^{3} (x_i - 2)^2$

   (g) $\sum_{i=1}^{3} x_i^2$

   (h) $\sum_{i=1}^{3} (x_i - 3)^2$

   (i) $\sum_{i=1}^{3} (x_i^2 - 6x_i + 9)$

2. Five measurements in a data set are $x_1 = 4$, $x_2 = 3$, $x_3 = 6$, $x_4 = 5$, $x_5 = 7$. Determine

   (a) $\sum_{i=1}^{5} x_i$

   (b) $\sum_{i=2}^{3} x_i$

   (c) $\sum_{i=1}^{5} 2$

   (d) $\sum_{i=1}^{5} (x_i - 6)$

   (e) $\sum_{i=1}^{5} (x_i - 6)^2$

   (f) $\sum_{i=1}^{4} (x_i - 5)^2$

## A1.2  SOME BASIC USES OF $\Sigma$ IN STATISTICS

Let us use the summation notation and its properties to verify some computational facts about the sample mean and variance.

### $\Sigma (x_i - \bar{x}) = 0$

The total of the deviations about the sample mean is always zero. Since $\bar{x} = (x_1 + x_2 + \cdots + x_n) / n$, we can write

$$\sum_{i=1}^{n} x_i = x_1 + x_2 + \cdots + x_n = n\bar{x}$$

Consequently, whatever the observations,

$$\sum_{i=1}^{n} (x_i - \bar{x}) = (x_1 - \bar{x}) + (x_2 - \bar{x}) + \cdots + (x_n - \bar{x})$$

$$= x_1 + x_2 + \cdots + x_n - n\bar{x}$$

$$= n\bar{x} - n\bar{x} = 0$$

We could also verify this directly with the second property for summation in A1.1, when $b = 1$ and $a = -\bar{x}$.

### ALTERNATIVE FORMULA FOR $s^2$

By the quadratic rule of algebra,

$$(x_i - \bar{x})^2 = x_i^2 - 2\bar{x}x_i + \bar{x}^2$$

Therefore,

$$\Sigma (x_i - \bar{x})^2 = \Sigma x_i^2 - \Sigma 2\bar{x}x_i + \Sigma \bar{x}^2$$

$$= \Sigma x_i^2 - 2\bar{x} \Sigma x_i + n\bar{x}^2$$

Using $(\Sigma x_i) / n$ in place of $\bar{x}$, we get

$$\Sigma (x_i - \bar{x})^2 = \Sigma x_i^2 - \frac{2(\Sigma x_i)^2}{n} + \frac{n(\Sigma x_i)^2}{n^2}$$

$$= \Sigma x_i^2 - \frac{2(\Sigma x_i)^2}{n} + \frac{(\Sigma x_i)^2}{n}$$

$$= \Sigma x_i^2 - \frac{(\Sigma x_i)^2}{n}$$

We could also verify this directly from the third property for summation, in A1.1, with $a = \bar{x}$.

This result establishes that

$$s^2 = \frac{\sum_{i=1}^{n} (x_i - \bar{x})^2}{n - 1} = \frac{\sum x_i^2 - (\sum x_i)^2 / n}{n - 1}$$

so the two forms of $s^2$ are equivalent.

## SAMPLE CORRELATION COEFFICIENT

The sample correlation coefficient and slope of the fitted regression line contain a term

$$S_{xy} = \sum_{i=1}^{n} (x_i - \bar{x})(y_i - \bar{y})$$

which is a sum of the products of the deviations. To obtain the alternative form, first note that

$$(x_i - \bar{x})(y_i - \bar{y}) = x_i y_i - x_i \bar{y} - \bar{x} y_i + \bar{x}\bar{y}$$

We treat $x_i y_i$ as a single number, with index $i$, and conclude that

$$\sum (x_i - \bar{x})(y_i - \bar{y}) = \sum x_i y_i - \sum x_i \bar{y} - \sum \bar{x} y_i + \sum \bar{x}\bar{y}$$
$$= \sum x_i y_i - \bar{y} \sum x_i - \bar{x} \sum y_i + n\bar{x}\bar{y}$$

Since $\bar{x} = (\sum x_i) / n$ and $\bar{y} = (\sum y_i) / n$,

$$\sum (x_i - \bar{x})(y_i - \bar{y}) = \sum x_i y_i - \frac{(\sum y_i)}{n} \sum x_i - \frac{(\sum x_i)}{n} \sum y_i$$
$$+ \frac{n}{n} \frac{(\sum x_i)}{n} \frac{(\sum y_i)}{n}$$
$$= \sum x_i y_i - \frac{(\sum x_i)\,(\sum y_i)}{n}$$

Consequently, either $\sum (x_i - \bar{x})(y_i - \bar{y})$ or $\sum x_i y_i - (\sum x_i)(\sum y_i)/n$ can be used for the calculation of $S_{xy}$ with similar choices for $S_{xx}$ and $S_{yy}$.

## SAMPLE MEAN AND VARIANCE OF $X - Y$ AND $X + Y$

Let $(x_i, y_i)$, for $i = 1, \ldots, n$ be the values of two characteristics on $n$ units.

The values $x_i - y_i$ for $X - Y$ have

sample mean $= \bar{x} - \bar{y}$

sample variance $= s_1^2 + s_2^2 - 2\frac{1}{n - 1} \sum (x_i - \bar{x})(y_i - \bar{y})$

The values $x_i + y_i$ for $X + Y$ have

> sample mean $= \bar{x} + \bar{y}$
>
> sample variance $= s_1^2 + s_2^2 + 2\dfrac{1}{n-1}\sum(x_i - \bar{x})(y_i - \bar{y})$

The term $s_{12} = \dfrac{1}{n-1}\sum(x_i - \bar{x})(y_i - \bar{y})$ is called the sample covariance.

According to the definition of summation, the $n$ values of the difference $x_i - y_i$ have sample mean

$$\frac{1}{n}\sum(x_i - y_i) = \frac{1}{n}\sum x_i + \frac{1}{n}\sum(-y_i) = \bar{x} - \bar{y}$$

and the squared deviation

$$
\begin{aligned}
(x_i - y_i - (\bar{x} - \bar{y}))^2 &= ((x_i - \bar{x}) - (y_i - \bar{y}))^2 \\
&= (x_i - \bar{x})^2 + (y_i - \bar{y})^2 \\
&\quad - 2(x_i - \bar{x})(y_i - \bar{y}))
\end{aligned}
$$

The result follows after summing and dividing by $n - 1$. The result for $x_i + y_i$ is obtained by replacing "$-$" with "$+$".

# A2

# Rules for Counting

Some basic rules for counting can help us calculate probabilities. We begin with an example. Brazil (B), United States (U.S.), Mexico (M), Canada (C), and Japan (J) are in the final round of a volleyball tournament. How many ways can a winner and second place finisher be selected?

To systematically address this problem, we create a tree diagram where the possibilities for first place are represented by the initial 5 branches. For each choice of a first place finisher, there are four choices for second place. These are represented by the sets of 4 branches in the second stage. For instance, the leftmost branch of the tree diagram depicts the case where Brazil is first and the United States second and so on.

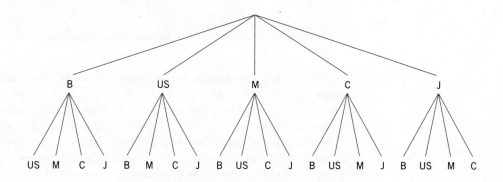

There are $5 \times 4 = 20$ terminal points or ways for these teams to finish first and second. Generalizing the tree diagram to $k$ choices followed by $m$ choices, we obtain the product rule for counting.

---

### Product Rule

An experiment consists of two parts where the first part has $k$ distinct possible outcomes and, for each outcome of the first part, there are $m$ distinct possible outcomes for the second part. Then, there are $k \times m$ distinct possible outcomes to the experiment.

---

The product rule is readily extended to experiments with more than two parts. Suppose we are interested in the winner, second, and third place finishers in the volleyball tournament. As above, by the product rule, there are $5 \times 4 = 20$ distinct possibilities for selecting a first and second place volleyball team. Again, by the product rule, each of these 20 outcomes must be paired with one of the 3 outcomes for a third place finisher so there are $20 \times 3 = 5 \times 4 \times 3 = 60$ possibilities for the first, second, and third place finishers.

The term permutation means an ordering or arrangement of objects. The number of different orderings or arrangements that can be formed with $r$ objects selected from a group of $n$ distinct objects is called the **number of permutations of $r$ out of $n$** and is denoted by $P^n_r$.

In the volleyball example, $P^5_2 = 5 \times 4 = 20$ for the first two places and $P^5_3 = 5 \times 4 \times 3 = 60$ for the first three places. When $r$ objects are selected, the first has $n$ possibilities and the last $n - r + 1$ possibilities.

---

### Rule of Permutations or Arrangements

The number of different arrangements of $r$ objects selected from a group of $n$ distinct objects is denoted by $P^n_r$ and

$$P^n_r = n(n - 1) \cdots (n - r + 1)$$

---

If all $n$ distinct objects are arranged in order, the number of possible arrangements is $P^n_n = n(n - 1) \times \cdots \times 2 \times 1$, which is the product $n$ times $n - 1$ on down through $2 \times 1$. This special product, of decreasing integers starting with $n$, is called $n$ **factorial** and is denoted by $n!$.

When order is not important, we speak about collections of objects. The number of different collections of $r$ objects selected from a group of $n$ distinct objects is called the **number of combinations of $r$ out of $n$**.

---

### Rule of Combinations

The number of combinations, or different collections, of $r$ objects selected from a group of $n$ distinct objects is denoted by $\binom{n}{r}$,

$$\binom{n}{r} = \frac{P^n_r}{r!} = \frac{n(n-1) \cdots (n-r+1)}{r!} = \frac{n!}{r!(n-r)!}$$

---

The general formula for combinations is obtained by the following correspondence and an application of the product rule.

$$\begin{bmatrix} \text{Arrange } r \text{ objects} \\ \text{selected from } n \end{bmatrix} \quad \begin{array}{c} \text{is the} \\ \text{same as} \end{array} \quad \begin{bmatrix} \text{first select } r \\ \text{objects from } n \end{bmatrix} \quad \begin{array}{c} \text{and} \\ \text{then} \end{array} \quad \begin{bmatrix} \text{arrange the } r \\ \text{selected objects} \end{bmatrix}$$

$$P^n_r \qquad = \qquad \binom{n}{r} \qquad \times \qquad r!$$

According to the last form, $\binom{n}{r}$ is symmetric in $r$ and $n - r$. Selecting $r$ objects to be in the collection is the same as choosing which $n - r$ to exclude.

If 2 of the 5 volleyball teams must be selected to play another match, so the order of selection is not important, there are 10 ways to select 2 to play or the 3 teams that will not play.

$$\binom{5}{2} = \frac{5 \times 4}{2 \times 1} = 10 = \frac{5 \times 4 \times 3 \times 2 \times 1}{2 \times 1 \times (3 \times 2 \times 1)} = \frac{5 \times 4 \times 3}{3 \times 2 \times 1} \text{ which equals } \binom{5}{3}$$

# A3
APPENDIX

# Expectation and Standard Deviation—Properties

The expected value of a discrete random variable is a summation of the products (value × probability). The key properties of expectations are then all inherited from the properties of summation. In this appendix, we indicate this development for some of the most useful properties of expectation and variance. The interested reader can consult Bhattacharyya and Johnson[1] for more details.

## A3.1 EXPECTED VALUE AND STANDARD DEVIATION OF $cX + b$

The units of the random variable $X$ may be changed by multiplying by a constant, for example,

$$X = \text{height in feet}, \qquad 12X = \text{height in inches}$$

or by adding a constant, for example,

$$X = \text{temperature } (^{\circ}F)$$
$$X - 32 = \text{degrees above freezing } (^{\circ}F)$$

[1]G. K. Bhattacharyya and R. A. Johnson, *Statistical concepts and methods*, New York: John Wiley & Sons, 1978.

The mean and standard deviation of the new random variables are related to $\mu = E(X)$ and $\sigma = sd(X)$.

If $X$ is multiplied by a constant $c$,

| Random Variable | Mean | sd |
|:---:|:---:|:---:|
| $cX$ | $c\mu$ | $|c|\sigma$ |

If a constant $b$ is added to $X$,

| Random Variable | Mean | sd |
|:---:|:---:|:---:|
| $X + b$ | $\mu + b$ | $\sigma$ (unchanged) |

Notice that adding a constant to a random variable leaves the standard deviation unchanged.

**Example**  Let $X$ have mean $= 3 = \mu$ and standard deviation $= 5 = \sigma$. Find the mean and sd of (a) $X + 4$, (b) $2X$, (c) $-X$, and (d) $\frac{1}{5}(X - 3)$.

SOLUTION  By the foregoing properties,

| Random Variable | Mean | sd |
|:---:|:---:|:---:|
| $X + 4$ | $3 + 4 = 7$ | $5$ |
| $2X$ | $2(3) = 6$ | $2(5) = 10$ |
| $-X = (-1)X$ | $(-1)3 = -3$ | $|-1|5 = 5$ |

Finally, $(X - 3)$ has mean $3 - 3 = 0$ and sd $= 5$, so $\frac{1}{5}(X - 3)$ has mean $= \frac{1}{5}(0) = 0$ and sd $= \frac{1}{5}(5) = 1$.

Any random variable having $E(X) = \mu$ and $Var(X) = \sigma^2$ can be converted to a

**Standardized variable**

$$Z = \frac{X - \mu}{\sigma}$$

The standardized variable $Z$ has mean $= 0$ and variance $= 1$. This was checked for $Z = \dfrac{X - 3}{5}$ in the example above.

## *VERIFICATION OF THE MEAN AND sd EXPRESSIONS FOR $cX + b$

Consider the random variable $cX + b$, which includes the two cases above. The choice $b = 0$ gives $cX$ and the choice $c = 1$ gives $X + b$. We restrict our verification to discrete random variables, where probability $f(x_i)$ is attached to $x_i$. Because $cX + b$ takes value $cx_i + b$ with probability $f(x_i)$,

$$(\text{value} \times \text{probability}) = (cx_i + b)f(x_i) = cx_i f(x_i) + bf(x_i)$$

and

$$
\begin{aligned}
\text{mean} &= \sum (\text{value} \times \text{probability}) \\
&= \sum c\, x_i f(x_i) + \sum bf(x_i) \\
&= c \sum x_i f(x_i) + b \sum f(x_i) = c\mu + b \cdot 1
\end{aligned}
$$

Next,

$$
\begin{aligned}
\text{deviation} &= (\text{random variable}) - (\text{mean}) \\
&= (cX + b) - (c\mu + b) = cX - c\mu = c(X - \mu)
\end{aligned}
$$

so

$$
\begin{aligned}
\text{variance} &= \sum (\text{deviation})^2 \times \text{probability} \\
&= \sum c^2 (x_i - \mu)^2 f(x_i) \\
&= c^2 \sum (x_i - \mu)^2 f(x_i) = c^2 \sigma^2
\end{aligned}
$$

Taking the positive square root yields $(cX + b) = |c|\sigma$.
Finally, taking $c = 1/\sigma$ and $b = \mu/\sigma$, we obtain

$$cX + b = \frac{1}{\sigma}X - \frac{\mu}{\sigma} = \frac{X - \mu}{\sigma} = Z$$

so the standardized variable $Z$ has

$$\text{mean} \quad c\mu + b = \frac{1}{\sigma}\mu - \frac{\mu}{\sigma} = 0$$

$$\text{sd} \quad c\sigma = \frac{1}{\sigma}\sigma = 1$$

## A3.2  ALTERNATIVE FORMULA FOR $\sigma^2$

An alternative formula for $\sigma^2$ often simplifies the numerical calculations. By definition,

$$\sigma^2 = \sum (\text{deviation})^2 (\text{probability}) = \sum (x_i - \mu)^2 f(x_i)$$

but $\sigma^2$ can also be expressed as

$$\sigma^2 = \sum x_i^2 f(x_i) - \mu^2$$

To deduce the second form, we first expand the square of the deviation:

$$(x_i - \mu)^2 = x_i^2 - 2\mu x_i + \mu^2$$

Then, multiply each term on the right-hand side by $f(x_i)$ and sum:

| | | | |
|---|---|---|---|
| First term | | $= \sum x_i^2 f(x_i)$ | |
| Second term | $-2\mu \sum x_i f(x_i) = -2\mu^2$ | since $\sum x_i f(x_i) = \mu$ | |
| Third term | $+\mu^2 \sum f(x_i) = \mu^2$ | since $\sum f(x_i) = 1$ | |
| Result: | | $\sigma^2 = \sum x_i^2 f(x_i) - \mu^2$ | |

**Example**  Calculate $\sigma^2$ by both formulas.

SOLUTION

| | | Calculation $\sum (x - \mu)^2 f(x)$ | | |
|---|---|---|---|---|
| $x$ | $f(x)$ | $xf(x)$ | $(x - 2)^2$ | $(x - 2)^2 f(x)$ |
| 1 | .4 | .4 | 1 | .4 |
| 2 | .3 | .6 | 0 | 0 |
| 3 | .2 | .6 | 1 | .2 |
| 4 | .1 | .4 | 4 | .4 |
| | | $2.0 = \mu$ | | $1.0 = \sigma^2$ |

| | | Calculation $\sum x^2 f(x) - \mu^2$ | |
|---|---|---|---|
| $x$ | $f(x)$ | $xf(x)$ | $x^2 f(x)$ |
| 1 | .4 | .4 | .4 |
| 2 | .3 | .6 | 1.2 |
| 3 | .2 | .6 | 1.8 |
| 4 | .1 | .4 | 1.6 |
| | | $2.0 = \mu$ | 5.0 |
| | | $\sigma^2 = 5 - 2^2 = 1$ | |

## A3.3  PROPERTIES OF EXPECTED VALUE FOR TWO RANDOM VARIABLES

The concept of expectation extends to two or more variables.

> With two random variables:
>
> Property 1.   $E(X + Y) = E(X) + E(Y)$
> (additivity or sum law of expectation).
> Property 2.   If $X$ and $Y$ are independent, then $E(XY) = E(X)E(Y)$.

*Remark:*  Property 1 holds quite generally, independence is not required.

## *DEMONSTRATION

We verify both Property 1 and Property 2 assuming independence. Independence implies that $P[X = x, Y = y] = P[X = x]P[Y = y]$ for all outcomes $(x, y)$. That is, the distribution of probability over the pairs of possible values $(x, y)$ is specified by the product $f_X(x)f_Y(y) = P[X = x, Y = y]$.

The expected value, $E(X + Y)$, is obtained by multiplying each possible value $(x + y)$ by the probability $f_X(x)f_Y(y)$ and summing

$$E(X + Y) = \sum_x \sum_y (x + y) f_X(x) f_Y(y)$$

$$= \sum_x \sum_y x f_X(x) f_Y(y) + \sum_x \sum_y y f_X(x) f_Y(y)$$

$$= \left( \sum_x x f_X(x) \right) \underbrace{\left( \sum_y f_Y(y) \right)}_{=1} + \underbrace{\left( \sum_x f_X(x) \right)}_{=1} \left( \sum_x y f_Y(y) \right)$$

$$= E(X) + E(Y)$$

Next,

$$E(XY) = \sum_x \sum_y x y f_X(x) f_Y(y)$$

$$= \sum_x x f_X(x) \left( \sum_y y f_Y(y) \right) = E(X)E(Y) \qquad \square$$

Under the provison that the random variables are independent, variances also add.

---

Property 3.  If $X$ and $Y$ are independent,

$$\text{Var}(X + Y) = \text{Var}(X) + \text{Var}(Y)$$

---

## *DEMONSTRATION

We set $\mu_1 = E(X)$ and $\mu_2 = E(Y)$, so by Property 1,

$$E(X + Y) = \mu_1 + \mu_2$$

Then, since variance is the expected value of (variable–mean)$^2$,

$$\text{Var}(X + Y) = E(X + Y - \mu_1 - \mu_2)^2$$
$$= E[(X - \mu_1)^2 + (Y - \mu_2)^2 + 2(X - \mu_1)(Y - \mu_2)]$$
$$= E(X - \mu_1)^2 + E(Y - \mu_2)^2 + 2E(X - \mu_1)(Y - \mu_2)$$

(by the sum law of expectation, property 1)

$$= \text{Var}(X) + \text{Var}(Y)$$

This last step follows since

$$E(X - \mu_1)(Y - \mu_2) = E(XY - \mu_1 Y - X\mu_2 + \mu_1\mu_2)$$
$$= E(XY) - \mu_1 E(Y) - \mu_2 E(X) + \mu_1\mu_2$$
$$= E(XY) - \mu_1\mu_2$$
$$= E(X)E(Y) - \mu_1\mu_2$$
$$= \mu_1\mu_2 - \mu_1\mu_2 = 0 \text{ by property 2} \qquad \square$$

## A3.4  PROPERTIES OF EXPECTED VALUE FOR SEVERAL RANDOM VARIABLES

The additivity of expected value stated in Property 1, and the variance in Property 3 of A3.3, clearly extend to any number of random variables.[2]

$$E(X) = E(X_1 + X_2 + \cdots + X_n)$$
$$= E(X_1) + E(X_2) + \cdots + E(X_n)$$

and, when the random variables are independent,

$$\text{Var}(X_1 + X_2 + \cdots + X_n)$$
$$= \text{Var}(X_1) + \text{Var}(X_2) + \cdots + \text{Var}(X_n)$$

**Example**  The Mean and Variance of the Binomial Distribution

Let $X$ be distributed as the binomial distribution proportion parameter $p$ and sample size $n$. Verify that the mean and variance of $X$ are given by

$$\mu = E(X) = np \qquad \text{and} \qquad \sigma^2 = \text{Var}(X) = np(1 - p)$$

[2]We avoid a formal proof by induction.

SOLUTION   The random variable $X$ can be expressed as

$$X = X_1 + X_2 + \cdots + X_n$$

where $X_1$ equals 1 if the first trial is a success and 0 otherwise, $X_2$ is the same for the second trial, and so on. That is, $X_i$ equals 1 if the $i$th trial is a success and 0 otherwise, for all $i = 1, 2, \ldots, n$. Further, $X_i$ has the same distribution as $X_1$ which has mean and variance

$$\mu_1 = E(X_1) = 0 \cdot (1 - p) + 1 \cdot p = p$$

$$\sigma_1^2 = \text{Var}(X_1) = E(X_1^2) - \mu_1^2 = 0^2 \cdot (1 - p) + 1^2 \cdot p - p^2$$
$$= p(1 - p)$$

Because the distributions of the $X_i$ are identical,

$$E(X) = E(X_1 + X_2 \cdots + X_n)$$
$$= E(X_1) + E(X_2) + \cdots + E(X_n)$$
$$= nE(X_1) = np$$

Also, the $X_i$ are independent so that

$$\text{Var}(X) = \text{Var}(X_1 + X_2 + \cdots X_n)$$
$$= \text{Var}(X_1) + \text{Var}(X_2) + \cdots + \text{Var}(X_n)$$
$$= n\sigma_1^2 = np(1 - p)$$

# The Expected Value and Standard Deviation of $\overline{X}$

Some basic properties of the sampling distribution of $\overline{X}$ can be expressed in terms of the population mean and variance when the observations form a random sample. Let

$$\mu = \text{population mean}$$
$$\sigma^2 = \text{population variance}$$

In a random sample, the random variables $X_1, \ldots, X_n$ are independent, and each has the distribution of the population. Consequently, each observation has mean $\mu$ and variance $\sigma^2$, or

$$E(X_1) = \cdots = E(X_n) = \mu$$
$$\text{Var}(X_1) = \cdots = \text{Var}(X_n) = \sigma^2$$

Next,

$$\overline{X} = \frac{1}{n}(X_1 + \cdots + X_n)$$

and $n$ is a constant. Using the additivity properties of expectation and variance discussed in Appendix A3.4, we obtain

$$E(\overline{X}) = \frac{1}{n} E(X_1 + \cdots + X_n)$$

$$= \frac{1}{n} [E(X_1) + \cdots + E(X_n)]$$

$$(\text{mean of sum} = \text{sum of means})$$

$$= \frac{1}{n} [\mu + \cdots + \mu] = \frac{n\mu}{n} = \mu$$

$$\text{Var}(\overline{X}) = \frac{1}{n^2} \text{Var}(X_1 + \cdots + X_n)$$

$$= \frac{1}{n^2} [\text{Var}(X_1) + \cdots + \text{Var}(X_n)]$$

$$(\text{variances add due to independence})$$

$$= \frac{1}{n^2} [\sigma^2 + \cdots + \sigma^2] = \frac{n\sigma^2}{n^2} = \frac{\sigma^2}{n}$$

Furthermore, taking the square root yields

$$\text{sd}(\overline{X}) = \sqrt{\text{Var}(\overline{X})} = \frac{\sigma}{\sqrt{n}}$$

# B

# Tables

**TABLE 1**  Random Digits

| Row | | | | | | | | | | |
|-----|------|------|------|------|------|------|------|------|------|------|
| 1   | 0695 | 7741 | 8254 | 4297 | 0000 | 5277 | 6563 | 9265 | 1023 | 5925 |
| 2   | 0437 | 5434 | 8503 | 3928 | 6979 | 9393 | 8936 | 9088 | 5744 | 4790 |
| 3   | 6242 | 2998 | 0205 | 5469 | 3365 | 7950 | 7256 | 3716 | 8385 | 0253 |
| 4   | 7090 | 4074 | 1257 | 7175 | 3310 | 0712 | 4748 | 4226 | 0604 | 3804 |
| 5   | 0683 | 6999 | 4828 | 7888 | 0087 | 9288 | 7855 | 2678 | 3315 | 6718 |
|     |      |      |      |      |      |      |      |      |      |      |
| 6   | 7013 | 4300 | 3768 | 2572 | 6473 | 2411 | 6285 | 0069 | 5422 | 6175 |
| 7   | 8808 | 2786 | 5369 | 9571 | 3412 | 2465 | 6419 | 3990 | 0294 | 0896 |
| 8   | 9876 | 3602 | 5812 | 0124 | 1997 | 6445 | 3176 | 2682 | 1259 | 1728 |
| 9   | 1873 | 1065 | 8976 | 1295 | 9434 | 3178 | 0602 | 0732 | 6616 | 7972 |
| 10  | 2581 | 3075 | 4622 | 2974 | 7069 | 5605 | 0420 | 2949 | 4387 | 7679 |
|     |      |      |      |      |      |      |      |      |      |      |
| 11  | 3785 | 6401 | 0540 | 5077 | 7132 | 4135 | 4646 | 3834 | 6753 | 1593 |
| 12  | 8626 | 4017 | 1544 | 4202 | 8986 | 1432 | 2810 | 2418 | 8052 | 2710 |
| 13  | 6253 | 0726 | 9483 | 6753 | 4732 | 2284 | 0421 | 3010 | 7885 | 8436 |
| 14  | 0113 | 4546 | 2212 | 9829 | 2351 | 1370 | 2707 | 3329 | 6574 | 7002 |
| 15  | 4646 | 6474 | 9983 | 8738 | 1603 | 8671 | 0489 | 9588 | 3309 | 5860 |
|     |      |      |      |      |      |      |      |      |      |      |
| 16  | 7873 | 7343 | 4432 | 2866 | 7973 | 3765 | 2888 | 5154 | 2250 | 4339 |
| 17  | 3756 | 9204 | 2590 | 6577 | 2409 | 8234 | 8656 | 2336 | 7948 | 7478 |
| 18  | 2673 | 7115 | 5526 | 0747 | 3952 | 6804 | 3671 | 7486 | 3024 | 9858 |
| 19  | 0187 | 7045 | 2711 | 0349 | 7734 | 4396 | 0988 | 4887 | 7682 | 8990 |
| 20  | 7976 | 3862 | 8323 | 5997 | 6904 | 4977 | 1056 | 6638 | 6398 | 4552 |

**TABLE 1**    *(Continued)*

| Row | | | | | | | | | | |
|-----|------|------|------|------|------|------|------|------|------|------|
| 21 | 5605 | 1819 | 8926 | 9557 | 2905 | 0802 | 7749 | 0845 | 1710 | 4125 |
| 22 | 2225 | 5556 | 2545 | 7480 | 8804 | 4161 | 0084 | 0787 | 2561 | 5113 |
| 23 | 2549 | 4166 | 1609 | 7570 | 4223 | 0032 | 4236 | 0169 | 4673 | 8034 |
| 24 | 6113 | 1312 | 5777 | 7058 | 2413 | 3932 | 5144 | 5998 | 7183 | 5210 |
| 25 | 2028 | 2537 | 9819 | 9215 | 9327 | 6640 | 5986 | 7935 | 2750 | 2981 |
| 26 | 7818 | 3655 | 5771 | 4026 | 5757 | 3171 | 6435 | 2990 | 1860 | 1796 |
| 27 | 9629 | 3383 | 1931 | 2631 | 5903 | 9372 | 1307 | 4061 | 5443 | 8663 |
| 28 | 6657 | 5967 | 3277 | 7141 | 3628 | 2588 | 9320 | 1972 | 7683 | 7544 |
| 29 | 4344 | 7388 | 2978 | 3945 | 0471 | 4882 | 1619 | 0093 | 2282 | 7024 |
| 30 | 3145 | 8720 | 2131 | 1614 | 1575 | 5239 | 0766 | 0404 | 4873 | 7986 |
| 31 | 1848 | 4094 | 9168 | 0903 | 6451 | 2823 | 7566 | 6644 | 1157 | 8889 |
| 32 | 0915 | 5578 | 0822 | 5887 | 5354 | 3632 | 4617 | 6016 | 8989 | 9482 |
| 33 | 1430 | 4755 | 7551 | 9019 | 8233 | 9625 | 6361 | 2589 | 2496 | 7268 |
| 34 | 3473 | 7966 | 7249 | 0555 | 6307 | 9524 | 4888 | 4939 | 1641 | 1573 |
| 35 | 3312 | 0773 | 6296 | 1348 | 5483 | 5824 | 3353 | 4587 | 1019 | 9677 |
| 36 | 6255 | 4204 | 5890 | 9273 | 0634 | 9992 | 3834 | 2283 | 1202 | 4849 |
| 37 | 0562 | 2546 | 8559 | 0480 | 9379 | 9282 | 8257 | 3054 | 4272 | 9311 |
| 38 | 1957 | 6783 | 4105 | 8976 | 8035 | 0883 | 8971 | 0017 | 6476 | 2895 |
| 39 | 7333 | 1083 | 0398 | 8841 | 0017 | 4135 | 4043 | 8157 | 4672 | 2424 |
| 40 | 4601 | 8908 | 1781 | 4287 | 2681 | 6223 | 0814 | 4477 | 3798 | 4437 |
| 41 | 2628 | 2233 | 0708 | 0900 | 1698 | 2818 | 3931 | 6930 | 9273 | 6749 |
| 42 | 5318 | 8865 | 6057 | 8422 | 6992 | 9697 | 0508 | 3370 | 5522 | 9250 |
| 43 | 6335 | 0852 | 8657 | 8374 | 0311 | 6012 | 9477 | 0112 | 8976 | 3312 |
| 44 | 0301 | 8333 | 0327 | 0467 | 6186 | 1770 | 4099 | 9588 | 5382 | 8958 |
| 45 | 1719 | 9775 | 1566 | 7020 | 4535 | 2850 | 0207 | 4792 | 6405 | 1472 |
| 46 | 8907 | 8226 | 4249 | 6340 | 9062 | 3572 | 7655 | 6707 | 3685 | 1282 |
| 47 | 6129 | 5927 | 3731 | 1125 | 0081 | 1241 | 2772 | 6458 | 9157 | 4543 |
| 48 | 7376 | 3150 | 8985 | 8318 | 8003 | 6106 | 4952 | 8492 | 2804 | 3867 |
| 49 | 9093 | 3407 | 4127 | 9258 | 3687 | 5631 | 5102 | 1546 | 2659 | 0831 |
| 50 | 1133 | 3086 | 9380 | 5431 | 8647 | 0910 | 6948 | 2257 | 0946 | 1245 |
| 51 | 4567 | 0910 | 8495 | 2410 | 1088 | 7067 | 8505 | 9083 | 4339 | 2440 |
| 52 | 6141 | 8380 | 2302 | 4608 | 7209 | 5738 | 9765 | 3435 | 9657 | 6061 |
| 53 | 1514 | 8309 | 8743 | 3096 | 0682 | 7902 | 8204 | 7508 | 8330 | 1681 |
| 54 | 7277 | 1634 | 7866 | 9883 | 0916 | 6363 | 5391 | 6184 | 8040 | 3135 |
| 55 | 4568 | 4758 | 0166 | 1509 | 2105 | 0976 | 0269 | 0278 | 7443 | 2431 |
| 56 | 9200 | 7599 | 7754 | 4534 | 4532 | 3102 | 6831 | 2387 | 4147 | 2455 |
| 57 | 3971 | 8149 | 4431 | 2345 | 6436 | 0627 | 0410 | 1348 | 6599 | 1296 |
| 58 | 2672 | 9661 | 2359 | 8477 | 3425 | 8150 | 6918 | 8883 | 1518 | 4708 |
| 59 | 1524 | 3268 | 3798 | 3360 | 2255 | 0371 | 7610 | 9114 | 9466 | 0901 |
| 60 | 6817 | 9007 | 5959 | 0767 | 1166 | 7317 | 7502 | 0274 | 6340 | 0427 |

**TABLE 1**    *(Continued)*

| Row | | | | | | | | | |
|-----|------|------|------|------|------|------|------|------|------|
| 61 | 6762 | 3502 | 9559 | 4279 | 9271 | 9595 | 3053 | 4918 | 7503 | 5169 |
| 62 | 5264 | 0075 | 6655 | 4563 | 7112 | 7264 | 3240 | 2150 | 8180 | 1361 |
| 63 | 5070 | 8428 | 5149 | 2137 | 8728 | 9110 | 2334 | 9709 | 8134 | 3925 |
| 64 | 1664 | 3379 | 5273 | 9367 | 6950 | 6828 | 1711 | 7082 | 4783 | 0147 |
| 65 | 6962 | 7141 | 1904 | 6648 | 7328 | 2901 | 6396 | 9949 | 6274 | 1672 |
| 66 | 7541 | 4289 | 4970 | 2922 | 6670 | 8540 | 9053 | 3219 | 8881 | 1897 |
| 67 | 5244 | 4651 | 2934 | 6700 | 8869 | 0926 | 4191 | 1364 | 0926 | 2874 |
| 68 | 2939 | 3890 | 0745 | 2577 | 7931 | 3913 | 7877 | 2837 | 2500 | 8774 |
| 69 | 4266 | 6207 | 8083 | 6564 | 5336 | 5303 | 7503 | 6627 | 6055 | 3606 |
| 70 | 7848 | 5477 | 5588 | 3490 | 0294 | 3609 | 1632 | 5684 | 1719 | 6162 |
| 71 | 3009 | 1879 | 0440 | 7916 | 6643 | 9723 | 5933 | 0574 | 2480 | 6893 |
| 72 | 9865 | 7813 | 7468 | 8493 | 3293 | 1071 | 7183 | 9462 | 2363 | 6529 |
| 73 | 1196 | 1251 | 2368 | 1262 | 5769 | 9450 | 7485 | 4039 | 4985 | 6612 |
| 74 | 1067 | 3716 | 8897 | 1970 | 8799 | 5718 | 4792 | 7292 | 4589 | 4554 |
| 75 | 5160 | 5563 | 6527 | 7861 | 3477 | 6735 | 7748 | 4913 | 6370 | 2258 |
| 76 | 4560 | 0094 | 8284 | 7604 | 1667 | 9286 | 2228 | 9507 | 1838 | 4646 |
| 77 | 7697 | 2151 | 4860 | 0739 | 4370 | 3992 | 8121 | 2502 | 7670 | 4470 |
| 78 | 8675 | 2997 | 9783 | 7306 | 4116 | 6432 | 7233 | 4611 | 7121 | 9412 |
| 79 | 3597 | 3520 | 5995 | 0892 | 3470 | 4581 | 1068 | 8801 | 1254 | 8607 |
| 80 | 4281 | 8802 | 5880 | 6212 | 6818 | 8162 | 0052 | 1755 | 7107 | 5197 |
| 81 | 0101 | 0907 | 9057 | 2263 | 0059 | 8553 | 7855 | 7758 | 1020 | 1264 |
| 82 | 8179 | 0109 | 4412 | 6044 | 7167 | 4209 | 5250 | 4570 | 1984 | 8276 |
| 83 | 8980 | 9662 | 9333 | 6598 | 2990 | 8173 | 1753 | 1135 | 1409 | 2042 |
| 84 | 3050 | 2450 | 9252 | 6724 | 2697 | 7933 | 9540 | 3700 | 6561 | 2790 |
| 85 | 4465 | 1307 | 8782 | 6763 | 9202 | 5594 | 7166 | 7050 | 4462 | 0426 |
| 86 | 1925 | 5402 | 1379 | 3556 | 5109 | 4846 | 9827 | 2881 | 5574 | 9027 |
| 87 | 8753 | 4602 | 1838 | 4624 | 4632 | 2512 | 2652 | 4804 | 1624 | 5116 |
| 88 | 2645 | 9197 | 4541 | 4822 | 7883 | 3352 | 3202 | 0906 | 3676 | 8141 |
| 89 | 4287 | 5473 | 4493 | 7086 | 4271 | 9140 | 3315 | 7073 | 4533 | 0653 |
| 90 | 5280 | 5426 | 7240 | 2154 | 7952 | 3804 | 8097 | 9328 | 8069 | 6894 |
| 91 | 9553 | 3136 | 2112 | 1369 | 5562 | 7360 | 5530 | 8074 | 6488 | 3682 |
| 92 | 2975 | 7924 | 0253 | 3503 | 9383 | 9454 | 3320 | 3234 | 9255 | 3527 |
| 93 | 2596 | 7274 | 8967 | 8138 | 6868 | 0385 | 4467 | 3792 | 3844 | 8700 |
| 94 | 4192 | 7440 | 6410 | 6064 | 4561 | 0411 | 9187 | 9940 | 2866 | 3345 |
| 95 | 3980 | 8594 | 9935 | 8560 | 0229 | 8778 | 2386 | 7852 | 4031 | 0627 |
| 96 | 1822 | 1177 | 6846 | 3997 | 5822 | 9188 | 2479 | 7951 | 3051 | 0110 |
| 97 | 8415 | 2623 | 2358 | 8895 | 5125 | 0173 | 3182 | 4151 | 4419 | 9049 |
| 98 | 2123 | 5798 | 5444 | 3282 | 8022 | 3931 | 4429 | 6028 | 5385 | 6845 |
| 99 | 1754 | 4076 | 3507 | 3705 | 7459 | 7544 | 6127 | 4820 | 3760 | 6476 |
| 100 | 3967 | 9997 | 0695 | 3562 | 9997 | 2934 | 8469 | 9706 | 4763 | 7132 |

**TABLE 1** *(Continued)*

| Row | | | | | | | | | | |
|-----|------|------|------|------|------|------|------|------|------|------|
| 101 | 7604 | 6645 | 6633 | 6288 | 5488 | 8355 | 9295 | 9637 | 5410 | 0452 |
| 102 | 6357 | 0216 | 1685 | 4308 | 0391 | 1517 | 1952 | 0108 | 1258 | 5498 |
| 103 | 5241 | 0554 | 6072 | 2412 | 1915 | 4451 | 0633 | 0449 | 9059 | 6873 |
| 104 | 9683 | 0618 | 2433 | 0154 | 0816 | 9885 | 3562 | 7392 | 4406 | 2994 |
| 105 | 8073 | 7718 | 9374 | 0965 | 8861 | 0018 | 2152 | 1736 | 5187 | 9347 |
| 106 | 3685 | 5901 | 6296 | 7748 | 6815 | 8033 | 5646 | 8691 | 3885 | 1550 |
| 107 | 9354 | 1854 | 1914 | 2592 | 9939 | 2468 | 0190 | 5882 | 3964 | 6938 |
| 108 | 2604 | 3040 | 9664 | 3962 | 4600 | 1314 | 8163 | 7869 | 2059 | 8203 |
| 109 | 9371 | 8390 | 6971 | 4931 | 1142 | 8588 | 2240 | 9256 | 7805 | 0153 |
| 110 | 5463 | 5569 | 1657 | 2797 | 9026 | 7754 | 8501 | 1953 | 1364 | 7787 |
| 111 | 5832 | 6510 | 1728 | 0531 | 9770 | 5790 | 8294 | 2702 | 4318 | 2494 |
| 112 | 6977 | 1478 | 4053 | 5836 | 5773 | 5706 | 8840 | 6575 | 6984 | 0196 |
| 113 | 6653 | 3177 | 7173 | 1053 | 8117 | 5818 | 2177 | 7524 | 3839 | 2438 |
| 114 | 2043 | 3329 | 3149 | 8591 | 8213 | 7941 | 0324 | 0275 | 2808 | 5787 |
| 115 | 1892 | 6495 | 7363 | 8840 | 6126 | 5749 | 5841 | 5564 | 3296 | 8176 |
| 116 | 4279 | 6686 | 2795 | 2572 | 6915 | 5770 | 0723 | 5003 | 6124 | 0041 |
| 117 | 9018 | 3226 | 1024 | 4455 | 4743 | 8634 | 7086 | 9462 | 5603 | 4961 |
| 118 | 6588 | 0445 | 5301 | 0442 | 7270 | 4287 | 9827 | 7666 | 4020 | 6061 |
| 119 | 3258 | 2829 | 5949 | 6280 | 9178 | 3614 | 8680 | 6705 | 1311 | 2408 |
| 120 | 9213 | 0161 | 4449 | 9084 | 8199 | 7330 | 4284 | 5061 | 1971 | 1008 |

**TABLE 2** Cumulative Binomial Probabilities $P[X \leq c] = \sum_{x=0}^{c} \binom{n}{x} p^x (1-p)^{n-x}$

| | | | | | | | $p$ | | | | | |
|--|--|------|------|------|------|------|------|------|------|------|------|------|
| | | .05 | .10 | .20 | .30 | .40 | .50 | .60 | .70 | .80 | .90 | .95 |
| | $c$ | | | | | | | | | | | |
| $n = 1$ | 0 | .950 | .900 | .800 | .700 | .600 | .500 | .400 | .300 | .200 | .100 | .050 |
| | 1 | 1.000 | 1.000 | 1.000 | 1.000 | 1.000 | 1.000 | 1.000 | 1.000 | 1.000 | 1.000 | 1.000 |
| $n = 2$ | 0 | .902 | .810 | .640 | .490 | .360 | .250 | .160 | .090 | .040 | .010 | .002 |
| | 1 | .997 | .990 | .960 | .910 | .840 | .750 | .640 | .510 | .360 | .190 | .097 |
| | 2 | 1.000 | 1.000 | 1.000 | 1.000 | 1.000 | 1.000 | 1.000 | 1.000 | 1.000 | 1.000 | 1.000 |
| $n = 3$ | 0 | .857 | .729 | .512 | .343 | .216 | .125 | .064 | .027 | .008 | .001 | .000 |
| | 1 | .993 | .972 | .896 | .784 | .648 | .500 | .352 | .216 | .104 | .028 | .007 |
| | 2 | 1.000 | .999 | .992 | .973 | .936 | .875 | .784 | .657 | .488 | .271 | .143 |
| | 3 | 1.000 | 1.000 | 1.000 | 1.000 | 1.000 | 1.000 | 1.000 | 1.000 | 1.000 | 1.000 | 1.000 |

**TABLE 2**    *(Continued)*

|  | | .05 | .10 | .20 | .30 | .40 | .50 | .60 | .70 | .80 | .90 | .95 |
|---|---|---|---|---|---|---|---|---|---|---|---|---|
| | | | | | | | $p$ | | | | | |
| $n = 4$ | $c$ 0 | .815 | .656 | .410 | .240 | .130 | .063 | .026 | .008 | .002 | .000 | .000 |
| | 1 | .986 | .948 | .819 | .652 | .475 | .313 | .179 | .084 | .027 | .004 | .000 |
| | 2 | 1.000 | .996 | .973 | .916 | .821 | .688 | .525 | .348 | .181 | .052 | .014 |
| | 3 | 1.000 | 1.000 | .998 | .992 | .974 | .938 | .870 | .760 | .590 | .344 | .185 |
| | 4 | 1.000 | 1.000 | 1.000 | 1.000 | 1.000 | 1.000 | 1.000 | 1.000 | 1.000 | 1.000 | 1.000 |
| $n = 5$ | 0 | .774 | .590 | .328 | .168 | .078 | .031 | .010 | .002 | .000 | .000 | .000 |
| | 1 | .977 | .919 | .737 | .528 | .337 | .188 | .087 | .031 | .007 | .000 | .000 |
| | 2 | .999 | .991 | .942 | .837 | .683 | .500 | .317 | .163 | .058 | .009 | .001 |
| | 3 | 1.000 | 1.000 | .993 | .969 | .913 | .813 | .663 | .472 | .263 | .081 | .023 |
| | 4 | 1.000 | 1.000 | 1.000 | .998 | .990 | .969 | .922 | .832 | .672 | .410 | .226 |
| | 5 | 1.000 | 1.000 | 1.000 | 1.000 | 1.000 | 1.000 | 1.000 | 1.000 | 1.000 | 1.000 | 1.000 |
| $n = 6$ | 0 | .735 | .531 | .262 | .118 | .047 | .016 | .004 | .001 | .000 | .000 | .000 |
| | 1 | .967 | .886 | .655 | .420 | .233 | .109 | .041 | .011 | .002 | .000 | .000 |
| | 2 | .998 | .984 | .901 | .744 | .544 | .344 | .179 | .070 | .017 | .001 | .000 |
| | 3 | 1.000 | .999 | .983 | .930 | .821 | .656 | .456 | .256 | .099 | .016 | .002 |
| | 4 | 1.000 | 1.000 | .998 | .989 | .959 | .891 | .767 | .580 | .345 | .114 | .033 |
| | 5 | 1.000 | 1.000 | 1.000 | .999 | .996 | .984 | .953 | .882 | .738 | .469 | .265 |
| | 6 | 1.000 | 1.000 | 1.000 | 1.000 | 1.000 | 1.000 | 1.000 | 1.000 | 1.000 | 1.000 | 1.000 |
| $n = 7$ | 0 | .698 | .478 | .210 | .082 | .028 | .008 | .002 | .000 | .000 | .000 | .000 |
| | 1 | .956 | .850 | .577 | .329 | .159 | .063 | .019 | .004 | .000 | .000 | .000 |
| | 2 | .996 | .974 | .852 | .647 | .420 | .227 | .096 | .029 | .005 | .000 | .000 |
| | 3 | 1.000 | .997 | .967 | .874 | .710 | .500 | .290 | .126 | .033 | .003 | .000 |
| | 4 | 1.000 | 1.000 | .995 | .971 | .904 | .773 | .580 | .353 | .148 | .026 | .004 |
| | 5 | 1.000 | 1.000 | 1.000 | .996 | .981 | .938 | .841 | .671 | .423 | .150 | .044 |
| | 6 | 1.000 | 1.000 | 1.000 | 1.000 | .998 | .992 | .972 | .918 | .790 | .522 | .302 |
| | 7 | 1.000 | 1.000 | 1.000 | 1.000 | 1.000 | 1.000 | 1.000 | 1.000 | 1.000 | 1.000 | 1.000 |
| $n = 8$ | 0 | .663 | .430 | .168 | .058 | .017 | .004 | .001 | .000 | .000 | .000. | .000 |
| | 1 | .943 | .813 | .503 | .255 | .106 | .035 | .009 | .001 | .000 | .000 | .000 |
| | 2 | .994 | .962 | .797 | .552 | .315 | .145 | .050 | .011 | .001 | .000 | .000 |
| | 3 | 1.000 | .995 | .944 | .806 | .594 | .363 | .174 | .058 | .010 | .000 | .000 |
| | 4 | 1.000 | 1.000 | .990 | .942 | .826 | .637 | .406 | .194 | .056 | .005 | .000 |
| | 5 | 1.000 | 1.000 | .999 | .989 | .950 | .855 | .685 | .448 | .203 | .038 | .006 |
| | 6 | 1.000 | 1.000 | 1.000 | .999 | .991 | .965 | .894 | .745 | .497 | .187 | .057 |
| | 7 | 1.000 | 1.000 | 1.000 | 1.000 | .999 | .996 | .983 | .942 | .832 | .570 | .337 |
| | 8 | 1.000 | 1.000 | 1.000 | 1.000 | 1.000 | 1.000 | 1.000 | 1.000 | 1.000 | 1.000 | 1.000 |
| $n = 9$ | 0 | .630 | .387 | .134 | .040 | .010 | .002 | .000 | .000 | .000 | .000 | .000 |
| | 1 | .929 | .775 | .436 | .196 | .071 | .020 | .004 | .000 | .000 | .000 | .000 |
| | 2 | .992 | .947 | .738 | .463 | .232 | .090 | .025 | .004 | .000 | .000 | .000 |
| | 3 | .999 | .992 | .914 | .730 | .483 | .254 | .099 | .025 | .003 | .000 | .000 |

**TABLE 2**    *(Continued)*

|  |  | | | | | | $p$ | | | | | |
|---|---|---|---|---|---|---|---|---|---|---|---|---|
|  |  | .05 | .10 | .20 | .30 | .40 | .50 | .60 | .70 | .80 | .90 | .95 |
|  | $c$ | | | | | | | | | | | |
|  | 4 | 1.000 | .999 | .980 | .901 | .733 | .500 | .267 | .099 | .020 | .001 | .000 |
|  | 5 | 1.000 | 1.000 | .997 | .975 | .901 | .746 | .517 | .270 | .086 | .008 | .001 |
|  | 6 | 1.000 | 1.000 | 1.000 | .996 | .975 | .910 | .768 | .537 | .262 | .053 | .008 |
|  | 7 | 1.000 | 1.000 | 1.000 | 1.000 | .996 | .980 | .929 | .804 | .564 | .225 | .071 |
|  | 8 | 1.000 | 1.000 | 1.000 | 1.000 | 1.000 | .998 | .990 | .960 | .866 | .613 | .370 |
|  | 9 | 1.000 | 1.000 | 1.000 | 1.000 | 1.000 | 1.000 | 1.000 | 1.000 | 1.000 | 1.000 | 1.000 |
| $n = 10$ | 0 | .599 | .349 | .107 | .028 | .006 | .001 | .000 | .000 | .000 | .000 | .000 |
|  | 1 | .914 | .736 | .376 | .149 | .046 | .011 | .002 | .000 | .000 | .000 | .000 |
|  | 2 | .988 | .930 | .678 | .383 | .167 | .055 | .012 | .002 | .000 | .000 | .000 |
|  | 3 | .999 | .987 | .879 | .650 | .382 | .172 | .055 | .011 | .001 | .000 | .000 |
|  | 4 | 1.000 | .998 | .967 | .850 | .633 | .377 | .166 | .047 | .006 | .000 | .000 |
|  | 5 | 1.000 | 1.000 | .994 | .953 | .834 | .623 | .367 | .150 | .033 | .002 | .000 |
|  | 6 | 1.000 | 1.000 | .999 | .989 | .945 | .828 | .618 | .350 | .121 | .013 | .001 |
|  | 7 | 1.000 | 1.000 | 1.000 | .998 | .988 | .945 | .833 | .617 | .322 | .070 | .012 |
|  | 8 | 1.000 | 1.000 | 1.000 | 1.000 | .998 | .989 | .954 | .851 | .624 | .264 | .086 |
|  | 9 | 1.000 | 1.000 | 1.000 | 1.000 | 1.000 | .999 | .994 | .972 | .893 | .651 | .401 |
|  | 10 | 1.000 | 1.000 | 1.000 | 1.000 | 1.000 | 1.000 | 1.000 | 1.000 | 1.000 | 1.000 | 1.000 |
| $n = 11$ | 0 | .569 | .314 | .086 | .020 | .004 | .000 | .000 | .000 | .000 | .000 | .000 |
|  | 1 | .898 | .697 | .322 | .113 | .030 | .006 | .001 | .000 | .000 | .000 | .000 |
|  | 2 | .985 | .910 | .617 | .313 | .119 | .033 | .006 | .001 | .000 | .000 | .000 |
|  | 3 | .998 | .981 | .839 | .570 | .296 | .113 | .029 | .004 | .000 | .000 | .000 |
|  | 4 | 1.000 | .997 | .950 | .790 | .533 | .274 | .099 | .022 | .002 | .000 | .000 |
|  | 5 | 1.000 | 1.000 | .988 | .922 | .753 | .500 | .247 | .078 | .012 | .000 | .000 |
|  | 6 | 1.000 | 1.000 | .998 | .978 | .901 | .726 | .467 | .210 | .050 | .003 | .000 |
|  | 7 | 1.000 | 1.000 | 1.000 | .996 | .971 | .887 | .704 | .430 | .161 | .019 | .002 |
|  | 8 | 1.000 | 1.000 | 1.000 | .999 | .994 | .967 | .881 | .687 | .383 | .090 | .015 |
|  | 9 | 1.000 | 1.000 | 1.000 | 1.000 | .999 | .994 | .970 | .887 | .678 | .303 | .102 |
|  | 10 | 1.000 | 1.000 | 1.000 | 1.000 | 1.000 | 1.000 | .996 | .980 | .914 | .686 | .431 |
|  | 11 | 1.000 | 1.000 | 1.000 | 1.000 | 1.000 | 1.000 | 1.000 | 1.000 | 1.000 | 1.000 | 1.000 |
| $n = 12$ | 0 | .540 | .282 | .069 | .014 | .002 | .000 | .000 | .000 | .000 | .000 | .000 |
|  | 1 | .882 | .659 | .275 | .085 | .020 | .003 | .000 | .000 | .000 | .000 | .000 |
|  | 2 | .980 | .889 | .558 | .253 | .083 | .019 | .003 | .000 | .000 | .000 | .000 |
|  | 3 | .998 | .974 | .795 | .493 | .225 | .073 | .015 | .002 | .000 | .000 | .000 |
|  | 4 | 1.000 | .996 | .927 | .724 | .438 | .194 | .057 | .009 | .001 | .000 | .000 |
|  | 5 | 1.000 | .999 | .981 | .882 | .665 | .387 | .158 | .039 | .004 | .000 | .000 |
|  | 6 | 1.000 | 1.000 | .996 | .961 | .842 | .613 | .335 | .118 | .019 | .001 | .000 |
|  | 7 | 1.000 | 1.000 | .999 | .991 | .943 | .806 | .562 | .276 | .073 | .004 | .000 |
|  | 8 | 1.000 | 1.000 | 1.000 | .998 | .985 | .927 | .775 | .507 | .205 | .026 | .002 |
|  | 9 | 1.000 | 1.000 | 1.000 | 1.000 | .997 | .981 | .917 | .747 | .442 | .111 | .020 |
|  | 10 | 1.000 | 1.000 | 1.000 | 1.000 | 1.000 | .997 | .980 | .915 | .725 | .341 | .118 |
|  | 11 | 1.000 | 1.000 | 1.000 | 1.000 | 1.000 | 1.000 | .998 | .986 | .931 | .718 | .460 |
|  | 12 | 1.000 | 1.000 | 1.000 | 1.000 | 1.000 | 1.000 | 1.000 | 1.000 | 1.000 | 1.000 | 1.000 |

**TABLE 2** *(Continued)*

| | | .05 | .10 | .20 | .30 | .40 | .50 | .60 | .70 | .80 | .90 | .95 |
|---|---|---|---|---|---|---|---|---|---|---|---|---|
| | | | | | | | $p$ | | | | | |
| | $c$ | | | | | | | | | | | |
| $n = 13$ | 0 | .513 | .254 | .055 | .010 | .001 | .000 | .000 | .000 | .000 | .000 | .000 |
| | 1 | .865 | .621 | .234 | .064 | .013 | .002 | .000 | .000 | .000 | .000 | .000 |
| | 2 | .975 | .866 | .502 | .202 | .058 | .011 | .001 | .000 | .000 | .000 | .000 |
| | 3 | .997 | .966 | .747 | .421 | .169 | .046 | .008 | .001 | .000 | .000 | .000 |
| | 4 | 1.000 | .994 | .901 | .654 | .353 | .133 | .032 | .004 | .000 | .000 | .000 |
| | 5 | 1.000 | .999 | .970 | .835 | .574 | .291 | .098 | .018 | .001 | .000 | .000 |
| | 6 | 1.000 | 1.000 | .993 | .938 | .771 | .500 | .229 | .062 | .007 | .000 | .000 |
| | 7 | 1.000 | 1.000 | .999 | .982 | .902 | .709 | .426 | .165 | .030 | .001 | .000 |
| | 8 | 1.000 | 1.000 | 1.000 | .996 | .968 | .867 | .647 | .346 | .099 | .006 | .000 |
| | 9 | 1.000 | 1.000 | 1.000 | .999 | .992 | .954 | .831 | .579 | .253 | .034 | .003 |
| | 10 | 1.000 | 1.000 | 1.000 | 1.000 | .999 | .989 | .942 | .798 | .498 | .134 | .025 |
| | 11 | 1.000 | 1.000 | 1.000 | 1.000 | 1.000 | .998 | .987 | .936 | .766 | .379 | .135 |
| | 12 | 1.000 | 1.000 | 1.000 | 1.000 | 1.000 | 1.000 | .999 | .990 | .945 | .746 | .487 |
| | 13 | 1.000 | 1.000 | 1.000 | 1.000 | 1.000 | 1.000 | 1.000 | 1.000 | 1.000 | 1.000 | 1.000 |
| $n = 14$ | 0 | .488 | .229 | .044 | .007 | .001 | .000 | .000 | .000 | .000 | .000 | .000 |
| | 1 | .847 | .585 | .198 | .047 | .008 | .001 | .000 | .000 | .000 | .000 | .000 |
| | 2 | .970 | .842 | .448 | .161 | .040 | .006 | .001 | .000 | .000 | .000 | .000 |
| | 3 | .996 | .956 | .698 | .355 | .124 | .029 | .004 | .000 | .000 | .000 | .000 |
| | 4 | 1.000 | .991 | .870 | .584 | .279 | .090 | .018 | .002 | .000 | .000 | .000 |
| | 5 | 1.000 | .999 | .956 | .781 | .486 | .212 | .058 | .008 | .000 | .000 | .000 |
| | 6 | 1.000 | 1.000 | .988 | .907 | .692 | .395 | .150 | .031 | .002 | .000 | .000 |
| | 7 | 1.000 | 1.000 | .998 | .969 | .850 | .605 | .308 | .093 | .012 | .000 | .000 |
| | 8 | 1.000 | 1.000 | 1.000 | .992 | .942 | .788 | .514 | .219 | .044 | .001 | .000 |
| | 9 | 1.000 | 1.000 | 1.000 | .998 | .982 | .910 | .721 | .416 | .130 | .009 | .000 |
| | 10 | 1.000 | 1.000 | 1.000 | 1.000 | .996 | .971 | .876 | .645 | .302 | .044 | .004 |
| | 11 | 1.000 | 1.000 | 1.000 | 1.000 | .999 | .994 | .960 | .839 | .552 | .158 | .030 |
| | 12 | 1.000 | 1.000 | 1.000 | 1.000 | 1.000 | .999 | .992 | .953 | .802 | .415 | .153 |
| | 13 | 1.000 | 1.000 | 1.000 | 1.000 | 1.000 | 1.000 | .999 | .993 | .956 | .771 | .512 |
| | 14 | 1.000 | 1.000 | 1.000 | 1.000 | 1.000 | 1.000 | 1.000 | 1.000 | 1.000 | 1.000 | 1.000 |
| $n = 15$ | 0 | .463 | .206 | .035 | .005 | .000 | .000 | .000 | .000 | .000 | .000 | .000 |
| | 1 | .829 | .549 | .167 | .035 | .005 | .000 | .000 | .000 | .000 | .000 | .000 |
| | 2 | .964 | .816 | .398 | .127 | .027 | .004 | .000 | .000 | .000 | .000 | .000 |
| | 3 | .995 | .944 | .648 | .297 | .091 | .018 | .002 | .000 | .000 | .000 | .000 |
| | 4 | .999 | .987 | .836 | .515 | .217 | .059 | .009 | .001 | .000 | .000 | .000 |
| | 5 | 1.000 | .998 | .939 | .722 | .403 | .151 | .034 | .004 | .000 | .000 | .000 |
| | 6 | 1.000 | 1.000 | .982 | .869 | .610 | .304 | .095 | .015 | .001 | .000 | .000 |
| | 7 | 1.000 | 1.000 | .996 | .950 | .787 | .500 | .213 | .050 | .004 | .000 | .000 |
| | 8 | 1.000 | 1.000 | .999 | .985 | .905 | .696 | .390 | .131 | .018 | .000 | .000 |
| | 9 | 1.000 | 1.000 | 1.000 | .996 | .966 | .849 | .597 | .278 | .061 | .002 | .000 |
| | 10 | 1.000 | 1.000 | 1.000 | .999 | .991 | .941 | .783 | .485 | .164 | .013 | .001 |
| | 11 | 1.000 | 1.000 | 1.000 | 1.000 | .998 | .982 | .909 | .703 | .352 | .056 | .005 |
| | 12 | 1.000 | 1.000 | 1.000 | 1.000 | 1.000 | .996 | .973 | .873 | .602 | .184 | .036 |

**TABLE 2**   *(Continued)*

| | | | | | | | *p* | | | | | |
|---|---|---|---|---|---|---|---|---|---|---|---|---|
| | | .05 | .10 | .20 | .30 | .40 | .50 | .60 | .70 | .80 | .90 | .95 |
| | *c* | | | | | | | | | | | |
| | 13 | 1.000 | 1.000 | 1.000 | 1.000 | 1.000 | 1.000 | .995 | .965 | .833 | .451 | .171 |
| | 14 | 1.000 | 1.000 | 1.000 | 1.000 | 1.000 | 1.000 | 1.000 | .995 | .965 | .794 | .537 |
| | 15 | 1.000 | 1.000 | 1.000 | 1.000 | 1.000 | 1.000 | 1.000 | 1.000 | 1.000 | 1.000 | 1.000 |
| *n* = 16 | 0 | .440 | .185 | .028 | .003 | .000 | .000 | .000 | .000 | .000 | .000 | .000 |
| | 1 | .811 | .515 | .141 | .026 | .003 | .000 | .000 | .000 | .000 | .000 | .000 |
| | 2 | .957 | .789 | .352 | .099 | .018 | .002 | .000 | .000 | .000 | .000 | .000 |
| | 3 | .993 | .932 | .598 | .246 | .065 | .011 | .001 | .000 | .000 | .000 | .000 |
| | 4 | .999 | .983 | .798 | .450 | .167 | .038 | .005 | .000 | .000 | .000 | .000 |
| | 5 | 1.000 | .997 | .918 | .660 | .329 | .105 | .019 | .002 | .000 | .000 | .000 |
| | 6 | 1.000 | .999 | .973 | .825 | .527 | .227 | .058 | .007 | .000 | .000 | .000 |
| | 7 | 1.000 | 1.000 | .993 | .926 | .716 | .402 | .142 | .026 | .001 | .000 | .000 |
| | 8 | 1.000 | 1.000 | .999 | .974 | .858 | .598 | .284 | .074 | .007 | .000 | .000 |
| | 9 | 1.000 | 1.000 | 1.000 | .993 | .942 | .773 | .473 | .175 | .027 | .001 | .000 |
| | 10 | 1.000 | 1.000 | 1.000 | .998 | .981 | .895 | .671 | .340 | .082 | .003 | .000 |
| | 11 | 1.000 | 1.000 | 1.000 | 1.000 | .995 | .962 | .833 | .550 | .202 | .017 | .001 |
| | 12 | 1.000 | 1.000 | 1.000 | 1.000 | .999 | .989 | .935 | .754 | .402 | .068 | .007 |
| | 13 | 1.000 | 1.000 | 1.000 | 1.000 | 1.000 | .998 | .982 | .901 | .648 | .211 | .043 |
| | 14 | 1.000 | 1.000 | 1.000 | 1.000 | 1.000 | 1.000 | .997 | .974 | .859 | .485 | .189 |
| | 15 | 1.000 | 1.000 | 1.000 | 1.000 | 1.000 | 1.000 | 1.000 | .997 | .972 | .815 | .560 |
| | 16 | 1.000 | 1.000 | 1.000 | 1.000 | 1.000 | 1.000 | 1.000 | 1.000 | 1.000 | 1.000 | 1.000 |
| *n* = 17 | 0 | .418 | .167 | .023 | .002 | .000 | .000 | .000 | .000 | .000 | .000 | .000 |
| | 1 | .792 | .482 | .118 | .019 | .002 | .000 | .000 | .000 | .000 | .000 | .000 |
| | 2 | .950 | .762 | .310 | .077 | .012 | .001 | .000 | .000 | .000 | .000 | .000 |
| | 3 | .991 | .917 | .549 | .202 | .046 | .006 | .000 | .000 | .000 | .000 | .000 |
| | 4 | .999 | .978 | .758 | .389 | .126 | .025 | .003 | .000 | .000 | .000 | .000 |
| | 5 | 1.000 | .995 | .894 | .597 | .264 | .072 | .011 | .001 | .000 | .000 | .000 |
| | 6 | 1.000 | .999 | .962 | .775 | .448 | .166 | .035 | .003 | .000 | .000 | .000 |
| | 7 | 1.000 | 1.000 | .989 | .895 | .641 | .315 | .092 | .013 | .000 | .000 | .000 |
| | 8 | 1.000 | 1.000 | .997 | .960 | .801 | .500 | .199 | .040 | .003 | .000 | .000 |
| | 9 | 1.000 | 1.000 | 1.000 | .987 | .908 | .685 | .359 | .105 | .011 | .000 | .000 |
| | 10 | 1.000 | 1.000 | 1.000 | .997 | .965 | .834 | .552 | .225 | .038 | .001 | .000 |
| | 11 | 1.000 | 1.000 | 1.000 | .999 | .989 | .928 | .736 | .403 | .106 | .005 | .000 |
| | 12 | 1.000 | 1.000 | 1.000 | 1.000 | .997 | .975 | .874 | .611 | .242 | .022 | .001 |
| | 13 | 1.000 | 1.000 | 1.000 | 1.000 | 1.000 | .994 | .954 | .798 | .451 | .083 | .009 |
| | 14 | 1.000 | 1.000 | 1.000 | 1.000 | 1.000 | .999 | .988 | .923 | .690 | .238 | .050 |
| | 15 | 1.000 | 1.000 | 1.000 | 1.000 | 1.000 | 1.000 | .998 | .981 | .882 | .518 | .208 |
| | 16 | 1.000 | 1.000 | 1.000 | 1.000 | 1.000 | 1.000 | 1.000 | .998 | .977 | .833 | .582 |
| | 17 | 1.000 | 1.000 | 1.000 | 1.000 | 1.000 | 1.000 | 1.000 | 1.000 | 1.000 | 1.000 | 1.000 |
| *n* = 18 | 0 | .397 | .150 | .018 | .002 | .000 | .000 | .000 | .000 | .000 | .000 | .000 |
| | 1 | .774 | .450 | .099 | .014 | .001 | .000 | .000 | .000 | .000 | .000 | .000 |
| | 2 | .942 | .734 | .271 | .060 | .008 | .001 | .000 | .000 | .000 | .000 | .000 |

**TABLE 2**   *(Continued)*

|  |  | .05 | .10 | .20 | .30 | .40 | .50 | .60 | .70 | .80 | .90 | .95 |
|---|---|---|---|---|---|---|---|---|---|---|---|---|
|  | c |  |  |  |  |  | *p* |  |  |  |  |  |
|  | 3 | .989 | .902 | .501 | .165 | .033 | .004 | .000 | .000 | .000 | .000 | .000 |
|  | 4 | .998 | .972 | .716 | .333 | .094 | .015 | .001 | .000 | .000 | .000 | .000 |
|  | 5 | 1.000 | .994 | .867 | .534 | .209 | .048 | .006 | .000 | .000 | .000 | .000 |
|  | 6 | 1.000 | .999 | .949 | .722 | .374 | .119 | .020 | .001 | .000 | .000 | .000 |
|  | 7 | 1.000 | 1.000 | .984 | .859 | .563 | .240 | .058 | .006 | .000 | .000 | .000 |
|  | 8 | 1.000 | 1.000 | .996 | .940 | .737 | .407 | .135 | .021 | .001 | .000 | .000 |
|  | 9 | 1.000 | 1.000 | .999 | .979 | .865 | .593 | .263 | .060 | .004 | .000 | .000 |
|  | 10 | 1.000 | 1.000 | 1.000 | .994 | .942 | .760 | .437 | .141 | .016 | .000 | .000 |
|  | 11 | 1.000 | 1.000 | 1.000 | .999 | .980 | .881 | .626 | .278 | .051 | .001 | .000 |
|  | 12 | 1.000 | 1.000 | 1.000 | 1.000 | .994 | .952 | .791 | .466 | .133 | .006 | .000 |
|  | 13 | 1.000 | 1.000 | 1.000 | 1.000 | .999 | .985 | .906 | .667 | .284 | .028 | .002 |
|  | 14 | 1.000 | 1.000 | 1.000 | 1.000 | 1.000 | .996 | .967 | .835 | .499 | .098 | .011 |
|  | 15 | 1.000 | 1.000 | 1.000 | 1.000 | 1.000 | .999 | .992 | .940 | .729 | .266 | .058 |
|  | 16 | 1.000 | 1.000 | 1.000 | 1.000 | 1.000 | 1.000 | .999 | .986 | .901 | .550 | .226 |
|  | 17 | 1.000 | 1.000 | 1.000 | 1.000 | 1.000 | 1.000 | 1.000 | .998 | .982 | .850 | .603 |
|  | 18 | 1.000 | 1.000 | 1.000 | 1.000 | 1.000 | 1.000 | 1.000 | 1.000 | 1.000 | 1.000 | 1.000 |
| $n = 19$ | 0 | .377 | .135 | .014 | .001 | .000 | .000 | .000 | .000 | .000 | .000 | .000 |
|  | 1 | .755 | .420 | .083 | .010 | .001 | .000 | .000 | .000 | .000 | .000 | .000 |
|  | 2 | .933 | .705 | .237 | .046 | .005 | .000 | .000 | .000 | .000 | .000 | .000 |
|  | 3 | .987 | .885 | .455 | .133 | .023 | .002 | .000 | .000 | .000 | .000 | .000 |
|  | 4 | .998 | .965 | .673 | .282 | .070 | .010 | .001 | .000 | .000 | .000 | .000 |
|  | 5 | 1.000 | .991 | .837 | .474 | .163 | .032 | .003 | .000 | .000 | .000 | .000 |
|  | 6 | 1.000 | .998 | .932 | .666 | .308 | .084 | .012 | .001 | .000 | .000 | .000 |
|  | 7 | 1.000 | 1.000 | .977 | .818 | .488 | .180 | .035 | .003 | .000 | .000 | .000 |
|  | 8 | 1.000 | 1.000 | .993 | .916 | .667 | .324 | .088 | .011 | .000 | .000 | .000 |
|  | 9 | 1.000 | 1.000 | .998 | .967 | .814 | .500 | .186 | .033 | .002 | .000 | .000 |
|  | 10 | 1.000 | 1.000 | 1.000 | .989 | .912 | .676 | .333 | .084 | .007 | .000 | .000 |
|  | 11 | 1.000 | 1.000 | 1.000 | .997 | .965 | .820 | .512 | .182 | .023 | .000 | .000 |
|  | 12 | 1.000 | 1.000 | 1.000 | .999 | .988 | .916 | .692 | .334 | .068 | .002 | .000 |
|  | 13 | 1.000 | 1.000 | 1.000 | 1.000 | .997 | .968 | .837 | .526 | .163 | .009 | .000 |
|  | 14 | 1.000 | 1.000 | 1.000 | 1.000 | .999 | .990 | .930 | .718 | .327 | .035 | .002 |
|  | 15 | 1.000 | 1.000 | 1.000 | 1.000 | 1.000 | .998 | .977 | .867 | .545 | .115 | .013 |
|  | 16 | 1.000 | 1.000 | 1.000 | 1.000 | 1.000 | 1.000 | .995 | .954 | .763 | .295 | .067 |
|  | 17 | 1.000 | 1.000 | 1.000 | 1.000 | 1.000 | 1.000 | .999 | .990 | .917 | .580 | .245 |
|  | 18 | 1.000 | 1.000 | 1.000 | 1.000 | 1.000 | 1.000 | 1.000 | .999 | .986 | .865 | .623 |
|  | 19 | 1.000 | 1.000 | 1.000 | 1.000 | 1.000 | 1.000 | 1.000 | 1.000 | 1.000 | 1.000 | 1.000 |
| $n = 20$ | 0 | .358 | .122 | .012 | .001 | .000 | .000 | .000 | .000 | .000 | .000 | .000 |
|  | 1 | .736 | .392 | .069 | .008 | .001 | .000 | .000 | .000 | .000 | .000 | .000 |
|  | 2 | .925 | .677 | .206 | .035 | .004 | .000 | .000 | .000 | .000 | .000 | .000 |
|  | 3 | .984 | .867 | .411 | .107 | .016 | .001 | .000 | .000 | .000 | .000 | .000 |
|  | 4 | .997 | .957 | .630 | .238 | .051 | .006 | .000 | .000 | .000 | .000 | .000 |
|  | 5 | 1.000 | .989 | .804 | .416 | .126 | .021 | .002 | .000 | .000 | .000 | .000 |

**TABLE 2** *(Continued)*

|  |  | $p$ | | | | | | | | | | |
|---|---|---|---|---|---|---|---|---|---|---|---|---|
|  |  | .05 | .10 | .20 | .30 | .40 | .50 | .60 | .70 | .80 | .90 | .95 |
|  | $c$ |  |  |  |  |  |  |  |  |  |  |  |
|  | 6 | 1.000 | .998 | .913 | .608 | .250 | .058 | .006 | .000 | .000 | .000 | .000 |
|  | 7 | 1.000 | 1.000 | .968 | .772 | .416 | .132 | .021 | .001 | .000 | .000 | .000 |
|  | 8 | 1.000 | 1.000 | .990 | .887 | .596 | .252 | .057 | .005 | .000 | .000 | .000 |
|  | 9 | 1.000 | 1.000 | .997 | .952 | .755 | .412 | .128 | .017 | .001 | .000 | .000 |
|  | 10 | 1.000 | 1.000 | .999 | .983 | .872 | .588 | .245 | .048 | .003 | .000 | .000 |
|  | 11 | 1.000 | 1.000 | 1.000 | .995 | .943 | .748 | .404 | .113 | .010 | .000 | .000 |
|  | 12 | 1.000 | 1.000 | 1.000 | .999 | .979 | .868 | .584 | .228 | .032 | .000 | .000 |
|  | 13 | 1.000 | 1.000 | 1.000 | 1.000 | .994 | .942 | .750 | .392 | .087 | .002 | .000 |
|  | 14 | 1.000 | 1.000 | 1.000 | 1.000 | .998 | .979 | .874 | .584 | .196 | .011 | .000 |
|  | 15 | 1.000 | 1.000 | 1.000 | 1.000 | 1.000 | .994 | .949 | .762 | .370 | .043 | .003 |
|  | 16 | 1.000 | 1.000 | 1.000 | 1.000 | 1.000 | .999 | .984 | .893 | .589 | .133 | .016 |
|  | 17 | 1.000 | 1.000 | 1.000 | 1.000 | 1.000 | 1.000 | .996 | .965 | .794 | .323 | .075 |
|  | 18 | 1.000 | 1.000 | 1.000 | 1.000 | 1.000 | 1.000 | .999 | .992 | .931 | .608 | .264 |
|  | 19 | 1.000 | 1.000 | 1.000 | 1.000 | 1.000 | 1.000 | 1.000 | .999 | .988 | .878 | .642 |
|  | 20 | 1.000 | 1.000 | 1.000 | 1.000 | 1.000 | 1.000 | 1.000 | 1.000 | 1.000 | 1.000 | 1.000 |
| $n = 25$ | 0 | .277 | .072 | .004 | .000 | .000 | .000 | .000 | .000 | .000 | .000 | .000 |
|  | 1 | .642 | .271 | .027 | .002 | .000 | .000 | .000 | .000 | .000 | .000 | .000 |
|  | 2 | .873 | .537 | .098 | .009 | .000 | .000 | .000 | .000 | .000 | .000 | .000 |
|  | 3 | .966 | .764 | .234 | .033 | .002 | .000 | .000 | .000 | .000 | .000 | .000 |
|  | 4 | .993 | .902 | .421 | .090 | .009 | .000 | .000 | .000 | .000 | .000 | .000 |
|  | 5 | .999 | .967 | .617 | .193 | .029 | .002 | .000 | .000 | .000 | .000 | .000 |
|  | 6 | 1.000 | .991 | .780 | .341 | .074 | .007 | .000 | .000 | .000 | .000 | .000 |
|  | 7 | 1.000 | .998 | .891 | .512 | .154 | .022 | .001 | .000 | .000 | .000 | .000 |
|  | 8 | 1.000 | 1.000 | .953 | .677 | .274 | .054 | .004 | .000 | .000 | .000 | .000 |
|  | 9 | 1.000 | 1.000 | .983 | .811 | .425 | .115 | .013 | .000 | .000 | .000 | .000 |
|  | 10 | 1.000 | 1.000 | .994 | .902 | .586 | .212 | .034 | .002 | .000 | .000 | .000 |
|  | 11 | 1.000 | 1.000 | .998 | .956 | .732 | .345 | .078 | .006 | .000 | .000 | .000 |
|  | 12 | 1.000 | 1.000 | 1.000 | .983 | .846 | .500 | .154 | .017 | .000 | .000 | .000 |
|  | 13 | 1.000 | 1.000 | 1.000 | .994 | .922 | .655 | .268 | .044 | .002 | .000 | .000 |
|  | 14 | 1.000 | 1.000 | 1.000 | .998 | .966 | .788 | .414 | .098 | .006 | .000 | .000 |
|  | 15 | 1.000 | 1.000 | 1.000 | 1.000 | .987 | .885 | .575 | .189 | .017 | .000 | .000 |
|  | 16 | 1.000 | 1.000 | 1.000 | 1.000 | .996 | .946 | .726 | .323 | .047 | .000 | .000 |
|  | 17 | 1.000 | 1.000 | 1.000 | 1.000 | .999 | .978 | .846 | .488 | .109 | .002 | .000 |
|  | 18 | 1.000 | 1.000 | 1.000 | 1.000 | 1.000 | .993 | .926 | .659 | .220 | .009 | .000 |
|  | 19 | 1.000 | 1.000 | 1.000 | 1.000 | 1.000 | .998 | .971 | .807 | .383 | .033 | .001 |
|  | 20 | 1.000 | 1.000 | 1.000 | 1.000 | 1.000 | 1.000 | .991 | .910 | .579 | .098 | .007 |
|  | 21 | 1.000 | 1.000 | 1.000 | 1.000 | 1.000 | 1.000 | .998 | .967 | .766 | .236 | .034 |
|  | 22 | 1.000 | 1.000 | 1.000 | 1.000 | 1.000 | 1.000 | 1.000 | .991 | .902 | .463 | .127 |
|  | 23 | 1.000 | 1.000 | 1.000 | 1.000 | 1.000 | 1.000 | 1.000 | .998 | .973 | .729 | .358 |
|  | 24 | 1.000 | 1.000 | 1.000 | 1.000 | 1.000 | 1.000 | 1.000 | 1.000 | .996 | .928 | .723 |
|  | 25 | 1.000 | 1.000 | 1.000 | 1.000 | 1.000 | 1.000 | 1.000 | 1.000 | 1.000 | 1.000 | 1.000 |

**TABLE 3** Cumulative Poisson Probabilities $P[X \leq c] = \sum_{x=0}^{c} \dfrac{e^{-m}m^x}{x!}$

| $m$ | 0 | 1 | 2 | 3 | 4 | 5 | 6 | 7 | 8 | 9 |
|------|------|------|------|------|------|------|------|------|------|---|
| .02 | .980 | 1.000 | | | | | | | | |
| .04 | .961 | .999 | 1.000 | | | | | | | |
| .06 | .942 | .998 | 1.000 | | | | | | | |
| .08 | .923 | .997 | 1.000 | | | | | | | |
| .10 | .905 | .995 | 1.000 | | | | | | | |
| .15 | .861 | .990 | .999 | 1.000 | | | | | | |
| .20 | .819 | .982 | .999 | 1.000 | | | | | | |
| .25 | .779 | .974 | .998 | 1.000 | | | | | | |
| .30 | .741 | .963 | .996 | 1.000 | | | | | | |
| .35 | .705 | .951 | .994 | 1.000 | | | | | | |
| .40 | .670 | .938 | .992 | .999 | 1.000 | | | | | |
| .45 | .638 | .925 | .989 | .999 | 1.000 | | | | | |
| .50 | .607 | .910 | .986 | .998 | 1.000 | | | | | |
| .55 | .577 | .894 | .982 | .998 | 1.000 | | | | | |
| .60 | .549 | .878 | .977 | .997 | 1.000 | | | | | |
| .65 | .522 | .861 | .972 | .996 | .999 | 1.000 | | | | |
| .70 | .497 | .844 | .966 | .994 | .999 | 1.000 | | | | |
| .75 | .472 | .827 | .959 | .993 | .999 | 1.000 | | | | |
| .80 | .449 | .809 | .953 | .991 | .999 | 1.000 | | | | |
| .85 | .427 | .791 | .945 | .989 | .998 | 1.000 | | | | |
| .90 | .407 | .772 | .937 | .987 | .998 | 1.000 | | | | |
| .95 | .387 | .754 | .929 | .984 | .997 | 1.000 | | | | |
| 1.00 | .368 | .736 | .920 | .981 | .996 | .999 | 1.000 | | | |
| 1.10 | .333 | .699 | .900 | .974 | .995 | .999 | 1.000 | | | |
| 1.20 | .301 | .663 | .879 | .966 | .992 | .998 | 1.000 | | | |
| 1.30 | .273 | .627 | .857 | .957 | .989 | .998 | 1.000 | | | |
| 1.40 | .247 | .592 | .833 | .946 | .986 | .997 | .999 | 1.000 | | |
| 1.50 | .223 | .558 | .809 | .934 | .981 | .996 | .999 | 1.000 | | |
| 1.60 | .202 | .525 | .783 | .921 | .976 | .994 | .999 | 1.000 | | |
| 1.70 | .183 | .493 | .757 | .907 | .970 | .992 | .998 | 1.000 | | |
| 1.80 | .165 | .463 | .731 | .891 | .964 | .990 | .997 | .999 | 1.000 | |
| 1.90 | .150 | .434 | .704 | .875 | .956 | .987 | .997 | .999 | 1.000 | |

**TABLE 3**    *(Continued)*

| | | | | | *x* | | | | | |
|---|---|---|---|---|---|---|---|---|---|---|
| *m* | 0 | 1 | 2 | 3 | 4 | 5 | 6 | 7 | 8 | 9 |
| 2.0 | .135 | .406 | .677 | .857 | .947 | .983 | .995 | .999 | 1.000 | |
| 2.2 | .111 | .355 | .623 | .819 | .928 | .975 | .993 | .998 | 1.000 | |
| 2.4 | .091 | .308 | .570 | .779 | .904 | .964 | .988 | .997 | .999 | 1.000 |
| 2.6 | .074 | .267 | .518 | .736 | .877 | .951 | .983 | .995 | .999 | 1.000 |
| | | | | | | | | | | |
| 2.8 | .061 | .231 | .469 | .692 | .848 | .935 | .976 | .992 | .998 | .999 |
| 3.0 | .050 | .199 | .423 | .647 | .815 | .916 | .966 | .988 | .996 | .999 |
| 3.2 | .041 | .171 | .380 | .603 | .781 | .895 | .955 | .983 | .994 | .998 |
| 3.4 | .033 | .147 | .340 | .558 | .744 | .871 | .942 | .977 | .992 | .997 |
| 3.6 | .027 | .126 | .303 | .515 | .706 | .844 | .927 | .969 | .988 | .996 |
| 3.8 | .022 | .107 | .269 | .473 | .668 | .816 | .909 | .960 | .984 | .994 |
| | | | | | | | | | | |
| 4.0 | .018 | .092 | .238 | .433 | .629 | .785 | .889 | .949 | .979 | .992 |
| 4.5 | .011 | .061 | .174 | .342 | .532 | .703 | .831 | .913 | .960 | .983 |
| 5.0 | .007 | .040 | .125 | .265 | .440 | .616 | .762 | .867 | .932 | .968 |
| 5.5 | .004 | .027 | .088 | .202 | .358 | .529 | .686 | .809 | .894 | .946 |
| 6.0 | .002 | .017 | .062 | .151 | .285 | .446 | .606 | .744 | .847 | .916 |
| 6.5 | .002 | .011 | .043 | .112 | .224 | .369 | .527 | .673 | .792 | .877 |
| 7.0 | .001 | .007 | .030 | .082 | .173 | .301 | .450 | .599 | .729 | .830 |

| | | | | | *x* | | | | | |
|---|---|---|---|---|---|---|---|---|---|---|
| *m* | 10 | 11 | 12 | 13 | 14 | 15 | 16 | 17 | 18 | 19 |
| 2.8 | 1.000 | | | | | | | | | |
| 3.0 | 1.000 | | | | | | | | | |
| 3.2 | 1.000 | | | | | | | | | |
| 3.4 | .999 | 1.000 | | | | | | | | |
| 3.6 | .999 | 1.000 | | | | | | | | |
| 3.8 | .998 | .999 | 1.000 | | | | | | | |
| | | | | | | | | | | |
| 4.0 | .997 | .999 | 1.000 | | | | | | | |
| 4.5 | .993 | .998 | .999 | 1.000 | | | | | | |
| 5.0 | .986 | .995 | .998 | .999 | 1.000 | | | | | |
| 5.5 | .975 | .989 | .996 | .998 | .999 | 1.000 | | | | |
| 6.0 | .957 | .980 | .991 | .996 | .999 | .999 | 1.000 | | | |
| 6.5 | .933 | .966 | .984 | .993 | .997 | .999 | 1.000 | | | |
| 7.0 | .901 | .947 | .973 | .987 | .994 | .998 | .999 | 1.000 | | |

**TABLE 4**   Standard Normal Probabilities

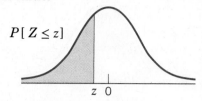

$P[\,Z \le z\,]$

| $z$ | .00 | .01 | .02 | .03 | .04 | .05 | .06 | .07 | .08 | .09 |
|---|---|---|---|---|---|---|---|---|---|---|
| −3.5 | .0002 | .0002 | .0002 | .0002 | .0002 | .0002 | .0002 | .0002 | .0002 | .0002 |
| −3.4 | .0003 | .0003 | .0003 | .0003 | .0003 | .0003 | .0003 | .0003 | .0003 | .0002 |
| −3.3 | .0005 | .0005 | .0005 | .0004 | .0004 | .0004 | .0004 | .0004 | .0004 | .0003 |
| −3.2 | .0007 | .0007 | .0006 | .0006 | .0006 | .0006 | .0006 | .0005 | .0005 | .0005 |
| −3.1 | .0010 | .0009 | .0009 | .0009 | .0008 | .0008 | .0008 | .0008 | .0007 | .0007 |
| −3.0 | .0013 | .0013 | .0013 | .0012 | .0012 | .0011 | .0011 | .0011 | .0010 | .0010 |
| −2.9 | .0019 | .0018 | .0018 | .0017 | .0016 | .0016 | .0015 | .0015 | .0014 | .0014 |
| −2.8 | .0026 | .0025 | .0024 | .0023 | .0023 | .0022 | .0021 | .0021 | .0020 | .0019 |
| −2.7 | .0035 | .0034 | .0033 | .0032 | .0031 | .0030 | .0029 | .0028 | .0027 | .0026 |
| −2.6 | .0047 | .0045 | .0044 | .0043 | .0041 | .0040 | .0039 | .0038 | .0037 | .0036 |
| −2.5 | .0062 | .0060 | .0059 | .0057 | .0055 | .0054 | .0052 | .0051 | .0049 | .0048 |
| −2.4 | .0082 | .0080 | .0078 | .0075 | .0073 | .0071 | .0069 | .0068 | .0066 | .0064 |
| −2.3 | .0107 | .0104 | .0102 | .0099 | .0096 | .0094 | .0091 | .0089 | .0087 | .0084 |
| −2.2 | .0139 | .0136 | .0132 | .0129 | .0125 | .0122 | .0119 | .0116 | .0113 | .0110 |
| −2.1 | .0179 | .0174 | .0170 | .0166 | .0162 | .0158 | .0154 | .0150 | .0146 | .0143 |
| −2.0 | .0228 | .0222 | .0217 | .0212 | .0207 | .0202 | .0197 | .0192 | .0188 | .0183 |
| −1.9 | .0287 | .0281 | .0274 | .0268 | .0262 | .0256 | .0250 | .0244 | .0239 | .0233 |
| −1.8 | .0359 | .0351 | .0344 | .0336 | .0329 | .0322 | .0314 | .0307 | .0301 | .0294 |
| −1.7 | .0446 | .0436 | .0427 | .0418 | .0409 | .0401 | .0392 | .0384 | .0375 | .0367 |
| −1.6 | .0548 | .0537 | .0526 | .0516 | .0505 | .0495 | .0485 | .0475 | .0465 | .0455 |
| −1.5 | .0668 | .0655 | .0643 | .0630 | .0618 | .0606 | .0594 | .0582 | .0571 | .0559 |
| −1.4 | .0808 | .0793 | .0778 | .0764 | .0749 | .0735 | .0721 | .0708 | .0694 | .0681 |
| −1.3 | .0968 | .0951 | .0934 | .0918 | .0901 | .0885 | .0869 | .0853 | .0838 | .0823 |
| −1.2 | .1151 | .1131 | .1112 | .1093 | .1075 | .1056 | .1038 | .1020 | .1003 | .0985 |
| −1.1 | .1357 | .1335 | .1314 | .1292 | .1271 | .1251 | .1230 | .1210 | .1190 | .1170 |
| −1.0 | .1587 | .1562 | .1539 | .1515 | .1492 | .1469 | .1446 | .1423 | .1401 | .1379 |
| −.9 | .1841 | .1814 | .1788 | .1762 | .1736 | .1711 | .1685 | .1660 | .1635 | .1611 |
| −.8 | .2119 | .2090 | .2061 | .2033 | .2005 | .1977 | .1949 | .1922 | .1894 | .1867 |
| −.7 | .2420 | .2389 | .2358 | .2327 | .2297 | .2266 | .2236 | .2206 | .2177 | .2148 |
| −.6 | .2743 | .2709 | .2676 | .2643 | .2611 | .2578 | .2546 | .2514 | .2483 | .2451 |
| −.5 | .3085 | .3050 | .3015 | .2981 | .2946 | .2912 | .2877 | .2843 | .2810 | .2776 |
| −.4 | .3446 | .3409 | .3372 | .3336 | .3300 | .3264 | .3228 | .3192 | .3156 | .3121 |
| −.3 | .3821 | .3783 | .3745 | .3707 | .3669 | .3632 | .3594 | .3557 | .3520 | .3483 |
| −.2 | .4207 | .4168 | .4129 | .4090 | .4052 | .4013 | .3974 | .3936 | .3897 | .3859 |
| −.1 | .4602 | .4562 | .4522 | .4483 | .4443 | .4404 | .4364 | .4325 | .4286 | .4247 |
| −.0 | .5000 | .4960 | .4920 | .4880 | .4840 | .4801 | .4761 | .4721 | .4681 | .4641 |

**TABLE 4**  *(Continued)*

$P[\,Z \leq z\,]$

| z | .00 | .01 | .02 | .03 | .04 | .05 | .06 | .07 | .08 | .09 |
|---|-----|-----|-----|-----|-----|-----|-----|-----|-----|-----|
| .0 | .5000 | .5040 | .5080 | .5120 | .5160 | .5199 | .5239 | .5279 | .5319 | .5359 |
| .1 | .5398 | .5438 | .5478 | .5517 | .5557 | .5596 | .5636 | .5675 | .5714 | .5753 |
| .2 | .5793 | .5832 | .5871 | .5910 | .5948 | .5987 | .6026 | .6064 | .6103 | .6141 |
| .3 | .6179 | .6217 | .6255 | .6293 | .6331 | .6368 | .6406 | .6443 | .6480 | .6517 |
| .4 | .6554 | .6591 | .6628 | .6664 | .6700 | .6736 | .6772 | .6808 | .6844 | .6879 |
| .5 | .6915 | .6950 | .6985 | .7019 | .7054 | .7088 | .7123 | .7157 | .7190 | .7224 |
| .6 | .7257 | .7291 | .7324 | .7357 | .7389 | .7422 | .7454 | .7486 | .7517 | .7549 |
| .7 | .7580 | .7611 | .7642 | .7673 | .7703 | .7734 | .7764 | .7794 | .7823 | .7852 |
| .8 | .7881 | .7910 | .7939 | .7967 | .7995 | .8023 | .8051 | .8078 | .8106 | .8133 |
| .9 | .8159 | .8186 | .8212 | .8238 | .8264 | .8289 | .8315 | .8340 | .8365 | .8389 |
| 1.0 | .8413 | .8438 | .8461 | .8485 | .8508 | .8531 | .8554 | .8577 | .8599 | .8621 |
| 1.1 | .8643 | .8665 | .8686 | .8708 | .8729 | .8749 | .8770 | .8790 | .8810 | .8830 |
| 1.2 | .8849 | .8869 | .8888 | .8907 | .8925 | .8944 | .8962 | .8980 | .8997 | .9015 |
| 1.3 | .9032 | .9049 | .9066 | .9082 | .9099 | .9115 | .9131 | .9147 | .9162 | .9177 |
| 1.4 | .9192 | .9207 | .9222 | .9236 | .9251 | .9265 | .9279 | .9292 | .9306 | .9319 |
| 1.5 | .9332 | .9345 | .9357 | .9370 | .9382 | .9394 | .9406 | .9418 | .9429 | .9441 |
| 1.6 | .9452 | .9463 | .9474 | .9484 | .9495 | .9505 | .9515 | .9525 | .9535 | .9545 |
| 1.7 | .9554 | .9564 | .9573 | .9582 | .9591 | .9599 | .9608 | .9616 | .9625 | .9633 |
| 1.8 | .9641 | .9649 | .9656 | .9664 | .9671 | .9678 | .9686 | .9693 | .9699 | .9706 |
| 1.9 | .9713 | .9719 | .9726 | .9732 | .9738 | .9744 | .9750 | .9756 | .9761 | .9767 |
| 2.0 | .9772 | .9778 | .9783 | .9788 | .9793 | .9798 | .9803 | .9808 | .9812 | .9817 |
| 2.1 | .9821 | .9826 | .9830 | .9834 | .9838 | .9842 | .9846 | .9850 | .9854 | .9857 |
| 2.2 | .9861 | .9864 | .9868 | .9871 | .9875 | .9878 | .9881 | .9884 | .9887 | .9890 |
| 2.3 | .9893 | .9896 | .9898 | .9901 | .9904 | .9906 | .9909 | .9911 | .9913 | .9916 |
| 2.4 | .9918 | .9920 | .9922 | .9925 | .9927 | .9929 | .9931 | .9932 | .9934 | .9936 |
| 2.5 | .9938 | .9940 | .9941 | .9943 | .9945 | .9946 | .9948 | .9949 | .9951 | .9952 |
| 2.6 | .9953 | .9955 | .9956 | .9957 | .9959 | .9960 | .9961 | .9962 | .9963 | .9964 |
| 2.7 | .9965 | .9966 | .9967 | .9968 | .9969 | .9970 | .9971 | .9972 | .9973 | .9974 |
| 2.8 | .9974 | .9975 | .9976 | .9977 | .9977 | .9978 | .9979 | .9979 | .9980 | .9981 |
| 2.9 | .9981 | .9982 | .9982 | .9983 | .9984 | .9984 | .9985 | .9985 | .9986 | .9986 |
| 3.0 | .9987 | .9987 | .9987 | .9988 | .9988 | .9989 | .9989 | .9989 | .9990 | .9990 |
| 3.1 | .9990 | .9991 | .9991 | .9991 | .9992 | .9992 | .9992 | .9992 | .9993 | .9993 |
| 3.2 | .9993 | .9993 | .9994 | .9994 | .9994 | .9994 | .9994 | .9995 | .9995 | .9995 |
| 3.3 | .9995 | .9995 | .9995 | .9996 | .9996 | .9996 | .9996 | .9996 | .9996 | .9997 |
| 3.4 | .9997 | .9997 | .9997 | .9997 | .9997 | .9997 | .9997 | .9997 | .9997 | .9998 |
| 3.5 | .9998 | .9998 | .9998 | .9998 | .9998 | .9998 | .9998 | .9998 | .9998 | .9998 |

**TABLE 5**  Percentage Points of $t_\alpha$ Distributions

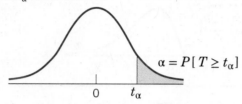

$\alpha = P[T \geq t_\alpha]$

| d.f. \ α | .25 | .10 | .05 | .025 | .01 | .00833 | .00625 | .005 |
|---|---|---|---|---|---|---|---|---|
| 1 | 1.000 | 3.078 | 6.314 | 12.706 | 31.821 | 38.204 | 50.923 | 63.657 |
| 2 | .816 | 1.886 | 2.920 | 4.303 | 6.965 | 7.649 | 8.860 | 9.925 |
| 3 | .765 | 1.638 | 2.353 | 3.182 | 4.541 | 4.857 | 5.392 | 5.841 |
| 4 | .741 | 1.533 | 2.132 | 2.776 | 3.747 | 3.961 | 4.315 | 4.604 |
| 5 | .727 | 1.476 | 2.015 | 2.571 | 3.365 | 3.534 | 3.810 | 4.032 |
| 6 | .718 | 1.440 | 1.943 | 2.447 | 3.143 | 3.287 | 3.521 | 3.707 |
| 7 | .711 | 1.415 | 1.895 | 2.365 | 2.998 | 3.128 | 3.335 | 3.499 |
| 8 | .706 | 1.397 | 1.860 | 2.306 | 2.896 | 3.016 | 3.206 | 3.355 |
| 9 | .703 | 1.383 | 1.833 | 2.262 | 2.821 | 2.933 | 3.111 | 3.250 |
| 10 | .700 | 1.372 | 1.812 | 2.228 | 2.764 | 2.870 | 3.038 | 3.169 |
| 11 | .697 | 1.363 | 1.796 | 2.201 | 2.718 | 2.820 | 2.981 | 3.106 |
| 12 | .695 | 1.356 | 1.782 | 2.179 | 2.681 | 2.779 | 2.934 | 3.055 |
| 13 | .694 | 1.350 | 1.771 | 2.160 | 2.650 | 2.746 | 2.896 | 3.012 |
| 14 | .692 | 1.345 | 1.761 | 2.145 | 2.624 | 2.718 | 2.864 | 2.977 |
| 15 | .691 | 1.341 | 1.753 | 2.131 | 2.602 | 2.694 | 2.837 | 2.947 |
| 16 | .690 | 1.337 | 1.746 | 2.120 | 2.583 | 2.673 | 2.813 | 2.921 |
| 17 | .689 | 1.333 | 1.740 | 2.110 | 2.567 | 2.655 | 2.793 | 2.898 |
| 18 | .688 | 1.330 | 1.734 | 2.101 | 2.552 | 2.639 | 2.775 | 2.878 |
| 19 | .688 | 1.328 | 1.729 | 2.093 | 2.539 | 2.625 | 2.759 | 2.861 |
| 20 | .687 | 1.325 | 1.725 | 2.086 | 2.528 | 2.613 | 2.744 | 2.845 |
| 21 | .686 | 1.323 | 1.721 | 2.080 | 2.518 | 2.601 | 2.732 | 2.831 |
| 22 | .686 | 1.321 | 1.717 | 2.074 | 2.508 | 2.591 | 2.720 | 2.819 |
| 23 | .685 | 1.319 | 1.714 | 2.069 | 2.500 | 2.582 | 2.710 | 2.807 |
| 24 | .685 | 1.318 | 1.711 | 2.064 | 2.492 | 2.574 | 2.700 | 2.797 |
| 25 | .684 | 1.316 | 1.708 | 2.060 | 2.485 | 2.566 | 2.692 | 2.787 |
| 26 | .684 | 1.315 | 1.706 | 2.056 | 2.479 | 2.559 | 2.684 | 2.779 |
| 27 | .684 | 1.314 | 1.703 | 2.052 | 2.473 | 2.552 | 2.676 | 2.771 |
| 28 | .683 | 1.313 | 1.701 | 2.048 | 2.467 | 2.546 | 2.669 | 2.763 |
| 29 | .683 | 1.311 | 1.699 | 2.045 | 2.462 | 2.541 | 2.663 | 2.756 |
| 30 | .683 | 1.310 | 1.697 | 2.042 | 2.457 | 2.536 | 2.657 | 2.750 |
| 40 | .681 | 1.303 | 1.684 | 2.021 | 2.423 | 2.499 | 2.616 | 2.704 |
| 60 | .679 | 1.296 | 1.671 | 2.000 | 2.390 | 2.463 | 2.575 | 2.660 |
| 120 | .677 | 1.289 | 1.658 | 1.980 | 2.358 | 2.428 | 2.536 | 2.617 |
| ∞ | .674 | 1.282 | 1.645 | 1.960 | 2.326 | 2.394 | 2.498 | 2.576 |

**TABLE 6**  Percentage Points of $\chi_\alpha^2$ Distributions

| $\alpha$ d.f. | .99 | .975 | .95 | .90 | .50 | .10 | .05 | .025 | .01 |
|---|---|---|---|---|---|---|---|---|---|
| 1 | .0002 | .001 | .004 | .02 | .45 | 2.71 | 3.84 | 5.02 | 6.63 |
| 2 | .02 | .05 | .10 | .21 | 1.39 | 4.61 | 5.99 | 7.38 | 9.21 |
| 3 | .11 | .22 | .35 | .58 | 2.37 | 6.25 | 7.81 | 9.35 | 11.34 |
| 4 | .30 | .48 | .71 | 1.06 | 3.36 | 7.78 | 9.49 | 11.14 | 13.28 |
| 5 | .55 | .83 | 1.15 | 1.61 | 4.35 | 9.24 | 11.07 | 12.83 | 15.09 |
| 6 | .87 | 1.24 | 1.64 | 2.20 | 5.35 | 10.64 | 12.59 | 14.45 | 16.81 |
| 7 | 1.24 | 1.69 | 2.17 | 2.83 | 6.35 | 12.02 | 14.07 | 16.01 | 18.48 |
| 8 | 1.65 | 2.18 | 2.73 | 3.49 | 7.34 | 13.36 | 15.51 | 17.53 | 20.09 |
| 9 | 2.09 | 2.70 | 3.33 | 4.17 | 8.34 | 14.68 | 16.92 | 19.02 | 21.67 |
| 10 | 2.56 | 3.24 | 3.94 | 4.87 | 9.34 | 15.99 | 18.31 | 20.48 | 23.21 |
| 11 | 3.05 | 3.81 | 4.57 | 5.58 | 10.34 | 17.28 | 19.68 | 21.92 | 24.72 |
| 12 | 3.57 | 4.40 | 5.23 | 6.30 | 11.34 | 18.55 | 21.03 | 23.34 | 26.22 |
| 13 | 4.11 | 5.01 | 5.89 | 7.04 | 12.34 | 19.81 | 22.36 | 24.74 | 27.69 |
| 14 | 4.66 | 5.62 | 6.57 | 7.79 | 13.34 | 21.06 | 23.68 | 26.12 | 29.14 |
| 15 | 5.23 | 6.26 | 7.26 | 8.55 | 14.34 | 22.31 | 25.00 | 27.49 | 30.58 |
| 16 | 5.81 | 6.90 | 7.96 | 9.31 | 15.34 | 23.54 | 26.30 | 28.85 | 32.00 |
| 17 | 6.41 | 7.56 | 8.67 | 10.09 | 16.34 | 24.77 | 27.59 | 30.19 | 33.41 |
| 18 | 7.01 | 8.23 | 9.39 | 10.86 | 17.34 | 25.99 | 28.87 | 31.53 | 34.81 |
| 19 | 7.63 | 8.90 | 10.12 | 11.65 | 18.34 | 27.20 | 30.14 | 32.85 | 36.19 |
| 20 | 8.26 | 9.59 | 10.85 | 12.44 | 19.34 | 28.41 | 31.41 | 34.17 | 37.57 |
| 21 | 8.90 | 10.28 | 11.59 | 13.24 | 20.34 | 29.62 | 32.67 | 35.48 | 38.93 |
| 22 | 9.54 | 10.98 | 12.34 | 14.04 | 21.34 | 30.81 | 33.92 | 36.78 | 40.29 |
| 23 | 10.20 | 11.69 | 13.09 | 14.85 | 22.34 | 32.01 | 35.17 | 38.08 | 41.64 |
| 24 | 10.86 | 12.40 | 13.85 | 15.66 | 23.34 | 33.20 | 36.42 | 39.36 | 42.98 |
| 25 | 11.52 | 13.11 | 14.61 | 16.47 | 24.34 | 34.38 | 37.65 | 40.65 | 44.31 |
| 26 | 12.20 | 13.84 | 15.38 | 17.29 | 25.34 | 35.56 | 38.89 | 41.92 | 45.64 |
| 27 | 12.88 | 14.57 | 16.15 | 18.11 | 26.34 | 36.74 | 40.11 | 43.19 | 46.96 |
| 28 | 13.56 | 15.30 | 16.93 | 18.94 | 27.34 | 37.92 | 41.34 | 44.46 | 48.28 |
| 29 | 14.26 | 16.04 | 17.71 | 19.77 | 28.34 | 39.09 | 42.56 | 45.72 | 49.59 |
| 30 | 14.95 | 16.78 | 18.49 | 20.60 | 29.34 | 40.26 | 43.77 | 46.98 | 50.89 |
| 40 | 22.16 | 24.42 | 26.51 | 29.05 | 39.34 | 51.81 | 55.76 | 59.34 | 63.69 |
| 50 | 29.71 | 32.35 | 34.76 | 37.69 | 49.33 | 63.17 | 67.50 | 71.42 | 76.15 |
| 60 | 37.48 | 40.47 | 43.19 | 46.46 | 59.33 | 74.40 | 79.08 | 83.30 | 88.38 |
| 70 | 45.44 | 48.75 | 51.74 | 55.33 | 69.33 | 85.53 | 90.53 | 95.02 | 100.43 |
| 80 | 53.54 | 57.15 | 60.39 | 64.28 | 79.33 | 96.58 | 101.88 | 106.63 | 112.33 |
| 90 | 61.75 | 65.64 | 69.13 | 73.29 | 89.33 | 107.57 | 113.15 | 118.14 | 124.12 |
| 100 | 70.06 | 74.22 | 77.93 | 82.36 | 99.33 | 118.50 | 124.34 | 129.56 | 135.81 |

**TABLE 7** Percentage Points of $F(v_1, v_2)$ Distributions

$$\alpha = .10$$

| $v_2 \backslash v_1$ | 1 | 2 | 3 | 4 | 5 | 6 | 7 | 8 | 9 | 10 | 12 | 15 | 20 | 25 | 30 | 40 | 60 |
|---|---|---|---|---|---|---|---|---|---|---|---|---|---|---|---|---|---|
| 1 | 39.86 | 49.50 | 53.59 | 55.83 | 57.24 | 58.20 | 58.91 | 59.44 | 59.86 | 60.19 | 60.71 | 61.22 | 61.74 | 62.05 | 62.26 | 62.53 | 62.79 |
| 2 | 8.53 | 9.00 | 9.16 | 9.24 | 9.29 | 9.33 | 9.35 | 9.37 | 9.38 | 9.39 | 9.41 | 9.42 | 9.44 | 9.45 | 9.46 | 9.47 | 9.47 |
| 3 | 5.54 | 5.46 | 5.39 | 5.34 | 5.31 | 5.28 | 5.27 | 5.25 | 5.24 | 5.23 | 5.22 | 5.20 | 5.18 | 5.17 | 5.17 | 5.16 | 5.15 |
| 4 | 4.54 | 4.32 | 4.19 | 4.11 | 4.05 | 4.01 | 3.98 | 3.95 | 3.94 | 3.92 | 3.90 | 3.87 | 3.84 | 3.83 | 3.82 | 3.80 | 3.79 |
| 5 | 4.06 | 3.78 | 3.62 | 3.52 | 3.45 | 3.40 | 3.37 | 3.34 | 3.32 | 3.30 | 3.27 | 3.24 | 3.21 | 3.19 | 3.17 | 3.16 | 3.14 |
| 6 | 3.78 | 3.46 | 3.29 | 3.18 | 3.11 | 3.05 | 3.01 | 2.98 | 2.96 | 2.94 | 2.90 | 2.87 | 2.84 | 2.81 | 2.80 | 2.78 | 2.76 |
| 7 | 3.59 | 3.26 | 3.07 | 2.96 | 2.88 | 2.83 | 2.78 | 2.75 | 2.72 | 2.70 | 2.67 | 2.63 | 2.59 | 2.57 | 2.56 | 2.54 | 2.51 |
| 8 | 3.46 | 3.11 | 2.92 | 2.81 | 2.73 | 2.67 | 2.62 | 2.59 | 2.56 | 2.54 | 2.50 | 2.46 | 2.42 | 2.40 | 2.38 | 2.36 | 2.34 |
| 9 | 3.36 | 3.01 | 2.81 | 2.69 | 2.61 | 2.55 | 2.51 | 2.47 | 2.44 | 2.42 | 2.38 | 2.34 | 2.30 | 2.27 | 2.25 | 2.23 | 2.21 |
| 10 | 3.29 | 2.92 | 2.73 | 2.61 | 2.52 | 2.46 | 2.41 | 2.38 | 2.35 | 2.32 | 2.28 | 2.24 | 2.20 | 2.17 | 2.16 | 2.13 | 2.11 |
| 11 | 3.23 | 2.86 | 2.66 | 2.54 | 2.45 | 2.39 | 2.34 | 2.30 | 2.27 | 2.25 | 2.21 | 2.17 | 2.12 | 2.10 | 2.08 | 2.05 | 2.03 |
| 12 | 3.18 | 2.81 | 2.61 | 2.48 | 2.39 | 2.33 | 2.28 | 2.24 | 2.21 | 2.19 | 2.15 | 2.10 | 2.06 | 2.03 | 2.01 | 1.99 | 1.96 |
| 13 | 3.14 | 2.76 | 2.56 | 2.43 | 2.35 | 2.28 | 2.23 | 2.20 | 2.16 | 2.14 | 2.10 | 2.05 | 2.01 | 1.98 | 1.96 | 1.93 | 1.90 |
| 14 | 3.10 | 2.73 | 2.52 | 2.39 | 2.31 | 2.24 | 2.19 | 2.15 | 2.12 | 2.10 | 2.05 | 2.01 | 1.96 | 1.93 | 1.91 | 1.89 | 1.86 |
| 15 | 3.07 | 2.70 | 2.49 | 2.36 | 2.27 | 2.21 | 2.16 | 2.12 | 2.09 | 2.06 | 2.02 | 1.97 | 1.92 | 1.89 | 1.87 | 1.85 | 1.82 |
| 16 | 3.05 | 2.67 | 2.46 | 2.33 | 2.24 | 2.18 | 2.13 | 2.09 | 2.06 | 2.03 | 1.99 | 1.94 | 1.89 | 1.86 | 1.84 | 1.81 | 1.78 |
| 17 | 3.03 | 2.64 | 2.44 | 2.31 | 2.22 | 2.15 | 2.10 | 2.06 | 2.03 | 2.00 | 1.96 | 1.91 | 1.86 | 1.83 | 1.81 | 1.78 | 1.75 |
| 18 | 3.01 | 2.62 | 2.42 | 2.29 | 2.20 | 2.13 | 2.08 | 2.04 | 2.00 | 1.98 | 1.93 | 1.89 | 1.84 | 1.80 | 1.78 | 1.75 | 1.72 |
| 19 | 2.99 | 2.61 | 2.40 | 2.27 | 2.18 | 2.11 | 2.06 | 2.02 | 1.98 | 1.96 | 1.91 | 1.86 | 1.81 | 1.78 | 1.76 | 1.73 | 1.70 |
| 20 | 2.97 | 2.59 | 2.38 | 2.25 | 2.16 | 2.09 | 2.04 | 2.00 | 1.96 | 1.94 | 1.89 | 1.84 | 1.79 | 1.76 | 1.74 | 1.71 | 1.68 |
| 21 | 2.96 | 2.57 | 2.36 | 2.23 | 2.14 | 2.08 | 2.02 | 1.98 | 1.95 | 1.92 | 1.87 | 1.83 | 1.78 | 1.74 | 1.72 | 1.69 | 1.66 |
| 22 | 2.95 | 2.56 | 2.35 | 2.22 | 2.13 | 2.06 | 2.01 | 1.97 | 1.93 | 1.90 | 1.86 | 1.81 | 1.76 | 1.73 | 1.70 | 1.67 | 1.64 |
| 23 | 2.94 | 2.55 | 2.34 | 2.21 | 2.11 | 2.05 | 1.99 | 1.95 | 1.92 | 1.89 | 1.84 | 1.80 | 1.74 | 1.71 | 1.69 | 1.66 | 1.62 |
| 24 | 2.93 | 2.54 | 2.33 | 2.19 | 2.10 | 2.04 | 1.98 | 1.94 | 1.91 | 1.88 | 1.83 | 1.78 | 1.73 | 1.70 | 1.67 | 1.64 | 1.61 |
| 25 | 2.92 | 2.53 | 2.32 | 2.18 | 2.09 | 2.02 | 1.97 | 1.93 | 1.89 | 1.87 | 1.82 | 1.77 | 1.72 | 1.68 | 1.66 | 1.63 | 1.59 |
| 26 | 2.91 | 2.52 | 2.31 | 2.17 | 2.08 | 2.01 | 1.96 | 1.92 | 1.88 | 1.86 | 1.81 | 1.76 | 1.71 | 1.67 | 1.65 | 1.61 | 1.58 |
| 27 | 2.90 | 2.51 | 2.30 | 2.17 | 2.07 | 2.00 | 1.95 | 1.91 | 1.87 | 1.85 | 1.80 | 1.75 | 1.70 | 1.66 | 1.64 | 1.60 | 1.57 |
| 28 | 2.89 | 2.50 | 2.29 | 2.16 | 2.06 | 2.00 | 1.94 | 1.90 | 1.87 | 1.84 | 1.79 | 1.74 | 1.69 | 1.65 | 1.63 | 1.59 | 1.56 |
| 29 | 2.89 | 2.50 | 2.28 | 2.15 | 2.06 | 1.99 | 1.93 | 1.89 | 1.86 | 1.83 | 1.78 | 1.73 | 1.68 | 1.64 | 1.62 | 1.58 | 1.55 |
| 30 | 2.88 | 2.49 | 2.28 | 2.14 | 2.05 | 1.98 | 1.93 | 1.88 | 1.85 | 1.82 | 1.77 | 1.72 | 1.67 | 1.63 | 1.61 | 1.57 | 1.54 |
| 40 | 2.84 | 2.44 | 2.23 | 2.09 | 2.00 | 1.93 | 1.87 | 1.83 | 1.79 | 1.76 | 1.71 | 1.66 | 1.61 | 1.57 | 1.54 | 1.51 | 1.47 |
| 60 | 2.79 | 2.39 | 2.18 | 2.04 | 1.95 | 1.87 | 1.82 | 1.77 | 1.74 | 1.71 | 1.66 | 1.60 | 1.54 | 1.50 | 1.48 | 1.44 | 1.40 |
| 120 | 2.75 | 2.35 | 2.13 | 1.99 | 1.90 | 1.82 | 1.77 | 1.72 | 1.68 | 1.65 | 1.60 | 1.55 | 1.48 | 1.45 | 1.41 | 1.37 | 1.32 |
| ∞ | 2.71 | 2.30 | 2.08 | 1.94 | 1.85 | 1.77 | 1.72 | 1.67 | 1.63 | 1.60 | 1.55 | 1.49 | 1.42 | 1.38 | 1.34 | 1.30 | 1.24 |

**TABLE 7**  (*Continued*)

$\alpha = .05$

$F_\alpha(v_1, v_2)$

| $v_2$ \ $v_1$ | 1 | 2 | 3 | 4 | 5 | 6 | 7 | 8 | 9 | 10 | 12 | 15 | 20 | 25 | 30 | 40 | 60 |
|---|---|---|---|---|---|---|---|---|---|---|---|---|---|---|---|---|---|
| 1 | 161.5 | 199.5 | 215.7 | 224.6 | 230.2 | 234.0 | 236.8 | 238.9 | 240.5 | 241.9 | 243.9 | 246.0 | 248.0 | 249.3 | 250.1 | 251.1 | 252.2 |
| 2 | 18.51 | 19.00 | 19.16 | 19.25 | 19.30 | 19.33 | 19.35 | 19.37 | 19.38 | 19.40 | 19.41 | 19.43 | 19.45 | 19.46 | 19.46 | 19.47 | 19.48 |
| 3 | 10.13 | 9.55 | 9.28 | 9.12 | 9.01 | 8.94 | 8.89 | 8.85 | 8.81 | 8.79 | 8.74 | 8.70 | 8.66 | 8.63 | 8.62 | 8.59 | 8.57 |
| 4 | 7.71 | 6.94 | 6.59 | 6.39 | 6.26 | 6.16 | 6.09 | 6.04 | 6.00 | 5.96 | 5.91 | 5.86 | 5.80 | 5.77 | 5.75 | 5.72 | 5.69 |
| 5 | 6.61 | 5.79 | 5.41 | 5.19 | 5.05 | 4.95 | 4.88 | 4.82 | 4.77 | 4.74 | 4.68 | 4.62 | 4.56 | 4.52 | 4.50 | 4.46 | 4.43 |
| 6 | 5.99 | 5.14 | 4.76 | 4.53 | 4.39 | 4.28 | 4.21 | 4.15 | 4.10 | 4.06 | 4.00 | 3.94 | 3.87 | 3.83 | 3.81 | 3.77 | 3.74 |
| 7 | 5.59 | 4.74 | 4.35 | 4.12 | 3.97 | 3.87 | 3.79 | 3.73 | 3.68 | 3.64 | 3.57 | 3.51 | 3.44 | 3.40 | 3.38 | 3.34 | 3.30 |
| 8 | 5.32 | 4.46 | 4.07 | 3.84 | 3.69 | 3.58 | 3.50 | 3.44 | 3.39 | 3.35 | 3.28 | 3.22 | 3.15 | 3.11 | 3.08 | 3.04 | 3.01 |
| 9 | 5.12 | 4.26 | 3.86 | 3.63 | 3.48 | 3.37 | 3.29 | 3.23 | 3.18 | 3.14 | 3.07 | 3.01 | 2.94 | 2.89 | 2.86 | 2.83 | 2.79 |
| 10 | 4.96 | 4.10 | 3.71 | 3.48 | 3.33 | 3.22 | 3.14 | 3.07 | 3.02 | 2.98 | 2.91 | 2.85 | 2.77 | 2.73 | 2.70 | 2.66 | 2.62 |
| 11 | 4.84 | 3.98 | 3.59 | 3.36 | 3.20 | 3.09 | 3.01 | 2.95 | 2.90 | 2.85 | 2.79 | 2.72 | 2.65 | 2.60 | 2.57 | 2.53 | 2.49 |
| 12 | 4.75 | 3.89 | 3.49 | 3.26 | 3.11 | 3.00 | 2.91 | 2.85 | 2.80 | 2.75 | 2.69 | 2.62 | 2.54 | 2.50 | 2.47 | 2.43 | 2.38 |
| 13 | 4.67 | 3.81 | 3.41 | 3.18 | 3.03 | 2.92 | 2.83 | 2.77 | 2.71 | 2.67 | 2.60 | 2.53 | 2.46 | 2.41 | 2.38 | 2.34 | 2.30 |
| 14 | 4.60 | 3.74 | 3.34 | 3.11 | 2.96 | 2.85 | 2.76 | 2.70 | 2.65 | 2.60 | 2.53 | 2.46 | 2.39 | 2.34 | 2.31 | 2.27 | 2.22 |
| 15 | 4.54 | 3.68 | 3.29 | 3.06 | 2.90 | 2.79 | 2.71 | 2.64 | 2.59 | 2.54 | 2.48 | 2.40 | 2.33 | 2.28 | 2.25 | 2.20 | 2.16 |
| 16 | 4.49 | 3.63 | 3.24 | 3.01 | 2.85 | 2.74 | 2.66 | 2.59 | 2.54 | 2.49 | 2.42 | 2.35 | 2.28 | 2.23 | 2.19 | 2.15 | 2.11 |
| 17 | 4.45 | 3.59 | 3.20 | 2.96 | 2.81 | 2.70 | 2.61 | 2.55 | 2.49 | 2.45 | 2.38 | 2.31 | 2.23 | 2.18 | 2.15 | 2.10 | 2.06 |
| 18 | 4.41 | 3.55 | 3.16 | 2.93 | 2.77 | 2.66 | 2.58 | 2.51 | 2.46 | 2.41 | 2.34 | 2.27 | 2.19 | 2.14 | 2.11 | 2.06 | 2.02 |
| 19 | 4.38 | 3.52 | 3.13 | 2.90 | 2.74 | 2.63 | 2.54 | 2.48 | 2.42 | 2.38 | 2.31 | 2.23 | 2.16 | 2.11 | 2.07 | 2.03 | 1.98 |
| 20 | 4.35 | 3.49 | 3.10 | 2.87 | 2.71 | 2.60 | 2.51 | 2.45 | 2.39 | 2.35 | 2.28 | 2.20 | 2.12 | 2.07 | 2.04 | 1.99 | 1.95 |
| 21 | 4.32 | 3.47 | 3.07 | 2.84 | 2.68 | 2.57 | 2.49 | 2.42 | 2.37 | 2.32 | 2.25 | 2.18 | 2.10 | 2.05 | 2.01 | 1.96 | 1.92 |
| 22 | 4.30 | 3.44 | 3.05 | 2.82 | 2.66 | 2.55 | 2.46 | 2.40 | 2.34 | 2.30 | 2.23 | 2.15 | 2.07 | 2.02 | 1.98 | 1.94 | 1.89 |
| 23 | 4.28 | 3.42 | 3.03 | 2.80 | 2.64 | 2.53 | 2.44 | 2.37 | 2.32 | 2.27 | 2.20 | 2.13 | 2.05 | 2.00 | 1.96 | 1.91 | 1.86 |
| 24 | 4.26 | 3.40 | 3.01 | 2.78 | 2.62 | 2.51 | 2.42 | 2.36 | 2.30 | 2.25 | 2.18 | 2.11 | 2.03 | 1.97 | 1.94 | 1.89 | 1.84 |
| 25 | 4.24 | 3.39 | 2.99 | 2.76 | 2.60 | 2.49 | 2.40 | 2.34 | 2.28 | 2.24 | 2.16 | 2.09 | 2.01 | 1.96 | 1.92 | 1.87 | 1.82 |
| 26 | 4.23 | 3.37 | 2.98 | 2.74 | 2.59 | 2.47 | 2.39 | 2.32 | 2.27 | 2.22 | 2.15 | 2.07 | 1.99 | 1.94 | 1.90 | 1.85 | 1.80 |
| 27 | 4.21 | 3.35 | 2.96 | 2.73 | 2.57 | 2.46 | 2.37 | 2.31 | 2.25 | 2.20 | 2.13 | 2.06 | 1.97 | 1.92 | 1.88 | 1.84 | 1.79 |
| 28 | 4.20 | 3.34 | 2.95 | 2.71 | 2.56 | 2.45 | 2.36 | 2.29 | 2.24 | 2.19 | 2.12 | 2.04 | 1.96 | 1.91 | 1.87 | 1.82 | 1.77 |
| 29 | 4.18 | 3.33 | 2.93 | 2.70 | 2.55 | 2.43 | 2.35 | 2.28 | 2.22 | 2.18 | 2.10 | 2.03 | 1.94 | 1.89 | 1.85 | 1.81 | 1.75 |
| 30 | 4.17 | 3.32 | 2.92 | 2.69 | 2.53 | 2.42 | 2.33 | 2.27 | 2.21 | 2.16 | 2.09 | 2.01 | 1.93 | 1.88 | 1.84 | 1.79 | 1.74 |
| 40 | 4.08 | 3.23 | 2.84 | 2.61 | 2.45 | 2.34 | 2.25 | 2.18 | 2.12 | 2.08 | 2.00 | 1.92 | 1.84 | 1.78 | 1.74 | 1.69 | 1.64 |
| 60 | 4.00 | 3.15 | 2.76 | 2.53 | 2.37 | 2.25 | 2.17 | 2.10 | 2.04 | 1.99 | 1.92 | 1.84 | 1.75 | 1.69 | 1.65 | 1.59 | 1.53 |
| 120 | 3.92 | 3.07 | 2.68 | 2.45 | 2.29 | 2.18 | 2.09 | 2.02 | 1.96 | 1.91 | 1.83 | 1.75 | 1.66 | 1.60 | 1.55 | 1.50 | 1.43 |
| $\infty$ | 3.84 | 3.00 | 2.61 | 2.37 | 2.21 | 2.10 | 2.01 | 1.94 | 1.88 | 1.83 | 1.75 | 1.67 | 1.57 | 1.51 | 1.46 | 1.39 | 1.32 |

**TABLE 8**   Selected Tail Probabilities for the Null Distribution of Wilcoxon's Rank-Sum Statistic

$$P = P[W_s \geq x] = P[W_s \leq x^*]$$

### Smaller Sample Size = 2

#### Larger Sample Size

| | 3 | | | 4 | | | 5 | | | 6 | |
|---|---|---|---|---|---|---|---|---|---|---|---|
| $x$ | $P$ | $x^*$ | $x$ | $P$ | $x^*$ | $x$ | $P$ | $x^*$ | $x$ | $P$ | $x^*$ |
| 8 | .200 | 4 | 10 | .133 | 4 | 11 | .190 | 5 | 13 | .143 | 5 |
| 9 | .100 | 3 | 11 | .067 | 3 | 12 | .095 | 4 | 14 | .071 | 4 |
| 10 | 0 | 2 | 12 | 0 | 2 | 13 | .048 | 3 | 15 | .036 | 3 |
| | | | | | | 14 | 0 | 2 | 16 | 0 | 2 |

| | 7 | | | 8 | | | 9 | | | 10 | |
|---|---|---|---|---|---|---|---|---|---|---|---|
| $x$ | $P$ | $x^*$ | $x$ | $P$ | $x^*$ | $x$ | $P$ | $x^*$ | $x$ | $P$ | $x^*$ |
| 15 | .111 | 5 | 16 | .133 | 6 | 18 | .109 | 6 | 19 | .136 | 7 |
| 16 | .056 | 4 | 17 | .089 | 5 | 19 | .073 | 5 | 20 | .091 | 6 |
| 17 | .028 | 3 | 18 | .044 | 4 | 20 | .036 | 4 | 21 | .061 | 5 |
| 18 | 0 | 2 | 19 | .022 | 3 | 21 | .018 | 3 | 22 | .030 | 4 |
| | | | 20 | 0 | 2 | 22 | 0 | 2 | 23 | .015 | 3 |

### Smaller Sample Size = 3

#### Larger Sample Size

| | 3 | | | 4 | | | 5 | | | 6 | |
|---|---|---|---|---|---|---|---|---|---|---|---|
| $x$ | $P$ | $x^*$ | $x$ | $P$ | $x^*$ | $x$ | $P$ | $x^*$ | $x$ | $P$ | $x^*$ |
| 13 | .200 | 8 | 16 | .114 | 8 | 18 | .125 | 9 | 20 | .131 | 10 |
| 14 | .100 | 7 | 17 | .057 | 7 | 19 | .071 | 8 | 21 | .083 | 9 |
| 15 | .050 | 6 | 18 | .029 | 6 | 20 | .036 | 7 | 22 | .048 | 8 |
| 16 | 0 | 5 | 19 | 0 | 5 | 21 | .018 | 6 | 23 | .024 | 7 |
| | | | | | | 22 | 0 | 5 | 24 | .012 | 6 |
| | | | | | | | | | 25 | 0 | 5 |

| | 7 | | | 8 | | | 9 | | | 10 | |
|---|---|---|---|---|---|---|---|---|---|---|---|
| $x$ | $P$ | $x^*$ | $x$ | $P$ | $x^*$ | $x$ | $P$ | $x^*$ | $x$ | $P$ | $x^*$ |
| 22 | .133 | 11 | 24 | .139 | 12 | 27 | .105 | 12 | 29 | .108 | 13 |
| 23 | .092 | 10 | 25 | .097 | 11 | 28 | .073 | 11 | 30 | .080 | 12 |
| 24 | .058 | 9 | 26 | .067 | 10 | 29 | .050 | 10 | 31 | .056 | 11 |
| 25 | .033 | 8 | 27 | .042 | 9 | 30 | .032 | 9 | 32 | .038 | 10 |
| 26 | .017 | 7 | 28 | .024 | 8 | 31 | .018 | 8 | 33 | .024 | 9 |
| 27 | .008 | 6 | 29 | .012 | 7 | 32 | .009 | 7 | 34 | .014 | 8 |
| 28 | 0 | 5 | 30 | .006 | 6 | | | | 35 | .007 | 7 |
| | | | 31 | 0 | 5 | | | | | | |

**TABLE 8** *(Continued)*

Smaller Sample Size = 4

Larger Sample Size

| | 4 | | | 5 | | | 6 | | | 7 | |
|---|---|---|---|---|---|---|---|---|---|---|---|
| $x$ | $P$ | $x^*$ | $x$ | $P$ | $x^*$ | $x$ | $P$ | $x^*$ | $x$ | $P$ | $x^*$ |
| 22 | .171 | 14 | 25 | .143 | 15 | 28 | .129 | 16 | 31 | .115 | 17 |
| 23 | .100 | 13 | 26 | .095 | 14 | 29 | .086 | 15 | 32 | .082 | 16 |
| 24 | .057 | 12 | 27 | .056 | 13 | 30 | .057 | 14 | 33 | .055 | 15 |
| 25 | .029 | 11 | 28 | .032 | 12 | 31 | .033 | 13 | 34 | .036 | 14 |
| 26 | .014 | 10 | 29 | .016 | 11 | 32 | .019 | 12 | 35 | .021 | 13 |
| 27 | 0 | 9 | 30 | .008 | 10 | 33 | .010 | 11 | 36 | .012 | 12 |
| | | | 31 | 0 | 9 | | | | 37 | .006 | 11 |

| | 8 | | | 9 | | | 10 | |
|---|---|---|---|---|---|---|---|---|
| $x$ | $P$ | $x^*$ | $x$ | $P$ | $x^*$ | $x$ | $P$ | $x^*$ |
| 34 | .107 | 18 | 36 | .130 | 20 | 39 | .120 | 21 |
| 35 | .077 | 17 | 37 | .099 | 19 | 40 | .094 | 20 |
| 36 | .055 | 16 | 38 | .074 | 18 | 41 | .071 | 19 |
| 37 | .036 | 15 | 39 | .053 | 17 | 42 | .053 | 18 |
| 38 | .024 | 14 | 40 | .038 | 16 | 43 | .038 | 17 |
| 39 | .014 | 13 | 41 | .025 | 15 | 44 | .027 | 16 |
| 40 | .008 | 12 | 42 | .017 | 14 | 45 | .018 | 15 |
| | | | 43 | .010 | 13 | 46 | .012 | 14 |
| | | | | | | 47 | .007 | 13 |

Smaller Sample Size = 5

Larger Sample Size

| | 5 | | | 6 | | | 7 | | | 8 | |
|---|---|---|---|---|---|---|---|---|---|---|---|
| $x$ | $P$ | $x^*$ | $x$ | $P$ | $x^*$ | $x$ | $P$ | $x^*$ | $x$ | $P$ | $x^*$ |
| 34 | .111 | 21 | 37 | .123 | 23 | 41 | .101 | 24 | 44 | .111 | 26 |
| 35 | .075 | 20 | 38 | .089 | 22 | 42 | .074 | 23 | 45 | .085 | 25 |
| 36 | .048 | 19 | 39 | .063 | 21 | 43 | .053 | 22 | 46 | .064 | 24 |
| 37 | .028 | 18 | 40 | .041 | 20 | 44 | .037 | 21 | 47 | .047 | 23 |
| 38 | .016 | 17 | 41 | .026 | 19 | 45 | .024 | 20 | 48 | .033 | 22 |
| 39 | .008 | 16 | 42 | .015 | 18 | 46 | .015 | 19 | 49 | .023 | 21 |
| | | | 43 | .009 | 17 | 47 | .009 | 18 | 50 | .015 | 20 |
| | | | | | | | | | 51 | .009 | 19 |

**TABLE 8**   *(Continued)*

|   | Larger Sample Size | | | | |
|---|---|---|---|---|---|
|   | 9 | | | 10 | |
| $x$ | $P$ | $x^*$ | $x$ | $P$ | $x^*$ |
| 47 | .120 | 28 | 51 | .103 | 29 |
| 48 | .095 | 27 | 52 | .082 | 28 |
| 49 | .073 | 26 | 53 | .065 | 27 |
| 50 | .056 | 25 | 54 | .050 | 26 |
| 51 | .041 | 24 | 55 | .038 | 25 |
| 52 | .030 | 23 | 56 | .028 | 24 |
| 53 | .021 | 22 | 57 | .020 | 23 |
| 54 | .014 | 21 | 58 | .014 | 22 |
| 55 | .009 | 20 | 59 | .010 | 21 |

Smaller Sample Size = 6

|   | Larger Sample Size | | | | | | | | | | |
|---|---|---|---|---|---|---|---|---|---|---|---|
|   | 6 | | | 7 | | | 8 | | | 9 | |
| $x$ | $P$ | $x^*$ | $x$ | $P$ | $x^*$ | $x$ | $P$ | $x^*$ | $x$ | $P$ | $x^*$ |
| 47 | .120 | 31 | 51 | .117 | 33 | 55 | .114 | 35 | 59 | .112 | 37 |
| 48 | .090 | 30 | 52 | .090 | 32 | 56 | .091 | 34 | 60 | .091 | 36 |
| 49 | .066 | 29 | 53 | .069 | 31 | 57 | .071 | 33 | 61 | .072 | 35 |
| 50 | .047 | 28 | 54 | .051 | 30 | 58 | .054 | 32 | 62 | .057 | 34 |
| 51 | .032 | 27 | 55 | .037 | 29 | 59 | .041 | 31 | 63 | .044 | 33 |
| 52 | .021 | 26 | 56 | .026 | 28 | 60 | .030 | 30 | 64 | .033 | 32 |
| 53 | .013 | 25 | 57 | .017 | 27 | 61 | .021 | 29 | 65 | .025 | 31 |
| 54 | .008 | 24 | 58 | .011 | 26 | 62 | .015 | 28 | 66 | .018 | 30 |
|   |   |   | 59 | .007 | 25 | 63 | .010 | 27 | 67 | .013 | 29 |
|   |   |   |   |   |   |   |   |   | 68 | .009 | 28 |

|   | 10 | |
|---|---|---|
| $x$ | $P$ | $x^*$ |
| 63 | .110 | 39 |
| 64 | .090 | 38 |
| 65 | .074 | 37 |
| 66 | .059 | 36 |
| 67 | .047 | 35 |
| 68 | .036 | 34 |
| 69 | .028 | 33 |
| 70 | .021 | 32 |
| 71 | .016 | 31 |
| 72 | .011 | 30 |
| 73 | .008 | 29 |

**TABLE 8**    *(Continued)*

Smaller Sample Size $= 7$

Larger Sample Size

| 7 | | | 8 | | | 9 | | | 10 | | |
|---|---|---|---|---|---|---|---|---|---|---|---|
| $x$ | $P$ | $x^*$ | $x$ | $P$ | $x^*$ | $x$ | $P$ | $x^*$ | $x$ | $P$ | $x^*$ |
| 63 | .104 | 42 | 67 | .116 | 45 | 72 | .105 | 47 | 76 | .115 | 50 |
| 64 | .082 | 41 | 68 | .095 | 44 | 73 | .087 | 46 | 77 | .097 | 49 |
| 65 | .064 | 40 | 69 | .076 | 43 | 74 | .071 | 45 | 78 | .081 | 48 |
| 66 | .049 | 39 | 70 | .060 | 42 | 75 | .057 | 44 | 79 | .067 | 47 |
| 67 | .036 | 38 | 71 | .047 | 41 | 76 | .045 | 43 | 80 | .054 | 46 |
| 68 | .027 | 37 | 72 | .036 | 40 | 77 | .036 | 42 | 81 | .044 | 45 |
| 69 | .019 | 36 | 73 | .027 | 39 | 78 | .027 | 41 | 82 | .035 | 44 |
| 70 | .013 | 35 | 74 | .020 | 38 | 79 | .021 | 40 | 83 | .028 | 43 |
| 71 | .009 | 34 | 75 | .014 | 37 | 80 | .016 | 39 | 84 | .022 | 42 |
| | | | 76 | .010 | 36 | 81 | .011 | 38 | 85 | .017 | 41 |
| | | | | | | 82 | .008 | 37 | 86 | .012 | 40 |
| | | | | | | | | | 87 | .009 | 39 |

Smaller Sample Size $= 8$

Larger Sample Size

| 8 | | | 9 | | | 10 | | |
|---|---|---|---|---|---|---|---|---|
| $x$ | $P$ | $x^*$ | $x$ | $P$ | $x^*$ | $x$ | $P$ | $x^*$ |
| 80 | .117 | 56 | 86 | .100 | 58 | 91 | .102 | 61 |
| 81 | .097 | 55 | 87 | .084 | 57 | 92 | .086 | 60 |
| 82 | .080 | 54 | 88 | .069 | 56 | 93 | .073 | 59 |
| 83 | .065 | 53 | 89 | .057 | 55 | 94 | .061 | 58 |
| 84 | .052 | 52 | 90 | .046 | 54 | 95 | .051 | 57 |
| 85 | .041 | 51 | 91 | .037 | 53 | 96 | .042 | 56 |
| 86 | .032 | 50 | 92 | .030 | 52 | 97 | .034 | 55 |
| 87 | .025 | 49 | 93 | .023 | 51 | 98 | .027 | 54 |
| 88 | .019 | 48 | 94 | .018 | 50 | 99 | .022 | 53 |
| 89 | .014 | 47 | 95 | .014 | 49 | 100 | .017 | 52 |
| 90 | .010 | 46 | 96 | .010 | 48 | 101 | .013 | 51 |
| | | | | | | 102 | .010 | 50 |

**TABLE 8**    *(Continued)*

| | Smaller Sample Size = 9 | | | | | | Smaller Sample Size = 10 | | |
| --- | --- | --- | --- | --- | --- | --- | --- | --- | --- |
| | Larger Sample Size | | | | | | | Larger Sample Size | |
| | 9 | | | | 10 | | | 10 | |
| $x$ | $P$ | $x^*$ | | $x$ | $P$ | $x^*$ | $x$ | $P$ | $x^*$ |
| 100 | .111 | 71 | | 106 | .106 | 74 | 122 | .109 | 88 |
| 101 | .095 | 70 | | 107 | .091 | 73 | 123 | .095 | 87 |
| 102 | .081 | 69 | | 108 | .078 | 72 | 124 | .083 | 86 |
| 103 | .068 | 68 | | 109 | .067 | 71 | 125 | .072 | 85 |
| 104 | .057 | 67 | | 110 | .056 | 70 | 126 | .062 | 84 |
| 105 | .047 | 66 | | 111 | .047 | 69 | 127 | .053 | 83 |
| 106 | .039 | 65 | | 112 | .039 | 68 | 128 | .045 | 82 |
| 107 | .031 | 64 | | 113 | .033 | 67 | 129 | .038 | 81 |
| 108 | .025 | 63 | | 114 | .027 | 66 | 130 | .032 | 80 |
| 109 | .020 | 62 | | 115 | .022 | 65 | 131 | .026 | 79 |
| 110 | .016 | 61 | | 116 | .017 | 64 | 132 | .022 | 78 |
| 111 | .012 | 60 | | 117 | .014 | 63 | 133 | .018 | 77 |
| 112 | .009 | 59 | | 118 | .011 | 62 | 134 | .014 | 76 |
| | | | | 119 | .009 | 61 | 135 | .012 | 75 |
| | | | | | | | 136 | .009 | 74 |

*Source*: Adapted from C. Kraft and C. van Eeden, *A Nonparametric Introduction to Statistics*, Macmillan, New York, 1968.

**TABLE 9** Selected Tail Probabilities for the Null Distribution of Wilcoxon's Signed-Rank Statistic

$$P = P[T^+ \geq x] = P[T^+ \leq x*]$$

| | $n = 3$ | | | $n = 4$ | | | $n = 5$ | | | $n = 6$ | |
|---|---|---|---|---|---|---|---|---|---|---|---|
| $x$ | $P$ | $x^*$ | $x$ | $P$ | $x^*$ | $x$ | $P$ | $x^*$ | $x$ | $P$ | $x^*$ |
| 5 | .250 | 1 | 8 | .188 | 2 | 12 | .156 | 3 | 17 | .109 | 4 |
| 6 | .125 | 0 | 9 | .125 | 1 | 13 | .094 | 2 | 18 | .078 | 3 |
| 7 | 0 | | 10 | .062 | 0 | 14 | .062 | 1 | 19 | .047 | 2 |
| | | | 11 | 0 | | 15 | .031 | 0 | 20 | .031 | 1 |
| | | | | | | 16 | 0 | | 21 | .016 | 0 |
| | | | | | | | | | 22 | 0 | |

| | $n = 7$ | | | $n = 8$ | | | $n = 9$ | | | $n = 10$ | |
|---|---|---|---|---|---|---|---|---|---|---|---|
| $x$ | $P$ | $x^*$ | $x$ | $P$ | $x^*$ | $x$ | $P$ | $x^*$ | $x$ | $P$ | $x^*$ |
| 22 | .109 | 6 | 27 | .125 | 9 | 34 | .102 | 11 | 40 | .116 | 15 |
| 23 | .078 | 5 | 28 | .098 | 8 | 35 | .082 | 10 | 41 | .097 | 14 |
| 24 | .055 | 4 | 29 | .074 | 7 | 36 | .064 | 9 | 42 | .080 | 13 |
| 25 | .039 | 3 | 30 | .055 | 6 | 37 | .049 | 8 | 43 | .065 | 12 |
| 26 | .023 | 2 | 31 | .039 | 5 | 38 | .037 | 7 | 44 | .053 | 11 |
| 27 | .016 | 1 | 32 | .027 | 4 | 39 | .027 | 6 | 45 | .042 | 10 |
| 28 | .008 | 0 | 33 | .020 | 3 | 40 | .020 | 5 | 46 | .032 | 9 |
| | | | 34 | .012 | 2 | 41 | .014 | 4 | 47 | .024 | 8 |
| | | | 35 | .008 | 1 | 42 | .010 | 3 | 48 | .019 | 7 |
| | | | | | | | | | 49 | .014 | 6 |
| | | | | | | | | | 50 | .010 | 5 |

| | $n = 11$ | | | $n = 12$ | | | $n = 13$ | | | $n = 14$ | |
|---|---|---|---|---|---|---|---|---|---|---|---|
| $x$ | $P$ | $x^*$ | $x$ | $P$ | $x^*$ | $x$ | $P$ | $x^*$ | $x$ | $P$ | $x^*$ |
| 48 | .103 | 18 | 56 | .102 | 22 | 64 | .108 | 27 | 73 | .108 | 32 |
| 49 | .087 | 17 | 57 | .088 | 21 | 65 | .095 | 26 | 74 | .097 | 31 |
| 50 | .074 | 16 | 58 | .076 | 20 | 66 | .084 | 25 | 75 | .086 | 30 |
| 51 | .062 | 15 | 59 | .065 | 19 | 67 | .073 | 24 | 76 | .077 | 29 |
| 52 | .051 | 14 | 60 | .055 | 18 | 68 | .064 | 23 | 77 | .068 | 28 |
| 53 | .042 | 13 | 61 | .046 | 17 | 69 | .055 | 22 | 78 | .059 | 27 |
| 54 | .034 | 12 | 62 | .039 | 16 | 70 | .047 | 21 | 79 | .052 | 26 |
| 55 | .027 | 11 | 63 | .032 | 15 | 71 | .040 | 20 | 80 | .045 | 25 |
| 56 | .021 | 10 | 64 | .026 | 14 | 72 | .034 | 19 | 81 | .039 | 24 |
| 57 | .016 | 9 | 65 | .021 | 13 | 73 | .029 | 18 | 82 | .034 | 23 |
| 58 | .012 | 8 | 66 | .017 | 12 | 74 | .024 | 17 | 83 | .029 | 22 |
| 59 | .009 | 7 | 67 | .013 | 11 | 75 | .020 | 16 | 84 | .025 | 21 |
| | | | 68 | .010 | 10 | 76 | .016 | 15 | 85 | .021 | 20 |
| | | | | | | 77 | .013 | 14 | 86 | .018 | 19 |
| | | | | | | 78 | .011 | 13 | 87 | .015 | 18 |
| | | | | | | 79 | .009 | 12 | 88 | .012 | 17 |
| | | | | | | | | | 89 | .010 | 16 |

**TABLE 9**    *(Continued)*

| $x$ | $P$ | $x^*$ |
|---|---|---|
| | $n = 15$ | |
| 83 | .104 | 37 |
| 84 | .094 | 36 |
| 85 | .084 | 35 |
| 86 | .076 | 34 |
| 87 | .068 | 33 |
| 88 | .060 | 32 |
| 89 | .053 | 31 |
| 90 | .047 | 30 |
| 91 | .042 | 29 |
| 92 | .036 | 28 |
| 93 | .032 | 27 |
| 94 | .028 | 26 |
| 95 | .024 | 25 |
| 96 | .021 | 24 |
| 97 | .018 | 23 |
| 98 | .015 | 22 |
| 99 | .013 | 21 |
| 100 | .011 | 20 |
| 101 | .009 | 19 |

*Source*: Adapted from C. Kraft and C. van Eeden, *A Nonparametric Introduction to Statistics*, Macmillan, New York, 1968.

**TABLE F1** General Formulas for Inferences about a Mean ($\mu$), Difference of Two Means ($\mu_1 - \mu_2$)

Confidence interval = Point estimator $\pm$ (Tabled value)(Estimated or true std. dev.)

$$\text{Test statistic} = \frac{\text{Point estimator} - \text{Parameter value at } H_0 \text{ (null hypothesis)}}{\text{(Estimated or true) std. dev. of point estimator}}$$

| | Ch. 8 | Ch. 9 | Ch. 10 | Independent Samples | | Matched Samples |
|---|---|---|---|---|---|---|
| Population(s) | General | Normal with unknown $\sigma$ | Normal $N(\mu_1,\sigma_1), N(\mu_2,\sigma_2)$ $\sigma_1 = \sigma_2 = \sigma$ | Normal $N(\mu_1,\sigma_1), N(\mu_2,\sigma_2)$ $\sigma_1 \neq \sigma_2$ | General | Normal for the difference $D_i = X_i - Y_i$ |
| Inference on | Mean $\mu$ | Mean $\mu$ | $\mu_1 - \mu_2$ | $\mu_1 - \mu_2$ | $\mu_1 - \mu_2$ | $\mu_D = \mu_1 - \mu_2$ |
| Sample(s) | $X_1, \ldots, X_n$ | $X_1, \ldots, X_n$ | $X_1, \ldots, X_{n_1}$ $Y_1, \ldots, Y_{n_2}$ | $X_1, \ldots, X_{n_1}$ $Y_1, \ldots, Y_{n_2}$ | $X_1, \ldots, X_{n_1}$ $Y_1, \ldots, Y_{n_2}$ | $X_1 \; Y_1 \;\; D_1 = X_1 - Y_1$ $X_2 \; Y_2 \;\; D_2 = X_2 - Y_2$ $\cdots$ $X_n \; Y_n \;\; D_n = X_n - Y_n$ |
| Sample size $n$ | Large $n > 30$ | $n \geq 2$ | $n_1 \geq 2$ $n_2 \geq 2$ | $n_1 \geq 2$ $n_2 \geq 2$ | $n_1 > 30$ $n_2 > 30$ | $n \geq 2$ |
| Point estimator | $\bar{X}$ | $\bar{X}$ | $\bar{X} - \bar{Y}$ | $\bar{X} - \bar{Y}$ | $\bar{X} - \bar{Y}$ | $\bar{D} = \bar{X} - \bar{Y}$ |
| Variance of point estimator | $\dfrac{\sigma^2}{n}$ | $\dfrac{\sigma^2}{n}$ | $\sigma^2\left(\dfrac{1}{n_1} + \dfrac{1}{n_2}\right)$ | $\dfrac{\sigma_1^2}{n_1} + \dfrac{\sigma_2^2}{n_2}$ | $\dfrac{\sigma_1^2}{n_1} + \dfrac{\sigma_2^2}{n_2}$ | $\dfrac{\sigma_D^2}{n}$ |
| Std. dev. of point estimator | $\dfrac{\sigma}{\sqrt{n}}$ | $\dfrac{\sigma}{\sqrt{n}}$ | $\sigma\sqrt{\dfrac{1}{n_1} + \dfrac{1}{n_2}}$ | $\sqrt{\dfrac{\sigma_1^2}{n_1} + \dfrac{\sigma_2^2}{n_2}}$ | $\sqrt{\dfrac{\sigma_1^2}{n_1} + \dfrac{\sigma_2^2}{n_2}}$ | $\dfrac{\sigma_D}{\sqrt{n}}$ |
| Estimated std. dev. | $\dfrac{S}{\sqrt{n}}$ | $\dfrac{S}{\sqrt{n}}$ | $S_{pooled}\sqrt{\dfrac{1}{n_1} + \dfrac{1}{n_2}}$ | $\sqrt{\dfrac{S_1^2}{n_1} + \dfrac{S_2^2}{n_2}}$ | $\sqrt{\dfrac{S_1^2}{n_1} + \dfrac{S_2^2}{n_2}}$ | $\dfrac{S_D}{\sqrt{n}}$ |
| Distribution | Normal | $t$ with d.f. $= n - 1$ | $t$ with d.f. $= n_1 + n_2 - 2$ | $t$ with estimated d.f.[a] | Normal | $t$ with d.f. $= n - 1$ |
| Test statistic | $\dfrac{\bar{X} - \mu_0}{S/\sqrt{n}}$ | $\dfrac{\bar{X} - \mu_0}{S/\sqrt{n}}$ | $\dfrac{(\bar{X} - \bar{Y}) - \delta_0}{S_{pooled}\sqrt{\dfrac{1}{n_1} + \dfrac{1}{n_2}}}$ | $\dfrac{(\bar{X} - \bar{Y}) - \delta_0}{\sqrt{\dfrac{S_1^2}{n_1} + \dfrac{S_2^2}{n_2}}}$ | $\dfrac{(\bar{X} - \bar{Y}) - \delta_0}{\sqrt{\dfrac{S_1^2}{n_1} + \dfrac{S_2^2}{n_2}}}$ | $\dfrac{\bar{D} - \mu_{D0}}{S_D/\sqrt{n}}$ |

$$S^2_{pooled} = \frac{(n_1 - 1)S_1^2 + (n_2 - 1)S_2^2}{n_1 + n_2 - 2}$$

$S_D$ = sample std. dev. of the $D_i$'s

[a] estimated by $\left(\dfrac{s_1^2}{n_1} + \dfrac{s_2^2}{n_2}\right)^2 \Big/ \left[\dfrac{(s_1^2/n_1)^2}{(n_1 - 1)} + \dfrac{(s_2^2/n_2)^2}{(n_2 - 1)}\right]$

**TABLE F2**   Inference About Proportions

| | Chapter 8 | Chapter 10 |
|---|---|---|
| Inference on | Proportion $p$ | $p_1 - p_2$ |
| Sample(s) | $X$ successes in $n$ trials | $X$ successes in $n_1$ trials <br> $Y$ successes in $n_2$ trials |
| Sample size(s) | $n$ large | $n_1$ and $n_2$ large |
| Point estimator | $\hat{p} = \dfrac{X}{n}$ | $\hat{p}_1 - \hat{p}_2 = \dfrac{X}{n_1} - \dfrac{Y}{n_2}$ |
| Large sample(s) Confidence Interval | $\hat{p} \pm z_{\alpha/2}\sqrt{\dfrac{\hat{p}(1-\hat{p})}{n}}$ | $\hat{p}_1 - \hat{p}_2 \pm z_{\alpha/2}\sqrt{\dfrac{\hat{p}_1(1-\hat{p}_1)}{n_1} + \dfrac{\hat{p}_2(1-\hat{p}_2)}{n_2}}$ |
| Null hypothesis | $H_0: p = p_0$ | $H_0: p_1 - p_2 = 0$ |
| Test Statistic | $\dfrac{\hat{p} - p_0}{\sqrt{\dfrac{p_0(1-p_0)}{n}}}$ | $\dfrac{\hat{p}_1 - \hat{p}_2}{\sqrt{\hat{p}(1-\hat{p})}\sqrt{\dfrac{1}{n_1} + \dfrac{1}{n_1}}}$ <br> where $\hat{p} = \dfrac{X + Y}{n_1 + n_2}$ |
| Distribution | $N(0, 1)$ | $N(0,1)$ |

# SUMMARY OF FORMULAS USEFUL FOR EXAMS

## CHAPTER 2 • ORGANIZATION AND DESCRIPTION OF DATA

The number of observations in an interval is called the **frequency** and

$$\textbf{Relative frequency} = \frac{\text{Frequency}}{\text{Total number of observations}}$$

$$\textbf{Sample mean } \bar{x} = \frac{\sum x}{n}$$

**Sample median** = middle most value of the ordered data set

$$s^2 = \frac{\sum (x - \bar{x})^2}{n - 1}$$

Also, $s^2 = \dfrac{1}{n - 1} \left[ \sum x^2 - \dfrac{(\sum x)^2}{n} \right]$

(convenient for hand calculation)

$$\text{Sample standard deviation } s = +\sqrt{s^2}$$

The **sample quartiles** divide the **ordered data** into nearly four equal parts. The **100$p$-th percentile** has least proportion $p$ at or below and proportion $1 - p$ at or above.

$$\textbf{Sample range} = \text{Largest observation} - \text{Smallest observation}$$

and

$$\textbf{Sample interquartile range} = \text{Third quartile} - \text{First quartile}$$

The **empirical rule for bell-shaped curves**

$\bar{x} \pm s$   includes approximately 68% of the data

$\bar{x} \pm 2s$   includes approximately 95% of the data

$\bar{x} \pm 3s$   includes approximately 99.7% of the data

## CHAPTER 3 • DESCRIPTIVE STUDY OF BIVARIATE DATA

**Qualitative data** refer to frequency counts in categories. These are summarized by calculating the

$$\textbf{Relative frequency} = \frac{\text{Frequency}}{\text{Total number of observations}}$$

for the individual categories.

For pairs of measurements $(x, y)$

$$\textbf{Sample correlation} \quad r = \frac{S_{xy}}{\sqrt{S_{xx}} \sqrt{S_{yy}}}$$

where $S_{xx} = \sum (x - \bar{x})^2$, $S_{yy} = \sum (y - \bar{y})^2$, and
$S_{xy} = \sum (x - \bar{x})(y - \bar{y})$.

$$\textbf{Fitted line} \qquad \hat{y} = \hat{\beta}_0 + \hat{\beta}_1 x$$

where

$$\textbf{Slope} \quad \hat{\beta}_1 = \frac{S_{xy}}{S_{xx}} \quad \text{and} \quad \textbf{Intercept} \quad \hat{\beta}_0 = \bar{y} - \hat{\beta}_1 \bar{x}$$

# CHAPTER 4 • PROBABILITY

If an experiment is repeated many times, the relative frequency of an event $A$

$$\frac{\text{Number of times } A \text{ occurs}}{\text{Number of times experiment is performed}}$$

provides an estimate of the probability of the event.

With a **uniform probability model,**

$$P(A) = \frac{\text{No. of } e \text{ in } A}{\text{No. of } e \text{ in } S}$$

Assign probabilities to the outcomes to satisfy

$$0 \le P(e_i) \quad \text{and} \quad \sum_{\text{all } e_i \text{ in } S} P(e_i) = 1$$

and then the probability of any event A is defined as

$$P(A) = \sum_{\text{all } e_i \text{ in } A} P(e_i)$$

**Law of complement** $P(A) = 1 - P(\bar{A})$
**Addition law** $P(A \cup B) = P(A) + P(B) - P(AB)$

When $A$ and $B$ are **incompatible** or **mutually exclusive,**

$$\textbf{Special addition law } P(A \cup B) = P(A) + P(B)$$

**Conditional probability** of $A$ given $B$

$$P(A \mid B) = \frac{P(AB)}{P(B)}$$

**Multiplication law**

$$P(AB) = P(B)P(A|B) = P(A)P(B|A)$$

$A$ and $B$ are **independent** if $P(A \mid B) = P(A)$ or if

$$P(AB) = P(A)P(B)$$

## CHAPTER 5 • PROBABILITY DISTRIBUTIONS

The **probability distribution** of a **random variable** $X$ is a list or formula of the pairs $x$ and $f(x) = P[X = x]$.

The **population mean**

$$\mu = \sum \text{Value} \times \text{Probability} = \sum x f(x)$$

is also called the **expected value** $E(X)$.

The **population variance**

$$\sigma^2 = \sum (\text{Deviation})^2 \times \text{Probability} = \sum (x - \mu)^2 f(x)$$
$$= \sum x^2 f(x) - \mu^2$$

The **population standard deviation** $\sigma$ is its square root.

**Bernoulli trials**

1.  Two outcomes: **success** (S) and **failure** (F)
2.  Constant probability of success
3.  Outcomes of trials are independent

The number of successes $X$, in a fixed number $n$ Bernoulli trials, has the **binomial distribution**

$$f(x) = \binom{n}{x} P^x (1 - p)^{n-x} \qquad x = 0, 1, 2, ..., n$$

whose mean $= np$ and variance $= np(1 - p)$.

## CHAPTER 6 • THE NORMAL DISTRIBUTION

If $X$ has mean $\mu$ and standard deviation $\sigma$, the **standardized variable**

$$Z = \frac{X - \mu}{\sigma}$$

has mean 0 and standard deviation 1. In addition, if $X$ is normally distributed, then $Z$ has a standard normal distribution.

The binomial distribution is approximately normal when $np, \; n(1 - p) \geq 15$ so treat as standard normal,

$$Z = \frac{X - p}{\sqrt{np(1 - p)}}$$

## CHAPTER 7 • SAMPLING DISTRIBUTIONS

When the underlying population has mean $\mu$ and standard deviation $\sigma$, the sampling distribution of $\overline{X}$ has

$$\text{Mean of } \overline{X} \;=\; \mu \;=\; \text{Population mean}$$

$$\text{Variance of } \overline{X} \;=\; \frac{\sigma^2}{n} \;=\; \frac{\text{Population variance}}{n}$$

If the sample size $n$ is large, $\overline{X}$ is approximately normal so $P[\,X \leq b\,]$ is approximately equal to the normal probability.

$$P\left[ Z \leq \frac{b - \mu}{\sigma} \right]$$

For normal populations, $\overline{X}$ is normal for all sample sizes.

## CHAPTER 8 • CHAPTER 9 • CHAPTER 10 • TESTS OF HYPOTHESES AND CONFIDENCE INTERVALS

Inference on a mean or two means. See Table F1.
  Inference on a proportion or two proportions. See Table F2.

## CHAPTER 11 • CHAPTER 12 • REGRESSION ANALYSIS

See Chapter 3 for definitions of $S_{xx}$, $S_{xy}$, and $S_{yy}$.
**Least squares estimators**

$$\widehat{\beta}_1 \;=\; \frac{S_{xy}}{S_{xx}} \qquad \widehat{\beta}_0 \;=\; \bar{y} \;-\; \widehat{\beta}_1 \, x$$

$$\text{Least squares fitted line } \widehat{y} \;=\; \widehat{\beta}_0 \;+\; \widehat{\beta}_1 \, x$$

$$\text{Residuals } \widehat{e}_i \;=\; y_i \;-\; \widehat{y}_i \;=\; y_i \;-\; \widehat{\beta}_0 \;-\; \widehat{\beta}_1 x_i$$

**Residual sum of squares**

$$\text{SSE} \;=\; \sum_{i=1}^{n} \widehat{e}_i^{\,2} \;=\; S_{xx} \;-\; \frac{S_{xy}^2}{S_{xx}}$$

$$\text{Estimate of variance } \sigma^2 \quad S^2 \;=\; \frac{\text{SSE}}{n - 2}$$

**Inferences**  All based on $t$ distributions.
**Slope $\beta_1$**

$$T \;=\; \frac{\widehat{\beta}_1 - \beta_1}{S / \sqrt{S_{xx}}}, \qquad \text{d.f.} \;=\; n - 2$$

A $100(1 - \alpha)\%$ confidence interval for $\beta_1$ is

$$\hat{\beta}_1 \pm t_{\alpha/2} \frac{S}{\sqrt{S_{xx}}}$$

**Intercept $\beta_0$**

$$T = \frac{\hat{\beta}_0 - \beta_0}{S\sqrt{\dfrac{1}{n} + \dfrac{\bar{x}^2}{S_{xx}}}}, \qquad \text{d.f.} = n - 2$$

A $100(1 - \alpha)\%$ confidence interval for $\beta_0$ is

$$\hat{\beta}_0 \pm t_{\alpha/2} \, S\sqrt{\frac{1}{n} + \frac{\bar{x}^2}{S_{xx}}}$$

**Expected response $\beta_0 + \beta_0 x^*$**
A $100(1 - \alpha)\%$ confidence interval for the expected response at $x^*$ is given by

$$\hat{\beta}_0 + \hat{\beta}_1 x^* \pm t_{\alpha/2} \, S\sqrt{\frac{1}{n} + \frac{(x^* - \bar{x})^2}{S_{xx}}}$$

**Single response at $x^*$**
A $100\,(1 - \alpha)\%$ prediction interval for a single response is

$$\hat{\beta}_0 + \hat{\beta}_1 x^* \pm t_{\alpha/2} \, S\sqrt{1 + \frac{1}{n} + \frac{(x^* - \bar{x})^2}{S_{xx}}}$$

**Decomposition of variability** $\quad S_{yy} = \dfrac{S_{xy}^2}{S_{xx}} + \text{SSE}$

**Sample correlation coefficient** $\quad r = \dfrac{S_{xy}}{\sqrt{S_{xx}S_{yy}}}$

**Proportion of variance explained $r^2$ or generally**

$$R^2 = \frac{\text{Regression SS}}{\text{Total SS}}$$

## CHAPTER 13 • ANALYSIS OF CATEGORICAL DATA

An $r \times c$ contingency table has $r$ rows and $c$ columns.
**Testing homogeneity** Independent samples from $r$ populations.

$$H_0: \quad \text{probability of each category is the} \atop \text{same for all populations}$$

**Test statistic**

$$\chi^2 = \sum_{\text{cells}} \frac{(O - E)^2}{E}, \qquad \text{d.f.} = (r - 1)(c - 1)$$

where for each cell

$$O = \text{observed cell frequency}$$

$$E = \frac{\text{row total} \times \text{column total}}{\text{grand total}}$$

$$\text{Rejection region} \quad \chi^2 \geq \chi_\alpha^2$$

**Testing independence** A single sample is cross-classified.

$$H_0 : \quad \text{the two classifications are independent}$$

Test statistic and rejection region same as when testing homogeneity.

# CHAPTER 14 • ANALYSIS OF VARIANCE (ANOVA)

Compare $k$ populations means.

**Completely randomized design (One-way ANOVA)**

Independent random samples.

$y_{ij}$ is the $j$-th observation from the $i$-th population (treatment).

$$\text{Total Sum of Squares} = \sum_{i=1}^{k} \sum_{j=1}^{n_i} (y_{ij} - \bar{y})^2$$

$$\text{Treatment Sum of Squares} \quad SS_T = \sum_{i=1}^{k} n_i (\bar{y}_i - \bar{y})^2$$

$$\text{Error Sum of Squares} \quad SSE = \sum_{i=1}^{k} \sum_{j=1}^{n_i} (y_{ij} - \bar{y}_i)^2$$

$$\text{Mean Square} = \frac{\text{Sum of squares}}{\text{Degrees of freedom}}$$

The $F$ − statistic
= ( Treatment Mean Square ) / ( Error Mean Square )
Multiple-$t$ confidence intervals for population mean differences $\mu_i - \mu_{i'}$ for all $m = k(k-1)/2$ statements.

$$\left( \bar{Y} - \bar{Y}_{i'} - t_{\alpha/2m} \sqrt{\frac{MSE}{n-k}}, \bar{Y} - \bar{Y}_{i'} + t_{\alpha/2m} \sqrt{\frac{MSE}{n-k}} \right)$$

**Randomized Block Experiment**
$y_{ij}$ is the observation on treatment $i$ in the $j$-th block.

**Treatment Sum of Squares**

$$SS_T = b \sum_{i=1}^{k} (\bar{y}_{i\cdot} - \bar{y}_{\cdot\cdot})^2$$

**Block Sum of Squares** $SS_B = k \sum_{j=1}^{b} (\bar{y}_{\cdot j} - \bar{y}_{\cdot\cdot})^2$

**Error Sum of Squares**

$$SSE = \sum_{i=1}^{k} \sum_{j=1}^{b} (y_{ij} - \bar{y}_{i\cdot} - \bar{y}_{\cdot j} + \bar{y}_{\cdot\cdot})^2$$

# CHAPTER 15 • NONPARAMETRIC INFERENCE

**Wilcoxon rank-sum statistic** Two independent samples

$W_A = $ sum of ranks of the $n_A$ observations from population $A$, among all $n_A + n_B$ observations

**Sign test statistic** Matched pairs sample

$$S = \text{No. of positive differences}$$

**Wilcoxon signed-rank statistic** Matched pairs sample

$T^+ = $ sum, over positive differences, of the ranks of their absolute values

**Spearmen's rank correlation coefficient**

$$r_{Sp} = \frac{\sum_{i=1}^{n} \left( R_i - \frac{n+1}{2} \right) \left( S_i - \frac{n+1}{2} \right)}{n(n^2 - 1)/12}$$

# Data Bank

The Jump River Electric Company serves several counties in northern Wisconsin. Much of the area is forest and lakes. The data on power outages from a recent summer include date, time, duration of outage (hours), and cause.

**TABLE D.1**   Power Outages

| Date | Time | Duration | Cause |
|------|------|----------|-------|
| 6/11 | 4:00 PM | 5.50 | Trees and limbs |
| 6/12 | 8:00 PM | 1.50 | Trees and limbs |
| 6/16 | 8:30 AM | 2.00 | Trees and limbs |
| 6/17 | 5:30 AM | 2.00 | Trees and limbs |
| 6/17 | 5:00 PM | 8.00 | Windstorm |
| 6/21 | 4:30 PM | 2.00 | Trees and limbs |
| 6/26 | 3:00 AM | 3.00 | Trees and limbs |
| 6/26 | 2:00 PM | 1.75 | Unknown |
| 6/26 | 12:00 AM | 2.00 | Lightning blew up transformer |
| 7/03 | 6:00 PM | 1.50 | Trees and limbs |
| 7/04 | 9:00 AM | 2.50 | Unknown |
| 7/04 | 5:00 PM | 1.50 | Trees and limbs |
| 7/05 | 5:00 AM | 1.50 | Trees and limbs |
| 7/08 | 7:00 AM | 3.50 | Lightning |
| 7/20 | 12:00 AM | 1.50 | Unknown |
| 7/21 | 9:00 AM | 1.50 | Animal |
| 7/28 | 12:00 PM | 1.00 | Animal |

**TABLE D.1**   *(Continued)*

| Date | Time | Duration | Cause |
|------|------|----------|-------|
| 7/30 | 7:00 PM | 1.00 | Squirrel on transformer |
| 7/31 | 7:30 AM | 0.50 | Squirrel on cutout |
| 8/04 | 6:00 AM | 2.50 | Trees and limbs |
| 8/06 | 8:00 PM | 2.00 | Beaver |
| 8/09 | 5:30 AM | 1.50 | Fuse-flying squirrel |
| 8/11 | 12:00 AM | 3.00 | Beaver-cut trees |
| 8/13 | 1:00 AM | 1.00 | Unknown |
| 8/15 | 12:30 AM | 1.50 | Animal |
| 8/18 | 8:00 AM | 1.50 | Trees and limbs |
| 8/20 | 4:00 AM | 2.00 | Transformer fuse |
| 8/25 | 9:00 AM | 1.00 | Animal |
| 8/26 | 2:00 AM | 10.00 | Lightning |
| 8/27 | 3:00 AM | 1.00 | Trees and limbs |

Madison recruits for the fire department need to complete a timed test that simulates working conditions. It includes placing a ladder against a building, pulling out a section of fire hose, dragging a weighted object, and crawling in a simulated attic environment. The times, in seconds, for recruits to complete the test for a Madison firefighter are shown in Table D.2.

**TABLE D.2**   Time to Complete Firefighters Physical Test (seconds)

| | | | | | | | | | | | |
|---|---|---|---|---|---|---|---|---|---|---|---|
| 425 | 389 | 380 | 421 | 438 | 331 | 368 | 417 | 403 | 416 | 385 | 315 |
| 427 | 417 | 386 | 386 | 378 | 300 | 321 | 286 | 269 | 225 | 268 | 317 |
| 287 | 256 | 334 | 342 | 269 | 226 | 291 | 280 | 221 | 283 | 302 | 308 |
| 296 | 266 | 238 | 286 | 317 | 276 | 254 | 278 | 247 | 336 | 296 | 259 |
| 270 | 302 | 281 | 228 | 317 | 312 | 327 | 288 | 395 | 240 | 264 | 246 |
| 294 | 254 | 222 | 285 | 254 | 264 | 277 | 266 | 228 | 347 | 322 | 232 |
| 365 | 356 | 261 | 293 | 354 | 236 | 285 | 303 | 275 | 403 | 268 | 250 |
| 279 | 400 | 370 | 399 | 438 | 287 | 363 | 350 | 278 | 278 | 234 | 266 |
| 319 | 276 | 291 | 352 | 313 | 262 | 289 | 273 | 317 | 328 | 292 | 279 |
| 289 | 312 | 334 | 294 | 297 | 304 | 240 | 303 | 255 | 305 | 252 | 286 |
| 297 | 353 | 350 | 276 | 333 | 285 | 317 | 296 | 276 | 247 | 339 | 328 |
| 267 | 305 | 291 | 269 | 386 | 264 | 299 | 261 | 284 | 302 | 342 | 304 |
| 336 | 291 | 294 | 323 | 320 | 289 | 339 | 292 | 373 | 410 | 257 | 406 |
| 374 | 268 | | | | | | | | | | |

Natural resource managers have attempted to use the Satellite Landsat Multispectral Scanner data for improved land-cover classification. The intensities of reflected light recorded on the near-infrared band of a thermatic mapper are given in Table D.3. Table D.3a gives readings from areas known to consist of forest and Table D.3b readings from urban areas.

**TABLE D.3a**  Near Infrared Light Reflected from Forest Areas

| | | | | | | | | | | | | | | | | | |
|---|---|---|---|---|---|---|---|---|---|---|---|---|---|---|---|---|---|
| 77 | 77 | 78 | 78 | 81 | 81 | 82 | 82 | 82 | 82 | 82 | 83 | 83 | 84 | 84 | 84 | 84 | 85 |
| 86 | 86 | 86 | 86 | 86 | 87 | 87 | 87 | 87 | 87 | 87 | 87 | 89 | 89 | 89 | 89 | 89 | 89 | 89 |
| 90 | 90 | 90 | 91 | 91 | 91 | 91 | 91 | 91 | 91 | 91 | 91 | 91 | 93 | 93 | 93 | 93 | 93 | 93 |
| 94 | 94 | 94 | 94 | 94 | 94 | 94 | 94 | 94 | 94 | 94 | 94 | 95 | 95 | 95 | 95 | 95 | 96 | 96 |
| 96 | 96 | 96 | 96 | 97 | 97 | 97 | 97 | 97 | 97 | 97 | 97 | 97 | 98 | 99 | 100 | 100 | 100 | 100 |
| 100 | 100 | 100 | 100 | 100 | 101 | 101 | 101 | 101 | 101 | 101 | 102 | 102 | 102 | 102 | 102 |
| 102 | 103 | 103 | 104 | 104 | 104 | 105 | 107 |

**TABLE D.3b**  Near Infrared Light Reflected from Urban Areas

| | | | | | | | | | | | | | | | | | | | |
|---|---|---|---|---|---|---|---|---|---|---|---|---|---|---|---|---|---|---|---|
| 71 | 72 | 73 | 74 | 75 | 77 | 78 | 79 | 79 | 79 | 79 | 80 | 80 | 80 | 81 | 81 | 81 | 82 | 82 | 82 | 82 |
| 84 | 84 | 84 | 84 | 84 | 84 | 85 | 85 | 85 | 85 | 85 | 85 | 86 | 86 | 87 | 88 | 90 | 91 | 94 |

Beginning accounting students need to learn to audit in a computerized environment. A sample of beginning accounting students took a test that is summarized by two scores shown in Table D.4: the Computer Attitude Scale (CAS), based on 20 questions, and the Computer Anxiety Rating Scale (CARS), based on 19 questions (courtesy of Douglas Stein). Males are coded as 1 and females as 0.

**TABLE D.4**  Computer Attitude and Anxiety Scores

| Gender | CAS | CARS | Gender | CAS | CARS |
|---|---|---|---|---|---|
| 0 | 2.85 | 2.90 | 1 | 3.30 | 3.47 |
| 1 | 2.60 | 2.32 | 1 | 2.90 | 3.05 |
| 0 | 2.20 | 1.00 | 1 | 2.60 | 2.68 |
| 1 | 2.65 | 2.58 | 0 | 2.25 | 1.90 |
| 1 | 2.60 | 2.58 | 0 | 1.90 | 1.84 |
| 1 | 3.20 | 3.05 | 1 | 2.20 | 1.74 |
| 1 | 3.65 | 3.74 | 0 | 2.30 | 2.58 |
| 0 | 2.55 | 1.90 | 0 | 1.80 | 1.58 |
| 1 | 3.15 | 3.32 | 1 | 3.05 | 2.47 |
| 1 | 2.80 | 2.74 | 1 | 3.15 | 3.32 |
| 0 | 2.40 | 2.37 | 0 | 2.80 | 2.90 |
| 1 | 3.20 | 3.11 | 0 | 2.35 | 2.42 |
| 0 | 3.05 | 3.32 | 1 | 3.70 | 3.47 |
| 1 | 2.60 | 2.79 | 1 | 2.60 | 4.00 |
| 1 | 3.35 | 2.95 | 0 | 3.50 | 3.42 |
| 0 | 3.75 | 3.79 | 0 | 2.95 | 2.53 |
| 0 | 3.00 | 3.26 | 1 | 2.80 | 2.68 |
| 1 | 2.80 | 3.21 | | | |

Data were collected on students taking the course Conditioning 1, designed to introduce students to a variety of training techniques to improve cardiorespiratory fitness, muscular strength, and flexibility (Table D.5); (courtesy of K. Baldridge).

$c_1$ Gender (1 = male, 2 = female)

$c_2$ Pretest percent body fat

$c_3$ Posttest percent body fat

$c_4$ Pretest time to run 1.5 miles (seconds)

$c_5$ Posttest time to run 1.5 miles (seconds)

$c_6$ Pretest time to row 2.5 kilometers (seconds)

$c_7$ Posttest time to row 2.5 kilometers (seconds)

$c_8$ Pretest number of situps completed in 1 minute

$c_9$ Posttest number of situps completed in 1 minute

**TABLE D.5**  Physical Fitness Improvement

| Gender | Pretest % Fat | Posttest % Fat | Pretest Run | Posttest Run | Pretest Row | Posttest Row | Pretest Situps | Posttest Situps |
|---|---|---|---|---|---|---|---|---|
| 1 | 15.1 | 12.4 | 575 | 480 | 621 | 559 | 60 | 67 |
| 1 | 17.1 | 18.5 | 766 | 672 | 698 | 595 | 42 | 45 |
| 2 | 25.5 | 15.0 | 900 | 750 | 840 | 725 | 32 | 36 |
| 2 | 19.5 | 17.0 | 715 | 610 | 855 | 753 | 28 | 38 |
| 2 | 21.7 | 20.6 | 705 | 585 | 846 | 738 | 46 | 54 |
| 1 | 17.7 | 18.5 | 820 | 670 | 630 | 648 | 18 | 41 |
| 2 | 22.7 | 17.2 | 880 | 745 | 860 | 788 | 22 | 33 |
| 2 | 26.6 | 22.4 | 840 | 725 | 785 | 745 | 29 | 39 |
| 2 | 36.4 | 32.5 | 1065 | 960 | 780 | 749 | 27 | 40 |
| 1 | 9.7 | 8.0 | 630 | 565 | 673 | 588 | 32 | 49 |
| 2 | 31.0 | 25.0 | 870 | 780 | 746 | 689 | 39 | 54 |
| 2 | 6.0 | 6.0 | 580 | 494 | 756 | 714 | 37 | 49 |
| 2 | 25.1 | 22.8 | 1080 | 806 | 852 | 838 | 43 | 38 |
| 1 | 15.1 | 13.4 | 720 | 596 | 674 | 576 | 48 | 62 |
| 2 | 23.0 | 21.1 | 780 | 718 | 846 | 783 | 30 | 38 |
| 2 | 9.7 | 9.7 | 945 | 700 | 890 | 823 | 38 | 44 |
| 1 | 7.0 | 6.1 | 706 | 657 | 652 | 521 | 52 | 63 |
| 1 | 16.6 | 15.1 | 650 | 567 | 740 | 615 | 38 | 47 |
| 2 | 21.7 | 19.5 | 686 | 662 | 762 | 732 | 29 | 47 |
| 2 | 23.7 | 21.7 | 758 | 718 | 830 | 719 | 35 | 42 |
| 2 | 24.7 | 21.7 | 870 | 705 | 754 | 734 | 24 | 37 |
| 1 | 11.6 | 8.9 | 480 | 460 | 640 | 587 | 55 | 60 |

**TABLE D.5**  *(Continued)*

| Gender | Pretest % Fat | Posttest % Fat | Pretest Run | Posttest Run | Pretest Row | Posttest Row | Pretest Situps | Posttest Situps |
|---|---|---|---|---|---|---|---|---|
| 2 | 19.5 | 16.0 | 715 | 655 | 703 | 731 | 36 | 44 |
| 1 | 4.8 | 4.2 | 545 | 530 | 625 | 571 | 40 | 45 |
| 2 | 31.1 | 21.6 | 840 | 790 | 745 | 728 | 30 | 36 |
| 1 | 23.2 | 19.9 | 617 | 622 | 637 | 620 | 35 | 32 |
| 1 | 12.5 | 12.2 | 635 | 600 | 805 | 736 | 37 | 52 |
| 2 | 29.3 | 21.7 | 790 | 715 | 821 | 704 | 41 | 42 |
| 2 | 19.5 | 16.0 | 750 | 702 | 1043 | 989 | 42 | 45 |
| 1 | 10.7 | 9.8 | 622 | 567 | 706 | 645 | 40 | 46 |
| 2 | 26.6 | 21.7 | 722 | 725 | 741 | 734 | 49 | 61 |
| 2 | 27.6 | 23.7 | 641 | 598 | 694 | 682 | 40 | 47 |
| 1 | 25.9 | 18.8 | 708 | 609 | 685 | 593 | 20 | 21 |
| 1 | 27.5 | 16.0 | 675 | 637 | 694 | 682 | 35 | 37 |
| 1 | 9.4 | 7.6 | 618 | 566 | 610 | 579 | 36 | 50 |
| 1 | 12.5 | 9.8 | 613 | 552 | 610 | 575 | 46 | 50 |
| 2 | 27.5 | 30.9 | 705 | 660 | 746 | 691 | 31 | 36 |
| 2 | 18.3 | 15.0 | 853 | 720 | 748 | 694 | 31 | 34 |
| 1 | 6.6 | 4.7 | 496 | 476 | 623 | 569 | 45 | 45 |
| 2 | 23.7 | 21.7 | 860 | 750 | 758 | 711 | 42 | 55 |
| 2 | 23.2 | 22.1 | 905 | 636 | 759 | 726 | 26 | 30 |
| 2 | 18.0 | 14.8 | 900 | 805 | 823 | 759 | 31 | 29 |
| 2 | 22.7 | 18.3 | 767 | 741 | 808 | 753 | 28 | 32 |
| 1 | 19.9 | 14.8 | 830 | 620 | 632 | 586 | 39 | 45 |
| 1 | 6.6 | 5.7 | 559 | 513 | 647 | 602 | 44 | 49 |
| 1 | 14.3 | 9.8 | 699 | 652 | 638 | 602 | 34 | 35 |
| 1 | 21.7 | 18.5 | 765 | 735 | 674 | 615 | 41 | 48 |
| 1 | 13.9 | 9.4 | 590 | 570 | 599 | 571 | 40 | 42 |
| 1 | 20.1 | 19.3 | 770 | 672 | 675 | 611 | 37 | 48 |
| 1 | 15.1 | 8.9 | 602 | 560 | 656 | 578 | 47 | 58 |
| 2 | 14.8 | 11.0 | 741 | 610 | 768 | 687 | 34 | 41 |
| 2 | 23.4 | 19.1 | 723 | 641 | 711 | 695 | 36 | 41 |
| 2 | 23.7 | 16.3 | 648 | 601 | 802 | 740 | 44 | 51 |
| 1 | 28.0 | 17.7 | 842 | 702 | 790 | 765 | 26 | 29 |
| 1 | 4.7 | 4.7 | 558 | 540 | 660 | 600 | 40 | 46 |
| 2 | 26.6 | 23.7 | 750 | 565 | 720 | 670 | 36 | 52 |
| 2 | 21.7 | 18.3 | 608 | 592 | 707 | 697 | 42 | 46 |
| 1 | 11.6 | 8.9 | 537 | 495 | 610 | 572 | 55 | 60 |
| 2 | 22.7 | 18.3 | 855 | 694 | 800 | 712 | 41 | 50 |
| 2 | 21.7 | 18.3 | 630 | 614 | 785 | 743 | 50 | 55 |

**TABLE D.5** *(Continued)*

| Gender | Pretest % Fat | Posttest % Fat | Pretest Run | Posttest Run | Pretest Row | Posttest Row | Pretest Situps | Posttest Situps |
|---|---|---|---|---|---|---|---|---|
| 2 | 25.9 | 21.9 | 902 | 820 | 771 | 717 | 34 | 38 |
| 1 | 28.4 | 21.7 | 780 | 664 | 756 | 703 | 57 | 64 |
| 2 | 21.9 | 20.8 | 665 | 670 | 673 | 667 | 41 | 44 |
| 2 | 27.5 | 22.7 | 675 | 646 | 689 | 674 | 49 | 50 |
| 1 | 5.7 | 3.8 | 473 | 472 | 551 | 546 | 53 | 53 |
| 2 | 22.7 | 22.0 | 715 | 682 | 678 | 672 | 40 | 43 |
| 2 | 33.2 | 25.7 | 795 | 740 | 817 | 721 | 30 | 31 |
| 2 | 16.0 | 13.6 | 688 | 615 | 811 | 705 | 45 | 49 |
| 1 | 14.3 | 11.6 | 530 | 497 | 589 | 570 | 39 | 50 |
| 1 | 22.7 | 19.5 | 840 | 705 | 788 | 780 | 29 | 32 |
| 2 | 25.0 | 18.3 | 690 | 618 | 816 | 701 | 42 | 50 |
| 1 | 4.2 | 3.2 | 545 | 527 | 577 | 543 | 55 | 59 |
| 2 | 25.7 | 21.7 | 760 | 727 | 849 | 724 | 39 | 44 |
| 1 | 13.9 | 9.4 | 620 | 515 | 689 | 580 | 32 | 41 |
| 1 | 12.2 | 7.6 | 605 | 564 | 661 | 614 | 35 | 38 |
| 1 | 21.7 | 20.6 | 688 | 625 | 750 | 686 | 39 | 32 |
| 1 | 7.0 | 5.1 | 590 | 529 | 631 | 619 | 60 | 65 |
| 2 | 34.6 | 27.5 | 720 | 694 | 690 | 698 | 34 | 45 |
| 1 | 3.2 | 3.2 | 500 | 459 | 644 | 599 | 35 | 37 |
| 1 | 6.1 | 5.1 | 540 | 492 | 579 | 546 | 56 | 60 |
| 2 | 28.4 | 21.7 | 885 | 825 | 804 | 733 | 33 | 36 |

Grizzly bears are magnificent animals that weigh up to several hundred pounds and can run over 30 mph in short distances. Males range over several miles but the females tend to stay closer to home. Grizzly bears usually keep to themselves. However, several grizzly bear attacks on humans are reported each year. Because of their large size, they can menace hikers and campers in some circumstances. Even so, it is important that a healthy grizzly bear population be maintained for future generations.

Harry Reynolds, has studied grizzly bears for over 10 years in an effort to protect them and learn what they need to survive. He and his colleagues, at the Alaska Game and Fish Department, typically spot bears from an airplane. Once a bear is located, they try to shoot it with a dart so that it falls asleep for a few minutes. During this brief period, they check its overall health and take measurements of the size of the bear. They also roll the bear onto a tarp, which is then lifted by a scale attached to a tripod. Age is given in years, weight in pounds, and the other size measurements in centimeters.

Occasionally, the researchers have to run to a vehicle for protection when the sedated bear awakes too quickly. These measurements are not easy to make!

**TABLE D.6** Grizzly Bear Data

| Bear | Sex | Age | Weight | Length | Neck | Girth | Head Length | Head Width |
|------|-----|-----|--------|--------|------|-------|-------------|------------|
| 1 | M | 9 | 245 | 199 | 84 | 135 | 24.5 | 39.0 |
| 2 | M | 9 | 200 | 198 | 83 | 129 | 22.5 | 39.3 |
| 3 | F | 4 | 102 | 174 | 65 | 104 | 20.3 | 33.4 |
| 4 | F | 5 | 180 | 183 | 57 | 101 | 17.9 | 32.7 |
| 5 | F | 5 | 225 | 170 | 57 | 100 | 18.6 | 33.3 |
| 6 | F | 5 | 230 | 188 | 62 | 103 | 19.3 | 32.9 |
| 7 | F | 13 | 315 | 185 | 62 | 118 | 22.1 | 36.9 |
| 8 | F | 19 | 280 | 193 | 63 | 114 | 22.2 | 35.5 |
| 9 | F | 5 | 235 | 177 | 56 | 93 | 19.4 | 33.4 |
| 10 | F | 2 | 70 | 152 | 46 | 80 | 16.8 | 31.3 |
| 11 | M | 2 | 91 | 157 | 58 | 102 | 17.3 | 30.6 |
| 12 | F | 6 | 280 | 181 | 61 | 100 | 19.4 | 34.5 |
| 13 | M | 11 | 465 | 204 | 84 | 132 | 23.7 | 38.8 |
| 14 | M | 4 | 125 | 191 | 60 | 97 | 20.1 | 33.4 |
| 15 | F | 6 | 265 | 189 | 59 | 100 | 20.4 | 33.3 |
| 16 | F | 3 | 107 | 156 | 45 | 81 | 16.3 | 28.1 |
| 17 | F | 2 | 68 | 158 | 43 | 77 | 16.0 | 29.9 |
| 18 | M | 2 | 75 | 163 | 45 | 77 | 17.8 | 29.7 |
| 19 | M | 2 | 82 | 164 | 48 | 80 | 16.7 | 32.3 |
| 20 | F | 6 | 245 | 176 | 66 | 104 | 20.5 | 34.5 |
| 21 | F | 4 | 215 | 176 | 55 | 103 | 19.1 | 32.0 |
| 22 | M | 2 | 84 | 169 | 53 | 89 | 19.9 | 31.5 |
| 23 | F | 17 | 109 | 176 | 62 | 112 | 19.0 | 32.3 |
| 24 | F | 5 | 240 | 185 | 61 | 103 | 20.5 | 33.6 |
| 25 | F | 5 | 247 | 175 | 57 | 102 | 18.5 | 33.3 |
| 26 | F | 3 | 130 | 148 | 45 | 77 | 13.6 | 27.4 |
| 27 | F | 3 | 130 | 166 | 53 | 86 | 17.5 | 30.0 |
| 28 | F | 7 | 280 | 177 | 57 | 106 | 18.7 | 33.1 |
| 29 | M | 11 | 620 | 224 | 95 | 164 | 24.2 | 42.0 |
| 30 | M | 3 | 150 | 151 | 49 | 85 | 18.0 | 32.4 |
| 31 | F | 1 | 115 | 134 | 45 | 78 | 14.4 | 26.5 |
| 32 | F | 1 | 115 | 137 | 43 | 77 | 14.0 | 26.8 |
| 33 | F | 4 | 285 | 188 | 64 | 110 | 19.6 | 34.0 |
| 34 | M | 4 | 415 | 199 | 77 | 137 | 23.0 | 38.5 |
| 35 | F | 1 | 120 | 135 | 49 | 77 | 14.0 | 26.3 |
| 36 | M | 6 | 578 | 219 | 90 | 134 | 24.8 | 38.6 |
| 37 | M | 2 | 140 | 148 | 56 | 88 | 17.0 | 29.9 |
| 38 | F | 2 | 125 | 131 | 48 | 80 | 16.3 | 27.2 |
| 39 | M | 4 | 265 | 178 | 63 | 102 | 18.8 | 32.2 |
| 40 | F | 2 | 135 | 153 | 49 | 83 | 16.2 | 28.7 |

**TABLE D.6** *(Continued)*

| Bear | Sex | Age | Weight | Length | Neck | Girth | Head Length | Head Width |
|------|-----|-----|--------|--------|------|-------|-------------|------------|
| 41 | M | 2 | 212 | 171 | 51 | 88 | 17.8 | 31.4 |
| 42 | F | 2 | 190 | 168 | 49 | 81 | 17.4 | 30.3 |
| 43 | M | 5 | 359 | 198 | 76 | 114 | 21.4 | 36.6 |
| 44 | F | 7 | 187 | 166 | 57 | 95 | 19.4 | 32.4 |
| 45 | M | 1 | 62 | 114 | 37 | 66 | 12.4 | 26.9 |
| 46 | F | 17 | 283 | 188 | 67 | 113 | 20.5 | 37.0 |
| 47 | M | 1 | 126 | 145 | 46 | 84 | 15.2 | 27.4 |
| 48 | F | 4 | 196 | 168 | 56 | 90 | 18.2 | 31.5 |
| 49 | M | 4 | 345 | 192 | 69 | 112 | 20.3 | 35.2 |
| 50 | M | 2 | 195 | 175 | 53 | 92 | 18.2 | 30.2 |
| 51 | M | 1 | 100 | 125 | 41 | 73 | 14.1 | 24.5 |
| 52 | F | 1 | 50 | 107 | 33 | 58 | 12.1 | 21.1 |
| 53 | M | 1 | 100 | 131 | 45 | 74 | 15.1 | 26.7 |
| 54 | F | 1 | 65 | 119 | 39 | 53 | 13.2 | 22.5 |
| 55 | F | 1 | 105 | 136 | 42 | 71 | 15.0 | 25.5 |
| 56 | M | 1 | 110 | 144 | 43 | 70 | 15.4 | 26.5 |
| 57 | F | 1 | 100 | 133 | 45 | 66 | 16.4 | 25.2 |
| 58 | M | 1 | 63 | 116 | 38 | 62 | 13.3 | 23.5 |
| 59 | F | 1 | 70 | 123 | 41 | 63 | 13.5 | 24.2 |
| 60 | F | 1 | 96 | 139 | 46 | 76 | 14.5 | 25.6 |
| 61 | M | 1 | 95 | 142 | 48 | 76 | 14.4 | 25.6 |

Salmon fisheries support a primary industry in Alaska and their management is of high priority. Salmon are born in freshwater rivers and streams, but then swim out into the ocean for a few years before returning to spawn and die. In order to identify the origins of mature fish and equably divide the catch of returning salmon between Alaska and the Canadian provinces, researchers have studied the growth of their scales. The growth the first year in freshwater is measured by the width of the growth rings for that period of life, and marine growth is measured by the width of the growth rings for the first year in the ocean environment. The scales are first magnified 100 times and then the measurements are made in hundredths of an inch. A set of these measurements, collected by the Alaska Department of Fish and Game, are given in Table D.7 (courtesy of K. Jensen and B. Van Alen).

**TABLE D.7**  Radius of Growth Zones for Freshwater and First Marine Year

| Males | | Females | |
|---|---|---|---|
| Freshwater Growth | First Year Marine Growth | Freshwater Growth | First Year Marine Growth |
| 147 | 444 | 131 | 405 |
| 139 | 446 | 113 | 422 |
| 160 | 438 | 137 | 428 |
| 99 | 437 | 121 | 469 |
| 120 | 405 | 139 | 424 |
| 151 | 435 | 144 | 402 |
| 115 | 394 | 161 | 440 |
| 121 | 406 | 107 | 410 |
| 109 | 440 | 129 | 366 |
| 119 | 414 | 123 | 422 |
| 130 | 444 | 148 | 410 |
| 110 | 465 | 129 | 352 |
| 127 | 457 | 119 | 414 |
| 100 | 498 | 134 | 396 |
| 115 | 452 | 139 | 473 |
| 117 | 418 | 140 | 398 |
| 112 | 502 | 126 | 434 |
| 116 | 478 | 116 | 395 |
| 98 | 500 | 112 | 334 |
| 98 | 589 | 117 | 455 |
| 83 | 480 | 97 | 439 |
| 85 | 424 | 134 | 511 |
| 88 | 455 | 88 | 432 |
| 98 | 439 | 99 | 381 |
| 74 | 423 | 105 | 418 |
| 58 | 411 | 112 | 475 |
| 114 | 484 | 98 | 436 |
| 88 | 447 | 80 | 431 |
| 77 | 448 | 139 | 515 |
| 86 | 450 | 97 | 508 |
| 86 | 493 | 103 | 429 |
| 65 | 495 | 93 | 420 |
| 127 | 470 | 85 | 424 |
| 91 | 454 | 60 | 456 |
| 76 | 430 | 115 | 491 |
| 44 | 448 | 113 | 474 |
| 42 | 512 | 91 | 421 |
| 50 | 417 | 109 | 451 |
| 57 | 466 | 122 | 442 |
| 42 | 496 | 68 | 363 |

The U. S. Department of Agriculture and State Agriculture Experiment Stations cooperate on the investigation of barleys for possible use in brewing processes. One year, the malt extract (%) was obtained for 40 different varieties from one experiment station.

**TABLE D.8**  Malt Extract

| | | | | | | | | | |
|---|---|---|---|---|---|---|---|---|---|
| 75.3 | 77.9 | 77.6 | 76.6 | 78.3 | 77.9 | 77.5 | 77.6 | 77.1 | 78.0 |
| 77.9 | 76.3 | 75.7 | 77.4 | 77.4 | 76.9 | 77.9 | 77.4 | 78.1 | 77.4 |
| 76.4 | 79.1 | 80.0 | 76.9 | 78.5 | 78.4 | 77.8 | 80.4 | 75.9 | 77.0 |
| 79.2 | 76.2 | 77.0 | 75.9 | 77.9 | 78.4 | 76.7 | 76.4 | 76.6 | 77.4 |

An ongoing study of wolves is being conducted at the Yukon-Charley Rivers National Preserve. Here are some of the physical characteristics of wolves that were observed (courtesy of John Burch National Park Service, Fairbanks, Alaska).

**TABLE D.9**  Physical Characteristics of Wolves

| Sex | Age | Weight (lb) | Body Length (cm) | Tail Length (cm) | Canine Length (mm) |
|---|---|---|---|---|---|
| M | 4 | 71 | 134 | 44 | 28.7 |
| F | 2 | 57 | 123 | 46 | 27.0 |
| F | 4 | 84 | 129 | 49 | 27.2 |
| M | 4 | 93 | 143 | 46 | 30.5 |
| M | 4 | 101 | 148 | 48 | 32.3 |
| M | 1 | 84 | 127 | 42 | 25.8 |
| M | 2 | 88 | 136 | 47 | 26.6 |
| M | 3 | 117 | 146 | 46 | 29.1 |
| F | 2 | 90 | 143 | 43 | 27.1 |
| M | 4 | 86 | 142 | 51 | 29.2 |
| F | 6 | 71 | 124 | 42 | 28.2 |
| F | 8 | 71 | 125 | 42 | 27.8 |
| M | 0 | 86 | 139 | 54 | 24.0 |
| M | 2 | 93 | 140 | 45 | 29.0 |
| M | 2 | 86 | 133 | 44 | 29.3 |
| F | 3 | 77 | 122 | 45 | 27.4 |
| F | 2 | 68 | 125 | 51 | 27.3 |
| M | 3 | 106 | 123 | 53 | 29.0 |
| F | 0 | 73 | 122 | 43 | 24.5 |

Researchers studying sleep disorders needed to obtain data from the general population to serve as a reference set. They randomly selected state employees who worked in certain divisions, and many agreed to spend a night in the sleep lab and be measured. Female = 0 and male = 1. The body weight index (BMI) is a person's weight (kg) divided by the square of his or her height (m). The percent of rapid eye movement (PREM) sleep is the percent of total sleep time that is spent in rapid eye movement sleep. The number of breathing pauses per hour (BPH) is the total number of breathing pauses divided by the total hours of sleep. The snoring frequency (SNORF) is a response to the survey question, "According to what others have told you, please estimate how often you snore?" The possible responses were (1) rarely, (2) sometimes, (3) irregular pattern but at least once a week, (4) three to five nights a week, or (5) every or almost every night (courtesy of T. Young).

**TABLE D.10**   Sleep Data

| Gender | Age | BMI | PREM | BPH | SNORF | Gender | Age | BMI | PREM | BPH | SNORF |
|--------|-----|-----|------|-----|-------|--------|-----|-----|------|-----|-------|
| 0 | 41 | 24.65 | 0.200 | 0.00 | 5 | 0 | 55 | 25.10 | 0.326 | 0.19 | 0 |
| 0 | 49 | 23.14 | 0.162 | 0.00 | 4 | 0 | 31 | 28.17 | 0.124 | 0.21 | 1 |
| 0 | 39 | 27.61 | 0.308 | 0.00 | 5 | 0 | 58 | 32.68 | 0.216 | 5.76 | 1 |
| 0 | 51 | 33.50 | 0.083 | 0.00 | 4 | 0 | 43 | 20.07 | 0.191 | 0.00 | 0 |
| 0 | 32 | 43.36 | 0.205 | 2.79 | 4 | 0 | 53 | 24.92 | 0.189 | 7.46 | 0 |
| 0 | 37 | 31.04 | 0.109 | 3.87 | 5 | 0 | 52 | 29.10 | 0.194 | 0.69 | 2 |
| 0 | 36 | 32.42 | 0.154 | 0.00 | 5 | 0 | 43 | 26.40 | 0.077 | 0.00 | 2 |
| 0 | 30 | 31.06 | 0.328 | 3.42 | 5 | 0 | 41 | 20.13 | 0.138 | 0.12 | 1 |
| 0 | 44 | 44.15 | 0.227 | 4.39 | 5 | 0 | 34 | 23.77 | 0.260 | 0.00 | 0 |
| 0 | 43 | 38.29 | 0.104 | 5.14 | 5 | 0 | 41 | 29.06 | 0.163 | 0.31 | 2 |
| 0 | 34 | 21.63 | 0.184 | 0.00 | 5 | 0 | 49 | 24.45 | 0.164 | 0.00 | 0 |
| 0 | 47 | 34.78 | 0.215 | 1.13 | 5 | 0 | 41 | 21.15 | 0.142 | 0.41 | 0 |
| 0 | 40 | 25.16 | 0.206 | 0.00 | 2 | 0 | 47 | 19.47 | 0.224 | 0.00 | 1 |
| 0 | 49 | 37.02 | 0.123 | 2.58 | 5 | 0 | 59 | 28.76 | 0.176 | 0.99 | 0 |
| 0 | 54 | 31.25 | 0.313 | 0.00 | 0 | 0 | 34 | 26.77 | 0.109 | 0.00 | 0 |
| 0 | 38 | 32.58 | 0.227 | 0.13 | 5 | 0 | 42 | 40.06 | 0.191 | 0.14 | 0 |
| 0 | 36 | 34.65 | 0.172 | 0.00 | 4 | 0 | 44 | 24.99 | 0.316 | 1.01 | 2 |
| 0 | 39 | 29.55 | 0.223 | 2.84 | 5 | 0 | 32 | 17.86 | 0.174 | 0.00 | 0 |
| 0 | 45 | 31.96 | 0.146 | 0.00 | 4 | 0 | 31 | 25.99 | 0.109 | 0.00 | 0 |
| 0 | 49 | 21.64 | 0.250 | 0.00 | 5 | 0 | 35 | 20.42 | 0.227 | 0.25 | 2 |
| 0 | 47 | 30.67 | 0.244 | 3.98 | 5 | 0 | 58 | 23.03 | 0.202 | 0.00 | 1 |
| 0 | 33 | 23.43 | 0.228 | 0.00 | 2 | 0 | 38 | 41.14 | 0.220 | 1.24 | 2 |
| 0 | 43 | 30.24 | 0.193 | 0.00 | 1 | 0 | 40 | 22.49 | 0.149 | 0.00 | 0 |
| 0 | 43 | 37.83 | 0.230 | 7.11 | 5 | 0 | 43 | 23.71 | 0.149 | 0.00 | 0 |
| 0 | 39 | 22.46 | 0.141 | 0.17 | 5 | 1 | 44 | 30.22 | 0.162 | 2.33 | 5 |
| 0 | 50 | 33.53 | 0.094 | 3.66 | 5 | 1 | 59 | 27.39 | 0.103 | 0.00 | 3 |
| 0 | 57 | 24.35 | 0.174 | 0.33 | 4 | 1 | 51 | 24.34 | 0.203 | 1.30 | 5 |
| 0 | 36 | 37.64 | 0.000 | 25.36 | 2 | 1 | 49 | 27.17 | 0.179 | 0.00 | 5 |
| 0 | 34 | 28.03 | 0.205 | 0.00 | 1 | 1 | 29 | 28.09 | 0.228 | 0.00 | 4 |
| 0 | 43 | 25.24 | 0.223 | 0.00 | 1 | 1 | 40 | 30.06 | 0.180 | 3.18 | 5 |
| 0 | 59 | 33.71 | 0.175 | 4.71 | 0 | 1 | 45 | 40.82 | 0.082 | 2.34 | 5 |
| 1 | 55 | 32.46 | 0.082 | 38.00 | 5 | 1 | 47 | 24.00 | 0.220 | 0.53 | 5 |
| 1 | 51 | 22.95 | 0.175 | 3.29 | 4 | 1 | 36 | 31.76 | 0.196 | 6.73 | 5 |

**TABLE D.10** *(Continued)*

| Gender | Age | BMI | PREM | BPH | SNORF | Gender | Age | BMI | PREM | BPH | SNORF |
|---|---|---|---|---|---|---|---|---|---|---|---|
| 1 | 38 | 21.97 | 0.123 | 0.00 | 5 | 1 | 59 | 27.12 | 0.255 | 6.18 | 5 |
| 1 | 36 | 25.57 | 0.179 | 0.00 | 4 | 1 | 42 | 24.34 | 0.137 | 1.46 | 5 |
| 1 | 39 | 26.67 | 0.216 | 0.24 | 5 | 1 | 59 | 35.14 | 0.160 | 3.56 | 5 |
| 1 | 38 | 28.95 | 0.231 | 0.55 | 5 | 1 | 37 | 26.53 | 0.122 | 3.80 | 4 |
| 1 | 42 | 35.65 | 0.139 | 0.20 | 5 | 1 | 55 | 31.28 | 0.170 | 0.58 | 4 |
| 1 | 34 | 43.65 | 0.131 | 47.52 | 4 | 1 | 31 | 35.61 | 0.066 | 19.60 | 4 |
| 1 | 55 | 42.70 | 0.191 | 5.41 | 5 | 1 | 53 | 27.55 | 0.083 | 8.01 | 4 |
| 1 | 42 | 25.04 | 0.242 | 0.00 | 5 | 1 | 41 | 34.92 | 0.207 | 15.28 | 4 |
| 1 | 47 | 31.28 | 0.147 | 7.26 | 5 | 1 | 45 | 33.20 | 0.228 | 4.37 | 5 |
| 1 | 46 | 31.02 | 0.124 | 22.55 | 5 | 1 | 41 | 29.94 | 0.134 | 2.42 | 5 |
| 1 | 39 | 25.43 | 0.169 | 10.35 | 5 | 1 | 50 | 26.26 | 0.179 | 0.16 | 5 |
| 1 | 37 | 26.29 | 0.178 | 7.05 | 5 | 1 | 40 | 34.94 | 0.174 | 8.76 | 5 |
| 1 | 41 | 29.59 | 0.259 | 1.37 | 5 | 1 | 54 | 28.60 | 0.197 | 0.00 | 4 |
| 1 | 58 | 29.24 | 0.170 | 3.89 | 5 | 1 | 39 | 31.55 | 0.135 | 0.81 | 5 |
| 1 | 57 | 28.72 | 0.152 | 2.33 | 5 | 1 | 49 | 22.84 | 0.109 | 1.53 | 2 |
| 1 | 33 | 28.56 | 0.126 | 1.09 | 5 | 1 | 46 | 18.89 | 0.114 | 0.00 | 0 |
| 1 | 58 | 30.04 | 0.160 | 0.18 | 4 | 1 | 39 | 25.70 | 0.233 | 0.00 | 0 |
| 1 | 54 | 35.43 | 0.117 | 1.24 | 5 | 1 | 41 | 26.89 | 0.247 | 1.94 | 1 |
| 1 | 46 | 34.78 | 0.151 | 19.47 | 5 | 1 | 46 | 31.29 | 0.230 | 0.00 | 2 |
| 1 | 46 | 25.66 | 0.223 | 0.00 | 3 | 1 | 40 | 29.76 | 0.000 | 0.71 | 0 |
| 1 | 55 | 31.80 | 0.087 | 4.58 | 5 | 1 | 44 | 26.71 | 0.163 | 0.39 | 2 |
| 1 | 56 | 29.74 | 0.162 | 18.77 | 5 | 1 | 41 | 24.69 | 0.209 | 0.16 | 2 |
| 1 | 35 | 24.43 | 0.423 | 0.16 | 5 | 1 | 37 | 26.58 | 0.074 | 0.00 | 0 |
| 1 | 49 | 23.33 | 0.196 | 21.58 | 4 | 1 | 44 | 23.61 | 0.234 | 0.00 | 0 |
| 1 | 50 | 26.45 | 0.199 | 0.00 | 4 | 1 | 44 | 29.86 | 0.147 | 5.13 | 2 |
| 1 | 42 | 27.38 | 0.155 | 2.63 | 5 | 1 | 41 | 28.41 | 0.097 | 0.19 | 1 |
| 1 | 46 | 27.06 | 0.123 | 0.00 | 5 | 1 | 46 | 23.12 | 0.156 | 0.19 | 0 |
| 1 | 37 | 24.82 | 0.175 | 0.18 | 4 | 1 | 34 | 29.88 | 0.195 | 0.00 | 0 |
| 1 | 38 | 30.64 | 0.190 | 70.82 | 5 | 1 | 41 | 18.02 | 0.149 | 0.00 | 0 |
| 1 | 44 | 37.61 | 0.254 | 7.61 | 5 | 1 | 44 | 20.90 | 0.181 | 0.21 | 0 |
| 1 | 48 | 26.85 | 0.219 | 0.42 | 4 | 1 | 39 | 22.39 | 0.223 | 0.00 | 1 |
| 1 | 53 | 28.23 | 0.103 | 0.33 | 4 | 1 | 35 | 29.23 | 0.152 | 0.00 | 2 |
| 1 | 55 | 29.48 | 0.150 | 4.57 | 5 | 1 | 38 | 23.83 | 0.163 | 0.00 | 2 |
| 1 | 56 | 27.76 | 0.185 | 7.82 | 5 | 1 | 47 | 22.23 | 0.129 | 0.76 | 1 |
| 1 | 57 | 32.08 | 0.205 | 28.26 | 5 | 1 | 47 | 28.04 | 0.146 | 2.53 | 0 |
| 1 | 58 | 27.55 | 0.128 | 26.37 | 5 | 1 | 50 | 32.82 | 0.183 | 12.86 | 0 |
| 1 | 37 | 31.98 | 0.241 | 5.38 | 1 | 1 | 48 | 20.09 | 0.083 | 1.59 | 2 |
| 1 | 30 | 24.09 | 0.188 | 0.68 | 5 | 1 | 38 | 26.63 | 0.229 | 1.81 | 2 |
| 1 | 51 | 33.80 | 0.169 | 0.90 | 5 | 1 | 37 | 30.72 | 0.203 | 2.13 | 2 |
| 1 | 44 | 28.08 | 0.130 | 0.53 | 5 | 1 | 50 | 24.73 | 0.188 | 0.00 | 2 |
| 1 | 45 | 30.83 | 0.105 | 6.86 | 4 | 1 | 37 | 33.35 | 0.274 | 17.78 | 2 |
| | | | | | | 1 | 44 | 24.70 | 0.170 | 0.42 | 2 |
| | | | | | | 1 | 49 | 18.24 | 0.223 | 0.00 | 1 |
| | | | | | | 1 | 41 | 24.86 | 0.216 | 2.31 | 0 |

Insecticides, including the long banned DDT, which imitate the human reproductive hormone estrogen, may cause serious health problems in humans and animals. Researchers examined the reproductive development of young alligators hatched from eggs taken from (1) Lake Apopka, which is adjacent to an EPA Superfund site, and (2) Lake Woodruff, which acted as a control. The contaminants at the first lake, including DDT, were thought to have caused reproductive disorders in animals living in and around the lake. The concentrations of the sex steroids estradiol and testosterone in the blood of alligators were determined by radioimmunoassay both at about six months of age and then again after the alligators were stimulated with LH, a pituitary hormone (courtesy of L. Guillette).

$x_1$ = 1 Lake Apopka, = 0 Lake Woodruff

$x_2$ = 1 male, = 0 female

$x_3$ = E2 estradiol concentration (pg/ml)

$x_4$ = T testosterone concentration (pg/ml)

$x_5$ = LHE2 estradiol concentration after receiving LH (pg/ml)

$x_6$ = LHT testosterone concentration after receiving LH (pg/ml)

**TABLE D.11**  Alligator Data

| Lake Apopka | | | | | | Lake Woodruff | | | | | |
|---|---|---|---|---|---|---|---|---|---|---|---|
| Lake | Sex | E2 | T | LHE2 | LHT | Lake | Sex | E2 | T | LHE2 | LHT |
| $x_1$ | $x_2$ | $x_3$ | $x_4$ | $x_5$ | $x_6$ | $x_1$ | $x_2$ | $x_3$ | $x_4$ | $x_5$ | $x_6$ |
| 1 | 1 | 38 | 22 | 134 | 15 | 0 | 1 | 29 | 47 | 46 | 10 |
| 1 | 1 | 23 | 24 | 109 | 28 | 0 | 1 | 64 | 20 | 82 | 76 |
| 1 | 1 | 53 | 8 | 16 | 12 | 0 | 1 | 19 | 60 | * | * |
| 1 | 1 | 37 | 6 | 220 | 13 | 0 | 1 | 36 | 75 | 19 | 72 |
| 1 | 1 | 30 | 7 | 114 | 11 | 0 | 1 | 27 | 12 | 118 | 95 |
| 1 | 0 | 60 | 19 | 184 | 7 | 0 | 1 | 16 | 54 | 33 | 64 |
| 1 | 0 | 73 | 23 | 143 | 13 | 0 | 1 | 15 | 33 | 99 | 19 |
| 1 | 0 | 64 | 16 | 228 | 13 | 0 | 1 | 72 | 53 | 29 | 20 |
| 1 | 0 | 101 | 8 | 163 | 10 | 0 | 1 | 85 | 100 | 72 | 0 |
| 1 | 0 | 137 | 9 | 83 | 7 | 0 | 0 | 139 | 20 | 82 | 2 |
| 1 | 0 | 88 | 7 | 200 | 12 | 0 | 0 | 74 | 4 | 170 | 75 |
| 1 | 0 | 73 | 19 | 220 | 21 | 0 | 0 | 83 | 18 | 125 | 45 |
| 1 | 0 | 257 | 8 | 194 | 37 | 0 | 0 | 35 | 43 | 19 | 76 |
| 1 | 0 | 138 | 10 | 221 | 3 | 0 | 0 | 106 | 9 | 142 | 5 |
| 1 | 0 | 220 | 15 | 101 | 5 | 0 | 0 | 47 | 52 | 24 | 62 |
| 1 | 0 | 88 | 10 | 141 | 7 | 0 | 0 | 38 | 8 | 68 | 20 |
| | | | | | | 0 | 0 | 65 | 15 | 32 | 50 |
| | | | | | | 0 | 0 | 21 | 7 | 140 | 4 |
| | | | | | | 0 | 0 | 68 | 16 | 110 | 3 |
| | | | | | | 0 | 0 | 70 | 16 | 58 | 18 |
| | | | | | | 0 | 0 | 112 | 14 | 78 | 5 |

* indicates missing data

Market researchers track attitudes and preferences of potential purchasers. An increasing number of brands are being linked to social causes and this study[1] was conducted to help determine if adding a social cause to products would cause a difference in consumers' preferences regarding credit cards. Questions about credit card use and preferences, as well as attitudes toward sharing wealth with others, were included.

Female $= 0$ and male $= 1$. The three age groups 19–24, 25–34, and 35–44 are coded as 1, 2, and 3, respectively. Charge is the average monthly amount charged on credit cards. The last three variables are measured on seven point rating scales. The frequency of credit card use, FreqCC, ranges from never (1) to several times a day (7), with weekly (4) in the middle. "Money given to charities goes for good causes" and "I consider myself a compassionate person" scored from strongly disagree (1) to strongly agree (7), (courtesy of Neeraj Arora).

**TABLE D.12**   Credit Card Data

| Row | Gender | Age | Charge | FreqCC | Charity | Compassion |
|-----|--------|-----|--------|--------|---------|------------|
| 1 | 0 | 3 | 100 | 1 | 6 | 6 |
| 2 | 1 | 3 | 150 | 3 | 5 | 7 |
| 3 | 0 | 3 | 500 | 3 | 7 | 7 |
| 4 | 1 | 3 | 0 | 1 | 5 | 7 |
| 5 | 1 | 3 | 0 | 2 | 6 | 6 |
| 6 | 0 | 1 | 100 | 4 | 5 | 5 |
| 7 | 1 | 3 | 1100 | 6 | 6 | 6 |
| 8 | 1 | 3 | 1000 | 5 | 5 | 7 |
| 9 | 0 | 3 | 200 | 6 | 7 | 7 |
| 10 | 1 | 3 | 1000 | 5 | 7 | 7 |
| 11 | 1 | 2 | 200 | 5 | 7 | 7 |
| 12 | 1 | 1 | 500 | 4 | 5 | 5 |
| 13 | 1 | 3 | 100 | 4 | 5 | 6 |
| 14 | 1 | 3 | 20 | 3 | 6 | 7 |
| 15 | 0 | 3 | 0 | 1 | 7 | 7 |
| 16 | 0 | 2 | 400 | 6 | 6 | 5 |
| 17 | 1 | 1 | 0 | 1 | 5 | 7 |
| 18 | 1 | 3 | 0 | 1 | 6 | 6 |
| 19 | 0 | 1 | 800 | 7 | 6 | 5 |
| 20 | 1 | 3 | 1000 | 7 | 7 | 7 |
| 21 | 1 | 1 | 100 | 4 | 5 | 4 |
| 22 | 1 | 3 | 500 | 4 | 5 | 5 |
| 23 | 1 | 3 | 0 | 4 | 4 | 5 |
| 24 | 1 | 2 | 300 | 4 | 5 | 5 |

[1]N. Arora and T. Henderson. "Embedded premium promotion: Why it works and how to make it more effective," *Marketing Science* **26**, (2007), pp. 514–531.

**TABLE D.12** *(Continued)*

| Row | Gender | Age | Charge | FreqCC | Charity | Compassion |
|-----|--------|-----|--------|--------|---------|------------|
| 25 | 1 | 2 | 400 | 5 | 5 | 5 |
| 26 | 1 | 1 | 50 | 4 | 6 | 7 |
| 27 | 1 | 2 | 100 | 4 | 6 | 6 |
| 28 | 1 | 2 | 200 | 5 | 4 | 7 |
| 29 | 0 | 2 | 1 | 6 | 5 | 5 |
| 30 | 1 | 2 | 2000 | 7 | 1 | 4 |
| 31 | 1 | 1 | 450 | 3 | 7 | 7 |
| 32 | 1 | 2 | 100 | 2 | 3 | 6 |
| 33 | 1 | 3 | 0 | 1 | 5 | 6 |
| 34 | 1 | 2 | 600 | 4 | 6 | 6 |
| 35 | 1 | 3 | 100 | 3 | 6 | 7 |
| 36 | 1 | 3 | 700 | 5 | 7 | 7 |
| 37 | 1 | 3 | 500 | 4 | 6 | 7 |
| 38 | 1 | 2 | 300 | 7 | 6 | 6 |
| 39 | 1 | 3 | 100 | 4 | 7 | 7 |
| 40 | 1 | 3 | 1500 | 7 | 7 | 5 |
| 41 | 0 | 3 | 600 | 6 | 6 | 6 |
| 42 | 1 | 1 | 40 | 2 | 6 | 7 |
| 43 | 1 | 2 | 1000 | 7 | 4 | 5 |
| 44 | 1 | 3 | 100 | 3 | 6 | 7 |
| 45 | 1 | 2 | 300 | 3 | 6 | 6 |
| 46 | 1 | 1 | 0 | 1 | 7 | 7 |
| 47 | 1 | 3 | 150 | 5 | 6 | 6 |
| 48 | 1 | 2 | 10 | 2 | 6 | 7 |
| 49 | 0 | 2 | 1000 | 5 | 7 | 7 |
| 50 | 1 | 2 | 500 | 5 | 7 | 7 |
| 51 | 0 | 3 | 0 | 2 | 4 | 6 |
| 52 | 0 | 1 | 500 | 4 | 5 | 4 |
| 53 | 1 | 2 | 450 | 4 | 5 | 5 |
| 54 | 1 | 2 | 200 | 5 | 6 | 7 |
| 55 | 1 | 2 | 25 | 3 | 2 | 6 |
| 56 | 1 | 2 | 500 | 5 | 5 | 6 |
| 57 | 1 | 3 | 70 | 4 | 2 | 4 |
| 58 | 0 | 3 | 300 | 4 | 5 | 1 |
| 59 | 1 | 3 | 300 | 5 | 4 | 6 |
| 60 | 1 | 1 | 0 | 2 | 4 | 6 |
| 61 | 1 | 1 | 0 | 1 | 5 | 6 |
| 62 | 1 | 2 | 200 | 5 | 5 | 5 |
| 63 | 0 | 2 | 2000 | 4 | 4 | 4 |
| 64 | 0 | 3 | 0 | 1 | 7 | 7 |
| 65 | 1 | 3 | 1500 | 7 | 4 | 5 |
| 66 | 1 | 2 | 0 | 1 | 3 | 3 |
| 67 | 1 | 1 | 60 | 3 | 6 | 6 |

**TABLE D.12** *(Continued)*

| Row | Gender | Age | Charge | FreqCC | Charity | Compassion |
|-----|--------|-----|--------|--------|---------|------------|
| 68 | 1 | 3 | 500 | 5 | 7 | 7 |
| 69 | 0 | 3 | 0 | 1 | 5 | 7 |
| 70 | 0 | 3 | 1200 | 5 | 5 | 6 |
| 71 | 0 | 3 | 2000 | 6 | 5 | 5 |
| 72 | 0 | 2 | 700 | 6 | 2 | 7 |
| 73 | 1 | 3 | 0 | 2 | 5 | 4 |
| 74 | 0 | 1 | 200 | 3 | 5 | 6 |
| 75 | 0 | 3 | 3000 | 6 | 6 | 6 |
| 76 | 1 | 3 | 50 | 5 | 5 | 3 |
| 77 | 0 | 3 | 300 | 4 | 6 | 6 |
| 78 | 1 | 3 | 3000 | 6 | 3 | 7 |
| 79 | 0 | 3 | 0 | 1 | 6 | 7 |
| 80 | 0 | 1 | 0 | 1 | 7 | 7 |
| 81 | 1 | 2 | 0 | 1 | 6 | 7 |
| 82 | 1 | 3 | 50 | 6 | 4 | 4 |
| 83 | 1 | 3 | 500 | 6 | 5 | 6 |
| 84 | 0 | 1 | 50 | 2 | 2 | 2 |
| 85 | 1 | 1 | 40 | 3 | 4 | 5 |
| 86 | 0 | 3 | 150 | 2 | 4 | 7 |
| 87 | 1 | 3 | 25 | 3 | 5 | 7 |
| 88 | 0 | 3 | 800 | 5 | 3 | 6 |
| 89 | 0 | 3 | 3500 | 7 | 5 | 6 |
| 90 | 1 | 2 | 100 | 5 | 3 | 7 |
| 91 | 0 | 2 | 200 | 3 | 1 | 7 |
| 92 | 0 | 3 | 400 | 5 | 5 | 6 |
| 93 | 1 | 1 | 300 | 5 | 3 | 3 |
| 94 | 1 | 2 | 350 | 5 | 6 | 6 |

Polar bears are among the world's largest land carnivores and males are typically larger than females. They can weigh over 1000 pounds and, when standing on their hind legs, be over 10 feet tall. Polar bears are well adapted to the harsh Arctic climate and they live on the islands and ice packs around the Arctic Circle. They do have to travel great distances on both land and water to find food, but they have no natural enemies.

The U. S. Geological Survey monitors the health of the population. Monitoring procedures include making morphometric measurements.

$x_1$ = age as determined from tooth growth layers

$x_2$ = total Length meaning the total body length from nose to tail (cm)

$x_3$ = girth taken on a sternally recumbent bear immediately behind the front legs (cm)

$x_4$ = skull length (cm)

$x_5$ = skull width (cm)

$x_6$ = weight or body mass (lb)

**TABLE D.13**   Measurements on Male Polar Bears

| Age | Total Length | Girth | Skull Length | Skull Width | Weight |
|-----|------|-------|------|------|------|
| 8 | 267.0 | 153.6 | 41.0 | 24.8 | 922 |
| 13 | 271.0 | 143.0 | 42.3 | 24.8 | 785 |
| 7 | 256.0 | 145.5 | 40.9 | 24.4 | 702 |
| 12 | 275.0 | 155.1 | 41.3 | 25.7 | 875 |
| 14 | 253.0 | 143.5 | 38.9 | 25.1 | 750 |
| 25 | 266.0 | 152.0 | 41.2 | 25.0 | 885 |
| 13 | 261.0 | 147.0 | 41.8 | 25.6 | 835 |
| 16 | 262.0 | 142.5 | 39.9 | 25.3 | 875 |
| 21 | 249.5 | 136.8 | 40.5 | 24.8 | 675 |
| 16 | 269.0 | 172.0 | 43.1 | 27.3 | 1060 |
| 14 | 255.0 | 148.0 | 40.7 | 27.1 | 985 |
| 9 | 248.0 | 150.3 | 41.8 | 26.1 | 770 |
| 12 | 266.0 | 158.0 | 42.6 | 25.7 | 988 |
| 14 | 262.0 | 165.0 | 43.7 | 25.2 | 995 |
| 9 | 239.0 | 161.0 | 40.8 | 27.0 | 925 |
| 9 | 270.0 | 155.5 | 40.3 | 26.3 | 998 |
| 7 | 246.5 | 140.5 | 38.8 | 23.5 | 662 |
| 11 | 271.0 | 154.3 | 42.4 | 26.1 | 925 |
| 13 | 251.5 | 150.6 | 39.8 | 26.7 | 887 |
| 8 | 249.8 | 123.4 | 38.8 | 22.4 | 630 |
| 9 | 252.0 | 142.0 | 39.6 | 23.3 | 741 |
| 12 | 262.0 | 165.2 | 41.6 | 27.1 | 956 |
| 6 | 251.0 | 143.5 | 37.8 | 23.0 | 719 |
| 17 | 256.0 | 166.5 | 40.1 | 27.0 | 996 |
| 11 | 234.5 | 125.0 | 37.0 | 23.0 | 547 |
| 10 | 240.0 | 157.0 | 41.5 | 27.6 | 885 |
| 9 | 259.0 | 141.0 | 39.1 | 24.1 | 696 |
| 10 | 260.7 | 135.2 | 40.1 | 24.2 | 747 |

*Source*: U. S. Geological Survey, Alaska Science Center.

# Answers to Selected Odd-Numbered Exercises

## CHAPTER 1

1.1 The statistical population is, for all males in US, the decision to buy flowers that day or day before, or, not buy at that time. The sample is the choices of the 451 males where 144 do purchase at that time.

1.3 Sample is the 20 numbers of close friends recorded. Statistical population is the collection of numbers from all college students.

1.5 The statistical population is collection of responses not more/more stressed for all 20-25 year olds in country. Sample is collection of 40 responses reported.

1.7 No. Likely to be among heavy drug users.

1.9 (a) Sample. (b) Anecdotal.

1.13 Comfortable not well defined. Could require majority of sample to rate as 5 on a five-point scale.

1.17 Lab, receptionist, and X-ray.

1.19 Population: Entire set of responses from all U.S. teens 13 to 17. Sample: Responses from the 1055 teens contacted.

1.21 (a) Person living in Chicago. (b) Whether a person is a registered voter (yes) or not (no). (c) Entire set of yes/no responses from all persons living in Chicago.

1.23 (a) Anyone who golfs. (b) The statistical population is the collection of estimated hole sizes, one for each golfer. The sample is the collection of 46 hole sizes the psychologists recorded.

1.25 It is likely that any person visiting the site does so because they have a problem with procrastination. The scores recorded would not be representative of the general population.

1.27 Population: The entire collection of responses, from all residents, of their favorite establishment. Sample: The responses from the persons who filled out the form. This is a self-selected sample, and not representative, as only those with strong opinions would fill out form.

1.29 (a) Anecdotal. (b) Sample based. (c) Sample based.

**1.31**  "Too long" is not well defined. Could, for instance, say half the time it takes longer than 5 minutes.

**1.37**  (b)   Including votes only from those with strong enough opinions to call in, like big dogs jumping above the fence, would produce unrepresentative results.

# CHAPTER 2

**2.1**  (a)  22.8%    (c)  28.5%, 42.4%, 77.2%

**2.3**  The frequencies are 7, 10, 13, 5, 2, 1, 1, 1.

**2.5**  (a)   The relative frequencies are Drive alone .625, Car pool .075, Ride bus .175, and Other .125.

**2.7**  3 belongs to two classes.

**2.9**  5 doesn't belong to any class.

**2.11**  (a)   Yes.     (d)  No.     (e)  No.

**2.13**  (c)   9 / 50  =  .18

**2.17**  (c)   1 / 15  =  .067

**2.19**  (a)   The frequency table

| Class | Frequency |
|---|---|
| 0.45–0.90 | 2 |
| 0.90–1.35 | 6 |
| 1.35–1.80 | 11 |
| 1.80–2.25 | 5 |
| 2.25–2.70 | 6 |

**2.21**  The frequencies are 12, 16, 10, 11, and 8. The classes are not of equal length.

**2.23**  The stem-and-leaf display:

| | |
|---|---|
| 0 | 6 |
| 1 | 2234455567777889 |
| 2 | 00000222445567799 |
| 3 | 022444566 |
| 4 | 1167 |
| 5 | 12 |

**2.25**  The double-stem display:

| 0 | 6 | 3 | 022444 |
|---|---|---|---|
| 1 | 22344 | 3 | 566 |
| 1 | 55567777889 | 4 | 11 |
| 2 | 00000022244 | 4 | 67 |
| 2 | 5567799 | 5 | 12 |

**2.27**  The five-stem display  Leaf Unit  =  1.0

| | |
|---|---|
| 20 | 0 |
| 20 | 2 |
| 20 | |
| 20 | |
| 20 | 89 |
| 21 | 011 |
| 21 | 2 |
| 21 | 4 |
| 21 | 7 |
| 21 | 99 |
| 22 | 0 |
| 22 | |
| 22 | 55 |
| 22 | |
| 22 | |
| 23 | 1 |
| 23 | 2333 |
| 23 | |
| 23 | |
| 23 | |
| 24 | |
| 24 | |
| 24 | 44 |
| 24 | |
| 24 | 8 |
| 25 | |
| 25 | 3 |

**2.29**  (a)   Median  =  3 and $\bar{x}$  =  3
(c)   Median  =  −1 and $\bar{x}$  =  −1

**2.31**  (a)  $\bar{x}$  =  254      (b)   Median since one large observation heavily influences the mean.

**2.33**  Claim ignores variability. July with 105 is hot.

**2.35**  (a)  $\bar{x}$  =  8.48      (b)   Either

2.37   Mean $=$ 118.05 is center where sum of positive deviations balance the sum of negative deviations. Median $=$ 117.00 has at least half of weights the same or smaller and at least half the same or larger.

2.39   (a)   2.60      (b)   2.00

2.43   181.5

2.45   (b)   $\bar{x} = 26.62$ sample median $= 24$
       (c)   $Q_1 = 22$, $Q_2 = 24$, and $Q_3 = 30$

2.47   (a)   Sample median $= 153$
       (b)   $Q_1 = 135.5$, and $Q_3 = 166.5$

2.49   Sample median $= 94$, $Q_1 = 73$, and $Q_3 = 105$

2.51   (a)   Median $= 110$, $Q_1 = 60$, $Q_3 = 340$
       (b)   90th percentile $= 400$

2.53   (b)   Mean $= 24.33°$C, median $= 24.44°$C

2.57   (b)   $s^2 = 7$ and $s = 2.646$

2.59   (b)   $s^2 = 18.667$ and $s = 4.32$

2.63   (a)   $s^2 = 1.30$      (c)   $s^2 = .286$

2.65   (a)   $s = 139.48$
       (b)   Reduces to almost half.

2.67   (a)   $s^2 = 155$, 226
       (b)   $s = 393.99$

2.71   (a)   Median $= 68.4$
       (b)   $\bar{x} = 68.343$      (c)   $s = 2.419$

2.73   (a)   $s^2 = 239.32$      (b)   25.00
       (c)   8 is nearly half as variable.

2.75   Interquartile range $= 31.0$

2.77   No. Extremes more variable than middle.

2.81   (a)   $\bar{x} = 2.600$ and $s = 1.303$
       (b)   Proportions .5, 1.0, and 1.0, respectively.

2.83   (a)   $-1.037$      (b)   1.677

2.87   (a)   Mean $= 21.92$, standard deviation $= 5.405$
       (b)   $22 / 24 = .917$

2.89   Mean $= 1.925$, standard deviation $= 1.607$

2.91   (a)   20.5      (b)   Maximum 63 and minimum $-$ 8 (gain) Bush.      (c) 71 Obama

2.95   Only the value 215 from the second period is outside of limits.

2.97   First an upward trend and then a downward trend. Control limits not appropriate.

2.99   (a)   The relative frequencies of the occupation groups are:

| Occupation Group | Relative Frequency | |
|---|---|---|
| | 2010 | 2000 |
| Goods producing | .117 | .161 |
| Service (private) | .730 | .702 |
| Government | .148 | .136 |
| Total | 1.000 | .999 |

2.103  (a)   Yes.      (b)   Yes.      (c)   No.

2.105  (a)   Mean $= 227.4$ pounds is center where sum of positive deviations balance the sum of negative deviations. Median $= 232.5$ has at least half of weights the same or smaller and at least half the same or larger.
       (b)   82.7      (c)   1.120

2.107  (a)   Median $= 9$
       (b)   $\bar{x} = 9.033$
       (c)   $s^2 = 3.895$

2.109  (a)   $\bar{x} = 7$, $s = 2$
       (c)   Mean $= -21$, standard deviation $= 6$.

2.113  (a)   $\bar{x} = 5.375$ and $s = 3.424$
       (b)   Median $= 5.0$
       (c)   Range $= 13$

2.115  (a)   Median $= 4.505$, $Q_1 = 4.30$, $Q_3 = 4.70$      (b)   4.935
       (c)   $\bar{x} = 4.5074$, $s = .368$

2.117  (a)   Median $= 6.7$, $Q_1 = 6.4$, $Q_3 = 7.1$
       (b)   $\bar{x} = 6.74$ and $s = .463$

2.119  (b)   Not reasonable with time trend.

2.121  Mean $= 80.30$ and standard deviation $= 39.30$

2.129  (b)   $\bar{x} = 109.1$, $s = 65.8$

# CHAPTER 3

3.1 (c) The pill seems to reduce the proportions of severe and moderate nausea.

3.3 The proportion who study more than 10 hours is highest for the physical sciences. It is .70 while social sciences have the lowest proportion .48.

3.7 Relative Frequencies

| | | Major | | | |
|---|---|---|---|---|---|
| | B | H | P | S | Total |
| Male | .245 | .082 | .102 | .286 | .715 |
| Female | .122 | 0 | .082 | .082 | .286 |
| Total | .367 | .082 | .184 | .368 | 1.001 |

3.11 (b) Research hospital best for either condition.

3.15 (b) Positive—better players are paid more.

3.17 No. Can only conclude there is not a straight line relationship.

3.19 (b) $r = -.415$

3.21 (a) Figure (c)　(b) Figure (b)
　　(c) Figure (a)

3.23 $r = -.578$

3.25 $r = .822$

3.27 (b) $r = .279$

3.33 (b) $r = .978$

3.37 (b) 16 or more

3.39 (c) $\hat{\beta}_0 = .52$　$\hat{\beta}_1 = .70$

3.41 (a) $\hat{\beta}_0 = 2.143$　$\hat{\beta}_1 = 1.085$

3.43 (b) $\hat{y} = -147.3 + 1.343x$

3.47 (c) Relative Frequencies

| 2-Wheel | 4-Wheel | Total |
|---|---|---|
| .343 | .657 | 1.000 |
| .444 | .556 | 1.000 |

3.49 (a) Negative. Only so much time.
　　(c) No relation.

3.51 No. Size of fire is a major lurking variable.

3.53 (b) $r = -.313$

3.55 (c) $\hat{y} = 1.15 + .37x$

3.57 (a) $x = $ road roughness and $y = $ gas consumption

3.59 (b) $r = .649$

3.61 (b) $r = .437$

# CHAPTER 4

4.1 (a) (iv)　(d) (vi)　(f) (i)

4.3 (a) (ii)　(b) (iii)

4.5 (a) $\{0, 1\}$
　(b) $\{0, 1, \ldots, 344\}$
　(c) $\{t: 90 < t < 425.4\}$

4.7 (a) $S = \{BJ, BL, JB, JL, LB, LJ, BS, JS, LS, SB, SJ, SL\}$
　(b) $A = \{LB, LJ, LS\}$
　$B = \{JL, LJ, JS, SJ, LS, SL\}$

4.9 .1

4.11 (a) Yes.
　(b) No, because the total is less than 1.
　(c) Yes.

4.13 .2

4.15 .804

4.17 (c) $\dfrac{3}{8}$

4.19 (a) $S = \{e_1, e_2, e_3\}$ where $e_i = $ [the ticket has number $i$]; $P(e_1) = \dfrac{2}{8}$, $P(e_2) = \dfrac{3}{8}$, $P(e_3) = \dfrac{3}{8}$
　(b) $\dfrac{5}{8}$

4.21 (c) $P(A) = \dfrac{5}{36}$, $P(B) = \dfrac{1}{6}$, $P(C) = \dfrac{1}{2}$, $P(D) = \dfrac{1}{6}$

4.23 (b) $\dfrac{4}{9}$

4.25 (a) $\dfrac{1}{15}$　(b) $\dfrac{11}{15}$　(c) $\dfrac{12}{15}$

4.27 $S = \{N, YN, YYN, YYYN, YYYYN, YYYYY\}$

4.29 (b) $\dfrac{1}{7}$

4.33 (b) $P(A) = \dfrac{1}{2}$, $P(B) = \dfrac{1}{3}$

4.35 (b) (i) $AB = \{e_6, e_7\}$
(iii) $A\bar{B} = \{e_4, e_5\}$

4.37 (b) $AB = \{e_2, e_6, e_7\}$,
$P(AB) = .41$      (d) $\bar{A}\,\bar{C} = \{e_3, e_4\}$, $P(\bar{A}\,\bar{C}) = .22$

4.39 (b) $A \cup B = \{e_1, e_3, e_4\}$, $AB = \{e_3\}$

4.41 (a) $P(A) = .30$ and $P(B) = .36$
(c) $P(A \cup B) = .40$

4.43 (b) $P(A\bar{B}) = .18, P(\bar{A}B) = .32$,
$P(\bar{A}\bar{B}) = .37$

4.45 (b) $P(\bar{A}B) = .36$
(c) $P(A \cup B) = .73$
(d) $.59$

4.47 (a) $P(\bar{A}) = .65$
(b) $P(A\bar{B}) = .15$

4.49 $\dfrac{7}{18}$

4.51 (a) $P(B) = .35$

4.55 Reasonable to assume $P(A|B) > P(A)$. Dependent.

4.57 $P(B|A) = .0099$, dependent.

4.61 (a) $P(A) = .4, P(B) = .4$
(b) Dependent because $P(AB) = .1 \neq P(A)P(B)$      (c) $P(A) = P(B) = .4$, independent.

4.63 (a) $.1686 = 2073/12299$
(b) $.1988 = 1032/5192$

4.65 (a) $688$      (c) $769$

4.67 (a) $\dfrac{2}{9}$      (b) $\dfrac{7}{12}$

4.69 (a) $BC, P(BC) = 0$
(c) $\bar{B}, P(\bar{B}) = .8$

4.71 $.99991$

4.73 (b) $.896$

4.75 (a) $.55$      (b) Yes.

4.77 (a) $P(B|A) = .320$

4.79 (b) $4.8\%$      (c) $.360$

4.81 (a) $.8426$      (b) $.4488$

4.83 (a) $.058$      (b) $.362$

4.85 (a) $210$      (b) $90$

4.87 (a) $462$      (b) $210$

4.91 (a) $.491$      (b) $.084$

4.93 No.

4.97 (a) $.018$      (b) $.393$

4.99 (a) $330$      (b) $.182$

4.105 (a) $A = \{23, 24\}$
(d) $A = \{t: 0 \leq t < 500.5\}$

4.107 $A = \{p: .10 < p < 100\}$, $p$ is percent.

4.109 (a) $.5$      (b) $.75$      (c) $.167$

4.111 $.5$

4.113 (a) No.      (b) Yes.      (c) No.

4.115 (a) Either a faulty transmission or faulty brakes or both.      (b) Transmission, brakes, and exhaust system all faulty.

4.117 $P(A) = .3, P(AB) = .10$,
$P(A \cup B) = .5$

4.121 (a) $.25$      (b) $.76$

4.123 (a) $.25$      (c) $.34$

4.125 $P(A|B) = .8$, they are independent.

4.127 (a) $P(A)P(C) = .15, P(AC) = .15$, independent.
(b) $P(A\bar{B})P(C) = .1$,
$P(A\bar{B}C) = .15$, dependent.

4.129 (a) $.44$      (b) $.3846$
(c) Dependent.

4.131 (a) $.00008$      (b) $.00002$

4.133 (a) $.001312$      (b) $.000719$
(c) $.000801$      (d) $.228$

4.135 (b) (ii) $.64$ (iii) $.49$

4.137 $\dfrac{1}{165} = .0061$

4.139 (a) $.108$      (b) $.515$

4.141 (a) $.008$

# CHAPTER 5

5.1 (a) Discrete.      (b) Continuous.
(c) Continuous.

5.3 (a)

| Possible Choices | $x$ |
|---|---|
| $\{1, 3\}$ | 2 |
| $\{1, 5\}$ | 4 |
| $\{1, 6\}$ | 5 |
| $\{1, 7\}$ | 6 |
| $\{3, 5\}$ | 2 |
| $\{3, 6\}$ | 3 |
| $\{3, 7\}$ | 4 |
| $\{5, 6\}$ | 1 |
| $\{5, 7\}$ | 2 |
| $\{6, 7\}$ | 1 |

(b)

| $x$ | 1 | 2 | 3 | 4 | 5 | 6 |
|---|---|---|---|---|---|---|
| $f(x)$ | .2 | .3 | .1 | .2 | .1 | .1 |

5.9

| $x$ | 4 | 5 | 6 | 7 | 8 |
|---|---|---|---|---|---|
| $f(x)$ | .1 | .4 | .1 | .2 | .2 |

5.13  (a)

| $x$ | 1 | 2 | 3 | 4 | Total |
|---|---|---|---|---|---|
| $f(x)$ | 0 | $\frac{1}{6}$ | $\frac{2}{6}$ | $\frac{3}{6}$ | 1 |

Yes, a probability distribution.

(b)

| $x$ | 2 | 3 | 4 | 5 |
|---|---|---|---|---|
| $f(x)$ | $-\frac{1}{3}$ | 0 | $\frac{1}{3}$ | $\frac{2}{3}$ |

Not a probability distribution; $f(2)$ is negative.

5.15

| $x$ | 0 | 1 | 2 |
|---|---|---|---|
| $f(x)$ | $\frac{1}{4}$ | $\frac{1}{2}$ | $\frac{1}{4}$ |

5.17

| $x$ | 0 | 1 | 2 | 3 | 4 |
|---|---|---|---|---|---|
| $f(x)$ | .4096 | .4096 | .1536 | .0256 | .0016 |

5.19  (b)  .5

5.21  $f(2) = .2$, $f(4) = .2$, $f(6) = .2$

5.25  (a)  .94      (b)  .47      (c)  .41

5.27  (a)  .45      (b)  .25
(c)  The capacity must be increased by 2 to a total of 4.

5.29  (b)  $E(X) = 3$, $\sigma = 1$

5.31  $931.20

5.33  $\mu = 2.4$, $\sigma = .980$

5.35  (c)  $228,500 with prep cost

5.37  (a)  $E(X) = 1.381$
(b)  $\sigma = 1.318$

5.41  (b)

| $x$ | 0 | 1 | 2 | 3 | 4 | Total |
|---|---|---|---|---|---|---|
| $f(x)$ | .2401 | .4116 | .2646 | .0756 | .0081 | 1.0000 |

(c)  $E(X) = 1.2$

5.43  Median $= 3$

5.45  (a)  Trials are dependent.
(c)  Bernoulli model plausible.

5.47  Identify $S =$ yellow, $F =$ other colors.
(a)  Yes, $p = .48$      (b)  No.
(c)  No.

5.49  (a)  Yes, $P(S) = .5$
(b)  Dependent.

5.51  (a)  Bernoulli trials model is not appropriate.    (b)  Appropriate.

5.53  (a)  .2401      (b)  .0081

5.55  (a)  .03456      (b)  .3456

5.57  (c)

| $x$ | 0 | 1 | 2 |
|---|---|---|---|
| $f(x)$ | $\frac{81}{256}$ | $\frac{126}{256}$ | $\frac{49}{256}$ |

5.59  (a)  Yes, $n = 10$, $p = \frac{1}{6}$
(b)  No, because the number of trials is not fixed.

5.61  (b)  .132

5.63  (a)  .9850      (b)  .1265
(c)  .3255

5.65  (a)  .7383      (b)  .6836      (c)  3

5.69  (a)  .233      (b)  .014      (c)  .014

5.71  (a)  .837      (b)  .901

5.73  (a)  Mean $= 15$ and sd $= 1.936$
(c)  Mean $= 12$ and sd $= 2.898$

5.75  .411

5.77  Mean $= 272.5$, sd $= 11.67$

5.79  (b)  Mean $= 12.308$, sd $= 2.919$

5.81  (a)  $P[X \leq 4$ or $X \geq 13] = .072$
(b)  $P[X \leq 3$ or $X \geq 12] = .07$

5.83  (c)  The proportions for all resturants are "in control".

5.85  (a)  .157      (b)  .109      (c)  .202

5.87  (a)  .217      (c)  .083

5.89  (a)  .101      (c)  .073

5.91  (a)  .5221      (b)  .478      (c)  .474

5.93  (b)

| $x$ | 0 | 1 | 2 |
|---|---|---|---|
| $f(x)$ | .400 | .467 | .133 |

5.95    (a)   The probability distribution is given in the first two columns.

| $x$ | $f(x)$ | $xf(x)$ | $x^2f(x)$ |
|---|---|---|---|
| 0 | $\dfrac{35}{120}$ | 0 | 0 |
| 1 | $\dfrac{63}{120}$ | .525 | .525 |
| 2 | $\dfrac{21}{120}$ | .350 | .7 |
| 3 | $\dfrac{1}{120}$ | .025 | .075 |
| Total | 1.00 | .9 | 1.3 |

  (b)   $\mu = .9$ and $\sigma^2 = .49$

5.97    (a)   $\mu = 1.0$   $\sigma^2 = .6$
5.99    (a)   .002      (b)   −$0.60
5.101  (a)   .30      (b)   $\sigma = 1.342$

5.103  (a)

| $x$ | 0 | 1 | 3 |
|---|---|---|---|
| $f(x)$ | $\dfrac{2}{6}$ | $\dfrac{3}{6}$ | $\dfrac{1}{6}$ |

  (b)   1.0

5.105  (a)

| $x$ | −15 | −5 | 5 | 15 |
|---|---|---|---|---|
| $f(x)$ | .1458 | .3936 | .3543 | .1063 |

  (b)   −$0.79      (c)   No.
5.107  (a)   $F(3) = .44, F(4) = .72,$
  $F(5) = .90, F(6) = 1.00$
5.109  (b)   Plausible.      (c)   Does not apply.
5.111  .125
5.113  .140
5.115  (a)   Binomial distribution with $n = 6, p = .4$      (b)   .821, .047, 2.4
5.117  (b)   1.759      (c)   .221
5.119  $P[X \leq 10] = .048$. This small probability casts doubt on the claim that $p = .7$
5.121  (d)   $E(X) = 2.0$ and $\text{Var}(X) = 1.2$
5.123  $P[X \geq 10] = .048,$
  $P[X = 10] = .031$
5.125  (a)   .598      (c)   $\mu = 3.2$ and $\sigma = 1.6$
5.127  (b)   .180      (c)   .156

# CHAPTER 6

6.1    (c)   Probability density function.
6.3    The interval 1.5 to 2 has higher probability.
6.5    Median = 1.414
6.9    (c)   $Z = \dfrac{X - 161}{5}$
6.11   (b)   $Z = \dfrac{X - 90}{12}$
6.13   (a)   .7357      (c)   .1515
6.15   (c)   .8907      (d)   .0510
6.17   (b)   .8164      (d)   .0695
6.19   (a)   $z = -.842$      (c)   $z = .97$
6.21   (a)   $z = -.93$      (d)   $z = .753$
6.23   (a)   .6293      (d)   $z = 1.881$
6.25   (a)   .3446      (c)   .2119
  (e)   .9093
6.27   (a)   $b = 67.64$
6.29   (a)   .0668      (c)   .9198
6.31   .8468
6.33   .2024
6.35   (a)   .0334
6.37   (a)   (iii) .3830      (b)   16.5 to 17.5
6.41   (a)   With continuity correction .2105.
  (c)   With continuity correction .8976.
6.43   No. $n(1 - p) = 2.4$ small.
6.47   (a)   .0923      (b)   .0728
6.49   .2784 using continuity correction
6.51   .600 no continuity correction
6.55   Large values of volume are too large for normal distribution.
6.57   Approximately a straight line pattern. Normal distribution reasonable.
6.59   (a)   Smallest values not small enough and largest a little too large for normal distribution.
  (b)   Normal distribution is reasonable for the fourth root.
6.61   (a)   0.5      (b)   First quartile = .25, second quartile = .5, and third quartile = .75
6.65   (a)   $z = -1.38$      (c)   $z = .64$
6.67   (a)   .2676      (c)   .0026

6.69 (a) .8371 (c) .8057
(e) .6096
6.71 (a) .2000 (c) .2047
6.73 .1056
6.75 (a) 0.5793
6.77 (b) 305.3 ounces
6.79 (b) (i) .1537 using continuity correction
6.81 (a) .0281 using continuity correction
6.83 (a) .0401 using continuity correction
6.85 (b) .0418 with continuity correction
6.87 Approximately a straight line pattern. Normal distribution reasonable for square root.

# CHAPTER 7

7.1 (c) Parameter. (d) Statistic.

7.5 (b)

| $\bar{x}$ | Probability $f(\bar{x})$ |
|-----------|--------------------------|
| 3 | $\frac{1}{9}$ |
| 4 | $\frac{2}{9}$ |
| 5 | $\frac{3}{9}$ |
| 6 | $\frac{2}{9}$ |
| 7 | $\frac{1}{9}$ |
| Total | 1 |

7.7 No, very best pictures become sample.
7.9 (a) $X = 4$ for 6 dots
7.11 Solid line is population density. Thicker tails and larger variance.
7.13 (a) $E(\bar{X}) = \mu = 83$, sd$(\bar{X}) = 19$
7.15 (a) .3098 (c) .0775
7.17 $E(\bar{X}) = 5$, sd$(\bar{X}) = \dfrac{2}{\sqrt{3}}$
7.19 (b) $E(\bar{X}) = 1.0$ (c) mean = 1.0, variance = $\dfrac{.6}{36} = .0167$

7.21 (a) $E(\bar{X}) = 115$
(b) sd$(\bar{X}) = 8.98$
(c) $N(115, 8.98)$
7.23 (a) .1587 (c) .0786
7.25 (a) $\bar{X}$ is approximately $N(51,000, 500)$. (b) .1587
7.29 .681
7.33 (b) $\sigma = 1.136$
7.35 (b) More variability in two months.
7.37 No. Means are different.
7.39 (b) Sampling distribution of $R$:

| Value of $R$ | Probability |
|--------------|-------------|
| 0 | 4 / 16 |
| 2 | 6 / 16 |
| 4 | 4 / 16 |
| 6 | 2 / 16 |
| Total | 1 |

7.41 (c) mean = 2 variance = $\dfrac{.4}{81} = .0049$
7.43 (b) Exactly a normal distribution.
(c) .7066
7.45 (a) $N(12.1, (3.2)/3)$
(c) About 26% smaller than 10.
7.47 (a) mean = 1.9, sd = .20 hour
(b) Approximately $N(1.9, .20)$
7.49 (a) $E(\bar{X}) = 4.660$ and sd$(\bar{X}) = .1289$
(b) Approximately $N(4.660, .1289)$
7.51 (b) $\sigma = .781$ (d) .095
7.53 (b) 305.29

# CHAPTER 8

8.1 (a) S.E. = 5.814, 98% error margin = 13.55
(c) S.E. = 8.165, 90% error margin = 13.43
8.3 $\bar{x} = 63.6$, estimated S.E. = 8.167
$\bar{x} = 79.4$, estimated S.E. = 6.745
8.5 $\bar{x} = 95.74$, 90% error margin = 5.73
8.7 (a) 2.143 (b) 3.525

8.9 (a) $n = 246$ (b) $n = 171$

8.13 $n = 68$

8.15 $(2.03, 3.14)$

8.19 $(28.94, 31.46)$ grams

8.21 $(17.1, 19.5)$ days

8.23 (a) $(3.604, 3.996)$

8.25 $(8.00, 9.20)$ miles

8.27 $(88.91, 102.57)$ dollars

8.29 $(-.009, .005)$

8.31 $(1.521, 1.909)$ centimeters

8.33 (b) In long run, covers with proportion .95.

8.35 (b) Yes, in the middle.
(d) No.

8.37 (a) $H_0: \mu = 14$, $H_1: \mu < 14$ days
(b) $H_0: \mu = 6.50$, $H_1: \mu > 6.50$ dollars

8.39 (a) Correct decision if $\mu = 14$. Wrong decision if $\mu < 14$, Type II error.
(b) Correct decision if $\mu = 6.50$ dollars. Wrong if $\mu > 6.50$, Type II error.

8.43 (a) $\alpha = .025$ (b) $c = 41.1$

8.45 (a) $Z = 2.26$., fail to reject $H_0$
(b) Fail to reject $H_0$ when $\mu \neq 2.0$.

8.47 (a) Reject $H_0$. (c) .01
(e) .005

8.49 (a) $Z = 1.58$, reject $H_0$.
(b) Falsely reject $H_0: \mu = 3.5$.

8.51 (a) $H_0: \mu = 183$, $H_1: \mu > 183$ days
(b) $Z = 1.88$, reject $H_0$
(c) Reject $H_0$ when $\mu = 183$.

8.53 $H_0: \mu = 3.0$, $H_1: \mu > 3.0$, $R: Z \geq 1.645$ Observed $z = 1.44$; $H_0$ is not rejected.

8.55 $H_0: \mu = 3000$, $H_1: \mu \neq 3000$, $R: |Z| \geq 1.96$ Observed $z = 2.13$; $H_0$ is rejected at $\alpha = .05$.

8.57 (a) (i) $p =$ proportion of adult population for which reading is a favorite leisure time activity. (ii) $\hat{p} = .299$ (iii) .029
(b) (i) $p =$ proportion of pet owners who buy their pets holiday presents. (ii) $\hat{p} = .666$ (iii) .044

8.59 (a) $n = 632$ (b) $n = 752$

8.61 (b) $(.49, .53)$.

8.63 (a) $\hat{p} = .287$ (b) S.E. $= .0467$
(c) $(.178, .396)$

8.65 (a) $H_0: p = .26$ $H_1: p > .26$

8.67 (c) (i) $H_0: p = .60$ versus $H_1: p \neq .60$ (ii) $\dfrac{\hat{p} - .6}{.0558}$
(iii) $R: |Z| \geq 2.33$

8.69 (a) $H_0: p = .30$ versus $H_1: p > .30$
(b) $R: Z = \dfrac{\hat{p} - .3}{\sqrt{.3 \times .7 / 60}} \geq 1.645$
(c) Fail to conclude that more than 30% have posters.
(d) Reject $H_0$.

8.71 (b) $P - $ value $= .0244$

8.73 $H_0: p = .75$ and $H_1: p > .75$
$Z = \dfrac{\hat{p} - .75}{\sqrt{.78 \times .22 / 5000}}$
$R: Z \geq 1.645$. Observed $z = 4.90$, reject $H_0$.
$P - $ value less than .0001.

8.75 (a) $(.47, .55)$ (b) $(3889, 4490)$

8.77 $(.58, .85)$

8.79 (a) $\bar{x} = 85.022$
(b) S.E. $= 1.201$
(c) 2.354

8.81 (b) A factor of 16.

8.83 (a) 2.8 (b) $(124.6, 129.2)$

8.85 (a) Correct

8.87 The alternative hypothesis $H_1$ is the assertion that is to be established; its opposite is the null hypothesis $H_0$.
(a) $H_0: \mu = 50$, $H_1: \mu < 50$
(d) $H_0: \mu = 16$, $H_1: \mu \neq 16$

8.89 (a) $\mu =$ mean number of days toothpaste lasts.
(c) $R: Z = \dfrac{\sqrt{n}\,(\bar{X} - 30.5)}{S} \geq c$
(d) $Z = 2.51$ reject $H_0$ (e) .006

8.91 (b) $P$–value $.014 / 2 = .007$ reject $H_0$.

8.93 (b) $\sigma = .781$ (c) $H_0: \mu = .7$ versus $H_1: \mu > .7$, $Z = \dfrac{\bar{X} - .7}{S / \sqrt{64}}$, $R: Z \geq 1.645$. Observed $z = 2.80$; $H_0$ is rejected at $\alpha = .05$.

8.95 (a) $\hat{p} = .7619$ (b) .091

8.97 (a) $(.190, .255)$

8.99  (a)  $H_0: p = .10$  $H_1: p > .10$

(b)  $z = \dfrac{\hat{p} - .10}{\sqrt{.10 \times .90 / 152}}$

(c)  $R: z \geq 1.645$    (e)  .033

8.101  (a)  $H_0: p = .8$  $H_1: p \neq .8$

$Z = \dfrac{\hat{p} - .8}{\sqrt{(.8 \times .2) / 200}}$,

$R: |Z| \geq 1.96$, Observed $z = -4.24$, $H_0$ is rejected at $\alpha = .05$.

$P$-value $P[|Z| \geq 4.24] < .0001$ is extremely small. The genetic model is strongly contradicted.

(b)  $(.62, .74)$

8.103  $(14.93, 23.51)$

8.105  (a)  Larger must cover more often, $(.260, .303)$.

(b)  Reject $H_0$ at level $\alpha = .05$.

8.107  (a)  $(1.55, 1.88)$    (b)  Observed $z = -2.46$, reject $H_0: \mu = 1.9$.

8.109  $(90.54, 106.16)$

8.111  (a)  $(2.63, 5.52)$

# CHAPTER 9

9.1  (a)  1.943    (b)  −2.228

9.3  (b)  3.365    (c)  −1.717

9.5  (a)  1.895    (b)  2.120

9.7  (b)  Between .025 and .05.

(c)  Between .05 and .10.

9.9  (a)  $(72.75, 79.45)$

(b)  Center = 76.1, length = 6.7.

(c)  Usually different since $S$ is random.

9.11  $(128.17, 147.03)$

9.13  $(3.2, 4.0)$ kg

9.15  (a)  $\bar{x} = 42.4$, $s = 5.98$

(b)  $(37.7, 47.1)$

9.17  $(5.372, 5.594)$

9.19  $H_0: \mu = 3.8$, $H_1: \mu < 3.8$

$T = \dfrac{\bar{X} - 3.8}{S / \sqrt{14}}$, d.f. = 13

$R: T \leq -1.771$. Observed $t = -1.86$, reject $H_0$ at $\alpha = .05$.

9.21  (a)  Cannot tell, mean unknown.

(b)  In the long run, 95% of the intervals obtained using this procedure will cover.

9.23  (b)  $(23.38, 29.86)$ feet

9.25  $H_0: \mu = 128$, $H_1: \mu > 128$

$T = \dfrac{\bar{X} - 128}{S / \sqrt{20}}$, d.f. = 19,

$t_{.05} = 1.729$, $R: T \geq 1.729$.

Observed $t = 2.13$, $H_0$ is rejected at $\alpha = .05$.

9.27  $H_0: \mu = 83$, $H_1: \mu \neq 83$ pounds.

$T = \dfrac{\bar{X} - 83}{S / \sqrt{8}}$, d.f. = 7, $t_{.025} = 2.365$, $R: T \leq -2.365$ or $T \geq 2.365$

Observed $t = -2.566$, $H_0$ is rejected at $\alpha = .05$.

9.29  (b)  $(83, 89)$

(c)  Normal population

9.31  (b)  Reject $H_0$ when it is true.

(d)  When $H_0$ prevails, $H_0$ will be rejected in about proportion .05 of the times.

9.33  (a)  $H_0$ is rejected at $\alpha = .10$.

9.35  (a)  $(1.28, 12.28)$    (c)  Observed $t = -3.53$, $H_0$ is rejected at $\alpha = .05$.

9.39  (a)  $\chi^2_{.10} = 22.31$

(b)  $\chi^2_{.95} = 13.85$

(c)  $\chi^2_{.10} = 51.81$

(d)  $\chi^2_{.95} = 2.73$

9.41  (a)  .773    (b)  $(.566, 1.220)$

9.43  $H_0: \sigma = .6$, $H_1: \sigma < .6$

$\chi^2 = \dfrac{(n - 1) S^2}{(.6)^2}$, d.f. = 39,

$R: \chi^2 \leq \chi^2_{.95} = 25.7$. Observed $\chi^2 = 24.4$, $H_0$ is rejected at $\alpha = .05$.

9.45  $(1.56, 4.15)$

9.47  (a)  $(15.99, 27.60)$

9.49  $(4.70, 10.83)$

9.51  (a)  1.895    (c)  −1.895

9.55  $(39.4, 54.6)$

9.57  (a)  $\bar{x} = 3.007$, 95% error margin = .243    (b)  $(2.75, 3.25)$

9.59 (a) $H_0: \mu = 42, H_1: \mu < 42$

(b) $T = \dfrac{\overline{X} - 42}{S/\sqrt{21}}$, d.f. $= 20$

(c) $R: T \leq -t_{.01} = -2.528$

(d) Observed $t = -3.23$,

$H_0$ is rejected at $\alpha = .01$.

9.61 $R: |T| \geq 2.681$, d.f. $= 12$

Observed $t = 2.74$, reject $H_0$ with

$\alpha = .02$.

9.63 (a) $H_0: \mu = 65, H_1: \mu > 65$

(b) $T = \dfrac{\overline{X} - 65}{S/\sqrt{9}}$, d.f. $= 8$,

$t_{.05} = 1.860, R: T \geq 1.860$.

(c) Observed $t = .18$, fail to reject

$H_0$. The claim is not demonstrated.

9.65 $H_0: \mu = 1500, H_1: \mu > 1500$

$T = \dfrac{\overline{X} - 1500}{S/\sqrt{5}}$, d.f. $= 4$,

$t_{.05} = 2.132, R: T \geq 2.132$.

Observed $t = \dfrac{1620 - 1500}{90/\sqrt{5}} = 2.98$;

$H_0$ is rejected at $\alpha = .05$.

9.67 (a) $(423.1, 456.1)$

(b) $t = -1.99$, reject $H_0$ at $\alpha = .10$

9.69 (a) $\chi^2_{.05} = 12.59$

(c) $\chi^2_{.95} = 1.64$

9.71 (a) $H_0: \mu = 55.0$ versus $H_1: \mu \neq$

$55.0$    (b) $T = \dfrac{\overline{X} - 55.0}{S/\sqrt{n}}$

(c) $R: T \geq c$ or $T \leq -c$

(d) $t = 2.75$ and $c = 2.228$ so reject

$H_0$ at level $\alpha = .05$.

9.73 (a) $\chi^2 = \dfrac{(n-1)S^2}{324}$, d.f. $= 9$,

$R: \chi^2 \leq 3.33$ Observed $\chi^2 = 3.25$;

$H_0$ is rejected.

9.75 (a) Cannot tell, mean unknown.

(b) In the long run, 95% of the intervals obtained using this procedure will cover.

9.77 $t = 2.25$, reject $H_0$.

# CHAPTER 10

10.1 (a) First group, using first letter of subject, $\{B, C\}, \{B, E\}, \{B, H\}, \{B, P\}$
$\{C, E\}, \{C, H\}, \{C, P\}$
$\{E, H\}, \{E, P\} \{H, P\}$

(b) 10

10.3 (a) Using first letter of name,
$\{(S, G), (T, E)\}\{(S, G), (T, R)\}$
$\{(S, G), (E, R)\}\{(S, J), (T, E)\}$
$\{(S, J), (T, R)\}\{(S, J), (E, R)\}$
$\{(J, G), (T, E)\}\{(J, G), (T, R)\}$
$\{(J, G), (E, R)\}$    (b) There are three sets each consisting of three pairs.

10.5 (a) (i) Yes.    (b) Yes.

10.7 (a) $\overline{x} - \overline{y} = 7$, (b) S.E. $= 2.48$

(c) $(2.1, 11.9)$

10.9 $H_0: \mu_1 - \mu_2 = 0, H_1: \mu_1 - \mu_2 \neq 0, R: |Z| \geq 1.96$. Observed $z = -4.17$; null hypothesis is rejected at $\alpha = .05$.

10.11 Let the population mean sentence length be $\mu_1$ for magazine 1 and $\mu_2$ for magazine 2.

(a) $H_0: \mu_1 - \mu_2 = .2, H_1: \mu_1 - \mu_2 > 2$

(b) $Z = \dfrac{\overline{X} - \overline{Y} - 2}{\sqrt{\dfrac{S_1^2}{n_1} + \dfrac{S_2^2}{n_2}}}$,

$R: Z \geq 1.645$

(c) Observed $z = 1.69, H_0$ is rejected, $P$-value $= .0455$.

10.13 $H_0$ is rejected at $\alpha = .05$.

10.15 $z = 2.945$ and $P$-value $= .002$

10.17 $H_0: \mu_1 - \mu_2 = 0, H_1: \mu_1 - \mu_2 > 0, Z = \dfrac{\overline{X} - \overline{Y}}{\sqrt{\dfrac{S_1^2}{n_1} + \dfrac{S_2^2}{n_2}}}$ Observed $z = 2.91, P$-value $= P[Z \geq 2.91] = .0018$. Strong evidence in support $H_1$.

10.19 (a) $s^2_{pooled} = 3.20$    (b) $1.789$

(c) $t = -1.46$

10.21 (a) $s^2_{pooled} = 2.400$    (b) $t =$ 3.35 and $P$-value $= .002$, reject $H_0: \mu_1 - \mu_2 = 0$
(c) $(.77, 3.23)$

10.23 (a) $t = -1.81$, d.f. $= 18$, reject $H_0: \mu_1 - \mu_2 = 0$
(c) $(-9.09, .69)$

10.25 (a) $Z$ test.    (b) $t$ test with pooling.
(d) $Z$ test.

10.27 (a) Diet and other exercise.

10.29 $(.94, 1.49)$

10.31 We drew slips with $\alpha, \beta, \tau$ so group 1 is { alpha, beta, tau }.

10.35 (a) $t = 2.07$    (b) d.f. $= 5$

10.37 (a) $t = 1.46$, the claim is not established at $\alpha = .05$.

10.39 (a) Observed $t = 2.87$, $H_0$ is rejected at $\alpha = .05$.    (b) $(1.52, 7.90)$.

10.41 $H_0: \mu_D = 0$, $H_1: \mu_D \neq 0$, $T = \dfrac{\bar{D}}{S_D / \sqrt{n}}$, d.f. $= 8$, $R: |T| \geq 2.306$
Observed $t = 1.49$, $H_0$ is not rejected. The difference is not significant.

10.43 (a) $t = 8.25$, d.f. $= 9$, $P$-value less than $.001$.    (b) $(60.27, 60.47)$

10.45 (a) $(.17, .36)$    (b) $z = 4.56$ reject $H_0$ at level $\alpha = .05$, $P$-value less than $.0001$.

10.47 $H_0: p_1 = p_2$, $H_1: p_1 < p_2$ (suffix 1 refers to the HRL group). Observed $z = -4.81$, the $P$-value less than $.0001$.

10.49 $H_0: p_1 = p_2$, $H_1: p_1 \neq p_2$ (suffix 1 refers to "with carbolic acid"), $R: |Z| \geq 1.96$. Observed $z = 2.91$; $H_0$ is rejected at $\alpha = .05$. $P$-value $= .0036$.

10.53 (a) $H_0: p_1 = p_2$, $H_1: p_1 > p_2$ (suffix 1 refers to "smokers"). Observed $z = 4.29$, $P$-value less than $.00001$. The conjecture is strongly supported.

10.55 (b) $(-.14, -.08)$

10.57 (a) $(.17, .38)$
(b) $(-.045, .103)$

10.59 (a) $(3.89, 6.11)$

10.61 (a) $H_0: \mu_1 - \mu_2 = -150$ versus $H_1: \mu_1 - \mu_2 < -150$    (c) $z = -1.79$, so reject $H_0$ at $\alpha = .05$, $P$-value $= .0367$.

10.63 (a) $(.13, 1.09)$    (b) $(4.31, 4.97)$

10.65 (a) $4.857$
(b) $t = .68$, d.f. $= 7$

10.67 $(-10.9, 42.9)$, d.f. $= 19$

10.69 (a) $H_0: \mu_1 - \mu_2 = 0$, $H_1: \mu_1 - \mu_2 \neq 0$ (suffix 1 refers to "city A"), $R: |Z| \geq 2.33$. Observed $z = -4.97$; reject $H_0$.
(b) $(-7.93, -2.87)$

10.71 (a) Do not reject $H_0$.
(b) $(-.026, .630)$

10.73 (a) $t = 1.27$, fail to reject $H_0$.

10.75 (a) $t = 2.00$, fail to reject $H_0$.
(b) $(.68, 18.65)$

10.77 Denote growth with additive $(x)$.
$$t = \frac{d}{s_d / \sqrt{n}} = 2.93 \quad \text{d.f.} = 9$$
greater growth established at $\alpha = .05$.

10.79 (a) $t = -1.95$, fail to reject $H_0$.

10.81 $H_0: p_1 = p_2$, $H_1: p_1 < p_2$ (suffix 1 refers to "seeded"), $R: Z \leq c$.
$$Z = \frac{\hat{p}_1 - \hat{p}_2}{\sqrt{\hat{p}\hat{q}}\sqrt{\dfrac{1}{n_1} + \dfrac{1}{n_2}}} \quad \text{Observed}$$
$z = -1.77$, $P$-value $= P[Z \leq -1.77] = .0384$. Fairly strong evidence in support of the conjecture.

10.83 $H_0: p_1 = p_2$, $H_1: p_1 > p_2$ (suffix 1 refers to the males). With $\alpha = .05$, $R: Z \leq -1.645$. Observed $z = -1.44$. Fail to reject $H_0$ at $\alpha = .05$

10.85 (a) Fail to reject $H_0$.
(b) $(-509, -3)$ Do not pool.

10.87 (a) Do not pool.
(b) $(-58.0, -16.0)$

10.89 $z = 2.961$, reject $H_0$.

10.91 $z = 13.13$, reject $H_0$.

# CHAPTER 11

11.1 Intercept $= 2$, slope $= 3$

11.3 (a) $x =$ duration of training, $y =$ measure of performance
(c) $x =$ level of humidity, $y =$ growth rate of fungus

11.5 $\beta_0 = 6, \beta_1 = -3$, and $\sigma = 3$

11.7 (a) $E(Y) = 1$, sd$(Y) = 4$

11.11 (b) No, only the mean is higher.

11.13 (d) The fitted line is $\hat{y} = 8.5 - 1.5x$.

11.15 (b) SSE $= 5.50$
(c) $s^2 = 1.375$

11.17 (b) $\hat{\beta}_1 = .962, \hat{\beta}_0 = 2.154$

11.19 (a) $\hat{y} = 3.07 + 12.0x$
(b) SSE $= 216.9$
(c) $s^2 = 18.1$

11.21 (a) $\hat{\beta}_1 = .5033, \hat{\beta}_0 = 89.61$
(c) $s^2 = 31.37$

11.25 (a) $\hat{\beta}_0 = 3.50, \hat{\beta}_1 = 2.50$, $s^2 = 3.833$
(b) $H_0: \beta_1 = 0, H_1: \beta_1 \neq 0$, $R: |T| \geq t_{.025} = 3.182$, d.f. $= 3$ Observed $t = 4.04$, reject $H_0$ at the 5% level.
(c) $(8.94, 13.06)$.

11.27 $(.530, 4.470)$

11.29 (b) $\hat{y} = 41.58 + .8694x$
(c) $(.72, 1.02)$

11.31 (a) $(.524, .595)$
(b) $(.426, .694)$

11.33 (a) $Y = \beta_0 + \beta_1 x + e$ and fit $\hat{y} = 994 + .10373x$ with sd $(e) = s = 299.4$. (b) Observed $t$-ratio $= 3.48$ with $P$-value .002. Reject $H_0: \beta_1 = 0$ at the 5% level.

11.35 (a) $Y = \beta_0 + \beta_1 x + e$ and fit $\hat{y} = .3381 + .83099x$ with sd $(e) = s = .1208$. (b) Observed $t$-ratio $= 9.55$ with $P$-value less than .001. Reject $H_0: \beta_1 = 0$ at 5% level.

11.37 (a) $\hat{y} = 26.31 + .5377x$
(c) $(27.67, 29.25)$

11.39 (a) $r^2 = .517$

11.41 (a) $r^2 = .868$ (b) same

11.43 (a) $r^2 = .956$ (b) $r = .978$

11.45 Proportion explained $= r^2 = .799$.

11.49 $x =$ (leaf length) $\times$ (leaf width) is the area of a rectangle containing leaf. Slope should be less than one.

11.51 (b) SSE $= 15.66$
(c) $s^2 = 2.24$

11.53 (a) $\hat{y} = 73.18 - 2.214x$
(b) $H_0: \beta_1 = -2, H_1: \beta < -2$, $R: T \leq t_{.05} = -1.771$, d.f. $= 13$. Observed $t = -.54$, fail to reject $H_0$.

11.55 (b) $H_0: \beta_1 = 0, H_1: \beta_1 > 0$, $R: T \geq t_{.05} = 1.943$, d.f. $= 6$. Observed $t = 7.44$, reject $H_0$.
(d) $(924.2, 1031.0)$ dollars

11.57 (b) $\hat{y} = 29.26 - 2.408x$
(c) $(-3.04, -1.77)$

11.59 $r = -.952$, proportion .906 explained.

11.61 (a) $r = .649, r^2 = .421$
(b) $r = .279, r^2 = .078$
(c) $r = .773, r^2 = .537$

11.63 (a) $\hat{y} = -87.17 + 1.2765x$
(b) Observed $t = 5.92$, d.f. $= 16$; reject $H_0: \beta_1 = 0$.

11.65 (b) MARINGRW $= 478 - 0.236 \times$ FRESHGRW

11.67 (a) $r^2 = .886$ (b) 115.4 mph

# CHAPTER 12

12.1 (b) $\hat{y} = 3.92 - .53x$ (c) $r^2 = .842$, but relation is not a straight line.

12.3 (a) $y' = \dfrac{1}{y^{1/3}}, x' = x$

12.5 (b) No. beetles $= 6.14 - .899 \times \log_e$ (distance)
(c) $(-1.52, -.28)$

12.7 $E(Y) = -13$

12.9 (a) $(-1.93, -1.19)$
(b) $R: T > t_{.05} = 1.68$, d.f. $= 48$ $(.0156 - .0125)/.00161 = 1.93$ reject $H_0$

12.11 (a) $\hat{\beta}_1 = -3.22, \hat{\beta}_2 = -.0207$, and $\hat{\beta}_0 = 45.3$

(c)  .795

(d)  11.58

12.13 (a)  $R: |T| > t_{.025} =$ 2.086,  d.f. $= 20$, $t =$ $-3.2243 / .4562 = -7.07$ reject $H_0$

(b)  $t = -2.41$, reject $H_0$

12.15 (b)  $\hat{y} = 46.55 - 11.77 \log_{10}(x)$, $r^2 = .749$

(c)  $(-17.66, -5.88)$

12.17 (a)  $\widehat{\log_{10} y} = -1.16 + .0305 x$, $r^2 = .988$

12.19 (a)  $(.235, .446)$      (b)  $R: |T| >$ $t_{.025} = 2.110$, d.f. $= 17$. $t = (.6532 - .5000) / .0632 =$ $2.42$ reject $H_0$

12.21 (b)  The residuals tend to increase over time. Errors dependent.

12.25 (a)  $\hat{y} = 81.2 + 0.998x +$ $0.00993 x^2$

(b)  .998

(c)  $R: T > t_{.05} = 1.860$ d.f. $= 8$ $t = .9983 / .1389 = 7.19$,  reject $H_0$

12.27 (a)  $C3 = -.167 + .237 C1$ $r^2 = .925$      (c)  $(9.67, 11.32)$

12.29 (a)  $\hat{y} = 50.4 + .1907 x_2$ and $r^2 = .03$

(c)  Even three variables do not predict well.

12.33 $\hat{y} = 9.999 + .155$ (gender) $+$ $.9015$ (initial no.) but $P$-value for gender is large.

# CHAPTER 13

13.1  Observed $\chi^2 = 13.90$, d.f. $= 5$, $\chi^2_{.05} = 11.07$. The model of a fair die is contradicted.

13.3  Observed $\chi^2 = 2.00$, d.f. $= 3$, $\chi^2_{.05} = 7.81$. The model is not contra-dicted.

13.5  Observed $\chi^2 = 17.336$, d.f. $= 3$, $\chi^2_{.025} = 9.35$. Strong evidence of mis-labeling.

13.7  Observed $\chi^2 = 12.44$, d.f. $= 5$, $\chi^2_{.05} = 11.07$. Reject $H_0$ at $\alpha = .05$.

13.9  (a)

| $x$ | 0 | 1 | 2 | 3 |
|-----|-----|-----|-----|-----|
| Probability | .216 | .432 | .288 | .064 |

(b)  Observed $\chi^2 = 1.098$, d.f. $= 2$, $\chi^2_{.05} = 5.99$. The binomial model is not contradicted.

13.11 (c)  With $\alpha = .05$ and d.f. $= 1$, $\chi^2_{.05} = 3.84$. Observed $\chi^2 = 9.298$, reject $H_0$.

13.13 Observed $\chi^2 = 5.580$, d.f. $= 3$, $\chi^2_{.05} = 7.81$. The two brands are not shown to be significantly different in quality.

13.15 (a)  Observed $\chi^2 = 1.848$, d.f. $= 1$, $\chi^2_{.05} = 3.84$. Fail to reject $H_0$.

(b)  No. Unable to prove different.

13.17 (a)  Observed $\chi^2 = 13.787$, d.f. $=$ $1$, $\chi^2_{.05} = 3.84$. Reject $H_0$. $P$-value less than .001.      (b)  Largest $\chi^2$ contribu-tion: large proportion of thin lenses that fail test.

13.21 Observed $\chi^2 = 15.734$, d.f. $= 2$, $\chi^2_{.05} = 5.99$. Reject $H_0$. The appeals decision and type of representation are dependent.

13.23 Observed $\chi^2 = 27.847$, d.f. $= 2$, $\chi^2_{.01} = 9.21$. Reject the null hypothesis of independence at $\alpha = .01$.

13.25 Observed $\chi^2 = 4.134$, d.f. $= 2$, $\chi^2_{.05} = 5.99$. Fail to reject the null hypothesis of independence.

13.27 (b)  Observed $\chi^2 = 16.238$, d.f. $= 1$, $\chi^2_{.01} = 6.25$. Reject $H_0$.

13.29 Observed $\chi^2 = 16.44$, d.f. $= 9$, $\chi^2_{.05} = 16.92$. Fail to reject $H_0$.

13.31 (b)  Observed $\chi^2 = 9.780$, d.f. $= 3$, $\chi^2_{.05} = 7.81$. Reject $H_0$.

(c)  $p_1: (.53, .79)$

13.33 Observed $\chi^2 = 48.242$, d.f. $= 1$, $\chi^2_{.01} = 6.63$. The null hypothesis of homogeneity is rejected even at $\alpha = .01$.

13.35 (a)  Observed $\chi^2 = 12.053$ statistic, d.f. $= 3$, $\chi^2_{.05} = 7.81$. Reject $H_0$.

Drugs are different.    (b)  Drug 1: ( .32, .55 ), Drug 3: ( .26, .48 )

13.37  (a)  Observed $\chi^2 = 0$. Incredible!, d.f. $= 1$, $\chi^2_{.01} = 6.63$. Fail to reject $H_0$ with $\alpha = .01$.    (b)  No. We only failed to prove dependence.

13.39  Observed $\chi^2 = 0.153$, d.f. $= 2$, $\chi^2_{.05} = 5.99$. Not significant.

13.41  Observed $\chi^2 = 2.764$, d.f. $= 1$, $\chi^2_{.05} = 3.84$. Fail to reject $H_0$.

13.43  Observed $\chi^2 = 18.338$, d.f. $= 1$, $\chi^2_{.01} = 6.63$. Reject $H_0$.

# CHAPTER 14

14.1    (b)–(d)

ANOVA Table

| Source | Sum of Squares | d.f. |
|---|---|---|
| Treatment | 28 | 3 |
| Error | 20 | 4 |
| Total | 48 | 7 |

14.3    (d)

ANOVA Table

| Source | Sum of Squares | d.f. |
|---|---|---|
| Treatment | 32 | 2 |
| Error | 22 | 8 |
| Total | 54 | 10 |

14.5

ANOVA Table

| Source | Sum of Squares | d.f. |
|---|---|---|
| Treatment | 30 | 3 |
| Error | 26 | 8 |
| Total | 56 | 11 |

14.7

ANOVA Table

| Source | Sum of Squares | d.f. |
|---|---|---|
| Magazines | 4414.80 | 2 |
| Error | 148196.39 | 57 |
| Total | 152611.19 | 59 |

14.9  (a)  $F_{.10}(3, 5) = 3.62$
(b)  $F_{.10}(3, 10) = 2.73$

14.11  $F_{.05}(5, 30) = 2.53$ and $F = 2.46$, fail to reject $H_0$.

14.13  $F_{.05}(2, 9) = 4.26$ and $F = 8.26$, reject $H_0$.

14.15  $F_{.05}(2, 98) = 3.09$ and $F = 5.63$, reject $H_0$. There is some difference in mean travel expenses.

14.17  (a)  With 26 d.f., $t_{.0083} = 2.559$
(b)  With d.f. $= 26$, $t_{.005} = 2.779$

14.19  (a)  $t_{.00833} = 2.45$ with 75 d.f. For $\mu_1 - \mu_2$: $t$ interval .90 $\pm$ .905,
(b)  multiple-$t$ interval .90 $\pm$ 1.33

14.21  $\dfrac{t_{.05}}{t_{.005}} = .595$ for 15 d.f.

14.25  (b)  Treatment SS $= 56$, Block SS $= 78$, Residual SS $= 64$    (c)  Treatment d.f. $= 2$, Block d.f. $= 3$, and Residual d.f. $= 2 \cdot 3 = 6$

14.27  Fail to reject equal treatment means. Blocks not significant.

| Source of Variation | Degrees of Freedom | Sum of Squares | Mean Square | F |
|---|---|---|---|---|
| Treatment | 2 | 56 | 28 | 2.62 |
| Blocks | 3 | 78 | 26 | 2.44 |
| Error | 6 | 64 | 10.667 | |
| Total | 11 | 198 | | |

14.29  (a)  Randomize the position of the loaves in the oven.
(b)  $F_{.05}(2, 8) = 4.46 > 3.92$, no significant treatment difference. $F_{.05}(4, 8) = 3.84 < 5.31$, block effects are significant.

14.31 (a) $F_{.05}(5, 25) = 2.603 <$ 106.79, treatments are highly significant. $F_{.05}(5, 25) < 6.11$, block effects are significant.

14.33 The residual or error array is

$$y_{ij} - \bar{y}_i$$

$$\begin{bmatrix} 0 & -1 & 2 & -1 & & \\ 3 & -2 & 0 & 1 & -2 & \\ -1 & 2 & 4 & -3 & 1 & -3 \end{bmatrix}$$

14.35 (a) $t_{.00833} = 2.44$

$\mu_T - \mu_B$: $(-7.56, -1.14)$

$\mu_T - \mu_N$: $(-6.29, -.07)$

$\mu_B - \mu_N$: $(-2.42, 4.76)$

(b) Yes. Signing at top reduces dishonest self-reporting.

14.37 $F = 20.85$ with $P$-value less than .0001.

14.39 (b) Treatment SS $= 56$, Block SS $= 138$, Residual SS $= 32$

(c) Treatment d.f. $= 2$, Block d.f. $= 3$, Residual d.f. $= 2 \cdot 3 = 6$

# CHAPTER 15

15.1 (a) Probability $= \dfrac{1}{15}$ for each rank collection.

(b) $P[W_B = 5] = \dfrac{2}{15}$

15.3 (a) $P[W_S \geq 39] = .063$

(c) $c = 66$

15.5 (a) $W_A = 2 + 4 + 5 = 11$

(b) $W_S = 1 + 3 = 4$

15.7 $W_S = 47$, reject $H_0$.

15.9 $W_S = 62$, fail to reject $H_0$.

15.11 (a) $BBBBBBBBBA$

(b) $P[W_A = 10] = .1$

15.13 .048

15.15 $S = 11$; with 7 ties, $n = 18$; do not reject $H_0$.

15.17 (a) .021    (c) $c = 31$

15.19 (b) $P[T^+ = t] = .125$ for $t = 0$, 1, 2 and 4, 5, 6

15.21 (a) $S = 11$; with $n = 13$, reject $H_0$.

(b) $T^+ = 105$, reject $H_0$.

15.23 $r_{Sp} = .80$

15.25 $z = -.60$, do not reject $H_0$ ($n = 10$ is not large).

15.27 $W_A = 14$

15.29 (a) Each triple has probability $\dfrac{1}{10}$.

(b) $P[W_A = 8] = .2$

15.31 (a) .098    (c) $c = 21$

15.33 $W_S = 51 \leq 69$, reject $H_0$.

15.35 (a) $r_{Sp} = -0.524$

(b) Do not reject independence.

(c) $\alpha = .05$ ($n = 8$ may be too small)

15.37 (a) $S = 1$    (b) $T^+ = 1$

# Index